SPECTROSCOPY AND STRUCTURE OF MOLECULAR COMPLEXES

SPECTROSCOPY AND STRUCTURE OF MOLECULAR COMPLEXES

Edited by

J. Yarwood

University of Durham,
Department of Chemistry,
Science Laboratories,
South Road,
Durham DH1 3LE
England

PLENUM PRESS • LONDON AND NEW YORK • 1973

Plenum Publishing Company Ltd
Davis House
8 Scrubs Lane
London NW10 6SE
Telephone 01-969-4727

U.S. Edition published by
Plenum Publishing Corporation
227 West 17th Street
New York, New York 10011

ISBN 0-306-30742-1
Library of Congress Catalog Card Number 73–77125

PRINTED IN GREAT BRITAIN BY
THE UNIVERSITIES PRESS, BELFAST, NORTHERN IRELAND

Contributors

D. A. Duddell	Chemistry Department, City of Leeds and Carnegie College, Beckett Park, Leeds LS6 3QS, England.
P. N. Gates	Chemistry Department, Royal Holloway College, Egham, Surrey TW20 0EX, England.
W. B. Person	Chemistry Department, University of Florida Gainesville, Florida 32611, U.S.A.
A. H. Price	Edward Davies Chemical Laboratories, University College of Wales, Aberystwyth SY23 1NE, Wales.
D. Steele	Chemistry Department, Royal Holloway College, Egham, Surrey TW20 0EX, England.
M. Tamres	Chemistry Department, University of Michigan, Ann Arbor 48104, Michigan, U.S.A.
J. L. Wood	Chemistry Department, Imperial College of Science and Technology, South Kensington, London SW7 2AY, England.
J. Yarwood	Chemistry Department, University of Durham, Science Laboratories, South Road, Durham City DH1 3LE, England.

Preface

This book is intended to be research text aimed principally at scientists beginning work in the field of molecular complexes. Since, however, much of the work described is only recently published, it should be useful also to more experienced workers as a source of the most recent ideas and interpretations. It should be emphasised that no attempt has been made to provide an up-to-date review of the literature. Such a review would clearly be impossible in a book of this size covering a wide range of molecular interactions. Rather, the treatment is illustrative. We have tried not to ignore the many difficulties (both experimental and interpretive) which arise in the process of spectroscopic study of many (especially weaker) complexes at the same time describing the value of such studies in our attempts to understand how molecules interact with one another.

It could be argued that such a text book ought to be written by a single author and, certainly, this would have had some advantages. However "production" of a multi-author text, with its inevitable variations in approach, style and presentation, has the important advantage of having each topic elucidated by an "expert" in that particular field. It is the editor's task to then ensure that differences of opinion and style do not destroy the book as an integrated text book. I hope that I have managed to perform this very difficult task with some degree of success.

The area of the most controversial difference of opinion among spectroscopists in the field at the present time should be mentioned specifically. It will become clear to the reader that the problem of deciding to what relative extents the various stabilizing forces (for example, electrostatic polarisation and charge-transfer forces) are important for any particular class of complex is a difficult one and is one which will, no doubt, provoke discussion for a long time to come. We have tried to indicate how difficult it is, for example, to differentiate (either experimentally or theoretically) cases where different types of forces predominate. This difficulty is severe, even for interactions for which the enthalpy change is much greater than **RT**. For "weaker" interactions the problem is made worse by the fact that one is not sure whether one is dealing with a stabilised ground state or with a system containing only "contacts" (and not "complexes"). Clearly, this is one of the most interesting areas of work, important for all types of complexes. These problems form a challenge, therefore, to both chemists and chemical physicists employing many different techniques.

We have placed major emphasis in this book on vibrational spectroscopy although magnetic resonance (and other) techniques are mentioned where especially appropriate. Only brief mention is made of electronic spectroscopy since this aspect has been covered fairly comprehensively in two recent publications (chapter 1, references 1 and 2). Molecular orbital calculations are rapidly becoming more important and these have been considered where appropriate. Finally a chapter has been included on the study of dielectric (microwave) absorption of molecular complexes and their dipole moments. Such studies have recently become important in the study of complex lifetimes and for testing the various complex "models".

We have employed SI units throughout but have retained the angstrom and wave number (cm^{-1}) for convenience. In order to minimise confusion during the transition to SI units we have included in the text the values of several less familiar parameters (*e.g.* transition moments, polarisabilities) in both c.g.s and SI systems. Conversion factors are included as footnotes to many of the tables and are summarised on page xii.

Finally, I should like to thank the contributors for their hard work and for their support and co-operation during the preparation of this book. Thanks are also due to Mrs R. Wadge and Mrs D. Armstrong who have typed (and retyped) various parts of the manuscript, to Miss L. Hall who helped to eliminate errors from the manuscript, and to Mrs K. Jones and Mr D. Hartshorn who prepared many of the diagrams. Grateful thanks are also due to the staff of the Plenum Publishing Co Ltd. for their helpful co-operation at all times. Individual acknowledgements are included, where appropriate, in the separate chapters.

J. Yarwood
Durham City
Sept. 1973

Glossary of Commonly used Symbols

A_i	The "true" absolute integrated intensity of the ith vibrational band (equation 2.5).
α	Molecular polarisability; Orbital exponent for Gaussian basis functions.
α_R	Damping constant for rotational diffusion (equation 2.21).
α_e	Vibrational-rotation interaction constant; The electronic polarisability.
α'	Polarisability derivative $-\partial\alpha/\partial Q_i$ or $\partial\alpha/\partial R_j$ (equation 5.1)
B_i	The "apparent" (observed) absolute integrated intensity of the ith vibrational band (equation 2.2).
B	Rotational constant.
β_1, β_0	Resonance integrals (equations 1.12)
β_v	Damping constant for vibrational relaxation (equation 2.17).
β_R	Damping constant for "free" rotation in the liquid phase.
C, c	Concentration, usually in mol dm^{-3}.
c	Velocity of light in a vacuum.
D_e	The equilibrium dissociation energy.
Γ_i	The absolute integrated intensity (equations 2.1 and 2.4)
Γ	The reducible representation.
E, ε	Energy
$\vec{E}$	Electrical field vector.
E_A^v	Vertical electron affinity of an acceptor (equation 1.10).
e	Electronic charge
ε	Molar extinction coefficient (equation 3.8).
$\boldsymbol{\epsilon}$	Electrical permittivity of vacuum.
ε'	Real part of the complex permittivity.
ε''	Imaginary part of the complex permittivity.
$\varepsilon^*(= \varepsilon' - i\varepsilon'')$	Complex permittivity (equation 7.24).
ε_0	Limiting permittivity at low frequency.
ε_∞	Limiting permittivity (debye) at high frequency.
F	Force constant (Urey-Bradley); The weight of the "no-bond" or "dative" states.
f	Force constant (valence force field).
f	Frequency (chapter 7).
g_i	Vibrational degeneracy.
ΔG	Difference between G_1, the stabilization energy of the "dative" state, and G_0, the corresponding stabilization energy of the "no-bond" state.

$\Delta G^{\ominus}$	Standard free energy change.
G^R, G^v	Rotational and vibrational components of the dipole autocorrelation function.
H	Force constant (Urey-Bradley).
$\Delta H^{\ominus}$	Standard enthalpy change.
$\hat{\mathscr{H}}$	The Hamiltonian operator.
h	The Planck constant.
η	Microscopic viscosity (equation 7.30).
θ	Molecular quadrupole moment (equation 7.43).
I_0, I	Incident and transmitted (true) radiation intensities.
I	Moment of inertia; spectral intensity (general).
I_D^v	Vertical ionisation potential of a donor (equation 1.10).
J_{ij}	Spin-spin coupling constant.
K	Force constant (Urey-Bradley)
k	Force constant (valence force field).
k	The Boltzmann constant.
K_c, K_a, K_{DA}	Equilibrium constant.
k	Rate coefficient
$k(\tilde{\nu})$	The absorption coefficient as a function of wave number (equation 2.10).
l, L	Vibrational eigenvector matrix elements (equation 1.52).
λ	Wavelength.
$M^{(i}$	The ith band moment (equation 2.10).
$\vec{M}_d$, $\vec{M}'_d$	Charge-transfer (vibronic) part of the total dipole moment derivative for a given vibrational mode equation 1.60).
μ	Reduced mass.
$\vec{\mu}$	Observed (*i.e.* measured) molecular dipole moment.
$\vec{\mu}_N$	Dipole moment of the ground state (Ψ_N) of a complex (equation 1.27).
$\vec{\mu}'$	The dipole moment derivative ($\partial\vec{\mu}/\partial Q$) or ($\partial\vec{\mu}/\partial R$) for a given vibrational mode—usually the total dipole derivative.
$\vec{\mu}'$	The induced dipole moment (equation 2.7).
m	Number of molecules per unit cell.
$\vec{m}'$	Dipole moment derivative for the "electrostatic" interactions ("general" solvent–solute interaction) (equation 1.72).
$\partial\vec{\mu}_N/\partial Q$ or $\partial\vec{\mu}_N/\partial R$	Dipole moment derivative due to electrostatic effects and due to changes in normal coordinates (equation 1.51).
n	Number of atoms in a molecule.
n	Real part of the complex refractive index.
N	The Avogadro constant.
ν	Vibrational mode; vibrational frequency.
$\tilde{\nu}$	Vibrational band position in cm^{-1}.
ν_s	A-H stretching mode (hydrogen-bonded complexes).

ν_b	In-plane bending mode (hydrogen-bonded complexes).
ν_d	Out-of-plane bending mode (hydrogen-bonded complexes).
ν_σ	Intermolecular stretching mode (hydrogen-bonded complexes).
ν_γ, ν_ϵ, ν_β, ν_δ	Intermolecular bending modes (hydrogen-bonded complexes).
ν_τ	Tunnelling mode (hydrogen-bonded complexes)
P	Molar polarisation (equation 7.1).
$\vec{p}^0$	Dipole moment of the "free" molecule.
$\vec{p}$	Calculated dipole moment (for example, from *MO* calculations).
$\partial\vec{p}/\partial Q$ or $\partial\vec{p}/\partial R$	Total calculated or observed dipole moment derivative from infrared data, or *MO* calculations (equation 1.46).
$\partial\vec{p}^0/\partial R$	Dipole moment derivative for "free" molecule (in absence of intermolecular forces).
ρ	Electron density at a particular atom.
ρ	Raman dipolarisation ratio
ρ	Density of a particular solution (equation 7.1).
q, Q	Normal vibrational coordinates.
q_i	Charge density on i^{th} atom.
r, R	Internuclear distance; Internal coordinate.
R	Molar refraction (equation 7.3).
S, S_{DA}, S_{da^-}, S_{01}	Overlap integrals (for example, equations 1.3, 1.7 and 1.17).
S	Symmetry coordinate.
$\Delta S^\ominus$	Standard entropy change.
σ'	Pauling covalent bond character.
T_0, T	Apparent (observed) incident and transmitted radiation intensities.
τ	Dielectric relaxation time (equation 7.26).
U	Interaction energy.
V	Intermolecular potential (equation 1.6).
v	Vibrational quantum number.
v	Specific volume.
$\phi(t)$	Transition dipole time autocorrelation function (equation 2.14).
ϕ	Wave function of individual atomic or molecular orbital.
χ	Slater-type or Gaussian basis function.
ψ	Unperturbed or modified donor or acceptor wave function (whole molecule) (equation 1.2).
Ψ	Ground or excited electronic wavefunction of complex (whole molecule) (equation 1.1).
Ψ°	Vibrational wave function.
ω	Angular frequency ($2\pi f$).
Z	Effective nuclear charge (equation 5.3).

Conversion Table Between c.g.s and SI Units

Physical quantity	SI unit	Conversion factor
mass	g or kg (as convenient)	—
length (interatomic distance)	nm (or Å)	1 Å = 0.1nm = 10^{-10}m
density	kg m^{-3}	1 gm cm^{-3} = 10^3kg m^{-3}
volume	m^3 (or dm^3)	1 litre = $10^{-3}m^3$ = 1dm^3
concentration	mol dm^{-3}	—
	mol m^{-3}	1 mol $litre^{-1}$ = 10^3mol m^{-3}
viscosity	kg $m^{-1}s^{-1}$	1gm cm^{-1} sec^{-1} = 10^{-1}kg $m^{-1}s^{-1}$
electrical polarisability	C m^2V^{-1}	1 $Å^3$ = $10^{-24}cm^3$ = 1.115 × 10^{-39} C m^2V^{-1}
molar polarisation	m^3mol^{-1}	1cm^3mole^{-1} = $10^{-6}m^3mol^{-1}$
molar refraction	m^3mol^{-1}	1cm^3mole^{-1} = $10^{-6}m^3mol^{-1}$
equilibrium constant (for 1 :1 interaction)	dm^3mol^{-1}	—
molar enthalpy	kJ mol^{-1}	1 kcal $mole^{-1}$ = 4.184kJ mol^{-1}
molar free energy	kJ mol^{-1}	1 kcal $mole^{-1}$ = 4.184kJ mol^{-1}
molar entropy	kJ $mol^{-1}K^{-1}$	1 kcal $mole^{-1}deg^{-1}$ = 4.184KJ $mol^{-1}K^{-1}$
molar extinction coefficient	m^2mol^{-1}	1 litre $mol^{-1}cm^{-1}$ = $10^{-1}m^2mol^{-1}$
integrated intensity	m mol^{-1} ($m^3mol^{-1}m^{-1}$)	1 litre $mol^{-1}cm^{-2}$ = 10m mol^{-1}
electrical dipole moment	C m	1 debye = 3.335 × 10^{-30}C m
electrical quadrupole moment	C m^2	1 esu cm^2 = 3.335 × 10^{-14}C m^2
dipole transition moment	C (*i.e.* C m m^{-1})	1 debye $Å^{-1}$ = 3.335 × 10^{-20}C
polarisability derivative	C m V^{-1}	1 $Å^2$ = 10^{-16} cm^2 = 1.115 × 10^{-30} C m V^{-1}
moment of inertia	kg m^2	1g cm^2 = 10^{-7}kg m^2
force constant	N m^{-1}	1md $Å^{-1}$ = 100N m^{-1}
wavelength	nm (ultraviolet and visible)	1mμ = 1nm = 10^{-9}m
	μm (infrared)	1μ = 1μm = 10^{-6}m
frequency	Hz	—

Contents

Chapter 1

Theoretical Aspects of the Study of Molecular Complexes

Willis B. Person

Department of Chemistry
University of Florida
Gainesville, Florida, 32601, USA

1.1. GENERAL THEORETICAL BACKGROUND

1.1.1. INTRODUCTION

In this chapter we should like to examine, from a theoretical viewpoint, the factors affecting the structure and spectra of molecular complexes. This subject has recently been thoroughly reviewed by

Mulliken and Person[1,2]; here we shall review it briefly, with emphasis on vibrational spectroscopy, and include some new results, in section 1.3, from approximate all-electron quantum mechanical calculations of whole-complex molecular orbitals. In general, we shall try to follow the nomenclature of Mulliken and Person[1], and we shall refer the reader to that text for historical discussion and for a more thorough treatment of some aspects.

A precise definition of the term *molecular complex* is not easy; Mulliken and Person attempted to provide one as follows:

> "A *molecular complex* between two molecules is an association, somewhat stronger than ordinary van der Waals associations, of definite stoichiometry . . . . The partners are often already *closed-shell* (saturated valence) electronic structures. In loose complexes the identities of the original molecules are to a large extent preserved."

Although it is not always easy to apply these criteria to a particular system, we believe this statement contains the key elements that characterise a molecular complex.

In particular the concept of "complex" refers to "weak" interactions in which the molecules preserve their properties, but in which they interact more strongly than in an "ordinary" association, such as those between different molecules in an ideal "non-interacting" solution. Hence, the electron distribution in the complex is expected to be quite similar to that for a pair of non-interacting molecules in juxtaposition, with only a small correction term. Nevertheless, the small correction term is important, because it is responsible for an interaction between a particular pair of molecules that is stronger than an ordinary van der Waals interaction.

Mulliken Resonance-Structure Theory

The resonance-structure theory developed by Mulliken[3–5] (see also references 1 and 2) is based on the following general ideas. The electronic wavefunction, Ψ_N, for the ground electronic state of a complex between two molecules is approximated by:

$$\Psi_N \simeq \underset{\text{no-bond}}{a\Psi_0(D, A)} + \underset{\text{dative}}{b\Psi_1(D^+\text{–}A^-)}. \qquad 1.1$$

Here D is the potential electron donor of the molecule pair and A is the potential acceptor. $\Psi_0(D, A)$, called the "no-bond" function with weighting coefficient a, is the wavefunction for the molecular pair if there were no special interaction, and $\Psi_1(D^+\text{–}A^-)$ is the "dative" wavefunction, occurring with coefficient b, describing the electronic configuration of a molecular pair after an electron has been transferred completely, usually from the highest filled molecular orbital of D to the lowest vacant orbital of A, followed by formation of a bond between D and A utilizing these two orbitals.

In writing the wavefunction in this way, one is assuming that the modification of Ψ_0 due to the interaction, although small (since b is expected to be small), is predominately toward that described by the dative wavefunction $\Psi_1(D^+\text{–}A^-)$. In this two-structure resonance theory description of the complex, other corrections to $\Psi_0(D, A)$ are completely ignored. As the interaction between the two molecules increases, it is quite possible that other distortion from $\Psi_0(D, A)$ will occur, so this two-structure description is expected to hold only for weak interactions.

The smaller distortions in the wavefunction that occur when two molecules (for example, two benzene molecules to form liquid benzene) are brought together in an "ordinary" van der Waals interaction to a configuration similar to that found at equilibrium in the "complex" may be included in $\Psi_0(D, A)$. Thus $\Psi_0(D, A)$ can be written as an antisymmetrized product of modified wavefunctions ψ_D^{mod} and ψ_A^{mod} of the individual D and A molecules in juxtaposition; because we do not know precisely how to write these distorted wavefunctions, one may in practice approximate them by the undistorted wavefunctions ψ_D^0 and ψ_A^0 for the free molecules. Thus we have:

$$\Psi_0(D, A) = \mathscr{A}\psi_D^{mod}\psi_A^{mod} \simeq \mathscr{A}\psi_D^0\psi_A^0. \qquad 1.2$$

(See reference 1, section 2.5, for a more complete discussion.)

Further details of the resonance-structure theory are given in section 1.1.2, below. In particular, we shall find that some resonance stabilization of the energy of the complex is expected as the dative wavefunction "mixes" with the no-bond wavefunction; this resonance stabilization is called the "charge-transfer" energy and *may* be responsible for the *extra* stabilization of a complex where it occurs. We emphasize again that in a *real* complex other modifications of Ψ_0 may occur to form Ψ_N, and other forces may be important in stabilizing the complex.

General Theory of Intermolecular Forces

The rigorous general theory of intermolecular interactions is indeed a complicated subject. It has recently been reviewed in some detail in a book edited by Hirschfelder[6]. In the first chapter, Hirschfelder and Meath divide the general problem into three different areas of increasing complication: namely, interactions occurring at large separations with essentially no overlap between the wavefunctions of the two molecules, interactions occurring at intermediate separations, where the overlap is small but non-zero, and interactions at small separations, where the overlap is larger.

When the overlap is large, at small separations, the two molecules involved in the interactions may lose their identity as their atoms rearrange to form an "activated complex" and then new products. For these small separations it would then be necessary to solve completely the quantum-mechanical problem for the entire "complex", using, for example, the variation procedure to obtain the total energy of the system. In fact, there have been attempts recently to carry out such calculations by approximate quantum-mechanical methods in order to map out the potential surface for chemical reactions[7]. One of the most complete studies of a chemical reaction is the exact (Hartree-Fock) quantum-mechanical study by Clementi and Popkie[8] of the reaction between Li^+ ion and H_2O to form the hydrated Li^+ ion. In the last section (1.3) of this chapter we shall describe such whole-complex molecular orbital methods as applied to complexes. However, the idea of the Mulliken resonance-structure theory of complexes is to avoid such complicated calculations, which still must be so approximate as to be of questionable value in predicting detailed properties of most complexes (although calculations such as that by Clementi and Popkie[8] are changing this picture rapidly).

In the general treatment of molecular forces the interactions between two molecules to form a weak complex are classified as "interactions at intermediate separations", and they are treated as a perturbation of the separated molecules. Thus, for example, the energy could be written in terms of the intermolecular separation, R, as:

$$E = E^{(0)} + \sum_{n} C_n/R^n,$$

where $E^{(0)}$ is the energy of the separated molecules and the coefficients, C_n, can be evaluated by a variation-perturbation technique. This procedure has been followed in detail for the interaction of two atoms (see reference 6, chapter 1), but it is not so useful for the long-range interaction between molecules as is the electrostatic perturbation treatment described below.

Electrostatic-Perturbation Calculations

The electrostatic-perturbation calculation is reviewed by Hirschfelder and Meath.[6] The procedure is described for the intermediate to long range values of R by Murrell, Randic and Williams.[9] In their procedure, the energy is expanded in a power series in the overlap integral:

$$S_{DA} = \int \psi_D^0 \psi_A^0 \, dv, \qquad 1.3$$

where ψ_D^0 is the unperturbed wavefunction of one molecule, D, of the pair undergoing interaction, and ψ_A^0 is the unperturbed wavefunction of the other molecule. When two molecules are just "touching" (say at the van der Waals distance), we can expect the value of this integral to be about 0.01 to 0.1. Such values can be estimated by writing approximate wavefunctions for ψ_D and ψ_A using Slater-type functions and then estimating S_{DA}. One can expect S_{DA} to be somewhat larger at the van der Waals distance for more realistic SCF wavefunctions (see reference 1). The perturbation theory of Murrell, Randic and Williams is expected to be valid for overlap values up to about $S_{DA} = 0.1$, so that it should apply to the weak interactions expected for molecular complexes.

In this perturbation treatment, the energy of interaction E_{DA}, is written as:

$$E_{DA} = E^{10} + E^{20} + E^{12} + E^{22} \dots \quad 1.4$$

Here the first superscript refers to the order of the perturbation in the potential energy and the second superscript refers to the order of the overlap perturbation. The first term, E^{10}, is the coulomb energy due only to electrostatic interaction. The second term contains the induction and London dispersion energies, the third term is the exchange-repulsion energy, and the fourth is the charge-transfer energy. The first two terms are exactly the long range interaction terms, and the last two terms are corrections for the effect of overlap at intermediate distances. Thus, we see the utility of this treatment in partitioning the interaction energy into parts with obvious physical significance, and which clearly become more important as the distance decreases (and as overlap increases).

In this treatment the wavefunction of the complex is:

$$\Psi_N = a_0\mathscr{A}\psi_D^0\psi_A^0 + \sum_{r,s} a_{rs}\mathscr{A}\psi_D^r\psi_A^s + \sum_{k,l} a_{kl}\mathscr{A}\psi_{D^+}^k\psi_{A^-}^l + \sum_{m,n} a_{mn}\mathscr{A}\psi_{D^-}^m\psi_{A^+}^n. \quad 1.5$$

Here the first term is just the antisymmetrized product of the ground state wavefunctions of the separated D and A molecules, with coefficient a_0. The second term is the sum over all the products of excited state functions of D and of A and describes the modification of ψ_D^0 and ψ_A^0 resulting when the molecules interact; the first two terms together are essentially the same (conceptually, at least) as Ψ_0 in equation 1.1. The third term is essentially the dative wavefunction Ψ_1 of equation 1.1, constructed with all excited D^+ and A^- wavefunctions, and not just with the lowest energy functions. The fourth term is an additional, *reverse* dative term representing the resonance structure D^-–A^+; it may be expected to have relatively small coefficients (a_{mn} in equation 1.5) for a complex formed between a neutral electron donor molecule, D, and a neutral acceptor, A.

If one writes out the perturbation-interaction potential, V_{DA}, as a sum of coulomb interaction terms between the electrons and nuclei on molecule D and those on A, then:

$$V_{DA} = -\sum_{j,A}\frac{Z_A e^2}{r_{Aj}} - \sum_{i,D}\frac{Z_D e^2}{r_{iD}} + \sum_{i,j}\frac{e^2}{r_{ij}} + \sum_{D,A}\frac{Z_A Z_D e^2}{R_{DA}}. \quad 1.6$$

Here the subscript A refers to a nucleus in the acceptor molecule with charge $Z_A e$, D refers to a nucleus on the donor with charge $Z_D e$, j refers to an electron on the donor and i refers to an electron on the acceptor. The summations are carried over all nuclei and electrons in D and in A.

Now, in principle, one could take accurately known wavefunctions (including excited state functions) for the separated molecules and for the D^+, A^- and D^-, A^+ ions, use the perturbation wavefunction (equation 1.5), the perturbation potential function (equation 1.6), and the procedure described by Hirschfelder and Meath[6] or by Murrell, Randic and Williams[9], to calculate exactly the several contributions to the energy shown in equation 1.4.

We see that this procedure is a refinement of the simplified Mulliken two-structure resonance theory described earlier, in that it would express Ψ_0 in terms of modified functions, ψ_D^{mod}, by including the second term of equation 1.5, and it would express Ψ_1, not simply as a product of the lowest energy functions $\psi^0_{D^+}$ and $\psi^0_{A^-}$, but including modifications to these functions as well. In addition the reverse dative structure functions $\Psi_2(D^-\text{–}A^+)$ are also included. The assumption made in using the simplified resonance-structure theory of equation 1.1 is that the additional refinements of equation 1.5 are not so important as the inclusion of just the simple dative function $\Psi_1(D^+\text{–}A^-)$.

In practice the refinements of equation 1.5 are still too difficult, so that approximations must be made in order to evaluate the terms in equation 1.4. Attempts to evaluate these terms are illustrated in the papers by Lippert, Hanna, and Trotter[10], and by Cook and Schug[11] (with a slightly different approach), who make good use of the approximate semiempirical expressions for the terms contributing to E_{DA} in equation 1.4, given by Hirschfelder, Curtiss and Bird[12].

Summary

The exact expression for the wavefunction, Ψ_N, and energy, E_N, for a weak complex D–A formed between a donor and an acceptor molecule must be found by solving the Schrödinger equation of the whole complex, as described in section 1.3. In attempting to avoid this complicated and expensive calculation, we can try to carry out the perturbation treatment using the wavefunctions in equation 1.5. Since the latter procedure is still too difficult to be practical for most systems, it is often useful to make the much more drastic approximation of equation 1.1 to a simplified two-structure resonance description of the wavefunction of the complex.

There is considerable discussion in the literature as to whether particular weak complexes are "electrostatic" or "charge transfer" in character. We see from the preceding discussion that this question is whether the charge-transfer term E^{22} is larger than the other terms in equation 1.4 and whether the coefficients a_{k1} are larger than the other coefficients (except a_0) in equation 1.5. As we shall see later, all the terms in equation 1.5 become larger as the D–A distance becomes smaller. Unless some experiment can be devised to distinguish between the two (or more) effects, or until the theory can accurately predict their uniqueness, such discussion is often largely semantic in nature.

At this point it is worth drawing attention to the "bond energy analysis" discussed by Clementi and Popkie[8]. They have defined a method for partitioning the total calculated energy for the $Li^+(H_2O)$ system, for example, into contributions from $\Delta\varepsilon_{1,0}(\text{Li})$ and $\Delta\varepsilon_{1,0}(\text{W})$, the polarization energies of the Li^+ ion by the water and of water by Li^+ ion, respectively; and $\varepsilon_1(\text{Li–W})$, the interaction energy. To some extent, the latter is related to the charge-transfer energy (approximately $E^{12} + E^{22}$ of equation 1.4) while the $\Delta\varepsilon_{1,0}$ terms are related to the electrostatic contributions ($E^{10} + E^{20}$ of equation 1.4). Unfortunately, we learned about this work[8] too late to make much use of it in this chapter, but this procedure will be of considerable interest in the future.

If we use the simplest term of the two-structure resonance theory of equation 1.1, then we can deduce the coefficient b of the dative structure from observed properties of the system—for example, from the apparent dipole moment of the benzene–iodine complex. If it is assumed that the two molecules in close juxtaposition have a dipole moment of zero, then the observed value of the dipole moment of the complex leads to the conclusion that $b \simeq 0.3$. On the other hand, it is reasonable that the quadrupole moment of benzene may induce a considerable dipole moment in the iodine, so that observed dipole moment of the complex may be due entirely to that effect, with b much smaller. For stronger interactions, however, it seems clear that both effects must be important.

In the next section we shall describe some attempts to evaluate the different terms contributing to the energy in equation 1.4.

1.1.2. SIMPLE RESONANCE-STRUCTURE THEORY AND ELECTROSTATIC-PERTURBATION THEORY

Before reviewing the attempts to calculate the relative magnitudes of the terms contributing (in equation 1.4) to the interaction energy, let us present, in very brief form, the details of the simple Mulliken resonance-structure theory. (See references 1 and 2 for a more detailed account.)

Simple Resonance-Structure Theory

The quantum-mechanical description of the complex in the ground state is given by the simplified two-structure wavefunction in equation 1.1.

We remind ourselves that the no-bond structure Ψ_0 *should* be expressed as an antisymmetrized product of *modified* functions (equation 1.2). These modified functions include all changes from the free-molecule wavefunction due to electrostatic interaction, induction, etc., represented in equation 1.5 by the second term sum over excited state wavefunctions. Since these modified functions are so difficult to evaluate, they are approximated by $\mathscr{A}\psi_D^0\psi_A^0$ in the usual simple theory, as described above. The wavefunction, Ψ_N, is normalized, so that

$$\int \Psi_N^2\, dv$$
$$= a^2 \int \Psi_0^2(D, A)\, dv + b^2 \int \Psi_1^2(D^+\text{–}A^-)\, dv + 2ab \int \Psi_0 \Psi_1\, dv = 1. \qquad 1.7$$

(Real wavefunctions are used throughout this chapter; if complex functions are used, we must, of course, use complex conjugates in the integrals.) Since Ψ_0 and Ψ_1 are individually normalized, and writing S_{01} for the overlap integral $\int \Psi_0 \Psi_1\, dv$ we have:

$$\int \Psi_N^2\, dv = a^2 + b^2 + 2abS_{01} = 1. \qquad 1.7a$$

The weight of the no-bond structure, F_{0N}, is defined by

$$F_{0N} = a^2 + abS_{01}, \qquad 1.8$$

and the weight of the dative structure, F_{1N}, is

$$F_{1N} = b^2 + abS_{01}. \qquad 1.9$$

As mentioned previously (also see reference 1) the value of F_{1N} is not expected to be very large for weak complexes, varying from 0.01 or so for the very weak benzene–halogen complexes to 0.3 or so for the stronger amine–halogen complexes. The weight of dative structure may be larger than this, varying up to 1.0 for reactions that form stable ion-pair (D^+–A^-) complexes (section 2.3.5). For the most part, however, if we are discussing "complexes", we are implicitly assuming relatively small values of F_{1N}.

Basically, the main reason that F_{1N} is small is that the hypothetical dative structure is usually a high-energy structure. Its energy W_1, relative to the ground state energy, W_0 is given (see reference 1, p. 118) by:

$$W_1 - W_0 = I_D^v - E_A^v + G_1 - G_0 = \Delta. \qquad 1.10$$

Here I_D^v is the vertical ionization potential of D and E_A^v is the vertical electron affinity of A. $W_0 + G_0$ is the energy of the no-bond structure, with W_0 the energy of the separated molecules and G_0 the energy of interaction between D and A when they are forced together in the configuration of the complex without further interaction. G_1 is the stabilization of the ion-pair dative state, due mainly to the coulomb energy of attraction as D^+ and A^- are brought together to the equilibrium separation of the complex, with a smaller contribution from the energy of formation of a D^+–A^- covalent bond.

As a result of the mixing between the no-bond and dative functions to form the gound state of the complex, the energy of the latter is lowered by a resonance interaction. The results from a variation method calculation of this energy are given in detail in section 2.2 of reference 1.

Usually for weak complexes the energy of the dative state, W_1, is expected to be considerably higher than that for the no-bond state, W_0 (*i.e.* Δ is large). In that case, the first-order perturbation expression for the energies and wavefunctions is expected to hold for this simplified resonance-structure theory, with

$$W_N = W_0 - \frac{\beta_0^2}{\Delta} = W_0 - X_0. \qquad 1.11$$

Here these symbols are as defined previously, with

$$(\Delta/2)^2 \gg \beta_0\beta_1; \quad W_{01} = \int \Psi_1 \hat{\mathscr{H}} \Psi_0 \, dv; \quad W_0 = \int \Psi_0 \hat{\mathscr{H}} \Psi_0 \, dv;$$

and

$$\begin{aligned} \beta_0 &= W_{01} - W_0 S_{01}; \qquad \beta_0, \beta_1 < 0; \\ \beta_1 &= W_{01} - W_1 S_{01}; \\ \beta_1 - \beta_0 &= -S_{01}\,\Delta. \end{aligned} \qquad 1.12$$

The coefficients a and b are also determined by these parameters, with

$$b/a = -\beta_0/\Delta \quad \text{and} \quad b^*/a^* = -\beta_1/\Delta. \qquad 1.13$$

Furthermore, the theory implies an excited electronic state, called the *charge-transfer state*, with energy W_V given approximately by:

$$W_V = W_1 + \frac{\beta_1^2}{\Delta} = W_1 + X_1, \qquad 1.14$$

with the wavefunction:

$$\Psi_V = -b^*\Psi_0(D, A) + a^*\Psi_1(D^+\text{–}A^-). \qquad 1.15$$

Because b* is expected to be considerably less than a* for weak complexes, the charge-transfer state is more nearly like the dative state than like the no-bond state. The electronic transition between W_N and W_V is expected to be quite strongly allowed, and the frequency, ν_{CT}, of this *charge-transfer band* is given approximately by:

$$\mathbf{h}\nu_{CT} = W_V - W_N = \Delta + \frac{\beta_0^2 + \beta_1^2}{\Delta} = \Delta + X_0 + X_1. \qquad 1.16$$

These quantities are defined in figure 1.1.

Measurements on complexes have been usually made in solution in an "inert" solvent (for example in *n*-heptane). There arises, therefore, an additional complication in this already over-simplified discussion due to the neglect of solvent effects. In figure 1.1 we have indicated that

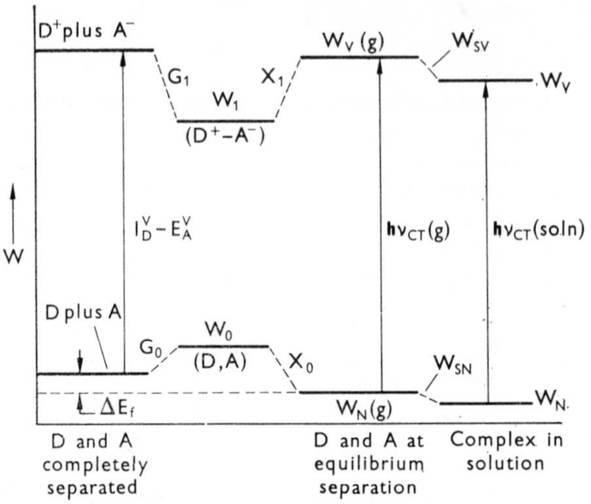

FIG 1.1 Energy diagram for a D–A complex, showing a convenient breakdown of terms for $h\nu_{CT}$. G_0, (shown here to be positive) may also be negative, especially in weak complexes. (Reproduced from Molecular Complexes; A Lecture and Reprint Volume, p 117, by permission of J. Wiley and Sons, Inc.)

interaction between solvent and complex causes a change in the energy levels by W_{SN} and by W_{SV}, respectively. These changes may be different for the two states; since they are unknown, we expect the charge-transfer frequency observed for a complex in solution to differ from the value for the same complex in the gas phase.

A detailed discussion is given in chapter 2 of reference 1 concerning the nature of Ψ_1, of the triplet charge-transfer state, and of a number of other points. We should, however, mention one or two of the most important of these.

First, we should emphasize again that S_{01} is the overlap integral between Ψ_0 and Ψ_1. It is related to another useful overlap integral, S_{da^-}, between the highest-energy filled molecular orbital ϕ_d on the donor and the lowest-energy empty molecular orbital ϕ_{a^-} on the acceptor anion (often approximated by the lowest-energy empty molecular orbital on the free molecule, ϕ_a) by the relation:

$$S_{01} \simeq (\sqrt{2}\, S_{da^-})/(1 + S_{da^-}^2)^{1/2}. \qquad 1.17$$

This overlap integral (S_{da^-}) is not to be confused with S_{DA} of equation 1.3, which is the overlap between the two antisymmetrized, molecular orbital, product functions ψ_D^0 and ψ_A^0. If ϕ_a is used to estimate S_{da^-}, and if the overlap between the highest filled donor orbital (ϕ_d) and the lowest empty orbital (ϕ_a) of the acceptor is the largest overlap term, then S_{da^-} may be approximately equal to S_{DA}.

The semiempirical evaluation of the parameters (Δ, β_0, S_{01}) that determine ν_{CT} and other properties of the complexes, if the simple resonance-structure theory is assumed correct, has been discussed in chapter 9 of reference 1. It is expected that a good electron donor has a relatively low ionization potential, and a good electron acceptor has a high electron affinity. Thus, Δ is expected to be relatively small (from equation 1.10) and the charge-transfer frequency, ν_{CT}, is often lower than electronic transitions to locally excited states of the donor or of the acceptor. Thus, complex formation involving charge transfer is often expected to be accompanied by a characteristic strong new ultraviolet (or visible) absorption band at lower frequencies than any absorption due to the component molecules.

Because of this apparently rather unique predicted behaviour, failure to observe a charge-transfer band has been used as an argument against charge transfer in attempting to establish the importance of this contribution in determining the properties of a complex (see section 4.4.3). This criticism may not always be valid, however, because the charge-transfer band is not *always* expected at lower frequencies than are the locally excited bands of the molecules.

Any pair of molecules bound together in a complex, *regardless of the primary force responsible for the interaction*, is expected to exhibit a charge-transfer (CT) absorption to one of the charge-transfer states of the "complex". If I_D^V is comparatively high or E_A^V low, the charge-transfer band may occur at frequencies *higher* than the locally-excited transitions of D or A, or even of the solvent, and thus not be observed.

This effect may well be responsible for the failure, generally, to observe charge-transfer bands in the ultraviolet absorption spectra of hydrogen-bonded systems. Thus, one may estimate that the *vertical* electron affinity of a typical HX acceptor would be approximately zero[13], or about 1–2 eV less than E_A^V for I_2, for example. For D–HX complexes the value of G_1 is expected to be a bit greater than for D–I_2, primarily because of the shorter D–A distance with the HX molecule. We estimate, by the reasoning given in chapter 9 of reference 1, that G_1 for HX complexes may be about 1 eV greater than for I_2 complexes. Thus $h\nu_{CT}$ is estimated for a D–HX complex from equation 1.16 to be about 1 eV (or 96.5 kJ mol^{-1}) greater than for I_2 complexes with that same donor. Thus we predict a blue shift of 8000 cm^{-1}, or a shift of a CT band from 330 nm for a benzene complex with I_2 to 240 nm for a benzene complex with HX, for example. Such an increase in ν_{CT} could be sufficient in that case, and in most cases, to shift the charge-transfer absorption until it overlaps that by the solvent or that by "free" donor. Hence, charge-transfer absorption may not always be observed, even though charge-transfer forces may be important in stabilizing the complex, or though effects of charge transfer may be important for other properties of the complex.

We note (see figure 1.1) that the energy of formation of the complex in the gas phase from the free D and A molecules is given by:

$$\Delta E_f = G_0 - X_0. \qquad 1.18$$

Thus, in terms of the electrostatic-perturbation treatment and equation 1.4 of the preceding section, $\Delta E_f = E_{DA}$, so that $G_0 \simeq E^{10} + E^{20} + E^{12}$, and $X_0 \simeq E^{22}$. A complex might properly be called a "charge-transfer complex" if the charge-transfer resonance term, X_0, is much larger than G_0. It is not easy to estimate reliably the relative magnitudes of X_0 and G_0, but it seems likely that the terms are often comparable. (see below for further discussion.) For solution studies the solvent interactions affect the energies W_N and W_V, as discussed above and in reference 1. Corrections (which are not easily estimated) must be made, as indicated in figure 1.1, and as discussed in the preceding section.

All of the equations presented in this section can be applied only when the perturbation of D by A is small, so that the wavefunction is given by equation 1.1 to a good approximation. For strong interactions, with resultant large perturbations, we expect that other correction terms (in addition to Ψ_1) should be added in equation 1.1. Thus, a reverse charge-transfer term, $\Psi_2(D^-–A^+)$ is often expected to become important. Excited charge-transfer states $\Psi_i(D^{+*}–A^-)$ are expected to be important. In fact, such states *may* be important even for such a weak complex as benzene–I_2. Mulliken has continually stressed the importance of considering these more complicated wavefunctions[1–5]; attempts to use the theory in its simplest two-structure resonance form can, indeed, lead to errors due to oversimplification.

Electrostatic Perturbation Theory

In an attempt to estimate the errors due to excessive reliance on the simplified resonance-structure theoretical description of complexes, let us go back to the expressions in equations 1.4–1.6 of the electrostatic perturbation theory, particularly to the energy expression of equation 1.4, and consider how each term may be evaluated.

The coulomb term, E^{10}, in equation 1.4, can be evaluated from classical electrostatic theory. At large separations the electrostatic energy is well represented by the multipole expansion of V_{DA} to give a sum of terms representing charge-charge, charge-dipole, dipole-dipole, quadrupole-dipole, *etc.*, interactions. Such expansions are discussed, for example, by Buckingham[14] and by Hirschfelder, Curtiss, and Bird[12], or by Lippert, Hanna and Trotter[10]. This E^{10} term could, in principle, be evaluated completely and accurately if the charge distributions in the free D and A molecules were well known, for example from accurate wavefunctions for the ground state, or from accurate experimental values of dipole, quadrupole, and higher moments. Cook and Schug[11] estimated E^{10} by numerical integration over charge distributions developed from Slater-type wavefunctions. When the multipole expansion is used, the centre of the expansion is sometimes difficult to define. For example in NH_3, is the point dipole centred at the centre of charge between the negative N atom and the positive H atoms, or is it centred in the middle of the N-atom lone-pair? Such decisions have considerable effect on the value calculated for E^{10} at intermediate separations, and hence upon the conclusion reached concerning the relative importance of the coulomb energy contribution to the interaction energy.

The second term, E^{20}, contains the induction energy and also the London dispersion energy. The former can be estimated, for dipole-induced dipole, quadrupole-induced dipole interactions, etc., from classical electrostatics[12,14] if the polarizability of D (and of A) is known, together with the charge distribution needed for the estimation of E^{10}. Again, estimates of the magnitude of this contribution depend critically upon decisions made concerning the location of the dipole moments, which polarizability is used (and it's precise value), *etc.* Because of the arbitrariness of some of these decisions, the otherwise simple nature of this calculation becomes complicated enough to lose some of its appeal.

The contribution from the dispersion energy to E^{20} can be calculated from the Drude model, given ionization potentials and polarisabilities for D and A. (See reference 12 for this procedure.)

The exchange repulsion energy, E^{12}, was estimated by Lippert, Hanna and Trotter[10] from

$$E^{12} = \sum_{i,j} c_{ij} S_{ij}^2 / r_{ij}. \qquad 1.19$$

Here r_{ij} is the distance between atom i on the donor and atom j on the acceptor, S_{ij} is the overlap integral between two orbitals, one on atom i, the other on atom j, and c_{ij} is a constant equal to c or to 2c depending on whether the two orbitals each contain one or two electrons. The empirical constant c was taken to be the same for all interaction pairs, and was evaluated by assuming that the total interaction energy, E_{DA}, is a minimum at the van der Waals distance. Cook and Schug[11] attempted to evaluate this term by a direct quantum-mechanical calculation in the *CNDO* approximation (see section 1.3 for further discussion of *CNDO* methods). They emphasize the critical importance of this term, since it *is* the *only* repulsive term considered, and so determines the equilibrium position and properties in a vital way.

Finally, the charge-transfer energy E^{22} is approximated[10] by:

$$E^{22} = \sum_{m,n} - [kS_{mn}]^2 / h\nu_{mn}. \qquad 1.20$$

Here S_{mn} is the overlap integral between molecular orbitals m, a filled orbital on the donor, and n, an empty orbital on the acceptor (presumably the most important term would be S_{da-}) and $h\nu_{mn}$ is the difference in energy between the orbitals. k is a proportionality constant between S_{mn} and the resonance integral, *i.e.* $H_{mn} \simeq -kS_{mn}$. Quite clearly E^{22} is the same as X_0 of equation 1.11, with $-kS_{ma} \simeq \beta_0$. Again, the choice of the parameters c and k is critical in estimating the magnitudes of E^{12} and E^{22} (see reference 9). Presumably the discussion of β_0 in chapters 9 and 10 of reference 1 should be helpful in choosing k.

At this point is is worth mentioning an alternative procedure for estimating the contribution to interaction energy that may be more practical than trying to determine each term as described above. The alternative has been used by O'Connell and Prausnitz to describe the interactions occurring in water vapour mixtures[15]. Their procedure is to represent the sum of the exchange repulsion energy, the symmetric part of the induction energy, and the London dispersion interaction energy by a realistic empirical interaction potential—in their case the Kihara potential:

$$V = 4U_0\left[\left(\frac{\sigma}{R_{DA} - 2a}\right)^{12} - \left(\frac{\sigma}{R_{DA} - 2a}\right)^{6}\right]. \qquad 1.21$$

Here a is the spherical core radius, R_{DA} is the D–A distance (between centers), and U_0 and σ are the energy and distance parameters, respectively, of the potential. These parameters can be estimated for D and A using "mixing" rules (see reference 12). The remaining electrostatic terms (E^{10} and the induction term) and the charge-transfer term then are calculated as described above. Thus, they assume that the energy expression 1.4 is given by:

$$E_{DA} = E^{10} + (E^{20})' + V + E^{22} = E_{DA}^{phys} + E^{22}. \qquad 1.22$$

Here E^{10} is the coulomb energy and is calculated as a sum of energies in an ordinary multipole expansion; $(E^{20})'$ is a smaller correction to the induction energy due to the asymmetric charge distribution in real molecules; and E^{22} is the charge-transfer energy. They have attempted to calculate the first three terms in equation 1.22; the difference between this calculated energy and the experimental energy is attributed[15] to the charge-transfer contribution.

Relative Magnitudes of Different Energy Terms in the Electrostatic Perturbation Treatment

In spite of the difficulty in fixing parameter values to be used in estimating the magnitudes of terms in equation 1.4, it is considered worthwhile examining some of these results. Since the exchange-repulsion and the charge-transfer terms are second order in the overlap, they are expected to vary especially rapidly as R_{DA} becomes smaller than the van der Waals distance. Thus comparisons (even of the relative magnitudes of the terms in equation 1.4) made for two calculations at different R_{DA} distances are expected to be misleading.

However, examination of table 1.1 does lead us to the conclusion that all four contributions to the intermolecular interaction energy are moderately large at the expected equilibrium distance for the complex. We note that E^{20} and E^{12} do tend to cancel approximately, leaving the energy of formation to be determined by the coulomb term, E^{10}, and the charge-transfer term E^{22}. It does make sense to combine E^{20} and E^{12} because their sum may be expected to be about the same at R_{DA}^{e} for *any* molecular pair, whereas the values of E^{10} and of E^{22} are expected to depend upon the *particular* molecular pairs involved. For the special pairs that show enough *extra* interaction to be called "complexes", the terms that are likely to be different enough to cause this special behaviour are just the coulomb attraction E^{10} and the charge-transfer attraction E^{22}. Of course, when this "extra" attraction is great enough, R_{DA}^{e} decreases and E^{12} will increase considerably, cancelling some of the "extra" stabilization.

The more extensive results tabulated by Lippert, Hanna and Trotter[10] and by Cook and Schug[11] (among others) in tables similar to those shown here have been used as arguments against the importance of charge-transfer in stabilizing the complex. As we can see by comparing the two sets of results for benzene–Cl_2 in table 1.1, the conclusion regarding the relative importance of the several different terms depends very much upon the assumptions made in estimating these parameters. It appears in table 1.1 that Cook and Schug have over-estimated the (positive) exchange-repulsion at each distance in their treatment, since their calculated equilibrium distance, $R_{DA}^{e} = 6.3$ Å, is too large. We believe Cook and Schug[11] may have underestimated the extent of both the charge-transfer (or E^{22}) and the coulomb energy (E^{10}) at shorter distances. We believe the magnitudes of both are comparable for benzene–Cl_2.

The result for the H_2O dimers shown in table 1.1 is very interesting. The magnitude of "charge transfer" here is estimated by taking the difference between the energy of the dimer calculated quantum-mechanically by Del Bene and Pople[16] and that calculated by Moore and O'Connell[17] assuming that only "physical" terms $E_{DA}^{phys} = E^{10} + V + (E^{20})'$ in equation 1.22 contribute and that they are adequately described by combining the E^{10} coulomb term (and the induction terms) with the Kihara potential. Thus $E^{22} = E_{DA}^{calc} - (E^{10} + (E^{20})' + V)$. The impressive thing about their calculation is that the "charge-transfer" energy calculated in this way, although almost surely over-estimated, is negative only for the few configurations (among the 50 considered by Del Bene and Pople[16]) for which the overlap is appropriate for charge-transfer action (for example $(H_2O)_2$, I in table 1.1 and in figure 1.2). For the other configurations $[E_{DA}^{calc} - E_{DA}^{phys}]$ is small and slightly positive, as indicated in table 1.1 for a typical different configuration, for example for $(H_2O)_2$ II. These two configurations are shown in figure 1.2.

It is quite instructive to examine the calculated interaction energy as a function of the D–A distance. It is possible to do so with the figures

TABLE 1.1

Estimated contributions in kJ mol^{-1} to the interaction energies (equation 1.4) of several D,A molecular pairs[a]

Complex, D–A	R_{DA}, Å	E^{10} oculomb	E^{20} induction	E^{20} dispersion	E^{12} exchange	E^{22} charge transfer	E_{DA}	ΔE^f (exp)[b]
Benzene–Cl_2^c	4.28	−13.64	−1.42	−3.85	+11.46	−6.48	−13.93	(−8)
Benzene–Cl_2^d	4.17	−7.38	−1.71	−4.70	+47.95	−1.37	+32.79	(−8)
Benzene–TCNE[e]	3.50	−16.74	−4.18	−15.06	+17.57	−7.53	−25.94	(−25)
$(H_2O)_2$ I[f]	2.8	−10.71		+0.50[g]		−13.76	−24.26	(−21)
$(H_2O)_2$ II[h]	2.8	−15.27		−1.13[g]		+4.60	−11.71	(?)

a. These values have been converted from kcal mol^{-1} by using 1 kcal mol^{-1} = 4.184 kJ mol^{-1}.
b. Experimental values for benzene–halogen complexes are from reference 1; values for the H_2O dimer are from Pimentel and McClellan, reference 28.
c. Values taken from Lippert, Hanna and Trotter, reference 10, with k = 11.5 eV, c = 1.891 a.u. They assumed axial geometry, and R_{DA} is between the centers of the molecules.
d. Values from Cook and Schug, reference 11. Their *calculated* equilibrium distance between the centres of the molecules is R_{DA}^e = 6.3 Å. They assumed axial geometry. The quoted values of E^{ij} were obtained from calculations using Slater-type functions.
e. TCNE = tetracyanoethylene. The results are from Lippert, Hanna and Trotter, reference 10.
f. Values taken from calculations by Moore and O'Connell (reference 17) for $(H_2O)_2$ I. This configuration is shown in figure 1.2. Here R_{DA} is the O—O distance. The value given here for the "charge-transfer" energy is estimated by taking the difference between the energy of interaction calculated for this configuration by Del Bene and Pople and the "physical" energy ($E^{10} + (E^{20})' + V$, equation 1.21) calculated by Moore and O'Connell.
g. This value is the "physical" energy of equation 1.21, evaluated by Moore and O'Connell.
h. This configuration is shown in figure 1.2. The values here are from the calculations by Moore and O'Connell (reference 17) combined with the results by Del Bene and Pople, made at an O—O distance of 2.8 Å.

given by Cook and Schug[11]. We have replotted their results in figure 1.3 to compare the calculated dependence of E_{DA} on R_{DA} for benzene–Br_2 with that for benzene–Kr; and for benzene–F_2, *vs.* benzene–Ne. We see from these comparisons that the predicted benzene–halogen interaction energy is very similar to that predicted *by the same kind of approximation* for benzene-rare gas interactions. We conclude, in agreement with Cook and Schug[11], that this calculation shows very little charge-transfer (or "extra") stabilization, at least over the range of R_{DA} for which their calculations are valid. However, we have already commented about the fact that their calculated equilibrium distance R_{DA}^e is too large, because of the overestimated repulsion energy, E^{12}.

The comparison in figure 1.4 of E_{DA}^{calc} and E_{DA}^{phys} for the two different water dimers of figure 1.2, shown as a function of R_{DA}, is also instructive. There we see that the calculated[15] E_{DA}^{phys} curve for $(H_2O)_2$ II is parallel to and only slightly different from the corresponding curve for

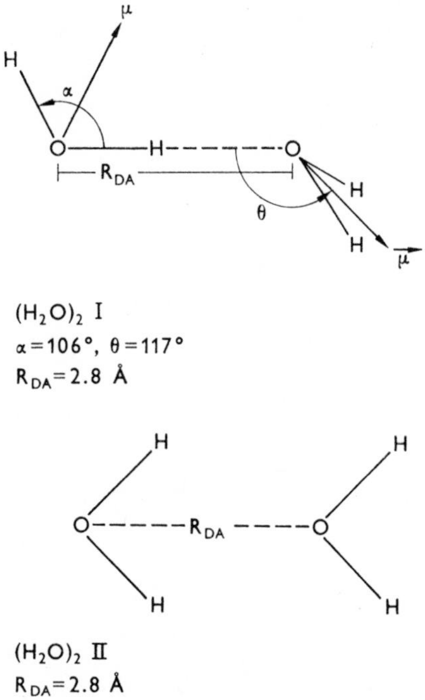

FIG 1.2 Configurations of H_2O dimers for results given in table 1.1 and in figure 1.4. (See Del Bene and Pople, reference 16, for detailed configurations. Our orientations I and II correspond to their configurations 50 and 11, respectively.)

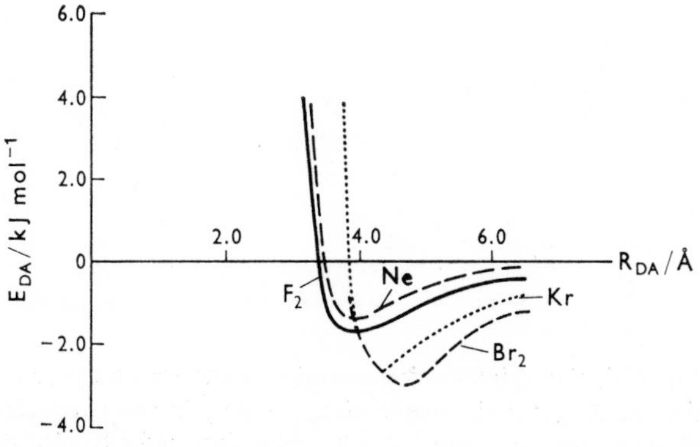

FIG 1.3 Calculated value of E_{DA} as a function of R_{DA} for benzene as the donor (D), and with Br_2, Kr, F_2 and Ne as acceptors (A).

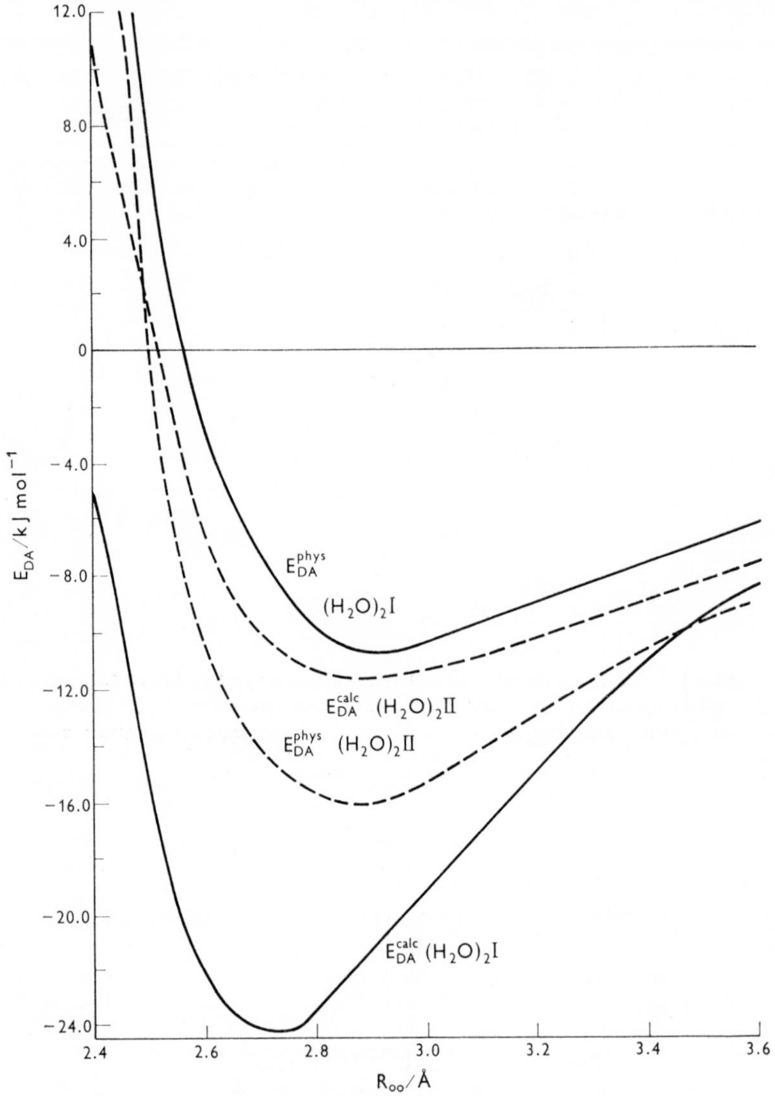

FIG 1.4 Calculated interaction energies (E_{DA}^{calc} and E_{DA}^{phys}) shown as a function of R_{DA} for the water dimer $(H_2O)_2$ in the two orientations (I and II) shown in figure 1.2. The solid lines refer to $(H_2O)_2$ I, and the dashed lines to $(H_2O)_2$ II.

E_{DA}^{calc} from the quantum-mechanical calculation by Del Bene and Pople[16]. On the other hand, the curve for E_{DA}^{calc} for the more stable $(H_2O)_2$ I is increasingly different from E_{DA}^{phys} as R_{DA} decreases, with a calculated minimum energy at shorter R_{DA} distance and considerably deeper than that predicted for "physical" interaction energy alone. Further discussion of the water dimer is given in section 1.3.3.

Finally we should point out that all of these calculations of interaction energies apply strictly only to the D,A pair in a gas-phase interaction. The experimental evidence for "complex" formation is almost all obtained from solution-phase studies. When the D–A complex forms in solution, it may be envisaged as resulting in the removal of two solvent molecules (S) (one from D and one from A), These D,S and A,S interactions are replaced by a D,A (and probably also an S,S) interaction. (See reference 1, section 7.4.) Thus we have

$$(\Delta E_f)_{soln} = (\Delta E_f)_{gas} - \Delta E_{SD} - \Delta E_{SA} + \Delta E_{SS}. \qquad 1.23$$

Here ΔE_f ($= E_{DA}$) is the gas-phase energy of formation of D–A, ΔE_{SD} is the energy of interaction between one solvent molecule and one donor molecule, and ΔE_{SA} and ΔE_{SS} are the corresponding interactions between solvent and acceptor, and solvent with itself. Thus, for D(S) + A(S) → D–A(S):

$$(\Delta E_f)_{soln} \simeq (E_{DA}^{22} - E_{DS}^{22} - E_{AS}^{22} + E_{SS}^{22}) + (E_{DA}^{10} - E_{DS}^{10} - E_{AS}^{10} + E_{SS}^{10}) \simeq E_{DA}^{22}. \qquad 1.24$$

Here we have assumed that $E^{20} + E^{12}$ for each pair cancels approximately, as in table 1.1 (or that the D,A and S,S interactions cancel the D,S and A,S interactions). Also we assume that E^{10} is approximately the same for any pair of interacting molecules, and that E^{22} is large only for a D,A pair that interacts by an electron donor-acceptor interaction and is small for D,S, D,A and S,S interactions.

Because the effect on stability due to the charge-transfer interaction *is* an *extra* effect, we see that it may be important for studies in solution, even though it is only one fairly small term contributing to the overall interaction energy between a molecular pair (see equation 1.4). Furthermore, we shall present arguments later which suggest that the effects on vibrational spectra resulting from complex formation are importantly, if not predominantly, determined by just this small charge-transfer contribution. Thus, even though the *stability* of the isolated complex is importantly (and even perhaps predominantly) determined by the coulomb interaction, it seems likely that the (possibly) small charge-transfer interaction still has important implications for the understanding of many properties of the complex. Such arguments are expected to apply to any electron donor-acceptor complex, including hydrogen-bonded complexes. The relative importance of charge-transfer effects in the latter is currently a subject of considerable debate and is discussed further in section 1.3, and in chapters 2 (section 2.3.4) and 3.

1.1.3. CLASSIFICATION OF DONORS, ACCEPTORS AND COMPLEXES

First of all, it is useful to classify complexes according to the number of molecules of each type. A complex of stoichiometry D_nA_m will be called an n:m complex. Many of the remarks earlier have been made on the assumption that we were discussing only interactions between one molecule of D and one of A, or 1:1 complexes. If the main force stabilizing a complex is charge transfer, then one would expect saturation effects, since the small positive charge acquired by the D molecule in the complex tends to discourage further action as an electron donor to another A molecule in the formation of DA_2. However, if the forces between D and A are electrostatic, there is no special reason that 1:2 complexes (DA_2), or 2:1 complexes (D_2A) for that matter, should not be formed. However, if the second A molecule approaches the D molecule from the side opposite to the first, then dipoles in D induced by the first A are cancelled by the approach of the second A molecule, resulting in an electrostatic energy for DA_2 which is smaller than twice the interaction energy of one (D,A) pair.

We shall restrict most of our discussion here to 1:1 complexes. The concepts can be extended to n:m complexes by (1) considering all the electrostatic interactions in the latter, generally resulting in smaller electrostatic stabilization, and (2) writing the wavefunction as:

$$\Psi_N = \sum_i a_i \Psi_{0i}(D, A) + \sum_i b_i \Psi_{1i}(D^+ - A^-) \cdots \qquad 1.25$$

Here $\Psi_{0i}(D,A)$ is the no-bond wavefunction between the *ith* pair of D,A molecules with weighting coefficient a_i, and $\Psi_{1i}(D^+-A^-)$ is the corresponding dative wavefunction for that *ith* pair of molecules with weighting coefficient b_i. The summations are carried out over all D,A pairs in the n:m complex. Additional resonance structure forms (*e.g.* reverse charge-transfer forms, *etc.*) may also be needed. This resonance-structure formulation of the structure of complicated m:n complexes is cumbersome and may not be especially useful, but it does permit us to extend, with caution, the general ideas developed here for 1:1 complexes.

Classification as Electrostatic or Charge-Transfer Complexes

It is tempting to attempt to classify complexes as "electrostatic" or "charge-transfer" complexes, depending upon which term dominates in its contribution to the interaction energy. As we have seen, it is not easy to make a clear-cut decision as to which term does predominate, and attempts to do so often lead to meaningless arguments when it is not possible to perform the necessary deciding experiments. However, some generalizations can be made.

Clearly, there will be a large electrostatic attractive energy when two oppositely charged ions come together. Clearly, also, there will be a large electrostatic attractive energy when two molecules with large dipole moments come together in a proper orientation. When such interactions occur for a pair of molecules for which the charge-transfer action is expected to be small (for example, between a D,A pair where I_D^v is large and E_A^v is small and where S_{da-}, and hence S_{01}, is also small) it seems quite logical that we should call the D,A pair an "electrostatic complex". However, the difficulty with this classification is that the situation is not always so obvious. Thus, benzene and iodine do not have dipole moments so one might expect little electrostatic stabilization, in contrast with a possibly important charge-transfer stabilization since I_D^v is relatively low and E_A^v relatively high. However, benzene has a sizable *quadrupole* moment, so that a sizable quadrupole-induced dipole interaction can be expected[18] with the highly polarizable I_2 molecule. On the other hand, it is not clear why benzene should not be expected to have almost as strong an electrostatic interaction with the highly polarizable carbon disulphide molecule (or even with the Xe atom). In fact we have estimated, using the methods described by Hanna[18], the electrostatic energy for an apparently non-existent benzene–CS_2 "complex" to be about half as large as the corresponding electrostatic energy of the well-known benzene–I_2 complex. It is just this difficulty in deciding relative magnitudes of the effects that makes it difficult to classify complexes as "electrostatic" or "charge-transfer" complexes.

Drago and Wayland[19a] have assigned empirically two quantities, E_D and C_D, to be characteristic of the ability (measured by E) of D to act electrostatically and the ability (measured by C) of D to form a "covalent" (or charge-transfer) bond. The energy of interaction between two molecules, D and A, is then written:

$$-\Delta H_f = E_A E_D + C_A C_D. \qquad 1.26$$

They used iodine as a reference acid and arbitrarily assigned the value of $E_A = 1.00$ and $C_A = 1.00$. Studying systems of amines with I_2 enabled them to determine E_D and C_D values (when they assumed that E_D was proportional to the dipole amount of the base and C_D to the distortion polarization) for a series of related amines, which would then be studied with different acids, *etc.* In this way they built up a fairly extensive set of parameter values for acids and bases[19a] which has since been extended[19b]. A few values of interest are shown in table 1.2.

These numbers can be used to classify complexes as electrostatic (large E_A and E_D, small C_A and C_D) or covalent (small E_A and E_D, large C_A and C_D). Similarly a given Lewis acid (A) or base (D) can be classified as one which prefers electrostatic or covalent action. Thus, we see from table 1.2 that amines tend to be covalent ($C_D > 5E_D$) while ethers are more electrostatic ($C_D < 5E_D$). Benzene and substituted

TABLE 1.2

Some E and C parameters[a] for selected donor and acceptor molecules. (These values,[b] when substituted into equation 1.26, give ΔH_f in kJ mol^{-1})

Donor	C_D	E_D
pyridine	13.09	2.39
trimethylamine	23.60	1.65
ammonia	7.08	2.78
acetonitrile	2.74	1.81
acetone	4.76	2.02
diethyl ether	6.65	1.97
p-dioxan	4.87	2.23
dimethyl sulphide	15.26	0.70
benzene	1.45	0.99
p-xylene	3.64	0.85
mesitylene	4.48	1.17
Acceptor	C_A	E_A
iodine	2.04[b]	2.04[b]
iodine monochloride	1.70	10.43
phenol	0.90	8.85
t-butyl alcohol	0.61	4.17
boron trifluoride	6.30	16.28
sulphur dioxide	1.65	1.88
tetracyanoethylene[c]	3.09	3.44

a. Values taken from Drago, Vogel and Needham, reference 19b.
b. E and C parameter values are converted to values that give ΔH in kJ mol^{-1} by multiplying the values from reference 19 by $\sqrt{4.184} = 2.045$. The arbitrary values chosen for E_A and C_A of I_2 then become 2.045.
c. Values taken from Drago and Wayland, reference 19a. Less reliable than other values.

benzenes are expected to be more covalent than electrostatic ($C_D < 5E_D$, like ethers) but weaker donors than the ethers or amines, since both C_D and E_D for benzenes are smaller than for ethers or amines. Since the E_A, C_A parameters for I_2 were arbitrarily fixed to be equal, the absolute numbers for a given acid or base are perhaps not as meaningful as are the relative values. We find, however, that the more polar ICl, with $E_A \simeq 5C_A$, should be classified as a more electrostatic acceptor than I_2. Alcohols and other electron acceptor systems active in hydrogen bonding are found to be quite electrostatic ($E_A \simeq 10C_A$). Tetracyanoethylene (TCNE) and iodine may be two of the few covalent acceptors ($E_A \simeq C_A$) according to this scheme.

Clearly this classification has limitations and is arbitrary in its evaluation of parameters. However, the classification of complexes as "electrostatic" or "charge transfer" by the E and C parameters appears to be in reasonable agreement with other attempts to classify them. Furthermore, these parameters appear to correlate well with the more qualitative, but conceptually useful, "hard" and "soft" classification suggested by Pearson.[20]

One of the goals of this chapter is to try to establish other experimental criteria for classification of complexes as "electrostatic" or "charge-transfer". We shall see that this problem is not simple or straight-forward, and one must realize that all such classification attempts are arbitrary, though often the different schemes are reasonably consistent.

Types of Electrostatic Action

We have already mentioned that charge-charge interactions between two ions are the strongest. Other large contributions to E^{10} of equation 1.4 result from dipole-dipole interactions between D and A molecules with permanent dipoles. If neither D nor A have a dipole moment, the first term which may be important in the coulomb contribution is the quadrupole-quadrupole term. If the quadrupole moments of both D and A are zero, higher order terms (octapole, hexadecapole, *etc.*) must be considered. Presumably the relative importance of E^{10} in equation 1.4 diminishes as the order of the terms increases.

In addition to the direct coulomb terms of E^{10}, we must consider also the induction interaction (E^{20} of equation 1.4) between the charge, dipole, quadrupole, *etc.* of one molecule and the induced dipole, *etc.* of the other. As we have mentioned, Hanna[18] stressed the importance of the quadrupole-induced dipole interaction to the stabilisation of benzene–halogen complexes. The calculation by Hanna[18] can serve as a useful model for similar estimates of the importance of terms of this type. In any attempt to evaluate the relative magnitudes of the different types of electrostatic interactions, however, it is important to try to distinguish between those types that occur between *any* molecular pair D,A, and those that occur *only* between a complex D–A pair.

Classification According to Charge-Transfer Action

If charge transfer dominates the properties of a complex, then it makes sense to classify complexes according to the mode of charge-transfer action. Even if charge transfer is less important to the stability of the complex, for example, than electrostatic forces, it is still useful to classify complexes so as to specify the mode of charge-transfer action, since the charge-transfer action may be important in controlling properties other than the stabilisation energy.

Mulliken[5] suggested a detailed classification of donors, acceptors, and complexes according to the type of donor or acceptor action; this

TABLE 1.3

Classification of donors, acceptors and complexes according to charge-transfer action (Reproduced, from Molecular Complexes; a Lecture and Reprint Volume, by permission of J. Wiley and Sons, Inc.)

Donors			Acceptors		
Number of electrons	Functional type	Structure type	Number of electrons	Functional type	Structure type
Odd	Radical	R	Odd	Radical	Q
Even	Increvalent	n	Even	Increvalent	v
Even	Sacrificial	$b\sigma$	Even	Sacrificial	$a\sigma$
Even	Sacrificial	$b\pi$	Even	Sacrificial	$a\pi$

Combinations of donors and acceptors

Donor type	Acceptor type: Q	v	$a\sigma$	$a\pi$
R	Compounds and $(TCNQ^-)_2$	Reaction intermediate	Reaction intermediate	Reaction intermediate
n	Reaction intermediate	$R_3N–BR_3$	$R_3N–I_2$	Often two-way
$b\sigma$	Reaction intermediate	$RX–AlX_3$	Contact	Contact
$b\pi$	Reaction intermediate	Often two-way	$Bz–I_2$	HMB–chloranil

scheme has been simplified and condensed in chapter 4 of reference 1 (p4 and p40), and is summarized in table 1.3. Donors with an even number of electrons are classified as n (for lone-pair), $b\sigma$ or $b\pi$ (for bonding σ and π), depending upon the type of molecular orbital from which the electron is donated; these are further classified as increvalent or sacrificial to emphasize that complex formation increases the total number of bonds, but that it may (or may not) be at the expense of the total net bonding in the donor. Even-electron acceptors are classified as v (for vacant orbital), $a\sigma$ or $a\pi$ (for antibonding σ or π-orbital) according to the type of acceptor orbital involved; they are also further classified as increvalent or sacrificial.

Typical n donors include amines (R_3N, H_3N), pyridine and its derivatives, sulphides, ethers, alcohols, water, *etc.* Typical $b\pi$ donors include benzene and other aromatic molecules, ethylene, *etc.* One expects less stable complexes with $b\pi$ donors than with n donors; the stability of complexes with $b\sigma$ donors is expected to be even less and it is not clear whether or not any examples of the latter occur. Possible $b\sigma$ donors include the aliphatic hydrocarbons, such as n-heptane, in weak interactions with halogens and oxygen characterised by "contact charge-transfer" absorption bands (see reference 1, p220).

Typical *v* acceptors include boron trifluoride, metal ions, *etc.* Typical *aσ* acceptors (which would be expected to form weaker complexes) include I_2, ICl and other halogens, the O—H bond in alcohols (or water) or any X—H bond. Typical *aπ* acceptors (which act with *bπ* donors to produce complexes comparable in stability with complexes between the same donor and an *aσ* acceptor—see section 2.3.5) include benzene or ethylene substituted by electron withdrawing groups, such as trinitrobenzene (TNB), tetracyanobenzene (TCNB), or tetracyanoethylene (TCNE), *etc.*

Complexes formed between donors and acceptors classified on this basis can be designated by two symbols—the first giving the type of donor action, the second the type of acceptor action. Thus a complex between R_3N and I_2 would be called an "*n–aσ*" complex, while the benzene–halogen complexes are "*bπ–aσ*" complexes and benzene-TCNE is a "*bπ–aπ*" complex. Possible combinations are listed in the second part of table 1.3.

Notice that we have not discussed, and do not plan to discuss in this chapter, complexes between odd-electron radical donors (R) and acceptors (Q). Extension of the concepts discussed here for complexes between even-electron donors and acceptors can be made to odd-electron systems if necessary[1].

Although we have seen that the electrostatic energy of interaction between D and A may be larger than the charge-transfer resonance energy of interaction for many of the D,A pairs, we have nevertheless classified the subject in the following chapters according to potential D–A action in terms of charge transfer. If desired, one might emphasize potential electrostatic action by adding a second set of symbols. Hence benzene–I_2 might be classified as a *bπ–aσ Q–Di* complex. Here "Q" is "quadrupole" and "Di" is for "induced–dipole". The water dimer might be called an *n–aσ*, *D–Di* complex, *etc.*

1.1.4. THE PRINCIPLE OF COMPROMISE GEOMETRY AND GROUND STATE PROPERTIES

The principle of compromise geometry states that "the bond lengths and bond angles (the "geometry") of a molecule (or complex) whose wavefunction (*e.g.* Ψ_N) can be written as a sum of two resonance wavefunctions (*e.g.* $a\Psi_0 + b\Psi_1$) are *weighted averages* of the values expected at equilibrium for each individual resonance structure". The "compromise" concept is applied to properties other than "geometry"; for example force constants, dipole moments, *etc.* The principle should not be accepted literally without question since, as pointed out in reference 2, the *formation energy* of a complex is most certainly *not* the weighted average of the energy of the two resonance structures, but is controlled by the off-diagonal matrix element, W_{01}. The value of this concept, however, is that it gives us a quick, intuitive, idea of the direction and magnitude of the correction from "no-bond" properties which is predicted from any charge-transfer action. In making such predictions we must realise, of course, that the properties of the molecular pair in the "no-bond" state are different from the properties of the isolated molecules. The "compromise" is between the properties of the "no-bond" pair and the "dative" pair.

Dipole Moments

It is instructive to consider application of the principle of compromise geometry to the dipole moment ($\vec{\mu}_N$) of the complex. The expectation value for the dipole moment is

$$\langle\vec{\mu}_N\rangle = \int \Psi_N \mu_{op} \Psi_N \, dv$$

$$= a^2 \int \Psi_0 \mu_{op} \Psi_0 \, dv + 2ab \int \Psi_1 \mu_{op} \Psi_0 \, dv + b^2 \int \Psi_1 \mu_{op} \Psi_1 \, dv. \quad 1.27$$

Thus,

$$\vec{\mu}_N = a^2 \vec{\mu}_0 + 2ab \vec{\mu}_{01} + b^2 \vec{\mu}_1.$$

Here $\vec{\mu}_0$ is the dipole moment of the no-bond structure, $\vec{\mu}_1$ is the dipole moment of the dative structure, and $\vec{\mu}_{01}$ is given approximately by $\frac{1}{2}S_{01}\vec{\mu}_1$ (see reference 1, p15, and p68). Thus we have

$$\vec{\mu}_N \simeq a^2 \vec{\mu}_0 + (b^2 + abS_{01})\vec{\mu}_1 \simeq F_{0N}\vec{\mu}_0 + F_{1N}\vec{\mu}_1. \quad 1.28$$

Application of equation 1.28 to experimentally determined dipole moments, using estimates for $\vec{\mu}_0$ that attempt to include the *induced* dipoles in the no-bond structure and estimates for $\vec{\mu}_1$ (usually just $\vec{\mu}_1 = \mathbf{e}\vec{R}_{DA}$) leads, in principle, to an evaluation of F_{1N} from the experimental dipole moments (see reference 1, p68).

We must emphasize here that $\vec{\mu}_0$ is the dipole moment of the no-bond state and so includes the dipole moments *induced in* D and A molecules as they come together to the equilibrium geometry of the complex. If these electrostatic contributions to the observed dipole moment are ignored, $\vec{\mu}_0$ will be underestimated and the magnitude of F_{1N} will be overestimated. The value of $\vec{\mu}_1$ is usually estimated from a simplified model for the electronic distribution in the dative state structure, with a positive charge of $+\mathbf{e}$ at the centre of the D molecule and a negative charge of $-\mathbf{e}$ at the centre of the A molecule, resulting in $\vec{\mu}_1 = \mathbf{e}\vec{R}_{DA}$. This model is grossly oversimplified and can lead to obvious errors in estimating F_{1N}. Nevertheless, the estimates of F_{1N} based on dipole moment values form one of the most reliable sources of such (semi-quantitative) information.

Nuclear Quadrupole Resonance Spectroscopy

The charges on the atoms in D and in A in the D–A complex, predicted by the principle of compromise geometry, can be checked against experimental measurements of the nuclear quadrupole resonance (n.q.r) spectrum of the complex. These measurements are made on solid samples, so the results may not apply to 1 : 1 complexes in solution. Nevertheless, they are of considerable interest.

The resonance frequency in an n.q.r experiment is determined by the field gradient at the nucleus due to the electrons. It is possible to deduce from the frequency the charge density on each atom for which resonance is observed[21]. Such observations for solid benzene–bromine have been interpreted[22] as indicating no charge transfer in that complex; a better statement might be that there is little charge transfer.[23] More illustrative n.q.r spectra for stronger halogen complexes[24] indicate a charge distribution: $D^{+(Q-q)}$ ------ X^{+q}—Y^{-Q}. Here q is less than Q and this interpretation suggests a fairly large electrostatic polarization of the XY halogen bond by the approach of the D molecule, followed by some charge transfer from D^+ to $(X\text{-}Y)^-$. (See sections 1.3.3 and 3.5.3 for more detailed discussion.)

Other Techniques

Other experimental techniques that may be related to the changes deduced from the compromise principle include infrared intensities (but these are complicated by other factors; see section 1.2), nuclear magnetic resonance (n.m.r.) measurements[25], and some recent measurements of the ^{129}I Mössbauer effect in I_2 complexes[26]. Further discussion of these effects is given later, especially in chapters 2 and 3.

We might mention here that geometry predictions from the principle of compromise geometry are also quite useful and reasonable. The decrease in D–A distance and increase in bond lengths in a sacrificial acceptor, for example, with increasing F_{1N}, appear to be in quite reasonable agreement with experimental results from X-ray diffraction studies of the crystals. This subject has been discussed in chapter 5 of reference 1.

Summary of Results Using Different Techniques

Finally, let us summarise the attempts to determine exactly the values of F_{1N} from experimental measurements of ground state properties, such as those mentioned here. As we have indicated above, such efforts can be expected to yield only semi-quantitative results. Even these depend upon our choice of model for the dative structure, and upon our somewhat arbitrary division of charge distribution into that from electrostatic interaction and that from charge transfer. We cannot expect to obtain exact values for F_{1N} with precise meanings. On the other hand, however, it is useful to have such semi-quantitative values available for the purpose of correlating a wide variety of experimental information about complexes. Thus, if we find from dipole moment data that the extent of charge transfer (F_{1N}) for one D–XY complex is 0.1 and for another, D′–XY, F_{1N} is 0.3, then we can predict with some confidence the relative values of other ground state properties of the two complexes. For example, the latter complex will be expected to be (thermodynamically) more stable; its XY bond will be weaker and will have a lower force constant (see sec. 1.2) and longer length. The n.q.r. and n.m.r. spectra will be expected to alter in a characteristic way and the frequencies and intensities of the two charge-transfer bond vibrations will be different. Table 1.4 lists some estimates for F_{1N} for different complexes that have been obtained from studies of several different properties. From this table we see that weak $b\pi$–$a\pi$ complexes are expected to have only a little charge transfer ($F_{1N} \sim 0.01 - 0.1$) while stronger n–$a\sigma$ complexes probably have values of F_{1N} as large as 0.25 – 0.4. Considering all the potential difficulties discussed above in the evaluation of F_{1N}, the values in table 1.4 from different techniques are in remarkable, and possibly partially fortuitous, agreement.

TABLE 1.4

Some estimates of F_{1N} ($= b^2 + abS_{01}$) which have been obtained for some donor-acceptor complexes by different techniques

Complex	Classification	F_{1N} (from $\vec{\mu}$[a])	F_{1N} (from $\Delta\bar{\nu}/\bar{\nu}$)	F_{1N} (other data)
benzene–I_2	$b\pi$–$a\sigma$	0.075[b]	0.02[b]	—
pyridine–I_2	n–$a\sigma$	0.25[b]	0.29[b]	—
triethylamine–I_2	n–$a\sigma$	0.28[b]	(0.4)[b]	—
trimethylamine–I_2	n–$a\sigma$	0.33[b]	0.41[b]	—
hexamethylene-tetramine–I_2	n–$a\sigma$	—	—	0.08[c]
3,5-dibromopyridine–Br_2	n–$a\sigma$	—	—	0.2[d]
3-bromopyridine–Br_2	n–$a\sigma$	—	0.27[e]	—
3,5-dibromopyridine–ICl	n–$a\sigma$	—	—	0.3[d]
pyridine–ICl	n–$a\sigma$	—	0.30[b]	0.26[f]

a. The values of F_{1N} from dipole moment measurements have, for the most part, ignored the induced dipole moments. Hence, these estimates are probably generally too high.
b. Values taken from the summary table 6.2 of Mulliken and Person, reference 1. (See section 1.2 of this chapter for a discussion of the results from the infrared frequency shift, $\Delta\bar{\nu}/\bar{\nu}$.) The value in parentheses is estimated by analogy with the trimethylamine–I_2 complex.
c. Values from Mössbauer studies by Ichiba *et al.*, reference 26.
d. Values from n.q.r. studies by Bowmaker and Hacobian, reference 24a.
e. Values from J. D'Hondt and T. Zeegers-Huyskens reference 45.
f. Values from n.q.r. studies by Fleming and Hanna, reference 24b.

1.2. VIBRATIONAL SPECTRA OF COMPLEXES (THEORETICAL BACKGROUND)

1.2.1. INTRODUCTION

One of the most direct applications of the compromise principle towards the qualitative prediction of experimental ground state properties of complexes is to the study of their vibrational spectra. For example, the stretching frequency of an *aσ* diatomic acceptor molecule, X-Y, is expected to decrease because of the sacrificial nature of its charge-transfer action. This decrease is expected to increase in magnitude as different donors of increasing donor strength are used with the XY molecule. At the same time, we expect changes in the vibrational frequencies of the D molecule corresponding to its charge-transfer action. The corresponding intensity changes are also related to the extent of D–A action. A decrease in frequency of the ν(X-Y) mode and a correspondingly large increase in its band intensity are characteristic of *aσ* acceptor action in complex formation. These spectral changes have received considerable attention ever since the first measurements of the Cl-Cl stretching vibration for chlorine dissolved in benzene[27]. This characteristic *decrease* in frequency and the corresponding *increase* in intensity of the XY stretching band are readily understandable in terms of the charge-transfer resonance-structure theory. They are considerably more difficult to explain quantitatively in terms of electrostatic interactions alone. Such changes are also characteristic of the O—H stretching frequency on hydrogen-bond formation[28], and can be interpreted as evidence for charge-transfer action in those complexes by the O—H bond as an *aσ* acceptor (see section 4.5.3 for a more detailed discussion of the ν_s(OH) band in hydrogen-bonded complexes).

Let us examine qualitatively the changes in vibrational spectrum of D + A as the two molecules are brought together to form the D–A complex. At large distances, the spectrum is just the superposition of those of the two free molecules. As they come to shorter R_{DA} distance, the electrostatic field of one molecule begins to influence the vibrations of the other, and *vice-versa*, causing small changes in frequency and intensity from the spectra of the free molecules. The magnitudes of these changes can be estimated by comparing the spectrum of a molecule in an "inert" solvent with that of the free molecule. Frequency shifts (generally, but not always, to lower frequency) of as much as 1% may be expected[29] when a molecule dissolves in an "inert" solvent, with some increase in intensity[30]. Sample calculations by Wiederkehr and Drickamer[31] indicate how frequency shifts from different electrostatic effects, from exchange repulsion, and from dispersion forces can be estimated[31]. These results are examined in somewhat more detail in the next section. It was the dramatically large frequency shifts and intensity changes associated with X-H stretching vibrations in hydrogen-bonding solvents (and the corresponding phenomena for halogen complexes) exceeding drastically all reasonable estimates of their magnitude from electrostatic considerations[31] (see also Hallam[29]) that led to the application of the charge-transfer model to these interesting systems.

Briefly, the observed spectral changes are attributed to changes in the internal electronic structure of the molecule (hence in force constants and in the charge distribution) which can most easily be described by considering the modification of the wavefunction of the complex Ψ_N from Ψ_0 towards Ψ_1 as the charge-transfer effect becomes stronger. If the charge-transfer action is sacrificial in A (as, for example, for I_2 as an acceptor) then we expect the internal force constants in A to decrease as the interaction becomes stronger. In addition smaller changes in frequency may occur in an internal vibration of D or A because the increasing strength of interaction between D and A results in low-frequency intermolecular D–A vibration which couples with one of the higher frequency internal vibrations of D or A, affecting the frequency of the latter.

Because the D (or A) molecule in the complex may have lower symmetry than it does when free, forbidden modes of vibration may become active due to mixing with other internal modes of D (or A) and small additional frequency shifts may occur due to this coupling. No one has seriously investigated this question, but such effects are probably quite important, for example, in explaining the changes in the spectrum of pyridine when it complexes with metal-containing acceptors. (See section 2.3.3 for further discussion of this point.)

In general, when the D and A molecules with $3n_D - 6$ and $3n_A - 6$ vibrational degrees of freedom (or $3n - 5$ for linear D or A molecules) from a complex with $3(n_D + n_A) - 6$ vibrational degrees of freedom, the result is that 6 (or 5 if D or A is linear) rotations and translations of the free molecular pair are converted to 6 (or 5) new vibrations of the tightly bound complex. These six new vibrations can be described, roughly in order of decreasing frequency, as: one D–A stretching vibration; two vibrations, perpendicular to each other and degenerate if the D–A bond has cylindrical symmetry, of D translating perpendicular to the D–A bond, called a D–A bending vibration; two D–A rocking vibrations perpendicular to each other (and again degenerate if there is cylindrical symmetry); and 1 internal rotation (torsion) of the D molecule about the D–A bond. (This vibration is missing if D or A is linear.)

All of these vibrations are expected to have low frequencies; for weak complexes they are of the order of lattice frequencies in pure solid D or A (*i.e.* 100 cm^{-1} or lower). For stronger interactions, the frequencies rise, so that the ν(D–A) vibration, especially, may appear in the range of 150–300 cm^{-1} or even higher. For such a strong interaction involving a halogen molecule (XY) as the acceptor, the ν(D–A) mode may then be expected to couple strongly with the ν(X-Y) vibration, since the latter may be expected in about the same spectral region (see section 2.3.2).

Only limited information is available even yet about the ν(D–A) mode for weaker complexes, partly because it *is* so difficult to sort out from the lattice motions. (See also section 3.4.3 for further discussion of this mode in "weak" halogen complexes.) Essentially no information is available, so far, about the other 4 or 5 new D–A vibrations. If the D–A interaction is only electrostatic, the new D–A vibrations are all expected to have very low frequencies; some charge-transfer (or covalent) character in the interaction is expected to increase the frequencies of all the vibrations appreciably.

Most of the information currently available is from infrared absorption spectra of the complexes; clearly the Raman spectrum should provide essentially the same information (and probably supplement the infrared spectral information because of the difference in selection rules). Here we should caution the enthusiast to be sure he knows just what species he is studying in solution; *e.g.* is iodine present in the complex, or has it reacted to give I_3^- [and possibly $(D–I)^+$] ions? Photochemical reactions are possible, and iodine solutions, for example, may be especially unstable when exposed to the intense Raman source. (See section 2.2 for a more complete discussion of the experimental difficulties involved.)

1.2.2. FREQUENCY SHIFTS ON COMPLEXATION

Let us now examine the effects leading to frequency shifts in somewhat more detail. The effects to be expected can be summarized as: (1) frequency shifts in D or A vibrations due to coupling of the internal vibrations of D or A with the new D–A stretching vibration; (2) frequency shifts in D or A vibrations due to coupling of these vibrations with other internal vibrations in D or A because of changes in the symmetry when complexation occurs; (3) frequency shifts in D or A vibrations due to structural changes in D or A when complexation occurs because of sacrificial or increvalent donor or acceptor action; and (4) changes in D or A vibrations due to electrostatic (or dispersion or repulsion) effects when the D and A molecules come together in the "no-bond" structure.

We can expect the theory to predict changes in *force constants*. In order to interpret frequency shifts that result from coupling of vibrations, it is necessary to go through a normal coordinate analysis (next section) before anything very quantitative emerges.

We shall examine the first effect in more detail in the next section, where we shall also discuss, briefly, the second effect. In both effects the normal coordinates of D or A are expected to be changed in the complex from those in the free molecules. For these effects it is difficult to discuss, in general terms, the frequency changes expected, since the shifts may be up or down, depending upon which internal D or A vibrations are mixed in the new normal coordinates of the complex.

The third effect is discussed in more detail in section 1.2.4 (see also chapter 6 of reference 1); however, we may summarize some qualitative considerations here. The following discussion refers to the classification given in table 1.3. The most important frequency decreases are to be expected when the donor or acceptor action is sacrificial and when the action is concentrated in one bond. The best examples are the *aσ* halogen (XY) acceptors, where the sacrificial action causes the X-Y bond to weaken and the frequency to shift. Since the X and Y atoms are quite heavy, the ν(X-Y) vibrational frequency is quite low, so that the magnitude of the frequency shift is fairly small, even for a large change in force constant. In measuring the magnitude of these effects, it is best to consider the *relative* frequency shift $\Delta\bar{\nu}/\bar{\nu}$. The relative shifts (see chapter 6 of reference 1) observed for the ν(X-Y) vibrations in halogen complexes are quite similar to those observed for the ν_s(X-H) vibration of hydrogen-bonded species, even though the absolute shifts are quite different.

The charge-transfer action is not expected to result in any very noticeable frequency shifts for vibrations of sacrificial donors or acceptors where the sacrificial action is spread out over the entire molecule, and not concentrated in a single bond. Thus, we do not expect to be able to observe any significant frequency shifts for benzene acting as a *bπ* donor nor for the vibrations of substituted benzenes acting as *aπ* acceptors. We do expect, however, that the internal vibrational frequencies of ethylene, when it acts as a *bπ* donor, may be significantly decreased. The ν(CC) vibration is expected to be most severely affected, since the *bπ* sacrificial action is localized in the CC double bond.

If the increvalent action from an *n* donor is indeed localized action from a true *lone*-pair, then we expect only small changes in the internal vibrations of the *n* donor, except those due to the coupling effects (1) and (2). Similarly, we do not expect significant effects for the vibrational spectrum of a *v* acceptor, except those due to changes in normal coordinates.

Electrostatic, repulsion, and dispersion effects are discussed in section 1.2.5. Here, however, we might examine the estimates made by Wiederkehr and Drickamer[31] for the magnitudes of these contributions to the frequency shift of the CN stretching vibration of acetonitrile (CH_3CN) in different solvents.

The results from their calculation[31] of these various contributions are shown in table 1.5. The value of the ν(CN) band frequency is predicted to decrease somewhat on going from gas to solution (in $CFCl_3$, for example). ($\Delta\bar{\nu}/\bar{\nu}$ is predicted to be 4.0×10^{-3}, compared to an observed relative solvent shift in $CFCl_3$, of 4.2×10^{-3}). This solvent

TABLE 1.5

Contribution to the relative spectral shifts $\Delta\bar{\nu}/\bar{\nu}$ of the ν(CN) vibrational band for CH_3CN dissolved in various solvents (calculated from table IV, Wiederkehr and Drickamer, reference 31)

	$10^3(\bar{\nu}_g - \bar{\nu}_s)/\bar{\nu}_g$[a]				
Solvent	Dispersion	Electrostatic	Induction	Repulsion	Total
$CFCl_3$	+4.93	+0.63	+1.65	−3.24	+3.99
CCl_4	6.12	0.28	2.02	−3.87	4.55
$CHCl_3$	5.33	1.38	1.84	−3.84	4.72
CH_2Cl_2	4.83	2.94	1.81	−4.12	5.47
CH_2Br_2	5.21	2.24	1.92	−4.27	5.11

a. $(\bar{\nu}_g - \bar{\nu}_s)/\bar{\nu}_g$ ($\equiv \Delta\bar{\nu}/\bar{\nu}$). Here $\bar{\nu}_g$ (or $\bar{\nu}_0$) is the wavenumber in the gas phase (2268 cm^{-1}) and $\bar{\nu}_s$ is the estimated wavenumber in solution in one of the "inert" solvents listed here.

shift is the resultant of a predicted *decrease* in $\Delta\bar{\nu}/\bar{\nu}$ due to repulsion ($-3.7 \pm 0.5 \times 10^{-3}$ for all the solvents considered) partially cancelling an *increase* from dispersion ($+5.5 \pm 0.7 \times 10^{-3}$), induction ($+1.8 \pm 0.2 \times 10^{-3}$) and a variable small increase in relative shift from the electrostatic interaction (from $+0.26$ to $+2.9 \times 10^{-3}$, depending on the solvent). Hence, we might conclude from this table that the major contribution to the relative solvent shift is an approximately constant value of $+3.5 \pm 0.9 \times 10^{-3}$ from the repulsion, dispersion, and induction, combined with an electrostatic contribution that varies from 0 to $+3 \times 10^{-3}$ from one solvent to another. If additional effects occurred in some solvent systems, for example due to a charge-transfer interaction between the solvent and the C–N bond acting as an $a\sigma$ acceptor, then we should expect an additional dramatic decrease in frequency. For the more probable action of the CN group as an n donor we should expect little change in ν(CN) frequency unless the coupling action caused it to increase slightly, or unless the electron pair in the n donor action is not strictly "lone", so that the donor action affects the C–N bond strength. (See sections 2.3.3 and 6.3 for more detailed discussion.)

1.2.3. NORMAL COORDINATE TREATMENTS FOR COMPLEXES

Consider now a specific D–A interaction between a donor such as pyridine and an acceptor such as I_2. The internal bonding in pyridine is strong, so that the vibration frequencies of pyridine are all above 375 cm^{-1}. Thus, for a low-frequency D–A stretching vibration near the equivalent of 150 cm^{-1} the atoms in the pyridine molecule move together more or less as a unit with an effective mass for D expected to be approximately equal to the total mass of the pyridine molecule ($M_D \simeq 75$ or 80 amu). The ν(I-I) band arises (Raman spectrum) at 215 cm^{-1} in the vapour, and so may be expected to interact rather strongly with the D–A stretching vibration, since both have the same symmetry and are expected at nearly the same frequencies for the pyridine–I_2 complex.

It is instructive to calculate the frequencies ν_2, of the "D–I stretch", and ν_1, of the "I-I stretch", for this system as a function of increasing value of the force constant k_{DI} ($= k_2$) for stretching the D–I bond. For this calculation we shall assume a "triatomic molecule" model for the complex with mass of D taken to be 75 amu and with k_{12}, the interaction constant joining Δr_{DI} and Δr_{II} in the potential, assumed to be zero.

$$\overset{k_2}{\mathrm{D} \cdots\cdots \mathrm{I}} \overset{k_1}{\cdots\cdots \mathrm{I}}.$$

$$\leftarrow r_{DI} \rightarrow \leftarrow r_{II} \rightarrow$$

This calculation has, first of all been carried out using the assumption that the force constant of the I-I bond in the complex does not change from its hypothetical value for the free molecule ($k_1 = 160$ N m^{-1}). The two stretching wavenumbers ($\bar{\nu}_1$ and $\bar{\nu}_2$) for the complex have been calculated using the formulae given by Herzberg[32]. The results are shown plotted in figure 1.5a. We see there that the result of this weak interaction ($k_2 < k_1$) on $\bar{\nu}_1$, for the "I-I stretch", is simply to increase

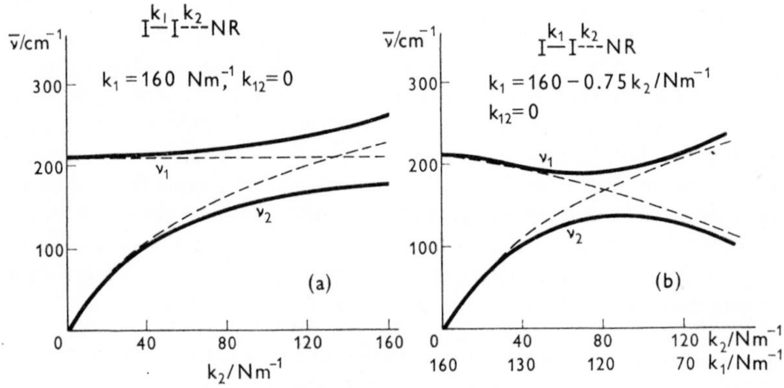

FIG 1.5 Predicted I-I stretching wave number ($\bar{\nu}_1$) and N–I stretching wave number ($\bar{\nu}_2$) for amine–I_2 complexes shown as a function of k_2, the N–I stretching force constant. In (a) the I-I force constant, k_1, is assumed to remain constant at 160 N m^{-1}; in (b) k_1 is assumed to decrease as k_2 increases ($k_1 = 160 - 0.75k_2$ N m^{-1}). For both (a) and (b) k_{12} is assumed to be zero, and the mass of the amine (D) is assumed to be 75 amu.

the I-I stretching frequency as a result of the mixing of the two stretching motions as the value of k_2 approaches k_1 (and hence $\bar{\nu}_2$ approaches $\bar{\nu}_1$). Here $\bar{\nu}_2$, the $\bar{\nu}$(D–I) wavenumber, increases proportionally to $\sqrt{k_2}$ until the two modes mix appreciably. From this calculation it is clear that formation of the D–I bond, without a corresponding decrease in the force constant of the I-I bond, will result in an *increase* in $\bar{\nu}_1$. Thus we conclude that an experimental observation of an appreciable decrease in $\bar{\nu}_1$ as a result of intermolecular interaction means that the I-I force constant has become weaker as a result of that interaction.

In figure 1.5b we show the frequencies calculated for our model system if we assume that k_1 decreases proportionally to the increase in k_2. There we see that the initial decrease in $\bar{\nu}_1$ is almost linearly proportional to k_2, and that the mixing interaction between the two modes occurs at smaller values of k_2 than in the previous case, as a result of this mixing, our description of the low frequency motion ($\bar{\nu}_2$) changes, of course, from a ν(D–I) mode on the left-hand side of figure 1.5b, to a ν(I-I) mode on the right-hand side of figure 1.5b.

The dashed lines in figure 1.5 show the values for $\bar{\nu}_1$ and $\bar{\nu}_2$ if no mixing occurred. When the D–I force constant is small, so that $\bar{\nu}_1$ has the value given by the dashed lines, the relationship between k_1 and $\bar{\nu}_1$ is given to a very good approximation by the "diatomic molecule" expression:

$$\bar{\nu}_1 = \frac{1}{2\pi c}\left(\frac{k_1}{\mu}\right)^{1/2}. \qquad 1.29$$

Here μ is the reduced mass of I_2, and $\bar{\nu}_1$ is given in cm^{-1}. When k_2 becomes large enough so that Δr_{II} and Δr_{DI} interact to give mixing, then the "triatomic approximation" must be used with the formulae from Herzberg[32] to obtain the correct relationship (given by the solid lines in figure 1.5) between $\bar{\nu}_1$, $\bar{\nu}_2$ and k_1, k_2.

For *n–aσ* complexes or for *bπ–aσ* complexes with halogen molecules acting as the *aσ* acceptor, we expect $\bar{\nu}_1$ to be given by the diatomic approximation (equation 1.29) for the weaker complexes (with small k_2) and also for complexes with the lighter halogens (with high $\bar{\nu}_1$). However, for stronger complexes between pyridine or trimethylamine or sulfur-containing donors and I_2 or IBr we expect important mixing between the two modes so that the "triatomic approximation" must be used. (There may be a small error in using the "diatomic" approximation for some ICl or Br_2 complexes.)

So far, we have assumed that the interaction constant k_{12} is zero. We may very well expect this constant to be non-zero; in fact, it may be expected to be quite large and to increase with increasing k_2. Referring to our model (D–I-I), we see that as the D–I bond increases in length for a relatively strong D–I interaction, the I-I bond is expected to become stronger (since there is less occupancy of the *aσ*-orbital) and *vice versa*. This tendency is expected to be reflected in a large positive value for k_{12}.

For symmetric trihalide ions a knowledge of both the symmetric (from the Raman spectrum) and antisymmetric (from the infrared spectrum) stretching frequencies permits the evaluation of k_1 (= k_2) and k_{12}. There k_{12} was found to be 40 N m^{-1}, compared to k_1 values of about 200 N m^{-1}[33].

However, k_{12} cannot be evaluated from the available experimental data on pyridine–I_2 complexes. For this complex there are three unknown force constants (k_1, k_2 and k_{12}) associated with experimental values for, at most, two wave numbers $\bar{\nu}_1$ and $\bar{\nu}_2$. We *may* be able to establish the value of k_{12} by obtaining more data, for example, the values of $\bar{\nu}_1$ and $\bar{\nu}_2$ for complexes with a perdeuterated donor molecule, by assuming that deuteration simply increases the mass of D. Lake and Thompson[34] attempted to do this by using data from I_2 complexes with methylated pyridines assuming that methylation changes only the mass of D and not the value of k_2. Gayles[35] made a serious effort to obtain force constants in this way, using trimethylamine and deuterated trimethylamine as D with I_2, IBr, and Br_2 as *aσ* acceptors. Unfortunately, he found serious coupling of $\bar{\nu}_1$ with other low-frequency modes of trimethylamine, so that D does not vibrate as a unit and even the "triatomic approximation" is apparently not valid for those particular complexes. It is not clear, however, whether his full normal coordinate treatment[35] applied to these data converged to the best set of force constants.

Recently Brownson and Yarwood[36] have obtained values of k_{12} for pyridine–IX complexes by studying the spectra in benzene solutions of the pyridine–IBr complex compared with the spectrum of pyridine-d_5–IBr under similar conditions. From this study they conclude that k_{12} is either +10 or +60 N m^{-1}. However, further arguments are needed to decide which choice is correct. The choice preferred by Brownson and Yarwood[36] (k_{12} = 10 N m^{-1}) is not definitive since experimental errors in the frequency data result in a considerable range of the values for the intersection points of the ellipses (hence the values of k_{12}, k_1 and k_2).

The ellipses mentioned above are the figures that arise when all possible values of k_1 and k_2 consistent with $\bar{\nu}_1$ and $\bar{\nu}_2$ for a D–I–I complex are plotted against k_{12}. Two ellipses result—one for k_1 and one for k_2—for each D molecule. The two ellipses for k_1 for pyridine and for pyridine-d_5 complexes are tilted slightly with respect to each other and intersect, hopefully at only one point, to give the value of k_1 and k_{12} for this complex (see section 2.3.2 and figure 2.7). The family of values of k_1 and k_2 are shown as a function of k_{12} for the pyridine–ICl complex[37] (in benzene solution) in figure 1.6. At each value of k_{12} the values of k_1 and k_2 that are consistent with k_{12} and with the frequency data are given by the ellipses. Thus, the possible values of k_{12} range from −10 to +120 N m^{-1}. For each value of k_{12} two values of k_1 and two values

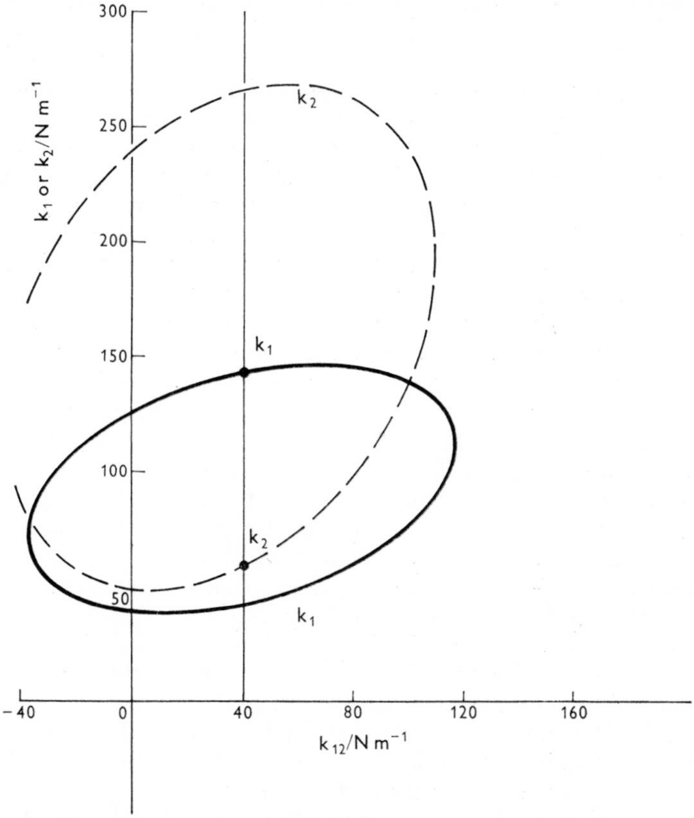

FIG 1.6 Force constant "ellipses" for the force constants k_1, k_2 and k_{12} of pyridine–ICl complexes in solution, treated as a triatomic D–ICl molecule. If $k_{12} = 40$ N m^{-1}, for example, the values for k_1 ($= k_{ICl}$) and k_2 ($= k_{NI}$) are marked as dots (●) at the intersection of the ellipses with the line for $k_{12} = 40$ N m^{-1}. (Data from Yarwood and Person, reference 37.)

of k_2 are obtained from the ellipses at that value of k_{12}. Physical arguments dictate the choice of the lower of the two values for k_2 and the higher one for k_1. However, without additional data to fix k_{12} we cannot tell which is the correct set of values for k_1, k_2 and k_{12} from the infinite choice shown in figure 1.6. Some model for the force field of the complex may help us chose which values of k_1, k_2 or k_{12} are "reasonable".

Person and Crawford[38] have suggested an alternative method of visualizing all possible force constants, normal coordinates, dipole moment derivatives, *etc*, for a D–I-I complex that would appear to be especially useful for problems of this type.

In our discussion so far we have been stressing a perturbation treatment (in the "diatomic" and "triatomic" approximations to the complex) of the normal coordinate problem to obtain force constants from frequency shifts and new frequencies of the complex. Thus, in the "triatomic" approximation for the pyridine–I_2 complex we assume that the high-frequency internal vibrations of D are essentially unchanged, and that the I-I stretching vibration and N–I stretching vibration interact and so are to be analyzed like a linear triatomic molecule in order to get force constants. We assume that the other new D–A vibrations are at very low frequencies, and so can be ignored. Since we know from experimental observation (chapters 2 and 3) that (small) frequency shifts *do* occur in internal vibrations of the D molecule, we know that these approximate treatments are not without error.

One may attempt to carry out a complete normal coordinate analysis on the complex (for example, see reference 35). When one attempts to do so, one finds that the usual problems of defining a force field for a molecule[39] are multiplied for the complex. At the very least, all the frequencies of the complex must be known, together with enough data from isotopically substituted complexes to fix the force constants. Because of the D–A interaction, the number of force constants needed to define the general valence force field is considerably greater than for the two isolated D and A molecules. Because of the D–A interaction, also, we cannot be sure that the internal force constants of D or of A remain the same as for the isolated molecules, even for the rare systems for which they are known for the molecules. Hence, it does not seem profitable, at present, to do a full normal coordinate treatment of a complex, except in very special cases. (See sections 2.3.2 and 3.4.4 for some preliminary results for the pyridine–ICl complex.)

On the other hand, it seems possible that a proper perturbation treatment of the vibrations of D in the complex can be made, utilizing the relationships given for example by Overend and Miyazawa[40], with the known normal coordinates and force constants of a D molecule. Thus, we can assume, for example, that the changes in the donor spectrum on complexing are not due to changes in the internal force constants but, rather, only to the change in geometry and to the D–A interaction force constants. The perturbation theory[40] can then be used to estimate change in vibrational frequencies of the internal vibrations of D, as a function of the unknown interaction constant, in order to determine whether or not the observed changes are consistent with "reasonable" values of the latter. So far no one, to our knowledge, has tried this apparently promising approach.

1.2.4. FORCE CONSTANT CHANGES DUE TO CHARGE TRANSFER

We have discussed in some detail in the preceding sections the qualitative changes expected in force constants as a result of charge transfer. A detailed discussion of the effects has been given in chapter 6 of reference 1. Here we shall try to summarize the main points of the latter. Let us again emphasize here that the theory can attempt some predictions for *force constants*, so that the observed *frequencies* must be analyzed to remove, for example, the effects of mixing with the internal modes that result in a different frequency behaviour on complexing from what is found for the force constants. Thus, it is necessary to perform some kind of normal coordinate analysis (either a "perturbation" treatment or a more complete treatment) to obtain the force constants, which may then be compared with theoretical predictions.

Force Constants Changes for $a\sigma$ Acceptors

As discussed previously, the largest changes in force constants of either D or A molecules are expected to be found for diatomic $a\sigma$ acceptors, such as I_2, ICl, Br_2, *etc.* (or for the OH bond in ROH hydrogen-bonding acceptors, for example). Friedrich and Person[13] attempted to derive a relationship between the relative change in force constant, $\Delta k_1/k_1^0$, and the weight of the dative structure, F_{1N}. This derivation is criticized in chapter 6 of reference 1. We may summarize by saying that the principle of compromise geometry strongly implies at least a semi-quantitative relationship of this sort. The simplest and most realistic estimate of the nature of this relationship is probably the following:

$$\Delta k_1/k_1^0 = (k_{XY}^0 - k_{XY})/k_{XY}^0 \simeq F_{1N}. \qquad 1.30$$

Here k_{XY}^0 is the stretching force constant for the X-Y bond in the "free" (*i.e.* solvated) A molecule, and k_{XY} is the corresponding force constant (k_1) for the A molecule in the complex. Similar expressions have been given also for the change in the O-H stretching force constant when hydrogen-bonded complexes form[41].

Equation 1.30 has been used to estimate F_{1N} to give the results labeled "from $\Delta\bar{\nu}/\bar{\nu}$" in table 1.4. The results are reasonably consistent with results from other studies, and so equation 1.30 seems to be semi-quantitatively correct. Part of the difficulty with its use appears to be in the choice, in practice, of the value for k_{XY}^0. Strictly speaking this reference force constant is the force constant for the XY stretch in the "no-bond" structure Ψ_0. Thus we should use the force constant for the "free" (gas-phase) molecule, corrected for all "solvent" effects (see section 1.2.2) except charge transfer. Possibly these other "solvent" effects, and particularly the effect of electrostatic action, have been underestimated in the past so that the k_{XY}^0 value chosen has been too large, thus overestimating F_{1N} from equation 1.30.

Note that if the perturbation by the complex is small so that the "diatomic approximation" of equation 1.29 is valid, then that equation can be differentiated to give:

$$2(\bar{\nu}_0 - \bar{\nu})/\bar{\nu}_0 = 2\Delta\bar{\nu}/\bar{\nu} = \Delta k_1/k_1^0 \qquad 1.31$$

This approximate equation is expected to be valid for small frequency shifts. Here $\Delta k_1/k_1^0$ is the same as that defined in equation 1.30, $\bar{\nu}_0$ is the reference wavenumber for the X-Y vibrational band in the "no-bond" structure, and $\bar{\nu}$ is the observed wavenumber of this band in the complex.

Finally, we should repeat that very little can be predicted for other force constants of *aσ* acceptors. In general, we expect *aσ* action to be localized in one bond. Thus, in ICN, for example, we expect the *aσ* action to be localized in the IC bond. The spectra of ICN in complexes change most significantly in the I-C stretching region, although some small changes do occur for both the CN stretching and ICN bending vibrations[42]. It seems likely that all the frequency changes may be due to a decrease in k_{IC} and an increase in k_{DI} force constants, with no important change in the other force constants of ICN. In general, we predict that the force constants for an *aσ* acceptor molecule in the complex are the same as for the free molecule except that a lower value is predicted for the bond involved in the *aσ* action. A bending vibration of an XYZ molecule, where XY is the *aσ* acceptor and where X is lighter than iodine, may be expected to show some change in frequency and the XYZ bending force constant may be expected to increase slightly. Internal vibrations within a polyatomic Z group would not be greatly affected by formation of the complex D–X-Y-Z.

Prediction of the D–X Stretching Force Constant, k_2

The other force constant about which something definite can be said is the intermolecular D–X stretching force constant, k_2. By reasoning similar to that leading to equation 1.30, we write[43],

$$k_2/k_{DX}^0 \equiv k_2/k_2^0 = F_{1N}. \qquad 1.32$$

Here k_{DX}^0 ($= k_2^0$) is a reference force constant for the D-I bond when it is fully formed. It seems that this reference force constant should be the force constant for the D-X bond in the dative $D^+ \cdots (X\text{-}Y\text{-}Z)^-$ structure; presumably this force constant is the same as that in a $(D\text{–}X)^+$ structure. For the *n–aσ* interaction between pyridine and ICl[37] and for pyridines with IBr[43], the N–I reference force constant was estimated approximately to be $k_{NI}^0 \simeq 250$ N m^{-1}. A similar application to the NBr stretching force constant for pyridine complexes with Br_2[44] estimates k_{NBr}^0 to be 150 N m^{-1}.

It would seem possible to estimate values of k_2^0 for other interacting species, and thus predict values of $\bar{\nu}_2$ for the D–X stretching vibration. For example, a reasonable estimate for k_{DI}^0 for the benzene–I_2 complex is $k_{DI}^0 \simeq 250$ N m^{-1}, the same as for the N–I interaction. If so, then equation 1.32 predicts $k_2 \simeq 0.02 \times 250 \simeq 5$ N m^{-1}. From this force constant a value of $\bar{\nu}_2$ can be estimated using the "diatomic model" (equation 1.29). If μ is 48 amu (for a benzene–I_2 "diatomic complex"), we estimate $\bar{\nu}_{DI}$ to be about 40 cm^{-1}. We may expect contributions from electrostatic forces to k_2 to be just as large or larger in a complex such as benzene–I_2 with so little charge-transfer character.

Similar estimates for the D–A stretching frequency in, for example, the benzene–TCNE, $b\pi$–$a\pi$ complex are expected to give similar estimates for k_2 and $\bar{\nu}_{DA}$. It would be of interest to extend this line of reasoning to other molecular complexes, including hydrogen-bonded systems.

Another possible use for the relationship suggested in equation 1.32 is to check the two force constants (k_1 and k_2) from a study of amine–IY complexes for consistency. Combining equations 1.30 and 1.32, we find,

$$\Delta k_1/k_1^0 = k_2/k_2^0. \tag{1.33}$$

Hence, for the pyridine–ICl complexes, with $k_1^0 = 227$ N m^{-1} and $k_2^0 = 250$ N m^{-1}, equation 1.33 predicts that $\Delta k_1/k_2$ must be 0.908. From the treatment of Brownson and Yarwood[36] k_2 is either 62 or 80 N m^{-1}; hence k_1 (from equation 1.33) must be either 165 or 147 N m^{-1}, respectively. Neither value fits exactly with the possible values of k_1 consistent with their data, since their values of k_1 corresponding to $k_2 = 62$ or 80 N m^{-1}, are 136 or 152 N m^{-1}, respectively. However, the set $k_1 = 152$, $k_2 = 80$ and $k_{12} = 60$ N m^{-1} seems more nearly consistent with equation 1.33 and with these values of k_1^0 and k_2^0.

Arguments About k_{12}

The value of the interaction constant k_{12} for D–XY complexes, where XY is an $a\sigma$ acceptor, is of considerable importance, as has been discussed in section 1.2.3. Because it was found to have a value of 40 N m^{-1} for the symmetric trihalide complexes where it can be evaluated,[33] this value (40 N m^{-1}) has been assumed to be the most likely value for that constant ever since. As discussed in the proceeding section, k_{12} *is* expected to be relatively large and positive. However, it is not clear what "relatively large" means. In the symmetric trihalides, k_{12} values actually ranged from 26 to 36 N m^{-1}; these values are to be compared to k_1 ($= k_2$) values in these ions which range from 90 to 130 N m^{-1}. Hence one might deduce that "relatively large" values for k_{12} might mean as large as one-fourth to one-third of the values of k_1 and k_2.

But suppose the values for k_1 and k_2 from the pyridine–ICl complex discussed above are approximately correct, so that k_1 is approximately 152 and k_2 is approximately 80 N m^{-1}. It would seem likely that k_{12} should not be expected to be greater than 20–25 N m^{-1} (or possibly k_{12} may be as high as 36 N m^{-1}, one-third the geometric mean of k_1 and k_2). It seems reasonable that k_{12} might be expected to increase approximately linearly with k_2, (assuming $k_1 > k_2$) reaching the "magic" value of $k_{12} = 40$ (or 60) N m^{-1} only when k_2 becomes as large as k_1. It is thus hard to justify such a large value of k_{12} as the value (60 N m^{-1}) suggested in the previous section.

The difficulty with this line of reasoning is that the values of k_1 associated with smaller values of k_{12} are smaller than seem reasonable. (See Brownson and Yarwood[36], Yarwood and Person[37] and table 1.4 of section 1.1.4.) The small values of k_1 imply larger values of F_{1N} by equation 1.30 than seems reasonable from other methods. It is possible that the difficulty here comes from the neglect of the effect on the band positions $\bar{\nu}_1$ and $\bar{\nu}_2$ from the interaction between Δr_{DI} and Δr_{IY} with the low-frequency internal modes of the pyridine molecule (compare with Gayles[35]). Another possibility is that the values of k_1^0 and k_2^0 used with equation 1.33 in the argument given in the preceding section are somewhat in error.

Other Force Constants

Much less can be said about changes in other force constants (*i.e.* besides k_1 in localized *aσ* acceptor bonds, k_2, the D–A force constant, and k_{12}, the interaction constant between these two bonds). We may summarize the qualitative expectations from the charge-transfer resonance structure theory in table 1.6.

The only prediction there that deserves further comment is the prediction about the decrease expected in the C═C stretching force constants for *bπ* donors or *aπ* acceptors. If several C═C bonds are involved in the donor or acceptor action, then the weakening effect is spread over all these bonds and any change in k_{CC} will be very difficult (if not impossible) to detect from frequency shifts in *D* (or *A*). On the other hand *bπ* action by ethylene as an electron donor in metal-ion (*v* acceptor) complexes is expected to cause the CC stretching force constant to decrease significantly. Any back-donation from the metal ion (as an *n* donor) to the ethylene as an *aπ* acceptor (see reference 1), is also expected to decrease the CC force constant. However, analysis of the observed frequencies of ethylene complexes is expected to be complicated by considerable mixing of the CC stretching coordinate with other internal coordinates in ethylene as well as with the new D–C vibrations, making it very difficult to obtain reliable information about even the qualitative prediction in table 1.6.

Other force constants associated with the new D–A vibrational degrees of freedom of the complex are predicted qualitatively to increase as k_2 increases. However, so little is known about these vibrations that it seems useless to try to make further prediction about them here.

Finally, let us emphasize that all of the discussion in this section has ignored the effect of electrostatic interactions. For "strong" complexes (F_{1N} large) the predictions in this section are expected to be in fairly good agreement with experiment unless the interaction is so strong that the additional coupling between the motions considered here and the internal motions of D or A must be considered. In that event a *full* normal coordinate analysis must be completed to obtain the force constants to be compared with the predictions made here for k_1, k_2

TABLE 1.6

Predicted changes in force constants for complexes, based on charge-transfer resonance-structure theory

<table>
<tr><th>Donor type[a]</th><th colspan="2">Comments[b]</th><th>Acceptor type[a]</th></tr>
<tr><td>n</td><td colspan="2">All force constants essentially unchanged from free D or A but important new D–A vibrations expected.</td><td>v</td></tr>
<tr><td>$b\sigma$</td><td>Too weak to be important.</td><td>Extensive changes. See previous discussion in this section.</td><td>$a\sigma$</td></tr>
<tr><td>$b\pi$</td><td colspan="2">This action is not localized in D or in A, so changes are very small. All C=C force constants in D or A are expected to be smaller.</td><td>$a\pi$</td></tr>
</table>

a. Classification from table 1.3.

b. General comments about expected changes in force constants. These comments apply when the intermolecular action between D and A can be classified strictly according to table 1.3. If, for example, a BX_3 molecule acts as an acceptor into an orbital that is not purely *v* (but has some bonding or antibonding character), then its force constants may be expected to change somewhat.

and k_{12}. For "weak" complexes electrostatic contributions to the force constants of the complexes are expected to be just as important as the charge-transfer contributions discussed in this section. The electrostatic contributions are discussed in the next section.

1.2.5. FORCE CONSTANT CHANGES DUE TO ELECTROSTATIC (COULOMBIC, REPULSIVE, DISPERSIVE, *ETC.*) INTERACTIONS

Because electrostatic interactions are expected to have an effect on the frequencies and force constants of molecules which is, qualitatively at least, similar to the effects caused by charge transfer, there has been an unnecessarily great confusion in the literature about the relative importance of these effects. It is essential, therefore, to review carefully the expected electrostatic effects and especially the magnitudes of the effects which can be expected. In this discussion we wish to consider not only purely electrostatic effects (coulombic forces), but also effects due to induction, dispersion and repulsion.

We have already given a preliminary discussion of these effects in section 1.2.2, in reviewing the frequency shifts calculated by Wiederkehr and Drickamer[31] (see table 1.5). Interest in the possibility that frequency shifts and intensity changes observed in the spectra of complexes could be explained simply by electrostatic interactions has recently been revived by a study of Hanna and Williams[45]. Let us now examine the treatment by Hanna and Williams [45] as applied to a benzene–I_2 complex in which we assume that the I_2 molecule is above the centre of the benzene ring, with its axis along the six-fold symmetry axis of benzene. For this complex there are 30 normal coordinates for the benzene molecule, q_i, 5 D–A normal coordinates (see section 1.2.1), q_α, and one I-I coordinate, Q. The potential energy for the complex can then be expressed in terms of these 36 normal coordinates as:

$$V(q_i, q_\alpha, Q) = V_I(q_i) + V_{II}(Q) + V_{I,II}(q_i, q_\alpha, Q). \qquad 1.34$$

Here $V_I(q_i)$ is the potential energy function for benzene alone, $V_{II}(Q)$ is the potential energy function for iodine alone, and $V_{I,II}$ is the intermolecular potential energy.

Hanna and Williams[45] did not consider the V_I terms but confined their attention only to the dependence on Q, and they assumed that $V_{I,II}$ was simply the electrostatic energy (V_{el}) of interaction between benzene and I_2, which they wrote as

$$V_{el} = -(\tfrac{1}{4})\alpha_{\|}(E_1^2 + E_2^2) - (\tfrac{1}{2})\theta_A E'_A. \qquad 1.35$$

Here $\alpha_{\|}$ is the polarizability of the iodine molecule parallel to its (z) axis, E_1 is the electric field strength (due to benzene) at the iodine atom closest to the benzene and E_2 is the corresponding electric field strength in the z direction at the other iodine atom. Finally, θ_A is the molecular quadrupole moment of the halogen and E'_A is the electric field gradient at the centre of the halogen molecule. Hanna and Williams[45] have plotted their calculated values of this electric field strength as a function of z and have showed that E_1 is expected to be about three or four times larger than E_2 for the benzene–I_2 complex. This expression for the electrostatic energy of interaction (equation 1.35) was suggested earlier by Hanna[18].

Continuing with the treatment from equation 1.34, we can write the potential function in a power series about any one of the normal coordinates, q_j:

$$\begin{aligned} V(q_i, q_\alpha, Q; q_j) &= V(q_i, q_\alpha, Q; 0) + (\partial V/\partial q_j)q_j + (\tfrac{1}{2})(\partial^2 V/\partial q_j^2)q_j^2 + \cdots \\ &= V_0 + (\partial V/\partial q_j)q_j + (\tfrac{1}{2})(\partial^2 V/\partial q_j^2)q_j^2 + \cdots. \end{aligned} \qquad 1.36$$

If q_j is the normal coordinate, Q, for the halogen stretching vibration, $V_I(q_i)$ is independent of Q so that:

$$\begin{aligned} V(q_i, q_\alpha, Q) &= V_0 + (\partial V_{I,II}/\partial Q)Q + (\tfrac{1}{2})[\partial^2 V_{II}/\partial Q^2 + \partial^2 V_{I,II}/\partial Q^2]Q^2 \cdots \\ &= V_0 + \tfrac{1}{2}k_0 Q^2 + aQ + bQ^2. \end{aligned} \qquad 1.37$$

Here the first term can be set to zero by the choice of energy zero. The interaction energy between the two molecules is not necessarily at a minimum with respect to the internal normal coordinates, so there is a non-zero contribution from $(\partial V_{I,II}/\partial Q)$ (= a). Hanna and Williams[45] then evaluated the difference between the energy levels for V_{II} represented by a Morse oscillator function and those for V(Q) represented by a Morse oscillator perturbed by the intermolecular interaction (equation 1.37). In this way they found that the linear term in Q from equation 1.37 caused a frequency shift for the $1 \leftarrow 0$ vibrational transition, and that there was an additional change predicted for this energy difference due to the change in the force constant from the $\partial^2 V_{I,II}/\partial Q^2$ term in equation 1.37.

It seems more straightforward just to utilize ordinary perturbation theory (for example, from Levine[46]) to determine the effect of the perturbation $(aQ + bQ^2)$ on the harmonic $(V_{II} = \frac{1}{2}k_0Q^2)$ energy levels of I_2. The energy levels, correct to second order perturbation theory, are given by[46]

$$E_v/\mathbf{hc} = \bar{\nu}_e(v + \tfrac{1}{2}) + (b/\alpha\mathbf{hc})(v + \tfrac{1}{2}) - a^2/2\alpha(\mathbf{hc})^2\bar{\nu}_e. \qquad 1.38$$

Here $\bar{\nu}_e$ is the harmonic wavenumber of I_2, α is a dimensionality constant $(\alpha = 4\pi^2\mathbf{c}\bar{\nu}_e\mu/\mathbf{h})$, μ is the reduced mass of I_2, v is a vibrational quantum number, and a and b are as defined in equation 1.37. Thus, we see that if the iodine vibrations are *harmonic*, then the linear term (aQ) contributes only a constant shift $(a^2/2\alpha(\mathbf{hc})^2\nu_e$—the same for all energy levels. If V_{II} is anharmonic, additional terms in the Taylor series expansion (equations 1.36, 1.37) may be included to give for the energy[46]

$$E_v/\mathbf{hc} = \bar{\nu}_e(v + \tfrac{1}{2}) - \bar{\nu}_e x_e(v + \tfrac{1}{2})^2 + (b/\alpha\mathbf{hc})(v + \tfrac{1}{2}) - [aU'''/2\alpha^2(\mathbf{hc})^2\bar{\nu}_e](v + \tfrac{1}{2}). \qquad 1.39$$

Here $\nu_e x_e$ is the same anharmonicity correction to ν_e as for the gas phase, U''' is the cubic anharmonicity term (*i.e.* $U''' = \partial^3 V/\partial Q^3$). Equation 1.39 gives an effective wavenumber $\bar{\nu}_{1\leftarrow 0}$ for the $1 \leftarrow 0$ vibrational transition of the complex $(E_1 - E_0)/\mathbf{hc}$:

$$\bar{\nu}_{1\leftarrow 0} = (\bar{\nu}_e - 2\bar{\nu}_e x_e) + b/\alpha\mathbf{hc} - aU'''/2\alpha^2(\mathbf{hc})^2\bar{\nu}_e. \qquad 1.40a$$

(Equation 1.39 omits higher order terms in b, a, U''', *etc.*) On substituting the definition of α:

$$\bar{\nu}_{1\leftarrow 0} = (\bar{\nu}_e - 2\bar{\nu}_e x_e) + b/4\pi^2\bar{\nu}_e\mu\mathbf{c}^2 - aU'''/2(4\pi^2\bar{\nu}_e\mu)^2\bar{\nu}_e\mathbf{c}^4. \qquad 1.40b$$

Writing

$$(\bar{\nu}_e - 2\bar{\nu}_e x_e) \approx \bar{\nu}_0 = \frac{1}{2\pi\mathbf{c}}\sqrt{k^0/\mu} \quad \text{and} \quad \bar{\nu}_{1\leftarrow 0} = \frac{1}{2\pi\mathbf{c}}\sqrt{k/\mu},$$

and assuming that $\bar{\nu}_e x_e$ is much smaller than $\bar{\nu}_e$, we find:

$$(k)^{1/2} = (k^0)^{1/2} + b/(k^0)^{1/2} - aU'''/2(k^0)^{3/2}. \qquad 1.41$$

Here $\bar{\nu}_0$ is the frequency of the transition for a gas-phase molecule, as contrasted with $\bar{\nu}_{1\leftarrow 0}$, for the transition in the complex.

Squaring both sides of equation 1.41 we have

$$k = k^0 + 2b - aU'''/k^0 + b^2/k^0 - abU'''/(k^0)^2 + a^2(U''')^2/4(k^0)^3.$$

Thus

$$\Delta k = k^0 - k = -2b + (aU''' - b^2)/k^0 + abU'''/(k^0)^2 - a^2(U''')^2/4(k^0)^3. \quad 1.42$$

Since the perturbation terms a and b are small compared to k^0, we may neglect the terms in $1/(k^0)^2$ and $1/(k^0)^3$ to obtain

$$\Delta k = k^0 - k = -2b + (aU''' - b^2)/k^0. \quad 1.43$$

Specific Application of the Hanna-Williams Model

The anharmonicity constant U''' for I_2 may be obtained from the vibration-rotation constant α_e for I_2 utilizing[46]:

$$\alpha_e = -(2B_e^2/\nu_e)[2B_e r_e^3/\mathbf{h}\nu_e^2 U''' + 3],$$

where B_e, D_e and ν_e are in s^{-1}(Hz). The centrifugal distortion constant De is

$$D_e = 4B_e^3/\nu_e^2.$$

(α_e should not be confused with α in the preceding section.) From the values given in Levine for I_2[46], we calculate:

$$U''' = -0.798 \times 10^{13}\ \text{J m}^{-3}.$$

Referring now to the model used by Hanna and Williams[45] for $V_{I,II}$ (equation 1.39) and using their values for $a = \partial V_{el}/\partial Q = -V_E$ (in their notation) $= -15.5 \times 10^{-11}$ J m^{-1} and $b = (\frac{1}{2})\ \partial^2 V_{el}/\partial Q^2 = (\frac{1}{2})\ \Delta k$ (in their notation) $= (\frac{1}{2})(-2.9)$ N m^{-1} we find, for benzene–I_2, from equation 1.43 that, $k^0 - k = 2.7 + 7.19 - 0.01 = 9.88$ N m^{-1}. Evaluating k for the benzene–I_2 complex with $\bar{\nu}_0 = 205$ cm^{-1}, k = 157 N m^{-1} and hence $\Delta k_{exp} = 15$ N m^{-1}. Or if k^0 is evaluated from $\bar{\nu}_e - 2\bar{\nu}_e x_e$ instead of from $\bar{\nu}_e$, it is 170 instead of 172 N m^{-1}, so that Δk_{exp} is 13 N m^{-1}. Hence this electrostatic model does account for more than two-thirds (9.88 compared to 13 or 15 N m^{-1}) of the change in force constant deduced from the observed (small) frequency shift of the ν(I-I) band for the benzene–I_2 complex.

Note that the derivation given above makes it quite clear that the reference wavenumber (to establish k^0) should be the gas-phase frequency, $\bar{\nu}_e - 2\bar{\nu}_e x_e$ (or approximately $\bar{\nu}_0$), in contrast with the discussion of this problem (k^0) by Hanna and Williams[45]. Furthermore, note that equations 1.39 and 1.42 are exact (to second order in the perturbation theory) and will give an exact prediction of force constant changes to be expected for I-I stretching force constant (on the assumption that the unperturbed I_2 potential V_{II} of equation 1.34 is the same for I_2 in the complex as it is in the gas phase) *if* we use an exact expression for the intermolecular perturbation potential, $V_{I,II}$. Hanna and Williams used a specific model for $V_{I,II}$ that included *only* electrostatic terms[45], so the results above cannot be a complete prediction. Finally, note that the results from the perturbation treatment here are the same as the results from the perturbed Morse oscillator treated by Hanna and Williams[45].

Comparison With "Solvent Effect" Theory

Before we rejoice too much at the apparent success of a "simple" electrostatic model in predicting the frequency shifts (or force constant changes) for the benzene–I_2 complex, let us examine just what we have done. Although it was not so apparent in the treatment given by Hanna and Williams,[45] the perturbation treatment given here, leading to equations 1.39 and 1.42, is readily seen to be identical to the treatment of "solvent effects" given by Buckingham[30d]. Our equation 1.39 is identical with Buckingham's equation 3.14[30d] and, in fact, with Wiederkehr and Dickamer's equation 1[31]. The only difference is that we have assumed here in applying equations 1.39 and 1.42 that $V_{I,II}$ is for a specific 1:1 interaction between benzene and iodine, and that it has the form $V_{I,II} = V_{el}$ of equation 1.35. In the general solvent theories[30,31] the interaction with solvent is thought to be more random so that the interaction potential of equation 1.35 is averaged over all orientations of D and A, and is averaged over all D molecules surrounding A.

Thus equation 1.43 may be used with equation 1.35, as in the preceding section, to predict the change in the I-I stretching force constant, on going from the gas phase to the 1:1 complex of benzene–I_2, in a particular configuration, assuming that there is no additional (for example, charge-transfer) effect to change k from gas phase to the complex in solution. However, a 1:1 interaction between I_2 and *any* solvent molecule may cause an effect approximately the same as that calculated from equations 1.43 and 1.35 for the benzene–I_2 system.

For I_2 dissolved in benzene solution the model of a 1:1 benzene–I_2 complex adapted by Hanna and Williams[45] may not be a very good one, since there will be *another* benzene molecule expected near the other end of the I_2 molecule (perhaps not in quite such close contact). The field from the second benzene will make the electric fields E_1 and E_2 in the treatment by Hanna and Williams more nearly the same, so that the contribution to the frequency shift (*i.e.* to *a* in equation 1.43) from $\partial(E_1^2 + E_2^2)/\partial Q$ would be overestimated. There *is* expected to be a solvent shift for I_2 dissolved in benzene, but it arises only from the $\partial\alpha/\partial Q$ term in the treatment by Hanna and Williams[45]. In fact we should note that the parameters appearing in the Hanna and Williams model for benzene–I_2 are quite similar to values that would be expected for a 1:1 interaction between benzene and carbon disulphide, in a similar model. Thus the calculated electrostatic force constant contributions for the I–I stretching force constant of benzene–I_2 are quite similar to those expected for the C-S stretching force constant for benzene–CS_2, or to those for any number of other pairs of molecules.

In fact, it is this depressing sameness of the estimates of effects due to electrostatic interaction which led to the realization that the large and specific observed frequency shifts of the ν(OH) band in solvents which formed hydrogen bonds (and later the shifts of the ν(XY) stretching band in halogen complexes involving $a\sigma$ acceptor action) must be due to some effect *in addition* to the electrostatic effects. The explanation may be that $V_{II}(Q)$ changes in the complex from that in the gas phase, due to the donor-acceptor charge-transfer action discussed in the previous section.

If one considers only a 1:1 interaction between benzene and iodine in a fixed configuration, then we might follow the treatment by Wiederkehr and Drickamer[31] to write the interaction potential energy $V_{I,II}$ in a form similar to equation 1.4 to include contributions from dispersion and repulsion as well as these from the electrostatic and induction forces. Thus we can write:

$$V_{I,II} = V^{el} + V^{ind} + V^{disp} + V^{rep} \qquad 1.44$$

The treatment by Hanna and Williams is equivalent to considering only the contribution from the "electrostatic and induction" terms in equation 1.44. The contribution to k is estimated for each term in equation 1.44, using the theory outlined above (for example, equations 1.34–1.43 to find Δk^{el}, and using corresponding expressions for V^{disp}, *etc.* from Wiederkehr and Drickamer[31] (see also section 1.1.2). One obtains:

$$\frac{\Delta k}{k^0} = \frac{k^0 - k}{k^0} = \frac{\Delta k^{el}}{k^0} + \frac{\Delta k^{ind}}{k^0} + \frac{\Delta k^{disp}}{k^0} + \frac{\Delta k^{rep}}{k^0}, \qquad 1.45$$

(Using equation 1.31, $\Delta k/k^0 \simeq 2\Delta\nu/\nu_0$.)

The results for the solvent shift of the ν(CN) band for acetonitrile, quoted in table 1.5, cannot be compared directly to the predictions above for the ν(I-I) mode made using the Hanna and Williams treatment since the electrostatic and inductive terms calculated by Wiederkehr and Drickamer[31] were for a random solvent interaction. Nevertheless, the results from table 1.5 show that $(\Delta k^{el} + \Delta k^{ind})/k^0$ for the ν(CN) mode in this random interaction is about 4.6 to 8.4×10^{-3}. If k^0 is approximately 1960 N m^{-1}, then $(\Delta k^{el} + \Delta k^{ind})$ is estimated to be from about 9 to 16 N m^{-1} for the ν(CN) vibration of acetonitrile dissolved in chlorinated methanes. This is a change in force constant of about the same order of magnitude estimated above for the ν(I-I) band in benzene–I_2. Presumably the change in CN stretching force constant predicted for a 1:1 axial interaction between benzene and the CN group would be even larger.

The results in table 1.5 suggest that, even for a specific 1:1 complex, the effect of repulsion in equation 1.45 will be to cancel some of the decrease in force constant predicted from electrostatic and induction effects alone.

An evaluation of $V_{I,II}$ by equation 1.44 using the methods given by Wiederkehr and Drickamer[31] would allow us to determine the contribution from these "electrostatic" interactions to all the intermolecular D–A force constants by evaluation of $\partial^2 V_{I,II}/\partial q_\alpha^2$. Another useful approach to this problem may be to attempt to determine, semi-empirically, a set of atom-atom force constants such as those given by Williams[48] for the interactions in solid hydrocarbons. This procedure would define a "physical" force constant, $k_{phys} = \partial^2 V^{phys}/\partial q_\alpha^2$ (*cf.* equation 1.22).

From the discussion here, it is obvious that a large part of the frequency shift for the ν(I-I) band, for example, from that for the isolated molecule in the gas phase to that for I_2 in a 1:1 benzene–I_2 complex is due to electrostatic (and induction, dispersion and repulsion) interactions. Because these interactions are similar to "solvent shifts" and, because we know that such solvent shifts are expected to be fairly similar in all solvents (except for the purely coulombic and inductive effects), we may be able to correct for them as follows. We measure the frequency for the I_2 molecule in an "inert" solvent (for example in carbon tetrachloride, $\nu_0 = 207\ cm^{-1}$, compared to 213 cm^{-1} in the gas phase). We then estimate (by the methods of Hanna and Williams[45]) the *difference* between the electrostatic shift induced in I_2 for a 1:1 interaction with carbon tetrachloride and that induced by the same kind of interaction with benzene. Since the latter interaction is expected to have a greater effect on k than the former, we may expect this calculation to predict an I-I stretching frequency in benzene less than that in carbon tetrachloride. If the observed frequency is still lower than this calculated frequency, it could presumably be explained by the charge-transfer interaction.

Raman Spectra of Iodine in Benzene Solutions

A study of the Raman spectrum of complexes would provide a valuable supplement to the information about vibrational energy levels from studies of the infrared spectrum. One system for which this has definite advantages is the benzene–I_2 system, since uncomplexed I_2 has a Raman spectrum (but not an infrared spectrum). Hence, in a mixture of I_2 and benzene in an inert solvent, such as *n*-hexane, one expects two Raman peaks—one due to uncomplexed I_2- the other due to the benzene–I_2 complex. As the concentration of benzene is increased, the intensity of scattering from the complexed I_2 is expected to increase while that from uncomplexed I_2 decreases. This situation is complicated by the fact that not all the I_2 is expected to be complexed even in pure benzene, and by the fact that the non-specific "solvent effects" on the band frequency of uncomplexed I_2 in benzene may cause the Raman band frequency to decrease from that for the solution in *n*-hexane even if no complex forms. A non-specific solvent shift is expected to vary smoothly with the change in solvent from *n*-hexane to benzene.

Rosen, Shen, and Stenman[48] have performed just this experiment. They did not observe two separate Raman lines in any solution, but only a smooth shift from 210.1 cm^{-1} for I_2 in *n*-hexane to 204.6 cm^{-1} for I_2 in benzene. They interpreted this result as providing conclusive evidence against the existence of a definite benzene–I_2 complex.

There are, however, at least two possible alternative explanations for their[48] observations. First, the life-time of the complex may be very short, so that the observed wavenumber of Raman scattering is a weighted average of the wavenumber from "free" I_2 and from complexed" I_2 (see section 3.4.3 for further discussion). The second possibility was suggested by Jao and Person[49]; namely, it is possible that the wavenumber for the "complexed" iodine is not very different (say 2–5 cm^{-1}) from the wavenumber of the "free" iodine. Since the Raman lines are quite broad, it is very likely that the two lines (from "free" and "complexed" I_2) would overlap so badly that they could not be resolved. Thus, the change in solvent from *n*-hexane to benzene may cause a smooth shift in ν(I-I) band wavenumber from 210.6 cm^{-1} to about 207 cm^{-1} for the free I_2 due to the expected larger averaged electrostatic effect of benzene. Then, when benzene complexes with the I_2 in a specific 1:1 complex, there may be either an increased electrostatic shift, or a charge-transfer effect, or both, of another 2–3 cm^{-1} to give the final observed wavenumber (204.6 cm^{-1}) for I_2 in benzene. Careful studies of this type may help sort out the ordinary generalized "solvent" effects from the specific effects of the one-to-one complex.

The discussion here illustrates the practical difficulties in defining the "solvent effect" or "electrostatic" effect empirically for these very weak complexes by determining the frequency in an "inert" solvent. It also emphasizes just how important it is to obtain more data of this type from additional studies of Raman spectra.

1.2.6. INTENSITY CHANGES ON COMPLEXATION

In addition to the frequency shifts which may be observed in the infrared spectra of molecules upon formation of complexes, one may usually expect even more dramatic changes in intensity. For *aσ* acceptors (XY) the striking *decrease* in frequency of the ν(XY) band is accompanied by an even more striking *increase* in the band intensity. This effect had been noted for hydrogen-bonded complexes[28]; the dramatic nature of the effect is illustrated by the observation by Collin and D'Or[27] of strong infrared absorption by chlorine in benzene, attributable to the Cl–Cl stretching vibration, which is of course *forbidden* in the gas phase. Hence, the donor-acceptor interaction is somehow responsible for some very large changes in intensity of molecular vibrations.

In discussing these intensity changes let us proceed along lines similar to the discussion of the frequency changes in sections 1.2.1 and 1.2.2.

There are, in general, three different mechanisms responsible for infrared intensity changes of the internal vibrations of D or A in the complex. Namely, (1) the intensity changes due to changes in local symmetry of D or A in the complex and to changes in the normal coordinates; (2) the intensity changes due to equilibrium electrostatic polarization effects; and (3) changes in intensity due to vibronic effects. We shall discuss each of these in turn, emphasizing primarily the intensity changes for internal vibrations of D or A, with only a short discussion of the intensities of the intermolecular D–A vibrations.

The normal coordinates of D or A may change from their values in the free molecule, causing intensity changes to occur. These changes in normal coordinates may be due to mixing of internal D (or A) coordinates with new D–A coordinates as discussed previously, or they may be due to mixing of internal D (or A) coordinates (together) if the local symmetry of the D (or A) molecule is different in the complex from that in the free molecule. Forbidden bands in the free molecule become allowed by the relaxed symmetry of the complex; they gain activity as their normal coordinates change in the complex, for example, from a mode where there is complete cancellation of the dipole moment derivative to one with only partial cancellation. The forbidden vibration is said to "mix" with an allowed band to gain intensity at the expense of the allowed band. Processes such as this, in which "intensity borrowing" occurs because the normal coordinates (for example, of D) change the extent of mixing of the set of internal coordinates, are expected to obey an intensity sum rule:

$$\sum_i (A_i)_D = K \sum_i (\partial\vec{p}/\partial Q_i)^2_D = K \sum_k (\partial\vec{p}/\partial q_k)^2_{D-A} = \sum_k (A_k)_{D-A} \qquad 1.46$$

Here Q_i is the set of internal normal coordinates in the "free" donor molecule, q_k is the corresponding set of normal coordinates for the complexed donor, A_i is the intensity of the vibration as defined below, and K is a proportionality constant; the summations are over the $(3n_D - 6)$ internal normal coordinates of the donor. If one vibration increases intensity, another decreases. Mixing of the q_i coordinates with the new D–A vibrational coordinates will, of course, cause further intensity changes that do not obey this simple sum rule.

In equation 1.46 we have made use of the well-known relationships[50] between the true integrated molar absorption coefficient, A_i (the "intensity"), and the transition dipole $\langle 1|\,\vec{p}\,|0\rangle$ (and hence the dipole moment derivative, with magnitude $\partial\vec{p}/\partial Q_i$), assuming that Q_i is harmonic, *viz.*:

$$A_i = \frac{8\pi^3 N\bar{\nu}_i}{3hc}\langle 1|\,\vec{p}\,|0\rangle^2 = \frac{N\pi g_i}{3c^2}\left(\frac{\partial\vec{p}}{\partial Q_i}\right)^2 = Kg_i\left(\frac{\partial\vec{p}}{\partial Q}\right)^2. \qquad 1.47$$

where **N** is the Avogadro number, g_i is the degeneracy of the ith normal vibration at wavenumber $\bar{\nu}_i$, and **c** is the velocity of light. The integrated molar absorption coefficient A_i is defined by:

$$A_i = \frac{1}{cl} \int \ln (I_0/I)\, d\bar{\nu}. \qquad 1.48$$

It is usual in practice to plot the absorbance [ln (I_0/I)] as a function of wavenumber (in cm^{-1}) and integrate numerically over the whole of the band to obtain the value of the integral in equation 1.48. Actually I and I_0 refer to the intensities of monochromatic light at wavenumber $\bar{\nu}_i$; for a real spectrometer we use the apparent values (T and T_0) given by the spectrometer. Integrating ln (T_0/T) over the band (in equation 1.48) gives an apparent integrated molar absorption coefficient B_i (see section 2.2). For reasonably careful work, $B_i \simeq A_i$ and we shall use these quantities interchangably in this chapter (however, see chapter 2 for further discussion).

If the concentration, c, in equation 1.48 is measured in mol dm^{-3} (≡ mol/liter) and the pathlength, l, is in cm, then A_i (or B_i) has units of $dm^3\ mol^{-1}\ cm^{-2}$. These units are the same as those of the "dark" (reference 50, p. 352–3) (*i.e.* liter $mol^{-1}\ cm^{-2}$, or cm $mmol^{-1}$). When using SI units the band area (= ln $(I_0/I)\, d\bar{\nu}$) is expressed in m^{-1}; the concentration is in mol m^{-3}, and the pathlength in m. The units of A or B are then $m^3\ mol^{-1}\ m^{-2}$ or m mol^{-1}. The conversion factor is clearly 1 "dark" = 1 $dm^3\ mol^{-1}\ cm^{-2}$ = 10 m mol^{-1}.

The value of A_i is related to the Γ_i (with units of $m^2\ mol^{-1}$, see equation 2.1) defined[50] by:

$$A_i \simeq \Gamma_i \bar{\nu}_i. \qquad 1.49$$

Here $\bar{\nu}_i$ is the wavenumber of the ith normal vibration in m^{-1}. The conversion factor for Γ_i from the usual units[50] ($cm^2\ mol^{-1}$) to SI units ($m^2\ mol^{-1}$) is then 1 $cm^2\ mol^{-1} = 10^{-4}\ m^2\ mol^{-1}$.

Since the dipole moment, $\vec{p}$, has dimensions of charge × distance, it is usually given in debye (D); in SI units, we should use C m, with 1D = 10^{-18} esu cm = 3.335×10^{-30} C m. The normal coordinate, Q is mass weighted, with units of Å $(amu)^{1/2}$ or Å $g^{1/2}$; the SI units are m $g^{1/2}$

$$1\ \text{Å}\ (amu)^{1/2} = N^{-1/2}\ \text{Å}\ g^{1/2} = 10^{-8}\ N^{-1/2}\ cm\ g^{1/2}$$
$$= 10^{-10}\ N^{-1/2}\ m\ g^{1/2},$$

where **N** is the Avogadro number. Thus, the units of $\partial\vec{p}/\partial Q$ are $DÅ^{-1}\ (amu)^{-1/2}$, with 1 $DÅ^{-1}\ (amu)^{-1/2} = (10^{-10}(N)^{1/2}$ esu $g^{-1/2}$ = $3.335 \times 10^{-20}\ N^{1/2}$ C $g^{-1/2} = (3.335 \times 10^{-2}\ N^{1/2}$ aC $g^{-1/2}$. Utilizing these relationships, and substituting values for the constants in equation 1.47 we have

$$\partial\vec{p}/\partial Q_i\ (esu\ g^{-1/2}) = 1.1936 \left[\frac{A_i\ (cm\ mmol^{-1})}{g_i}\right]^{1/2}$$

or

$$\partial\vec{p}/\partial Q_i\ (D\mathring{A}^{-1}\ amu^{-1/2}) = 1.1936(A_i/g_i)^{1/2}(10^{10}/N^{1/2})$$
$$= 1.537 \times 10^{-2}\left[\frac{A_i(\text{cm mmol}^{-1})}{g_i}\right]^{1/2}$$
$$= 1.537 \times 10^{-2}\left[\frac{A_i'(\text{m mol}^{-1})}{10g_i}\right]^{1/2}. \qquad 1.50$$

(*N.B.* the degeneracy, g_i, should not be confused with the unit of mass, g.) Conversion of $\partial\vec{p}/\partial Q_i$ to C $g^{-1/2}$ is then achieved as indicated above.

Referring again to equation 1.46, we summarize that result, using the numerical values and conversion factors given above (from equations 1.46 to 1.50) by noting that the sum of the infrared intensities of the internal donor normal vibrations for the complexed donor molecule is expected to be the same as the corresponding sum for the "free" donor molecule, assuming that there is no appreciable mixing of donor coordinates with the new D–A coordinates, and assuming that there are no contributions to the intensity changes of the donor molecule from electrostatic polarization or vibronic effects. Of course, we may expect these latter effects usually to be very important, and to dominate the intensity changes observed.

Let us now consider the second reason (mentioned above) that the intensity of a vibration of D or A might change on complexing; namely, that the D or A molecule may be polarized in the complex so that the equilibrium charge distribution is different from that in the free molecule. This electrostatic polarization will result in two effects on the infrared intensity of vibrations of D (or A) in the complex. First, the different equilibrium charge distribution in D results in a change of the dipole moment, $\vec{p}$, and of the dipole moment derivative $\partial\vec{p}/\partial Q_i$, even for vibrations of D in which Q_i does not change from the form it has in the free molecule. Second, and probably more important, the extent of polarization of D by the A molecule in the complex may change as the D molecule vibrates along the coordinate Q_i. If so, then there may be a sizable *electrostatic vibronic* contribution to $\partial\vec{p}/\partial Q_i$.

For a non-specific random orientation of A molecules about D (or *vice versa*) these effects are essentially just the "solvent effects" on intensities (see reference 30). In general, such effects are small, so that the intensities in solution (or for complexed D) differ from the intensities of free D by less than a factor of two. One method of examining the magnitude of these effects, both on band intensities and also on their frequencies, is to measure the infrared spectrum of a pure solid compound and compare it to the spectrum in the gas phase. Since it has been suggested[45] that benzene is unique in the kind of electric field it produces along the six-fold axis, it seems reasonable to choose benzene for such a comparison. Results from measurements of the absorption intensities for solid benzene[51–53] are compared in table 1.7 with values

TABLE 1.7

Infrared intensities ($10^{-4}B_i$/m mol^{-1}) of benzene fundamental vibrations in different phases[a]

Mode (band position)	Gas	Solid	Bz–HCl[b]	$\Delta B_{g-s}/B_s$[c]
ν_{18}, $(\nu_8 + \nu_9)$, ν_{13} (ν(CH) region, 3000 cm^{-1})	6.0	2.1 ± 0.1	1.25 ± 0.3	1.8
ν_{19} (1475 cm^{-1})	1.3	3.6 ± 0.3	3.0 ± 0.3	−0.64
ν_{20} (1037 cm^{-1})	0.88	1.6 ± 0.15	0.7 ± 0.1	−0.45
ν_{11} (679 cm^{-1})	8.8	9.2 ± 0.8	8.6 ± 0.5	−0.04
ν_{10} (860 cm^{-1})	0	0	0.1	—

a. Summary table taken from Szczepaniak and Person (from reference 53).
b. Intensity of benzene in an HCl "matrix' with an HCl:Bz ratio of 13:1 (from reference 53).
c. $(B_{gas} - B_{solid})/B_{solid}$.

for the gas phase (quoted by Overend[50]). Although some questions have been raised recently[54] concerning the absolute accuracy of absorption intensities in the solid phase, their order of magnitude (in table 1.7) is believed to be correct.

We see in table 1.7 that the effect of the surrounding benzene molecules on the intensity of the "free" gas-phase-allowed benzene fundamentals is to change the band intensity (on going from gas to solid) by less than a factor of two. Some additional small changes occur in the intensities of these benzene fundamentals as the environment is changed from pure benzene to an HCl "matrix" (HCl:benzene ratio is 13:1). This comparison may show the order of magnitude that can be expected for the intensity changes of vibrations of a donor molecule as a result of the "electrostatic" effects which occur when D is placed in an environment of more-or-less randomly oriented A molecules. The results for the out-of-plane bending vibration, ν_{10}, are also given in table 1.7 to illustrate the magnitude of changes found in the intensities of gas-phase-forbidden bands. The changes in the intensities of these bands are an order of magnitude smaller than the changes found for the intensities of the allowed fundamental bands. Probably many of these forbidden bands become active in the crystal (or in the complex) because of changing normal coordinates instead of polarization effects; at any rate they do not usually ever acquire much intensity.

In summary, then, we may expect intensity of vibrations of D (or A) to change by a factor of two from the "free" molecule to the molecule surrounded by randomly non-specific oriented A molecules because of "ordinary electrostatic" (also dispersion, induction and repulsion) effects that polarise the molecule. On the other hand, the change from "free" D to a specific 1:1 D–A complex may be considerably larger than the factor of two. The "solvent effect theories" (see reference 30) may be of help in estimating the smaller non-specific effects, and an extension of the treatment by Buckingham[30d], using models such as that used for frequencies by Wiederkehr and Drickamer[31] may be profitable. Treatment of the electrostatic effects for a specific 1:1 complex is considered later in section 1.2.7.

Finally, we come to the third reason for expecting intensity changes in vibrations of D (or A), on going from the free molecule to the complex; namely, "*vibronic coupling*". The idea is that motion during a particular vibration (say the X-Y stretching vibration of an $a\sigma$ acceptor) may cause extensive reorientation of electrons as the electronic wavefunction changes with the changing nuclear configuration. Such effects are not uncommon in the vibrations of isolated molecules; the very strong infrared absorption of carbon dioxide (and carbon disulphide) by ν_3, the asymmetric stretching vibration, is explained[55] as a vibronic effect resulting in a change in the wavefunction of carbon dioxide from one with equal contributions at equilibrium from the resonance structures $O^-—C\equiv O^+$ and $O^+\equiv C—O^-$ to a wavefunction at one extreme of the vibration that favours one of these highly polar structures.

A similar effect is expected for some of the vibrations of the molecule in the complex. Thus, the length of the X-Y bond of the halogen molecule (XY) acting as an $a\sigma$ acceptor is expected to affect the vertical electron affinity of the acceptor. The importance of the dative structure, as measured by F_{1N}, is thus expected to change as the X-Y bond distance changes in the X-Y stretching vibration. This vibronic charge-transfer effect has been discussed in detail in chapter 6 of reference 1, and it has a very large influence on the observed intensity of this X-Y stretching vibration. This effect is responsible for at least part of the dramatic intensification found to be characteristic of the O-H stretching vibration on forming hydrogen bonds[28]; for example, although the intensities of the benzene vibrations in the "HCl matrix" of table 1.7 are quite similar to the intensities in the free molecule, the intensity of the HCl stretching vibration changes from 3.9 to 34 × 10^4 m mol^{-1} as the environment of the HCl molecule changes from the gas phase to a benzene "matrix"[53]. Such *very* large intensity changes are characteristic of "vibronic" effects; they cannot be explained by changes in the equilibrium electrostatic polarization described above as the "second effect". However, we can also expect large *electrostatic vibronic* effects (mentioned above) for specific D–XY complexes to contribute importantly to the intensification of the ν(X-Y) vibrational band.

Our discussion so far has emphasized changes in the intensities of D (or A) internal vibrations. Let us now consider briefly the intensities of the new D–A stretching vibrations. Even though D and A may be non-polar molecules, the D–A stretching vibration is expected to have some infrared activity. In part this activity comes because of the electrostatic polarization effects; D and A become polarized in the molecule so that stretching the D–A bond causes a dipole moment change. Also the polarization is expected to change rather sharply with changing D–A distance, causing a further possibly large electrostatic vibronic contribution to $\partial\vec{p}/\partial q_2$. Furthermore, the extent of charge-transfer (F_{1N}) might change as the D–A distance changes, resulting in a vibronic charge-transfer contribution to $\partial\vec{p}/\partial q_2$. The bending vibrations of D against A may also be expected to have non-zero intensities as a result of these effects, although there does not seem to be so much reason to expect these bending vibrations to be very intense.

1.2.7. INTENSITY CHANGES DUE TO VIBRONIC EFFECTS

As mentioned previously, this subject has been reviewed (for vibronic charge-transfer effects) in chapter 6 of reference 1. The problem is to evaluate the transition dipole moment, $\langle 1|\ \vec{p}\ 0|\rangle$ of equation 1.47, allowing for the fact that the coefficients a and b of the wavefunction Ψ_N for the ground electronic state (equation 1.1) may be dependent upon the normal coordinate Q_i for the particular vibration under consideration. (For the electrostatic vibronic effect the variation with Q_i of the a_{rs} coefficients of equation 1.5 causes the vibronic effect). Although this evaluation of the vibronic charge-transfer effect is, in general, complicated, it reduces (reference 1, p. 71) to the following simple expression relating $\partial\vec{p}/\partial Q_i$ (from the relationship between $\partial\vec{p}/\partial Q_i$ and $\langle 1|\ \vec{p}\ |0\rangle$ in equation 1.47) to the coefficients in equation 1.1, using the assumption that the complex is fairly weak and that the overlap (S_{01}) is small compared to the other terms:

$$\partial\vec{p}/\partial Q_i \simeq \partial\vec{\mu}_N/\partial Q_i + 2b(\partial b/\partial Q_i)\ |\vec{\mu}_1 - \vec{\mu}_0|. \qquad 1.51$$

Here $\vec{\mu}_N$ is the dipole moment of the complex, so that $\partial\vec{p}_N/\partial Q_i$ represents the magnitude of change in this moment when the Q_i coordinate changes: $|\vec{\mu}_1 - \vec{\mu}_0|$ is the difference between the dipole moment in the dative state, $\vec{\mu}_1$, and that in the no-bond state, $\vec{\mu}_0$, and is approximately equal to $\vec{\mu}_{VN}$, the electronic transition moment for the charge-transfer absorption band.

The first term in equation 1.51 is the ordinary term responsible for the intensity of the Q_i vibration if the coefficients in equation 1.1 do not change with Q_i; the second term is the "vibronic charge-transfer" contribution. The first term will be different from the corresponding derivative for the free molecule, $(\partial\vec{p}^0/\partial Q_i^\circ)$. (where $\vec{p}^0$ is the dipole moment of the "free" A (or D) molecule, and Q_i° is the normal coordinate for this particular vibration in the free A (or D) molecule). The difference (between $\partial\vec{\mu}_N/\partial Q_i$ and $\partial\vec{p}^0/\partial Q_i^\circ$) arises because of the possible change in Q_i° and because $\vec{\mu}_N$ may differ from $\vec{p}^0$ due to the "electrostatic effect" (both the equilibrium charge effect and the vibronic effect) discussed in the previous section and also below. As usual, attempts to estimate the magnitude of this change are controversial, and quantitative discussion, especially for weak complexes, is almost impossible. We shall consider this problem again later in this section.

The second term in equation 1.51 gives the vibronic charge-transfer effect. It is expected that some vibrations will cause b to change so this second term ($\partial b/\partial Q_i$) may be expected to be different from zero. If so, then equation 1.51 implies a large contribution of $\partial \vec{p}/\partial Q_i$ from the vibronic charge-transfer effect, because $|\vec{\mu}_1 - \vec{\mu}_0|$ is expected to be so large. In essence some of the very large transition moment ($\vec{\mu}_{VN}$) of the charge-transfer absorption is "borrowed" by the vibration, Q_i.

In the discussion of vibronic effects it is convenient to change coordinates from the normal coordinates, Q_i, to internal coordinates, r_j, because the latter can always be defined, whereas the former depend on the masses and the extent of mixing of internal coordinates. The transformation equation is given in matrix form as[39]:

$$\mathbf{R} = \mathbf{LQ}; \qquad \mathbf{Q} = \mathbf{L}^{-1}\mathbf{R}. \tag{1.52}$$

Here $\mathbf{R}$ is the $(3n - 6) \times 1$ column vector of the r_j internal coordinates, $\mathbf{Q}$ is the corresponding column vector of the normal coordinates Q_i, and $\mathbf{L}$ (or its inverse, $\mathbf{L}^{-1}$) is the $(3n - 6) \times (3n - 6)$ normal coordinate transformation matrix. For the isolated D and A molecules, the number of coordinates is related to n_D (or n_A)—the number of atoms in the D molecule; the normal coordinates are q_i^0 for D and Q_i^0 for A. For the complex, as discussed above, there are $[3(n_A + n_D) - 6]$ normal coordinates, in general; $3n_D - 6$ q_i's are expected to be similar to the q_i^0 coordinates, $3n_A - 6$ Q_i's are expected to be similar to the Q_i^0 coordinates, and there are six new q_α coordinates for the new D–A vibrations. These coordinates are related to the same $3n_D - 6$ r_j internal coordinates for D as were used for the free molecule; to the same $3n_A - 6$ R_j internal coordinates used to describe the free A molecule; and to 6 new r_α internal coordinates for the new D–A degrees of freedom described in section 1.2.1. As we shall see in the next section, a particular normal coordinate of the free D, say q_k^0, may change in form to q_k as D forms the complex. This new coordinate q_k may still be described in terms of the same internal coordinates (r_j) but with different coefficients ($L_{ij} \neq L_{ij}^0$). If this q_k coordinate is formed by mixing q_k^0 with a vibration of A or with a new D–A motion, then the description of q_k must be given in terms of some of the internal coordinates of A (R_j) or in terms of the new D–A coordinates (r_α).

Now consider the dipole moment derivative ($\partial \vec{p}/\partial Q_i$) obtained, for example, for a vibration of complexed A from a measurement of the infrared intensity using equations 1.47 and 1.48. This derivative is related to the derivatives with respect to the internal coordinates by:

$$\begin{aligned} \partial \vec{p}/\partial Q_i &= \sum_j (\partial \vec{p}/\partial R_j)(\partial R_j/\partial Q_i) \\ &= \sum_j L_{ji}(\partial \vec{p}/\partial R_j). \end{aligned} \tag{1.53}$$

Or, conversely

$$\partial\vec{p}/\partial R_j = \sum_i L_{ij}^{-1}(\partial\vec{p}/\partial Q_i). \qquad 1.54$$

Here L_{ji} is the ji coefficient from the normal coordinate transformation given in equation 1.52; L_{ji}^{-1} is the corresponding element from the inverse transformation. The sum over i and j is over all the internal coordinates that participate in Q_i (see reference 50). For simplicity, let us consider these internal coordinates to be symmetry coordinates;[39] the summation is then over all coordinates of the same symmetry.

Now in considering the vibronic effects the change of the coefficients in the wavefunctions (equations 1.1 or 1.5) with a change an *internal* coordinate, for example R_j, is given, by analogy with equation 1.51, by:

$$\partial\vec{p}/\partial R_j = \partial\vec{\mu}_N/\partial R_j + 2b(\partial b/\partial R_j)\,|\vec{\mu}_1 - \vec{\mu}_0|. \qquad 1.55$$

The theory is most conveniently applied in this form since R_j can be well-defined (for example for a series of D–X-Y complexes with constant X-Y but changing D), while Q_i is expected to change from one complex to another. In order to compare the theoretical prediction of $\partial\vec{p}/\partial R_j$ with experimental values of $\partial\vec{p}/\partial Q_i$, however, we must evaluate (or assume values for) the coefficients L_{ji} in equation 1.53.

Now let us go on to evaluate (from chapter 6 of reference 1) the vibronic charge-transfer effect for these derivatives. It is expected that $\partial b/\partial R_j$ will be different from zero for R_j coordinates that change either Δ (equation 1.10) or β_0 (equation 1.12), since (using equation 1.13)

$$\partial b/\partial R_j \simeq \partial(-a\beta_0/\Delta)/\partial R_j.$$

Thus

$$\partial b/\partial R_j \simeq (b/a)(\partial a/\partial R_j) - (a/\Delta)\,\partial\beta_0/\partial R_j - (b/\Delta)\,\partial\Delta/\partial R_j. \qquad 1.56$$

Rearranging, using the normalization condition (equations 1.7a, 1.8 and 1.9) we find:

$$(1 + F_{1N}/F_{0N})(\partial b/\partial R_j) = -(a/\Delta)\,\partial\beta_0/\partial R_j - (b/\Delta)\,\partial\Delta/\partial R_j. \qquad 1.57$$

Since $F_{1N} \ll F_{0N}$ for weak complexes, we may wish to neglect F_{1N}/F_{0N} in equation 1.57 for that special case.

Examining equation 1.57 for the special case of $R_j = R_1$ the stretching coordinate of the X-Y bond of a complexed halogen molecule (I_2, for example) we see that $\partial b/\partial R_1$ may be expected to be different from zero, since stretching this bond is expected[1] to cause a considerable increase in the vertical electron affinity of I_2. This change is expected[1] to dominate the value of $\partial\Delta/\partial R_1$ so that:

$$-(b/\Delta)(\partial\Delta/\partial R_1) \simeq -(b/\Delta)(\partial E_A^V/R_1).$$

If this term is more important for the X-Y stretching vibration than $\partial\beta_0/\partial R_1$, then the vibronic charge-transfer contribution to the dipole moment derivative in equation 1.55 is

$$(\partial\vec{p}/\partial R_1) - (\partial\vec{\mu}_N/\partial R_1) \simeq -2b\left(\frac{F_{0N}}{F_{0N}+F_{1N}}\right)\left(\frac{b}{\Delta}\right)\left(\frac{\partial E_A^V}{\partial R_1}\right)|\vec{\mu}_1 - \vec{\mu}_0|. \tag{1.58}$$

Or, since $F_{1N} \simeq b^2$, for weak complexes, and since $F_{0N} + F_{1N} = 1$, by the normalization condition, we write:

$$\vec{M}_d' = (\partial\vec{p}/\partial R_1) - (\partial\vec{\mu}_N/\partial R_1) \simeq -(2F_{0N}F_{1N})\frac{\partial E_A^V/\partial R_1}{\Delta}|\vec{\mu}_1 - \vec{\mu}_0|. \tag{1.59}$$

This equation is a slightly modified version of equation 6.15 of reference 1. Note that $\partial E_A^V/\partial R_1$ is negative so that the vibronic term on the right hand side of equation 1.55 is positive. The vibronic term on the right hand side of equation 1.55 defines the *delocalization moment*, M_d', of reference 1, p. 74–5. (Properly, of course, M_d' should be called a delocalization *charge*.)

The value of $\partial E_A^V/\partial R_1$ may be estimated from potential curves for the XY and $(XY)^-$ molecules, the value of $|\vec{\mu}_1 - \vec{\mu}_0|$ can be estimated or the experimental value of $\vec{\mu}_{VN}$ can be used, together with a value of F_{1N}, perhaps from a measurement of the $\bar{\nu}$(X-Y) band shift (see sections 1.2.2 and 1.2.4), to predict the value of M_d', and hence of the intensification expected for the X-Y stretching vibration in the complex (using equation 1.53) *if* all these assumptions, plus those in deriving equation 1.59, are correct.

The results of the calculation are compared with experimental values for a few halogen complexes in table 1.8. There we present the results estimated for the vibronic charge-transfer contribution ($\vec{M}_d'$) to $\partial\vec{p}/\partial R_j$ for these complexes. The values in this table are taken from table 6-3 of reference 1. In order to compare these values with experimental measurements of the intensity of the X-Y stretching vibration, we must make additional assumptions concerning the normal coordinates. Because these complexes are weak (except for Pyr–I_2) we can assume that $Q_1 \simeq Q_1^0 = (m_r)^{1/2} R_1$, where m_r is the reduced mass of the XY molecule $[m_r = m_X m_Y/(m_X + m_Y)]$. If so, then the vibronic contribution to $\partial\vec{p}/\partial Q_1$ is:

$$\vec{M}_d = \partial\vec{p}/\partial Q_1 - \partial\vec{\mu}_N/\partial Q_1 = \vec{M}_d'/(m_r)^{1/2} \tag{1.60}$$

Here $\vec{M}_d$ has units of D Å^{-1} amu$^{-1/2}$ or C g$^{-1/2}$ and $\vec{M}_d'$ has units of D Å^{-1} or C. (Conversion to SI units has been described above.) The "observed values" of M_d' listed in table 1.8 are calculated using equation 1.60 from the experimental values of $\partial\vec{p}/\partial Q_1$ obtained[1] using equation 1.50, assuming that $\partial\vec{\mu}_N/\partial R_1$ is the same as the $\partial\vec{p}^0/\partial R_1$ obtained from the gas-phase intensity of the ν(X-Y) vibrational band.

For the pyr–I_2 complex we have assumed the "triatomic" approximation of section 1.2.3, so that $Q_1 = L_{11}^{-1} R_1 + L_{12}^{-1} r_2$, where r_2 is the

TABLE 1.8

Comparison of calculated values of the vibronic charge-transfer contribution to the observed values of $\vec{M}'_d$[a,b] (recalculated from table 6.3 of reference 1)

	Observed		Calculated[e]
Complex[c]	$\partial\vec{p}^0/\partial R_1$[d]	$\vec{M}'_d$	$\vec{M}'_d$
bz–I_2	0	5.0	1.7
bz–Br_2	0	6.3	3.3
bz–Cl_2	0	2.7	5.3
bz–ICl	9.0	3.3	10.0
tol–ICl	9.0	4.0	9.7
p-xy–ICl	9.0	2.3	10.3
pyr–I_2	0	31.7 ± 1.7	25.7

a. Here $\vec{M}'_d = (\partial\vec{p}/\partial R_1) - (\partial\vec{\mu}_N/\partial R_1)$. To obtain the observed value of $\vec{M}'_d$, the experimental value of $\partial\vec{p}/\partial Q_1$ is converted $\partial\vec{p}/\partial R_1$ by multiplying by $(m_r)^{1/2}$ as described in the text. Furthermore, it is assumed that $\partial\vec{\mu}_N/\partial R_1 = \partial\vec{p}^0/\partial R_1$. Values of $(m_r)^{1/2}(\text{amu}^{-1/2})$ are 7.97, 6.32, 4.31 and 5.27 for I_2, Br_2, Cl_2 and ICl, respectively.

b. The units of $\vec{M}'_d$ and $\partial\vec{p}^0/\partial R_1$, are 10^{-20} C. They were converted from D Å^{-1} using,

$$1 \text{ D Å}^{-1} = 3.335 \times 10^{-20} \text{ C}.$$

The charge on one electron, **e**, is 16.02×10^{-20} C.

c. Abbreviations: bz is benzene; tol is toluene; *p*-xy is *p*-xylene; and pyr is pyridine.

d. The dipole derivative from the gas-phase intensity, using equations 1.50 and 1.60.

e. Value of $\vec{M}'_d$ calculated from equation 1.59, using parameter values given in reference 1, chapter 6.

D–I stretching coordinate of that section. The normal coordinates were calculated by Yarwood and Person[36] assuming $k_{12} = 40$ N m^{-1}. From these and from experimental measurements of $\partial\vec{p}/\partial Q_1$ and $\partial\vec{p}/\partial Q_2$ (the derivative with respect to the D–I normal coordinate) they calculated[37] the experimental values of $\partial\vec{p}/\partial R_1$ reported in table 1.8.

We see from the comparison in table 1.8 that the estimated vibronic charge-transfer contributions to the intensity are large enough to account for most of the observed intensification of the X-Y stretching vibration in these halogen complexes. The calculated vibronic charge-transfer effects are not expected to agree quantitatively with the experimental values of $\vec{M}'_d$ since we have ignored the "electrostatic effects" discussed in the previous section. These effects are discussed later.

Application to Vibrations of D–$a\sigma$ Complexes

Before considering further the magnitude of these "electrostatic effects" let us examine in more detail the delocalization moment, $\vec{M}'_d$, for the specific case of the vibration for an X-Y bond acting as a localized $a\sigma$ acceptor, continuing the discussion begun in the preceding section.

Consider the benzene–I_2 complex. If the I_2 is perpendicular to the benzene ring, along the six-fold axis of benzene (axial geometry[1]) then the vibronic charge-transfer term $\vec{M}'_d$, increases the dipole moment in the sense $A^- D^+$ along the six-fold axis as the I-I bond is stretched. If the acceptor is ICl (D–ICl), with $\partial\vec{\mu}_N/\partial R_{ICl} \neq 0$, then the vibronic term, $\vec{M}'_d$, is in the same direction as the intrinsic moment change ($\partial\vec{\mu}_N/\partial R_{ICl}$) and the intensity of the ICl stretch increases. If the acceptor is a HO bond (D–H-O-R), and if $\partial\vec{\mu}_N/\partial R_{OH}$ is such that stretching the HO bond causes the O^-H^+ bond dipole moment to decrease as r_{OH} increases, then the vibronic moment change $\vec{M}'_d$ is opposed to $\partial\vec{\mu}_N/\partial R_{OH}$. Hence, the intensity of the ν_s(O-H) band is expected first to decrease and then to increase as $\vec{M}'_d$ increases from zero to a large value as F_{1N} increases for a series of donor molecules with HOR. This example may seem artificial for the OH bond, but the calculated dipole derivatives for CH bonds *are* negative[56], and there is some evidence (see reference 19 of chapter 2) that weak interaction may cause an intensity *decrease* before stronger interaction with the CH bond causes a larger intensity increase which is observed in hydrogen bonding with CH bonds. In fact this effect is believed to be the cause of the *decrease* in the intensity of the ν(CH) band in benzene from gas to solid (see table 1.7).

If the benzene–I_2 complex does have axial geometry, then some contribution to M'_d may be expected because $\partial\beta_0/\partial R_1$ (equation 1.57) may not be zero [1].However, it seems quite clear that stretching the I-I bond, while keeping the D–I bond constant, will not affect the value of S_{01} nor of β_0. Thus, $\partial\beta_0/\partial R_1 = 0$, and the *only* contribution to $\partial b/\partial R_1$ from equation 1.57 arises from $\partial\Delta/\partial R_1$. Furthermore, the only term in Δ that can depend on R_1 is E^V_A, so that equation 1.59 gives the only vibronic charge-transfer contribution that can be expected for the I-I stretching coordinate.

Similarly, if we consider r_2, the D–X stretching coordinate, it is just as clear that Δ does not depend upon r_2. The only term contributing to $\partial b/\partial r_2$ in equation 1.57 is $\partial\beta_0/\partial r_2$. To evaluate this term, refer to the discussion of β_0 in section 3.5 of reference 1; β_0 is proportional to the overlap integral S_{da^-}:

$$\beta_0 = (2)^{1/2}S_{da^-}\int[(\phi_d\phi_{a^-}/S_{da^-}) - \phi^2]V_A\,dv = \sqrt{2}\,kS_{da^-} \simeq kS_{01}. \qquad 1.61$$

Here k is the proportionality constant between S_{01} and β_0 (see also equation 1.20 and discussion immediately following). From the discussion in chapter 10 of reference 1 of these parameters for $b\pi$–I_2 complexes, it seems that k may be between -10 and -20 eV, depending upon which of the various estimates is accepted as correct. A reasonable value for k may be about -15 eV for these complexes. For amine–I_2 complexes (or for complexes with ethers and alcohols) k may be about -5 to -6 eV. Thus $\partial\beta_0/\partial r_2 = \sqrt{2}\,k\,\partial S_{da^-}/\partial r_2$. We may evaluate $\partial S_{da^-}/\partial r_2$ by examining the graph of this overlap integral shown in figure 9.4 of reference 1, to obtain the slope of S_{da^-} for SCF wave-functions. The value of $\partial S_{da^-}/\partial r_2$ is about -0.1 Å^{-1} for D–X distances from about 2 to 4 Å, decreasing to smaller values at either extreme of this range of distances.

Thus:
$$-(a/\Delta)\,\partial\beta_0/\partial r_2 = -(a/\Delta)((2)^{1/2}k)(\partial S_{da^-}/\partial r_2). \qquad 1.62$$

Substituting into equation 1.55, using equation 1.59:
$$\vec{M}_d'(r_2) = -(2(2)^{1/2}F_{0N}F_{1N}ka/b\Delta)(\partial S_{da^-}/\partial r_2)\,|\vec{\mu}_1 - \vec{\mu}_0|. \qquad 1.63$$

Hence, using the approximate values discussed above for the parameters, we predict for $b\pi$–I_2 complexes:

$$\begin{aligned}\vec{M}_d'(r_2) &\simeq [\vec{M}_d'(R_1)]((2)^{1/2}ak/b)[(\partial S_{da^-}/\partial r_2)/\partial E_A^V/\partial R_1]\\ &\simeq -0.499(a/b)[\vec{M}_d'(R_1)]. \end{aligned} \qquad 1.64$$

For a weak complex, such as benzene–I_2, a/b $\approx$ 10, so $\vec{M}_d'(r_2) \simeq -5\,\vec{M}_d'(R_1)$. However, for the stronger n donor complexes with I_2 the value of k is expected to be only about one-third the value for $b\pi$–$a\sigma$ complexes, so we predict $\vec{M}_d'(r_2) \approx -0.166\,(a/b)\,\vec{M}_d'(R_1)$. For these stronger complexes, a/b is nearly 2, so that $\vec{M}_d'(r_2) \simeq -0.3\,\vec{M}_d'(R_1)$ for n–$a\sigma$ complexes with I_2. For ICl complexes with amine donors, we also predict $\vec{M}_d'(r_2) \simeq -0.30\vec{M}_d'(R_1)$, assuming a/b = 0.76/0.41, the value for pyridine–ICl.

The experimental values of the ratios of $\vec{M}_d'(r_2)/\vec{M}_d'(R_1)$ from the study of Brownson and Yarwood[36] are near these predicted values if the interaction constant, k_{12}, *is* very large. For $k_{12} \simeq 60$ N m^{-1}, the experimental value[36] for $\vec{M}_d'(r_2)/\vec{M}_d'(R_1)$ for pyr–ICl is -0.40 compared with -0.30 predicted above, while for pyr–I_2 the experimental ratio[36] is -0.30, in exact agreement with the predicted value of -0.30.

Summary of Predicted Vibronic Charge Transfer Effects

Since both $\partial\vec{p}/\partial R_1$ and $\partial\vec{p}/\partial r_2$ contribute to $\partial\vec{p}/\partial Q_1$ and $\partial\vec{p}/\partial Q_2$ (equation 1.53) and thus to the intensities of both the ν(X-Y) and ν(D–X) vibrational bands, the intensities of both bands will be changed by the vibronic charge-transfer effect in D–$a\sigma$ complexes. In fact, the intensities of both vibrations are expected to increase drastically as the D–X interaction increases. Let us consider whether the intensities of any of the other vibrations of the acceptor molecule may also be expected to change. To a first approximation, we would expect that this would not be the case, since changing the Y-Z distance in an X-Y-Z acceptor is not expected to affect Δ, so that $\partial\Delta/\partial R_j \simeq 0 \simeq \partial\beta_0/\partial R_j \simeq \vec{M}'_d(R_j)$; (*i.e.* for $R_j \neq R_1$).

Let us consider vibronic effects for distortion of the donor. A totally symmetric deformation of the donor is expected to change the ionization potential of the donor, so that $\partial\Delta/\partial r_j \simeq \partial I_D^v/\partial r_j$. It is possible that other non-symmetric deformations may also affect the ionization potential. However, in contrast with the very large change expected[1] for the vertical electron affinity of most *aσ* XY acceptors when the X-Y distance changes, the value of $\partial I_D^v/\partial r_j$ is expected to be quite small for *bπ* donors. Again, the delocalized nature of *bπ* action would make such effects relatively smaller than for localized *aσ* action. A conservative estimate would be that this term ($\partial I_D^v/\partial r_j$) is generally less than the corresponding $\partial E_A^v/\partial r_j$ by one or two orders of magnitude, so that the contribution to $\vec{M}'_d$ (r_j) (r_j being a coordinate of the donor) is expected to be only 0.01 to 0.1 as large as the corresponding contribution for the ν(X-Y) mode of an *aσ* acceptor.

For *n* donors, such ammonia, however, one might expect a fairly large change in ionization potential associated with the "umbrella bending" coordinate. Similarly the hydridization of the N atom may be expected to change as the N-H bonds stretch, so that there may also be a non-negligible contribution to $\partial I_D^v/\partial r_{NH}$ for the symmetric stretching

TABLE 1.9

Summary of predicted vibronic charge-transfer intensity changes [$\vec{M}'_d(R_j)$] for complexes, based on resonance-structure theory

Donor type[a]	Comments[b]		Acceptor type[a]
n	Possibly some fairly large contributions to $\vec{M}'_d(r_j)$ for symmetric vibrations due to expected fairly large values of $\partial I_D^v/\partial r_j$ (or $\partial E_A^v/\partial R_j$). Little effect for other vibrations.		*v*
bσ	to weak to be important	Large effects on X-Y stretching vibrations and on D–X stretching mode (see text).	*aσ*
b*π*	The action is not localized, so effects are small; furthermore, $\partial I_D^v/\partial r_j (\partial E_A^v/\partial R_j)$ expected to be fairly small, so very small effects on intensities.		*aπ*

a. classification from table 1.3.
b. general comments about expected values for vibronic term, $\vec{M}'_d(R_j)$.

coordinate. This question would appear to deserve further experimental study.

We summarize our predictions for intensity changes in table 1.9, similar to the summary of predictions for force constants in table 1.6.

Evaluation of $\partial\vec{\mu}_N/\partial R_j$. The Electrostatic Vibronic Effect

Let us now consider further the value of $\partial\vec{\mu}_N/\partial R_j$ of equation 1.55 (or $\partial\vec{\mu}_N/\partial Q_i$ of equation 1.51). So far we have assumed that it was just $\partial\vec{p}^0/\partial R_j$, the derivative for the free molecule, while realizing that it is probally quite different. We have mentioned two factors that may cause this derivative ($\partial\vec{\mu}_N/\partial Q_i$) to change in the complex. The change in the normal coordinates (from Q_i^0 for the free molecule to Q_i in the complex) is a change in the coefficients L_{ij} of the normal coordinate transformation. Such a change does not affect R_j, so that $\partial\vec{\mu}_N/\partial R_j$ would be equal to $\partial\vec{p}^0/\partial R_j$, if only this factor needed consideration. However, there is also a change due to the "electrostatic effects" discussed previously; both to the change in equilibrium charge and to the electrostatic vibronic effect.

Furthermore, there is an additional charge-transfer effect, since, $\vec{\mu}'_N$ is expected to be different from $\vec{\mu}_0$. From the principle of compromise geometry, $\partial\vec{\mu}'_N/\partial R_j$ is expected to be given by analogy with equation 1.28, by:

$$\frac{\partial\vec{\mu}'_N}{\partial R_j} = F_{0N}(\partial\vec{\mu}_0/\partial R_j) + F_{1N}(\partial\vec{\mu}_1/\partial R_j). \qquad 1.65$$

This effect is expected to be important in that derivatives that are zero by symmetry in a free donor molecule, for example, may become nonzero in the dative structure. This question has not been thoroughly explored, however, so that it is not really possible to make very meaningful predictions about the magnitude of these effects. Changes of this sort may be as important in determining intensities of benzene vibrational bands in complexes as are the vibronic changes due to $\partial I_D^v/\partial r_j$ terms.

The importance of the "electrostatic" contribution to the change in $\partial\vec{\mu}_N/\partial R_j$ from the value in the *free* molecule ($\partial\vec{p}^0/\partial R_j$) was emphasized in the discussion of the benzene–I_2 complex by Hanna and Williams[45] (their model was described in section 1.2.5). In order to estimate the contribution from the electrostatic effects to the intensity of the ν(X-Y) vibrational band in the complex (D–X-Y), Hanna and Williams evaluated the change in the *induced dipole* in XY ($\vec{\mu}_i$) during the vibration.

The induced dipole arises from the interaction of the field along the *z* axis from the benzene molecule with the polarizable halogen molecule:

$$\vec{\mu}_i = +(\tfrac{1}{2})\alpha_{\parallel}(\vec{E}_1 + \vec{E}_2). \qquad 1.66$$

Here $\alpha_{\parallel}$ is the polarizability of the halogen parallel to its axis (in the z direction); $\vec{E}_1$ is the field from the benzene (in the axial model[1]) at the nearest halogen atom (X), and $\vec{E}_2$ is the field at the halogen atom (Y) further away from the benzene. The induced dipole is directed from Y^- to X^+. When the XY bond is stretched, the induced dipole is expected to change. Re-writing Hanna and Williams' equation 4 in terms of the internal coordinate, $\Delta R_1[\equiv (\Delta z_2 - \Delta z_1)]$ for the halogen stretching motion, we find that

$$(\partial\vec{\mu}_i/\partial R_1) = +(\tfrac{1}{2})(\partial\alpha_{\parallel}/\partial R_1)(\vec{E}_1 + \vec{E}_2) + (\tfrac{1}{4})\alpha_{\parallel}[(\partial\vec{E}_1/\partial z)_{z=z_2} - (\partial\vec{E}_2/\partial z)_{z=z_1}.] \quad 1.67$$

Hence equation 1.67 predicts an electrostatic vibronic contribution.

The relation between the induced moment and $\vec{\mu}_N$ is:

$$\vec{\mu}_N = \vec{\mu}'_N + \vec{\mu}_i, \quad 1.68$$

where $\vec{\mu}'_N$ is the dipole moment for D and A placed together in the configuration for the complex so that the charge distribution is modified as shown in equation 1.65. If only electrostatic interactions are important:

$$\partial\vec{\mu}_N/\partial R_j - \partial\vec{\mu}'_N/\partial R_j = \partial\vec{\mu}_i/\partial R_j. \quad 1.69$$

Hence, $\partial\vec{\mu}_i/\partial R_j$ is the electrostatic vibronic contribution to the derivative. For the particular case of $R_j = R_1$, the X-Y stretching vibration of a halogen *aσ* acceptor, $\partial\vec{\mu}_i/\partial R_j$ is given by equation 1.67.

From the values given by Hanna and Williams[45] in their table I, the values of $\partial\vec{\mu}_i/\partial R_1$ are calculated for the halogen complexes and listed in table 1.10. Comparing the calculated values in the next to last column with the observed values in the last column, we see that apparently *all* of the observed intensification of the halogen-halogen stretching vibration can be attributed to this change in the induced dipole moment from the simple electrostatic interaction. Apparently no vibronic charge transfer is required to explain the observed intensity change.

Since the dipole moment derivative predicted from this electrostatic effect has the same direction as that predicted in the previous discussion for the vibronic charge transfer effect, the magnitude of the total dipole derivative from equations 1.69 and 1.55 is

$$\partial\vec{p}/\partial R_1 = \frac{\partial\vec{\mu}'_N}{\partial R_1} + \partial\vec{\mu}_i/\partial R_1 + \vec{M}'_d(R_1). \quad 1.70$$

In practice we would probably ignore the effect from equation 1.65 and replace $\vec{\mu}'_N$ by $\vec{p}^0$, the dipole moment of the non-interacting molecule pair in the gas phase. For benzene–I_2, in a one-to-one axial configuration with a Bz–I distance of 3.5 Å (approximately equal to the van der Waals distance), the values from tables 1.10 and 1.8 in equation 1.70 predict:

$$\partial\vec{p}/\partial R_1 = 0 + \partial\vec{\mu}_i/\partial R_1 + \vec{M}'_d = 6.7 + 1.7 = 8.4 \times 10^{-20}\ \text{C}.$$

This predicted value is almost twice the experimental value of 5.0×10^{-20} C. For this example, the vibronic charge-transfer effect is

TABLE 1.10

Calculated values contributing to the electrostatic vibronic term $(\partial \vec{\mu}'_d / \partial R_1)$ for an increase in the X-Y coordinate R_1 for some axial benzene–halogen complexes (see table I of reference 45)

Halogen	$+\frac{1}{2}(\partial \alpha_{\parallel}/\partial R_1) \times (\vec{E}_1 + \vec{E}_2)$ $(10^{-20}$ C)[a]	$+\frac{1}{2}\alpha_{\parallel} \times (\partial \vec{E}/\partial z)_{z_2}$ $(10^{-20}$ C)[a]	$-\frac{1}{2}\alpha_{\parallel} \times (\partial \vec{E}/\partial z)_{z_1}$ $(10^{-20}$ C)[a]	Calculated $\partial \vec{\mu}_i/\partial R_1$[b] $(10^{-20}$ C)[a]	Observed $\lvert \vec{M}'_d \rvert$[c] $(10^{-20}$ C)[a]
Cl_2	−3.6	+0.3	−1.2	−4.1	2.7
Br_2	−4.3	+0.3	−1.8	−5.1	6.3
I_2	−5.4	+0.4	−2.9	−6.7	5.0
ICl	−3.9	+0.4	−2.3	−4.9	3.3

a. Converted from D/Å using 1 D $Å^{-1}$ = 3.335 × 10^{-20} C (see table 1.8).
b. Derivative of induced dipole, calculated using equation 1.67. The negative sign means that the dipole moment becomes larger in the sense $D\text{-}X^{+}\text{-}Y^{-}$ as the X-Y bond length increases.
c. The experimental absolute value of the vibronic term for the halogen in the complex, taken from table 1.8 (see chapter 6 of reference 1).

expected to contribute only about one-fifth of the total change in dipole derivative, with the major effect coming from the electrostatic effect. For other halogen complexes with benzene, the results in tables 1.8 and 1.10 suggest that the vibronic charge transfer contribution to $\partial \vec{p}/\partial R_1$ may vary from $\frac{1}{6}$ to $\frac{2}{3}$ of the total predicted change. This fluctuation in estimates for the different halogens with benzene probably measures the reliability of the estimates rather than any significantly different behaviour from one halogen molecule to another.

As we go from complexes of I_2 with benzene to complexes with stronger donors, we expect F_{0N} to increase, so that the vibronic charge-transfer contribution (equation 1.59) is expected to increase considerably. Thus, in table 1.8, that contribution for the pyridine–I_2 complex is predicted to be about 10 times as large as for benzene–I_2. On the other hand, the predicted electrostatic contribution $(\partial \vec{\mu}_i/\partial R_1)$ would be about the same for pyridine–I_2 as for benzene–I_2. For the latter contribution the only change between the two complexes is in the electric field ($\vec{E}_1$ and $\vec{E}_2$ as well as $\partial \vec{E}/\partial z$). This change at the I_2 molecule is expected to be fairly small on going from an axial benzene complex to the pyridine–I_2 complex. The field at I_2 due to pyridine for a planar symmetric C_5H_5N–I-I complex (see reference 1, chapter 5, for the structure) is primarily due to the lone-pair of electrons, somewhat modified by the charges on the other atoms. Thus it is negative, and similar in magnitude to the field estimated for benzene.

Hence, we may observe the vibronic charge-transfer effect by studying the intensities of the ν(I-I) band for complexes with a series of electron donors of increasing strength (hence increasing F_{0N} value). The observed changing intensification is due primarily to the changing vibronic contribution. The initial intensification for very weak complex, such as benzene–I_2, may be due primarily to the electrostatic effect.

1.2.8. INTENSITY CHANGES DUE TO VIBRATIONAL MIXING

As we have discussed in the preceding section and in section 1.2.3, the normal coordinates of the complex (Q_i) are expected to be mixtures of symmetry (or internal) coordinates (r_j) of D among themselves and also with those of A. Let us consider how the normal coordinate transformation coefficients L_{ij} change for the Q_1 and Q_2 normal coordinates of a D–XY *n-aσ* (or *bπ-aσ*) complex as a function of the strength of interaction between D and A, in a calculation similar to the one made in section 1.2.3 (see figure 1.5) for the force constants.

In the notation of section 1.2.3, assuming that only R_1, and r_2 mix to form the normal coordinates Q_1 (the "X-Y stretching" mode) and Q_2 (the "D-X stretching" mode), we write out equation 1.52 to obtain:

$$\begin{aligned} R_1 &= L_{11}Q_1 + L_{12}Q_2 \\ r_2 &= L_{21}Q_1 + L_{22}Q_2. \end{aligned} \qquad 1.71$$

In figure 1.7 we have plotted the normal coordinates (L_{ij}) as a function of k_2, the D–I stretching force constant for an amine–I_2 complex, as the value of k_2 varies from 0 to 160 N m^{-1}, the latter being the assumed value for the I-I force constant. In figure 1.7a we have assumed that the I-I force constant does not change as k_2 increases. In figure 1.7b we assume that k_1 decreases as k_2 increases ($k_1 = 160 - 0.75k_2$). In both calculations we assume $k_{12} = 0$. We see in both figures that there is considerable change in the normal coordinates as a result of changing k_2. The second assumption (k_1 decreases as k_2 increases) accelerates the changes in L_{ij}.

As a result of these changes in the form of the normal coordinates, there will be a considerable change also in the intensities of the ν_1 [$A_1 = K(\partial\vec{p}/\partial Q_1)^2$, equation 1.47] and ν_2 [$A_2 = K(\partial\vec{p}/\partial Q_2)^2$] bands. Let us assume, for an example, that $\partial\vec{p}/\partial R_1 = \varepsilon$ and that $\partial\vec{p}/\partial r_2 = -\varepsilon$ coulomb. (There is no special significance to this particular choice of parameters. As we have seen in section 1.2.7, the vibronic charge-transfer effect predicted for benzene–I_2 that $\partial\vec{p}/\partial R_1 \simeq -0.2\partial\vec{p}/\partial r_2$, and that for other complexes $\partial\vec{p}/\partial R_1$ was predicted to be about $-5\partial\vec{p}/\partial r_2$, so an assumption that $\partial\vec{p}/\partial R_1 = -\partial\vec{p}/\partial r_2$ is just somewhere in between these extremes. Arguments can also be made that suggest that the electrostatic vibronic effect would also predict a sign for $\partial\vec{p}/\partial R_1$ that is opposite to that for $\partial\vec{p}/\partial R_2$.) Under this assumption, we can use the normal coordinates from figure 1.7 with equation 1.53 to calculate $\partial\vec{p}/\partial Q_1$ and $\partial\vec{p}/\partial Q_2$ as a function of k_2. The results are shown in figure 1.8a computed on the assumption that k_1 does not change, and figure 1.8b shows the results computed on the assumption that k_1 decreases as k_2 increases.

For both assumptions the qualitative behaviour predicted for the intensities is similar, but the intensity changes with changing k_2 are

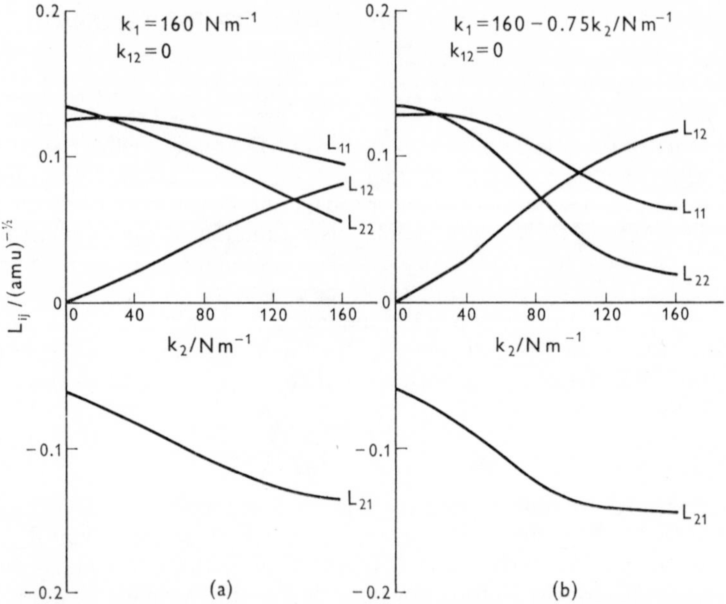

FIG 1.7 Normal coordinate transformation matrix elements, L_{ij} (equation 1.71) for an amine–I_2 "triatomic" molecule (D–II) plotted as a function of k_2, the force constant for stretching the N—I bond. In (a) the value of k_1, the force constant for stretching the I—I bond, is assumed to be constant ($k_1 = 160$ N m^{-1}); in (b) the value of k_1 is assumed to decrease as k_2 increases ($k_1 = 160 - 0.75k_2$ N m^{-1}). For both (a) and (b) k_{12} is assumed to be zero and the mass of D is assumed to be 75 amu. Three other solutions exist, depending on choice of signs for Q_j and R_i: one with negative signs for both L_{11} and L_{12}, one with both signs of L_{21} and L_{22} changed, and one with all four signs changed.

accelerated if k_1 decreases. The qualitative prediction is that $\partial\vec{p}/\partial Q_1$ is approximately equal to $-(2)^{1/2}\,\partial\vec{p}/\partial Q_2$ (so that the intensity of the "ν(I-I)" band is about twice that of the "ν(D–I)" band when $k_2 = 0$). The intensities of both the ν_1 and ν_2 bands are expected to be fairly small, depending on the magnitude of ε. We may assume that when $k_2 = 0$, there is no interaction between D and A but that the two molecules simply touch each other with only electrostatic effects operating. As k_2 increases from zero, the intensity of the "ν(I-I)" band (proportional to the square of $\partial\vec{p}/\partial Q_1$, as above) is expected (from figure 1.8b) to increase by about 50% as k_2 increases to 80 N m^{-1}, while the intensity of the "ν(D–I)" band is expected to decrease to zero. Upon increasing interaction ($k_2 > 80$ N m^{-1}) the intensity of the "ν(D–I)" band is expected to increase again at the expense of the "ν(I-I)" band. Of course, if the $\partial\vec{p}/\partial R_j$ values change as a function of changing k_2, as would be expected if a "vibronic contribution" is

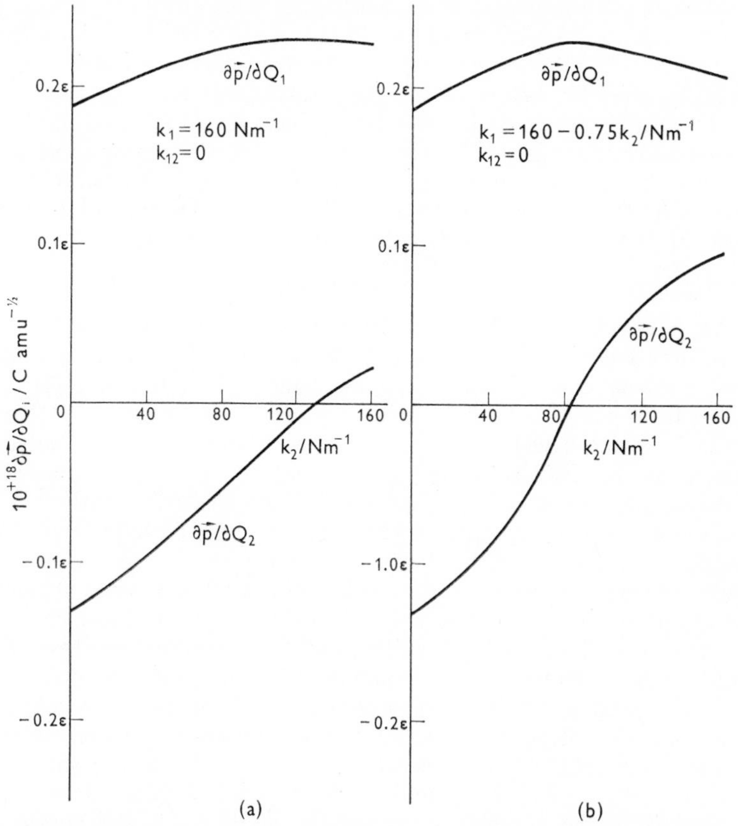

FIG 1.8 The dipole moment derivatives $\partial\vec{p}/\partial Q_i$ with respect to the "I-I stretching" normal coordinate (Q_1) and the "N–I stretching" normal coordinate (Q_2) shown as a function of k_2, using the normal coordinate transformation elements (L_{ij}) shown in figure 1.7. We assume that $\partial\vec{p}/\partial R_1 = \varepsilon = -\partial\vec{p}/\partial R_2$. In (a) it is assumed that k_1 is constant; in (b) k_1 is assumed to decrease as k_2 increases ($k_1 = 160 - 0.75k_2$ N m^{-1}).

important, then the intensity predictions here must be modified to allow ε to vary with k_2.

Figure 1.8 illustrates the intensity sum rules of equation 1.46 in this situation. They hold under this assumption that $\partial\vec{p}/\partial R_1$ and $\partial\vec{p}/\partial r_2$ do not change as the strength of interaction (measured by k_2) changes.

We see from this example that it is necessary to carry out a proper normal coordinate analysis in order to obtain values of $\partial\vec{p}/\partial R_i$ from the observed intensities. Figure 1.8b indicates that use of the "diatomic approximation" $[\partial\vec{p}/\partial R_1 \approx (m_r)^{1/2}(\partial\vec{p}/\partial Q_1)]$ (see section 1.2.7) may lead to an overestimate of the experimental value of $\partial\vec{p}/\partial R_1$ if $k_2 > 0$.

As we have discussed in section 1.2.3, we expected the major effects due to vibrational mixing in the complex to occur for the X-Y stretch in an *aσ* acceptor, as a result of strong mixing with the D–X stretching vibrations. However, similar effects may be expected to occur, perhaps to a smaller extent, for all of the vibrations of the complex, depending upon the extent with which each mixes with the others. Because the mixing *is* generally expected to be small, these effects have usually been ignored, but they should always be investigated rather carefully.

1.2.9. SOLVENT EFFECTS IN VIBRATIONAL SPECTRA

We have referred to "solvent effects" to some extent in several of the preceding sections. These "solvent effects" refer to frequency shifts and intensity changes for vibrational bands of A, for example, on going from gas to solution. Let us consider briefly here the "solvent effects" on the new D–A vibrations in a complex and also "solvent effects" on the X-Y stretching vibration of an *aσ* acceptor molecule, both complexed with D and when XY is uncomplexed in solution.

Consider a D–A complex formed in the vapour phase. We have considered the forces acting between D and A in sections 1.1 and 1.2 and we have noted there that the energy of interaction between D and A measured in solution is expected to be decreased from that measured for the isolated gas-phase complex because of competition from the D,S and A,S interactions (see equations 1.23 and 1.24).

As discussed in section 1.1.2, we consider both the free donor and the free acceptor molecules in solution to be surrounded by solvent molecules. When the D and A molecules encounter each other in solution, a D,A interaction may replace a D,S and an S,A interaction. The properties of the complex measured in solution give the *difference* between the D,A interaction and the D,S, A,S interactions, as in equation 1.25 for the energy. If we examine the far-infrared spectrum of a D–A complex in solution, we expect to see an absorption spectrum characteristic of the D–A vibrations against a background of D,S and A,S vibrations. For a weak complex, we may have much higher concentrations of the latter species than we do of D–A, so that we can only expect to see absorption by D–A vibrations of the complex if they are sufficiently different (much more intense, or much higher in frequency) from the D,S or A,S vibrational spectrum.

Thus, we have information about the D–I stretching vibration in amine–I_2 complexes because the interaction is sufficiently strong so that its frequency rises above 150 cm^{-1} and its intensity increases so that it emerges from the general D,S and A,S background. Similarly, we have information about the behaviour of the I-I (or I-Cl, O-H, *etc.*) stretching vibration of *aσ* acceptors in complexes because that behaviour is sufficiently different from ordinary solvent behaviour (D,S or A,S) to be noticeable and interpretable.

Random "Solvent Effects" *vs.* Specific One-to-One Interaction Effects

Suppose that we have in a benzene solution two different kinds of I_2 molecules: one that is complexed and held in close contact with a single benzene molecule, for example by charge-transfer forces, and another that is randomly surrounded by benzene molecules—no one of which is in close, one-to-one contact. Clearly, there will be no contribution to $\partial\vec{\mu}_i/\partial R_1$ for the second molecule, so the ν(I-I) mode is still expected to have no intensity in the infrared spectrum for that molecule (assuming $R_1 \simeq Q_i$). If a benzene molecule near one end of that second I_2 molecule were oriented properly for the electrostatic contribution (equation 1.67) to $\partial\vec{\mu}_i/\partial R_1$ to be appreciable, we might also expect a second benzene molecule to be at the same orientation on the other end of the I_2 molecule. The change in field at the I_2 from the second benzene molecule would cancel the change in field from the first benzene molecule. Thus, it seems quite unlikely that the second term in equation 1.67 would contribute to $\partial\vec{\mu}_i/\partial R_j$ for this randomly surrounded I_2 molecule. Thus, we may ignore the contribution to $\partial\vec{\mu}_i/\partial R_1$ from columns 3 and 4 in table 1.10 for this I_2 molecule.

However, the contribution $\partial\vec{\mu}_i/\partial R_1$ from the first term in equation 1.67 (column 2 of table 1.10) may still be important. For the I_2 molecule with benzene molecules properly oriented on both ends, the magnitude of this first term (column 2 in table 1.10) must be re-examined. Clearly, $\vec{E}$ in equations 1.66 and 1.67 is a vector quantity, so that $\partial\vec{\mu}_i/\partial R_1$ in equation 1.67 has a sign depending upon whether the induced moment in the $+z$ direction increases or decreases as R_1 increases. It is clear that the induced moment from the benzene on one end cancels the induced moment from the benzene on the other end (and the same for the derivatives, $\partial\vec{\mu}_i/\partial R_1$), so that the intensification predicted in table 1.10 for the I-I stretching vibration simply will not be observed for the I_2 molecule with one oriented benzene on each end. (If the two benzene molecules are not attached to the I_2 molecule in exactly the same manner, then there *may* be some residual non-zero contribution to $\partial\vec{\mu}_i/\partial R_j$, but it will clearly be much less than expected from equation 1.67 and listed in table 1.10.)

For this randomly surrounded I_2 molecule, then, we may expect no intensification of the ν(I-I) band upon dissolving in pure benzene, even if partial orientation of the benzene molecules occurs. Similarly, we can expect essentially no effect for I_2 dissolved in a solution of excess benzene in n-heptane, if the interaction is random. We shall, of course, expect some intensification in the latter solution because 1:1 complexes are expected to form. However, for the randomly oriented acceptor molecule, the "electrostatic contributions" to the intensity are given by the Buckingham[30d] theory of "solvent effects" on intensity. For a vibrating molecule placed a spherical cavity within a continuum of solvent the dipole derivative $\vec{m}' = \partial\vec{p}/\partial R_j$ ($R_j \simeq Q_i$) is:

$$\vec{m}' = \vec{\mu}' + (2/a^3)\left\{\frac{\varepsilon - 1}{2\varepsilon + 1}\vec{\mu}_e\alpha' + \frac{n^2 - 1}{2n^2 + 1}\vec{\mu}'\alpha_e\right\} + O(a^{-6}). \quad 1.72$$

Here a is the radius of the spherical cavity, ε and n are the permittivity and refractive index of the medium, $\vec{\mu}'$ and α' are the dipole derivative and polarizability derivative for the isolated molecule, $\vec{\mu}_e$ and α_e are the equilibrium dipole moment and polarizability of the molecule, and $O(a^{-6})$ are terms which depend on the inverse sixth power of a. We see that Buckingham has included the contribution to the intensity from α'. Since $\vec{\mu}_e = 0$ for the molecules we have considered here, the contribution suggested by Hanna and Williams is included in the $O(a^{-6})$ term. Again we should remind ourselves that it was the failure of theories, such as that given by Buckingham,[30d] to explain the very large intensity changes observed for the X-Y stretching vibrational bands of *aσ* acceptors that led to the realization of the unusual nature of the latter and of the necessity (in order to explain them) for involving some effect, such as charge-transfer and the vibronic effects associated with it.

Now let us consider the other kind of I_2 molecule in the benzene solution; namely, the I_2 in the 1:1 complex with a single benzene molecule. This molecule is held in the axial configuration by the resultant of all the forces of interaction, and it is thought to interact much more strongly with this one benzene molecule than with the others surrounding it. Of course, the effect of the surrounding molecules is partly to cancel the terms contributing in equation 1.67 to $\partial\vec{\mu}_i/\partial R_j$, but that equation may still give a reasonable estimate of the upper limit for $\partial\vec{\mu}_i/\partial R_1$. However, as we have mentioned in the discussion above, the estimated value of $\partial\vec{p}/\partial R_1$ for the benzene–I_2 complex is about twice as large as the experimental value. It seems quite possible that the interaction between the 1:1 benzene–I_2 complex and the surrounding benzene molecules in solution *would* reduce the value of $\partial\vec{\mu}_i/\partial R_1$ considerably.

If only randomly oriented *A* (or I_2) molecules are present then there will be only one absorption band observed for I_2 in a mixed solution of *D* (benzene) in *n*-heptane, for example. As the benzene-to-heptane ratio is varied there will be a smooth shift in the frequency from that for ν_1 in heptane to ν_1 in benzene. However, as discussed in section 1.2.5, observation of a single absorption band smoothly shifting in frequency as *D* increases is no guarantee that complexing between *D* and *A* does not occur. If the complex I_2 molecules are in rapid equilibrium with the randomly surrounded I_2 molecules, then only one I-I absorption peak is expected with a frequency and intensity between those for random I_2 and complexed I_2. If the rate of the exchange between the complex and randomly surrounded I_2 is less than the difference in frequencies [ν_1 (random) $-$ ν_1 (complex)], then we expect to observe two peaks; if the exchange rate is greater than [ν_1 (random) − ν_1 (complex)], we expect to see only one peak (see reference 46, p. 328 ff.). The frequency and intensity are averages weighted by the relative concentrations of the random and complexed molecules. Thus, for benzene–I_2, the measured intensity calculated from equation 1.49 using the total I_2 concentration for c must be multiplied by the fraction of I_2 molecules that are complexed. This fraction, f, is given by:

$$f = \frac{K_c[bz]}{1 + K_c[bz]} = \frac{[bz - I_2]}{[I_2]_0} \qquad 1.73$$

The observed values of $\partial\vec{p}/\partial R_1$ listed in tables 1.8 and 1.10 have been obtained in this way.

"Solvent Effects" on Frequencies of D–A Complexes

There is also expected to be a relatively small solvent effect on frequencies measured for a stable 1:1 D–A complex in solution compared to those measured in the vapour. These effects have been discussed at great length[29]; they are generally quite small. These frequency effects were first predicted by Kirkwood[57] and by Bauer and Magat[58]; the prediction is that:

$$\Delta\bar{\nu}/\bar{\nu} = (\bar{\nu}_g - \bar{\nu}_s)/\bar{\nu}_g = C(\varepsilon - 1)/(\varepsilon + 2)$$
$$\approx C(n^2 - 1)/(n^2 + 2). \qquad 1.74$$

Here $\bar{\nu}_g$ is the wavenumber in the gas phase; $\bar{\nu}_s$ is the wavenumber in solution; ε is the permittivity of the solvent and n is its refractive index.

It is important to realize that C in equation 1.74 (called the Kirkwood-Bauer-Magat (or KBM) equation) is proportional to the intensity of the vibrational band. Hence for the very strong absorptions due to D–A stretching vibrations or to ν(X-Y) vibrations of $a\sigma$ acceptors, C is quite large. The frequencies of the latter are quite low (say equivalent to 200 cm^{-1}) so that if $\Delta\bar{\nu}/\bar{\nu}$ is on the order of 20×10^{-3}, then $\Delta\bar{\nu} \simeq$ 4 cm^{-1}. As we go from one solvent to another, n might change from 1.4 to 1.6; the predicted difference (from equation 1.74) in wavenumber is only 2 or 3 cm^{-1}.

It is quite clear that the rather large frequency shifts observed for these vibrations of complexes in different solvents (see section 3.4) cannot be due to this KBM effect, even when modified as described by Hallam[29], based on Buckingham's treatment[30], for example. One possible explanation may be that the exchange between specific 1:1 complexes and randomly surrounded complexes occurs very rapidly, as discussed in the previous section, so that the wavenumber recorded for the vibration of the D or A (or D–A) molecule in solution is the weighted average of the wavenumber of complexed D and of randomly solvated D. This explanation may possibly account for the very large frequency shifts observed, for example, in pyridine–ICl complexes from solution in benzene to solution in pyridine (see reference 37, for example). This possibility needs to be explored further. It does suggest that caution is needed in the choice of solvents if one is interested in properties of a 1:1 complex.

1.2.10. SUMMARY

We have seen in the previous sections the various factors contributing to the infrared spectrum of a complex. For weak complexes the new D–A vibration frequencies are low and the frequency shifts of D or A vibrations are small. Corresponding intensity changes are also expected to be small. For these weak complexes it may be very difficult indeed to decide whether or not charge-transfer is more important than electrostatic effects, for example, or to tell which electrostatic effect is most important, *etc.*

However, we can arrange different D molecules in expected order of electron donor ability, and study the changes in the spectrum of A (or of D–A vibrations) as we go from one D–A complex to another D′–A complex. To a somewhat rough approximation we can say that the equilibrium "electrostatic effects" may be expected to be the same for all of these D molecules, so that a trend in vibrational properties may be identified either with charge-transfer action or with electrostatic vibronic effects. By controlling the types of D molecules, it might be possible to distinguish between the two effects. As we have seen above, it will generally be necessary to carry out a normal coordinate analysis especially for stronger complexes, in order to be able to interpret the observed spectral changes.

Although we expect this approach (varying D systematically) to be generally applicable to all complexes, it has not been attempted for very many types of complexes. We have seen that charge-transfer effects on frequencies and intensities of vibrations for $b\pi$ donors and for $a\pi$ acceptors are expected to be small so that systematic variation of D might not produce changes large enough to be interpretable. Many of the small changes observed in the spectra of such molecules on complexing are probably due just to mixing of normal coordinates in the new complexed molecule, or are due to electrostatic effects. Some interesting effects are expected in the spectra of n donors and of v acceptors; the new D–A vibrations are also expected to be at high frequencies and so will have interesting and complicated interactions with the vibrations of D and A, making a good normal coordinate analysis necessary.

Thus, we may expect the largest easily interpretable charge-transfer effects to be observed in the vibrational spectra of $a\sigma$ acceptors, especially those with n donors, since those interactions tend to be the strongest. The charge-transfer effects on the ν(X-Y) frequency of a localized $a\sigma$ acceptor bond and on the new ν(D–X) band frequency are expected to compete with the electrostatic vibronic effects for dominance in the large spectral changes observed for these two vibrations.

It is important to stress the relationship between intensity changes and frequency shifts for these vibrations as D is changed systematically. The characteristic decrease in frequency and corresponding increase in intensity of the ν(X-Y) vibrational band as the D–A interaction increases has been stressed several times in the preceding discussion. We should stress that this correlation can be made more quantitative by some correlation plots. Huggins and Pimentel[59] plotted the intensity *vs.* the ν(O-H) frequency shift for hydrogen-bonded complexes. Later Person, Humphrey and Popov[42] and then Person, Erickson and Buckles[60] found empirically that $[(\partial\vec{p}/\partial R_1 - \partial\vec{p}^0/\partial R_1)]$ plotted *vs.* $\Delta k_1/k_1^0$ gave an approximate straight line for X-Y vibrations of complexes with ICl, Br_2, Cl_2, ICN and HOR $a\sigma$ acceptors. This is shown in figure 1.9. The values of $\Delta k_1/k_1^0$ and of $\partial\vec{p}/\partial R_1$ were obtained using the "diatomic approximation" discussed in sections 1.2.3 and 1.2.8; the general trend is clear. (These correlation graphs for hydrogen-bonded complexes are discussed more fully in section 2.3.4.)

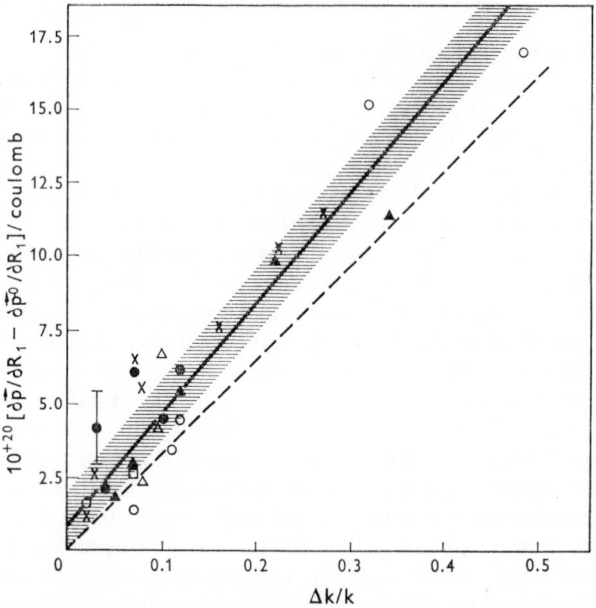

FIG 1.9 The observed change in dipole derivative $[\partial\vec{p}/\partial R_1 - \partial\vec{p}^0/\partial R_1]$ for the X-Y stretching vibration of halogen and hydrogen-bonded $a\sigma$ acceptors, plotted *vs* the relative change in force constant $\Delta k_1/k_1^0$. Here ● and ○ signify ICl complexes from two different studies; Δ signifies Br_2 complexes; □ signifies Cl_2 complexes; ▲ signifies ICN complexes (the ν(I–C) vibration) and × signifies data for the ν(OH) mode of a hydrogen-bonded complex. (Reproduced from *J. Amer. Chem. Soc.*, **82,** 20 (1960), by permission of the American Chemical Society.)

The reason for this linear relationship in figure 1.9 is not so clear. Since $\Delta k_1/k_1^0$ is expected from equation 1.30 to be approximately equal to F_{1N} and since the vibronic charge-transfer contribution to $[\vec{M}'_d(R_1)]$ to $\partial\vec{p}/\partial R_1$ is expected from equation 1.59 to be linearly proportional to F_{1N}, it is reasonable to expect $\vec{M}'_d$ also to be linearly proportional to $\Delta k_1/k_1^0$. However, equation 1.59 suggests that it is only coincidence that the data for all *aσ* acceptors fall on the same straight line plot. It is quite possible that the stronger interaction between D and I_2 also results in a larger electric field from D at the I_2 molecule that the electrostatic vibronic term (equation 1.67) may also increase linearly with $\Delta k_1/k_1$. At any rate the empirical behaviour is characteristic of the effect of this kind of interaction on the X-Y stretching vibration of *aσ* acceptors.

1.3 WHOLE-COMPLEX-MOLECULAR-ORBITAL CALCULATIONS OF VIBRATIONAL PROPERTIES

1.3.1. INTRODUCTION

The previous discussion (sections 1.1 and 1.2) illustrates both the power and the disadvantage of a resonance-structure theory. The advantage of such a theory is that it provides a quick and easy intuitive prediction of qualitative behaviour, that can be made at least semi-quantitative by careful study. For complicated systems, as for most molecular complexes, it has not been, and still may not be, possible to carry out more accurate quantum-mechanical calculations. Hence, this kind of simple resonance-structure theory has been, and continues to be, useful to the chemist. However, we find that when we try to force a resonance-structure theory to make exact quantitative predictions, there are many difficulties. Have we included all important resonance structures in our thinking? Exactly what are the properties of the "dative structure"? For that matter, what are the properties of the "no-bond structure"? Can we define the difference between "electrostatic effects" and "charge-transfer effects"? We have seen in the previous section numerous examples illustrating the difficulties with these and other questions. For these reasons there is a definite pressure forcing us to attempt to obtain accurate wavefunctions, Ψ_N, for the complex using standard *ab initio* quantum-mechanical methods. This pressure exists even though enough approximations still have to be made in such calculations for these very complicated systems that the results may not be entirely reliable.

(However, this situation is changing rapidly, and *ab initio* Hartree-Fock calculations for these large molecular complexes may soon be "routine". In a recent review Clementi[61] describes a computational procedure that reduces computer time (on an IBM 360/195 computer) to 80 minutes for calculation of the 0.8×10^9 integrals needed to obtain the energy and wavefunctions for the cytosine-guanine complex. In a private communication he reports that further improvements reduce the required computer time by another factor of two. Such achievements will surely enhance the importance of molecular orbital (*MO*) calculations for complexes in the near future.)

A qualitative general description of whole-complex-molecular-orbital methods applied to complexes has been given in chapter 12 of reference 1. In general, we expect the results of such calculations for 1:1 complexes in the absence of solvation effects to be better than might be expected from similar calculations for single molecules. This expectation is because we are considering the change in properties of the complex from the properties of the individual molecules; since the latter are closed-shell systems, we expect a single-electron-configuration *MO* description of the complex to have proper behaviour on dissociation.

In the whole-complex *MO* method, a geometric structure is assumed for the complex and the Schrödinger equation is solved for the entire complex treated as a single molecule. Using the variation method, the ground-state wavefunction, Ψ_N, and an estimate of the energy E_N are obtained. Other properties of the complex can be computed; for example, variation of the geometry maps out the energy as a function of the configuration of the nuclei. If the changes in geometry are chosen to correspond to changes in internal coordinates, we can fit a potential curve to the energy calculated at several values of the coordinate near equilibrium to evaluate the force constant associated with that coordinate. From the wavefunctions at each configuration $[\Psi_N(R)]$ we can calculate the dipole moment $\vec{\mu}_N(R)$ at each configuration (designated by R) using equation 1.27. From the results of this calculation of $\vec{\mu}_N(R)$ at different values of an internal coordinate R_j, we can compute,

$$\Delta\vec{\mu}/\Delta R_j = [\vec{\mu}_N(R) - \vec{\mu}_N(R_e)]/(R - R_e) \simeq \partial\vec{\mu}_N/\partial R_j. \qquad 1.75$$

Comparison of these calculated properties with the corresponding calculated properties of the free D and A molecule make it possible to predict the changes in these properties.

Levels of Approximation

Since the Schrödinger equation for a problem with as many electrons as in a complex can never be solved exactly, approximate methods are used. The wavefunction may be approximated by:

$$\Psi_N \approx \mathscr{A}[\phi_1(1)\phi_2(2)\phi_3(3)\cdots]. \qquad 1.76$$

Here the ϕ_i are spinorbitals from the ground state electronic configuration for the complex. They are given approximately by:

$$\phi_i = \sum_m a_{im}\chi_m$$

where the χ_m are "basis" functions—part of a complete, orthonormal set spanning the same function space as do the ϕ_i functions. The χ_m

may be "Slater-type-functions" (see reference 1) or Gaussian orbitals of the form $N x^l y^m z^n e^{-\alpha r^2}$, where N is the normalization constant and l, m, n are integers. They are specified by "orbital exponents" α which define the spread of the function in space and which may be varied in a variation method calculation to obtain the lowest energy, or which may be specified by some formula that attempts to predict values close to those that would give the lowest energy:

$$E_N = \int \Psi_N \hat{\mathscr{H}} \Psi_N \, dv. \tag{1.77}$$

In almost all procedures the a_{im} coefficients would be chosen by the variation method to give minimum energy. The number and kind of basis functions used in the *ab initio* calculation (since the set can never be *complete*) determines the quality of the calculation.

Early approximate procedures for this calculation used an approximate Hamiltonian operator–for example, the Hückel calculations ignored all electrons in a complicated molecule but the π-electrons, and the integrals in equation 1.77 were evaluated empirically. Such "semi-empirical" methods gave only questionable results for free molecules. The predictions for complexes based on such calculations may well be worse than those from the simpler two-structure resonance theory. (See the discussion in sections 12.3 and 12.4 of reference 1.)

More recently it has been possible to compute energies and wavefunctions without ignoring any of the electronic terms in the Hamiltonian operator using *S*elf-*C*onsistent-*F*ield *L*inear *C*ombination of *A*tomic *O*rbital *M*olecular *O*rbital (*SCF LCAO MO*) theory. No assumptions are made in these calculations, but all integrals are evaluated using the very large computers required for such calculations. These *ab initio* methods and results for molecules have been described recently in several places[62,63]. Reading such reviews should convince even the skeptical experimentalist that such carefully done quantum-mechanical theory has much to offer the spectroscopist. The problem now is that such *ab initio* calculations, if done with large enough basis sets (*i.e.* enough different χ_i functions) to be reliable, are still expensive in computer time for most large molecules or small complexes. However, such calculations will be of increasing importance in the future.[16]

If a more accurate representation of Ψ_N is required we can write

$$\Psi_N = \sum_i c_i \Phi_i,$$

where each Φ_i is a determinantal wave function of the form written for Ψ_N in equation 1.76. The coefficients c_i may be found in a "configurtion interaction" (*CI*) calculation. We shall not consider further this additional refinement for complexes, except to realize that the electron distribution, even for an accurate *SCF LCAO MO* determination of the wavefunction in Eq. 1.76, is not expected to be entirely correct if *CI* has been ignored. Such *CI* calculations are still too difficult to be practical for complexes.

For weak complexes between stable molecules, calculations ignoring *CI* do not properly give all the "polarisation" of the molecules as they are brought together. In particular the dispersion energy (E^{20} of equation 1.4) is not given, and the charge polarisation of the molecules as a result of this interaction will not be given unless *CI* is included. However, the examples discussed in the following parts of this section show a very considerable charge shift (which we shall call "polarization") which occurs due to the coulomb repulsions (and attractions) by the electrons and nuclei of D and A as they come together. Estimation of these large charge shifts requires a knowledge of the charge distribution in D and in A, given quite well by the single-configuration *SCF* calculation. These charge shift effects are expected to be the most important "polarisation" effects, at least for our considerations here.

In the following examples we shall not attempt, by any means, a complete review of the *MO* calculations of complexes, but instead attempt to illustrate some of the factors to be considered in trying to understand the results from such calculations.

H_3N–HCl: An Example of an *Ab Initio* Calculation

An example of the kind of *ab initio* calculation which can be made for complexes is provided by the NH_3, HCl system studied at different intermolecular distances by Clementi[64]. This theoretical exploration of the potential surface for the reaction between these two molecules to form the $NH_4^+Cl^-$ ion pair still serves as the model for this kind of calculation. This particular example may not be so useful for the understanding of *weak* complexes since the reaction goes all the way to the formation of the hydrogen-bonded ion pair. Still, the study of the way the wavefunctions and energies change as a function of the R_{DA} distance is very instructive. This calculation was not designed to give information about vibrational properties of the system; however, it is not difficult to see how it might be modified in the future to obtain directly the vibrational properties, as well as the wavefunctions and energies of the complex. Even so, the analysis of the calculated frequencies by Clementi and Gayles[64c] is indicative of what may be expected for complex, if still somewhat confusing.

From our point of view, it is of interest to examine first the calculations labeled by Clementi[64a] as *A-1*, *A-2*, and *A-3*. For these calculations, the HCl lies along the three-fold symmetry axis of the NH_3 molecule, H_3N–HCl, with the N-Cl distance fixed at 6.037 Å (11.40873 au). Calculations were then made at different values of the HCl distance. This particular *A* set is of interest to us because the HCl is far enough away from the NH_3 molecule (since the van der Waals N–Cl distance is expected to be about 4 Å) so that it maintains its identity, but it is still close enough to show some effects of the interaction and so is indicative of effects that may be expected for weak complexes.

For the *A* set the HCl equilibrium distance and vibrational wavenumber are calculated to be[64c] 1.319 Å and 2711 ± 260 cm^{-1}, respectively, compared to calculated values of 1.301 Å and 2882 ± 25 cm^{-1}, respectively, for the free HCl molecule. Thus the interaction at this N–Cl distance is predicted to result in a decrease in the HCl frequency of 171 cm^{-1}, corresponding approximately to a $\Delta k_1/k_1^0$ value of 0.12, with an associated increase of 0.018 Å in the equilibrium distance of the HCl molecule in the "complex" (with, however, considerable uncertainty in the calculated frequency shift and $\Delta k/k$ value). From the Mulliken[65] gross atomic population analysis of the *A*-set wavefunctions[64b] (*A-2*) we find that the charge distribution near equilibrium in the "complex" is:

$$\underset{(D)}{\overset{(+0.0013)}{[(H^{+0.3624})_3N^{-1.0859}]}} \cdots \underset{(A)}{\overset{(-0.0013)}{[H^{+0.2455}\,Cl^{-0.2467}]}}.$$

Here the total charge on each molecule in the complex is indicated by the upper figure in parentheses, indicating the extent of charge transferred (F_{1N}) toward formation of D^+–A^-. The charges are given in multiples of **e**, the electronic charge (**e** = 16.02 × 10^{-20}C).

This distribution in the complex is to be contrasted with charge distributions of:

$$\underset{(D)}{(H^{+0.3592})_3N^{-1.077}} \quad \text{and} \quad \underset{(A)}{H^{+0.1761}\,Cl^{-0.1761}}$$

for the free molecules ("case *F*")[64b], which are, of course, neutral. We see from this comparison that the *A* calculation predicts a net charge transfer from H_3N to HCl of 0.0013 **e** at this N–Cl distance, with a considerable change predicted in the charge distribution of the molecules in the "complex", compared to the free molecules. In addition we see a strong charge shift or "polarization" in each molecule in the D–A complex compared to the free D and A molecules. The charge distribution in H_3N in the complex may be interpreted as that for the free molecule with a large charge shift in the H_3N molecule due to the presence of the H^+-Cl^- dipole, resulting in an extra induced charge of +0.0032 **e** on each H atom (with, of course, a corresponding change in the charge on the N atom). We must conclude that an appreciable change from the equilibrium dipole is induced in H_3N as a result of the electrostatic interaction with HCl. A similar additional charge (+0.0694 **e**) is induced on the H atom of the HCl, due to its interaction with the $^+H_3N^-$ dipole (complicated by the charge transfer).

This result can be compared with the *C* set of calculations[64], made at an N–Cl distance of 3.920 Å (approximately the van der Waals distance). The calculated HCl equilibrium distance and vibrational wavenumber for that set were 1.319 Å and 2745 cm^{-1}, respectively, or about the same as for the *A* calculation at the longer N–Cl distance. The charge distribution (for calculation *C-2*) is:

$$[(\overset{(+0.0179)}{H^{+0.3726})_3N^{-1.0999}}] \cdots [\overset{(-0.0177)}{H^{+0.2305}\,Cl^{-0.2482}}].$$

Thus, we see that at the van der Waals distance the extent of electrostatic polarisation has increased slightly (from that occurring at $R_{NCl} =$ 6.04 Å) so that the positive charge on each H atom on H_3N, for example, increases by 0.01 **e**. In addition there is also an increase in the extent of charge transfer from H_3N to HCl to almost 0.02 **e**, predicting a value of F_{1N} about 15 times greater than that calculated at a N–Cl distance only 2 Å larger. We believe that the calculated large increase in charge-transfer effects with decreasing D–A distance is characteristic, as is the more gradual change predicted for the electrostatic effects when the D–A distance decreases.

We see that the calculated decrease in ν(HCl) wavenumber of 137 cm^{-1} (corresponding to $\Delta k_1/k_1^0$ of 0.095) is larger than the arguments in section 1.2 have led us to expect for the small extent of charge transfer. The calculation is for an isolated H_3N, HCl molecular pair, however, and so includes a sizable contribution from the "electrostatic effects" discussed in section 1.2.5. The large uncertainty in the calculated wavenumber of HCl in the complex suggests that no conclusion can yet be reached about these calculated values of $\Delta k_1/k_1$.

Clementi[64] goes on to discuss the charge rearrangements that occur at shorter N–Cl distances as the H atom changes from an attachment to the Cl atom to form the hydrogen-bonded NH_4^+ ion in the $NH_4^+Cl^-$ ion pair at equilibrium. However, our examination of the properties of the H_3N, HCl molecular pair at the van der Waals distance is indicative of the information from this calculation that is applicable to properties of weak complexes.

Before leaving these results, let us compare the charge distribution from the *C-1* calculation (with $R_{NCl} = 3.920$ Å, but $R_{HCl} = 1.195$ Å instead of 1.275 Å in *C-2*) with that given for the near-equilibrium distance charge distribution quoted earlier (for *C*-2). For the distorted HCl (C-1) the charge distribution is:

$$[(\overset{(+0.0155)}{H^{+0.3711})_3N^{-1.0978}}] \cdots [\overset{(-0.0156)}{H^{+0.1906}\,Cl^{-0.2062}}].$$

From this result we can estimate the change in the dipole derivative $[\partial\vec{p}/\partial R_1 - \partial\vec{p}^0/\partial R_1] = \vec{M}'_d(R_1)$ for the HCl molecule in contact with the H_3N molecule near the van der Waals N–Cl distance. To do this, we use the dipole derivative from the *ab initio* calculation by Cade and Huo[67] for the free HCl molecule, $\partial\vec{p}^0/\partial R_1 = -4.74 \times 10^{-20}$ C. (Here the negative sign means that the dipole increases in the H^+–Cl^- sense when the HCl bond length increases.)

Since the population analysis of the calculated wavefunctions gives us a "charge" q_i on each atom in the complex, it is convenient to write the corresponding "dipole moment" as:

$$\vec{p} = \sum_i q_i \vec{r}_i. \qquad 1.78$$

Here the origin of coordinates can be chosen at any convenient position, since the dipole moment is independent of this choice. For this complex, we chose the N atom to be the origin so that:

$$\vec{p} = \left[\sum_{i=1}^{3} q_{H_i} \vec{r}_{H_i}\right] + q_{H'} \vec{r}_{H'} + q_{Cl} \vec{r}_{Cl}.$$

Here the first sum can be replaced by $q_H \, d_{NH}(-\vec{k})$; $\vec{r}_{H'}$, the vector to the H atom in HCl, by $d_{NH'}\vec{k}$; and $\vec{r}_{Cl'}$ by $d_{NCl}\vec{k}$. Substituting, and evaluating the z-component (along the N-HCl direction) of the derivative, $\partial\vec{p}/\partial r_1$, we find, since $d_{NH'} = d_{NCl} - r_{HCl}$ (and $r_{HCl} = R_1$):

$$\partial\vec{p}/\partial r_1 = -(\partial q_H/\partial R_1)\, d_{NH} + (\partial q_{H'}/\partial R_1)(d_{NCl} - R_1) + (\partial q_{Cl}/\partial R_1)\, d_{NCl} - q_{H'}. \qquad 1.79$$

The changes in charge $(\partial q_i/\partial R)\,\Delta R_1$ when the HCl bond is increased by 0.08 Å are calculated from the two previously given charge distributions to be:

$$\overset{(+0.0024)}{[(H^{+0.0015})_3 N^{-0.0021}]} \cdots \overset{(-0.0021)}{[H^{+0.0399}\, Cl^{-0.0420}]}$$

Part of the change in charge on each atom is due to charge transfer, but most of the change on Cl, for example, is due to the change in polarisation. For Cl, we estimate that the latter is -0.0410 **e**, while the former only -0.00110 **e**.

Calculating the dipole moment derivative for the complex from equation 1.79 with this "change in charge distribution" gives:

$$\begin{aligned}\partial\vec{p}/\partial R_1 &\\ &= -\left(\frac{0.0015}{0.08}\right)1.01 + \left(\frac{0.040}{0.08}\right)[3.92 - 1.32] - \left(\frac{0.042}{0.08}\right)3.92 - 0.23\,\mathbf{e}.\\ &= -0.02 + 1.30 - 2.06 - 0.23\,\mathbf{e} = -1.01\,\mathbf{e}\\ &= -16.18 \times 10^{-20}\,\mathrm{C}.\end{aligned}$$

Hence, $[\partial\vec{p}/\partial R_1 - \partial\vec{p}^0/\partial R_1]$ for the H_3N, HCl pair at the N–Cl van der Waals distance is calculated to be $-16.18 + 4.74 = -11.44 \times 10^{-20}$ C. Of this, the contribution from charge transfer ($\vec{M}'_d(R_1)$ of equation 1.70) may be estimated by multiplying the charge transferred (0.0022 **e**) from the centre of the H_3N molecule to the centre of the

HCl molecule by the distance transferred to get Δp, and then dividing by ΔR_1, or:

$$\vec{M}_d'(r_1) = \frac{\Delta q}{\Delta R_1}(R_{DA}) = \left(\frac{-0.0022}{0.08}\right)(0.50 + 2.60 + 0.66)\,\mathbf{e}$$
$$= -0.10\,\mathbf{e} = -1.6 \times 10^{-20}\,\mathrm{C}.$$

From this comparison we see that for the H_3N, HCl pair at the van der Waals distance a sizable increase is predicted for the derivative $\partial\vec{p}/\partial R_1$. However, at that distance only about 10% of this increase is due to the effect of charge transfer; the remainder is due to the electrostatic vibronic effect [*e.g.*, to the terms containing $\partial q_{H'}/\partial r_1$ and $\partial q_{Cl}/\partial r_1$ (which here do also include vibronic charge transfer)].

This calculation can be expected to give only a rough guide to changes expected in dipole derivatives as the interaction progresses, since it was not designed for that purpose and so uses too great a change in the HCl bond length. However, these results indicate the potential value of such calculations. Certainly it will be important to investigate other model systems of this type to see whether it is always true that the electrostatic vibronic effect dominates the increase in $\partial\vec{p}/\partial R_1$, even at distances less than the van der Waals distance, as might occur for a real, weak, D–A complex. Some of the more approximate calculations cited in later parts of this section suggest that the charge transfer contribution may then dominate. The importance of these calculations as an aid in our understanding of these matters is clearly indicated by the analysis given here.

1.3.2. APPROXIMATE METHODS

Most complexes are complicated enough so that it is still (in 1972) not practical to carry out *ab initio* calculations. Thus, approximate quantum-mechanical methods, simpler than *ab initio* methods, but accurate enough to give useful information about complexes, are still of interest. One such procedure is the *C*omplete *N*eglect of *D*ifferential *O*verlap method (*CNDO*/2) developed by Pople and co-workers[68,69]. This method and another modification called the *I*ntermediate *N*eglect of *D*ifferential *O*verlap (*INDO*) method[70] are very nicely described in the book by Pople and Beveridge[69]. These approximate methods are based on *ab initio SCF* molecular orbital theory and are applied to all of the valence-shell electrons in the molecule. However, the rather drastic zero-differential overlap assumption is made for all products of different atomic orbitals, $\phi_\mu\phi_\nu$; thus the two-electron integrals appearing in the Fock matrix in the *SCF* method[69] are simplified considerably. Further simplification is achieved by an additional assumption that the remaining two-electron integrals depend only on the nature of the atoms A and B to which ϕ_μ and ϕ_ν belong, and not on the type of orbital[69]. These integrals (γ_{AB}) and the two-electron resonance integrals ($\beta_{\mu\nu} = \beta^0_{AB} S_{\mu\nu}$) are parameterized by choosing values for A and B atoms so that the calculated *CNDO* wavefunctions and energies for diatomic molecules agree fairly well with the values calculated by *ab initio* methods. (See reference 69, chapter 3.) The details of the parameterization and of the approximation differ for several existing different "*NDO*" procedures. The two most commonly used methods are *CNDO*/2 and *INDO*, thanks partly to the fact that computer programs are available from Quantum Chemistry Program Exchange[71]. Hence, many experimentalists have found it possible to use these programs to calculate properties of direct personal interest.

Although these calculations *are* approximate, and although the parameterization has not been completed for most of the atoms in the periodic table (only H through Cl), these relatively easy calculations are inexpensive enough to be done even for "complexes" involving these atoms. The results are believed to be correct enough so that they provide a valuable supplement to the simple resonance-structure theory (section 1.1.2) as a guide to behaviour of complexes. We shall describe a few such model calculations in the remaining section of this chapter.

1.3.3. SOME EXAMPLES OF *CNDO/INDO* CALCULATIONS

Let us now examine some results of *CNDO/2* or *INDO* calculations for some simple intermolecular interactions. There have been a relatively large number of such calculations (and also of *ab initio* calculations) for hydrogen-bonded systems, particularly for the water dimer (H_2O)[16,72] and for other simple hydrogen-bonded dimers involving HF, H_2O and NH_3[72a,73]. The *ab initio* study of $Li^+(H_2O)$ by Clementi and Popkie[8] (and additional, as yet unpublished results, on related systems by the same authors) provides an especially interesting example. This literature provides a good, if sometimes confusing, basis for a real understanding of the nature of the hydrogen-bond interaction. Much less work has been reported on *ab initio* or *CNDO/INDO* calculations for other *n–aσ* "complexes". However, an *INDO* calculation has been reported by Carreira and Person[74] for the interaction of H_3N with several different diatomic molecules: H_2, N_2, F_2, HF and Cl_2. The results from these exploratory calculations are indicative of the future of the theory of complexes.

Before examining a few of these calculations in detail, let us again stress one general problem in their interpretation. In the preceding sections (especially in sections 1.2.5, 1.2.7 and 1.2.9) we have emphasized that much of our knowledge about properties of complexes, and particularly vibrational properties, comes from experimental studies of these complexes in solution. The theoretical predictions are generally for interactions between isolated D,A pairs of molecules and so correspond to properties of such pairs in the gas phase. (However,

Clementi and Popkie[8] indicate how properties of isolated hydrated ions would be modified by the presence of additional water molecules, indicating how calculations may be made in the future to include solvation effects.) As we have suggested previously, some of the changes predicted from free D,A pairs to the one-to-one D–A complex may cancel in solution because of the effect of the solvent. Thus, comparison of calculated results (for example, the relative importance of electrostatic and charge-transfer effects) with experimental results from studies in solution must be made with caution.

Ab Initio and *CNDO* Results for the Water Dimer: $(H_2O)_2$

Let us now examine some of the calculated results for the water dimer[16,72]. The several *ab initio SCF* calculations differ in the size of the basis set used in the calculation. All of them use Gaussian basis functions and, in principle, the calculations with larger basis sets should give more accurate results. Some of the predicted properties of molecules (notably, the dipole moment) depend rather strongly on the choice of basis set, so some caution must be used in interpreting results from these calculations. An idea of the variation in properties with different basis sets can be obtained by comparing the predicted values for some one-electron properties of the H_2O monomer for different basis sets, shown in table III of reference 72e. However, the calculations are in semiquantitative agreement for many of the properties of the dimer.

It is generally agreed that the structure of the water dimer is the "linear form" indicated by $(H_2O)_2$, I, in figure 1.2. There is some debate about the exact angle θ between the bisector of the HOH angle in the molecule on the RHS of figure 1.2 with the O-H—O bond, —calculated results vary from 122°[16] to 140°[72e]—but there is good agreement on the general structure of the dimer. The different treatments give slightly different estimates of the O—O distance at equilibrium, ranging from 2.6 to 3.0 Å, with some indication that the larger values are better estimates. Incidentally, earlier discussion about the structure of this dimer was somewhat confused by the tentative conclusion from infrared spectral studies of the spectrum of $(H_2O)_2$ isolated in an nitrogen matrix that the structure of the dimer was cyclic[75] [$(H_2O)_2$, II in figure 1.2]. This conclusion has been reversed by a more recent study of that system, and it is now clear that the infrared spectrum is consistent with the linear structure for the dimer[76].

Calculated values for the energy of formation of the dimer vary from -10.75 to -52.72 kJ mol^{-1}, with the smaller values being preferred. In comparing results from previous calculations with their own, Hankins, Moskowitz and Stillinger[72e] point out that the small-basis-set calculations have large errors in the calculation of energy, compared to the Hartree-Fock limit. These errors are such that the calculation tends to predict smaller equilibrium distances for a molecular D,A pair and a larger energy of interaction. This general tendency of a calculation to exaggerate properties of complexes must be considered in evaluating calculated properties of complexes. However, the calculated energy of interaction, predicted equilibrium angular geometry, and O—O distance for the water dimer seem to be in quite reasonable agreement with what is known from experiment about these properties[17,28].

The charge distribution in the water dimer calculated from Mulliken gross atomic population analysis[65] of the molecular wavefunctions is of considerable interest. The results (in multiples of the electronic charge, **e**) from Hankins, Moskowitz and Stillinger[72e] at their computed equilibrium O—O distance (3.0 Å) are:

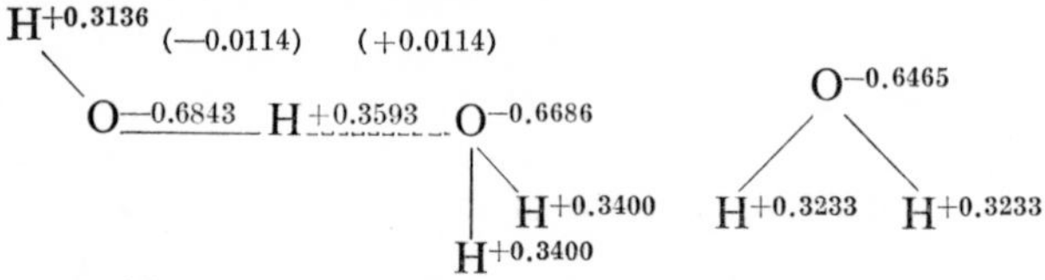

(A) (D) (monomer)

The extent of charge transferred from D to A is calculated to be 0.0114 **e** at this O—O distance. (The charge distribution on the monomer shown here[72e] is somewhat different than that given by Clementi and Popkie[8].) The results are quite similar to those of Kollman and Allen[72b] who found that the extent of charge transfer was only 0.0064 **e**.

Kollman and Allen[72b] have specifically considered the vibrational properties of the water dimer. They calculated the energy for the monomer at several configurations in which they stretched the OH bond and held the other constant. The force constant they calculate is then f_r (= k_1) the coefficient of the internal coordinate r_{OH} in the expression for the potential energy. For monomeric H_2O they calculated f_r to be 1246 N m^{-1}, compared with an experimental value of 845 N m^{-1} reported by Mills[39]. Because this calculated value is in error, the significance of the value they[72b] calculate in the same way for the OH bond of the acceptor in the dimer (f_r = 996 N m^{-1}) must be questioned. The prediction that f_r is smaller in the dimer than in the monomer may be correct; the magnitude of the predicted change ($\Delta k_1/k_1 = 0.20$) is probably too large.

In the better (*ab initio*) calculation by Hankins, Moskowitz and Stillinger[72e], the calculated force constant in the H_2O monomer is f_r = 836 N m^{-1}, in excellent agreement with the experimental value (845 N m^{-1}). Their calculated bending force constant is f_α = 49 N m^{-1}, compared to 70 N m^{-1} given by Mills[39]. The latter result can be compared to the value of f_α = 95 N m^{-1} calculated by *CNDO/2* methods[69]. The latter gives a stretching force constant of 1702 N m^{-1} (1639 N m^{-1} from *INDO* calculations) for the diatomic OH radical, compared to an experimental value of 780 N m^{-1}. These comparisons show that accurate quantum-mechanical calculations can predict values for stretching force constants, in excellent agreement with experiment, but that results from less accurate calculations do have to be treated cautiously.

Kollman and Allen[72b] have also calculated the change in dipole moment ($\partial\vec{p}/\partial r_1$) for stretching the OH bond of A in the O-H—O part of the dimer, compared to stretching a single OH bond in the monomer ($\partial\vec{p}^0/\partial r_1$),

$$(\partial\vec{p}/\partial r_1 - \partial\vec{p}^0/\partial r_1) = 3.2 - 0.7 = 2.5\ \mathrm{D\,Å^{-1}} = 8.34 \times 10^{-20}\ \mathrm{C},$$

at their calculated equilibrium O—O distance of 2.75 Å. No experimental values are available, but the intensity for the OH stretching normal coordinate in the dimer does not appear[75,76] to be quite so large as predicted here, possibly because the calculation is made for an O—O distance that may be too short. Hankins, Moskowitz and Stillinger[72c] did not make a similar calculation for comparison.

Because many of these properties (including $\partial\vec{p}/\partial r_1$) do depend rather strongly on the O—O distance, it is of some interest to estimate that variation. For this purpose the admittedly less accurate *CNDO/INDO* calculations can still be of considerable interest. Kollman and Allen[72a] have shown that the results at equilibrium [geometries, except for θ (see figure 1.2); energies of formation; charge distributions] from *CNDO* calculations for dimers are in surprisingly good agreement with these calculated by *ab initio* methods.

Figures 1.10 and 1.11 show a plot of some properties of the water dimer, calculated by the *INDO* method[77], as a function of O—O distance. At each value of the O—O distance shown by a point on the curve, the energy and wavefunctions were calculated for the dimer, assuming the linear geometry of figure 1.2. Then the distances of the OH bonds in the acceptor water molecule were varied according to the symmetric [$R_1 = (1/\sqrt{2})(\Delta r_1 + \Delta r_2)$] or antisymmetric [$R_3 = [(1/\sqrt{2} \times (\Delta r_1 - \Delta r_2)$] symmetry stretching coordinates and the energies and dipole moments evaluated. Since these two symmetry coordinates are almost identical with the corresponding normal coordinates (after mass-weighting), the results of the calculation can be fitted to a potential curve to obtain $\bar{\nu}_1$ and $\bar{\nu}_2$ (the symmetric and antisymmetric stretching wave numbers) and also the value of $\partial\vec{p}/\partial R_1$, $\partial\vec{p}/\partial R_3$ and $\partial\vec{p}^0/\partial R_1$, $\partial\vec{p}^0/\partial R_3$. As we have seen above, the *INDO* calculation is not expected to give very good agreement for stretching force constants or stretching frequencies. Hence, the calculated OH stretching frequency for the water monomer was multiplied by a scaling factor to force it to agree with the observed harmonic frequency[39]. The frequencies calculated for the dimer were then all multiplied by the same scaling factor.

In figures 1.10 and 1.11 the energy of formation of the dimer from the *INDO* calculation is -54 kJ mol^{-1}, much larger than the better estimates of about -20 kJ mol^{-1}, and the calculated equilibrium

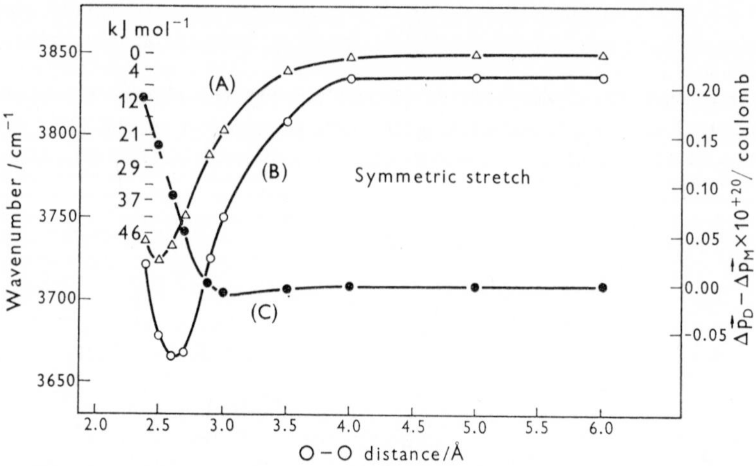

FIG 1.10 Some properties calculated by INDO methods[77] for the water dimer, shown as a function of the assumed O—O distance, for configuration I of figure 1.2. Curve A shows the calculated potential energy, read from the small inner scale on the left side of the figure. Curves B and C give, respectively, the calculated frequency and dipole moment change on stretching, for the symmetric stretching mode (R_1) in the acceptor H_2O molecule. The scale for B is on the left side and that for C on the right side of the figure.

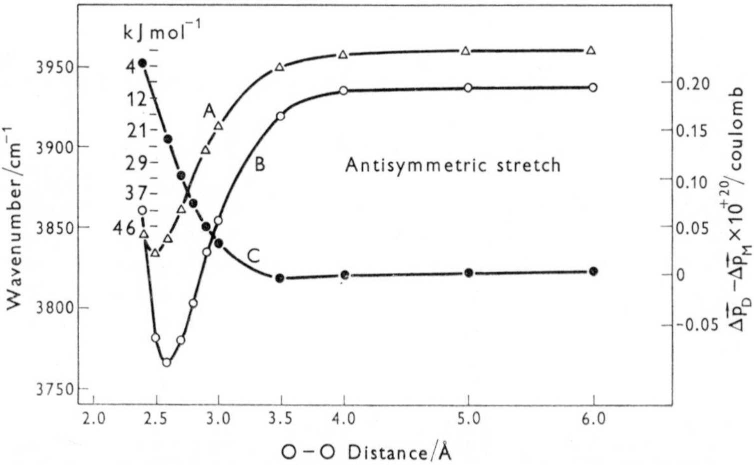

FIG 1.11 Some additional properties calculated by *INDO* methods[77] for the water dimer. The scales are the same as for figure 1.10. These calculations for frequency changes (Curve B) and dipole moment changes (Curve C) are for the antisymmetric stretch (R_3).

O—O bond distance (2.6 Å) is much too short. However, properties of the dimer calculated by *CNDO/2* or *INDO* methods at the *experimental* O—O distance may be expected to be reasonably close to the values to be expected from an experimental study[72a,74]. At an O—O distance of 3 Å, $\Delta\bar{\nu}$ is predicted to be about 100 cm^{-1} for both frequencies of the dimer, a bit larger than the observed shifts (10–28 cm^{-1} for the symmetric stretch or 7–85 cm^{-1} for the antisymmetric stretch) for the dimer in a nitrogen matrix. At that O—O distance only a slight increase in $\partial\vec{p}/\partial R_3$ is predicted: $(\partial\vec{p}/\partial R_3 - \partial\vec{p}^0/\partial R_3 \simeq 0.7$ D Å^{-1} = 2.33×10^{-20} C, and essentially no change is predicted for $\partial\vec{p}/\partial R_1$.

Note that both frequencies are predicted to continue to drop, and that the dipole derivatives are predicted to increase, as the O—O distance becomes shorter. These figures show clearly that the values calculated for these properties depend strongly on the value assumed for the O—O distance.

Finally, figure 1.12 shows the change in the extent of the charge transfer predicted from the INDO calculation[77] as the O—O distance is decreased. At 3.0 Å, the extent of charge transfer is predicted to be 0.01 e, in excellent agreement with Hankins, Moskowitz and Stillinger[72e] (diagram, p. 86), but the extent of charge transfer is predicted to increase drastically as the O—O distance decreases.

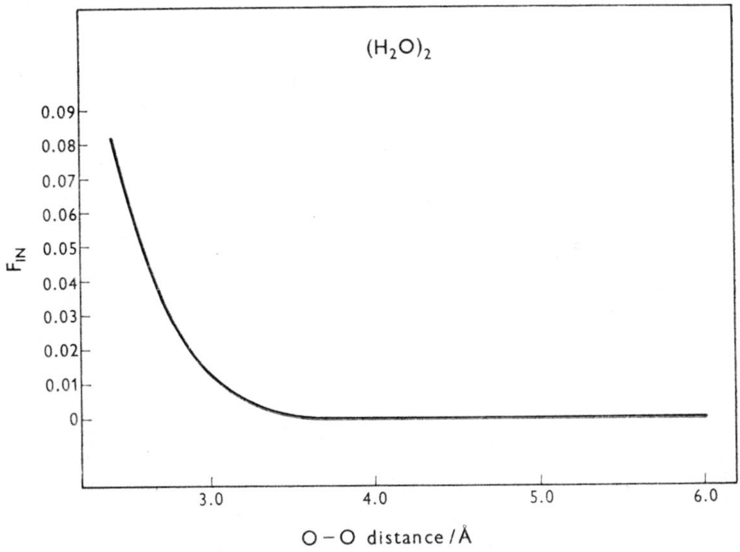

FIG 1.12 Extent of charge transfer (F_{1N}) from the donor H_2O molecule to the acceptor in the water dimer, calculated by INDO methods[77], as a function of O—O distance.

Returning to the charge distribution for the water dimer given above, the biggest effect in the dimer is a sizable charge shift (or electrostatic polarisation) from the charge distribution of the isolated molecule. Kollman and Allen[73] have previously stressed the importance of this "charge shift" in hydrogen bonding, in general, and in the water dimer in particular. In terms of resonance-structure theory[78,79], (see section 1.1.1) the wavefunction for the hydrogen-bonded complex is (by analogy with equation 1.1) a combination of the following wavefunctions from valence bond structures: Ψ_0(X-H, Y) (no-bond); Ψ_2(X^--H^+, Y) (X-H polarised); and Ψ_1 (X^-, H-Y^+) (dative). (For the water dimer X is an OH radical, and Y is the electron donor H_2O molecule.) The *ab initio* calculations for water dimer may be interpreted in these terms leading to the conclusion that the importance of Ψ_2 (which also contributes to the wavefunction for free H_2O) is enhanced in the dimer. Kollman and Allen[73] stress that this result is general for all hydrogen-bonded complexes.

Figure 1.13 summarizes the *changes* in the charges shown above for

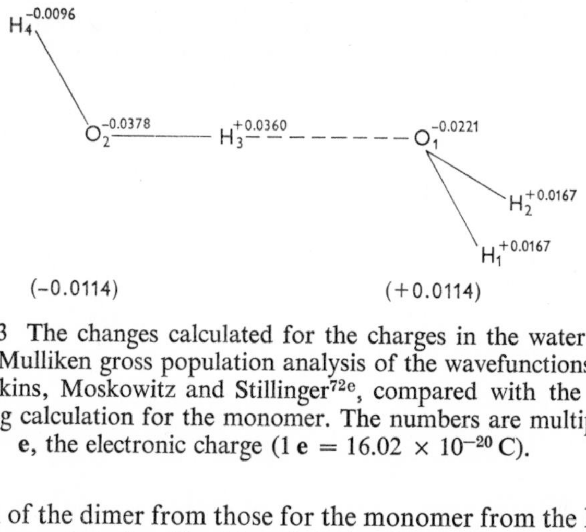

FIG 1.13 The changes calculated for the charges in the water dimer from a Mulliken gross population analysis of the wavefunctions given by Hankins, Moskowitz and Stillinger[72e], compared with the corresponding calculation for the monomer. The numbers are multiples of **e**, the electronic charge (1 **e** = 16.02 × 10^{-20} C).

each atom of the dimer from those for the monomer from the Mulliken population analysis by Hankins, Moskowitz and Stillinger[72e]. The charge rearrangement can be broken down into steps: (1) 0.0114 **e** is transferred from the O_1 atom of D to the O_2 atom of A; (2) the O_2H_3 bond in A is polarised by the negative O_1 atom neighbour increasing q_3 (the charge on H_3) by 0.0360 **e** and decreasing q_{O_2} by the same; (3) the resulting polarised O_2H_3 bond induces some polarisation in the O_2H_4 bond (decreasing q_4 by 0.0096 **e** and increasing q_{O_2} the same); (4) finally, the positive H_3 atom neighbour induces polarisation of the O_1H_1 and O_1H_2 bonds of D (increasing q_1 and q_2 by 0.0167 **e**). From the changes in charge distribution given in figure 1.13, the increase in dipole moment $\Delta\vec{p}$ from that for two oriented non-interacting monomers can be estimated using equation 1.78 (based on an origin at the O_2 atom):

$$\begin{aligned}\Delta\vec{p} = {} & \Delta q_4 d_{OH}[\sin(180^\circ - \alpha)\vec{i} - \cos(180 - \alpha)\vec{k}] \\ & + \Delta q_3 d_{OH}\vec{k} + \Delta q_{O_1} d_{OO}\vec{k} \\ & + 2\,\Delta q_1\{[-d_{OH}\cos(\alpha/2) \times \sin(180^\circ - \theta)]\vec{i} \\ & + [d_{OO} + d_{OH}\cos(\alpha/2)\cos(180 - \theta)]\vec{k}\}\end{aligned} \qquad 1.80$$

Substituting values from figure 1.13 into equation 1.80, we calculate $|\Delta\vec{p}| = 0.09063\ \mathbf{e}\,\text{Å} = 1.45 \times 10^{-30}$ C m. We may estimate the charge-transfer contribution to this increased dipole moment of the dimer to be the charge transferred (0.0114 **e**) multiplied by the distance of transfer (3.0 Å) or $\Delta\vec{p}_{CT} = 0.0342\ \mathbf{e}\,\text{Å} = 0.55 \times 10^{-30}$ C m. Thus the contribution from charge transfer to the calculated increase in dipole moment is only about 38% of the total.

The magnitude ($0.0360\ \mathbf{e}\,\text{Å} = 0.12 \times 10^{-30}$ C m) of $\Delta\vec{\mu}_2$ the dipole induced in the O_2H_3 bond in step (2) above, is about what would be estimated electrostatically for the dipole induced in that bond from the neighbouring $^-O_1H_1^+H_2^+$ dipole if the polarizability along the O_2H_3 bond ($\alpha_{\parallel}$) is about 1.9×10^{-40} C m^2 V^{-1}, a value that seems fairly reasonable. The apparent success of this calculation and its evaluation of the parameter ($\alpha_{\parallel}$) needed for the application of the Hanna-Williams electrostatic theory (section 1.2.7) to the dipole derivative ($\partial\vec{p}/\partial r_1$) for this OH bond in the dimer suggests that we may attempt to estimate the change in $\partial\vec{p}/\partial r_1$ from monomer to dimer due to the electrostatic effects, using these calculated charges for the dimer.

If population analyses for wavefunctions of the water dimer at different values of the O_2H_3 distance were available, then $\partial\vec{p}/\partial r_1$ for the dimer could be estimated, as for the H_3N, HCl system. Since such results are not available, it is worthwhile to estimate the change in $\partial\vec{p}/\partial r_1$, from monomer to dimer using the Hanna-Williams ideas[45], and then to compare them with the value calculated by Kollman and Allen[72b].

The change in dipole moment for the water dimer may be estimated by analogy with equation 1.79. In that equation the first term gives the vibronic (mostly electrostatic) contribution due to the change in charge for the hydrogen atom of A not involved in the hydrogen bond; here there will be a similar term for H_1 and H_2, the atoms on D. These terms are expected to be small—perhaps the total is less than half the other vibronic term described below. The major vibronic contribution is contained in the next two terms of equation 1.79, which include both the electrostatic vibronic term and the vibronic charge-transfer contributions. The former may be estimated by calculating the Hanna-Williams[45] induced dipole $\Delta\vec{\mu}_i$, of equations 1.66–1.69 dividing by

Δr_1; the result is expected to equal approximately the electrostatic vibronic contribution of the second two terms of equation 1.79.

To estimate $\Delta\vec{\mu}_i/\Delta r_i$ from the Hanna-Williams[45] theory, we use the previously estimated value of $\alpha_{\|}$ (1.9×10^{-40} C m^2 V^{-1}, or about $\frac{1}{4}\alpha_{\|}$ for Cl_2), estimating $\partial\alpha_{\|}/\partial r_1$ to be about $\frac{1}{10}$ the value of Cl_2, the field due to O_1 to be about half that of benzene, and the field gradient at H_3 to be about the same as for benzene–Cl_2. From these, the electrostatic vibronic contribution to $\partial\vec{p}/\partial r_1$ of the dimer is about $+1.00 \times 10^{-20}$ C. About half this value may be just from the change in q_3 from monomer to dimer (an increase of about $+0.53 \times 10^{-20}$ C).

In addition, $\partial\vec{p}/\partial r_1$ may be expected to change in the dimer because the extent of charge transferred from D to A changes as the O_2H_3 bond length changes ($\vec{M}'_d(r_1)$ in equation 1.70). Comparing our estimate above for the electrostatic vibronic contribution to $\partial\vec{p}/\partial r_1$ with the total change calculated by Kollman and Allen[72b] at $R_{OO} = 2.6$ Å, and quoted above [*e.g.* $(\partial\vec{p}/\partial r_1 - \partial\vec{p}^0/\partial r_1) = 8.33 \times 10^{-20}$ C], the total value of this derivative is either drastically overestimated or else the contribution from the vibronic charge-transfer effect for the water dimer is much larger than the electrostatic effect estimated above. Even if the parameter values for the Hanna-Williams theory have been underestimated here, it seems unlikely that the electrostatic vibronic contribution to $[\partial\vec{p}/\partial r_1 - \partial\vec{p}^0/\partial r_1]$ for the O_2H_3 bond in the water dimer could be greater than those given in table 1.10 for the halogen complexes (or about 2–6×10^{-20} C). Assuming that the total change in this derivative calculated by Kollman and Allen[72b] is overestimated by a factor of two, then the electrostatic vibronic contribution may be from $\frac{1}{8}$ to $\frac{1}{2}$ of the total, with the vibronic charge-transfer term contributing the remainder. Of course, a calculation similar to that by Kollman and Allen, but made at $R_{OO} = 3.0$ Å might predict a considerably smaller vibronic charge-transfer contribution at the longer distance.

(Note that the intensity of the shifted "symmetric and antisymmetric OH stretching vibrations" of the complexed acceptor water molecule will be determined not only by $\partial\vec{p}/\partial r_{O_2H_3}$ but also by an essentially unchanged (from the monomer) $\partial\vec{p}/\partial r_{O_2H_4}$, weighted by the normal coordinate coefficients. Hence, the change in intensity observed will be less than that expected from the large change predicted above for $\partial\vec{p}/\partial r_1$.)

Kollman and Allen[72a,73] report calculations for a large number of hydrogen-bonded complexes. The charge distributions predicted at equilibrium are similar to those for the water dimer, with small charge transfer effects and fairly large charge shifts. Since the calculated amount of charge transfer does depend very critically on the assumed value of the O—O hydrogen bond distance (see figure 1.12) and since these distances are generally not known accurately, either from experimental work or from calculations, no real conclusion can be drawn at the present time concerning the actual magnitude of charge transfer in these hydrogen-bonded complexes. A small amount of charge transfer probably occurs, with an important effect on the static dipole moment of the complex, and also on the intensity and frequency of the OH stretching vibration.

Interaction of NH_3 with Some Diatomic Molecules

In an attempt to examine further the questions raised in the preceding discussion, let us consider the results from some *INDO* calculations for NH_3 paired with different X_2 molecules;[74] first for H_3N,F_2, and then compare those with the results for the H_3N,N_2 system. For the latter molecular pair we do not expect charge transfer, to be important; in the former, it may have some effect. It would be more interesting to calculate the interaction between H_3N and I_2 (instead of F_2); we cannot do so because the *CNDO*/2 or *INDO* parameters have not yet been well determined even for Cl_2. However, the *CNDO*/2 calculation was made for NH_3 with Cl_2[74], using the available parameter values. That system is quite similar in its calculated behaviour to the H_3N,F_2 system, and we shall not discuss it further.

The variation in several different properties of the H_3N, XY pairs as the X-Y molecule approaches the N atom along the three-fold symmetry axis (and also as it approaches along one of the NH bonds) are shown in figures 2–9 of reference 74. Here, let us concentrate on some of these results for just two systems: H_3N,N_2 and H_3N,F_2, with three-fold symmetry, in order to contrast the *INDO* predictions for these two systems to try to detect any difference due to the expected difference in extent of charge transfer.

In this comparison we must recognize the tendency for the *INDO* calculation to underestimate the repulsion energy (see reference 74), so that the *calculated* equilibrium N–X distance is expected to be too small, the calculated energy at equilibrium too large, etc. Furthermore, we may keep in mind that calculations for some of the systems (for example, H_2 approaching along an N-H bond) predicted *no* energy minimum, so that the calculations do, in fact, distinguish between different kinds of behaviour.

The calculated energy, extent of charge transfer (F_{1N}), frequency shift of the ν(XY) band ($\Delta\bar{\nu}_1/\bar{\nu}_1$), and dipole moment derivative ($\partial\vec{p}/\partial r_1$) for stretching the XY bond are given in figure 1.14 for the C_{3v} pairs H_3N,N_2 and H_3N,F_2 as the function of NX distance. We note that the *INDO* calculation does predict a significant (if exaggerated, by underestimating repulsion) stabilization of both the H_3N,N_2 and H_3N, F_2 molecular pairs. The van der Waals radius of F is 1.35 Å, that of N is 1.5 Å, so that one expects the van der Waals distance for H_3N,N_2 to be $R_{NN} = 3.0$ Å; for H_3N,F_2 the van der Waals distance is $R_{NF} = 2.85$ Å. The calculated energy minima in figure 1.14a came at 1.8 and 1.6 Å, for R_{NN} and R_{NF} respectively, with corresponding calculated stabilization energies of -28.87 kJ mol^{-1} and -43.51 kJ mol^{-1}. These calculated equilibrium NX distances may be compared with the

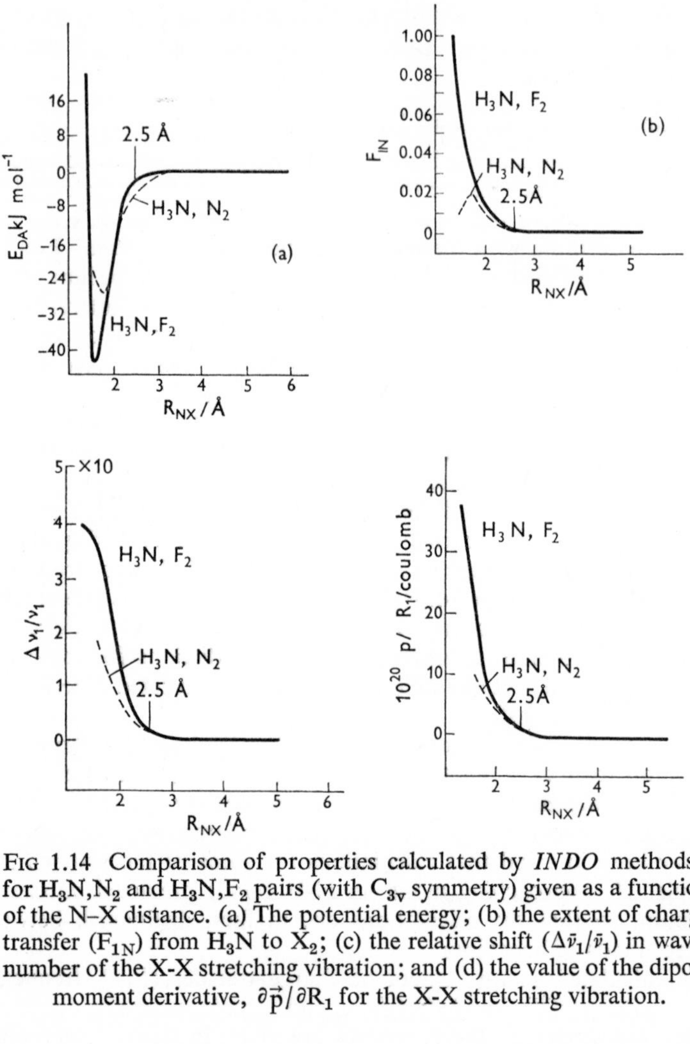

Fig 1.14 Comparison of properties calculated by *INDO* methods[74] for H_3N,N_2 and H_3N,F_2 pairs (with C_{3v} symmetry) given as a function of the N–X distance. (a) The potential energy; (b) the extent of charge transfer (F_{1N}) from H_3N to X_2; (c) the relative shift $(\Delta\tilde{\nu}_1/\tilde{\nu}_1)$ in wavenumber of the X-X stretching vibration; and (d) the value of the dipole moment derivative, $\partial\vec{p}/\partial R_1$ for the X-X stretching vibration.

sum of covalent radii; $R^{cov}_{NN} = 1.40$ Å; $R^{cov}_{NF} = 1.34$ Å. For strong complexes of halogens with *n* donors the DX distance has been found to be about 0.25 Å larger than the sum of covalent radii; for weaker complexes with dioxan, the O–I distance was 0.58 Å larger than the sum of covalent radii for dioxan–ICl, rising to an increase of 0.91 Å for the O–Br distance in dioxan–Br_2 and to 1.02 Å for the O–Cl distance in dioxan–Cl_2, where the distance is 0.53 Å less than the sum of van der Waals radii. (See chapter 5 of reference 1.) We conclude that the true equilibrium distances for H_3N,N_2 and H_3N,F_2 molecular pairs may be between 2.4 and 3.0 Å for the former and between 2.3 and

2.8 Å for the latter. Probably the experimental distance for H_3N,N_2 may be about 2.7 Å, while that for H_3N,F_2 might be about 2.5 Å.

Since it is found empirically that the properties calculated for molecules or for "complexes" by *INDO* (or *CNDO*) methods at the *experimental* equilibrium distances are often in quite good agreement with *ab initio* calculations [*cf.* $(H_2O)_2$], and also with experimental properties, let us examine figure 1.14 closely at $R_{NN} = 2.7 \pm 0.3$ Å and $R_{NF} = 2.5 \pm 0.3$ Å in order to get a better idea of the probable experimental behaviour of these two molecular pairs. The properties (in figure 1.14) at the shorter calculated equilibrium distances of these pairs are interesting as the (probably) limiting values for these molecular properties at very short NX distances.

From figure 1.14a, the calculated energy of stabilization for the H_3N,N_2 pair at 2.7 Å (and also for the H_3N,F_2 pair at 2.5 Å) is −2.09 kJ mol^{-1}. From figure 1.14b, F_{1N} at the *calculated* equilibrium distances is 0.018 for H_3N,N_2 and 0.078 for H_3N,F_2; at 2.7 Å $F_{1N} \simeq 0$ for H_3N,N_2 and $F_{1N} \simeq 0.003$ for H_3N,F_2 at 2.5 Å. From figure 1.14c the relative frequency shifts at the *calculated* equilibrium distances are 0.012 for the N_2 pair and 0.041 for the F_2 pair; at 2.7 Å for the N_2 pair and also at 2.5 Å for the F_2 pair, $\Delta\bar{\nu}/\bar{\nu} \simeq 0.001$. Finally, from figure 1.14d we find that at the *calculated* equilibrium distances, $\partial\vec{p}/\partial R_{XY} = \partial\vec{p}/\partial R_1 = 7.00 \times 10^{-20}$ C for N_2 and 20.68×10^{-20} C for the F_2 system. At 2.7 and 2.5 Å, respectively, the values are 0.0 for N_2 and about 0.10×10^{-20} C for the H_3N,F_2 pair.

From these comparisons, we see that the two different molecular pairs are expected to have almost identical properties at the probable experimental NX distances of 2.5–2.7 Å. Only the extent of charge transfer is calculated to be *slightly* different for the two systems. However, as the NX distance decreases for each pair the value of F_{1N} for H_3N,F_2 becomes considerably greater than does that for H_3N,N_2, as does the predicted stabilization energy, and the frequency shift and intensification of the XY stretching vibration. Hence, if we have slightly overestimated the experimental distances, charge transfer could contribute to significant differences in the behaviour of H_3N,F_2 compared to H_3N,N_2.

These comparisons illustrate the concept (see section 1.1.1) that the charge transfer is an "extra" or "special" effect. The extent of charge transfer in the H_3N,F_2 pair cannot be very great at the experimental equilibrium distance; however, it can have an important effect on the properties of the D,A pair where it occurs, as illustrated here in figure 1.14 at shorter NF distances, as contrasted with the behaviour predicted for the H_3N,N_2 pair.

Let us now examine the calculated charge distribution for the H_3N,F_2 pair. Figure 1.15 shows the charge distribution calculated near the *calculated* equilibrium distance. As we have seen in the discussion above, the extent of charge transfer at this distance is expected to be

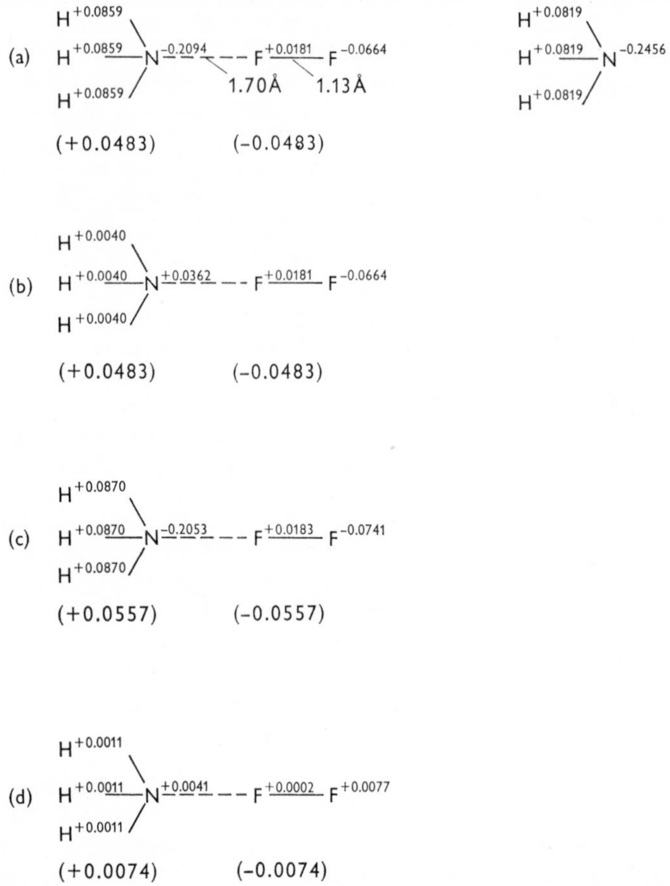

FIG 1.15 The charge densities on each atom from the INDO calculation[74] of H_3N,F_2, for a configuration near the *calculated* equilibrium N–F distance (1.70 Å) and at the *calculated* F-F equilibrium distance (1.13 Å). (a) The charge densities on each atom of the complex compared with those for free H_3N. (For free F_2 the charges are, of course, zero.) (b) The *difference* in charge densities for atoms in the complex from those for the atoms in the free molecules. (c) The charge densities for the complex in (a) after stretching the F-F bond by 0.02 Å. (d) The *difference* in charge for the complex with the F-F bond stretched by 0.02 Å from the charges given in (a) for the complex with the F-F bond at its equilibrium value. The numbers give the charges as multiples of **e**, the electronic charge, and the numbers in parenthesis give the charge transferred from the H_3N molecule to the F_2 molecule.

greater than for the pair at its true equilibrium distance. However, Figure 1.15 does show quite clearly *what* the limiting charge-transfer effects may be, and gives an idea (also much exaggerated) of how they compare with the electrostatic polarization effects expected for this molecular pair.

Figure 1.15a summarizes the charge densities from the *INDO* calculation for the pair at the near-calculated-equilibrium NF distance of 1.70 Å. There is a charge transfer from H_3N to F_2 of 0.0483 **e**; if the charge transfer is assumed to be into the symmetrical $a\sigma$ orbital of F_2, half of the charge (0.0242 **e**) is transferred to each F atom. The negative dipole of the H_3N molecule induces an electrostatic polarisation charge of +0.0423 **e** in the F atom closest to N and a corresponding negative charge (−0.0423 **e**) in the other F atom. The resulting induced polarisation moment $\Delta\vec{\mu}_1$ from the *INDO* calculation (0.76×10^{-30} C m) compares favourably with the value of 0.54×10^{-30} C m estimated for this induced dipole for an interaction between a point dipole, $^+H_3N^-$, of 6.30×10^{-30} C m at the N atom with an F_2 molecule with a polarisability ($\alpha_{\|}$) of 1.115×10^{-40} C m^2 V^{-1}. This induced dipole in F_2 in turn induces a dipole ($\Delta\vec{\mu}_2$) in each NH bond of NH_3; the value of $\Delta\vec{\mu}_2$ from the INDO calculation (0.064×10^{-30} C m) is somewhat smaller than that calculated from a point dipole ($\Delta\vec{\mu}_1$) of 0.76×10^{-30} C m centred in the F_2 molecule interacting with an NH_3 molecule with $\alpha_{\|} = 2.70 \times 10^{-40}$ C m^2 V^{-1}. The total change in the static dipole moment for the complex from that of the component molecules may be calculated (using equation 1.78) from the changes in charges summarised in figure 1.15b to be 0.1611**e** Å $= 2.58 \times 10^{-30}$ C m, of which about two-thirds comes from the charge transfer, and the remaining one-third from the polarisation (or charge shift).

We may now compare the charge distribution in figure 1.15a,b for the near-equilibrium configuration with that given in figures 1.15c and 1.15d for a H_3N,F_2 pair at the same NF distance but with the F-F distance increased by 0.02 Å. The charge difference from the near-equilibrium pairs (figure 1.15a) is given for the pair with the stretched F-F bond in figure 1.15d. Using these charge differences (for $\Delta R_1 = +0.02$ Å) in equation 1.78, we compute for $\partial\vec{p}/\partial R_1$ of the H_3N,F_2 pair (by analogy with equation 1.79, with the N atom at the origin):

$$\partial\vec{p}/\partial r_1 = +\left(\frac{0.0011}{0.02}\right)1.0 - \left(\frac{0.0002}{0.02}\right) \times 1.70 + \left(\frac{0.0077}{0.02}\right)2.85 + 0.0664\,\mathbf{e}$$

$$= +1.201\,\mathbf{e} = +19.24 \times 10^{-20}\,\mathrm{C}.$$

Here the plus sign means that the dipole moment increases in the $^+H_3N,F\text{-}F^-$ sense when the F-F bond length increases.

This change may be compared with the charge-transfer contribution $\vec{M}'_d(R_1)$ given by the product of the increased charge transferred

(0.0074 e) multiplied by the distance transferred [1.70 + (1.15/2) = 2.27 Å] and divided by ΔR_1 (0.02 Å): $\vec{M}'_d(R_1) = 0.84\,e = +13.45 \times 10^{-20}$ C. Hence, for this molecular pair, at this N-F distance, the INDO calculation indicates that the charge-transfer vibronic contribution is 70% of $\partial\vec{p}/\partial R_1$, with only 30% from the electrostatic contributions. Of the latter, only about one-sixth comes from the purely static charge on the F atom (analogous to $q_{H'}$ in equation 17 9).

It is of some interest to use the Hanna-Williams theory[45], expressed in equations 1.66–1.69, to try to estimate independently the electrostatic contribution to $\partial\vec{p}/\partial R_1$ for the H_3N,F_2 pair with geometry and charges given in figures 1.15a, b. We estimate $\partial\alpha_{\|}/\partial r_1$ for F_2 to be about one half that for Cl_2, the field (and field gradient) at F′ and F to be about the same as for chlorine from benzene, and the polarizability of F_2 to be about one-sixth that of chlorine (see above). From these estimates (using table 1.10) we predict a change in the dipole derivative of about $+2.00 \times 10^{-20}$ C due to the electrostatic vibronic term. This term is thus predicted to contribute about 10% of the total value of $\partial\vec{p}/\partial R_1$, in fair agreement with the predicted 25% contribution from the estimate given above from the difference between the calculated $\partial\vec{p}/\partial R_1$ and the vibronic charge-transfer term.

We see that the simple electrostatic arguments of Hanna and Williams[45] with reasonable guesses for parameter values do give estimates for the electrostatic vibronic contribution to $\partial\vec{p}/\partial R_1$ that are in reasonable agreement with those from the *INDO* calculation. From equations 1.67 the magnitude of this contribution is expected to be smaller at longer NF distances (say at 2.5 Å). On the other hand, it is quite clear that the vibronic charge-transfer contribution to $\partial\vec{p}/\partial R_1$, while dominant at an NF distance of 1.70 Å, will be considerably smaller at longer NF distances. Indeed, since the field drops as the square of the distance, the magnitude of $[-\frac{1}{4}(\vec{E}_1 + \vec{E}_2)(\partial\alpha_{\|}/\partial R)]$ in equation 1.67 may be expected to decrease only by a factor of two when the NF distance changes from 1.7 to 2.5 Å; if the electrostatic contribution to $\partial\vec{p}/\partial R_1$ is, for example, actually as much as 2.0×10^{-20} C at an NF distance of 1.7 Å, it may decrease to about 1×10^{-20} C at 2.5 Å. Since the calculated value of $\partial\vec{p}/\partial R_1$ (figure 1.14d) is only about 0.67×10^{-20} C, we must conclude that the electrostatic vibronic term quite possibly dominates $\partial\vec{p}/\partial R_1$ at this longer NF distance.

Referring again to the H_3N,N_2 pair, we realize that the parameters for the Hanna-Williams[45] theory applied to the ν(NN) vibration may very probably have nearly the same values as for the H_3N,F_2 pair. Figure 1.14b indicates that there is nearly the same extent of charge transfer (F_{1N}) at values of R_{NX} down to about 2.3 Å for both pairs; below that value F_{1N} is much greater for the H_3N,F_2 pair. At the longer NX distances it seems likely that the electrostatic terms dominate $\partial\vec{p}/\partial R_1$, and that their contribution is the same for both systems. Thus, for example, the calculated values of $\partial\vec{p}/\partial R_1$ are the same for both pairs for R_{NX} values down to about 2.3 Å, where the calculated value for $\partial\vec{p}/\partial R_1$ of the H_3N,F_2 pair increases sharply over the value calculated for the H_3N,N_2 pair, presumably because charge transfer becomes important in the H_3N,F_2 pair at shorter NX distance.

Before leaving this subject, let us look again at the empirical correlation found between $[\partial\vec{p}/\partial R_1 - \partial\vec{p}^0/\partial R_1]$ and $\Delta k_1/k_1$, shown in figure 1.9. Carreira and Person[74] plotted the value of $\Delta\bar{\nu}/\bar{\nu}$ calculated for the ν(XY) vibration of the several complexes vs. the calculated value of F_{1N}. (Both values were those from the *calculated* equilibrium NX distance.) They found (see figure 11 of reference 74) that $\Delta k_1/k_1$ ($\simeq 2\Delta\bar{\nu}/\bar{\nu}$, equation 1.31) is approximately proportional to F_{1N} as suggested by equation 1.30. However the slope is about 0.3 instead of 1.0, predicted by equation 1.30. Furthermore, examination of the empirical graph of figure 1.9 indicates that the slope is about one-tenth that predicted[74] for the corresponding figure[74] giving the correlation between $\partial\vec{p}/\partial R_1$ values and $\Delta k_1/k_1$ values calculated by *INDO* methods.

The strong dependence of the calculated values of $\partial\vec{p}/\partial R_1$, F_{1N} and $\Delta\bar{\nu}_1/\bar{\nu}_1$ on the NX distance suggests that the correlation given by Carreira and Person (figure 11 of reference 74) is too naive. The arguments given above suggest clearly that both the electrostatic and the charge-transfer contributions to $\partial\vec{p}/\partial R_1$ will change when we study the vibration of an XY acceptor with donor molecules of increasing strength. It is not clear that either of these changes (or both in combination) should be expected to be linearly related to the corresponding force constant change $\Delta k_1/k_1$, as found empirically in figure 1.9. At the present time we believe that the electrostatic contributions to both $\Delta k_1/k_1$, and $[\partial\vec{p}/\partial R_1 - \partial\vec{p}^0/\partial R_1]$ should be nearly constant or only slowly varying for a series of complexes between XY and different donor molecules (especially for a series of closely related donors). Hence, we still believe that the relatively large changes of figure 1.9 must be explained by the vibronic charge-transfer effect. The results in this section, however, from both *ab initio* and approximate calculations, suggest very strongly that the electrostatic contributions to properties of weak complexes may often be more important than charge-transfer effects. Certainly, attempts to separate these two effects for a new system, for example by making plots of $\partial\vec{p}/\partial R_1$ *vs.* $\Delta k_1/k_1$, and comparing the results with figure 1.9, do not appear to be definitive tests. (Further discussion of the possible interpretation of these plots is given in section 2.3.4.)

It is hoped that this discussion of the H_3N,N_2 and H_3N,F_2 systems may illustrate some of the concepts and techniques from quantum-mechanical calculations that promise to be helpful in interpreting complexes. We believe these ideas help to build an intuitive model to interpret some of the changes observed in the vibrational spectra of complexes. It is expected that the more exact *ab initio* calculations with fairly large basis sets which are beginning now to be possible will provide an even more detailed understanding of the properties of complexes.

1.3.4. SUMMARY

As we have seen, there have been only enough preliminary quantum-mechanical calculations to suggest their coming importance. The results at present are tantalizing, in that they suggest a number of intriguing possibilities, but do not yet differentiate strongly between them. We can use these results, however, to summarize some concepts in connection with complexes between *n* donor molecules and *aσ* acceptors.

We have seen in the preceding section that the charge distribution calculated for the D–A molecule in the complex differs from that in the free D and A molecule because of two effects: (1) charge transfer, and (2) polarisation (or "charge shift). We have seen[74] that both effects occur specifically for certain orientations of D with respect to A. Both effects increase rather strongly as D and A come closer together, although the charge transfer effect is expected to increase more strongly at shorter distances, *if* it occurs. Because of the tendency for *INDO* calculations to underestimate the repulsion effects, these calculations tend to exaggerate the importance of charge transfer at the too-small calculated equilibrium D–A distance. Nevertheless, it is clear that a little bit of charge transfer may be expected at the true equilibrium D–A distance for many *n–aσ* interactions, and that a little bit may go a long way in its effect on many phenomena, particularly on the dipole moment and on $\partial\vec{p}/\partial R_1$.

It is also quite clear that the calculations give, at best, the effect on the properties of the D,A pair due to their interaction in the gas phase. The effect on these properties of surrounding the D,A pair by even an inert solvent is not at all clear. The *INDO* results from reference 74 can help in a discussion of solvent effects for the NH_3,N_2 pair. The interaction energy between N_2 and NH_3 is calculated to be about the same for approach along the C_3 axis as for approach along the N-H axis.[74] If we consider all possible orientations of NH_3 molecules about an N_2 molecule, then, we may expect them all to be approximately equally probable. (This statement is a considerable extrapolation from the fact that H_3N,N_2 and H_2N-H,N_2 interactions have the same energy.) The properties for N_2 dissolved in liquid ammonia might then be expected to be the average of the calculated values of these properties taken over all possible orientations of H_3N molecules about the N_2.

Consider the value of $\partial\vec{p}/\partial R_1$, for example. From figure 6 of reference 74, no change is predicted in $\partial p/\partial R_1$ for approach of an NH bond to the N_2, whereas there is a considerable effect predicted for $\partial\vec{p}/\partial R_1$ for the approach of the N atom (from H_3N) along the axis of the N_2

molecule. We might expect the approach of each H atom (from H_3N) to the N′ atom of N_2 to be equally probable, and equal in probability to the approach of the N atom itself to the N′ atom. Thus, we might expect the observed increase in $\partial\vec{p}/\partial R_1$ for N_2 in a random one-to-one solution in liquid H_3N, for example, to be about one-fourth of the value predicted for the one-to-one pair in the gas phase for the approach along the C_3 axis.

Suppose now that two NH_3 molecules interact with the N_2 molecule; one at each end. If both approach the N_2 molecule with the N atom on the N-N axis, then the fields from H_3N cancel each other so that the electrostatic contribution to $\partial\vec{p}/\partial R_1$ is expected partially to cancel. It is not clear whether or not some charge-transfer effect remains. It seems possible that the argument can go either way. However, one H_3N molecule may approach the N_2 molecule with the N atom first, and the other with an H atom first, *viz.*, H_3N, N-N, HNH_2. The observed value of $\partial\vec{p}/\partial R_1$ may then be expected to be about the same as the calculated value for the one-to-one pair with C_{3v} geometry. This latter configuration might be expected on a random basis to occur three times as often as the symmetric H_3N, N-N, NH_3 configuration, and so a non-zero value of $\partial\vec{p}/\partial R_1$ *may* be expected for N_2 in liquid ammonia, even if there is *no* charge transfer.

If the arguments in the preceding section are correct, then it is quite clear that the observed value of $\partial\vec{p}/\partial R_1$ may *not* be used to measure the importance of charge transfer in a particular complex, unless a correction is first made for the electrostatic effects. We have seen how difficult it may be to make such corrections, in general.

We have, of course, oversimplified this discussion by assuming that the encounters are completely random in orientation. The average of the property being discussed must be an energy-weighted average. If the energy of interaction is much greater for one D,A orientation, we may expect that orientation to dominate in D,A mixtures, so that the properties of the pair in the mixture will be approximately those predicted for this orientation of the one-to-one gas-phase pair. In fact, the interaction between two H_2O molecules does appear to give one such predominant configuration. It is not yet clear for H_3N, XY pairs whether that situation is also often true there, but it is probably not true for the particular case of the H_3N,N_2 pair discussed above.

In conclusion, the calculations presented in this section suggest that the interaction between D and A, in a weak one-to-one interaction in the gas phase, is often to be described as an approximately equal mixture of electrostatic effects and charge transfer effects. Since both effects are increased as the interaction forces the D,A pair together, it will be difficult to find experimental criteria for separating these effects. We do not believe it is any wiser to ignore charge–transfer effects in *n–aσ* complexes than it is to ignore "electrostatic effects". However, since the latter *can* be estimated from classical electrostatics, as indicated by Hanna and Williams[45], we can *try* to estimate their magnitudes and search for *extra* charge-transfer effects.

ACKNOWLEDGEMENT

It is a pleasure to acknowledge my indebtedness to friendly discussions with many people, including especially R. S. Mulliken, J. Yarwood, E. Clementi, and M. Tamres, during the preparation of this manuscript. I am grateful also to the National Science Foundation for encouragement and some financial support.

REFERENCES

1. R. S. Mulliken and W. B. Person, Molecular Complexes, (John Wiley and Sons, New York, 1969)
2. R. S. Mulliken and W. B. Person (Eds, H. Eyring, D. Henderson, and W. Jost) Physical Chemistry, Vol III (Academic Press, New York, 1969), chapter 10
3. R. S. Mulliken, *J. Amer. Chem. Soc.*, **72,** 600 (1950)
4. R. S. Mulliken, *J. Amer. Chem. Soc.*, **74,** 811 (1952)
5. R. S. Mulliken, *J. Phys. Chem.*, **56,** 801 (1952)
6. J. O. Hirschfelder, Ed., Intermolecular Forces, (Adv. in Chem. Phys., Vol. 12) (Interscience Publishers, New York, 1967)
7. For example, see J. P. Lowe, *J. Amer. Chem. Soc.*, **93,** 301 (1971)
8. E. Clementi and H. Popkie, *J. Chem. Phys.*, **57,** 1078 (1972)
9. J. N. Murrell, M. Randic and D. R. Williams, *Proc. Roy. Soc.*, **A284,** 566 (1965) Also see J. N. Murrell and G. Shaw, *J. Chem. Phys.*, **46,** 1768 (1967)
10. J. L. Lippert, M. W. Hanna and P. J. Trotter, *J. Amer. Chem. Soc.*, **91,** 4035 (1969)
11. E. G. Cook, Jr. and J. C. Schug, *J. Chem. Phys.*, **53,** 723 (1970); See also J. C. Schug, W. M. Chang and M. C. Dyson, *Spectrochim. Acta*, **28A,** 1157 (1972)
12. J. O. Hirschfelder, C. E. Curtiss and R. B. Bird, Molecular Theory of Gases and Liquids, (John Wiley and Sons, New York, 1954)
13. H. B. Friedrich and W. B. Person, *J. Chem. Phys.*, **44,** 2161 (1966)
14. A. D. Buckingham, Chapter 2 of reference 6
15. J. P. O'Connell and J. M. Prausnitz, *Ind. Eng. Chem. Fundam.*, **8,** 453 (1969); *ibid.*, **9,** 579 (1970); see also reference 17
16. J. Del Bene and J. A. Pople, *J. Chem. Phys.*, **52,** 4858 (1970)
17. L. S. Moore and J. P. O'Connell, *J. Chem. Phys.*, **55,** 2605 (1971)
18. M. W. Hanna, *J. Amer. Chem. Soc.*, **90,** 285 (1968)
19. (a) R. S. Drago and B. B. Wayland, *J. Amer. Chem. Soc.*, **87,** 3571 (1965); (b) R. S. Drago, G. C. Vogel, and T. E. Needham, *J. Amer. Chem. Soc.*, **93,** 6014 (1971)
20. R. G. Pearson, *J. Amer. Chem. Soc.*, **85,** 3533 (1963); see also S. Ahrland, J. Chatt and N. R. Davies, *Quart. Rev.*, **12,** 265 (1958)
21. C. H. Townes and B. P. Dailey, *J. Chem. Phys.*, **17,** 782 (1949); see also T. P. Das and E. L. Hahn, Solid State Phys., Suppl. No. 1 (1958)
22. H. O. Hooper, *J. Chem. Phys.*, **41,** 599 (1964); see also D. F. R. Gilson and C. T. O'Konski, *ibid.*, **48,** 2767 (1968), and R. A. Bennett and H. O. Hooper, *ibid.*, **47,** 4855 (1967)
23. R. S. Mulliken and W. B. Person, *J. Amer. Chem. Soc.*, **91,** 3409 (1969)
24. (a) G. A. Bowmaker and S. Hacobian, *Aust. J. Chem.*, **22,** 2047 (1969); (b) H. Creswell-Fleming and M. W. Hanna, *J. Amer. Chem. Soc.*, **93,** 5030 (1971)

25. For example, see J. Yarwood, *Spectrochim. Acta*, **26A,** 2099 (1970); also J. P. Larkindale and D. J. Simkin, *J. Chem. Phys.*, **55,** 5048 (1971)
26. S. Ichiba, H. Sakai, H. Negita and Y. Maeda, *J. Chem. Phys.*, **54,** 1627 (1971)
27. J. Collin and L. D'Or, *J. Chem. Phys.*, **23,** 397 (1955)
28. G. C. Pimentel and A. L. McClellan, The Hydrogen Bond, (W. H. Freeman and Co., San Francisco, 1960)
29. See the discussion of solvent effects by H. E. Hallam, in Infrared Spectroscopy and Molecular Structure, (Ed., M. Davies), (Elsevier Publishing Co., Amsterdam, 1963) Chapter XII. Dr. Hallam stresses the large solvent effects associated with hydrogen bonding, but he gives a good general review of the theory of "ordinary" solvent effects in his section 10
30. (a) H. E. Hallam, section 11 of reference 29; (b) S. R. Polo and M. K. Wilson, *J. Chem. Phys.*, **23,** 2376 (1955); (c) W. B. Person, *J. Chem. Phys.*, **28,** 319 (1958); (d) A. D. Buckingham, *Proc. Roy. Soc.*, **A248,** 169 (1958), *ibid.*, **A255,** 32 (1960); (e) A. D. Buckingham, *Trans. Faraday Soc.*, **56,** 753 (1960)
31. R. R. Wiederkehr and H. G. Drickamer, *J. Chem. Phys.*, **28,** 311 (1958)
32. G. Herzberg, Molecular Spectra and Molecular Structure. II Infrared and Raman Spectra of Polyatomic Molecules) (D. Van Nostrand Co., Inc., New York, 1945).
33. See W. B. Person, G. R. Anderson, J. N. Fordemwalt, H. Stammreich and R. Forneris, *J. Chem. Phys.*, **35,** 908 (1961)
34. R. F. Lake and H. W. Thompson, *Proc. Roy. Soc.*, **A297,** 440 (1967)
35. J. N. Gayles, *J. Chem. Phys.*, **49,** 1840 (1968)
36. G. W. Brownson and J. Yarwood, *J. Mol. Struct.*, **10,** 147 (1971)
37. J. Yarwood and W. B. Person, *J. Amer. Chem. Soc.*, **90,** 3930 (1968); *ibid.*, **90,** 594 (1968)
38. W. B. Person and B. L. Crawford, Jr., *J. Chem. Phys.*, **26,** 1295 (1957)
39. I. M. Mills, chapter 5 of reference 29.
40. T. Miyazawa and J. Overend, *Bull. Chem. Soc. Jap.*, **39,** 1410 (1966)
41. For example, see K. Szczepaniak and A. Tramer, *J. Phys. Chem.*, **71,** 3035 (1967); also see P. G. Puranik and V. Kumar, *Proc. Indian Acad. Sci.*, **A58,** 29, 327 (1963)
42. W. B. Person, R. E. Humphrey and A. I. Popov, *J. Amer. Chem. Soc.*, **81,** 273 (1959)
43. Y. Yagi, W. B. Person and A. I. Popov, *J. Phys. Chem.*, **71,** 2439 (1967); see also reference 36.
44. J. D'Hondt and Th. Zeegers-Huyskens, *J. Mol. Structure*, **10,** 135 (1971)
45. M. W. Hanna and D. E. Williams, *J. Amer. Chem. Soc.*, **90,** 5358 (1968)
46. I. N. Levine, Quantum Chemistry. II. Molecular Spectroscopy, (Allyn and Bacon, Boston, 1970) p. 145 *ff.*
47. D. E. Williams, *J. Chem. Phys.*, **45,** 3770 (1966)
48. H. Rosen, Y. R. Shen and F. Stenman, *Mol. Physics*, **22,** 33 (1971)
49. T.-C. Jao and W. B. Person, (unpublished results)
50. J. Overend, chapter 10 of reference 29.
51. J. L. Hollenberg and D. A. Dows, *J. Chem. Phys.*, **37,** 1300 (1962)
52. H. Yamada and W. B. Person, *J. Chem. Phys.*, **38,** 1253 (1963)
53. K. Szczepaniak and W. B. Person, *Spectrochim. Acta*, **28A,** 15 (1972)
54. A. J. Zelano and W. T. King, *J. Chem. Phys.*, **53,** 4444 (1970)
55. W. B. Person and L. C. Hall, *Spectrochim. Acta*, **20,** 771 (1964)
56. (a) G. A. Segal and M. L. Klein, *J. Chem. Phys.*, **47,** 4236 (1967); (b) R. E. Bruns and W. B. Person, *J. Chem. Phys.*, **57,** 324 (1972), and other unpublished results; (c) T. P. Lewis and I. W. Levin, *Theoret. Chim. Acta*, **19,** 55 (1970)
57. J. G. Kirkwood, quoted by W. West and R. T. Edwards, *J. Chem. Phys.*, **5,** 14 (1937)
58. E. Bauer and M. Magat, *J. Phys. Radium*, **9,** 319 (1938)
59. C. M. Huggins and G. C. Pimentel, *J. Phys. Chem.*, **60,** 1615 (1956)

60. W. B. Person, R. E. Erickson and R. E. Buckles, *J. Amer. Chem. Soc.*, **82,** 29 (1960)
61. E. Clementi, *Proc. Nat. Acad. Sci. USA*, **69,** 2942 (1972)
62. See, for example, H. F. Schaefer, III, The Electronic Structure of Atoms and Molecules, (Addison-Wesley Publishing Co., Reading, Mass., 1972)
63. E. Clementi, (a) *Chem. Rev.*, **68,** 341 (1968); (b) Physics of Electronic and Atomic Collisions, VII ICPEAC, 1971 (North-Holland, 1972), pp. 399–426; (c) B. J. Duke, *Ann. Rept.*, **68A,** 3 (1971)
64. (a) E. Clementi, *J. Chem. Phys.*, **46,** 3851 (1967); (b) *ibid.*, **47,** 2323 (1967); (c) E. Clementi and J. N. Gayles, *J. Chem. Phys.*, **47,** 3837 (1967)
65. R. S. Mulliken, *J. Chem. Phys.*, **23,** 1833, 1841, 2338, 2343 (1955)
66. R. G. Body, D. S. McClure and E. Clementi, *J. Chem. Phys.*, **49,** 4916 (1968)
67. P. Cade and W. M. Huo (private communication).
68. (a) J. A. Pople, D. P. Santry and G. A. Segal, *J. Chem. Phys.*, **43,** S129 (1965); (b) J. A. Pople and G. A. Segal, *J. Chem. Phys.*, **43,** S136 (1965); (c) J. A. Pople and G. A. Segal, *J. Chem. Phys.*, **44,** 3289 (1966)
69. J. A. Pople and D. L. Beveridge, Approximate Molecular Orbital Theory (M Graw-Hill Book Co., New York, 1970).
70. J. A. Pople, D. L. Beveridge and P. A. Dobosh, *J. Chem. Phys.*, **47,** 2026 (1967)
71. G. A. Segal, QCPE 91 *CNDO/2*—Molecular Calculations with Complete Neglect of Differential Overlap, and P. A. Dobosh, QCPE 141 *CNINDO*, *CNDO* and *INDO* Molecular Orbital Program (FORTRAN IV). (Available from Quantum Chemistry Program Exchange, University of Indiana, Bloomington, Indiana)
72. There have been a great many calculations for $(H_2O)_2$. See (a) P. A. Kollman and L. C. Allen, *J. Amer. Chem. Soc.*, **92,** 753 (1970), for a review of literature and for a comparison of their *CNDO/2* results with *ab initio* calculations; and (b) P. A. Kollman and L. C. Allen, *J. Chem. Phys.*, **51,** 3286 (1969) for a particularly interesting (for its treatment of vibrational properties) *ab initio* calculation; also see *ab initio* calculations by (c) K. Morokuma and L. Pedersen, *J. Chem. Phys.*, **48,** 3275 (1968); (d) K. Morokuma and J. R. Winick, *J. Chem. Phys.*, **52,** 1301 (1970); (e) D. Hankins, J. W. Moskowitz and F. H. Stillinger, *J. Chem. Phys.* **53,** 4544 (1970); and reference 15.
73. P. A. Kollman and L. C. Allen, *J. Amer. Chem. Soc.*, **93,** 4991 (1971)
74. L. A. Carreira and W. B. Person, *J. Amer. Chem. Soc.*, **94,** 1485 (1972)
75. M. van Thiel, E. D. Becker, and G. C. Pimentel, *J. Chem. Phys.*, **27,** 486 (1957)
76. A. J. Tursi and E. R. Nixon, *J. Chem. Phys.*, **52,** 1521 (1970)
77. L. A. Carreira and W. B. Person, unpublished results.
78. H. Tsubomura, *Bull. Chem. Soc. Jap.*, **27,** 445 (1954); *J. Chem. Phys.*, **24,** 927 (1956)
79. C. A. Coulson (Ed. D. Hadzi), Hydrogen Bonding, (Pergamon Press, New York, 1959), p. 339

Chapter 2

The Measurement and Interpretation of Vibrational Spectra of Molecular Complexes

J. Yarwood

Chemistry Department, University of Durham, Durham DH1 3LE, *England*

2.1 INTRODUCTION

This chapter is intended to be a link between the expected spectral changes outlined in some detail in chapter 1 and the bulk of experimental data presented in chapters 3–7. We have attempted here to discuss the relative merits of the various experimental techniques available, and the general type of results which are obtained. We have also tried to show how the data may be interpreted in the light of theoretically based predictions. To do this we have chosen only a few of the better-known complexes for which there is a relatively large amount of data or for which particularly definitive data are available.

Emphasis has been placed on vibrational studies (including infrared intensities and low frequency studies using far-infrared and Raman spectroscopy) since complex studies using electronic spectroscopy have been covered in some considerable detail in other recent works.[1–3]

Some consideration is made of the magnetic resonance techniques (which are becoming increasingly used in the field) and of other lesser known methods where they are important (dielectric absorption and dipole moment studies being covered fully in chapter 7). However, it is emphasised that the maximum information about a particular interaction will only be achieved by using *all* the techniques available for the system, and that quantitative data (where they can be measured) are likely to yield a more complete and accurate interpretation.

Some effort has also been made (particularly in section 2.3) to correlate, compare and contrast data obtained for different types of complexes covered in chapters 3 to 7. In doing this it has been necessary to try to avoid serious overlapping with individual chapters. It was nevertheless thought worthwhile, for example, to point out that, while some spectral features may be common to many modes of interaction, others may be peculiar to a specific complex or type of complex. It is hoped that, in this way, a brief overall view of the field may be presented.

2.2. EXPERIMENTAL TECHNIQUES

2.2.1. SURVEY OF TECHNIQUES AVAILABLE

In the early days of molecular complex studies the only spectroscopic techniques readily available were electronic (200–1000 nm) and (conventional) vibrational spectroscopy (4000–650 cm^{-1}). Although a considerable amount of useful information was (and still is) obtained by these methods, several other techniques of importance have, more recently, yielded additional important data. These "new" techniques include nuclear magnetic resonance (particularly proton, fluorine and boron n.m.r. and halogen n.q.r.), electron spin resonance, Mössbauer, far-infrared and Raman methods (the latter vastly improved recently by advances in source technology and instrument design), dielectric absorption, and dipole moment studies. Such studies have already made a considerable impact on complex investigations and there is little doubt that they will continue to do so. These "newer" techniques provide information of the utmost importance in trying to understand the nature of molecular interactions. For example, low frequency vibrational spectroscopy often provides a direct measure of the strength of the intermolecular bond. Magnetic resonance spectroscopy and dipole moment studies give some idea of the extent and nature of electron re-distribution and polarisation occurring on complexation. Dielectric loss studies in the microwave region provide some idea of the lifetime of the complex in solution (also available in principle by other methods—see below). Such information is not available directly from conventional spectroscopy in the infrared, visible or ultraviolet regions and is of considerable importance therefore in supplementing the data obtained by these better-known methods.

For complexes studied in solution (usually in a "non-complexing" solvent or the pure donor) it is clear that the complex will be breaking up and forming at a rate which depends on its lifetime. The exact frequency of this relaxation depends, of course, on the overall "strength" of donor-acceptor interaction (which in turn may depend on several factors). If the interaction is "strong" relaxation will occur relatively slowly and the complex is said to have a long lifetime. On the other hand, if interaction is "weak" relaxation will occur more quickly due to the complex having a much shorter lifetime. The lifetime of a complex species in solution is important in assessing the relative merits of the experimental techniques available. The different spectroscopic experiments are carried out at different frequencies and different effects are likely to be seen depending on the relative frequencies of the relaxation process and the experiment concerned.

Thus, when using n.m.r. spectroscopy it is clear that signals corresponding to complexed and uncomplexed molecules will only be observed[4] if the rate of relaxation is less than the frequency of separation of the two signals (in the absence of relaxation or exchange). This frequency separation is likely to be of the order of 10–1000 s^{-1} and it follows that separate signals will only be observed if the lifetime is greater than about 10^{-3} s. Even if separate signals are not observed it is, in principle, possible to estimate limiting values of complex lifetimes from the broadening of signals which occur due to the "exchange" process[4]. Such estimates have been made for some iodine complexes[5] and for boron trifluoride complexes[6] and *maximum* lifetimes are in the region of 10^{-3}–10^{-4} s. These results do not agree very well with the data obtained for a range of complexes using microwave methods[7,8] (which indicate lifetimes of 10^{-10} to 10^{-12} s) although it should be emphasised that the "strengths" of complexes studied by the latter method were, in most cases, very "weak" (as measured by thermodynamic methods). There are quite serious problems associated with trying to study "strong" complex lifetimes by microwave methods. These arise from the fact that relaxation always occurs by the *faster* process so that if the decomposition process is *slow* compared with the relaxation process of the "rigid" complex (relaxation of the complex dipoles in the electric field) then only the latter process will be observed and no estimate of the complex lifetime will be obtained. (These difficulties are discussed more fully in chapter 7.) There have been several reports of n.m.r. signals due to both complexed and uncomplexed species being observed simultaneously.[9–13] These results have not usually been interpreted in terms of complex lifetimes but calculations from the published data[9] show that the *minimum* lifetimes are of the order of 10^{-3} s. In view of the rather large K_c values observed for pyridine complexes with iodine and iodine monochloride[14,15] it is somewhat surprising that separate n.m.r. signals are not observed.[14] These complexes are at least as "strong" (thermodynamically) as the nitrile–BX_3 complexes[9] for which separate

^{19}F signals *are* observed. (There are, however, complications in the pyridine–halogen system due to nitrogen and halogen quadrupole broadening and much higher temperatures—see section 3.5.3).

On the other hand, it should be possible, in the infrared and Raman spectra, to observe bands due to complexed and uncomplexed molecules whatever the complex "strength". Vibrational frequencies range from 10^{14} s^{-1} to 10^{12} s^{-1} and in the vast majority of cases the rate of relaxation is considerably slower than this (even accepting the lowest lifetimes reported at 10^{-12} s). Thus many vibrations may take place before the complex can break up. Interestingly, if the estimates of complex lifetime for "weak" complexes are correct[7,8] it could be that low frequency vibrations (for example the halogen–halogen stretching frequency) may give rise only to averaged absorption bands as in the n.m.r. case. No definite identification of this phenomenon has so far been made, although there have been a number of attempts[16–18] to observe separate "complex" and "free" ν(X-X) bands using Raman spectroscopy. Klaeboe[16] and Rosen *et al.*[17] have observed two Raman bands in the ν(Br-Br) and ν(I-I) regions for *some* aromatic donors (but *not* with benzene itself). This information serves to place an approximate lower limit on the complex lifetime. (This work and its implications in discussing complex stability are discussed further in sections 1.2.5 and 3.4.3.)

Several interesting changes in infrared band shape have been observed on complexation. While some pyridine vibrational bands in "strong" interhalogen complexes show a drastic decrease in half-band width[19] there are several cases where considerable band broadening occurs. This phenomenon has been observed for the a_{2u} mode of C_6F_6 in C_6H_6–C_6F_6[20], for chloroform bands in C_6H_6–$CHCl_3$[21] and for both the SO_2, b_1 mode, and several pyridine, a_1 modes in C_5H_5N–SO_2[22]. It is, in principle, possible to calculate[20,23] (from measurements of band widths) transition dipole correlation times[24] and estimates of half-lives and dissociation rate constants for complex decomposition assuming that such broadening is caused by vibrational degradation when the complex undergoes chemical relaxation. Such calculated rate constants are in the region of 10^{11} dm^3 mol^{-1} s^{-1} and lend support to the lifetimes estimated for weak complexes from dielectric measurements[7,8] (see chapter 7). Band width decreases on complexation can be then interpreted as being due to a decrease in diffusional rotation of the molecules in the complex. It is probable that, for a given complex, both these effects are operative and that the observed band widths and shapes reflect the relative importance of the two mechanisms as controlled by the lifetime of the complex in solution. (Studies of band shape changes due to molecular interaction are described in more detail in section 2.3.6.)

In the solid phase it has now become almost a routine matter to determine the crystal structures of molecular complexes[25] provided always that good crystals may be obtained and that the complex is chemically stable. Recently there has also been an increase in the number of n.q.r. studies particularly on halogen complexes[26–32]. The principal reason for this revival of interest in halogen n.q.r. spectroscopy is that the technique offers a method of finding, in a direct way, what changes in electron density occur (at each halogen) on going from free molecule to complex[33] (chapter 3, p. 293). This, in principle, should provide very important information about the relative contributions of charge-transfer and classical polarisation forces to the stability of halogen complexes—a subject about which there has been, and still is, considerable controversy[34–37] (see also chapter 1). The method is complicated by several factors, including crystal field effects[26,28,29] and the apparent difficulty, on occasions, of detecting all the relevant signals[25,26]. Further, it is necessary to know the crystal structure of the complex before starting to interpret the data and to be sure that one has the *same* crystalline sample as was employed for the crystal structure determination (see section 2.2.2 for problems involved with the elucidation of species present in the system). Despite the difficulties and uncertainties a certain amount of success has been achieved in detecting the shifted halogen signals and interpreting them in terms of various complex models. (These measurements and their interpretation are discussed more fully in sections 1.1.4 and 3.5.3.) It seems likely that this technique will be used more fully in the future especially for systems where well-defined crystalline samples may be prepared.

It should be clear, that, if we are to gain the maximum information about a particular complex, it is desirable to employ as wide a range of experimental techniques as is practicable. Only by making measurements, for example, of interaction "strength" (say $-\Delta H_f^{\ominus}$ or $\Delta G_f^{\ominus}$), complex lifetime, and electronic redistribution for a given complex by *different* methods will one be in a position to assess the relative merits of each technique and be able to give any interpretation maximum credibility.

2.2.2. EXPERIMENTAL DIFFICULTIES

The experimental difficulties which arise in the spectroscopic study of molecular complexes may be divided into two broad areas,

A. Difficulties arising from the chemistry of complex systems.
B. Difficulties arising from inadequacies and inaccuracies inherent in spectrometer systems.

In (A) the difficulties are ones which, in the main, are peculiar to the study of complex-containing systems when two or more compounds (often quite reactive ones) are mixed together. In (B) the experimental uncertainties are inherent in the spectroscopic study of any system although they may well be more important when studying complexes. Such instrumental uncertainties are, of course, different, in both nature and extent, for different spectroscopic techniques, whereas the chemical difficulties are a function of the chemical system studies in the majority of cases. In this section we shall discuss in some detail the difficulties involved with vibrational spectroscopic studies with brief reference to other techniques. Some emphasis will also be placed on the determination of intensity data—which are desirable if maximum information is likely to be gained about a particular complex. The discussion is confined to studies in the liquid phase (usually in solution in an "inert" solvent). For a discussion of gas phase studies on halogen complexes see section 3.2.

(A) Difficulties Arising from the Chemistry of the System

On placing certain initial concentrations of donor and acceptor into a suitable solvent one needs to consider carefully several factors before making any spectral measurements. One needs to be sure, for example,

(a) that any chemical reactions—other than complex formation—are negligibly small. It is often possible to (partly) avoid such complications by making measurements soon after solution preparation since some reactions are at least, in the initial stages, fairly slow. Any chemical reaction will tend to give rise to unexpected bands and spectral assignment may be made ambiguous or impossible. Such bands can usually be picked out from the time-dependence of their intensities but great care must be taken, especially in quantitative work, since any concentrations calculated will be in error unless it is possible to estimate quantitatively how far side-reactions have proceeded.

(b) that the complex species in solution is "well-defined" (1:1, 2:1, *etc.*) or that the nature of the species formed (if more than one) is *known*. This is especially important, again in quantitative work, where it is essential to know the relevant equilibrium constant(s) if complex concentrations are to be calculated. A good deal of spectroscopic work (ultraviolet, visible, infrared and n.m.r.) is aimed at the determination of these constants (see chapter 3 and references 1 and 2). In this work it is essential to know how many complex species are formed[2,38,39] if accurate and consistent data are to be obtained. The effect of competing reactions could, of course, be disastrous and, since reactions are more likely to occur in more concentrated solutions, it is best to use the lowest concentrations practicable.

(c) that the effect (or otherwise) of the solvent is carefully considered and investigated. Solvent competition with either donor or acceptor will be expected to produce (at best) poor quantitative data or (at worst) meaningless results. (See section 3.2 for a detailed discussion of likely effects of solvent competition.)

Such competition is unlikely to be detected spectroscopically since the solvent is usually present in large excess. Any spectral changes occurring as a result of solvent interaction will therefore be masked by the bulk of uncomplexed solvent molecules. This problem can of course, be avoided by using pure donor (or acceptor) if it is a liquid at normal temperatures. However, comparison of the spectra with those obtained for the "free" acceptor (or donor) (in a different "solvent") is difficult because of the varying effect of bulk dielectric medium[40–43] (especially on the intensity data).

In addition to these major sources of error there are a number of less serious difficulties which, nevertheless, place some restrictions on complex studies, especially in solution. It is often the case that, when using "non-complexing" or "inert" solvents (hopefully heptane, hexane, carbon tetrachloride, *etc*), the solubility of one of the components is insufficient to make infrared or Raman studies feasible. There is a limit on the pathlength which one can employ successfully because solvent absorptions result in almost complete absorption over the frequency range of interest. This is especially the case between 4000 and 500 cm^{-1} where solvent absorption is severe at pathlengths above about 500 μm (solvent compensation techniques can help in some cases). At lower frequencies, however, pathlengths up to 15 000 μm may be employed with a non-polar solvent, so the problem is less serious. Many electron acceptors (*e.g.* iodine, tetracyanoethylene, metal halides) are only slightly soluble in organic, non-polar or non-complexing solvents* so that studies on donor vibrations, requiring quite a high concentration of the acceptor (even if the K_c value is fairly large†) are quite difficult from this point of view. It is then necessary to either use a more reactive solvent (*e.g.* dichloromethane, chloroform, dioxan or tetrahydrofuran) and risk error due to solvent competition or to increase the pathlength to its limit and work with low transmitted energies—neither of which will help in attempts to obtain accurate data. Further, error may occur if one or both of the components is highly volatile and although loss of material can be avoided, in principle, by working in a closed system, a fairly complex arrangement is necessary to transfer the sample to a spectroscopic cell. Using a closed system is also probably the best way of avoiding unwanted moisture getting into the system. In some cases it has been clearly shown[44] that working in a dry atmosphere has virtually no effect on the spectra but on the other hand inconsistencies in the data for some systems (*e.g.* pyridine–SO_2[22]) *could* be attributed to the instability of the

* This situation can also lead to weighing errors during quantitative work.

† For a K_c value of 10 $dm^3\ mol^{-1}$ if the concentration of acceptor were 0.1 mol dm^{-3} only 50% of the donor would be complexed even assuming that the acceptor were in large excess. For equal concentrations (at 0.1 mol dm^{-3}) only 38% of the donor would be complexed.

complex towards water vapour. In section 2.3 some of these difficulties are illustrated by reference to specific systems.

(B) Difficulties Arising from Instrumental Errors and Inadequacies

In order to make frequency measurements in the infrared using modern commercial equipment[45–47] it is only necessary to ensure that the frequency (or wavelength) calibration is accurate and that the resolution (which is a function of spectral slit width[46]) is adequate to give reasonable separation of the bands (in solution a resolution of 2–3 cm^{-1} is quite adequate in most, *but not all*, instances). If, however, it is desired to make quantitative transmission measurements in order to obtain accurate extinction coefficients or integrated intensities then several additional instrument conditions need to be fulfilled.

It is necessary to ensure (as far as possible) that the bands are not distorted or broadened by the instrument recording system. In order to achieve a "true" band shape it is necessary to arrange that the spectral slit, width, s, is small compared with the apparent half-band width $\Delta\nu^a_{1/2}$[46,48]. The rule used normally is that quoted by Ramsay[48]. That is, if $s \lesssim 0.2\ \Delta\nu^a_{1/2}$ then the integrated intensity value is close as possible to the "true" value. Ramsay, however, determined his intensities by the peak height/half-band width procedure, valid only for Lorentzian bands (see section 2.3.6). It has been found in our laboratory that integrated areas (obtained by numerical integration) are by no means as sensitive to the $s/\Delta\nu^a_{1/2}$ ratio and that often the intensity does not change significantly over a considerable range of $s/\Delta\nu^a_{1/2}$ provided it is less than 0.5. This is illustrated by the figures shown in table 2.1 for one of the pyridine, a_1 bands in the pyridine–ICl complex. It can be seen that, within the experimental error, for values of $s/\Delta\nu^a_{1/2} < 0.33$, the intensity remains essentially the same although the peak absorbance continues to increase to a maximum value at $s \simeq 0.8\ cm^{-1}$. It is important to notice that the spectral resolution *is* important if $s/\Delta\nu^a_{1/2}$ gets larger than 0.5 in which case the intensity decreases and the band-width increases considerably (table 2.1). If band shape analyses[20,23,24] are attempted (section 2.3.6) then excessive instrumental distortion will obviously give erroneous results. This distortion is illustrated in figure 2.1 which compares the band profile of the same band using 3.0 cm^{-1} and 1.0 cm^{-1} slits.

The recommended procedure[48] is to decrease the slit width manually until the peak height of the band under consideration reaches a *maximum* showing that s is lower than $\Delta\nu^a_{1/2}$. It is generally found that as the peak height increases the band width decreases (see table 2.1) so that for these slit conditions the band has least instrumental broadening. As the slits are closed it will be necessary, of course, to increase the amplifier gain and so often a compromise to give reasonably small s and acceptable noise level will be necessary. The data shown in table 2.1 at $s = 0.6\ cm^{-1}$ are subject to errors caused by a high noise level since the instrument response is very poor at lower gain settings.

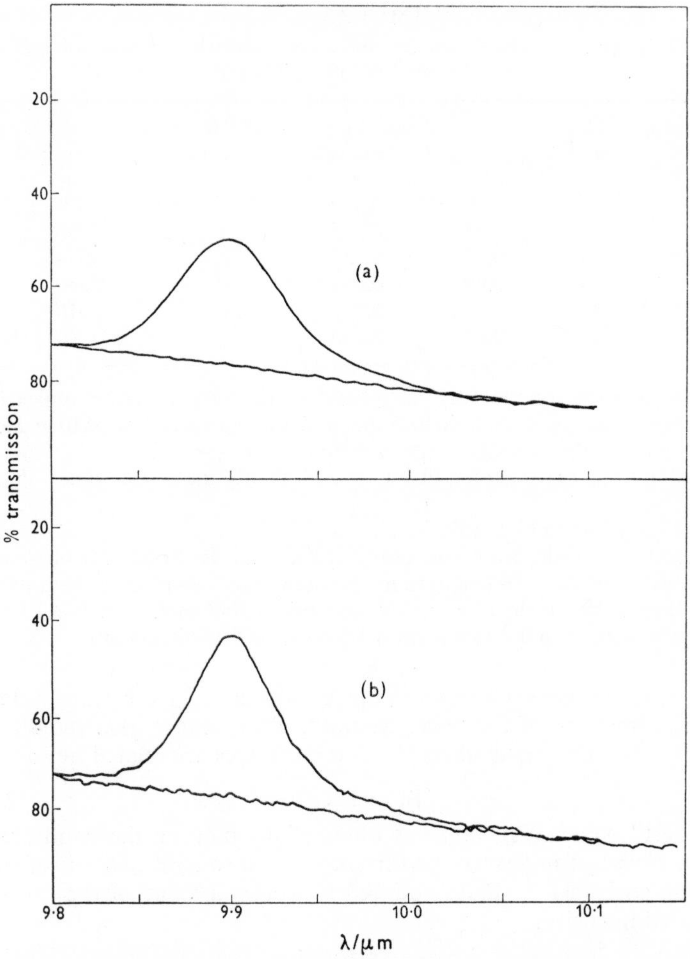

FIG 2.1 Pyridine–ICl in chloroform, the ν_9 band at 1012 cm^{-1}. Effect of resolution on the shape and intensity of an infrared absorption band. (a) slit width, 3.0 cm^{-1} (b) slit width, 1.0 cm^{-1}. The apparent half-bands width of this band is 3.5 cm^{-1}.

Once the slit conditions have been investigated and set at the appropriate value the apparent integrated intensity can be obtained from:

$$\Gamma_i^a(\text{apparent}) = \frac{1}{cl}\int_{\text{band}} \ln\left(\frac{T_0}{T}\right)_{\ln \nu} d(\ln \nu) \qquad 2.1$$

or

$$B_i(\text{apparent}) = \frac{1}{cl}\int_{\text{band}} \ln\left(\frac{T_0}{T}\right)_{\nu} d\bar{\nu} \qquad 2.2$$

TABLE 2.1

Effect of spectral slit width on the maximum absorbance and integrated area of a narrow infrared band[a]

Slit width/ cm^{-1}	$\Delta\nu^{a}_{1/2}$/ cm^{-1}	$s/\Delta\nu^{a}_{1/2}$	$\log (T_0/T)_{\bar{\nu}}$ (at 1012 cm^{-1})	10^{-2} Band area/ m^{-1}	$10^{-1}B_i$[d]/ m mol^{-1}
3.0	4.6	0.66	0.495	2.75 ± 0.05[b]	2720 ± 50
1.4	4.0	0.35	0.535	2.78	2750
1.2	3.6	0.33	0.657	3.06	3030
1.0	3.4	0.29	0.691	2.89	2860
0.8	3.2	0.25	0.710	3.00	2970
0.6	3.5	0.17	0.638	3.24 ± 0.1[c]	3210 ± 100

a. These data are for the 1012 cm^{-1} band of the pyridine–ICl complex in chloroform. The pathlength is 97.5 μm and the concentration 0.1036 mol dm^{-3}. A Grubb-Parsons GS2A spectrometer was employed.
b. Error quoted is estimated from several bands measured under same conditions.
c. Error large due to high noise level.
d. Integrated intensity from equation 2.2. This unit is the SI unit obtained by conversion of the path length to m, the area to m^{-1} and the concentration into mol m^{-3}. It can then be readily seen that 1 dm^3 mol^{-1} cm^{-2} = 10 m^3 mol^{-1} m^{-2} = 10 m mol^{-1} (see section 1.2.6 for further discussion).

where c is the concentration, l the pathlength, T_0 the transmitted radiation intensity of the "background", T the transmitted radiation intensity of sample, and where the two intensities are related by[49]:

$$B_i = \Gamma^a_i \bar{\nu}_i \qquad 2.3$$

The "best" value of B_i or Γ^a_i is obtained by plotting the band area against cl and, if necessary, performing a least squares analysis of the resulting straight line. If it is desired to obtain a value of the "true" integrated intensity:

$$\Gamma^t_i = \frac{1}{cl} \int_{\text{band}} \ln \left(\frac{I_0}{I}\right)_{\ln \nu} d(\ln \bar{\nu}) \qquad 2.4$$

or

$$A_i = \frac{1}{cl} \int_{\text{band}} \ln \left(\frac{I_0}{I}\right)_{\bar{\nu}} d\bar{\nu} \qquad 2.5$$

(where I_0 is the incident radiation intensity and I the transmitted radiation intensity) then it is necessary to plot B_i or Γ^a_i against cl and extrapolate to $cl = 0$[49,50]. This extrapolation, it should be noted, is valid only for constant incident radiation and resolving power over the width of the band[49]. (This condition is never strictly fulfilled in practice since, if the spectral slit is kept constant, then the incident intensity will fall off with decreasing frequency (below 5000 cm^{-1})[45,46] due to the variation of source intensity with frequency.) For pure liquids a considerable variation of B_i with l was observed[51] (c cannot be varied) and extrapolation was necessary. However, with solutions of complexes and other solutes studied so far[14,19,52] the values of B_i have been essentially constant with variations in cl and have shown only random variations caused mainly by base-line errors. The change of B_i for pure liquid studies may be due to the variation in reflection losses[23,53] when relatively thin cells are employed.

In order to numerically integrate a particular absorption curve and call this an "absolute" band area (with any confidence) one has to be sure that the transmission (T, T_0) or absorbance [log (T_0/T)] read off the chart recorder (or the data tape in the case of automatic digital recording) are close to the "true" values for the particular sample. Whether or not "true" transmission values are recorded depends on several instrumental factors. These are the transmission linearity, the amount of "stray" light, the grating or prism drive speed and the value of the electrical time constant. The first two factors control the deviation from "true" transmission values actually read off the chart scale while the latter two are the variable factors which control the extent of "lagging" of the (mechanical) pen recorder behind the (electronic) detector and amplification system (thereby causing an error in transmission reading at any particular frequency). The scale linearity may be checked using sectored discs[54] designed to transmit given percentages of the incident radiation. For an "optical null" instrument (double beam) any non-linearity is due to non-linearity of the attenuator comb in the reference beam[46] and is usually small. The amount of "stray" light may be checked by placing in the sample beam a compound which is completely absorbing (for example, 100 μm of carbon tetrachloride between 780 and 900 cm^{-1}). The recorder should draw a straight line along the %T = 0 (or absorbance = ∞) line. Any deviation towards higher transmission values is the result of "stray" light entering the monochromator from sources other than the spectral slits. In the "conventional" infrared region down to 500 cm^{-1} there is usually very little "stray" light. At lower frequencies stray light is a real problem since effective scattering of the "stray" radiation is much more difficult and since the detector sensitivity is relatively low. This problem has now been largely eliminated by the careful use of interferometer systems (see below).

In order to check that the time constant, gain and drive speed settings have been adjusted to their optimum values it is wise to check for "tracking error" as follows; At a point where the transmission of the sample is changing rapidly (*i.e.* on the side of a band) both the grating drive and the chart drive are turned off. Observation is made of *how far* the pen moves before it comes to a stop, (this is the "tracking error"). It should ideally be zero. If there is a large tracking error

(> 1%) then the drive speed is probably too high for the given time constant and gain settings. It is obvious that the necessary drive speed will depend on the band shape but a scanning speed of 1 cm^{-1}/min is *not* too slow in most cases. Indeed Wilson and Shimozawa[55] in their measurements on the "standard" band at 878 cm^{-1} in 1,2,4,5-tetrachlorobenzene used a scanning speed as low as 3 min/cm^{-1}.

One of the major sources of error involved in all quantitative work in the infrared (or Raman) is the position of the "base" line. A displacement of 1% in the position of the T_0 curve can result in a change in a band area of between 3 and 10%[19,56]—the change depending, of course, on the total band area—so that this is probably the most serious source of systematic error in band intensity measurements. For solution work it is normal to employ the pure solvent to obtain a T_0 curve but it is sometimes found that this does not "fit" the T curve in the "wings" of the band. The reasons for this are not easy to discover but some of the factors involved are probably (a) absorption by the solute in the wings due to other bands in the same region, (b) changes in solvent "concentration" when the solute is removed and (c) small changes in the optical properties (*e.g.* refractive index), which lead to differences in reflective properties, when the solute is removed[53]. Nevertheless, the situation in solution is a good deal more straightforward than it is in the pure solid or liquid phase where there are a number of additional complications. In neither case is there an easy way to obtain a T_0 curve in the absence of the absorbing sample. In the case of a liquid, reflection losses in the empty cell[51,57,58] make it unsuitable for a background determination, while refractive index differences between different compounds make the use of a "non-absorbing" liquid undesirable[23,53,57]. In the case of a solid film background problems again arise—due to similar changes in refractive index and in the number of reflective surfaces when the sample is removed[59–62]. For solids there is also the problem of the cooled sample acting as an energy "sink", if the radiation is chopped between sample and detector[60]. This problem can, however, be overcome if the chopping is carried out between source and sample; and if the zero of energy is chosen carefully[60,62]. These problems usually result in the use of a base-line, T_0 which is intuitively *drawn* between the "wings" of the absorption band. An analysis of the error involved in doing this for the solid phase has been given by Yamada and Person[62]. Few attempts have been made so far to obtain solid state intensity data on molecular complexes but where they have been made[63–65] base lines *have* been drawn in.

For many systems, the accurate determination of intensity data may be hampered by the fact that overlapping of bands occurs. This is especially true in complex systems where there are two or more components which absorb and where one would like to make intensity data comparison for a particular band in the free donor (or acceptor) with

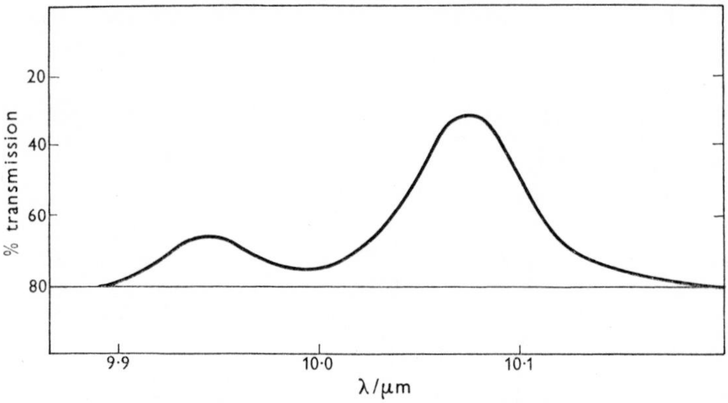

FIG 2.2 Pyridine–I_2 complex in carbon tetrachloride. Showing the "free" and "complexed" pyridine ν_9 bands at 991 and 1006 cm^{-1} respectively.

that in the complexed molecule. This situation is illustrated in figure 2.2 for one of the pyridine bands in a solution containing pyridine–I_2. Often, however, the frequency shift is insufficient for separate bands to be observed and manual separation of the bands is necessary. We have recently found[22] that such separation is achieved quite readily and accurately using a Du Pont curve resolver[66]. A total of ten Gaussian or Lorentzian functions is available and a reproducibility of 1% is obtained for a given band. This treatment of overlapping bands is usually necessary if association constants are to be obtained from infrared[2,67,68] or Raman data[69,70].

So far no explicit mention has been made of the effect of the signal-to-noise ratio for a given spectrometer and sample system (although the noise level is controlled, of course, by the amplifier gain, time constant and slit conditions used). In the near infrared (4000–600 cm^{-1}) there is usually no problem with excessive noise levels for the order of slit widths necessary to make reasonable intensity measurements. However, below 600 cm^{-1} the source intensity begins to drop off rapidly and increased amplification is required. This reduces the signal-to-noise ratio and leads, in general, to spectra of poor quality. Further, despite the introduction of filters, stray light is a problem if accurate intensity data are required. The lack of a high energy source for this region led, in 1956, to the first far-infrared Michelson interferometer[71]. The interferometric technique has the principal advantage[72–74] of high energy throughput for any desired optical resolution. This is because all frequencies in the range of interest fall on the detector simultaneously and the result is a corresponding improvement in signal-to-noise ratio. The normal source is a mercury lamp (which has a relatively high output

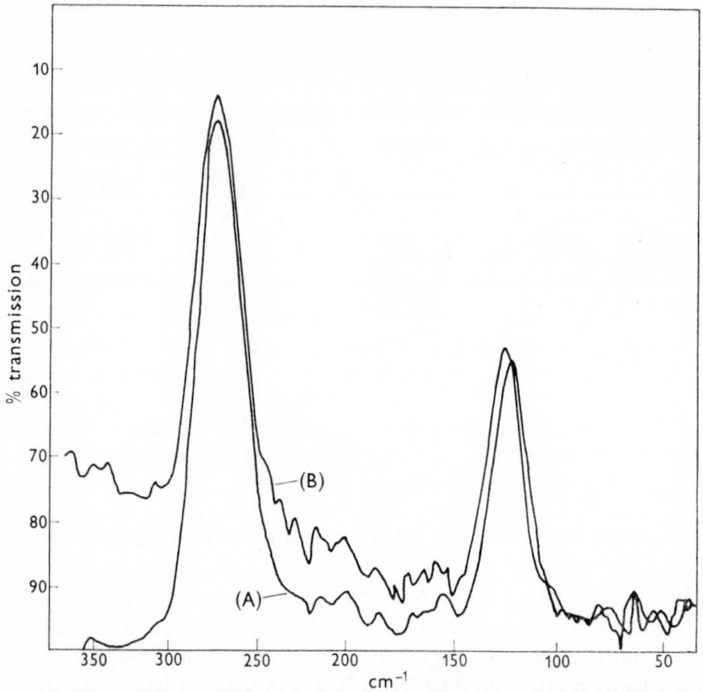

FIG 2.3 Effect of phase error on the computed spectrum using a Beckman-RIIC FS720 interferometer (A) with phase error correction (B) without phase error correction. (Reproduced, by permission, from reference 75.)

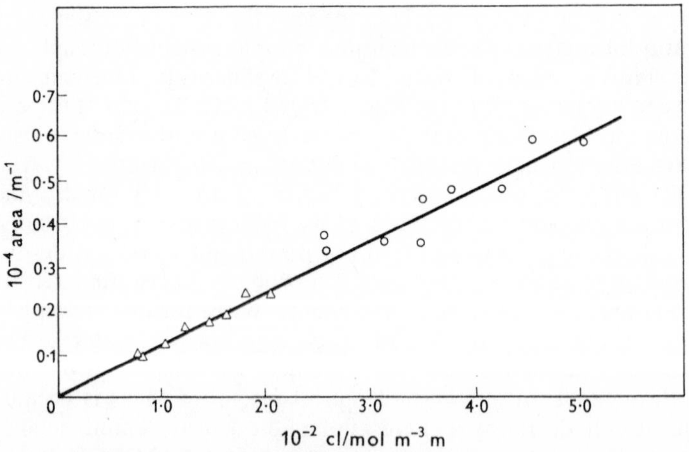

FIG 2.4 The Beer-Lambert Law plot for the ν(I-Cl) band of pyridine–ICl in benzene. △, obtained using Beckman IR-11 spectrometer, ⊙ obtained using FS270 interferometer. (Reproduced, by permission, from reference 75.)

in the region 10–500 cm^{-1}) which is used in conjunction with a Golay detector[45]. The detector has a relatively large receiving area so that only a cheap condensing system (polyethylene lens and simple light tube) is necessary. The principal disadvantage of the interferometer is that it is a single beam system. It is particularly difficult to make meaningful intensity measurements with such a system for the following reasons (a) The d.c. part of detector signal (which consists of a constant d.c. voltage with the interferogram modulation superimposed upon it) is a function of source intensity. Any variation with time (over the period of obtaining an interferogram) or between the sample and background "runs" will cause errors in the calculated transmission. (b) Different gain settings must be employed for sample and background since in general they have differing transmission properties. We have carried out a series of careful investigations[75] in an attempt to evaluate the reliability of absolute intensity data obtained using our Beckman-RIIC, FS720 far-infrared interferometer. We have found that the principal factor affecting the computed transmission data is the phase correction which is applied to the (one-sided) interferogram before Fourier transformation. The effect of lack of phase correction[76–78] (in this case deliberately caused[75] by employing a high electrical time constant) is shown in figure 2.3. The transmission curve, obtained without correction, is seen to be grossly distorted especially at high frequencies. Other factors which affect the transmission data slightly are the resolution employed and the source aperture. The spectra are very noisy if no apodisation is employed (see reference 75 for a full discussion of these factors). Otherwise, provided that "background" and "sample" interferograms are obtained close to one another† and provided an adequate "gain" correction is applied we have had no great difficulty obtaining what we believe to be accurate values of T_0/T for solutions (estimated accuracy $\pm$ 2–3%). We have, however, noticed that the ratio T_0/T does not always reach a value of 1.0 in the wings of a band; maximum values are sometimes as low as 0.8–0.9. This is probably due to a combination of two factors (a) normalisation to peak noise level—which may be different for sample and background, (b) residual phase errors—since it is known that phase errors produce this sort of transmittance error (see figure 2.3). We have nevertheless been able to get good agreement between intensity data (for iodine monochloride complexes) obtained using the FS720 and a Beckman IR-11 filter/grating spectrometer[14b]. Figure 2.4, which shows the Beer-Lambert law plot for the ν(I–Cl) band in the pyridine–ICl complex, clearly demonstrates this agreement.

In conclusion, it is clear that considerable care is necessary if reliable and quantitative studies are to be made on complex-containing systems. It is demonstrated, however, in chapters 3–7 that a wide range of molecular complexes *has* been very successfully studied using vibrational (and other) spectroscopic techniques.

† Within about one hour of each other.

2.3. RESULTS OF SPECTRAL MEASUREMENTS AND THEIR INTERPRETATION

2.3.1. INTRODUCTION

There are three principal objectives in writing this section. The first is to illustrate the *kind* of spectroscopic data which has been obtained for molecular complexes taking specific examples from a range of "types" of complex which are discussed individually in chapters 3–7. The second is to illustrate (and attempt to correlate) the ways in which the data have been interpreted. Finally, we shall try to illustrate some of the experimental difficulties discussed in section 2.2 for real systems and to point out how they may be minimised or overcome. No attempt is made here to give *exhaustive* lists of spectroscopic properties of a large number of complexes. Rather, an attempt will be made to discuss in some detail a *few well-chosen cases* of each "type" of interaction in order to show how spectroscopic techniques are of value for the elucidation of their structures.

It seems clear that if, as we might expect, the different complex "types" (*e.g.* *bπ-aπ*, *n-aσ*, *bπ-aσ*, *n-v*, *etc.*) are all stabilised by a combination of the same kinds of forces (coulombic, inductive, repulsive, charge transfer, *etc.*) then we might expect *some* correlation of their spectroscopic properties. It should, however, be remembered that the spectroscopic effects due to each kind of force are superimposed and that it is likely to be difficult to "separate" these effects quantitatively from the experimental data alone. Nevertheless, in favourable cases, differences in spectral behaviour in going from one complex to another or one complex "type" to another *might* be interpreted as being due to differing relative effects of the various stabilising (or destabilising) forces, or to a differing overall extent of interaction. Although only vibrational spectra are considered in depth, we have mentioned other spectroscopic studies especially where they lend support to (or appear to contradict) the interpretation of vibrational data, or where they throw some light on the problem of distinguishing relative stabilising or destabilising influences.

The vibrational changes (section 1.2) which occur on complexation lead to frequency and intensity changes (the former usually quite small and the latter sometimes very large) and the observation of "new" bands. Such "new" bands usually arise either from modes made "active" by complexation or from "intermolecular" modes, for example, the ν(D-A) stretching mode between the two component molecules. Weak bands may also arise from coupling of the donor or acceptor modes with modes such as ν(D-A). In the event of a "new" band being observed, care must be taken to ensure that the band does not arise from some impurity nor that it is a frequency-shifted band of donor or acceptor (the "free" donor or acceptor band being observed also (in solution) since, in general, an equilibrium mixture is present). (See figure 2.2.)

Having assigned the observed bands to expected modes of vibration of the complex one faces the interpretation problem. So far such data, coupled with band intensity data when available, have been interpreted broadly in the following ways.

(a) Frequency shifts have been used to compare interaction "strengths" within a *closely related* series of complexes[14,79–82] where normal coordinate changes might be expected to be similar in each case (sections 1.2.3 and 1.2.8). The direction and specificity of frequency shifts, and intensity perturbations have been used to provide information about the mode of interaction between the molecules (see chapters 3–7) (*i.e.* about the site of interaction and the possible electronic changes which take place).
(b) For some relatively simple complexes frequency data have been used to calculate force constants and normal coordinates for the whole complex molecule[52,83,84,85]. Comparison within a series of complexes is, of course, made on more justifiable grounds if force constants are available even for simplified models[14,82,86,87,88]. This is emphasised in section 1.2.2 where it is pointed out that normal coordinate changes can produce frequency shifts without any force constant change.
(c) Frequency and intensity data have been combined to obtain transition moments ($\partial\vec{p}/\partial R$) for some vibrations[81,82,86,87,52] usually on the basis of a simplified model. Such computations enable estimates to be made of the electronic redistribution occurring during complexation.
(d) The experimental data (including intensity and force constant changes, solvent effects and dipole moments) have been used to test the predictions of the various theories of molecular donor-acceptor interaction[34–36,81,82,87,89,90]. So far, however, such tests have been applied only for a few complexes (see later).

There are several major problems associated with testing the various complex "models". The ideal way to go about doing this is to compare the experimental data (energies of formation, force constant and intensity changes, dipole moments, *etc*) with values of these parameters calculated using the best available wave functions (see section 1.3). There are, however, despite recent advances in computer technology and in computational techniques, limitations on the size of the molecule which can be treated "*ab initio*" while *CNDO*/*INDO MO* calculations are less reliable (section 1.3). In particular, they tend to overemphasise the charge-transfer effects due to inadequate treatment of exchange-repulsion. There are also two serious problems which arise on the experimental side (apart from the superimposition of the effects of classical electrostatic and charge-transfer forces) (i) In general it may be not known precisely how the different forces, say charge-transfer and polarisation forces, are likely to differ in their effect on the spectroscopic parameters (some semiquantitative predictions have been made, however, in section 1.2) (ii) There may be, as pointed out by Mulliken (reference 1, R14), distinct parallels between, say charge-transfer donor action and classical electrostatic donor action (or similar parallels for acceptor action). In other words, donors which have low ionisation potential may be the ones for which greatest classical electrostatic interaction (through coulombic or polarisation forces) might be expected. It may well be therefore that in order to draw definite conclusions about predominant stabilising forces one needs to examine cases in which such parallelisms are known not to apply (the example given by Mulliken, reference 1, p. 452 relates to the donor properties and charge distributions for RX molecules).

We shall now proceed to consider how solutions to some of these problems have been sought for different types of complex.

2.3.2. COMPLEXES OF THE HALOGENS AND RELATED MOLECULES

Apart from hydrogen-bonded complexes, by far the most commonly studied $a\sigma$ acceptors are the halogens, interhalogens, and pseuohalogens (*e.g.* XCN, X = Cl, Br, I) which form a range of complexes varying widely in their "strength" and spectral properties. Large amounts of data are available for complexes with both $b\pi$ and n donors and there have been quite determined attempts made to investigate their structures and the differences between them. These are described in detail in chapters 1 and 3.

There are several difficulties associated with the study of halogen complexes which should be noted. Since the complexes *are* of widely varying "strengths" it seems likely that the forces stabilising them probably change considerably *both* in nature and extent on going from one complex to another. Since the contributions to complex stability are continuously changing in this way they are that much more difficult to separate. Thus, although there are cases, for example benzene–I_2, where collision-induced electrostatic forces seem likely to predominate (section 3.4.3) (but not to the *total* exclusion of a charge-transfer contribution) and cases, for example trimethylamine–I_2, where charge-transfer forces seem likely to predominate, there are also "intermediate" cases, for example pyridine–I_2, where we now believe that *both* charge-transfer and polarisation forces play an important role. Many of the most interesting halogen complexes are "stable" only in solution and have not so far been isolated (including both benzene and pyridine complexes with I_2). This means that crystallographic techniques (section 3.5.2) cannot be employed for structure determinations. Attempts to isolate solid 1:1 complexes[52] usually result in the formation of polyhalide ions (*e.g.* $Pyr_2I^+I_7^-$ and analogous compounds). Even in solution the ions ICl_2^-, I_3^-, IBr_2^- are readily detected[52] if a highly polar solvent[79–82,86–88], large excesses of a donor such as pyridine,[16,79] or salt plate windows[91] are employed. Any quantitative work is of course, ruined by complex decomposition of this kind.

We shall now consider the two main types of halogen complex separately.

Complexes with $b\pi$ Donors

The classical example in this category is, of course, the controversial benzene–I_2 complex which was one of the first complexes of iodine to be examined using spectroscopic techniques. We shall examine in some detail all the results available for this complex and the analogous benzene–XY complexes and attempt to discover whether or not there is a single "model" which is consistent with all the data.

The association constant for the benzene–I_2 complex is so small (0.18 $dm^3\ mol^{-1}$ at 27°C in carbon tetrachloride[92]—table 3.1) that there has been considerable discussion as to whether this system should be regarded as a specific "complex" or not (see sections 1.2 and 3.4.3). This value of the K_c value is, of course, obtained from a study of the electronic charge-transfer spectrum and does not necessarily mean that the "complex" ground state is stabilised. Further, it has been said[93] that if only random collisions occur an "association constant" of between 0.15 and 0.3 $dm^3\ mol^{-1}$ might be expected (see section 3.3.4 for a further discussion of this point). There is, however, considerable spectroscopic evidence that benzene and iodine *do* form specific complexes. Several important changes occur in the vibrational spectrum. Two of the vibrations which are infrared inactive in the "free" molecule give rise to bands in the spectrum of liquid benzene when iodine (or bromine) is added. These are the $\nu_1(a_{1g})$ mode with a band at 992 cm^{-1} and the $\nu_{10}(e_{1g})$ mode with a band at 850 cm^{-1} [89,94,95]. Their intensities have been measured by Ferguson and Chang[95] both in the presence of iodine and on adding other solvents. It was found that, although *all* the Raman active modes have bands which show perturbations when a solvent was added, the ν_1 and ν_{10} bands show *specific* enhancement when halogens were added (see table 2.2 for a summary of data). There is also a perturbation of the ν(I-I) mode which, although inactive in the "free" halogen, gives an easily-seen absorption band at ~205 cm^{-1} (see figure 2.5) when the iodine is dissolved in benzene or other $b\pi$ donors[86] (as shown in table 2.2 this phenomenon also occurs for the other homonuclear halogens). Attempts at studying the infrared spectra of interhalogens in benzene and other aromatic hydrocarbons have been hampered by reactions which occur between donor and acceptor[22] (producing halogenated hydrocarbons of unknown halogen content). We *have* succeeded[52] in studying both IBr and ICl in benzene (at low concentrations) in the far-infrared region and find that both ν(I-Cl) and ν(I-Br) band intensities and frequencies are considerably

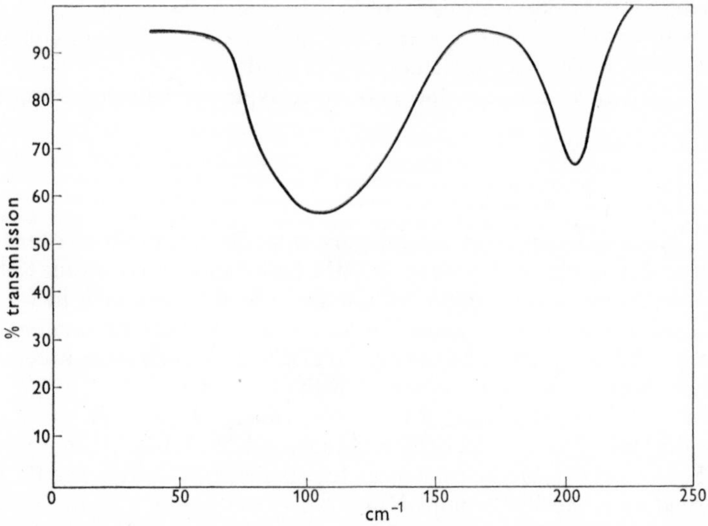

FIG 2.5 Far-infrared spectrum of iodine in benzene at 10°C. Bands at 205 and 100 cm^{-1}. Absorption due to "excess" benzene has been removed. The concentration is 0.2 mol dm^{-3} and the pathlength 2 mm.

TABLE 2.2

Perturbed bands due to complexation of aromatic π donors with halogen molecules (all data in pure donor)

Complex	Band position/ cm^{-1}	Assignment	$10^{-1}B_i$/ m mol^{-1}	Reference
benzene–iodine	992	$\nu_1(a_{1g})$	190 ± 20	94, 95
	850	$\nu_{10}(e_{1g})$	640 ± 200	94, 95
	205	ν(I-I)	160 ± 20	86
benzene-d_6–iodine	994	$\nu_1(a_{1g})$	—	94
	662	$\nu_{10}(e_{1g})$	—	94
benzene–bromine	992	$\nu_1(a_{1g})$	—	94
	850	$\nu_{10}(e_{1g})$	—	94
	307	ν(Br-Br)	—	104
benzene–chlorine	531	ν(Cl-Cl)	—	91, 106
toluene–iodine	1002	a_1	130 ± 20	89, 95
	1210	a_1	95 ± 20	89, 95
	842	a_2	—	89, 95
	204	ν(I-I)	—	16
mesitylene–iodine	998	a_1'	80 ± 50	89, 95
	1300	a_1'	350 ± 30	89, 95
	880	e''	700 ± 200	89, 95
	575	a_1'	—	89, 95
p-xylene–iodine	836	a_{1g}	120 ± 70	89, 95
	1204	a_{1g}	250 ± 40	89, 95
	943	b_{2g}	90 ± 20	89, 95

TABLE 2.3

Summary of band position and intensity data for ICl and IBr complexes in a range of solvents

	Solvent	$\bar{\nu}_{max}$/ cm^{-1}	$10^{-1}B_i$/ m mol^{-1}	$\Delta\bar{\nu}_{1/2}$/ cm^{-1}	Reference
ICl	*n*-heptane	375	—	10	52, 104
	carbon tetrachloride	375	1000	8	104
	carbon disulphide	361	1400	12	104
	benzene	355	1900 ± 200	15	104
	benzene	353	2000 ± 200	15	52
IBr	*n*-heptane	261	40[a]	—	88, 52
	carbon disulphide	253	—	—	88
	benzene	249	—	—	88
	benzene	249	90 ± 10	13	52

a. Intensity an estimate only due to extreme weakness of band.

perturbed from their values in heptane or hexane (the data are summarised in table 2.3). As mentioned in section 3.4.3, we have now observed the "free" ν(I-X) band in benzene (a "complexing" solvent) when it is mixed with cyclohexane (see figure 3.12). Such a "free" halogen stretching band has not been positively identified for benzene solutions of X_2 halogens but it is suspected to arise in the Raman spectrum of bromine in benzene (chapter 1, reference 49) and has been definitely identified in the Raman spectra of solutions of iodine in mesitylene[17]. It is thus unclear as to whether the halogens from "specific" complexes with benzene but there is now little doubt that iodine does so with mesitylene and that ICl does so with benzene. (More work is obviously required on the infrared spectra of the interhalogens with these aromatic donors.) The Raman spectrum does not, of course, tell us what *kind* of forces are responsible for complex formation. One parameter which might give us some idea of whether charge transfer is important is the intensity of the ν(I-I) band. Charge-transfer interaction is likely to give rise to a large intensity perturbation (section 1.2.7) and this perturbation should increase as the donor power of the hydrocarbon increases (as I_D^v decreases), We have examined[22] the ν(I-I) band intensity data obtained for iodine with durene and hexamethylbenzene in cyclohexane. The values are summarised in table 2.4 and it can be seen that a very large increase in the band intensity occurs (these measurements are subject to relatively large errors due to the quite high absorption of the hydrocarbons themselves in this region). Evidence that charge-transfer effects are present is therefore provided in the case of methylated benzenes, since the first ionization potential falls considerably on going from benzene to hexamethylbenzene (reference 2, p. 45). If activity of this band were due solely to quadrupole-induced dipole effects then one

TABLE 2.4

Intensity perturbation of the ν(I-I) band in the presence of aromatic hydrocarbons

Donor	Solvent	$\bar{\nu}$(I-I)/ cm^{-1}	$\Delta\bar{\nu}_{1/2}$/ cm^{-1}	$10^{-1}B_i^a$/ m mol^{-1}
benzene	benzene	207	12	160 ± 20
1,2,4,5-tetramethylbenzene	cyclohexane	201	10	900 ± 200
hexamethylbenzene	cyclohexane	198–200	10	3000 ± 500

a. Based on equilibrium constant data quoted by Keefer and Andrews (reference 96).

would expect the intensity to be proportional to square of the quadrupole moment of the hydrocarbon[97,98]. This would mean (table 2.4) that the quadrupole moments of durene and hexamethylbenzene would need to be approximately twice and four times that of benzene (respectively)[98,99]. It seems that the quadrupole moments of these hydrocarbons have not yet been determined but they must clearly be similar in value to that of benzene since the contribution from π-quadrupoles is expected to predominate[36,100]. It therefore seems unlikely that the large intensity increase is accounted for by electrostatic forces, although for benzene itself these forces could be responsible for at least *some* of the enhancement[36].

The idea that collision-induced electrostatic interaction is a relevant factor to be considered in connection with these "weak" complexes has been recently supported by the observation[98,101] of a very broad absorption band in the 70–100 cm^{-1} region when the halogens or interhalogens are dissolved in benzene (and other solvents such as xylene, ethyl benzene, dioxan and carbon disulphide). The origin of these bands (shown in figure 2.5, p. 124) is discussed more fully in section 3.4.3 but it is important to notice here that the reciprocal frequency of the experiment is now $\sim 10^{-12}$ s and the lifetimes (see below and chapter 7) of these complexes are therefore also expected to be of this order. We would thus not expect evidence for "stable" complex formation at these frequencies. This could be why no separation of "free" and "complexed" ν(I-I) bands occurs for benzene–I_2 at 200 cm^{-1} even though at higher frequencies (equivalent to 10^{13} s^{-1}) one might find evidence for "stable" complex formation. Some effort has been made to study other properties of $b\pi$-halogen complexes which might be expected to reflect the extent of charge-transfer in the ground state. There have been, for example, attempts to detect changes in the n.m.r. and n.q.r. spectra on going from free donor (or acceptor) to a potential donor/acceptor mixture. The n.q.r. data[26,32] for the solid benzene–Br_2 mixture shows an increase in ^{81}Br frequency (compared with Br_2) rather than a decrease as might be expected if charge transfer is important. However, as pointed out in section 3.5.3, crystal field effects[24] (producing frequency increases usually) are operative and, since their magnitude is difficult to estimate accurately, some frequency decrease due to transfer of charge may be obtained and masked by crystal effects. It is interesting that there is some disagreement between the two measurements reported on the benzene–Br_2 system[26,32] (although both values show an increase in frequency from that of pure bromine). This may be due to differences in the crystallinity of the two samples and shows that the precise crystal structure of the sample examined needs to be known before unambiguous interpretation is possible for these "weak" complexes. Studies on the proton resonance spectrum of benzene with added iodine have been equally disappointing from the point of view of achieving unambiguous results. No change in the benzene signal has been observed[15,22] but, with such a low association constant[92] it is very difficult to complex more than about 10% of the benzene in solution since iodine is relatively insoluble in benzene or other common organic solvents ($\sim$0.05 mol dm^{-3} is the upper limit in most cases). Even using methyl-substituted benzenes[102,103] where the K_c values are higher[96] only mesitylene shows any down field shift of the proton signal when iodine is added. The n.m.r. studies using better acceptors (ICl, IBr or Br_2) are again hampered by reactions which occur, especially at high halogen concentrations. Dielectric absorption studies have also been reported[7] for iodine in benzene the data yielding values of 3×10^{-12} s and 2.3×10^{-30} C m for the relaxation time and dipole moment respectively. This relaxation time has now been shown to be invalid (chapter 7) but a complex lifetime of about 10^{-12} s is still indicated (see above).

Let us now summarise the situation for "weak" halogen complexes. Although electrostatic forces can make a considerable contribution to their energy of formation (section 1.2.5), can account for their dipole moments[98] and can explain some of their low frequency spectral properties (section 3.5.3) it seems clear that some of the observed spectral changes cannot be properly explained without considering the charge-transfer model. It has been shown (section 1.2.5) that the electrostatic effects calculated by Hanna and his co-workers[34–36] are probably overestimates. Improved theoretical calculations are therefore necessary in order to establish the validity (or otherwise) of the varying interpretations currently being employed. However, the experimental data we have discussed in this section are all consistent with a benzene–I_2 which *is* stabilised by a combination of polarisation and charge-transfer forces but which has a lifetime of the order of 10^{-12} s. As the "strength" of interaction increases (on going, for example, to mesitylene, durene or hexamethylbenzene complexes) the charge-transfer (and possibly polarisation) forces increase and the complex lifetime also increases. All it is then necessary to accept is that a *small* amount of charge transfer can produce relatively large spectral changes as is indicated in section 1.2.

Complexes with *n* Donors

Although many complexes with "lone-pair" donors have been studied, (e.g. see ref. 105) by far the most thoroughly studied donor is pyridine and we shall consider only pyridine–XY complexes in detail in this section. Interaction between pyridine and the halogens is thermodynamically "strong"—the association constants being in the range 10^3–10^5 dm^3 mol^{-1} and the $\Delta H^{\ominus}$ values being about 30 kJ mol^{-1} (table 3.1). In several cases (with ICl and IBr, for example) pyridine forms well-characterised solid 1:1 complexes[14b,87]. These are among the most easily studied complexes since they dissolve in organic solvents and problems of band overlap due to "free" and "complexed" donor or acceptor do not occur. The "strength" of these complexes is reflected in the large spectral changes which occur for both donor and acceptor (see below and section 3.4) but the unambiguous interpretation of these changes is by no means straightforward. As pointed out in section 3.4 there are at least three contributory factors which may lead to vibrational frequency and intensity changes. These are:

1. Vibronic effects which arise as a result of charge-transfer interaction. These are expected (section 1.2.4) to give rise to small frequency (*i.e.* force constant) changes for a molecule such as pyridine but their effects on the intensities could be quite large (section 1.2.7).
2. Classical electrostatic effects, due mainly in this case to dipole-dipole and dipole-induced dipole interaction. It is not entirely clear how large these effects may be but (as shown in section 1.2.6) their effect on band intensities is likely to be considerably less than those due to vibronic effects. Their effect on the force field of the pyridine molecule is again expected to be small (section 1.2.5).
3. Vibrational coupling effects due to changes of normal coordinate on going from the "free" pyridine and halogen to the complex. The sizes of the changes expected from such vibrational mixing are again not known precisely but they can be estimated using an approximate normal coordinate treatment (sections 1.2.3 and 1.2.8).

We shall now examine the data for pyridine–IX complexes (X = Cl, Br, I, CN) and consider how they compare with the expectations of section 1.2. The frequency and intensity data which have been determined for pyridine–IX complexes in the far-infrared spectrum are given in table 3.7. The frequency changes in the pyridine spectrum are given in table 3.13 and the intensity changes observed in benzene and chloroform solutions for pyridine–ICl are summarised in table 2.5. The magnitude of some of the changes is emphasised by pointing out the six-fold increase in intensity (tables 2.3 and 3.7) and the $\sim 25\%$ frequency shift of the ν(I-Cl) band for pyridine–ICl[87]; also the six-fold intensity increase in many of the pyridine vibrational bands for the same complex[14b,19]. (The intensity changes for the pyridine molecule in pyridine–IBr and pyridine–I_2 are similar to those for pyridine–ICl (see table 2.8).)

TABLE 2.5

Summary of the infrared intensities B_i of pyridine and pyridine–ICl bands in chloroform and benzene (Reproduced from Trans. Faraday Soc., **65**, 934 (1969), by permission of the Chemical Society).

No.	Class	$\bar{\nu}$/cm^{-1}	$\Delta\bar{\nu}_{1/2}$/ cm^{-1}	$10^{-4}B_i$/ m mol^{-1}	$\bar{\nu}$/cm^{-1}	$\Delta\bar{\nu}_{1/2}$/ cm^{-1}	$10^{-4}B_i$/ m mol^{-1}
		pyridine in chloroform			pyridine–ICl in chloroform		
4	a_1, b_1	1586	11.0	2.81 ± 0.2[a]	1601	4.2	2.95 ± 0.3
5	a_1	1484	4.5	0.34 ± 0.01	—	—	0.0[b]
6	a_1	1220	5.0	0.46	1212	3.5	1.8
7	a_1	1069	5.2	0.46 ± 0.04	1067	2.5	3.35 ± 0.4
8	a_1	1031	4.8	0.70 ± 0.07	1034	2.8	0.75 ± 0.15
9	a_1	991	5.3	0.72 ± 0.08	1011	3.5	3.43 ± 0.5
10	a_1	605	5.9	0.45 ± 0.03	631	4.0	1.94 ± 0.5
13	b_1	1572	—	—[c]	—	—	0.0[b]
14	b_1	1440	4.5	3.07 ± 0.20	1451	4.2	4.30 ± 0.4
15	b_1	1375	—	—	1353	—	0.25
16	b_1	1217	5.0	0.46	1212	3.5	1.8
17	b_1	1147	5.4	0.32 ± 0.03	1155	5.0	0.45 ± 0.05
18	b_1	1069	5.2	0.46 ± 0.04	1067	2.5	3.35 ± 0.4
		pyridine in benzene			pyridine–ICl in benzene		
4	a_1, b_1	1581	12.0	3.1[a]	1599	4.2	3.2
5	a_1	—[d]			—[d]		
6	a_1	1217	6.0	0.42	1210	3.0	1.6
7	a_1	1068	9.0	0.43	1066	5.0	2.9
8	a_1	—[d]			—[d]		
9	a_1	991	5.7	0.63	1010	3.0	2.5
10	a_1	602	6.0	0.39	629	5.0	2.5
13	b_1	1575		—[c]			~0.0[b]
14	b_1	1438	8.0	3.1	1450	3.9	4.5
15	b_1				1350		
16	b_1	1217	6.0	0.42	1210	3.0	1.6
17	b_1	1145	6.0	0.33	1154		~0.2
18	b_1	1068	9.0	0.43	1066	5.0	2.9

a. Intensity of combined triplet of bands near 1600 cm^{-1}.
b. Band not present in spectrum of complex solution—intensity very low.
c. Intensity included in value given for ν_4.
d. Band masked by solvent absorption.

For such strong complexes it is clear that vibrational mixing must occur to some extent and so a normal coordinate analysis is essential to properly interpret the frequency and intensity changes. This is especially true for the two low frequency modes approximately described as ν(I-I) and ν(D–I) which have the same symmetry and which are expected to mix strongly. It has been usual to treat the complex as a linear triatomic molecule[107] (section 1.2.3). (It should be noticed here that, although previously the geometry of pyridine complexes in solution had been assumed to be the *same* as that in the solid, data which confirm that this *is* the case for pyridine–I_2 have recently been obtained[108].) Difficulty is experienced, however, even using such a simple model since we have three force constants defined by the general valence force field. The potential energy for in-plane motion is:

$$2V = k_{IX}\,\Delta r_{IX}^2 + k_{DI}\,\Delta r_{DI}^2 + k_{12}\,\Delta r_{IX}\,\Delta r_{DI} \qquad 2.6$$

(the parameters in equation 2.6 being defined in figure 2.6). Since only two frequencies can be measured it is impossible to determine k_{12} using a single isotopic species (despite claims[109] to the contrary).

It is, however, simple to set up the normal coordinate calculations[81,86,87] so that one can obtain k_{IX} and k_{DI} *as a function* of k_{12}

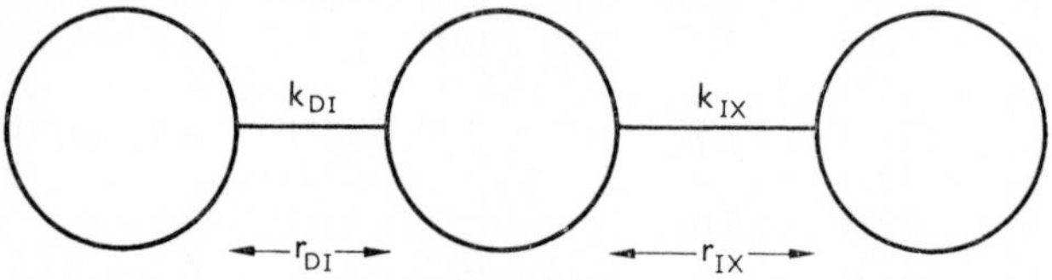

FIG 2.6 Definition of force constants and internal coordinates for the D–IX linear triatomic model.

which is allowed to vary-within limits considered reasonable for such an interaction constant from previous results (mainly for trihalide anions[110,111]). This procedure is described in detail in section 1.2.3 where the effects of mixing of these two vibrations (with and without a change of k_{IX}) as k_{DI} increases is examined. Having decided by some means on the most reasonable value of k_{12} it is a simple matter, knowing the intensities of the two bands, to calculate the dipole derivatives $\partial\vec{p}/\partial R_j$ given by equation 1.54, p. 57. The values of these dipole derivatives, along with the data necessary to compute them, are given in table 2.6. We have recently[107] obtained extra frequency data for the pyridine-d_5–IBr complex in benzene solution and so have been able to calculate unique values of k_{12}, these corresponding to the crossing points of the pyridine–IBr and pyridine-d_5–IBr force constant ellipses[87,107], which are shown in figure 2.7. The two values obtained are 10 and 60 N m^{-1} and the corresponding force constants are underlined in table 2.6 for the pyridine complexes. It should be emphasised that there is no special reason why the constant k_{12} should be the same for all complexes even with the same donor. However, we have been unable to detect sufficient frequency shifts for the pyridine-d_5–ICl and pyridine-d_5–I_2 systems for separation of the force constant ellipses. So, for the purposes of calculating the dipole derivatives, we have assumed that the value of k_{12} remains the same. The values of $\partial\vec{p}/\partial R_1$ and $\partial\vec{p}/\partial R_2$ for the possible values of k_{12} are given in table 2.6 and summarised in

TABLE 2.6

Normal co-ordinate and observed transition moment data for the two low frequency vibrations of pyridine–IX complexes (reference 107)

Complex	$\bar{\nu}$(IX)/ cm^{-1}	$\bar{\nu}$(D–I)/ cm^{-1}	$10^{-1}B^c$/ m mol^{-1}	$10^{-1}B^c$/ m mol^{-1}	k_{12}/ N m^{-1}	k_{IX}/ N m^{-1}	k_{DI}/ N m^{-1}	L_{11}^{-1}[a]	L_{21}^{-1}[a]	L_{21}^{-1}[a]	L_{22}^{-1}[a]	$10^{20}\partial\vec{p}/\partial R_{IX}$/ coulomb	$10^{20}\partial\vec{p}/\partial R_{DI}$/ coulomb
pyridine–ICl (benzene)	292	140	11900 ± 200	3700 ± 200	0	136	63	6.57	−0.88	2.68	9.36	22.0	−26.4
					10	141	62	6.74	−0.22	2.22	9.40	24.2	−23.1
					20	145	62	6.87	0.41	1.76	9.39	25.6	−20.8
					30	148	64	6.97	1.02	1.31	9.34	27.0	−18.2
					40	151	68	7.04	1.62	0.85	9.25	28.9	−15.5
					50	152	73	7.08	2.21	0.40	9.13	29.8	−12.5
					60	152	80	7.10	2.82	−0.08	8.96	31.0	−9.1
					10							16.0	−22.7
					60							22.1	−13.3
pyridine–IBr (benzene)	206 (205)[b]	134 (127)[b]	6500 ± 100	2800 ± 100	0	100	73	6.75	−3.92	7.08	8.93	6.8	−31.4
					10	118	62	8.16	−1.84	5.40	9.57	14.8	−26.0
					20	128	60	8.87	−0.39	4.13	9.74	19.7	−21.4
					30	136	62	9.31	0.79	3.03	9.72	23.5	−17.9
					40	140	63	9.61	2.03	1.80	9.53	27.1	−13.6
					50	143	68	9.76	3.17	0.62	9.21	30.2	−9.2
					60	143	76	9.75	4.35	−0.66	8.72	32.6	−4.4
					10							12.7	−25.4
					60							37.0	−5.5
pyridine–I_2 (cyclohexane)	183	93	2500 ± 40	1950 ± 60	0	113	34	9.12	−1.91	6.87	9.82	6.1	−21.0
					10	127	31	9.99	−0.61	5.62	9.94	10.0	−18.7
					20	137	31	10.51	+0.50	4.49	9.99	13.0	−16.6
					30	145	33	10.90	1.52	3.38	9.89	15.7	−14.4
					40	150	36	11.19	2.51	2.26	9.69	18.6	−12.0
					50	154	41	11.37	3.45	1.15	9.39	20.6	−9.6
					60	155	48	11.41	4.43	−0.07	8.98	22.8	−7.0

a. units are 10^{-12} $g^{1/2}$.
b. data for pyridine-d_5–IBr.
c. errors are the standard errors obtained from Beer-Lambert law graphs (see section 2.2).
d. "corrected" values—the transition moment for the "free" interhalogen ν(IX) band has been removed (see also table 3.8).

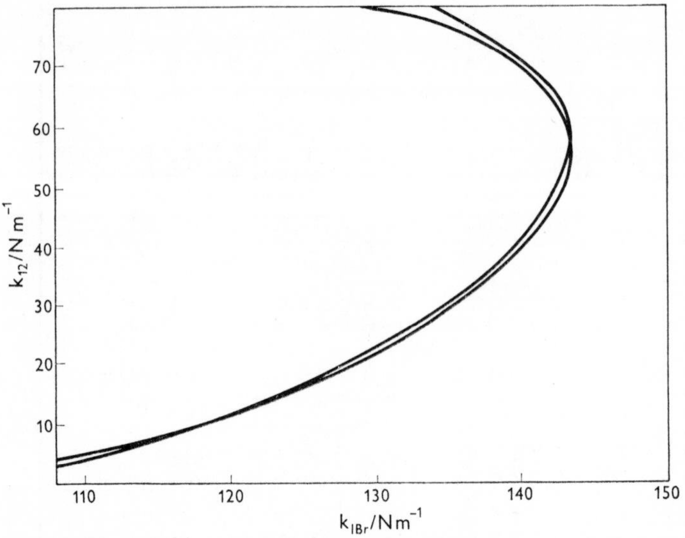

FIG 2.7 Force constant ellipses for pyridine–IBr and pyridine-d_5–IBr using the linear triatomic model (Reproduced, by permission, from *J. Mol. Struct.* **10,** 147 (1971).

table 3.8 (the symmetry coordinates used being $\Delta R_1 = \Delta R_{IX}$ and $\Delta R_2{}^{\dagger} = \Delta R_{DI}$). It can be seen that $\partial\vec{p}/\partial R_j$ for the possible k_{12} values are quite different in both absolute and relative magnitude. (It should be noted that these values of $\partial\vec{p}/\partial R_j$ represent one of the two possible sets which arise because of the sign ambiguity in $\partial\vec{p}/\partial Q_i$ values (see equation 1.54).) It has been argued[87] that the two dipole moment derivatives should have opposite signs and, if this is correct, then the relative sign combination of the $\partial\vec{p}/\partial Q_i$ values is fixed. Notice that this argument is based on the premise that the ν(D-I) band intensity is controlled by a change in overlap of the component molecules during vibration (hence giving a negative $\partial\vec{p}/\partial R_{DI}$ when the D–A distance increases because the overlap and $\vec{p}$ both decrease), and that the ν(I-X) band intensity is controlled by a change in electron affinity during vibration (hence giving a positive $\partial\vec{p}/\partial R_{IX}$ when the I-X distance increases because the electron affinity and $\vec{p}$ both increase). The argument is therefore valid for a charge-transfer model and so in choosing the values of $\partial\vec{p}/\partial R_j$ which are given in table 2.6 we have assumed this model to be a reasonable one. It should be noticed, that *if* this model *is* a valid one we might expect the stretching-stretching

† This coordinate is labelled Δr_2 in chapter 1 in accordance with the more general nomenclature employed there.

interaction force constant, k_{12}, to be positive, by arguments similar to those employed by Mills[112,113] and Coulson[114] since, as the D–I bond stretches, the I-X bond ought to become shorter (*i.e.* nearer to its length in the free I-X molecule). It is therefore interesting that, while we get positive k_{12} values using pyridine-d_5 isotope shifts and a linear triatomic model, Gayles[83], using a somewhat more sophisticated model, which includes the CNC "skeleton", for trimethylamine–XY and trimethylamine-d_9–XY complexes, obtains a best fit of force constants to the frequency data when $k_{12} < 0$ and so concludes that electrostatic interactions are important in the N-X-Y region of the complex molecule. Further, Gayles finds that if $k_{12} < 0$ then $\partial\vec{p}/\partial R_{XY}$ and $\partial\vec{p}/\partial R_{DX}$ must be of the same sign. His normal coordinates show that the *lower* of the two a_1 bands in the 150–250 cm^{-1} region arises due to vibration which consists essentially of X-Y stretching motion coupled to other skeletal modes. The data shown in table 2.6 are obtained, of course, *assuming* that the I-X band has the highest of the two force constants in the corresponding pyridine complexes. There are two reasons why, having decided on the use of a linear model, the choice of frequency assignment (table 2.6) is the most reasonable. Firstly, we find that it is the lower frequency vibration which shows a shift on deuteration. Notice that this is the opposite effect to that observed by Gayles[83] and indicates that the lower frequency band should be assigned to ν(D–I). Furthermore, if the alternative assignment is accepted then no crossing of the force constant ellipses is observed so no unique values of k_{12} are obtained.

It is interesting to see what happens when we treat the trimethylamine complexes using the simpler linear triatomic model in an attempt to discover how a change of model affects the value of k_{12}. On calculating force constants using a linear model for the TMA–IBr and TMA–I_2 complexes it was found that all the real solutions for k_{IX} and k_{DI} correspond to positive values of k_{12} and that crossing of the force constant ellipses for different isotopic molecules occurred in the region of 40–60 N m^{-1}. (No intensity data are available so no transition moments can be obtained.) It seems clear, therefore, that different models may give rise to significantly different force constants. It is, however, likely that the simple model will be a better one for the pyridine complexes than for the trimethylamine complexes. This is because the lowest in-plane skeletal mode of the pyridine molecule has a frequency corresponding to 405 cm^{-1} [19] and so coupling of the two low frequency stretching modes with the pyridine vibrations will probably be quite small. Work on a more sophisticated whole-complex model for the pyridine–IX complexes has shown (as indicated in section 3.4) that there is some mixing between the ν(D–I) mode and the lower frequency a_1 modes of the pyridine. The mixing of the ν(I-X) mode with the pyridine modes is very small.

From the values given in table 2.6 it is seen that the order of complex strength using $\partial\vec{p}/\partial R_{IX}$ as a measure is,

for $k_{12} = 10\ N\ m^{-1}$ $ICl > IBr > I_2$
for $k_{12} = 60\ N\ m^{-1}$ $IBr > I_2 > ICl$

and using $\partial\vec{p}/\partial R_{DI}$ it is,

for $k_{12} = 10\ N\ m^{-1}$ $IBr > ICl > I_2$
for $k_{12} = 60\ N\ m^{-1}$ $ICl > I_2 > IBr$

We therefore considered, using this simple model, that an interaction constant of $10\ N\ m^{-1}$ was to be preferred since the data are more consistent and since, from the thermodynamic data, it is clear that the iodine complex is at least an order of magnitude less stable than the other two complexes. This choice has been questioned (section 1.2) on the grounds that the ratio $\Delta k_{ICl}/k_{DI}$ for the pyridine–ICl complex (equation 1.33, p. 41) is more consistent with the values at $k_{12} = 60\ N\ m^{-1}$. One possibility, which we did not examine previously, is that the band assigned to a ν(D–I) mode of the complex may not be entirely due to such an internal mode. Evidence is presented in section 3.4.3 that this band may be partly due to a residual rotation of the complex dipole in the liquid phase. This would mean that the values of $\partial\vec{p}/\partial R_{DI}$ (in table 2.6) would be considerably smaller (the values of $\partial\vec{p}/\partial R_{IX}$ would also decrease but not by so much—see equation 1.54). It is impossible to say whether or not the orders of complex "strength" shown above would be changed (at, say, $60\ N\ m^{-1}$) by such a phenomenon since the contribution due to the Poley-Hill[115,116] mechanism is not known—and is difficult to calculate for a dilute solution[101,117] (see equation 3.29). However, if the contribution of Poley-Hill absorption were roughly constant for all three complexes then the values of $\partial\vec{p}/\partial R_j$ *would* decrease relative to those of the other two complexes. On the other hand, if the value of $k_{12} = 60\ N\ m^{-1}$ *is* correct the lack of a sensible "order" of values of $\partial\vec{p}/\partial R_j$ could simply be due to errors introduced by the use of a simplified model.

Having made a normal coordinate calculation of this kind on the D–I-X system it should be possible to say something about the nature and extent of bonding between the molecules. Firstly, the force constant of the D–I bond is of the order of 30–60 $N\ m^{-1}$ which means that this bond may be thought of as about $\frac{1}{3}$rd to $\frac{1}{5}$th of a single covalent N-I bond (reference 112, p. 192). This is, of course, a considerable degree of interaction. Secondly, the transition moments for stretching of the bonds are very high (the intensities of these bands are very high). This means either a high degree of delocalisation of charge during vibration or vibration of atoms carrying large charges (*i.e.* vibration of highly polar bonds). Some information has recently been obtained on the electronic structure of these complexes using other techniques. It was

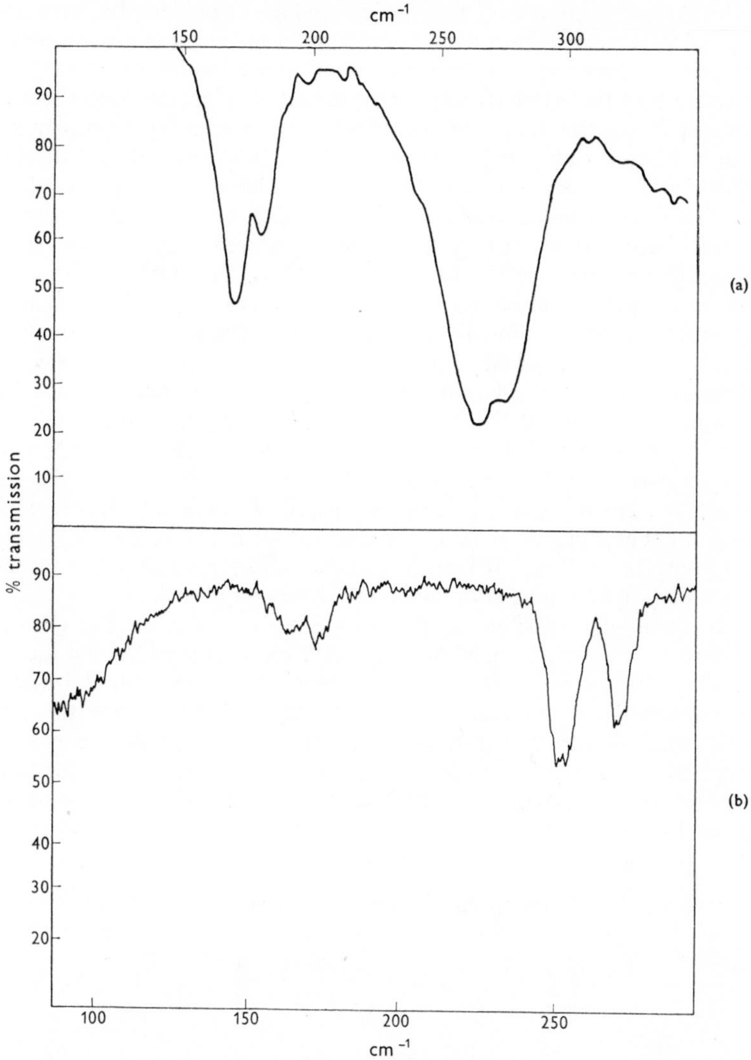

FIG 2.8 The infrared (a) and Raman (b) spectra of the solid pyridine–ICl complex.

noticed, for example, that in the Raman spectra[118] of solid pyridine–ICl and pyridine–IBr complexes that the ratio of the intensities of the two low frequency bands change considerably from that observed in the infrared spectrum. From figure 2.8 it can be seen that the Raman intensity of the band assigned to ν(D–I) has decreased relative to the

ν(I-Cl) band as compared with the infrared data (incidentally there is no sign of the low frequency band in the Raman spectrum of the complex in solution in benzene, but the background scattering level is too high to be sure that the band is completely missing). This low polarisability change during the stretching of the D–I bond indicates quite clearly that the N–I bond is a highly polar one. The bond between the two molecules therefore has a large dipolar contribution. (There are, of course, plenty of data for very polar bands (for example, C-F bonds) where, although intense infrared stretching bands are observed, the Raman active stretching modes give rise to very weak lines.) This information needs to be carefully considered in the light of the n.q.r. data obtained recently for the pyridine–IX complexes[29,30] in the solid phase (sections 1.1.4 and 3.5.3). These data, when reduced to calculated electron densities, show considerable polarisation of the halogen molecule along with about 20% of charge transfer. The calculated densities[30] are shown on p. 293 for the ICl and the complex. Clearly, the N and I atoms carry considerably greater charges than they do in the free molecule. However, the polarity of the ICl bond is also increased so one might expect a decrease of the Raman intensity of the ν(I-Cl) band, on going from "free" ICl to the complex. So far no absolute Raman intensity data have been obtained for these complexes.

It should be possible using a very simple model for the classical electrostatic interaction of pyridine with iodine to gain some idea of whether the observed transition moments shown in table 2.6 can be accounted for using such a model. In figure 2.9 we show the total calculated dipole moment, $\vec{p}$, obtained as a result of the gradual approach of a pyridine molecule towards the iodine, which leads to an increasing induced dipole on the latter. This induced dipole moment is given by:

$$\vec{\mu}' = \alpha\vec{E} \qquad 2.7$$

where α is the iodine molecule average polarisability[34–36] and where $\vec{E}$ is the field produced at the centre of the iodine molecule by a 7.35×10^{-30} C m dipole at distance r. This field is given by[119]:

$$\vec{E} = 2\vec{\mu}_{pyr}/r^3 \qquad 2.8$$

The total dipole moment is obtained by simple addition of $\vec{\mu}'$ to the pyridine dipole moment of 7.35×10^{-30} C m. By measuring the slope of this curve one can, of course, estimate a very rough value of $\partial\vec{p}/\partial R_{DI}$ for the complex. It can be seen from table 2.7 that for intermolecular distances similar to those measured[25] for crystalline complexes of this kind (*i.e.* about 35 nm or 3.5 Å) the total transition moment is in the region of 5.0×10^{-20} C which is considerably less than the observed value (even if $k_{12} = 60$ N m^{-1}) although the observed (static) dipole moment (15.1×10^{-30} C m) is seen to be accounted for very well by

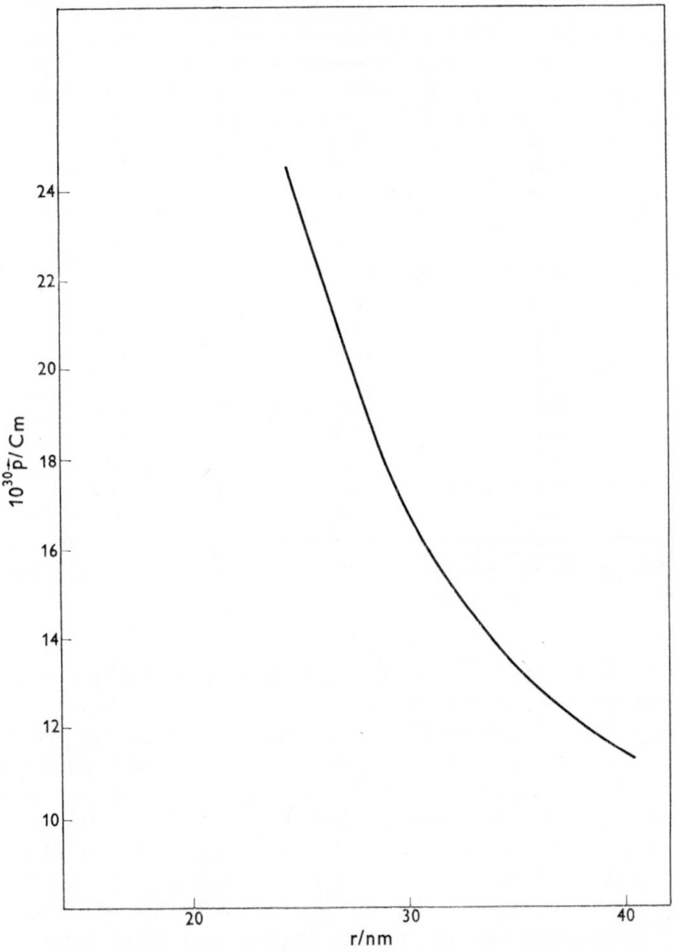

FIG 2.9 Total dipole moment of the pyridine–I_2 complex assuming only dipole-induced dipole forces between the molecules.

this calculation. The model is a crude one, of course, but in chapter 7 it is shown that such simple calculations of this kind often give results indistinguishable from apparently more sophisticated ones. The indications are that, although the dipole moment of the complex can be accounted for by electrostatic interaction, the infrared intensity data are too large to be entirely due to this effect. It should be noted that repulsive effects have not been taken into account and *may* effect the intensity data[35]. In addition, as pointed out above and in section

TABLE 2.7

Calculated induced and total dipole moments for the pyridine–I_2 complex

r/nm	$10^{30}\ \mu_{ind}$/C m	10^{30} total μ/C m[a]	$10^{20}\ \partial\mu/\partial R_{DI}$/coulomb
20	32.2	39.6	
25	16.4	23.8	
26	14.6	22.0	
29	10.6	18.0	
30	9.5	16.9	10.4
31	8.5	15.9	9.4
32	7.8	15.2	7.7
33	7.2	14.6	6.7
34	6.5	13.9	6.4
35	6.0	13.4	5.0
36	5.5	12.9	5.0
37	5.1	12.5	4.3
38	4.7	12.1	4.0
39	4.3	11.7	3.3
40	4.0	11.4	3.3

a. Obtained by adding the dipole moment of pyridine (7.4×10^{-30} C m).

3.4.3, some of the band intensity may arise from the hindered rotation of the complex dipole of the liquid.

A similar situation arises with regard to the vibrations of the pyridine moiety in that the large intensity changes (table 2.5) are unlikely to occur as a result of electrostatic interactions or changes in the normal coordinates (*i.e.* by vibrational mixing). As shown in section 1.2.7, electrostatic forces are likely to lead to only small (factor of two) intensity perturbations. Thus, if the n.q.r. data give a reasonable measure of the charge transfer for pyridine–ICl, then the data in table 2.5 show how sensitive intensity changes may be to even small amounts of charge transfer. From the comparison of the intensities of the two most "sensitive" pyridine bands for the three complexes pyridine–IX (X = Cl, Br, I)—shown in table 2.8—it may be seen that the degree of perturbation varies very little through the series. Although the frequency and intensity changes of these two bands (especially the one near 600 cm^{-1}) are probably controlled partly by changes in normal coordinate (section 3.4.4) the amount of charge transfer as measured by the intensity increase, seems similar in each case.

Some of the most interesting, but least studied, phenomena associated with complex formation and with complex environmental changes are the changes of vibrational band shape[20] which occur. These manifest themselves most obviously by a change (often quite drastic) in the half-band width. This effect is best illustrated for pyridine complexes

TABLE 2.8

Comparison of band width and intensity data for the most 'sensitive' pyridine vibrational bands for pyridine–IX complexes

		$\tilde{\nu}_9(a_1)$			$\tilde{\nu}_{10}(a_1)$		
Complex	Solvent	$\tilde{\nu}$/ cm^{-1}	$\Delta\nu_{1/2}$/ cm^{-1}	10^{-4}B[a]	$\tilde{\nu}$/ cm^{-1}	$\Delta\nu_{1/2}$/ cm^{-1}	$10^{-4}B_i$[a]
(pyridine)	chloroform	991	5.3	0.72	605	5.9	0.45
(pyridine)	carbon tetrachloride	991	5.5	0.53	605	5.0	0.48
pyridine–ICl	chloroform	1011	3.5	3.43	631	4.0	7.94
pyridine–IBr	chloroform	1010	4.0	3.20	627	5.4	3.20
pyridine–I_2	carbon tetrachloride	1006	5.0	3.0	620	7.0	2.6

a. Intensity units are m mol^{-1}, the scatter of values is usually about $\pm 10\%$.

by reference to tables 2.5 and 3.9 and is discussed further in section 2.3.6. It should be noted here that both increases and decreases in band width have been observed on complexation and that such band shape changes can (potentially) provide a great deal of information[23,24] about the molecular interactions involved. Unfortunately band widths and their changes from one system to another are rarely quoted in the literature.

2.3.3. COMPLEXES WITH METAL-CONTAINING ACCEPTORS

Introduction

Even though a vast amount of vibrational data has been collected on donor–metal complexes over the last 20 years, their unambiguous interpretation is still a very difficult problem. This is principally due to the fact that the complex molecules are relatively large and so do not easily lend themselves to force constant and molecular orbital calculations. However, some progress has been made recently in doing both types of calculation on organometallic complexes (see chapter 5), and these will be among the principal "handles" which will be used to interpret the experimental data in the future.

The vibrational spectra of organometallic complexes have been used in attempts to obtain several kinds of information.

(a) to determine the complex geometry and stereochemistry (in the solid state and in solution) by a study of the infrared and Raman selection rules for the possible point groups (the stoichiometry having been determined by other methods). One difficulty with this approach is that the transition moments of infrared or Raman active modes are often low and the relevant "active" bands may be missed. There may also be problems with overlapping bands in regions of dense spectral activity. This means that in order to be certain of predicting geometries from vibrational data alone one must be able to make a complete and unambiguous assignment. In practice this is rarely possible (see below). The situation is improved somewhat if *both* the vibrational spectra and crystal structures for related complexes have been determined.

(b) to assign a band or bands in the spectrum to the "metal–donor stretching" vibration and hence to deduce information about the nature of the metal–donor bonding. Again one is faced with the same problems as in (a) and, in addition, the vibration concerned will rarely, if ever, be a pure "M–D stretching" mode. Mixing will occur with other low frequency modes of the complex (of the same symmetry) and "metal–donor stretching" may be involved in the normal coordinates associated with *several* observed bands. It is mainly for this reason that correlations between "$\bar{\nu}$(M–D)" and other measures of donor-acceptor interaction or between "$\bar{\nu}$(M–D)" and mass and chemical factors for related donors are significantly lacking. This probably means that the vibrational mixing changes in going from one complex to another.

(c) to determine the extent and nature of perturbation of the donor using the frequency shifts observed on going to the complex. Attempts have been made to discuss such frequency shifts in terms of (i) the extent of π bonding between the metal and the donor molecule; *i.e.* the extent of the so-called "back-bonding" from metal to donor by dπ-pπ interaction (see section 5.1), (ii) the site of interaction in complexes where more than one donor site is possible and the nature of the interaction (*i.e.* whether the ligand acts as a $b\pi$ or n donor, for example), (iii) the extent of interaction, usually using empirical methods based on frequency shifts observed in a series of related complexes. Usually, in studying the complexed donor spectrum, the assignment problem is by no means as severe mainly because frequency shifts are relatively small since changes in internal normal coordinates of the donor on complexation are small. One might expect much greater (and therefore more useful) intensity changes on complexation. Very few such data have been so far obtained for reasons outlined in section 2.2 (see below for a discussion of the results which *have* been obtained).

In order to improve the chances of making a correct assignment (especially of the low frequency bands) it has recently been popular to use approximate force constant calculations to check that a "reasonable" set of force constants does in fact give rise to a set of frequencies which agrees fairly well with the observed data. At the same time it is possible to estimate the degree of vibrational coupling between the various low frequency modes. Complete potential energy distributions have been calculated in some cases. Although such calculations are useful, one should remember that the force constants can usually be obtained only by trial and error. Even using simplified models based on the complex "skeleton" with donor molecules treated as point masses, the number of force constants is usually larger than the number of observed frequencies. The set of force constants may therefore not be unique and any calculation of vibrational coupling is expected to be approximate (as pointed out for halogen complexes in section 2.3.2). Nevertheless, such calculations do establish that such coupling exists and allow *estimates* of the most important force constants to be obtained. Some of the more detailed information derived from these force fields may be less reliable.

We shall now illustrate some of these methods and problems using examples chosen from the literature. We shall also attempt to correlate spectral properties (and their interpretation) of selected donors with metal compounds and other acceptors. In doing so we shall refer to chapters 5 and 6 for more detailed comments and try to avoid serious overlap with those chapters.

Nitrile Complexes

Complexes of simple nitriles have been studied with many metal containing acceptors. An excellent review of the data obtained up to 1965 is available[120], but much of the most important data has been obtained since then (excellent sources of recent data are the specialist reports[121] published by the Chemical Society). Perhaps the major feature which is apparent in the vibrational spectra of simple nitrile complexes (not only with metals but other acceptors) is the increase in both frequency and intensity of the band usually assigned to the $\nu(C{\equiv}N)$ vibration. However, intensity measurements are still relatively few probably due to low complex stability (particularly towards moisture which makes solution work, in particular, difficult) and low solubility in organic solvents. Acetonitrile complexes with metal halides have been most frequently studied. This work not only illustrates many of the difficulties associated with studies on such systems, but also has provided more complete structural information than has been obtained for other nitriles. Many metal halides and nitrates dissolve fairly easily in acetonitrile and both Raman and infrared studies on the species present (either in solution or in solid precipitates) have been made[122–124]. It is found that several of the vibrations of acetonitrile are increased in frequency when nitrates of zinc, cadmium and mercury are dissolved. Raman intensity studies[123] have been used in the zinc nitrate case to show that the principal species present is $[Zn(CH_3CN)_2](NO_3)_2$. It is emphasised, however, that the nitrile complexes nCH_3CN–$SnCl_4$ where n = 1[125], n = 2[126], n = 3[127], and n = 4[125] have all been studied. In solution, especially, it is not always easy to establish which stoichiometries are present and in what amounts. Table 2.9 gives a summary of the changes observed in the acetonitrile spectrum due to complexation with various metal salts. Included for comparison are the data for CH_3CN–ICl[129]. There is some controversy[123,124,128] over the assignment of bands which arise in the "C≡N stretching" region in the CH_3CN–$SnCl_4$ complexes. This arises mainly because of the strong Fermi resonance between the $\nu(C{\equiv}N)$ mode and the combination mode $(\nu_3 + \nu_4)$ and illustrates yet another problem which may arise;

TABLE 2.9

Perturbation of acetonitrile due to complexation with metal-containing Lewis acids[e] (all data in cm^{-1})

Acceptor	Stoichiometry (if known)	ν_1, a_1 $\bar{\nu}(CH_3)$	ν_2, a_1 $\bar{\nu}(C{\equiv}N)$	$\bar{\nu}_3 + \bar{\nu}_4$	ν_3, a_1 $b(CH_3)$	ν_4, a_1 $\bar{\nu}(C—C)$	ν_6,e $\delta(CH_3)$	ν_8,e b(C—C—N)	Reference
("free" CH_3CN)		2944	2251	2289	1376	918	1443	375	123, 130
$ZnCl_2$[a]		2941	2314	2283 (R)	1368	938	1443[d]	395 (R)	128
$ZnCl_2$ and $Zn(NO_3)_2$[b]	1:2		2291	2310	1376	943	1422	398	123
CdI_2 and $Cd(NO_3)_2$[b]			2275	2304	1377	932	1470	386	123
$Hg(NO_3)_2$[b]			2275	2294	1372	918[d]	1443[d]	376[d]	123
$AlCl_3$[c]	1:1	2934	2330		1363	957		399 (R)	125, 131
Zn^{2+}	1:4	2941	2306		1368	939			125, 131
BF_3[c]	1:1	2935	2355		1362	965			125, 131
$SnCl_4$[f]	1:1	2954	2302		1351	940			125
$SnCl_4$[c]	1:2		2307 (R)	2285 (R)	1354 (R)	939 (R)			126
ICl	1:1	2954	2268		1371	920			125, 129

a. in solution in aqueous acetonitrile. b. in solution in anhydrous acetonitrile. c. in solid. d. no shift observed. f. solution in benzene. e. infrared spectra unless marked (R).

namely that band assignments are not always unambiguous and, as in this case, may lead to difficulty in assessing the true change of frequency.

The problem of accounting for the change in the $\nu(C{\equiv}N)$ frequency on complexation is discussed in section 6.3.4. An increase in frequency has in the past[120,125] been attributed mainly to an increase in the force constant of the C≡N bond. This implied shortening of the bond but the crystallographic data available at that time for $CH_3CN–BF_3$[132] showed a shortening which was comparable with the error involved in the determination. More precise data for $CH_3CN–BX_3$[133] and $CH_3CN–SnCl_4$[126] show that the r(C≡N) value drops from 1.157 Å in the free nitrile[123] to 1.135 Å (BF_3 complex), to 1.122 Å (BCl_3 complex) or to 1.088 or 1.111 Å ($SnCl_4$ complex). The measured shortening, which is about 0.02 Å (*i.e.* three times the standard error in the data), is regarded as a significant change. As can be seen from table 2.10, for simple complexes of mono-nitriles the $\nu(C{\equiv}N)$ frequency shift is equivalent to between 5 and 100 cm^{-1}, and is accompanied by a 5–10-fold intensity increase. It has been shown (section 6.3.4) that the frequency increase in most cases is too large to be caused purely by vibrational coupling and so most of the increase in frequency is attributable to a change in the

TABLE 2.10

Position and intensity changes for the $\nu(C{\equiv}N)$ band in simple nitrile complexes

Donor	Acceptor	Stoichiometry donor:acceptor	$\tilde{\nu}(C{\equiv}N)$/ cm^{-1}	$\Delta\tilde{\nu}$/ cm^{-1}	ΔB/B[a]	Reference
CH_3CN	$SnCl_4$	2:1	2303	47	—	134
C_6H_5CN	$SnCl_4$	1:1	2258	27	12.0	134
p-$NO_2C_6H_4CN$	$SnCl_4$	1:1	2263	28	—	134
p-FC_2H_4CN	$SnCl_4$	1:1	2258	25	9.3	134
p-$CH_3C_6H_4CN$	$SnCl_4$	1:1	2254	24	12.5	134
C_2H_5CN	$SnCl_4$	2:1	2278	34	11.0[b]	135
CH_3CN	BF_3	1:1	2359	111		135
C_3H_7CN	BF_3	1:1	2337	93	6.8[b]	135
C_6H_5CN	BF_3	1:1	2336	111	3.5[b]	135
CH_3CN	BCl_3	1:1	2325	69		136
HCN	BCl_3	1:1	2189	92		138
C_6H_5CN	BCl_3	1:1	2304	75	9.0	136
p-$NO_2C_6H_4CN$	BCl_3	1:1	2313	76	6.0	136
ClCN	$TiCl_4$	2:1	2255	36		139
CH_3CN	$TiCl_4$	2:1	2304	48		134
CH_2CHCN	$TiCl_4$	2:1	2273	45		134, 140
CH_2CHCN	$NbCl_5$	1:1	2264	36		140
CH_2CHCN	$FeCl_2$	1:1	2257	29		137
CH_2CHCN	$CoCl_2$	1:1	2260	32		137
CH_2CHCN	$ZnCl_2$	1:1	2268	40		137
CH_3CN	ICl	1:1	2268	12	5.4	129
CH_2CHCN	ICl	1:1	2235	4	6.4	129
C_6H_5CN	ICl	1:1	2240	8	5.3	129

a. fractional ratio of intensities of "free" and "complexed" nitrile bands.
b. ratio of maximum extinction coefficients of "free" and "complexed" nitriles.

electronic structure of the nitrile bond due to complexation. The explanation accepted at present is that donation of lone pair electron density on the nitrogen leads to a less polar C≡N bond. This leads to a decrease in the electron-withdrawing effect of the nitrogen (see section 6.4.3, p. 495). Since electron density withdrawn by the nitrogen must come from the bonding orbitals the bond gets stronger when complexation occurs. A similar explanation[136] couched in slightly different language was provided some years ago. If the free nitrile is considered to have a structure made up of contributions from resonance form I and II,

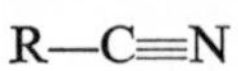

$$R-C\equiv N \qquad R-\overset{+}{C}=\overset{-}{N}$$

$$\text{I} \qquad\qquad\qquad \text{II}$$

then, when complexation occurs with a metal, the two extreme forms are III and IV,

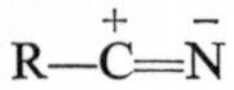

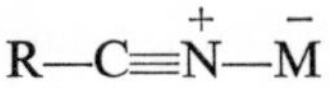

$$R-C\equiv\overset{+}{N}-\overset{-}{M} \qquad\qquad R-\overset{+}{C}=N$$
$$\diagdown$$
$$M^-$$

$$\text{III} \qquad\qquad\qquad\qquad \text{IV}$$

Since resonance between III and IV involves a change of hybridisation at the nitrogen atom it is very unlikely and the complex is expected to be composed entirely of form III. It can be seen that this leads to an increase in the formal bond order (from that found for the "free" nitrile by resonance of I and II) of the nitrile bond and explains the increases in frequency and force constant which have been found (see table 6.4, p. 484, for example). The vibrational spectral changes given in table 2.9 have been explained using a force field calculated by Purcell and Drago[125]. These authors have also attempted to explain the intensity increase of the $\nu(C\equiv N)$ band in terms of vibrational mixing of the $\nu(C\equiv N)$ and $\nu(N-M)$ modes. The metal atom carries a large negative charge, whose oscillation is thought to cause a perturbation of the $\nu(C\equiv N)$ bond. It should, however, be noted that vibration of the N-M bond will give rise to changes of electron density at the nitrogen atom also—giving another contribution to the $\nu(C\equiv N)$ band intensity. One of the striking features of tables 2.9 and 2.10 is the apparent difference in donor strength of acetonitrile towards metal acceptors and halogen acceptors. The relatively small frequency shift on complexation with ICl is reflected by the smaller increase in CN force constant[125]. However, the intensity perturbation for the ICl complex is no less severe and since vibrational coupling in this case is expected to be small (no one has yet positively identified a "$\nu(D-I)$" band for this complex—see section 3.5.3) any mechanism depending on vibrational mixing must be used with caution. (Normal coordinate calculations for nitrile complexes are discussed further in section 6.3.4.)

Recently, there have been a number of cases documented where the band shifts to lower wavenumber[141-144]. These interesting complexes

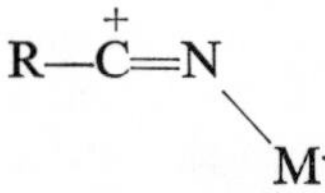

are ones in which the nitrile molecule is coordinated to the metal via the π-electrons of the CN triple bond[143,144] (thereby causing the bond to weaken) or where metal-nitrogen overlap is such that electron density from the metal is donated to the antibonding orbitals of the CN group (again causing a weakening of the triple bond). This effect depends, as would be expected, on the metal oxidation state[142a]. As can be seen from table 2.11 the frequency decrease caused by π complexation of RCN with the metal is much greater than that produced by the "back-bonding effect" on the σ-bonded nitrile group. In some circumstances, therefore, RCN acts as a strong π donor and in others as rather

TABLE 2.11

Selected data for nitrile complexes showing a drop in the nitrile stretching frequency

Donor	Acceptor	Stoichiometry (donor:acceptor)	$\tilde{\nu}$(C≡N)/ cm^{-1}	$\Delta\tilde{\nu}$/ cm^{-1}	Reference
CH_3CN	$Co(H)(PPh_3)_3$	1:1	2210	−40	141
CH_3CN	$Ru(NH_3)_5X_2$	1:1	2239	−11	142
C_6H_5CN	$Ru(NH_3)_5X_2$	1:1	2194	−37	142
$(C_2H_5)_2NCH_2CN$	$SnCl_4$	1:1	2180	−40	143
$(C_2H_5)_2NCH_2CN$	$TiCl_4$	1:1	2150	−70	143
$CNCH_2CN$	$Mn(CO)_3Cl$	1:1	2066	−212	144
$CNCH_2CH_2CN$	$Mn(CO)_3Cl$	1:1	2070	−187	144

weak *n* donor. On the other hand, it has been shown[143] that amino-nitriles (which have potential chelating properties) act as rather weak π complexes (from the nitrile site), and that cyano-pyridines[142a] co-ordinate via the nitrile lone pair rather than via the pyridyl lone pair (although this is probably due to steric considerations when the metal is already complexed to five ammonia molecules).

A considerable amount of work has also been done recently in trying to establish the geometry and stereochemistry of nitrile–metal complexes and in seeking low frequency bands which can be attributed to metal–donor stretching modes (such vibrations being, of course, coupled with other internal modes of the complex—see below). This work is illustrated by reference to the $2CH_3CN$–$SnCl_4$ complex since it has been one of the most thoroughly studied complexes and because solid phase spectroscopic data can be checked against the known crystal structure[126]. In principle, at least, it is possible to distinguish between *cis* and *trans* configurations of such an octohedral complex using the different infrared and Raman selection rules. These are summarised for ML_2X_4 complexes in table 2.12. It can be seen that for the *cis* configuration *four* ν(Sn-Cl) stretching modes are expected to give rise to bands in the 300–360 cm^{-1} region. These four modes are both infrared and Raman active. For the *trans* configuration one infrared and two Raman bands are expected in this region. The infrared spectrum[145] of this complex,

TABLE 2.12

Number and activities M–L and M–X stretching modes for octahedral complexes, ML_2X_4

Approximate description	*cis* (C_{2v})	*trans* (D_{4h})
ν(M–L)	a_1(I.R., R) + b_1(I.R., R)	a_{1g}(R) + a_{2u}(I.R.)
ν(M–X)	$2a_1$(I.R., R) + $2b_2$(I.R., R)	a_{1g}(R) + b_{1g}(R) + e_u(I.R.)
b(MLX)	$3a_1$(I.R., R) + $2a_2$(R) + $2b_1$(I.R., R) + $2b_2$(I.R., R)	b_{2g}(R) + b_{2u}(i.a) + e_u(I.R.)
Totals	13 I.R. + 15 R	5 I.R. + 6 R

both for the solid and solution, shows three bands which can be assigned to ν(Sn-Cl) modes and two which can be assigned to ν(Sn–N) modes. However, the Raman spectrum originally obtained[145] showed only one band in the 300–360 cm^{-1} region and so, on the basis of the Raman spectrum alone, the molecule may well have been thought to have a *trans* (D_{4h}) configuration. The Raman spectrum[126] obtained more recently for the solid complex shows (figure 2.10(A)) that there are four ν(Sn-Cl) bands as expected for a *cis*-octohedral complex. Three of the bands are, however, very weak and were previously unobserved. It is clear that great care must be taken in drawing conclusions from such vibrational spectroscopic data and it has been pointed out[146] that assignment of stereochemistry, in some cases, may not be possible from vibrational data alone.

Another problem which arises is that of assignment of bands especially in the low frequency region where the normal modes are likely to be made up of mixtures of different types of motion. This problem is apparent in trying to assign bands to the two expected ν(M–N) modes. In the $2CH_3CN$–$SnCl_4$ case two bands at $\sim$200 and $\sim$220 cm^{-1} were chosen as belonging to Sn–N stretching modes and this assignment has recently been confirmed[126] using deuterated donors. The ν(Sn–N) frequencies are expected to decrease on deuteration and the two bands at 219 and 206 cm^{-1} (for $2CH_3CN$–$SnCl_4$ solid complex) do shift by approximately 8 cm^{-1}. By contrast, the bands assigned by ν(Sn-Cl) modes, between 300 and 360 cm^{-1}, and those assigned to b($SnCl_2$) modes, between 100 and 150 cm^{-1}, show little, if any, shift on deuteration. Table 2.13 gives a list of bands which have been assigned to metal–N stretching modes of nitrile complexes and the force constants estimated for these bonds. From this table it is apparent that such modes give rise to bands at widely differing frequencies. However, this is only partly a reflection of differing strengths of bonding between the molecules since vibrational mixing is almost certain to contribute to

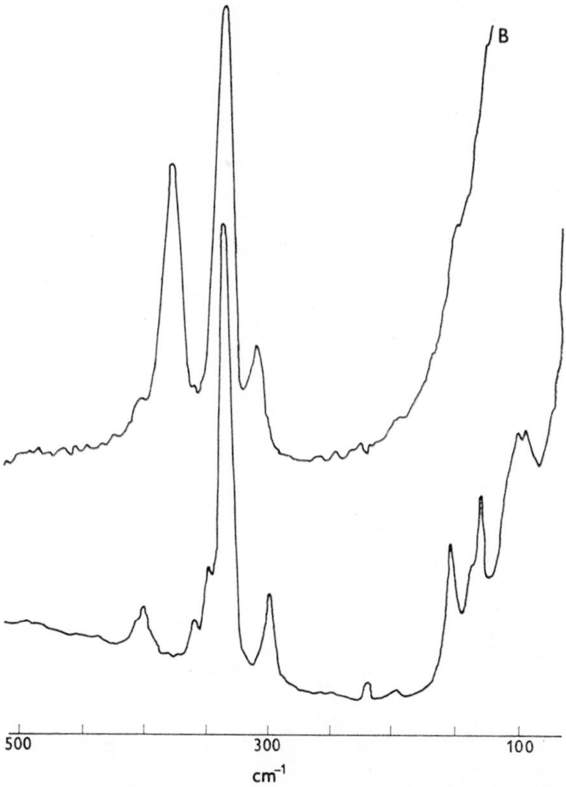

FIG 2.10 The Raman spectrum of (A) solid $2CH_3CN–SnCl_4$ complex (B) a saturated solution of $SnCl_4$ in CH_3CN (reproduced, by permission, from *J. Chem. Soc.*, 2443 (1969).

frequency shifts on going from one complex to another[153]. It can be seen from table 2.13 that the range of force constants is mainly between 100 and 300 N m^{-1}. As an illustration of the problems involved in calculating valid force constants and normal coordinates for complexes of this kind, let us consider the $CH_3CN–BX_3$ complexes which have been studied in some detail recently by independent groups[149,152]. Although the vibrational frequencies are well known for several isotopic modifications of these complexes, and despite the fact that both sets of authors claim that their force fields reproduce closely these vibrational frequencies, the two sets of force constants vary significantly. Table 2.14 gives a summary of the principal force constants obtained by the two groups. It may be seen that, for the BF_3 complex, the B-F stretching valence force constants are quite different for the E symmetry species although good agreement is observed for the A_1 species (the complex molecule belongs to the C_{3v} point group). The decrease in the B-F force

TABLE 2.13

Metal–nitrogen stretching band positions and force constants for simple nitrile complexes

Donor	Acceptor	Stoichiometry donor:acceptor	Stereochemistry (if known)	$\tilde{\nu}$(M–N)/cm^{-1}	f_{mn}/N m^{-1}	Reference
CH_3CN	$SnCl_4$	2:1	*cis*-octahedral	222, 207		147
				219, 206		126
CD_3CN	$SnCl_4$	2:1	*cis*-octahedral	211, 197		126
CH_3CN	$PtCl_2$	2:1	*cis*-square planar	120, 100		148
CH_3CN	$PdCl_2$	2:1	*trans*-square planar	125		148
CH_3CN	$NbBr_5$ and $TaBr_5$	1:1	octahedral	~220	40–120	151
CH_3CN	$ZnCl_2$	2:1	tetrahedral	174	50–300	124
CH_3CN	BF_3	1:1	tetrahedral	646	250	149a, 152
CH_3CN	BCl_3	1:1	tetrahedral	712	340	149b
CH_3CN	BBr_3	1:1	tetrahedral	706	350	149b
CH_3CN	$Mo(CO)_3$	3:1	tetrahedral	533, 481		150
CH_3CN	$Cr(CO)_3$	3:1	tetrahedral	552, 495		150
HCN	BCl_3	1:1	tetrahedral	750	140	139
HCN	$TiCl_4$	2:1	*trans*-octahedral	281, 252		139
ClCN	$TiCl_4$	2:1	*trans*-octahedral	285		138
BrCN	$TiCl_4$	2:1	*trans*-octahedral	317		138
ClCN	$AlCl_3$	1:1	tetrahedral	465	200	138
BrCN	$AlCl_3$	1:1	tetrahedral	525	280	138

TABLE 2.14

Principal force constants for the $CH_3CN–BX_3$ complexes (all values in N m^{-1})

	CH_3CN^a or BF_3^b	$CH_3CN–BF_3^c$	$CH_3CN–BF_3^d$	$CH_3CN–BCl_3^d$	$CH_3CN–BBr_3^d$
f_{CN}	1740	1890	1880	1870	1860
f_{CC}	530	530	530	510	510
f_{BN}	—	250	250	340	350
F^e_{BF}, a_1	882	653	649	332	294
F^e_{BF}, e	782	277	385	324	245

a. reference 125. b. reference 155. c. reference 152. d. reference 149. e. these are valence force constants, reference 149b.

constant on going to the complex is therefore different using the two different calculations. This may well be a situation in which multiple solutions to the force constant problem arise[154] and it emphasises the need for great care in interpreting force constants calculated by a least squares fit to the observed frequencies, especially if the problem is under-determined. The potential energy distributions calculated by Shriver and Swanson[149] for the BX_3 complexes show that the band assigned to the "ν(B–N)" normal mode has, in fact, contributions from at least six of the symmetry coordinates and that these contributions do vary on going from one complex to another as mentioned above (table 2.15). However, it seems clear that the interaction strength does increase in the order $BBr_3 > BCl_3 > BF_3$ as indicated by the force constants of the B–N bonds (table 2.13) and, in these cases, by the frequencies of the bands observed (see section 6.3.5 for further discussion).

Complexes of Pyridine and Related Donors

In this section we shall illustrate the kind of results which have been obtained using the complexes of simple amines and heterocyclic nitrogen bases (especially pyridine and its derivatives). These donors usually have (formally) a single lone pair which may, however, be to a certain

TABLE 2.15

Potential energy distribution for ν(B–N) normal mode in $CH_3CN–BX_3$ complexes (reference 149)

X	$\tilde{\nu}_{max}/cm^{-1}$	$V_{44}{}^a$	V_{55}	V_{66}	V_{77}	V_{56}	V_{57}	V_{67}
F	866		1.00	0.22	0.35	−0.17	−0.18	−0.22
Cl	712.4	0.17	0.69	0.20	0.65	−0.17	−0.39	−0.21
Br	706.6	0.21	0.69	0.27	0.31		−0.38	−0.11

a. subscripts identify V_{kl} with the symmetry coordinates defined in reference 149.

extent delocalised within the molecule by intramolecular electronic effects (reference 1, chapter 18). The whole or part of the lone-pair electron density may then be available for intermolecular bonding to a metal centre. The vibrational spectroscopic effects observed will again depend on the geometrical configuration of the complex and the degree of vibrational coupling, *besides* the extent of the interaction. Often, the changes observed in the donor vibrational spectrum are confined to small frequency shifts (section 1.2.2). Although this allows relatively easy assignment of the bands in the complex spectrum, it makes the problem of interpretation more difficult (since well-defined, systematic changes on going from one complex to another are rare). The most studied donor is pyridine itself. About five or six of the pyridine modes show systematic frequency shifts when the molecule is complexed to a metal. Table 2.16 shows a selection of the data obtained for pyridine–metal complexes. Comparison with table 3.13 (p. 282) shows that the changes are similar to those observed for pyridine–halogen complexes. Some of the other pyridine bands do show small shifts (for example, the 745 cm^{-1} (a_2) band shows a shift of $\sim$20 cm^{-1} for some complexes)[160] but the six bands shown are the ones which are consistently sensitive to metal complex formation. Although extent of the shifts varies from complex to complex, the direction of the shifts remain the same. One exception to this is the *low* frequency shift of the 1030 cm^{-1} band in the *cis*-octohedral $(pyr)_4$–$PtCl_2$ complex[160] although it should be noted that there are a number of weak, unassigned bands in the spectra of these complexes, and that the choice of bands made is not always obvious. If table 2.16 and table 3.13 (p. 282) are compared it is readily seen that the frequency shifts are very similar whether the acceptor is a metal or a halogen molecule. The relatively small perturbation of the force field of the pyridine molecule when complexed to a metal compared with that on protonation[167] has been used[158] as evidence for metal–donor π-"back-bonding" in pyridine–metal complexes. If such a deduction is valid then "back-bonding" from a halogen molecule seems likely. Such a phenomenon has been tenuously suggested but never established. However, the argument is a poor one[158] in view of the very small frequency shifts between pyridine–platinum halide complexes and pyridine–BX_3 (where π-"back-bonding" is unlikely). Our normal coordinate calculation for pyridine–IX complexes (section 2.3.2) would appear to indicate that a considerable part of the frequency perturbation is caused, not by electronic effects, but by vibrational mixing. The same may be true for metal complexes (see below). On the other hand, it has recently been shown[22] that the intensity perturbation of the pyridine molecule in a metal complex is different from that for a halogen complex. Intensity measurements are difficult to make for metal complexes but we have obtained some data for the soluble $(pyr)_2MX_2$ tetrahedral complexes (M = Zn, Ni, Co). Table 2.17 shows a comparison of band intensities for "sensitive" modes of the donor. It seems clear therefore

TABLE 2.16

Changes in the pyridine spectrum[a] caused by complexation to a metal (all data in cm^{-1})

Complex	$\bar{\nu}_4$, a_1	$\bar{\nu}_{14}$, b_1	$\bar{\nu}_8$, a_1	$\bar{\nu}_9$, a_1	$\bar{\nu}_{10}$, a_1	$\bar{\nu}_{27}$, b_2	Reference
pyridine (liquid)	1580	1439	1029	992	605	405	164–6
pyridine–BF_3	1631	1464	1038	1026	619	—	156
pyridine–BF_3[b]	1629	1464	—	—	619	—	156
2 pyridine–$ZnCl_2$	1610	1452	1046	1017	637	417	158
2 pyridine–ZnI_2	1602	1444	1042	1014	635	420	158
pyridine–$HgCl_2$	1607 1597		1033 1041	1018 1015	630	—	157
2 pyridine–$CuCl_2$	1603 1596	1449	1042	1016 1009	645	444	159
2 pyridine–$NiCl_2$	1603	1447	1042	1017 1013	632	437	159
2 pyridine–$PtCl_2$ (*cis*)	1605 1596	1447	1050	1018 1010	643	467	160
2 pyridine–$PtBr_2$ (*cis*)	1605 1595	1448	1049	1017 1006	643	464	160
2 pyridine–$PtBr_2$ (*trans*)	1605 1597	1449	1052	1016 1009	646	478	160
2 pyridine–$PdCl_2$ (*cis*)	1603	1448	1049	1017	653	419	160
2 pyridine–NbF_4					620	410	161
2 pyridine–$TiCl_4$	1605	1444	1044	1015	645	436	162
2 pyridine–$ZrCl_4$	1607	1446	1059	1010	635		162
pyridine–$AlCl_3$	1620	1455	—	1020	—	—	131
2 pyridine–$SnCl_2$	1600	1460	1035	1010 1005	625 620	410	163
pyridine–$SnCl_2$	1605	1460	1040	1015 1012	632 620	415	163

a. solid phase except where otherwise stated. The double splittings observed are probably due to crystal field effects.
b. solution in benzene.

that there *are* differences in electronic redistribution for the two kinds of complex. For the protonated species the electronic perturbation is probably much more severe since the $C_5H_5NH^+$ species is isoelectronic with benzene.

Since many solid pyridine complexes are octohedral in structure (of general formula ML_2X_4 or ML_4X_2) it is possible, assuming that to a first approximation the ligand molecules may be treated as point

TABLE 2.17

Comparison of the intensity[a] perturbation of the pyridine molecule in halogen and metal halide complexes

Compound	ν_4, a_1	ν_5, a_1	ν_{14}, b_1	ν_8, a_1	ν_9, a_1	ν_{10}, a_1	ν_{27}, b_2
pyridine in chloroform	2800	340	3070	700	720	450	190
pyridine–ICl in chloroform	2950	0[b]	4300	750	3430	1940	200
2 pyridine–$ZnBr_2$ in chloroform	—	700	4000	1340	700	980	900
2 pyridine–$CoCl_2$ in chloroform	2000	—	—	950	360	650	700

a. all intensities are quoted as 10^{-1} B_i/m mol^{-1}.
b. intensity very small—band disappears.

masses, to examine the low frequency spectrum and use table 2.12 to distinguish *cis* and *trans* configurations. Notice that molecules with a *trans*-configuration are expected to show no coincidence of infrared and Raman bands (since the molecule has a centre of symmetry). Attempts to distinguish stereochemistry in this way have been partly successful. For example, the pyridine complex of niobium tetrafluoride was shown unambiguously to have *trans*-octahedral stereochemistry[161]. In the region 180–600 cm^{-1} there is a complete lack of coincidence of infrared and Raman bands (see figure 2.11). In addition the number of

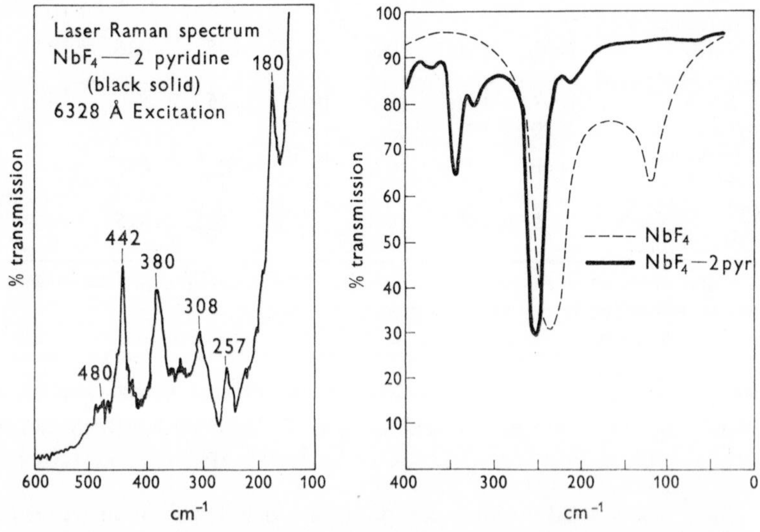

FIG 2.11 The infrared and Raman spectra of the 2pyr–NbF_4 complex (Reproduced, by permission, from *Spectrochim. Acta*, **25A**, 1875 (1969)).

TABLE 2.18

Infrared and Raman spectra of 2pyridine–NbF_4 (Reproduced, by permission, from *Spectrochim. Acta*, **25A,** 1879 (1969))

Infrared	Raman	Symmetry class	Assignment
	180		
212 w		E_u	Nb—N bend (in-plane)
252 s		A_{2u}	Nb—F bend
	257	B_{2g}	Nb—F bend (in-plane)
	308	A_{1g}	Nb—N sym. str.
320 w		E_u	Nb—F bend
343 m		A_{2u}	Nb—N asy.
367 w			
	380		(Ligand)
410 w			(Ligand)
	442	B_{1g}	Nb—F sym. str.
	480	A_{1g}	Nb—F sym. str.
505 w			(Ligand)
525 w			(Ligand)
554 w			(Ligand)
590 vs		E_u^*	Nb—F asy. str.
610 vs			
	620		(Ligand)

* Band is split, probably due to a combination with a lattice vibration.

bands fits almost exactly the expectations of table 2.12 (see table 2.18).

On the other hand, there have been a number of cases where it has proved impossible to deduce stereochemistry even when complete vibrational data have been obtained. A typical example is the 2 pyridine–$SiCl_4$[146] complex for which deuterated pyridine complex spectra have also been obtained. Although good spectral evidence for the *trans* configuration of the similar complex 2trimethylamine–$SnCl_4$ was obtained it was found that the bands observed for 2pyridine–$SiCl_4$ were not consistent with either *cis* or *trans* selection rules. This work, it should be noted, was carried out *after* an X-ray crystallographic study[167a] had shown the complex to have *trans*-octahedral geometry. A previous assignment[168] of the three bands in the 400 cm^{-1} region to SiCl stretching vibrations had led to the belief that the complex had a *cis* configuration (although it was later predicted that the *trans* configuration was correct). Force constant calculations[146], based on the octahedral model using an N–Si stretching force constant of 140 N m^{-1}, show that one of these bands arises from a mode which is partly SiCl out-of-plane bending and partly N–Si stretching. The other two bands are claimed to arise from two components of the, e_u, SiCl antisymmetric stretching mode, the degeneracy being split by reduction of the site symmetry from D_{4h} to D_{2h}. Thus, in this case, distortion of the octahedral arrangement is sufficient to cause a wrong deduction to be made. This work also illustrates the importance of vibrational mixing in determining the appearance of the far-infrared and Raman spectra especially if the symmetry is reduced by complex formation. In addition there are sometimes severe problems due to the overlapping of "skeletal" bands of the complex with bands due to internal vibrations of the ligand.

Factors which Affect Metal–Ligand Bonding in Simple Organometal Complexes

Since by measuring the frequency of the "metal–donor stretching" vibration one might expect to gain a direct measure of the extent of interaction between donor and acceptor, it is tempting to try to correlate this with quantities such as donor mass, position and electronic effect of donor substituents, basicity of donor, *etc.* This has been attempted on a number of occasions[157,159,169–171] usually for complexes of related stereochemistry and with a given metal oxidation state. There are several reasons why, even under the most favourable conditions, these attempts may be abortive. One of the most difficult problems is that of reliable assignment, which is related to the way in which low frequency vibrations of the same symmetry are coupled together. Since frequency shifts may occur due to changes in vibrational mixing on going from one donor (or acceptor) to another it is necessary to check on assignments in each individual case. The methods which have been used to pick out ν(metal–donor) bands have been summarised by Nakamoto[172]. They include (a) Using isotopic substitution of the metal[172] or the donor[126,160,173] to cause the ν(donor–metal) band to shift without affecting the other modes too much. The extent of shifts of the bands due to other low frequency modes will depend on the extent of changes in their normal coordinates produced by isotopic modification. Donor band frequency shifts in the low frequency region may cause complications in some cases. (b) Changing the metal in a series of complexes of the *same structure* which should lead to a systematic change in ν(donor–metal) band frequency provided that the vibration is reasonably well "isolated" or the degree of coupling remains the same on going from one complex to another. This technique has been shown to work well on some occasions. Systematic shifts may also be caused by using several closely related donors[174]. In both cases one must be careful to ensure that other bands in the same general region (*e.g.* due to metal-halide stretching modes) do not show the same systematic shifts (c) changing the halogen in a metal halide complex. This produces ν(M–X) band frequency shifts but should leave the ν(donor–metal) band(s) virtually unchanged except possibly for a small shift due to the mass effect[149,173,175–177] or due to small changes in π-bonding (see below).

Several of the attempts at correlating the ν(donor–metal) band position, with electronic, steric and mass properties of donor ligands *have* been successful to a certain degree and they are mentioned here to illustrate the difficulties and limitations of such procedures. Burgess[170], for example, has attempted to separate the mass and electronic effects by using substituted pyridines of approximately the same mass but with

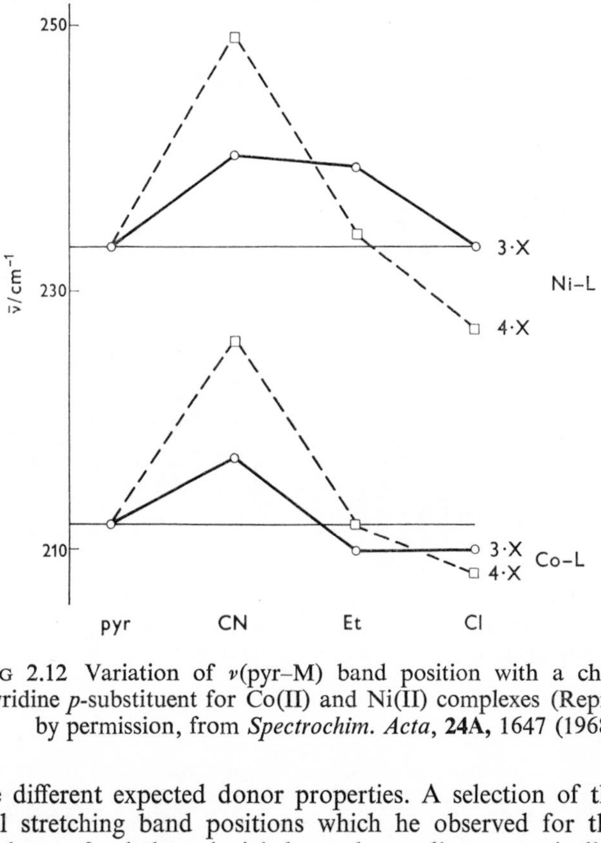

FIG 2.12 Variation of ν(pyr–M) band position with a change of pyridine *p*-substituent for Co(II) and Ni(II) complexes (Reproduced, by permission, from *Spectrochim. Acta*, **24A,** 1647 (1968)).

quite different expected donor properties. A selection of the donor–metal stretching band positions which he observed for thiocyanate complexes of cobalt and nickel are shown diagrammatically in figure 2.12. Since no substituents were present at the 2- and 6-positions[2,159] it was considered that no serious steric hindrance to complexation was likely. The results show a considerable increase in $\bar{\nu}$(M–N) for the 4-cyanopyridine complexes, as compared with the pyridine complexes, when M = Co or Ni. This is thought to be due to increased metal–donor π-bonding caused by the strong π-acceptor property of the cyano group. (This is reinforced by electron release from the NCS group to the metal.) This increase does not occur if M = Cu since in this case the complex has a distorted *trans*-octohedral structure[159] (as distinct from

TABLE 2.19

Band positions and stretching force constants for the M–D bond in n donor-metal complexes

Donor	Acceptor	Stoichiometry donor:acceptor	Stereochemistry	$\bar{\nu}$(M–D)/cm^{-1}	f_{MD}/N m^{-1}	Reference
NH_3	$AlCl_3$	1:1	tetrahedral	579	191	177
NH_3	BF_3	1:1	tetrahedral	742	397	181
NH_3	$ZnCl_2$	2:1	tetrahedral	432, 429	142	173
NH_3	$ZnBr_2$	2:1	tetrahedral	428, 421	138	173
NH_3	$PtCl_2$	2:1	*trans*-square planar	505	192	173
NH_3	$PdCl_2$	2:1	*trans*-square planar	496	160	173
$(CH_3)_3N$	SiF_4	2:1	*trans*-octahedral	224	150	182
$(CH_3)_3N$	$AlCl_3$	1:1	tetrahedral	582	150	183
$(CH_3)_3N$	$AlCl_3$	1:1	tetrahedral	565	232	177
$(CH_3)_3N$	$SnCl_4$	2:1	octahedral	194	140–160	146
$(CH_3)_3N$	BF_3	1:1	tetrahedral	927, 694	300	184
$(CH_3)_3N$	BF_3	1:1	tetrahedral	695	336	177
$(CH_3)_3N$	BCl_3	1:1	tetrahedral	745	307	177
$(CH_3)_3N$	BBr_3	1:1	tetrahedral	727	308	177
C_5H_5N	$SiCl_4$	2:1	octahedral	387, 240	130	146
C_5H_5N	BF_3	1:1	tetrahedral	694	—	156
C_5H_5N	$CuCl_2$	2:1	polymeric octahedral	268	158	174
4-$CH_2OHC_5H_4N$	$CuCl_2$	2:1	polymeric octahedral	276	174	174
4-$CH_3COC_5H_4N$	$CuCl_2$	2:1	polymeric octahedral	265	150	174
4-$COOCH_3C_5H_4N$	$CuCl_2$	2:1	polymeric octahedral	264	149	174
4-CNC_5H_4N	$CuCl_2$	2:1	polymeric octahedral	243	105	174
4-$CH_3C_5H_4N$	$CuCl_2$	2:1	polymeric octahedral	285	186	174
4-$CH_3C_5H_4N$	$NiCl_2$	4:1	polymeric octahedral	260	139	174
4-$CH_3C_5H_4N$	$ZnCl_2$	2:1	tetrahedral	238	108	174
4-$CH_3C_5H_4N$	$CoCl_2$	2:1	tetrahedral	235	93	174
4-$CH_3C_5H_4N$	$MnCl_2$	2:1	polymeric octahedral	230	80	174
$HCOOCH_3$	BF_3	1:1	tetrahedral	643	260–320	185
CH_3COOCH_3	BF_3	1:1	tetrahedral	712		185
CH_3COCH_3	BF_3	1:1	tetrahedral	493, 623	250	84
CH_3SOCH_3	BF_3	1:1	tetrahedral	385, 720	250	85
$(CH_3)_2S$	BF_3	1:1	tetrahedral	310	150	84
$(CH_3)_2Se$	BF_3	1:1	tetrahedral	256	150	84
$(CH_3)_2O$	$AlCl_3$	1:1	tetrahedral	325, 552	160	186
$(CH_3)_2O$	$AlCl_3$	1:1	tetrahedral	315, 547	150	186

cis-octohedral when M = Co or Ni) and the spatial distribution the metal $d\pi(T_{2g})$ orbitals prevents transmission of π-electron density from NCS, via the metal, into vacant CN $a\pi(\pi^*)$ orbitals. The ν(Cu–N) band frequencies show only a small mass effect on going from pyridine to 4-XC_5H_4N (X = CN, Et, Cl). Consistent with this model is the observed decrease in $\bar{\nu}$(M–N) for 4-methylpyridine complexes when M = Co or Ni. The methyl group is a π-releasing group which, it is thought, causes metal($d\pi$)–pyridine(π^*) electron transfer to decrease. The extent of π-bonding therefore decreases. For M = Cu the mass effect is the only one observed. For M = Zn the complexes are tetrahedral and the 4-alkylpyridine complexes have a $\bar{\nu}$(Zn–N) value which is lower than that for the pyridine complex even though the thermodynamic stabilities[178] of such alkyl-substituted complexes are greater than that of the pyridine complex. This means that *both* π- and σ-electron releasing properties of the methyl group must be important since a decrease in the metal–donor π-bonding cannot alone be responsible for the apparently weaker bonding.

The situation becomes even more complicated when 2- and 6-substituted pyridines are used[146,159,169,178] since steric effects are now observed. These effects are even strong enough to change the stereochemistry so that halogen bridging (essential for an octohedral structure) no longer occurs.[159] The $\bar{\nu}$(Cu–N) values[159] follow the trend pyr > 2-methyl > quinoline > 2-ethyl > 2,6-dimethyl apparently showing the effect of steric repulsion. A similar result was obtained by Frank and Rogers[169]. In neither case was there any attempt to check on the extent to which vibrational mixing occurs. There is, however, a strong indication in this work that π-bonding between metal and donor may to some extent be important. The $\sim 10\%$ *increase* in ν(M–N) band frequency for 4-cyanopyridine complexes (M = Co, Ni) shows that the model proposed by Burgess has a good chance of being a reasonable one. Such a large shift is unlikely[153] to be caused by changes in vibrational mixing in a series of such closely related complexes.

Obviously it is desirable wherever possible to compare f_{MD} force constants in order to try to get some indication of the donor–metal interaction strength. Even so it is unlikely that, at the present level of sophistication in force constant calculations, one would be able to separate σ- and π-effects (as pointed out in section 5.1). Without such comparison one can only assume that the normal coordinates remain the same on going from one complex to another. Table 2.19 gives a selection of donor–metal complexes for which f_{MD} force constants have been obtained. It can be clearly seen from this table that the f_{MD} often do not correlate very well with the frequencies of the observed "ν(M–D)" bands even within a series of related complexes. The value of reliable force constant data is emphasised by the data for 4-methylpyridine complexes with MX_2 halides[174]. Using an empirical measure of the metal–donor π-bonding (based rather arbitrarily on the frequency shift of a skeletal donor vibration) it was suggested that the π-bonding tendency increased as the 1st ionisation potential of the metal[179] decreased (*i.e.* Ni(II) > Cu(II) > Co(II) > Mn(II) > Zn(II)). However, the force constants, f_{MD}, obtained assuming C_{2v} symmetry (table 2.19), are in the order Cu(II) > Ni(II) > Zn(II) > Co(II) > Mn(II) which is identical with the order of stability of complexes of these metals predicted by Irving and Williams[180]. (See section 5.2.4 for further discussion of this work.) Assuming that the force constants are valid, this indicates that the σ-contribution to metal–ligand bonding is the predominant one.

It is interesting to compare the f(pyridine–metal) force constants (table 2.19) with those obtained for pyridine–halogen complexes (section 2.3.2). The k_{DI} values (table 2.6) are not very different from those for the pyridine–metal complexes when one considers that the highest $\bar{\nu}$(pyridine–IX) value is $\sim$140 cm^{-1}. There seems no reason (from these results) to suppose that there is any difference in the nature of the bonding between pyridine and the two types of acceptor although some metals are, of course, considerably better acceptors than the halogens. In several cases where reasonably "good" normal coordinate calculations have been made it has been clearly shown how vibrational mixing, if undetected, can lead to erroneous conclusions. Two of these are outlined here to illustrate the dangers of neglecting the effects of such mixing.

The distribution of potential energy among the various modes of a given symmetry class has been obtained[84,85] for acetone, trimethylamine and dimethyl sulphoxide complexes with boron trifluoride and for the $(CH_3)_2X$–BF_3 system (X = S, Se). The 1:1 complexes have been treated using a pseudotetrahedral framework with C_{3v} symmetry. The vibrations of the BF_3D skeleton then comprise 3e modes (each doubly degenerate) and 3a_1 modes. All the vibrations are infrared and Raman active. Initial force constants were taken from the work of Sowondy and Goubeau[177] and from that of Amster and Taylor[187] for amine–BF_3 complexes and the BD stretching force constant and some of the interaction constants were varied until a good observed/calculated frequency fit was achieved. The calculated frequencies of the three a_1 vibrations have been plotted against the BD stretching force constant and a value chosen from the best frequency fit (figure 2.13). The potential energy distribution among these three modes varies, of course, as f_{BD} is varied. However, two things stand out quite clearly. Firstly, the three modes are considerably mixed and almost all of the observed bands arise from vibrational modes which contain some of the character of symmetric stretching and bending of the BF_3 group and also some B–D stretching motion. It is thus impossible to assign any of the observed bands to a given individual "pure" mode. Secondly, the extent to which mixing occurs goes down as the mass of the donor increases (see table 2.20), and this correlates with a decrease in the calculated BD stretching

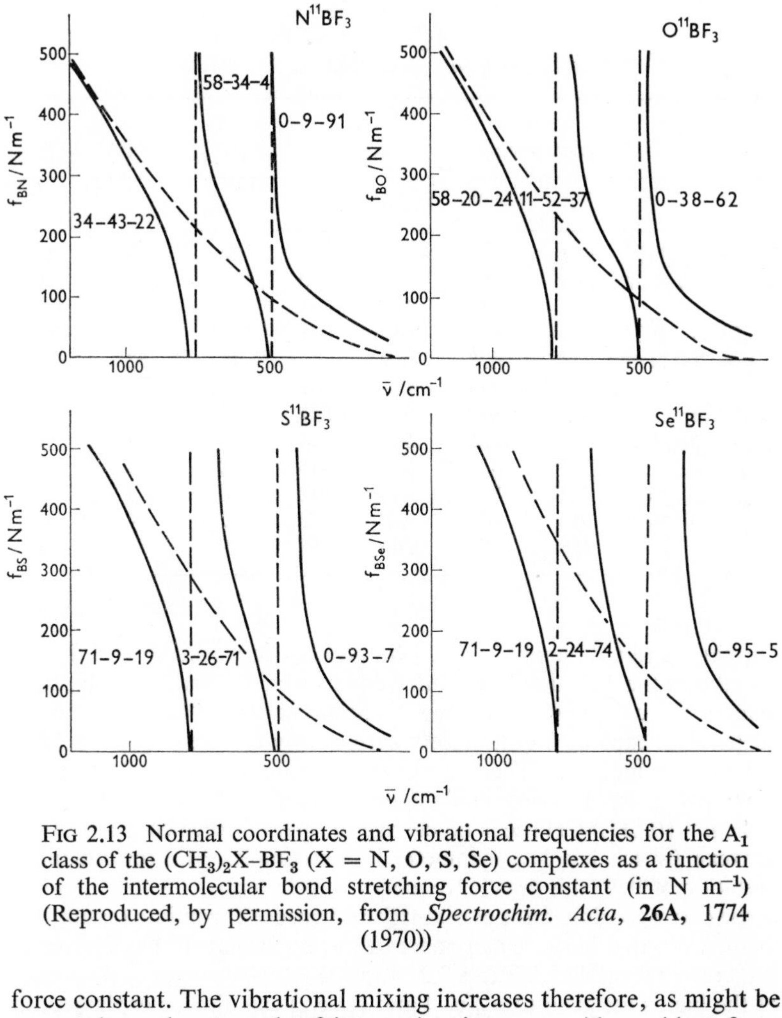

FIG 2.13 Normal coordinates and vibrational frequencies for the A_1 class of the $(CH_3)_2X–BF_3$ (X = N, O, S, Se) complexes as a function of the intermolecular bond stretching force constant (in N m^{-1}) (Reproduced, by permission, from *Spectrochim. Acta*, **26A**, 1774 (1970))

force constant. The vibrational mixing increases therefore, as might be expected, as the strength of interaction increases. Also evident from table 2.20 is the gradual change in character of the band at lowest frequency. In going from $(CH_3)_3N$ to $(CH_3)_2Se$ this band changes its character from 91% $\delta(BF_3)$ to 95% ν(BD), there being an intermediate stage when this band arises from a 60%/40% mixture of the $\delta(BF_3)$ and ν(BD) modes. Obviously, any conclusion drawn from a consideration of the *frequency changes* of this band assuming it to be a ν(metal–donor) band would be of little significance.

The second example chosen to underline this point is that of the $(C_6H_6)_2M$ complexes which have subjected to complete 25 particle

TABLE 2.20

Infrared band positions (cm^{-1}) and calculated potential energy distributions for $(CH_3)_3N$–BF_3 and $(CH_3)_2X$–BF_3 complexes (Reproduced, by permission, from *Spectrochim. Acta*, **26A**, 1773 (1970))

	$\bar{\nu}$ (experimental)		$\bar{\nu}$ (calculated)		$\nu_s(BF_3)$	$\nu(BD)$	$\delta_s(BF_3)$
	^{11}B	^{10}B	^{11}B	^{10}B			
$BF_3[(CH_3)_3N]$	927	952	946	979	34	43	22
$f_{BN} = 300\ N\ m^{-1}$	694	696	699	700	58	38	4
	458	459	489	493	0	9	91
$BF_3[(CH_3)_2CO]$	858	874	850	875	58	20	24
$f_{BO} = 200\ N\ m^{-1}$	623	628	619	627	11	52	37
	493	495	460	461	0	38	62
$BF_3[(CH_3)_2S]$	828	837	823	842	70	9	19
$f_{BS} = 150\ N\ m^{-1}$	620	635	564	574	3	26	71
	310	310	336	336	0	93	7
$BF_3[(CH_3)_2Se]$	812	825	823	842	70	9	19
$f_{BSe} = 100\ N m^{-1}$	618	630	562	573	2	24	74
	256	256	266	266	0	95	5

normal coordinate analysis by Cyvin and his co-workers[188–190]. It was found that the direction of certain frequency shifts (and, roughly, their magnitude) on going from "free" benzene to these "sandwich" complexes were explained quite nicely *without* any change of force field of the benzene molecule. The results are summarised in table 2.21. Thus, the $\nu_{11}(a_{2u})$ band which shifts from 671 cm^{-1} ("free" benzene) to 794 cm^{-1} ($(C_6H_6)_2Cr$) needs no change of internal force constant of the benzene (which has D_{6h} symmetry in the complex[191]). Previous interpretation of[190,192] this drastic frequency shift in terms of a 40% increase in the corresponding force constant for dibenzene chromium (assuming, incorrectly, that benzene can be treated as an "isolated" D_{6h} molecule) is therefore invalidated. Table 2.21 shows that the changes of frequency of the bands assigned to $\nu_{23}(a_{2u})$ and $\nu_{25}(e_{1u})$ can also be explained, at least qualitatively, by differences in normal coordinates on going from one complex to another since the calculations were performed[188] allowing *only* the metal mass to change from one complex to another. The coupling is caused by the non-vanishing off-diagonal element in the appropriate **G** matrix block (reference 112, p. 173). It thus appears that the bonding between benzene and the metal atom in these complexes is controlled more by the *mass* of the central atom than by its electronic configuration. Attempts to interpret these frequency changes in terms of differing strengths of interaction are therefore likely to be abortive. It should be noted, finally, that the frequency shifts caused by

TABLE 2.21

Infrared band positions (cm^{-1}) for metal–benzene complexes (Reproduced, by permission, from *Acta Chem. Scand.*, **24**, 3420 (1970))

	$\bar{\nu}_{11}$		$\bar{\nu}_{25}$		$\bar{\nu}_{23}$	
Complex	obs.	calc.	obs.	calc.	obs.	calc.
$Cr(C_6H_6)_2$	794	794	459	459	490	490
$Cr(C_6H_6)_2^+$	795	794	415	459	466	490
$V(C_6H_6)_2$	742	796	429	463	478	493
$Mo(C_6H_6)_2$	773	760	362	372	424	415
$Mo(C_6H_6)_2^+$	—	760	333	372	410	415
$W(C_6H_6)_2$	798	745	331	313	386	355
$W(C_6H_6)_2^+$	—	745	306	313	378	355
$Tc(C_6H_6)_2^+$	—	759	358	369	431	411
$Re(C_6H_6)_2^+$	—	745	336	312	396	354
C_6H_6	671	671	—	—	—	—

normal coordinate changes are much greater than the 2% "norm" suggested by Murrell[153].

Infrared Intensity Studies on Metal Complexes

There are several problems encountered when attempts are made to study the band intensities of inorganic or organometallic complexes of the types we have considered in this section. The problems of measurements are principally (a) lack of solubility in suitable solvents and (b) overlapping of bands in the spectra of relatively large molecules. These experimental problems are, however, dwarfed by the problems of interpretation. Normal coordinate analyses for such large molecules are difficult without severe approximations and molecular orbital calculations on such molecules are still in their infancy. The separation of σ and π-electronic effects is therefore difficult even assuming that vibrational coupling is successfully estimated or justifiably ignored (see section 5.1 for a more detailed discussion). The problems and methods of interpretation have recently been excellently reviewed by Kettle and Paul[193] and by Larsson[194]. From these reviews it is clear that, despite the difficulties, a large amount of very important information about the bonding in inorganic complexes can be obtained in return for the extra effort involved in measuring relative or absolute band intensities. Metal carbonyls have been by far the best studied class of complexes but there have also been important studies on N_2, NO and SCN complexes (see later), in addition to the studies on pyridine–metal complexes[22]—see table 2.17. (Notice that the study of Raman intensities has been covered in chapter 5.) We shall consider here the individual types of complex which have been studied and show how intensity data are interpreted to give structural information.

(i) *Metal Carbonyls*

Virtually all the studies made[193] have been on the bands due to carbonyl stretching modes in the 1800–2200 cm^{-1} region. There are, of course, usually several infrared bands depending on the symmetry of the complex[195], and for comparison purposes it is usual to measure the perturbation of intensity using the value of the "specific intensity"[193,196], *i.e.* the intensity of the absorption observed *per* carbonyl group in the molecule. This figure does not, of course, take any account of vibrational mixing of modes belonging to the same symmetry species (which can be accompanied by intensity redistribution—see section 1.2). However, it does suffice to demonstrate (see reference 193, tables 2–6 and reference 197, table 5) that the increase of intensity of the $\nu(C{=}O)$ vibrational band from that in "free" carbon monoxide is by a factor of between 5 and 100[198]. (Notice, however, that the specific intensity can be somewhat misleading in the sense that some modes—for example of pentacarbonyl complexes[199,204,205]—can give rise to quite weak bands and, under some circumstances[200], decreases in intensity of the "perturbed" CO molecule can occur). The generally observed increase in intensity (table 2.22) is accompanied by a decrease in energy of the $\nu(CO)$ vibration of between 20 and 250 cm^{-1} [196,197,200] (reference 1, section 17.8). Both frequency drop and intensity increase are thought to be caused by the fact that CO is a very weak *n* donor (the

TABLE 2.22

Summary of some typical infrared spectra data and f_{CO} force constants for carbon monoxide–metal complexes

Compound	Solvent	$\tilde{\nu}(CO)/cm^{-1}$ (symmetry)	10^{-3} B_i/ m mol^{-1}	$\Delta\nu_{1/2}$/ cm^{-1}	f_{CO}/N m^{-1}	Reference
CO	(gas)	2143.3	54.0	—	1900	1, 202
BH_3–CO	(gas)	2165 (a′)	—	—	1897	201
$Mn(CO)_6^+$	THF[a]	2090 (t_{1u})	3480	12.2		203
$Re(CO)_6^+$	THF[a]	2078 (t_{1u})	2920	13.1		203
$Ni(CO)_4$	CCl_4	2944 (t_2)	3200	10.5	1760	203
	THF[a]	2040 (t_2)	4560	23.1		
$Cr(CO)_6$	THF[a]	1981 (t_{1u})	8430	16.0		203
	CCl_4	1986 (t_{1u})	7070	10.0		203
$W(CO)_6$	CCl_4	1983 (t_{1u})	9140	10.6		203
	THF[a]	1977 (t_{1u})	10710	16.6		203
$V(CO)_6$	CH_2Cl_2	1973 (t_{1u})	9000	38.0		203
$Mo(CO)_6$	C_6H_{12}	1990 (t_{1u})	7360	3.7		204
$Co(CO)_4^-$	diglyme	1888 (t_2)	6910	23.4	1440	203
		1885 (t_2)	4900	28		196
$V(CO)_6^-$	THF[a]	1859 (t_{1u})	13080	14.8		203

a. tetrahydrofuran.

ionisation potential for removal of the carbon 5σ electrons is 14.01 eV—reference 1, p 278) but a very good $a\pi$ acceptor. The structure of the CO molecule has been outlined by Mulliken and Person (reference 1, p 278). The σ- and π-contributions to bonding in the free molecule are indicated by the following polarisation diagrams:

$$\overset{\delta-}{\mathrm{C}}\text{—}\overset{\delta+}{\mathrm{O}} \quad \text{bonding } \sigma \text{ part}$$

$$\overset{\delta+}{\mathrm{C}}\text{—}\overset{\delta-}{\mathrm{O}} \quad \text{bonding } \pi \text{ part}$$

$$\overset{25\%}{\mathrm{C}}\text{—}\overset{75\%}{\mathrm{O}} \quad \text{antibonding } \pi^* \text{ (or } a\pi\text{) part}$$

The overall dipole moment of the molecule[200] is very small (0.37×10^{-30} C m) and its transition moment ($\partial\vec{p}/\partial R$) is also small[198]. *n* donor action alone by the CO molecule through the carbon lone pair would be expected to lead to a smaller $\partial\vec{p}/\partial R$ value (since the C atom becomes more positive) while π acceptor action would lead to a higher charge separation and a larger $\partial\vec{p}/\partial R$. However, when *n* donor action occurs to the metal the carbon atom becomes more positively charged and metal, ($d\pi$)–CO, ($p\pi^*$) "back-bonding" can occur. This weakens the CO bond leading to a decrease in f_{CO} force constant. The intensity perturbation arises from a variation of the $d\pi$–$p\pi^*$ interaction during the vibration of the carbon monoxide. Lengthening of the CO bond leads to a lowering of the $p\pi^*$-orbital energy so that the amount of charge-transfer from metal $d\pi$-orbitals increases. This leads to a large delocalisation moment (or vibronic effect) similar to those operative for halogen complexes (sections 2.3.2 and 3.4).

This model may be tested in two ways. Firstly, one can study co-ordinated carbon monoxide in a situation where little or no π-"back-bonding" is expected. For example, the results of Bethke and Wilson[201] for BH_3–CO show that the CO force constant is only marginally decreased from that of "free" CO (1897 N m^{-1} compared with 1900 N m^{-1}—see table 2.22). Since BH_3 is expected to be a very poor π donor the result seems consistent with this expectation and with the proposed model. *n* donor action (alone) by the CO is expected to give rise to an *increase* in ν(CO) band frequency (*c.f.* the nitrile group) and a decrease in intensity (since the bond will approach more closely to 'homo-polarity'—reference 1, p. 278). Such spectral changes have been observed for CO weakly adsorbed on zinc oxides[200,202] (see table 2.23). Secondly, it is possible to observe the effect of metal oxidation state and other ligands on the carbonyl band intensities. For the octohedral hexacarbonyls only one ν(CO) mode (t_{1u}) is infrared active. Table 2.22 includes the data for a series of neutral and ionic hexacarbonyls[203] from which it is clearly seen that, as the metal nuclear charge decreases, the ν(CO) band intensity increases and the frequency (and force constant) decreases. This is presumably due to the increase in *d*-orbital

TABLE 2.23

Band positions ($\bar{\nu}$(CO)) and intensities for carbon monoxide adsorbed on metal oxides (from reference 202)

Adsorbing surface	$\bar{\nu}$(CO)/cm^{-1}	10^{-3} B_i/m mol^{-1}
CO in N_2 matrix at 78 K	2139.5	119
CO in A matrix at 78 K	2145	94
ZnO	2212	90
	2200	56
	2187	18
ZnO(Ga_2O_3)	2212	90

energy and size as the positive charge on the nucleus decreases. On the other hand, introduction of a good π acceptor directly bonded to the metal is expected to increase the metal nuclear charge and produce a weakening of the M–CO bonding by decreasing the dπ-orbital energy. This ought to lead to an increase in CO force constant and a decrease in intensity. Table 2.24 shows that the specific intensity does decrease when a halogen is directly bonded to the metal in pentacarbonyls[196,200,204]. The inductive effect of the halogen is apparently also important and the weakening effect decreases in the order I < Br < Cl. For an entirely σ bonding ligand, for example pyridine or an amine[205], the dπ-orbital on the metal is again raised in energy and enhanced dπ–pπ* bonding occurs with the carbonyl groups.

In order to interpret quantitatively these changes in band intensity it is necessary to compute the transition moments $\partial\vec{p}/\partial R_{CO}$ usually

TABLE 2.24

Data for pentacarbonyl complexes showing the effect of a π- or σ-bonding substituent (in chloroform solution unless otherwise stated)

Compound	$\bar{\nu}$(CO)/cm^{-1}	10^{-3} × specific intensity[b]/m mol^{-1}	Reference
$Mn(CO)_5Cl$	2125, 2044, 2007	705	204
$Mn(CO)_5Br$	2133, 2049, 2004	867	204
$Mn(CO)_5I$	2138, 2053, 2001	975	204
$Re(CO)_5Br$[a]	2150, 2044, 1985	980	204
$Re(CO)_5I$[a]	2146, 2042, 1990	966	204
$Mo(CO)_5C_5H_{10}NH$[a,c]	2074, 1941, 1922	1420	200
$Mo(CO)_5C_5H_5N$[a,c]	2074, 1943, 1922	1380	200

a. these data may be compared with those for the 'parent' hexacarbonyl given in table 2.22.
b. as defined in the text.
c. in cyclohexane solution.

referred to as the dipole moment derivatives $\vec{\mu}'_{CO}$. These parameters are obtained from the absolute intensities with the aid of a set of normal coordinate matrices **L** which can only be obtained if a force field is available (section 1.2). Many studies on metal carbonyls have used the Cotton-Kraihanzel[206,207] method of obtaining a force field for the CO stretching motion. This method employs several simplifying assumptions[193], aimed at reducing the number of interaction force constants to be determined. It is also assumed that zero vibrational coupling (*i.e.* mechanical interaction) occurs between any two carbonyl groups. (The CO oscillators are thus treated separately from the rest of the molecule, the metal atom remaining at rest.) If, *in addition*, all the carbonyl groups are assumed to have the *same* dipole moment derivative associated with them, then approximate relative intensities can be predicted[193] and the process can be used as an assignment aid when complicated spectra are encountered. The procedure has since been modified to allow[197,200,205,208,209] more interaction constants to be employed and to investigate the validity of the assumptions previously made with respect to equal dipole derivatives, zero transverse dipole derivatives and the overwhelming importance of π-bonding in determining the ν(CO) band intensities. We shall illustrate the usefulness of absolute intensity data in testing the theories of metal–CO interations by reference to some of the work on pentacarbonyls, $ML(CO)_5$, which have essentially octohedral geometry and which (in the absence of distortion) belong to the point group C_{4v}. There are three active CO stretching of modes and one (b_1) inactive mode since:

$$\Gamma_{CO} = 2a_1 + e + b_1$$

The forms of the a_1 and e stretching modes are shown in figure 2.14. For most complexes $ML(CO)_5$ the $a_1^{(1)}$ mode is essentially an "axial" CO stretch while the $a_1^{(2)}$ mode (giving rise to the band at highest frequency)[205] is a CO stretching mode with the five CO groups moving in phase. One might expect from simple symmetry considerations that the e, ν(CO) mode would have the highest intensity followed by the $a_1^{(1)}$ mode. This is often (but not always) found to be the case (see table 2.25). However, mixing of the two a_1 modes is expected and so attempts have been made to calculate $\partial\vec{p}/\partial R$ values (usually denoted by $\vec{\mu}'_{CO}$ or $\vec{\mu}'_{MCO}$ by the original authors) assuming a modified Cotton-Kraihanzel force-field and **L** matrix elements based on a 2×2 secular determinant for these two vibrations. Notice that in only one paper[200] published so far has any account of mixing between the a_1, ν(CO) and ν(MC) modes has been taken. The effect of making assumptions[206,207] about the values of the $f_{CO.CO}$ interaction constants and of ignoring of metal–carbon bonding has, however, been investigated by Darensbourg and Brown[205] and found to be negligible as far as the values of the **L** matrix elements are concerned. From the summary of intensity data and $\partial\vec{p}/\partial R$ values given in table 2.25 it can be seen that the values are very high as expected

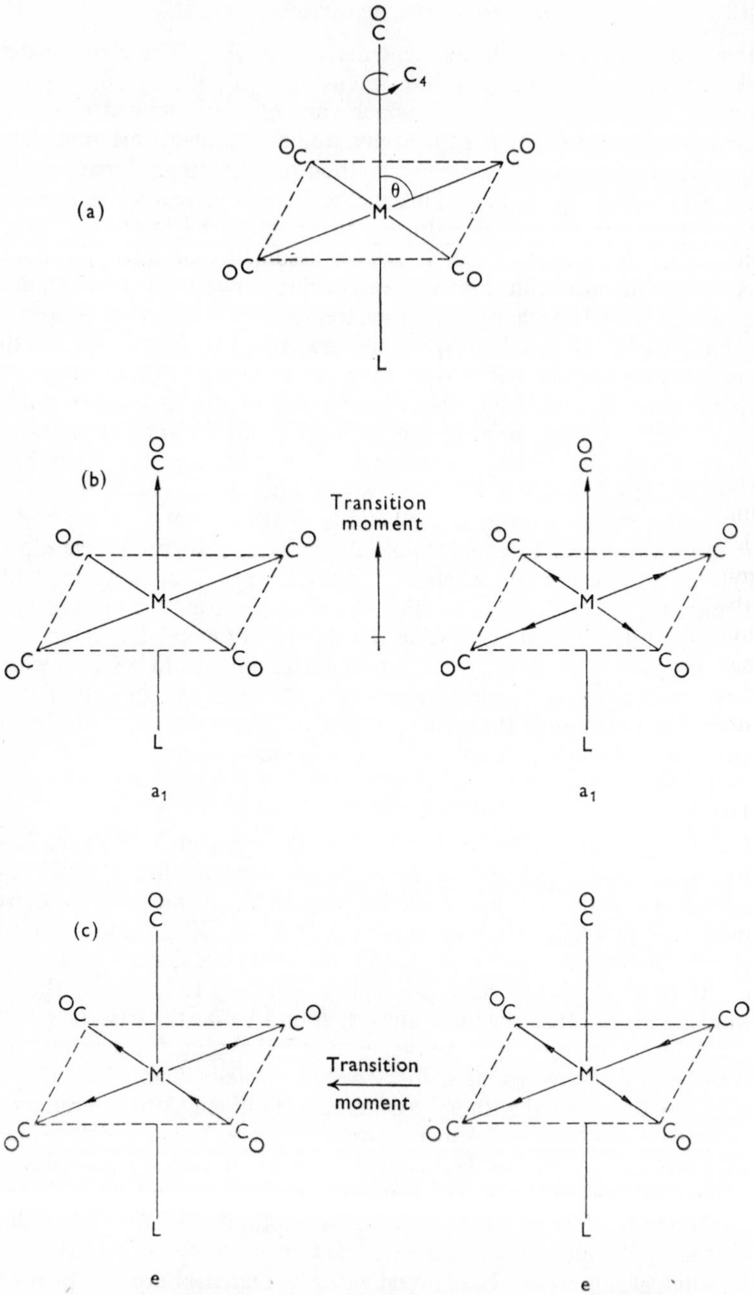

FIG 2.14 (a) $ML(CO)_5$ complex geometry and angle θ between "axial" and "radial" carbonyl groups (b) forms of the symmetry coordinates for the a_1, $\nu(CO)$ vibrations (c) forms of the symmetry coordinates for the degenerate, e, $\nu(CO)$ vibrations.

TABLE 2.25

Summary of band intensities, transition dipole derivatives and bond angles for pentacarbonyl complexes of molybdenum (reference 205)

Compound	10^{-3} B_i/m mol^{-1}			10^{20} $\partial\vec{p}/\partial R$/ coulomb			
	$a_1^{(1)}$	$a_1^{(2)}$	e	$\vec{\mu}_{MCO}(a_1)$	$\vec{\mu}_{MCO}(e)$	Ratio[d]	θ^c
$Mo(CO)_5C_5H_{10}NH$	1250	94.5	5720	49.4[b]	50.9	0.97	90.3
$Mo(CO)_5C_5H_5N$	1220	119	5540	49.3[b]	50.2	0.98	89.5
$Mo(CO)_5As(C_6H_5)_3$	1340	347	4780	54.5[b]	46.1	1.18	85.3
$Mo(CO)_5P(C_2H_5O)_3$	1410	183	4730	53.8[b]	46.3	1.16	90.4
$Mo(CO)_5P(CH_3O)_3$	1390	208	4700	53.9[b]	46.2	1.17	89.8
$Mo(CO)_5PCl_3$	1210	330	4470	51.9[b]	45.2	1.16	87.5

b. value calculated assuming $\vec{\mu}_{MCO}(a_1^{(1)}) = \vec{\mu}_{MCO}(a_2^{(2)})$ (see text).
c. calculated values—observed values all within a few degrees of 90°—see reference 205.
d. the ratio $\vec{\mu}_{MCO}(a_1)/\vec{\mu}_{MCO}(e)$.

(the dipole derivative for 'free' CO is $\sim 10 \times 10^{-20}$ C). This data has been used to provide structural information in the following ways. (i) The intensity data and **L** matrix elements have been used[193,205,209] to calculate the angle θ (figure 2.14a) between the axial and radial carbonyl groups. Comparison with observed angles gives some idea of the validity of the (simplified) analysis. Many of the bond angles calculated are different from 90° although few direct experimental determinations show appreciable deviation from planarity of the radial CO groups. Although there are experimental errors involved in the measurement of this angle using intensity data[193,209] it is felt[209] that the disagreement (by more than 15° in the case of $Re_2(CO)_{10}$ [209]) between the calculated and observed experimental value is due charge movement along the C_4 axis during vibration of the radial CO groups. However, the effect of a transverse bond moment derivative (normal to the bond direction) has been shown to be small[193,211]. Other effects—possibly insufficiently accurate **L** matrix elements or solvent effects[193]—are therefore operative in determining the calculated bond angles. For most of the pentacarbonyls studied, however, the calculated and observed θ values are in good agreement. The intensity ratios are thought[205] to be controlled to some extent by deviations from C_{4v} symmetry, and in support of this assertion, the b_1 ("inactive") vibrational mode is weakly active for some of these complexes. Clearly, however, this method of testing a particular model must be used with caution.

(ii) The ratios of the dipole moment derivatives ($\vec{\mu}'_{CO}$ or $\vec{\mu}'_{MCO}$) have also been used to test the assumptions made about the complex model. Braterman *et al.*[209], for example, have assumed that there are only two different $\vec{\mu}'_{CO}$ values, $\vec{\mu}'_{CO}$ (radial) and $\vec{\mu}'_{CO}$ (axial). (Strictly, there are

three values $\vec{\mu}'_{CO}(1)$, $\vec{\mu}'_{CO}(2)$ and $\vec{\mu}'_{CO}(3)$—one corresponding to each of the three active ν(CO) stretching modes—figure 2.14b-c; They have assumed $\vec{\mu}'_{CO}(2) = \vec{\mu}'_{CO}(3)$ since only two values can be determined[209] along with the angle θ, from three intensity values). The ratio of $\vec{\mu}'_{CO}(a)/\vec{\mu}'_{CO}(r)$ is then used to test the validity of this assumption. The values of this ratio are generally >1.0 but do approach 1.0, for example, if L = H. Darensbourg and Brown[205] have made the alternative assumption that there is a unique $\vec{\mu}'_{MCO}$ value for each symmetry species; they have assumed $\vec{\mu}'_{MCO}(1) = \vec{\mu}'_{MCO}(2)$ since, it is claimed[205], the $\vec{\mu}'_{MCO}(3)$ value is likely to be greater than the other two values (see above). The ratio $\vec{\mu}'_{MCO}(1)/\vec{\mu}'_{MCO}(3)$ varies between about 0.97 and 1.18 for a series of ligands L (table 2.25) and the differences are related to the electronic (σ and π) effects of the particular ligands.
(iii) The relative values of the intensities of the three ν(CO) bands (and their corresponding force constants and $\vec{\mu}'_{CO}$ derivatives) have been discussed at some length[200,204,205,209,210] in terms of the expected electronic effects of the ligand L. For example when L = halogen[209,210] the following trends are observed on going from L = Cl to L = I (see table 2.25)

$$\text{Cl} \rightarrow \text{Br} \rightarrow \text{I}$$

(a) $\rightarrow \vec{\mu}'_{CO}(a)$ and $\vec{\mu}'_{CO}(r)$ increase $\rightarrow$
(b) $\rightarrow f_{CO}$ (axial) increases (weaker bonding) $\rightarrow$
(c) $\rightarrow f_{CO}$ (radial) decreases (stronger bonding) $\rightarrow$

The increase in $\vec{\mu}'_{CO}(a)$ along the series is greater than that of $\vec{\mu}'_{CO}(r)$. This is attributed to the increasing polarisability of the halogens which leads to an increased pπ-dπ halogen–metal interaction as the CO bond vibrates. This effect is more pronounced in the axial than the radial dipole derivative since the $\vec{\mu}'_{CO}(a)$ vector is along the M–L direction. On the other hand, the decreasing polarity along the series leads to an increase in the electron donating ability of the metal and the force constant (f_{CO}) drops on going from Cl to Br to I. However, the most polar halogens also have the greatest π donor ability so the opposite trend is expected, f_{CO} decreasing from Cl to Br to I. This effect is sufficient to reverse the electronegativity-controlled force constant shift only for the axial f_{CO} value since the axial CO group is *trans* to the ligand (figure 2.14a).

The effect of σ- and π-bonding ligands on the CO groups in $ML(CO)_5$ complexes has been considered by Darensbourg and Brown[205]. As mentioned above the ratio $\vec{\mu}'_{MCO}(1)/\vec{\mu}'_{MCO}(3)$ increases considerably on going from an essentially σ-bonding ligand (pyridine or an amine) to a ligand with π-bonding ability (a phosphine or phosphite derivative). This is attributed to an increased $\vec{\mu}'_{MCO}(1)$ and a decreased $\vec{\mu}'_{MCO}(3)$ on going from L = pyridine to L = phosphine because the effect of π-π interaction between the phosphine and metal cannot be transmitted through to the essentially "radial" dipole derivative $\vec{\mu}'_{MCO}(3)$ but only to the "axial" mode. On the other hand, the entirely σ-bonding ligand produces a large negative charge on the metal and dπ "back-bonding" to the CO groups increases. These dipole derivative changes appear to support the idea of a predominant π-bonding effect in these complexes. However, it is pointed out[205] that the negligible increase in $\vec{\mu}'_{MCO}(3)$ on going from $M(CO)_6$ to $ML(CO)_5$ (L = π-ligand) indicates that there is very little increase in π-bonding to the other five CO groups. The fact that such an increase in $\vec{\mu}'_{MCO}$ *does* occur for L = pyridine shows that not all the change of force constant of the CO bond is due to π-bonding effects. Since an increase of electron density on the metal (from a σ-bonding ligand) will lead to a decrease in the (*n*-*v*) metal–carbon σ-bond strength, this alone will lead to a weakening of the CO bond. Any model based entirely on π-bonding considerations is therefore likely to be only an approximate one.

(ii) *Nitrosyl and Nitrogen Complexes*

A number of intensity studies have been made on nitrogen and nitrosyl complexes, sometimes in the presence of CO group(s) attached to the same metal. The data have been used in much the same way[212,213] as described for carbonyl complexes. Studies[212] on $Co(CO)_2NOL$, for example, show that while changing the ligand increases the ν(CO) band specific intensity the ν(NO) band is considerably weakened. However, no correlation could be found between the π or σ-bonding abilities of the ligands and the intensity changes observed. It is suspected that some of the more bulky ligands may give rise to steric effects. Interestingly the half-band width of both ν(CO) and ν(NO) bands increase considerably from those of the parent $Co(CO)_3NO$ when one ligand is introduced. This has been likened to the solvent effect observed for these bands[203] which is, however, much more drastic than that observed due to substitution (see section 2.3.6 for further discussion of band shapes). Much of the work involving molecular nitrogen complexes has centred around the comparison of the σ donor and π acceptor abilities of the CO and N_2 molecules[214,215]. This comparison is discussed in detail in section 5.4.1 and so is mentioned only briefly here. It is clear that infrared intensity measurements, in contrast to frequency shifts[216], do give an unambiguous indication that the CO molecule is both a better σ donor and a better π acceptor than N_2. The (localised) group dipole moment derivatives demonstrating this for a series of osmium complexes[214] are shown in table 5.24 (p. 444) along with the vibrational data from which they were derived. From the consideration of the relative distributions of the aπ-orbitals on CO and N_2 Darensbourg[214] predicts that the contribution to the dipole derivative due to π-charge transfer from the metal should be greater for the N_2 complex than to the CO complex (assuming equal π acceptor abilities). Since

$\vec{\mu}'_{CO}$ is considerably greater than $\vec{\mu}'_{N_2}$ it follows that the π acceptor ability of the CO molecule is much higher than that of N_2. There is also evidence that N_2 is a poorer σ donor in that N_2 does not, like CO, form a complex with BH_3 (see table 2.22). These conclusions have been supported by theoretical considerations[217–219] and by independent experimental studies.[217] They are not altered by the fact that vibrational mixing of the ν(CO) and ν(NN) modes is ignored since stronger coupling of the CO (due to lower $\bar{\nu}$(CO) values) would lead to a *underestimated* value of $\vec{\mu}'_{CO}$ relative to that of $\vec{\mu}'_{N_2}$. However, it should be noticed that no allowance was made in comparing $\vec{\mu}'_{CO}$ and $\vec{\mu}'_{N_2}$ for the intensity of the "free" ν(CO) band.

(iii) *Thiocyanate Complexes*

Larsson[194] has shown that infrared intensity data are extremely useful in distinguishing linkage isomers in thiocyanate and *iso*-thiocyanate complexes. This was achieved by observing that the "$\bar{\nu}$(CN)" value (between 2000 and 2150 cm^{-1}) was respectively unchanged, increased and decreased in intensity for the three "classes" of metal ions (see below). (The results for a few typical cases are summarised in table 2.26). The first class of metals comprised bivalent first row transition metals to which the SCN^- anion is thought to be electrostatically bound (without anion-metal electron exchange). The second class of metal ions, comprising tetrahedral Co^{2+}, UO_2^{2+}, Fe^{3+}, *etc*, are thought to have a metal–N covalent bond. This would lead to an increase in the

TABLE 2.26

Infrared data thiocyanate complexes of transition metal ions (Reproduced, by permission, from *Acta Chem. Scand.*, **16**, 1447 (1962))

Complex system	$\bar{\nu}/cm^{-1}$	$\Delta\bar{\nu}_{1/2}/cm^{-1}$	10^{-5} B_i^a/m mol^{-1}
SCN^-	2066	37	1.99
Mn^{2+}/SCN^-	2093	31	2.03
Zn^{2+}/SCN^-	2109	34	2.32
Co^{2+}/SCN^-	2112	29	2.02
Ni^{2+}/SCN^-	2119	32	2.05
Co^{2+}/SCN^-	2075	29	4.9
UO_2^{2+}/SCN^-	2066	40	4.4
Fe^{3+}/SCN^-	2045	49	6.4
Cr^{3+}/SCN^-	2088	39	3.7
Pt^{4+}/SCN^-	2126	15	0.4
Hg^{2+}/SCN^-	2112	27	1.1

a. Obtained from the product of maximum extinction coefficient and half-band width.

importance of resonance structures (b) and (c) of the thiocyanate ion

$$\underset{(a)}{N\equiv C-S^-} \qquad \underset{(b)}{^-N=C=S} \qquad \underset{(c)}{^{2-}N-C\equiv S^+}$$

and consequently to an increase in the CN bond moment (and presumably the ν(CN) mode transition moment). The third class of metals ions (Hg^{2+}, Pt^{4+}) are thought to have a M–S covalent bond which would enhance the importance of resonance-structure (a) and decrease the ν(CN) mode transition moment. Although this rationalisation appears to work well in explaining the intensity changes there does not seem any complete explanation of the frequency shifts, $\Delta\nu$(CN), for the covalently bonded SCN groups. Both positive and negative shifts are observed and since it is thought[194] that mechanical coupling is responsible for the increase of $\bar{\nu}$(CN) for class 1 metal complexes, it seems likely that normal coordinate effects could be operative for classes 2 and 3 also. In the case of class 3 metal ions the quite large increase in $\bar{\nu}$(CN) could be due to the formal increase in CN bond order (resonance structure (a)) (compare the nitrile case, p. 144). Smaller increases or decreases for class 2 metal ions could then be indicative of opposed normal coordinate and electronic effects. Clearly, intensity data are far superior to frequencies in attempting to distinguish these different types of interaction.

2.3.4. HYDROGEN-BONDED COMPLEXES

Introduction

In chapter 4, J. L. Wood has pointed out quite clearly that, despite the very large volume of spectroscopic and theoretical work on hydrogen-bonded systems (which has been exhaustively reviewed[221–228], there are a number of fundamental problems which still remain to be solved. We shall concentrate in this section on high-lighting some of these problems and on showing what efforts have been made, in attempting to solve them.

The Nature and Extent of Forces Stabilizing Hydrogen-Bonded Complexes

Perhaps the most important, and certainly the most controversial, problem which attracts attention is that of defining quantitatively the nature of the forces which stabilise hydrogen-bonded complexes. The relative magnitudes of the various stabilising forces have not, except in a few simple cases (see chapter 4, p. 377) been unambiguously determined. The problem is the same one as that which arises for any molecular complex but it is emphasised here since it is for hydrogen-bonded complexes that opinions among spectroscopists seem to be most divergent on the matter. As for other complexes, the most obvious basic method of tackling the problem is to compare the values of experimental parameters (force constant changes, intensity changes, heats of formation, *etc.*) with those calculated using the best available wave functions (section 1.3). Since heavy atoms are likely to be absent in simple hydrogen-bonded complexes reasonably accurate *MO* calculations can be made (sections 1.3 and 4.7). In addition, changes in the vibrational spectra of the component molecules are quite large and, for the most part relatively easily measured. Nevertheless, at the present time, only a few complexes have been extensively studied by *both* methods and much effort has been directed at attempting to "separate" and quantify the various stabilising forces by consideration of *either* experimental data *or* a *MO* calculation. As pointed out in 2.3.1 there are major difficulties with both the experimental and theoretical approaches.

Although the hydrogen bond has been, and is still,[37] thought to be stabilised predominantly by electrostatic forces (*i.e.* coulombic and polarisation terms) there have been a number of workers who have, following Mulliken and Person[1,3,91], claimed that charge-transfer (electron migration) forces are *partly* responsible for complex stabilisation (see chapter 1 for further discussion). Such forces were thought to have some importance mainly because the force constant change for the AH bond (usually in the "diatomic" approximation, section 1.2.7, p. 59) and the ν_s(AH) band intensity increase correlated[91] in a very similar way to those parameters for halogen complexes (see section 1.2.10, p. 75). The latter were at the time, and indeed some still are, thought to be predominantly stabilised by charge-transfer forces (see chapter 3). Even allowing for the fact that some weak halogen complexes are now thought[34–36] to be stabilised mainly by classical forces it remains true that the very large increase in intensity observed for ν_s(AH) cannot be easily explained using an entirely electrostatic model[225,228,229].

Let us first examine the experimental evidence that charge-transfer effects are important in determining the intensity of ν_s(AH) band in the complex. It has been shown,[230,231] for example, that if the effects of charge-transfer are predominant in the hydrogen-bond formation of a particular proton donor with a series of bases (proton acceptors) then one might expect linear relations between (a) the base ionisation potential, I_D^v, and $[\Delta\bar{\nu}/\bar{\nu}_0]^{-1/2}$ for the AH stretching mode (b) I_D^v and $(\partial\vec{p}/\partial R)^{-1/2}$—effectively the *increase* in intensity of ν_s(AH) band when interaction with the base occurs using the diatomic molecule approximation[1,91,104]. There should therefore also be a linear relationship between $\Delta\bar{\nu}/\bar{\nu}_0$ (or $\Delta k/k_0$) and $\partial\vec{p}/\partial R$ for the ν(AH) mode. It should be emphasised that this charge-transfer approach to hydrogen bonding assumes that the *entire* force-constant and intensity changes observed are due to vibronic (delocalisation) effects. Even if charge transfer is important it is unlikely to be the only effect present. Further, it should be pointed out that the equations relating the frequency shift and intensity parameters contain "constants" such as overlap integrals, AH bond polarities and ion pair dipole moments which may not strictly be constant and will only be approximately so for complexes of a given AH molecule with a closely related series of bases B.

Linear relations *have* been obtained between these parameters for a number of proton donors. Szczepaniak and Tramer[231] have shown that quite good linear relationships are obtained for several proton donors (including *t*-butyl alcohol, pyrrole, and hydrogen chloride) with a series of methylated benzenes but over a rather limited range of I_D^v. Basila *et al.*[233] have also studied *t*-butyl alcohol with a similar set of aromatics and have obtained a good I_D^v vs $[\Delta\bar{\nu}/\bar{\nu}]^{-1/2}$ linear plot. However, a plot of $\partial\vec{p}/\partial R$ against $\Delta k/k$ showed considerable scatter from linearity. Yoshida and Osawa[234] have determined free energy changes for complexes of aromatic hydrocarbons with phenol but have not tested the relationships predicted by the charge-transfer approach. It is interesting that in neither case[233,234] was there a good correlation between $\Delta G^{\ominus}$ and the observed frequency shift. Since, for these relatively weak complexes, the frequency shifts are small it follows that intensity measurements (section 2.2) are likely to be difficult and this could account for the poor correlation[233] of $\partial\vec{p}/\partial R$ and $\Delta k/k$. In connection with these hydrogen-bonded complexes with $b\pi$ electron donors it should also be remembered that, as Mulliken[1,232] has pointed out, the negative charge on the ring carbons increases as the number of methyl groups increases. Thus the classical electrostatic attraction for a nearby proton will also be enhanced.

On the other hand, examination of the charge-transfer theory has been carried out for other "stronger" proton acceptors[235] where the A—B distance is likely to be smaller and where charge-transfer effects should be enhanced (reference 225, p. 221). Examination was made of the intensity change of the ν(NH) band in diphenylamine and ν(OH) band in 2,4,6-trimethylphenol when complexation occurs with a series of substituted pyridines. $\partial\vec{p}/\partial R$ values were computed using the diatomic molecule approximation and the relationship between these two parameters was found to be distinctly non-linear in both cases. A comparison of the dipole moments of these complexes, calculated using the simple charge-transfer model, with the experimental values lends support to the idea that charge-transfer forces are not predominant in stabilising these complexes. At any rate, such forces are not needed to account for complex dipole moment (see also section 1.3 and chapter 7). This does not mean, however, that electron delocalisation may not be significant in the generation of a large $\partial\vec{p}/\partial R$ for the complex ν_s mode during vibration. This work underlines the fact that the experimental data are likely to result from a superimposition of several effects which are difficult to separate.

It seems clear that infrared intensity measurements ought to play a more important role in the investigation of hydrogen bonding. This parameter is the most sensitive one to hydrogen-bond formation and it is a parameter reasonably easily calculated using a "good" molecular orbital calculation. However, the amount published intensity data for the perturbed ν_s(AH) band is relatively small compared with the frequency shifts observed since reliable measurements are so much more difficult to make. Nevertheless, it seems worthwhile examining the data

TABLE 2.27
Summary of the OH bond force constant and transition moment data for hydrogen-bonded complexes of *t*-butanol

Base	$\tilde{\nu}_s$(OH)	k_{OH}[b]/ N m^{-1}	$\Delta k/k_0$	10^{-4} B_i/ m mol^{-1}	$\partial\vec{p}/\partial R_{OH}$[c]	Reference
toluene	3583	717	0.019	5.3	1.00	233
p-xylene	3579	715	0.022	5.7	1.20	233
mesitylene	3569	711	0.027	6.0	1.27	233
durene	3568	710	0.028	5.8	1.20	233
pentamethylbenzene	3560	708	0.031	8.0	1.87	233
hexamethylbenzene	3549	703	0.038	7.7	1.80	233
acetonitrile	3552	703	0.038	4.6	1.87[d]	233
benzonitrile	3554	705	0.036	4.5	1.83[d]	243
trimethylacetonitrile	3548	702	0.039	4.2	1.73[d]	243
chloroacetonitrile	3568	711	0.027	3.6	1.46[d]	243
trichloroacetonitrile	3594	721	0.015	1.8	0.56[d]	243
acetone	3517	691	0.055	25.1	5.23	245
acetophenone	3541	699	0.043	19.1	4.27	245
dimethyl sulphoxide	3434	658	0.100	14.3	4.47[d]	243
dibenzyl sulphoxide	3476	674	0.078	13.5	4.27[d]	243
diethyl ether	3498	683	0.066	26.0	5.40	e
dioxan	3500	684	0.064	21.8	4.74	245
dimethyl formamide	3475	674	0.078	36.6	6.83	245
pyridine	3366	632	0.135	48.0	8.27	245
pyridine	3387	639	0.125	46.8	8.14	247
triethylamine	3277	599	0.181	46.0	8.04	e

a. for *t*-butanol $\tilde{\nu}$(OH) is 3618 cm^{-1}, k_0 is 7.31 N m^{-1}, B_0 is 2.7×10^4 m mol^{-1}.
b. "diatomic" molecule approximation.
c. corrected value of the removal of $\partial\vec{p}_0/\partial R_{OH}$ due to "free" alcohol band, units are 10^{-20} C.
d. based on a B_0 value of 0.88×10^4 m mol^{-1} (reference 243).
e. data from G. M. Barrow, *J. Chem. Phys.*, 1955, **23**, 896.

which is available in an attempt to see what overall picture (if any) emerges. A selection of the results extracted from the literature[236–248] is given in table 2.27. This serves to indicate the range of intensity perturbation observed for a typical proton donor using different bases and to demonstrate the degree of agreement to be expected from different laboratories. We have combined as much data as we could find in order to make the $\Delta k/k$ against $\partial\vec{p}/\partial R$ plots shown in figures 2.15–2.18. (Note that we have plotted $[\partial\vec{p}/\partial R - \partial\vec{p}^0/\partial R]$ in these diagrams—see caption to figure 1.9, p. 75). Examination of the graphs leads to two principal conclusions. Firstly, if the same base (electron donor) or very similar bases are employed then the plots obtained

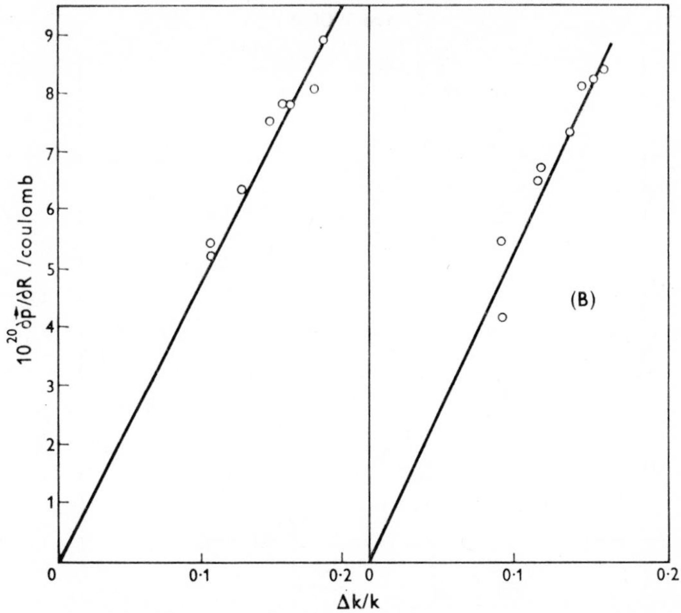

FIG 2.15 Plot of Δk/k against $\partial\vec{p}/\partial R$ for hydrogen-bonded complexes of (A) methanol (B) *iso*-propanol with a series of substituted pyridines (data from reference 246).

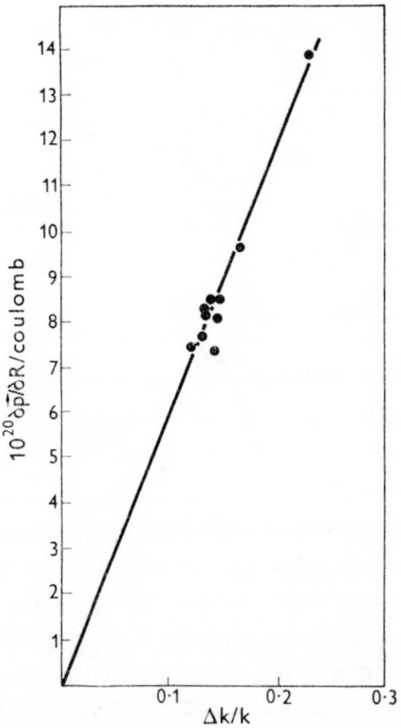

FIG 2.16 Plot of Δk/k against $\partial\vec{p}/\partial R$ for hydrogen-bonded complexes of pyridine with a series of alcohols.

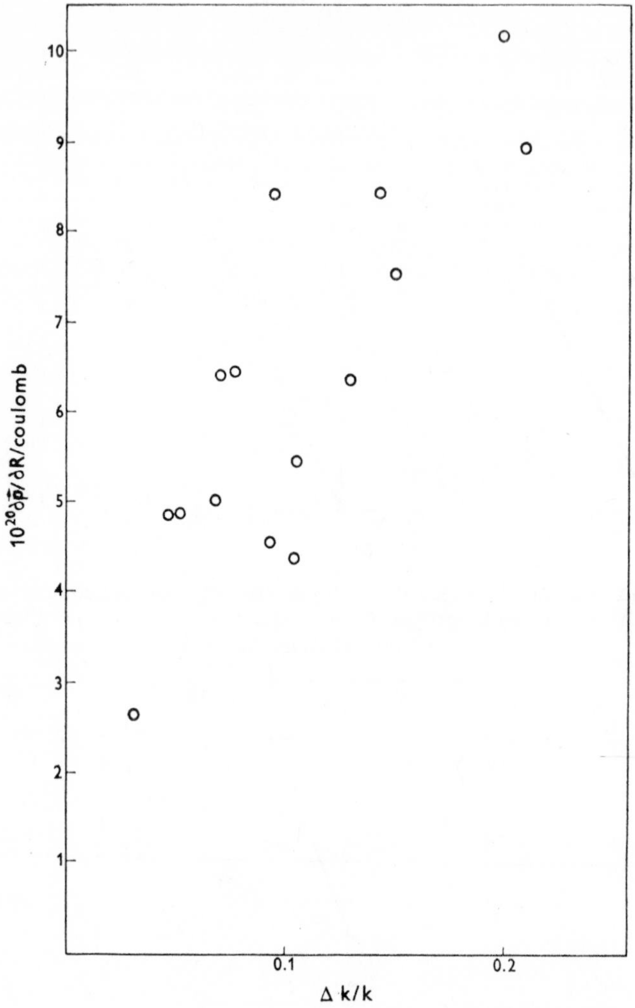

FIG 2.17 Plot of Δk/k against $\partial\vec{p}/\partial R$ for hydrogen-bonded complexes of methanol with a range of bases of different chemical type.

(figures 2.15 and 2.16) are reasonably linear. Deviations could be due to errors in the intensity measurements or solvent effects (see below). Thus the correlation of $\Delta k/k_0$ and $\partial\vec{p}/\partial R$ for a series of alcohols with pyridine[245–247] and for a single alcohol with pyridine derivatives[248] is quite good. Secondly, if a single alcohol is considered complexed to a series of *different* bases then the correlation is very poor (figures 2.17 and 2.18). It is interesting that the correlation is also poor for *t*-butanol

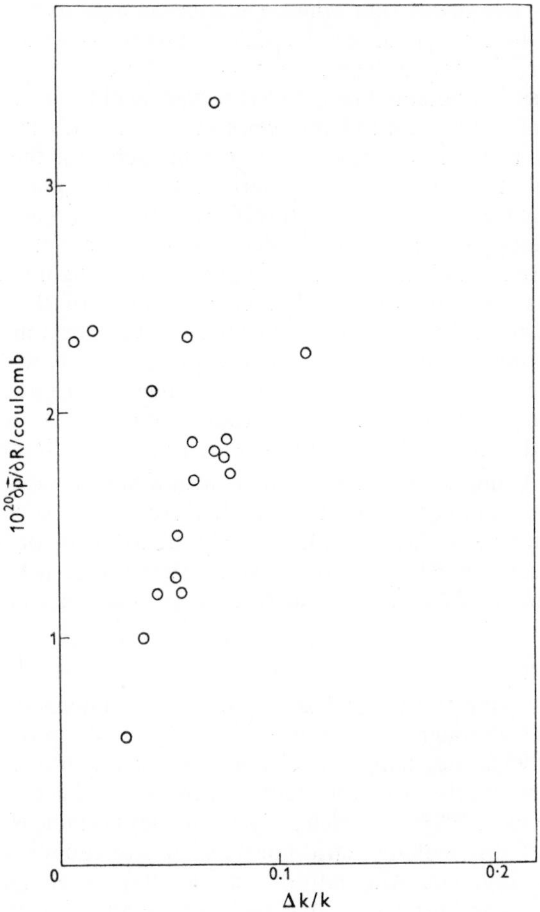

FIG 2.18 Plot of Δk/k against $\partial\vec{p}/\partial R$ for hydrogen-bonded complexes of *t*-butanol with aromatic bases, ethers, sulphides, ketones and nitriles (data from table 2.27).

with closely related aromatic hydrocarbons and nitriles (figure 2.18). Since, for these weaker donors, charge transfer is expected to be less important (and electrostatic forces relatively more important) than for, say, pyridine derivatives, one might expect greater scatter. (Notice also, however, that the accuracy of intensity measurements for weak complexes is expected to be low due to overlapping of "free" and "complexed" ν_s(AH) bands). Thus it seems clear that, if a particular base (or series of similar bases) is used, then the intensity perturbation fits, reasonably well, the charge-transfer model. This *may*, however, be because the electrostatic contribution to the intensity change (whose magnitude is unknown, (but which *could* be as high as 50% of the perturbation in some hydrogen-bonded complexes—see section 1.3, p. 92) is approximately constant and the vibronic contribution varies with the ionisation potential or the base of the "acidity" of the proton donor (only alcohols seem to have been tested sufficiently well so far). On the other hand, if the base is different in each case the "donor" orbital is likely to lead to different polarisation forces between the acid and base. Any trend in the charge-transfer contribution would then be masked. This appears to be clearly demonstrated for methanol complexes by figure 2.17. If this is a correct interpretation of figures 2.15–2.18 then the data support the perturbation mechanism of Boobyer and Orville-Thomas[249]. They regard the intensity perturbation as due to a variation of polarisation† of the highly polarisable base lone-pair orbital as the AH bond vibrates against it. This contribution to the intensity would be expected to remain roughly constant for a series of closely related bases (or for the same base). The lack of linear correlation of $\partial\vec{p}/\partial R$ and $\Delta k/k_0$ for an alcohol with *different* bases parallels that for other complex parameters. Thus there is a lack of $\Delta\tilde{\nu}/\Delta H$ and $\Delta\tilde{\nu}/K_b$ linear correlations unless closely related acids or bases are used, as is expected[222], since for a given strength of interaction a parameter such as $\Delta\tilde{\nu}$ is likely to depend on both donor and acceptor molecule.

Solvent Effects

The ν_s(AH) vibrational band shows some very interesting solvent effects which, although not yet fully understood, do provide some indication of what may happen to a molecule when hydrogen bonding occurs. Solvent effects have been studied in two ways. There have been a number of studies[250–253] in which a given complex is studied in a range of solvents of different dielectric constant, proton donor ability or π-bonding ability, *etc.* Alternatively, studies have been made[253–257] by varying the concentration of the (polar) base added to the proton donor in an "inert" solvent such as cyclohexane or carbon tetrachloride. In each case it has been found that the ν_s(AH) band shifts to lower frequency as the concentration of base is increased or as the proton donor property of the solvent increases (see, for example, reference 252). Effectively, therefore, the hydrogen-bond strength is increased by the solvent effect due to an increase in the acidity of the AH proton.

There have been two principal approaches in the attempted interpretation of these frequency shifts. Bellamy[251,252] and his co-workers have sought to rationalise the effects in terms of specific complex-solvent

S—H·········A—H———B   A—H———A—H′·········S

I   II

† This mechanism is similar to the one described as the electrostatic vibronic effect in section 1.2.7.

interactions of the kind I or II, where S or SH is a solvent molecule. Type I interaction can, of course, occur only if the solvent molecule can act as a proton donor. Type II interaction occurs for hydrogen-bonded self-dimers and now S may be a proton acceptor. This interaction is caused by an increase in the basicity of the A atom (oxygen or nitrogen, for example) or an increase in the acidity of the H′ atom when hydrogen bonding occurs. This then accounts for the fact that the "free" ν(AH) band shows no "solvent" effect of this kind. This mechanism is supported by the following observations.

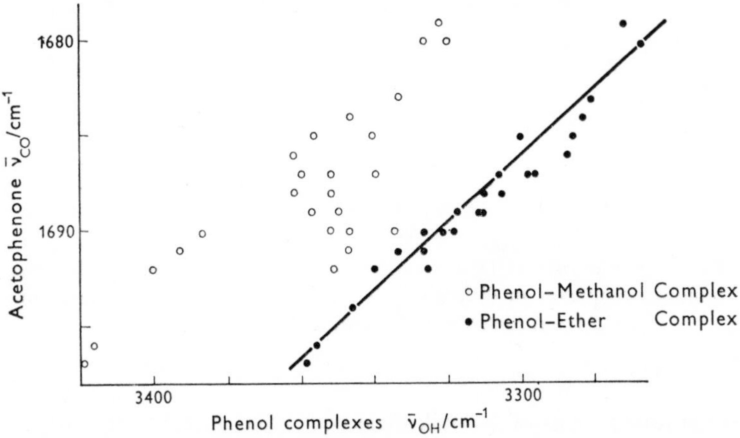

FIG 2.19 Solvent shifts relative to ν(CO) of acetophenone for complexes with and without free OH group (Reproduced, by permission, from *Spectrochim. Acta*, **22,** 535 (1966)).

(i) Whereas alcohol dimers show such solvent shifts in both proton donor and proton acceptor solvents, complexes of the phenol–ether type have a ν_s(OH) band which is sensitive only to protic solvents[251]. This is clearly illustrated[251] by plotting the phenol $\tilde{\nu}$(OH) shift in a range of solvents against the $\tilde{\nu}$(CO) shift for acetophenone in the same solvent. Solvent dipolar interaction at the alcohol OH oxygen for the phenol–ether complex is the same as that thought to occur at the carbonyl oxygen in acetophenone. It is seen (figure 2.19) that there is a reasonably good linear correlation between the two shifts. However, for the phenol-methanol "dimer" complex a wide scatter of shifts is observed since now solvent interaction with proton acceptor solvents can also occur (through the free OH group).

(ii) It has been shown that, for complexes of carbonyl compounds with phenol[252], the solvent effect on the ν(CO) band is much greater than that on the ν_s(OH) band. Figure 2.20 shows the correlation of $\Delta\tilde{\nu}$(CO) and $\Delta\tilde{\nu}$(OH) for these complexes when the solvent is changed. The initial $\Delta\tilde{\nu}$(CO):$\Delta\tilde{\nu}$(OH) ratio due to complex formation is about 0.1. Clearly there is a specific solvent interaction—probably at the carbonyl

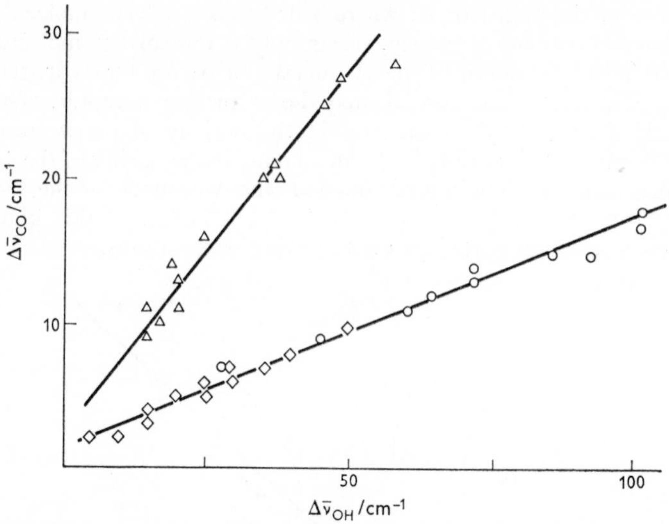

FIG 2.20   Plot of $\Delta\bar{\nu}(CO)$ against $\Delta\bar{\nu}(OH)$ for ketone–phenol hydrogen-bonded complexes in different solvents (Reproduced, by permission, from *Spectrochim. Acta*, **27A**, 705 (1971).

oxygen lone pair since a "bulk dielectric" effect (see below) would be expected to affect both $\nu(OH)$ and $\nu(CO)$ bands in the same way. These results strongly suggest that the additional solvent-complex interaction, III is the most likely for these complexes although interaction of type I may also occur.

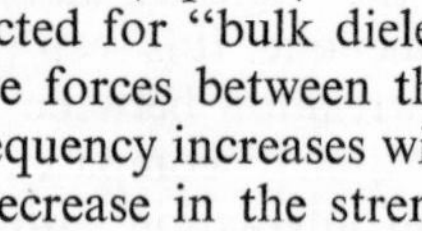

$$
\begin{array}{c}
S \\
\vdots \\
A\!-\!H\!-\!\!-\!\!-\!O\!\!=\!\!CR \\
III
\end{array}
$$

(iii) The effect of a temperature variation has been shown[250] to be consistent with the solvent (dipolar) interaction mechanism and contrary to what is expected for "bulk dielectric" effects which arise from long-range inductive forces between the molecules. Figure 2.21 shows how the $\nu_s(OH)$ frequency increases with a drop in temperature. This is attributed to a decrease in the strength of the dipolar interactions leading to a decrease in the acidity of the OH proton and a consequent weakening of the hydrogen bond. No temperature effect is observed on the $\nu(OH)$ band in the gas phase and very little effect is observed for the "free" $\nu(OH)$ band in pure solvent (see figure 2.21). It should be noticed, however, that the temperature effects may be partly caused by changes in solvent density (which would change the average interacting molecule–molecule distances). The authors also

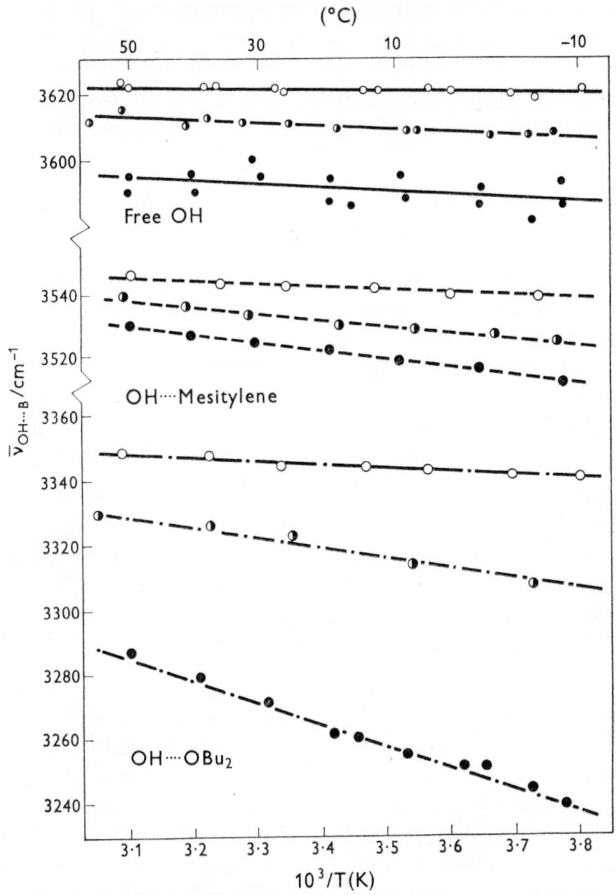

FIG 2.21 Effect of temperature on the OH stretching frequencies of phenol in solutions of di-*n*-butyl ether (— · — ·) and mesitylene (- - - -) in *n*-hexane (○), carbon tetrachloride (◑), and chloroform (●). Initial concn: phenol, ~0.02M; di-*n*-butyl ether, 0.24M (in *n*-hexane), 0.90M (in CCl_4), 0.06–0.21M (in $CHCl_3$); mesitylene, 0.57M (in *n*-hexane), 0.93M (in CCl_4), 0.3M (in $CHCl_3$) (Reproduced, by permission, from *Spectrochim. Acta*, **23A**, 2029 (1967)).

regard their interpretation as tentative in view of the fact that the model[258] describing the interaction of two dipoles *i.e.*

$$E = -\text{const.}\frac{\vec{\mu}_1^2\vec{\mu}_2^2}{T}$$

applies strictly only to the vapour phase.

The alternative approach[253,255,256] is to attempt to account for the solvent frequency shifts in terms of a general "bulk dielectric" or "reaction field" effect which is caused by changing the refractive index and permittivity of the medium. The application of this model to the determination of band frequency and intensity changes is described in references 41–43. In particular, Buckingham has shown[41] that the frequency shift in a medium of refractive index n and permittivity ε is given by,

$$\tilde{\nu} - \tilde{\nu}_0 = C_1 \frac{(\varepsilon - 1)}{(2\varepsilon + 1)} - C_2 \frac{(n^2 - 1)}{(2n^2 + 1)} \qquad 2.9$$

where C_1 and C_2 are constants related to the Onsager molecular model[259] which is used to describe the "reaction field". Horak *et al.*[255,256] have fitted their data for phenol–nitrile complexes (in solutions with varying amounts of nitrile) to an equation of this type and were able to deduce a value for $\tilde{\nu}_0$, the position of the ν_s(OH) band in the "gas-phase" complex. The relevant $\tilde{\nu}$(OH) values for the (typical) phenol–acetonitrile complex were as follows half-band widths shown in brackets),

phenol, $\tilde{\nu}$(OH), gas phase, 3655 cm^{-1}
phenol "free" $\tilde{\nu}$(OH), carbon tetrachloride, 3611 cm^{-1} (20 cm^{-1})
phenol $\tilde{\nu}$(OH) in neat acetonitrile, completely "solvated" complex, 3409 cm^{-1} (138 cm^{-1})
phenol-acetonitrile $\tilde{\nu}$(OH) extrapolated to "gas-phase", 3540 cm^{-1} (15 cm^{-1})

This means that the frequency shift due to complexation is equivalent to 115 cm^{-1} and the solvent shift is equivalent to 131 cm^{-1}—a result which is completely out of line with what is expected for such a "reaction-field" solvent effect (see section 1.2) since the formation of a specific complex is expected to lead to much the largest shift. Further, the extrapolated half-band width of 15 cm^{-1} for the isolated complex ($\tilde{\nu}_0$(OH)) is *less* than that of the phenol in carbon tetrachloride (referred to as a "contact" complex—see section 1.2.10). Although these data appear to fit the Buckingham equation quite well the authors do point out that the extrapolation to $\tilde{\nu}_0$ is a long one so that the values may be somewhat unreliable. They also point out that the data for some phenol–ether and phenol–nitrile complexes in solvents such as chloroform and toluene do not fit the relation well and specific solvent-complex interactions are then said to be operative, (for chloroform type I interaction would seem likely). However, it is not entirely clear how the frequency shift of ν_s(OH) band occurs for the phenol–acetonitrile complex in a mixed solvent of carbon tetrachloride and acetonitrile as the concentration of acetonitrile is increased (see table 2.28). It is unlikely that type I interaction occurs since the acetonitrile protons are not very acidic. Type II interaction in this case would mean formation of 2:1 phenol–acetonitrile complexes due to the increased basicity of the phenol oxygen atom. If a cyclic system, IV, were formed this would certainly lead to a downward shift of $\tilde{\nu}_s$(OH) (see chapter 4, p. 325).

TABLE 2.28

Positions and half-band widths (in cm^{-1}) of the ν_s(O–H) band in phenol/acetonitrile/carbon tetrachloride mixtures (Reprinted from *Collect. Czech. Chem. Comm.*, **36,** 2757 (1971), by permission of the publishing house of the Czechoslovak Academy of Sciences)

Conc. CH_3CN in CCl_4/mol %	10^2 × phenol conc./ mol dm^{-3}	"free" $\bar{\nu}_s$(OH) band		"complexed" $\bar{\nu}_s$(OH) band	
		$\bar{\nu}$	$\Delta\bar{\nu}_{1/2}$	$\bar{\nu}$	$\Delta\bar{\nu}_{1/2}$
100	3.004			3409	138
38.2	3.060	3608	24	3422	121
25.0	3.062	3609	22	3428	117
17.2	3.003	3609	22	3431	116
10.5	2.989	3610	23	3437	109
3.6	2.974	3611	22	3450	106
1.8	3.098	3611	20	3456	100
0.78	2.025	3611	20	3454	93
0.39	2.009	3611	20	3460	90
0.19	2.038	3611	20	3460	90

Alternatively dipolar interaction may occur between the complexed phenol oxygen lone pair (now more basic) and the "excess" nitrile molecule dipoles leading to type III interaction where the solvent

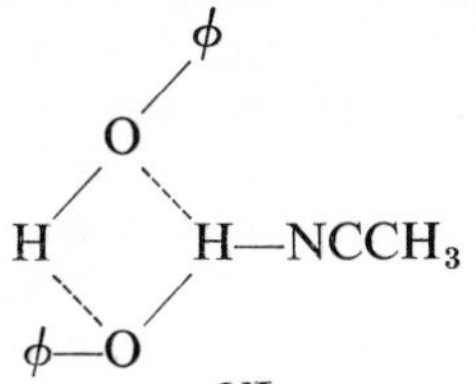

VI

molecule S (acetonitrile) interacts with the A end of the complex rather than the B end. More work is needed on this very interesting system before one can be certain which mechanism is most likely and whether or not "reaction-field" effects are operative.

There are some possibly significant similarities between the band broadening effects observed for hydrogen-bonded complexes (see table 2.28) and those observed for other complexes. For example, the two low frequency bands of the pyridine–I_2 complex in cyclohexane show a quite drastic broadening when varying amounts of excess pyridine are added[260] (see section 3.4.2). Such broadening effects have also been observed[193] for the ν(CO) bands of metal carbonyl complexes when polar solvents are used. It appears likely, in view of the conclusions reached for hydrogen-bonded complexes, that such effects may be, at least partly, due to the formation of ternary (and more complex) species due to specific interaction with the solvent molecules. (See section 2.3.6 for further discussion).

2.3.5. COMPLEXES BETWEEN bπ DONORS AND aπ ACCEPTORS

Introduction

A wide range of molecules including simple olefines, aromatic hydrocarbons, aromatic amines, heterocyclic aromatics and condensed ring systems, can be regarded as potential bπ donors. The π-orbitals concerned have a range of first ionisation potentials (reference 2, table 13.4) sometimes as low as 6.5 eV. If the electron affinity is high, as in the case of chloranil or tetracyanoethylene, for example (reference 2, table 13.5), then the formation of an "inner complex" ($D^{\dot{+}}A^{\dot{-}}$), by complete electron transfer is by no means unexpected (reference 1, chapter 16). On the other hand, a relatively "weak" bπ donor such as benzene (or one of its methylated derivatives) might be expected to form a complex with an essentially neutral ground state. Since the formation of an "inner" complex produces a pair of radical ions the extensive use of electron spin resonance[261–63] and conductivity measurements[264] (reference 2, chapter 9) has been made to study such interaction. Electronic spectroscopy[261–265] has also been used to provide information about the nature of the complex ground state. In this section we shall be mainly concerned with the vibrational perturbation which arises as a result of complex formation and with the deductions which can be made about the nature of the bonding between the component molecules of a bπ-aπ "outer" complex. One advantage of bπ-aπ complexes, as compared with those of the halogens for example, is that often the complex readily crystallises in such a way that the geometry may be determined by crystallographic methods[266,267] (reference 2, chapter 8). The crystals are often of sufficiently high quality for "orientated-crystal" vibrational spectra to be determined and extra information is thereby obtained (see below). Nevertheless, although the charge-transfer model for such complexes has been shown[34–36] to be somewhat inadequate in many ways, it is still not clear exactly how the interaction ought to be envisaged, and the charge-transfer model is still widely employed.

Perturbation of the Component Molecule Spectra

By comparison with the complexes described in sections 2.3.2 to 2.3.4, the amount of vibrational spectroscopic work published on bπ-aπ complexes is extremely small. The main reason for this is that, in line with the expectations (see sections 1.2.4 and 1.2.6), the vibrational perturbation of the component molecules (for "outer" complexes) is small. There are at least two reasons for this. The first, and most important, reason is that the interaction for neutral (1:1) complexes of this type are relatively "weak" so that severe changes of force constants (or normal coordinates) are not expected. The second is that the electronic effects of interaction are unlikely to be localised (for example, for an aromatic bπ donor) except in the case of olefin-type donors or acceptors where the π-electrons are more or less localised near the CC double bond (such interactions are then expected to lead to a decrease of the ν(CC) band frequency).

Vibrational spectroscopy has principally been used in order to distinguish a neutral "outer" complex, D–A, from the radical ion pair formed by electron transfer. This distinction is easily made by a comparison of the spectra of the "free" components, the complex and the alkali metal salt of the acceptor; the latter being regarded as a

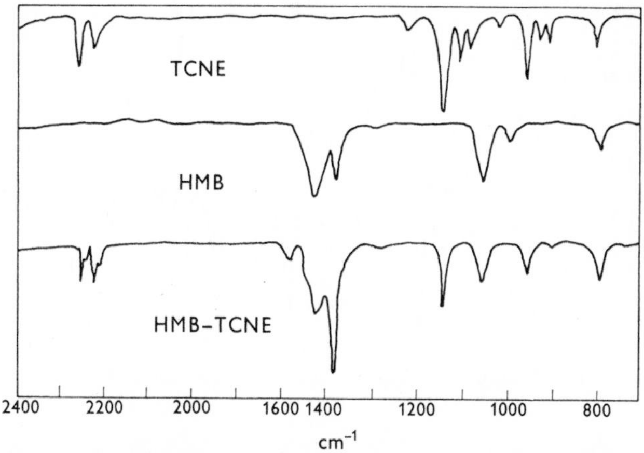

FIG 2.22 Infrared spectra of TCNE, HMB and HMB–TCNE in KBr pellets (Reproduced from *J. Phys. Chem.*, **70,** 2011 (1966), by permission of the American Chemical Society).

'model' for the situation in which complete electron transfer has taken place (*i.e.* 100% charge transfer). Thus, it has been clearly shown that arene complexes of halogenated benzoquinones[262,268,269] and of tetracyanoethylene (TCNE)[270–274] are "outer" complexes with little, if any, contribution from charge transfer. On the other hand, there is good evidence that crystalline complexes of chloranil with aromatic amines[264,268] (such as aniline and *p*-phenylenediamine derivatives) are composed entirely of radical ions. These contrasting situations are illustrated by figures 2.22 and 2.23. In figure 2.22 the spectrum[271] of hexamethylbenzene–TCNE is seen to be a superposition (with some perturbation—see below) of the component "free" molecule spectra. In figure 2.23 the spectrum[264] of *p*-phenylenediamine–*p*-chloranil is seen to resemble more closely that of the lithium salt of the corresponding semiquinone. The presence (or absence) of radical ions in these differing situations has been confirmed by electronic[265] and e.s.r. spectroscopy[262,263]. It should be noted that neutral complexes between

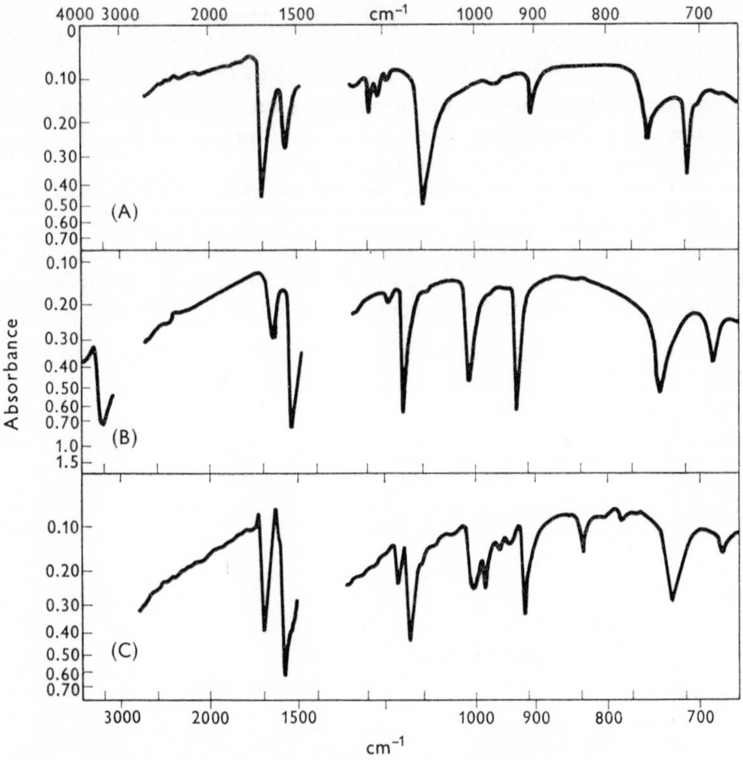

FIG 2.23 Infrared spectra of (A) *p*-chloranil, (B) the lithium salt of the semiquinone and the (C) *p*-chloranil–*p*-phenylenediamine complex (all in the solid state) (Reproduced from *J. Chem. Phys.*, **41**, 1609 (1964), by permission of the American Institute of Physics).

pyridines (and other heterocyclicdonors) and the substituted benzoquinones $C_6X_4O_2$ (X = halogen) or TCNE *have* been detected in a non-polar or weakly polar medium (carbon tetrachloride or dichloromethane, for example) using electronic spectroscopy[276,277]. The concentrations were very low (typically 10^{-5} mol dm^{-3}) and the spectra of the radical ions were then only observed in polar solvents. For some "weaker" bπ donors and aπ acceptors 1:1 complexes may be observed[275], even in polar solvents. For a more detailed discussion of the interaction (and reaction) of π donors and acceptors in solution see reference 2 and Kosower's excellent review[278].

Since, for bπ-aπ complexes, there exists a unique opportunity to study the spectra of one of the species (*i.e.* $A^{\overline{\cdot}}$) resulting from complete electron transfer, there is a great temptation to attempt to estimate the extent of charge transfer occurring for 1:1 neutral complexes by a direct comparison of the vibrational perturbation of A in the complex with that of $A^{\overline{\cdot}}$ in the $M^+A^{\overline{\cdot}}$ salt. Such a comparison is part of the extensive infrared study of the hexamethylbenzene–tetracyanoethylene (HMB–TCNE) crystalline complex by Devlin *et al.*[270–272] and others[273,274]. The vibrational spectrum of $Na^+TCNE^{\overline{\cdot}}$ is significantly different from that of TCNE (see figure 2.22). In particular, only (normally) infrared inactive (a_g) modes give rise to bands in the spectrum of $TCNE^{\overline{\cdot}}$ and most of the infrared active (b_{1u} and b_{2u}) modes have bands of vanishingly small intensity. Bands at about 1380 and 2190 cm^{-1} are assigned to the totally symmetric ν(C=C) and ν(CN) modes; these being, as expected[270], at considerably longer wavelengths than the corresponding TCNE bands. The red shifts of these bands (table 2.29)—taken along with the blue shift of the b_{2u}, ν(C-C) band—when the electron affinity of the cation is increased (from Cs to Li) seems good evidence that the vibrational spectrum of the radical anion, $TCNE^{\overline{\cdot}}$, depends on the degree of charge transfer. In view of this result the shift of the ν(C=C) band on going from TCNE to the HMB–TCNE complex (10 cm^{-1}) is compared with that ($\sim$180 cm^{-1}) on going $TCNE^{\overline{\cdot}}$ and it is deduced that $\sim$5–10% of charge transfer occurs for the 1:1 complex. This evidence for a small, but significant, charge-transfer contribution is supported by the intensity of the b_{2u}, ν(C-C) stretching mode which decreases by about 25% on going from TCNE to the complex (table 2.29). This interpretation, it should be noted, involves the assumption (which is supported by a simple quantum mechanical calculation) that the very severe perturbation of the TCNE vibrational modes on going to $TCNE^{\overline{\cdot}}$ is due to a "vibronic" effect in which the transition moments of the b_{1u} and b_{2u} modes are very nearly cancelled by the transition moment due to "electron vibration" of the radical electron (see section 2.3.2 for a more detailed description of this mechanism). The extent of this cancellation then depends on the extent of charge transfer for the HMB–TCNE complex. Although this mechanism is supported by the fact that at least two of the HMB fundamental bands are also perturbed in intensity (table 2.29), no allowance is made for the fact that classical electrostatic forces between the molecules may be responsible for part of the observed frequency and intensity changes (see section 1.2). Thus the estimated extent of charge transfer could be in error. It is, in any case, small even assuming an exclusively charge-transfer "model" and, by comparison with halogen complexes (section 3.4.2), it is likely to decrease on going from the solid phase into solution. A similar mechanism is invoked to explain the infrared activity of the totally symmetric, a_g, modes of the TCNE, in the complex and in its salts. This perturbation is envisaged as due a change of scalar polarisability of the TCNE molecule in the complex or salt as the "ν(C=C)" or "ν(CN)" vibration takes place. The resulting electronic redistribution (leading to a finite value of the transition moment) may be the result of enhanced charge transfer between $D(M^+)$ and $A(A^{\overline{\cdot}})$ or, in the case of the salts, of a changing cation—anion electrostatic interaction. Changes in the spectrum due to changes

TABLE 2.29

Vibrational band positions (cm^{-1}) and intensities ($m\ mol^{-1}$) of HMB, TCNE, HMB–TCNE and $M^{+}TCNE^{\overline{\cdot}}$ (all data for the solid phase (Reproduced from *J. Phys. Chem.*, **75**, 325 (1971), by permission of the American Chemical Society)

Mode (symmetry)	HMB $\bar{\nu}$	HMB 10^{-1} B_i	TCNE $\bar{\nu}$	TCNE 10^{-1} B_i	HMB-TCNE $\bar{\nu}$	HMB-TCNE 10^{-1} B_i	$Li^{+}TCNE^{\overline{\cdot}}$ $\bar{\nu}$	$Na^{+}TCNE^{\overline{\cdot}}$ $\bar{\nu}$	$K^{+}TCNE^{\overline{\cdot}}$ $\bar{\nu}$	$Cs^{+}TCNE^{\overline{\cdot}}$ $\bar{\nu}$
$\nu(CN)(a_g)$			2236		2230		2198	2197	2190	2187
$\nu(C{=}C)(a_g)$			1570		1560		1387	1390	1371	1358
$\nu(C{-}C)(b_{2u})$			1155	1140	1155	856	1184	1187	1187	1190
$d(CH_3)$	1380	1785			1390	3995				
$w(CH_3)$	1460	12250			1460	7425				

in normal coordinate are thought[270] to be small since there is evidence that the geometry of the TCNE$^{\overline{\cdot}}$ ion is similar to that of the neutral molecule[279].

The 1:1 HMB–TCNE complex is an excellent one with which to illustrate the usefulness of "orientated crystal" infrared spectra[271,272]. Since the component molecules are in "stacked" in linear chains (–D–A–D–A–D–), with their molecular planes roughly parallel to each other, the complex crystallises in needles which may be used to obtain spectra with the radiation polarised both parallel and perpendicular to the molecular planes (*i.e.* perpendicular and parallel, respectively, to the needle axis). This immediately distinguishes in-plane and out-of-plane modes which are active only if the radiation is polarised parallel to the particular transition moment. Further, since vibrations rendered active by the vibronic mechanism of Ferguson[89] and Person and Friedrich[90] are expected to have a "delocalisation" (charge-oscillation) moment perpendicular to the molecular plane their activity with the radiation vector parallel to the needle axis makes them easily identified. The band at 1560 cm^{-1} in the spectrum of HMB–TCNE is identified in this way as the totally symmetric ν(C=C) mode of TCNE (corresponding to the Raman band at 1569 cm^{-1}). This observed infrared dichroism is, of course, interpreted as evidence that the "vibronic interaction" mechanism is in fact operative for this complex, at least in the solid phase.

Intermolecular Vibrations for bπ-aπ Complexes

If the interaction between the component molecules of the complex is strong enough (or, in solution, if the complex lifetime is long enough—see section 3.4.3) then one ought to be able to detect a band due to an intermolecular vibration similar to that observed for pyridine–halogen, metal–donor or hydrogen-bonded complexes. Although a number of reports of the detection (mainly indirectly) of a low frequency band attributable to this mode have been made, no completely unambiguous assignment has yet been possible.

The indirect detection of such a vibration is, in theory possible, by a study of the vibrational structure of the electronic (charge-transfer) bands of the complex. Vibrational modes which involve a large electronic perturbation of the molecule are expected to show progressions in the electronic spectrum of the molecule. One might therefore expect the charge-transfer band to show a progression in the intermolecular vibrational mode (see reference 1, chapter 8). Progressions with a band separation of 180–250 cm^{-1} have been observed[280,281] for complexes of arenes with tetrahalo-*p*-benzoquinone in chloroform but, surprisingly, not for other very similar bπ-aπ complexes. As pointed out by Mulliken and Person (reference 1, p113) this vibrational structure corresponds to the separation of vibrational levels in the *excited* electronic state and therefore may not correspond to the frequency expected for

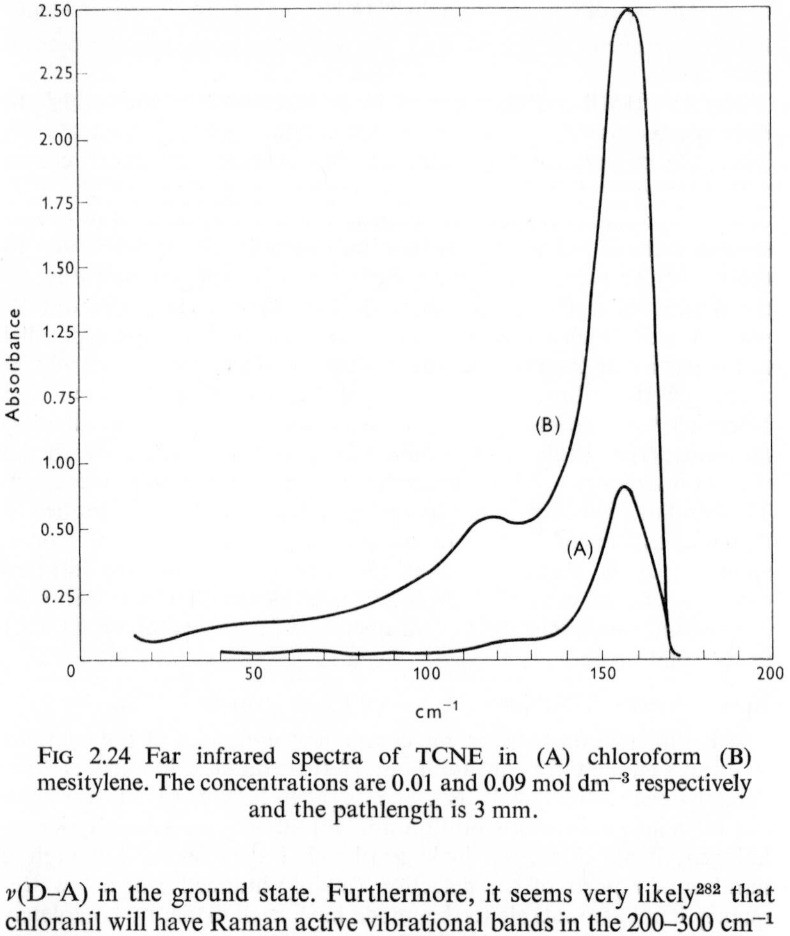

FIG 2.24 Far infrared spectra of TCNE in (A) chloroform (B) mesitylene. The concentrations are 0.01 and 0.09 mol dm^{-3} respectively and the pathlength is 3 mm.

ν(D–A) in the ground state. Furthermore, it seems very likely[282] that chloranil will have Raman active vibrational bands in the 200–300 cm^{-1} region and so the assignment of this vibrational fine structure must be regarded as tentative especially in the absence of similar data for other acceptors.

The only published attempt to detect the ν(D–A) band directly for these complexes is that of Larkindale and Simpkin[283]. They report a band at 115 cm^{-1} observed for a 0.09 mol dm^{-3} solution of TCNE in mesitylene (which, they claim, is not present in a 0.01 mold m^{-3} chloroform solution). This band is assigned to a ν(D–A) mode despite the fact that TCNE has a weak infrared fundamental band at 119 cm^{-1} and a Raman active mode which gives rise to a band at 127 cm^{-1}.[283a]. Since either of these bands could be perturbed on complexation (*i.e.* in mesitylene solution) it is clearly necessary to regard the assignment with some skepticism until the behaviour of this "new" band shown in figure 2.24 has been investigated further.

Since the mean D–A distance for crystalline bπ-aπ complexes[266,267] is about 3.1–3.3 Å, as compared with 3.2–2.8 Å for halogen complexes, (see table 3.15, p. 288) it is perhaps not surprising that a band due to the ν(D–A) vibration is elusive especially since searches have normally taken place in solution (crystal spectra being complicated by lattice vibrational bands) where the forces between the molecules are almost certainly smaller. It is pertinent to ask whether, in the supposed absence of a ν(D–A) band, one should expect a band due to collision-induced dipole fluctuations in the liquid phase (see section 3.4.3). From dielectric studies it is seen (table 7.13, p. 567) that the estimated complex lifetime is of the order of 10^{-9} s for TCNE in mesitylene (as compared with an estimate lifetime of 10^{-12} s for the benzene–I_2 complex). It is therefore clear that collision-induced absorption would not arise in this region although absorption due to the Poley-Hill mechanism (section 3.4.3) may occur (though only weakly since the dipole moment is small and the moment of inertia large). It can be seen from figure 2.24 that there is, so far, little evidence that any absorption is present in the 20–100 cm^{-1} region but more work is needed to establish whether or not this is the case.

2.3.6. THE STUDY OF INTERMOLECULAR INTERACTIONS USING VIBRATIONAL BAND SHAPES

Introduction and Theoretical Background

It has recently been realised that the "shape" or "profile" of an infrared or Raman absorption band contains a great deal of information about the interaction of the absorbing molecules with their environment. Such information, which until recently has been lost, can provide some insight about the molecular relaxation processes which occur, especially in the liquid phase. The overall band profile is usually determined by a combination of vibrational and rotational mechanisms[284–287] each of which is affected by the *intermolecular* forces operative in the system. The effect of such forces can therefore, in principle, be determined by a complete analysis of the band profile.

A convenient, but crude, measure of the band shape is provided by the *overall* half-band width, often quoted, but rarely interpreted in any depth, by vibrational spectroscopists. Definitive interpretation is facilitated if the whole band profile is used to calculate the band moments[23,288] and/or time autocorrelation functions[23,24,289,290] (or relaxation functions) which may be directly related to parameters describing the molecular dynamics and interactions of the system.

The truncated band moments are given by:

$$M^{(n)} = \int_{\text{band}} (\bar{\nu} - \bar{\nu}_0)^n k(\bar{\nu})\, d\bar{\nu} \Big/ \int_{\text{band}} k(\bar{\nu})\, d\bar{\nu} \qquad 2.10$$

where $\bar{\nu}_0$ is a suitably chosen band origin and where $k(\bar{\nu})$ is the absorption coefficient, $\log(T_0/T)/4\pi\bar{\nu}l$, given as a function of wave number which defines the band profile. Jones *et al.*[291] use moments defined by:

$$M_j^{(n)} = \frac{1}{b^n}\int_{-j}^{+j}(\bar{\nu}-\bar{\nu}_0)^n k(\bar{\nu})\,d\bar{\nu}\bigg/\int_{-j}^{+j}k(\bar{\nu})\,d\bar{\nu} \qquad 2.11$$

where the half-band width is 2b and where $j = (\bar{\nu} - \bar{\nu}_0)/b$. The band moments calculated this way are pure numbers and are obtained as a function of j. They thus show how the band accumulates and contributes to the moments as one gets further away from $\bar{\nu}_0$.

For totally symmetric functions (such as Gaussian or Cauchy (*i.e.* Lorentzian) functions) *odd* moments are zero[23] and the third moment $M^{(3)}$ may be conveniently used as a measure of band asymmetry. The first moment, $M^{(1)}$, can be shown[288] to be useful for the study of solvent-shifted band origins. The second moment, $M^{(2)}$, can be shown[288] to depend, in the classical limit, only on the moment of inertia and the temperature. For example, for the parallel band of a linear molecule[288c]:

$$M^{(2)} \approx 2\mathbf{k}T/I \qquad 2.12$$

giving a measure of the average rotational kinetic energy of the molecule. The 4th moment, $M^{(4)}$, is related to the intermolecular torque acting on the molecule. Again for a linear molecule we have[288c]:

$$M^{(4)} \approx [(8\mathbf{k}T)^2 + (OV)^2]/I^2 \qquad 2.13$$

where OV is the mean square torque acting on the rotating molecules and I is the moment of inertia (corresponding values for symmetric top molecule are given in reference 295). Hindering of the rotation, for example by increased intermolecular forces, can be shown to lead to an increase of $M^{(3)}$ and $M^{(4)}$. It should be noted that these relationships are valid only in the first approximation but quantum corrections have been shown[288a] to be small.

It is clear that, in principle, much useful information may be obtained from computed band moments and the theory for linear molecules has been tested on a number of occasions (see below). As yet the theory has not been developed for larger molecules but Wilson *et al.*[55] have shown how the basic principles may be applied to the study of substituted benzenes. It should be noticed, however, that the truncated moments are extremely sensitive to small variations in the experimental absorption data. Very accurate absorbance data are therefore required if meaningful results are to be obtained.

The time autocorrelation function[20,23,24,292] for the transition moment vector of a vibrational transition is defined by:

$$\phi(t) = \langle\vec{u}(0)\cdot\vec{u}(t)\rangle = \int_{-\infty}^{+\infty} k(\omega)\exp[i(\omega-\omega_0)t]\,d\omega \qquad 2.14$$

The integral is of course, calculable only over the region of the absorption band and so some problems may occur due to truncation of the data. It is usual to calculate the *real* part of the normalised autocorrelation function (the imaginary part being very small[20]). Converting angular frequency, ω, to wave number, $\bar{\nu}$, and normalising we have:

$$\phi(t) = \langle \vec{u}(0) \cdot \vec{u}(t) \rangle = \int_{\text{band}} \frac{k(\bar{\nu}) \cos [2\pi c(\bar{\nu} - \bar{\nu}_0)t]\, d\bar{\nu}}{\int_{\text{band}} k(\bar{\nu})\, d\bar{\nu}} \qquad 2.15$$

where $\vec{u}$ is a unit vector pointing along the direction of the transition moment. The direction cosine $\langle \vec{u}(0) \cdot \vec{u}(t) \rangle$ then represents the average change of angle between the dipole transition moment at times $t = 0$ and $t = t$. The value of $\phi(t)$ therefore enables one to "follow" the molecular motion of the molecules during the vibrational transition period (~0–3 ps). This molecular "motion" is principally a result of translation and rotational diffusion of the molecules in the medium but the form of $\phi(t)$ is *also* determined by vibrational relaxation (damping) caused by coupling of the excited vibrational with other degrees of freedom in the system (for example, by collision with the surround molecules by translational diffusion[284–7].) The total dipole autocorrelation function can thus be written[284]:

$$\phi(t) = G^R(t)G^v(t) \qquad 2.16$$

where G^R and G^v are the rotational and vibrational component autocorrelation functions. More specifically Morowitz and Eisenthal[287] show that one can write,

$$\phi(t) = G^R(t) \exp(-\beta_v t). \qquad 2.17$$

where β_v is the vibrational damping constant. It should be noticed that this treatment is based on the assumption that collisions leading to the two processes (rotational diffusion and vibrational relaxation) are independent. The two processes are therefore assumed to be uncorrelated.

When $\phi(t)$ is plotted as a function of time two distinct regions are usually observed. At short times ($<$ ~0.5 ps) the molecules execute essentially "free" rotational motion and the curve is Gaussian in shape. It should be noticed (equation 2.15) that at small t values $(\bar{\nu} - \bar{\nu}_0)$ is large and these values of $\phi(t)$ are derived from experimental information in the wings of the absorption band. For a *diatomic* molecule Bratos *et al.*[284] show that the band profile has the form shown in table 2.30. In this table ω_c is the shifted band origin[284] and the parameters β_v, β_R and α_R are the damping constants for vibrational relaxation, "free" rotation and rotational diffusion respectively. ξ is fraction of molecules with energy less than the intermolecular potential V_R). The motion during this initial time period is principally controlled

TABLE 2.30

Vibrational bandprofiles for a diatomic molecule assuming that one relaxation mechanism is predominant (Reproduced from *J. Chem. Phys.*, **52,** 439 (1970), by permission of the American Institute of Physics)

Relaxation mechanism	Profile $I(\omega - \omega_c) = I(\Delta\omega)$
Translation diffusion	$[(\pi/\beta_v)]^{1/2} \exp[-(\Delta\omega^2/4\beta_v)]$
Free rotation	$(\pi/2\beta_R)\,|\Delta\omega| \exp[-(\Delta\omega^2/4\beta_R)]$
Rotational diffusion	$2\alpha_R/(\alpha_R^2 + \Delta\omega^2)$
Intermediate type reorientation	$(1 - \xi)\dfrac{\pi}{2\beta_R}\exp\left(-\dfrac{\Delta\omega^2}{4\beta_R}\right) + \xi\dfrac{2\alpha_R}{\alpha_R^2 + \Delta\omega^2}$

by the molecular moment of inertia since intermolecular torques only blur the rotational structure. For short times, in the classical limit[287];

$$G^R(t) = 1 - at^2 + bt^4 \qquad 2.18$$

where

$$a = \mathbf{k}T/I$$

and

$$b = \tfrac{1}{3}(\mathbf{k}t/I)^2 + (24I^2)^{-1}\langle(OV)^2\rangle. \qquad 2.19$$

Taking into account vibrational relaxation one has new coefficients of t^2 and t^4 in equation 2.18 which are:

$$a' = a + (\beta_v^2/2)$$
$$b' = b + a(\beta_v^2/2) + (\beta_v^2/4). \qquad 2.20$$

At longer times intermolecular collisions, leading to rotational diffusion, cause the correlation function to decay exponentially since process leads to a band shape (table 2.30) which is Lorentzian (this part of the $\phi(t)$ curve being derived from information near $\bar{\nu} = \bar{\nu}_0$). The log $\phi(t)$ – t plot therefore becomes linear. The slope of this linear part may be related to a sum of damping constants:

$$\log \phi(t) = -(\alpha_R + \beta_v)t. \qquad 2.21$$

Since bands in solids are very narrow it is sometimes assumed (see below) that $\alpha_R \gg \beta_v$ but it should be noticed that this is only true to a first approximation. In general, the infrared band profile is determined by *both* rotational diffusion *and* by vibrational relaxation. For Raman lines[293] it is possible in principle to separate these effects by a study of both the isotropic and anisotropic parts of the scattered light. However, little Raman band shape work on the study of intermolecular interactions has so far been published.

Bratos *et al.*[284,285] have suitably summarised the situation for diatomic molecules for which, as noted above, vibrational relaxation is thought to occur via translational diffusion. The band profiles and corresponding different relaxation processes are given in table 2.30 and shown in figure 2.25. It can be clearly seen that if "free" rotation occurs the band profile resembles that in the gas phase spectrum with somewhat blurred P and R branch lines (case b). The band half-width $\Delta\bar{\nu}_{1/2}$ is of the order of $(4\mathbf{k}T/I)^{1/2}$. If reorientational motion is predominant in determining $\phi(t)$ then the band profile is Lorentzian near the centre and exponential in the wings (case c). The half-band width, $\Delta\bar{\nu}_{1/2}$, which is related to the rotational diffusion coefficient, D, by $\Delta\bar{\nu}_{1/2} = 2D$, is strongly temperature dependent. In the event of the predominant relaxation being a reorientational motion intermediate between "free" rotation and rotational diffusion then the band profile has the appearance of d in (figure 2.25A) with a central maximum and side bands, the relative intensity of which depend on the potential barrier to rotation of the molecules and therefore on the intermolecular potential (V_R). For large V_R the central peak is intense and the side bands weak. Finally, in the case of a predominant contribution from vibrational relaxation (via translational diffusion in this case), the band profile is seen to be essentially Gaussian (case a) and if rotational relaxation is reduced (or eliminated) this type of broadened profile might be expected. It should be noticed that the band half-width in this case is of the order of $\beta_v^{1/2}$ and is strongly dependent on the upper state vibrational quantum number. Since rotational effects are expected to be essentially the same for, say, $1 \leftarrow 0$ and $2 \leftarrow 0$ transitions, this enables one to test for the importance of vibrational relaxation effects. In practice, of course, two or more of these processes may be operative in determining the band profile.

Figure 2.25B shows, in the diatomic case, the types of profile and $\phi(t)$ functions expected under different circumstances when more than one type of relaxation is involved. It will be noticed that profiles similar to case b are often observed in infrared spectra but there is also sometimes evidence for a profile of the type shown in case c (for example, for the bands due to e modes of acetonitrile). However, it is often observed that real infrared band profiles show considerable asymmetry which, again, can be introduced by vibrational or rotational effects. The vibrational asymmetry results from large solute—solvent interaction[305,306] while rotational asymmetry increases with decreasing moment of inertia[284]. Both effects are quantum mechanical in origin[306a] and follow from a non-vanishing imaginary component of the correction function $\phi(t)$ (equation 2.15). Figure 2.25C shows the band profiles for the same combination of relaxation processes as in figure 2.25B but with the inclusion of a parameter n which is related to the number of effective "neighbours" for a given "solute" molecule.

Although the theory on which the diagrams in figure 2.25 depend is

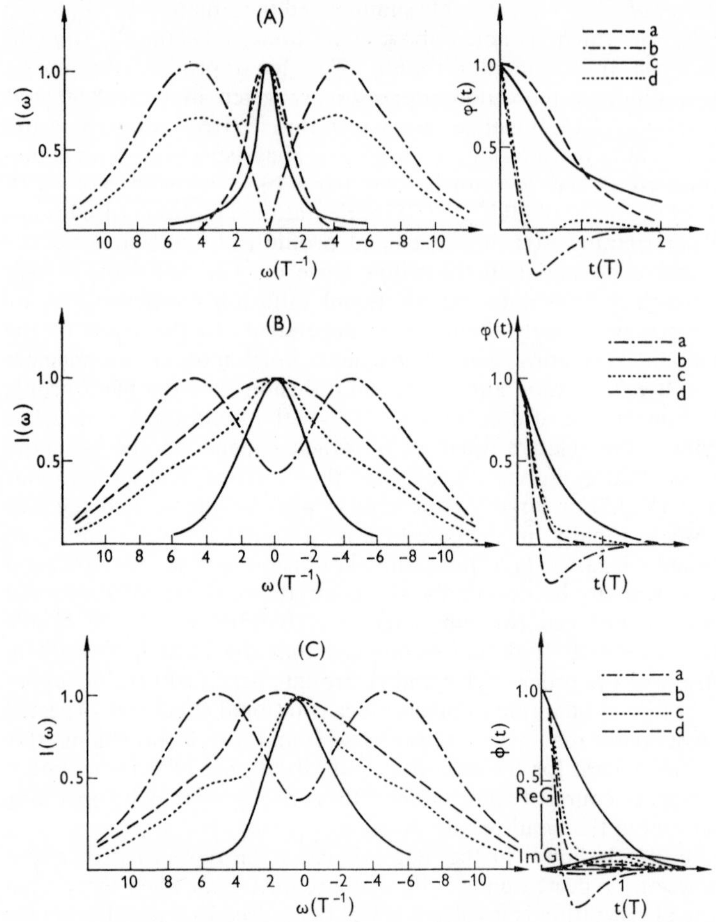

FIG 2.25 Calculated band profiles and dipole autocorrelation functions for a diatomic molecule (A) assuming one relaxational process is dominant (B) allowing for two simultaneous relaxational processes (C) allowing for asymmetry in the band profile due to solute-solvent interaction. For different individual cases see text (Reproduced from *J. Chem. Phys.*, **52**, 439 (1970), by permission of the American Institute of Physics).

only strictly valid for a diatomic molecule, many of the general ideas will also apply to larger molecules. It is clear, however, that *instrumental* band broadening and distortion (due mainly to the use of finite spectrometer slits) must be carefully considered before attempts are made to interpret experimental band profiles.

A set of computer programmes (in FORTAN IV) for the analysis of vibrational band shapes using the techniques described above has been produced at the National Research Council of Canada in Ottawa. These are contained in N.R.C. Bulletins 11, 12 and 13 published by Dr. R. N. Jones and his co-workers.

Results and Discussion

In this section we shall attempt to illustrate how vibrational band profiles may be used to study molecular interactions in the liquid phase using the techniques outlined above.

There have been several attempts to study the intermolecular torques acting on linear and symmetric top molecules using the truncated moment method[293–6]. In general, it is shown [295] that the intermolecular torque (equation 2.13) increases on going from the gas phase to solution (as expected) and that solvents such as CCl_4 and CH_2Cl_2 show considerably greater interaction[296] with the absorbing molecules than do hydrocarbon solvents such as heptane or hexane. Indeed, the intermolecular torques acting on the methyl halides seem to be greater in carbon tetrachloride solution than in the pure liquid[295]. A selection of the results obtained for methyl halides is included in table 2.31. It is interesting that, except where free rotation is expected (*i.e.* in the gas phase) the second moment obtained is generally greater than the calculated value (equation 6, reference 295). In the similar work[296] on N_2O in a series of solvents, on the other hand, the second moment is *lower* than the calculated value when the intermolecular torque is high, and this is attributed solute-solvent interaction (table 5, reference 296). However, since Gordon[288a] has shown that, in the classical limit (expected to be valid in these cases), the $M^{(2)}$ value is independent of intermolecular potential it would appear that other factors are important in determining the band profile. For the methyl halides[295] the $M^{(2)}$ (expt.) $-$ $M^{(2)}$ (calc.) value does increase as the solute mass increases (on going from CH_3Cl to CH_3Br or CH_3I) so that the importance of the "shift fluctuation" contribution[288c] (*i.e.* the non-classical contribution) to $M^{(2)}$ may increase relative to the (classical) dynamic term which is mass dependent—see equation 2.12). However, it is also possible that (in both cases) vibrational relaxation effects are important especially since these effects are said to increase as the mass increases (molecular vibrational frequencies decrease)[284,288d]. It should be noted here, however, that no attempt to determine the experimental error in the determination of $M^{(2)}$ is reported.

There have also been a number of studies[298–302] recently on the temperature dependence of vibrational band widths. The general decrease in overall half-band width with decreasing temperature has been correlated with viscosity measurements[298] and barriers to (Brownian) rotational diffusion in the liquid phase of about 4-12 kJ mol^{-1} have been determined[298]. It was found that this potential barrier (determined using an Arrhenius-type equation) depended only on the

TABLE 2.31

Intermolecular torques acting on the methyl halides[a] and nitrous oxide[b] in different environments (references 295 and 296)

Molecule (solvent)	$M^{(2)}$ (calc.)[c]/ cm^{-2}	$M^{(2)}$ (expt.)/ cm^{-2}	$10^{-2}\langle(OV)^2\rangle^{1/2}$/ cm^{-1} rd^{-1}
$CHCl_3$ (gas)	92	89	~0
$CHCl_3(CS_2)$	92	75	4.37
$CHCl_3$ (liquid)	92	71	3.57
$CHCl_3(CCl_4)$	92	177	10.71
$CHBr_3$ (gas)	35	43	2.62
$CHBr_3$ (liquid)	35	70	12.74
$CHBr_3(CCl_4)$	35	229	29.13
$CHI_3(CS_2)$	17	39	9.19
$CHI_3(CCl_4)$	17	182	34.39
$N_2O(n\text{-}C_6H_{14})$	15175	16666	393.2
$N_2O(CCl_4)$	15175	18777	546.6
$N_2O(CHBr_3)$	15175	17000	591.9

a. torques quoted are those opposing rotation about an axis perpendicular to the C_3 axis (*i.e.* they are derived from information for the ν(CH) vibration which has its transition moment essentially along the C_3 axis).
b. using information on the ν_3 (parallel) band so rotation is observed perpendicular to the molecular axis.
c. calculated using equation 2.13 or the symmetric top equivalent (reference 295).

solvent and *not* on the nature of the solute nor on the band which was chosen for study. Parallel and perpendicular bands of the symmetric top molecule, CH_3SCN, gave the same potential barrier. Since the *energy* barriers to rotational diffusion about these two axes are likely to be different it is clear that there is a considerable difference in the *entropy* terms controlling the two different motions of the molecule.

By choosing different bands of a simple, symmetrical molecule it is often possible to relate the overall band profile to differences in rotational diffusion about different axes[294,301]. It is, however, preferable to compute autocorrelation functions if possible to investigate the broadening effects of other processes such as vibrational relaxation. Although some work along these lines has been carried out for simple molecules[303–306], vibrational effects were not specifically taken into account. Nevertheless, the results, based on the assumption that the Lorentzian part of $\phi(t)$ is caused by rotational diffusion, for the methyl halides, symmetrically substituted benzenes and acetonitrile seem quite reasonable. Thus the diffusion coefficients (table 2.32) show that rotation about the symmetric top (unique) axis is considerably easier than rotation about the non-unique axis and that rotational diffusion is considerably slower for the highly polar acetonitrile than for the

TABLE 2.32
Comparison of rotational diffusion coefficients[a] for acetronitrile and methyl iodide (Reproduced from *Spectroscopy Letters*, **5**, 193 (1972), by permission of Marcel Dekker Inc.)

Molecule	$10^{12}\alpha_R(a_1)/s^{-1}$	$10^{12}\alpha_R(e)/s^{-1}$	$10^{12}D_x/s^{-1}$	$10^{12}D_z/s^{-1}$	$10^{47}I_x$/ kg m^2	$10^{47}I_z$/ kg m^2	Reference
Acetonitrile (liq.)	0.73 ± 0.04	1.66 ± 0.10	1.29 ± 0.10	0.37 ± 0.02	5.3	91.2	304
Acetonitrile (CCl_4 soln)	0.71 ± 0.04	1.96 ± 0.10	1.61 ± 0.10	0.34 ± 0.02	5.3	91.2	304
Acetonitrile-d_3 (liq)			1.2[b]	0.14	10.6	106.5	c
Methyl Iodide (liq)	1.06	2.5	1.97	0.53	5.5	111.5	303

a. data from reference 304 were obtained at ~298 K.
b. values as low as 0.9 × 10^{12} s^{-1} are possible.
c. values from T. T. Bopp, *J. Chem. Phys.*, **47**, 3621. (1967).

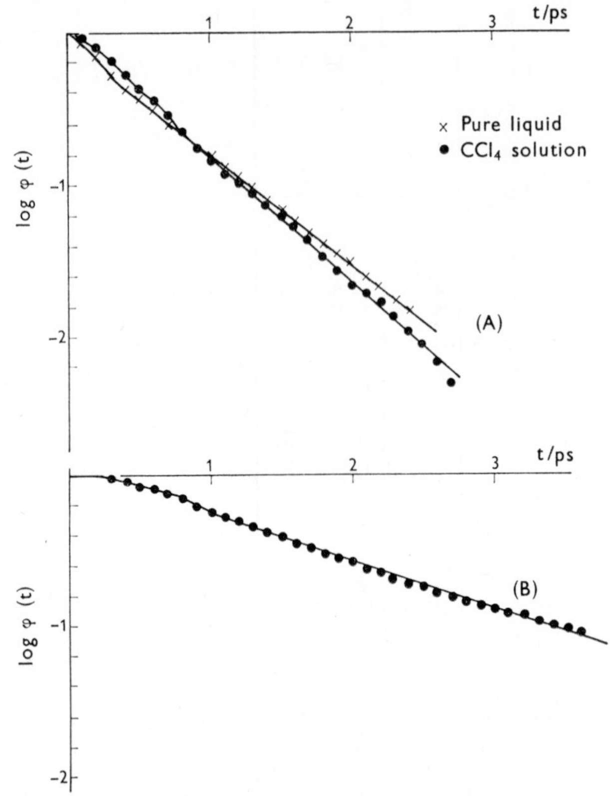

FIG 2.26 Dipole autocorrelation functions for acetonitrile (A) the ν_7(e) band at 1042 cm^{-1} (B) the ν_3(a$_1$) band at 920 cm^{-1}. In case (B) the curves for pure liquid and carbon tetrachloride solution could not be distinguished (Reproduced from *Spectroscopy Letters*, **5**, 193 (1972), by permission of Marcel Dekker Inc.).

methyl halides—presumably due to dipole-dipole intermolecular forces between the molecules. Further, as may be seen from figure 2.26, these forces are only diminished to a small extent when acetonitrile is diluted ten times in carbon tetrachloride solution. It is possible, however, that part of the intermolecular torque in solution could be provided by the solvent in view of the results mentioned above for the methyl halides[295].

A vivid illustration of the effects on band shape which may be caused by intermolecular forces is provided by the pyridine–ICl complex[22]. The bands due to vibrations of the pyridine moiety in the complex show a considerable decrease in overall band width from those of the "free" pyridine molecule in the same solvent[19]. This effect is shown for the in-plane, a$_1$, CCC skeletal stretching band at 991 (or 1012 cm^{-1}) in figure 2.27. The decrease in rotational diffusion coefficient (*i.e.* the slope of the log ϕ(t) -t curve for long t) on going to the complex is greater than the decrease in the overall band width and a change of the relative contributions of Lorentzian and Gaussian components to the total profile is indicated. For example, it appears from the log ϕ(t) — t plots (figure 2.27) that the complex band may be considerably more Gaussian in character than that of the "free" molecule (see reference 23, figure 7). Thus it seems likely that the band profile of the complex may be controlled essentially by vibrational relaxation effects since the rotation of any transition dipole parallel to (or almost parallel to) the C_2 axis is drastically hindered by the attachment of the very heavy ICl molecule. This is presumably what causes the drastic decrease in overall band width. It is interesting that on going to the band at 290 cm^{-1}, essentially due to ICl stretching in the complex, it is found that an *increase* in overall band width is obtained (from 15 cm^{-1} to 21 cm^{-1}, see table 3.6). Figure 2.28 shows a comparison of the autocorrelation curves for the "free" and complexed ν(ICl) bands. Since rotation of this transition moment must be hindered in the same way as those for the pyridine molecule vibrations, it seems likely that vibrational relaxation effects are even more important. However, the data obtained so far are for benzene solutions. Since benzene and ICl are known to form a complex (section 3.4) it might be thought possible that the ν(I-Cl) band is narrowed in benzene by the effects of intermolecular forces. In fact, the ν(I-Cl) half-band width in carbon tetrachloride is only 8 cm^{-1} (table 3.6).

Vibrational relaxation effects are also thought to be relevant to the interpretation of the changes of band width which occur for the ν(I-I) band of the pyridine–I_2 complex when the concentration of excess pyridine is increased[260] (see table 3.9, p. 267). The log ϕ(t) -t plots are shown in figure 2.29 for increasing pyridine concentration. Curve 4 shows the log ϕ(t) function for the complex in benzene solution. Originally it was thought that these changes were due to increased solvation of the complex by the increasingly polar medium (see section 3.4.2) but the result in benzene solution, along with the fact that the band *increases* in width as the amount of pyridine increases (*i.e.* the apparent rotational diffusion coefficient increases rather than decreasing as is expected when intermolecular forces are operative) tend to discredit this explanation. Band broadening could be due to the formation of multimeric species (*i.e.* n:1 complexes, n = 2, 3, . . .) but, in view of the "Gaussian" shapes of these log ϕ(t) — t curves over the 0-2 ps region, it seems likely that vibrational effects are important. It should be noted that vibrational relaxation is also thought[20] to be responsible for broadening of a band at 215 cm^{-1} in the spectrum of C_6F_6 when benzene is added. It thus seems that low frequency vibrational bands are more severely affected, as expected if the effect is proportional to $1/\bar{\nu}_{max}$[284]. However, more detailed work on the analysis of these

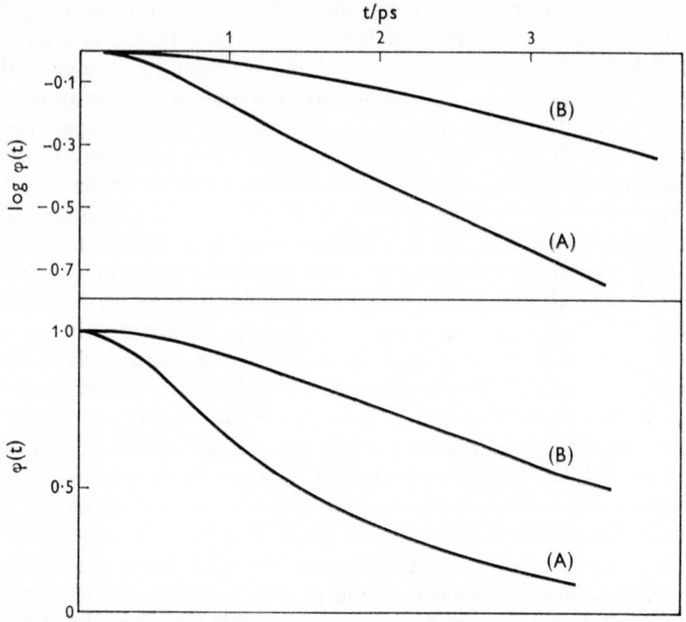

FIG 2.27 The dipole autocorrelation function (and log $\phi(t)$) as a function of time for the ν_9, a_1 band of pyridine; (A) pyridine in chloroform, (B) pyridine–ICl in chloroform.

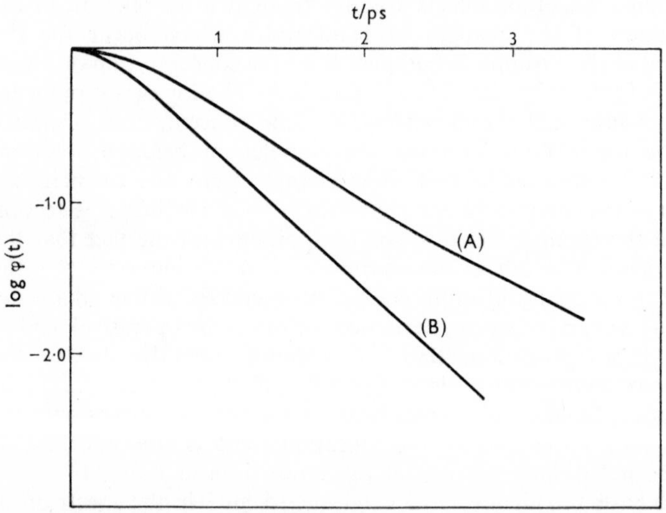

FIG 2.28 Comparison of log $\phi(t)$-t graphs for the ν(I-Cl) band; (A) for ICl in benzene, $\bar{\nu} = 355\ cm^{-1}$. (B) for pyridine–ICl in benzene, $\bar{\nu} = 292\ cm^{-1}$ (room temperature).

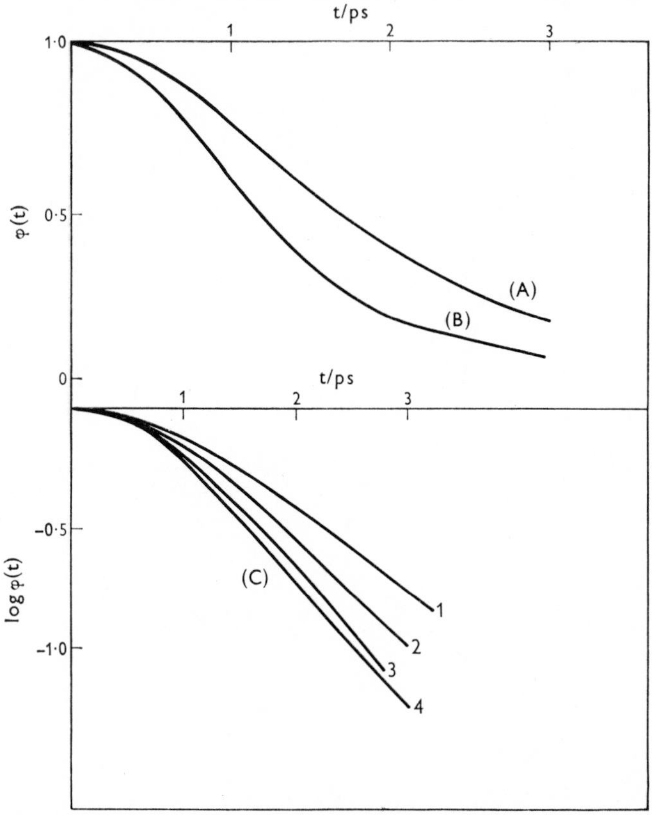

FIG 2.29 Autocorrelation functions for the ν(I–I) band the pyridine–I_2 complex in cyclohexane; (A) low concentration (0.17 mol dm^{-3}) of pyridine, $\bar{\nu}$ = 183 cm^{-1}; (B) high concentration (1.2 mol dm^{-3}) of pyridine, $\bar{\nu}$ = 174 cm^{-1}; (C) the effect of a variation of pyridine concentration (1) 0.17 mol dm^{-3} (2) 0.62 mol dm^{-3} (3) 1.21 mol dm^{-3} (4) 0.24 mol dm^{-3} (solution 4 is in benzene solution). The iodine concentration is constant at 0.02 mol dm^{-3} (room temperature) (Reproduced, by permission, from *Advances in Mol. Relaxation Processes*, **6,** 1 (1973)

band profiles is required before we shall be able to present a more complete interpretation.

A detailed study of the weak interaction between benzene and chloroform has been made using the time autocorrelation approach by Rothschild[9,307,308]. It is shown, from a study of the parallel (ν_3) band of $CDCl_3$ and the perpendicular (ν_4) band of $CHCl_3$ in benzene solution, that the rotational diffusion process (obtained by assuming that broadening due to vibrational relaxation is negligible) is largely

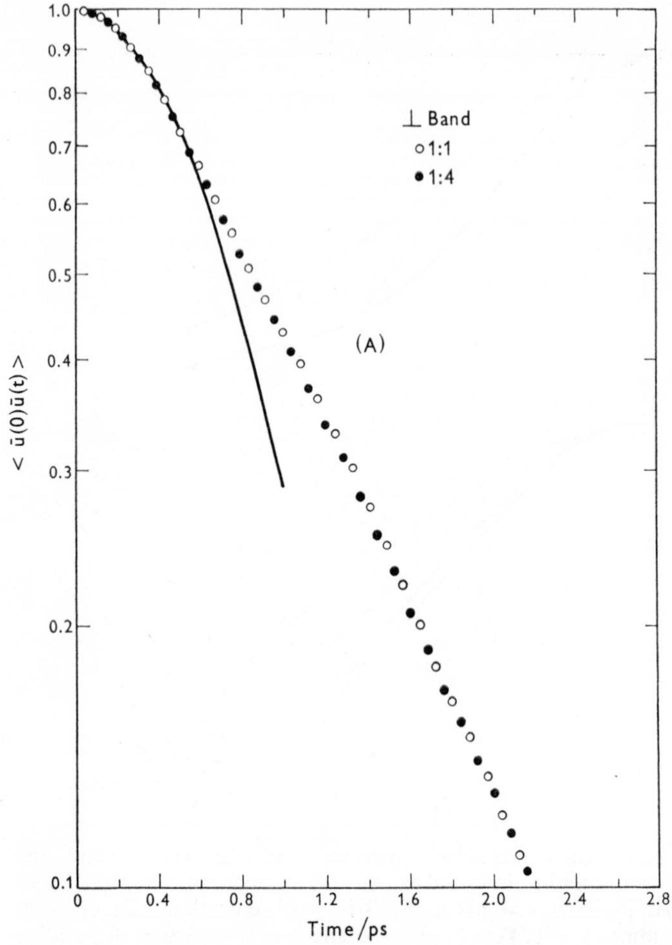

FIG 2.30A The dipole autocorrelation functions for the ν_4, perpendicular band of $CHCl_3$ in benzene solution (rotation about the C_3 axis and the other perpendicular axis). Open points; equimolar mixture, solid points; 1:4 molar $CHCl_3$:C_6H_6 mixture. The solid line represents "freely" rotating $CHCl_3$ molecules (Reproduced from *J. Chem. Phys.*, **55,** 1405 (1971), by permission of the American Institute of Physics).

independent of the chloroform concentration (see figure 2.30A) and that the intermolecular torques (derived from the band moment analysis) opposing rotational reorientation are little different from those obtained[12] for liquid chloroform and for chloroform in carbon disulphide—see table 2.33. In addition, a study of the translational motion of benzene in chloroform using a spin-echo technique shows

TABLE 2.33
Intermolecular torques for chloroform in various solvents

Solvent	Intermolecular torque/cm^{-1} rad^{-1}	Reference
pure liquid	357	295
carbon disulphide	437	295
benzene	470, 290[a]	307
carbon tetrachloride	1071	295

a. for rotation about C_3 axis, other data for rotation perpendicular to the C_3 axis.

that the translational trapping time[307] (the time between translational "jumps") is also the same as for neat liquid benzene. These results indicate that weak intermolecular forces between the benzene and chloroform molecules have little effect on the rotational and translational motions of the individual components. This conclusion is supported by the calculation of the autocorrelation function (for the ν_3 band) corresponding to "free" rotation of the "rigid" $CDCl_3$–C_6H_6 "complex" in the presence of uncomplexed chloroform. This function is a weighted sum of two Gaussian functions. For a 1:4 mixture of $CDCl_3$:C_6H_6, using the published association constant, approximately 50% of the chloroform molecules are complexed and the function for "free" rotation about the C_3 axis of chloroform is:

$$\phi(t)(\text{“free,” } C_3\,\text{axis}) = 0.53\exp[-(2\pi c/\mathbf{h}(B_x + B_y)\mathbf{k}Tt^2] + 0.47\exp[-2\pi c/\mathbf{h}(B'_x + B'_y)\mathbf{k}Tt^2 \quad 2.22$$

where B_x, B_y, B'_x and B'_y are the rotational constants of the $CHCl_3$ and $CDCl_3$–C_6H_6 molecules (respectively) about axes perpendicular to the C_3 axis of the chloroform (along which the $CDCl_3$–C_6H_6 "bond" is presumed to lie). This function is shown in figure 2.30B and it may be seen that it lies *above* the experimental $\phi(t) - t$ curve. Since this calculated function represents the fastest (*i.e.* "free") rotation of the chloroform-containing species in solution, and since any intermolecular forces would be expected to give rise to *hindering* of the decay of $\phi(t)$, it follows that the postulated "rigid" $CDCl_3$–C_6H_6 complex does not exist. It is concluded that the weak forces between the two molecules are largely independent of the relative orientations of independent components.

From these examples it may be clearly seen that the study of band profiles (and their temperature and concentration dependence) may be extremely useful for the elucidation of problems associated with the nature and strengths of molecular interactions. It is expected that these

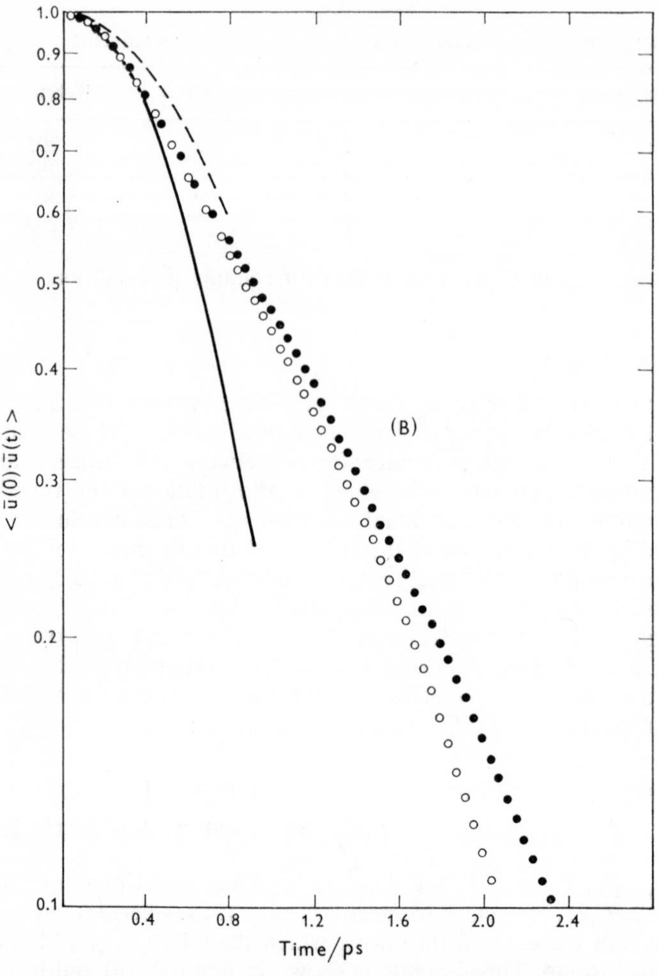

FIG 2.30B the ν_3, parallel, band of $CDCl_3$ in benzene (rotation about axes perpendicular to the C_3 axis). Solid and open points, and solid line represent the same as in A. The broken line represents "free" rotation of $CDCl_3$ in an equilibrium mixture of uncomplexed $CDCl_3$ and rigidly complexed C_6H_6–$CDCl_3$ (Reproduced, by permission, from *Chem. Phys. Lett.*, **9**, 149 (1971)).

techniques will make an important contribution to future research in this field.

Future Work

It is clear that many more band shape studies, especially on small symmetrical molecules, are needed before one will be able to unambiguously separate the effects of the various relaxational contributions (especially those of rotational diffusion and vibrational relaxation) for larger molecules of more direct interest to the chemist. The current upsurge of interest in Raman spectroscopy ought to, at least, partly satisfy this desire. Indeed, several very interesting accounts of the effect of intermolecular forces on Raman band shapes have been published[309–311] and it is anticipated that more will follow.

To summarise, it seems that there are several, so far relatively unexplored, areas where the study of band shapes may be useful. One is in the study of "specific" solvent effects on the vibrational spectra of dissolved molecules. Polar solvents are known, for example, to cause considerable broadening[193] of the carbonyl stretching bands of metal carbonyls, and it has been claimed[193] that the Gaussian component is enhanced. Such broadening in solvents of high dielectric constant is said[313] to be due to a fluctuation of the frequency shift. If there is indeed a significant change in band shape (as opposed to a simple broadening effect) then the techniques described here ought to detect the change and help to substantiate its origin. Band shape techniques may also be useful for the study of hydrogen-bonded complexes. Part, but not all, of the extreme width of the ν_s(A-H) band can be explained by dissociative relaxation (see chapter 4, p. 323). It is felt that detailed band-shape studies may well help to disentangle the various dynamic phenomena which contribute to the profiles of these very interesting bands. Examples of the possibilities in this direction for the water molecule have already been published[314,315]. Finally, it would be extremely useful to investigate further, by these methods, the changes of band shape which occur when relatively strong intermolecular forces are known to occur between component molecules in a mixture (*i.e.* when a "well-defined" complex is formed). The overall band broadening which usually occurs (see above and in section 3.4) is an extremely interesting phenomenon which may be significantly similar to the "solvent" effect mentioned above. A detailed study of any differences between predominant dynamic processes for different situations could lead to valuable information about the nature and (or) extent of the intermolecular forces involved.

ACKNOWLEDGEMENT

It is a pleasure to acknowledge valuable discussions with Prof. M. Tanres, Prof. W. B. Person, Prof. S. Bratos, Prof. M. M. Davies, Dr. D. Steele, Dr. G. Kohnstam, Dr. P. Young and Dr. A. H. Price during the preparation of this chapter.

REFERENCES

1. R. S. Mulliken and W. B. Person, Molecular Complexes—A Lecture and Reprint Volume. (J. Wiley and Sons, New York, 1969)

2. R. Foster, Organic Charge-Transfer Complexes, (Academic Press, London, 1969)
3. R. S. Mulliken and W. B. Person, (Eds., H. Eyring, D. Henderson and W. Jost), Physical Chemistry, Vol. 3. (Academic Press, New York, 1969), chapter 10
4. J. A. Pople, W. G. Schneider and H. J. Bernstein, High Resolution Nuclear Magnetic Resonance, (McGraw-Hill, New York and London, 1959), chapter 10
5. D. W. Larsen and A. L. Allred, *J. Amer. Chem. Soc.*, **87,** 1216, 1219 (1965)
6. R. A. Craig and R. E. Richards, *Trans. Faraday Soc.*, **59,** 1962 (1963)
7. R. A. Crump and A. H. Price, *Trans. Faraday Soc.*, **66,** 92 (1970)
8. A. H. Price and V. L. Brownsell, Molecular Relaxation Process, *Chem. Soc. Spec. Pub.*, No 20, 1966, 83
9. R. W. Taft and J. W. Carten, *J. Amer. Chem. Soc.*, **86,** 4199 (1964)
10. A Fratiello, R. Schuster and D. P. Miller, *Mol. Phys.*, **11,** 597 (1966)
11. A. Fratiello and R. Schuster, *J. Phys. Chem.*, **71,** 1948 (1967)
12. J. H. Swinehart and H. Taube, *J. Chem. Phys.*, **37,** 1579 (1962)
13. S. Thomas and W. L. Reynolds, *J. Chem. Phys.*, **44,** 3148 (1966)
14. J. Yarwood, (a) *J. Chem. Soc.*, D, 809 (1967); (b) *Spectrochim. Acta*, **26A,** 2099 (1970)
15. A. Fratiello, *J. Chem. Phys.*, **41,** 2204 (1964)
16. P. Klaeboe, *J. Amer. Chem. Soc.*, **89,** 3667 (1967)
17. H. Rosen, Y. R. Shen and F. Stenman, *Mol. Phys.*, **22,** 33 (1971)
18. T.-C. Jao and W. B. Person, University of Florida, unpublished data.
19. J. Yarwood, *Trans. Faraday Soc.*, **65,** 934 (1969)
20. D. Steele, Theory of Vibrational Spectroscopy, (W. B. Saunders, London, 1971), chapter 11
21. D. H. Whiffen, *Trans. Faraday Soc.*, **49,** 878 (1953)
22. J. Yarwood, University of Durham, unpublished data
23. P. Young and R. N. Jones, *Chem. Rev.*, **71,** 219 (1971)
24. R. G. Gordon, *J. Chem. Phys.*, **43,** 1307 (1965); H. Shimizu, *J. Chem. Phys.*, **43,** 2453 (1965)
25. O. Hassel and Chr. Rømming, *Quart. Rev.*, **16,** 1 (1962)
26. H. O. Hooper, *J. Chem. Phys.*, **41,** 599 (1964)
27. C. D. Cornwell and R. S. Yamasaki, *J. Chem. Phys.*, **27,** 1060 (1957)
28. D. C. Douglass, *J. Chem. Phys.*, **32,** 1882 (1960)
29. G. A. Bowmaker and S. Hacobian, (a) *Aust. J. Chem.*, **21,** 551 (1968); (b) *ibid.*, **22,** 2047 (1969); Private communication from Dr. Bowmaker
30. H. Creswell-Fleming and M. W. Hanna, *J. Amer. Chem. Soc.*, **93,** 5030 (1971)
31. C. T. O'Konski and D. F. R. Gilson, *J. Chem. Phys.*, **48,** 2767 (1968)
32. P. Cornil, M. Read, J. Duchesne and R. Cahay, *Bull. Classe. Sci. Acad. Roy. Belg.*, **50,** 235 (1964)
33. C. H. Townes and B. P. Dailey, *J. Chem. Phys.*, **17,** 782 (1949); *ibid*, **20,** 35 (1952); *ibid*, **23,** 118 (1955)
34. M. W. Hanna, *J. Amer. Chem. Soc.*, **90,** 285 (1968)
35. M. W. Hanna and D. E. Williams, *J. Amer. Chem. Soc.*, **90,** 5358 (1968)
36. J. L. Lippert, M. W. Hanna and P. J. Trotter, *J. Amer. Chem. Soc.*, **91,** 4035 (1969)
37. R. S. Mulliken and W. B. Person, *J. Amer. Chem. Soc.*, **91,** 3409 (1969)
38. G. D. Johnson and R. E. Bowen, *J. Amer. Chem. Soc.*, **87,** 1655 (1965)
39. R. Foster and D. R. Twiselton, *Rec. Trav. Chim. Pay-Bas*, **89,** 325, 1020 (1970)
40. H. E. Hallam (Ed., M. M. Davies), Infrared Spectroscopy and Molecular Structure, (Elsevier Ltd., Amsterdam and London, 1963), chapter 12
41. A. D. Buckingham, (a) *Proc. Roy. Soc.*, **A248,** 169 (1958); (b) *Trans. Faraday Soc.*, **56,** 753 (1960)
42. S. R. Polo and M. Kent Wilson, *J. Chem. Phys.*, **23,** 2376 (1955)
43. R. R. Wiederkehr and H. G. Drickamer, *J. Chem. Phys.*, **28,** 311 (1958)

44. H. W. Thompson, D. L. Glusker and R. S. Mulliken, *J. Chem. Phys.*, **21,** 1407 (1953)
45. A. E. Martin (Ed., M. M. Davies), Infrared Spectroscopy and Molecular Structure (Elsevier Ltd., Amsterdam and London, 1963), chapter 2
46. W. J. Potts, Chemical Infrared Spectroscopy, Vol. 1, (J. Wiley and Sons, London, 1963), chapter 6
47. G. W. Chantry, Submillimetre Spectroscopy, (Academic Press, London and New York, 1971), chapter 4
48. D. A. Ramsay, *J. Amer. Chem. Soc.*, **74,** 72 (1952)
49. J. Overend (Ed. M. M. Davies) Infrared Spectroscopy and Molecular Structure, (Elsevier Ltd., Amsterdam and London, 1963), chapter 10
50. E. Bright Wilson and A. J. Wells, *J. Chem. Phys.*, **14,** 578 (1946)
51. J. Yarwood, Ph.D. Thesis, University of Wales, 1964
52. G. Brownson and J. Yarwood, University of Durham, unpublished data
53. S. Maeda and P. N. Schatz, *J. Chem. Phys.*, **35,** 1617 (1961)
54. J. E. Stewart, *Appl. Opt.*, **1,** 75 (1962)
55. M. K. Wilson and J. T. Shimozawa, *Spectrochim. Acta*, **22,** 1591 (1966)
56. G. J. Boobyer, *Spectrochim. Acta*, **23A,** 335 (1967)
57. R. P. Young and R. N. Jones, *Spectrochim. Acta* (in press)
58. T. Fujiyama, J. Herrin and B. L. Crawford, *Appl. Spectrosc.*, (in press)
59. O. S. Heavens, Optical Properties of Thin Solid Films, (Dover Pub. Inc., New York, 1965)
60. A. Bandy, S. Rudys and W. B. Person, Technical manual, University of Florida, 1968
61. S. Maeda, G. Thyagarajan and P. N. Schatz, *J. Chem. Phys.*, **39,** 3474 (1963)
62. H. Yamada and W. B. Person, *J. Chem. Phys.*, **40,** 309 (1964)
63. W. B. Person, C. F. Cook and H. B. Friedrich, *J. Chem. Phys.*, **46,** 2521 (1967)
64. W. B. Person and K. Szczepaniak, *Spectrochim. Acta*, **28A,** 15 (1972)
65. W. B. Person and J. N. Gales, unpublished data on pyridine–Br_2 and pyridine–I_2 solid complexes
66. E. I. du Pont de Nemours and Co. Inc., Model 310. (Thanks are due to Dr. D. T. Clark for the use of this equipment)
67. H. Yamada and K. Kozima, *J. Amer. Chem. Soc.*, **82,** 1543 (1960)
68. G. Roland, *Spectrochim. Acta*, **25A,** 1135 (1969)
69. G. Michel and G. Duyckaerts, *Spectrochim. Acta*, **21,** 279 (1965)
70. G. Leclere and G. Duyckaerts, *Spectrochim. Acta*, **22,** 403 (1966)
71. H. A. Gebbie and G. A. Vanasse, *Nature*, **178,** 432 (1956)
72. A. Finch, P. N. Gates, K. Radcliffe, F. N. Dickson and F. F. Bentley, Chemical Applications of Far-infrared Spectroscopy, (Academic Press, London and New York, 1970)
73. W. J. Hurley, *J. Chem. Ed.*, **43,** 236 (1966)
74. P. Jacquinot, *Appl. Opt.*, **8,** 497 (1969)
75. J. Yarwood, Absolute Intensity Measurements in the region 10-600 cm^{-1} (Beckman-RIIC. Ltd., London, 1971)
76. M. L. Forman, W. H. Steel and G. A. Vanasse, *J. Opt. Soc. Amer.*, **56,** 59 (1966)
77. D. Neale and L. Thorpe, Beckman-RIIC Ltd., (private communication)
78. M. L. Forman, *J. Opt. Soc. America*, **56,** 978 (1966); J. W. Cooley and J. W. Tukey, *Math. Computation*, **19,** 296 (1965)
79. S. G. W. Ginn and J. L. Wood, *Trans. Faraday Soc.*, **62,** 777 (1966)
80. S. G. W. Ginn, I. Haque and J. L. Wood (a) *Spectrochim. Acta*, **23A,** 959 (1967); (b) *Spectrochim. Acta*, **24A,** 1531 (1968)
81. R. F. Lake and H. W. Thompson, *Proc. Roy. Soc.*, **A297,** 440 (1967)
82. R. F. Lake and H. W. Thompson, *Spectrochim. Acta*, **24A,** 1321 (1968)
83. J. N. Gayles, *J. Chem. Phys.*, **49,** 1840 (1968)
84. M-T. Forel, M. Fouassier and M. Tranquille, *Spectrochim. Acta*, **26A,** 1761 (1970)

85. M-T. Forel, M. Tranquille and M. Fouassier, *Spectrochim. Acta*, **26A,** 1777 (1970)
86. J. Yarwood and W. B. Person, *J. Amer. Chem. Soc.*, **90,** 594 (1968)
87. J. Yarwood and W. B. Person, *J. Amer. Chem. Soc.*, **90,** 3930 (1968)
88. Y. Yagi, A. I. Popov and W. B. Person, *J. Phys. Chem.*, **71,** 2439 (1967)
89. E. E. Ferguson, *J. Chim. Phys.*, **61,** 257 (1964)
90. H. B. Friedrich and W. B. Person, *J. Chem. Phys.*, **44,** 2161 (1966)
91. W. B. Person, R. E. Erickson and R. E. Buckles, *J. Amer. Chem. Soc.*, **82,** 29 (1960)
92. B. B. Bhowmik, *Spectrochim. Acta*, **27A,** 321 (1971)
93. J. E. Prue, Chemical Society Aniversary meeting, Exeter, April 1967, Paper C14; *J. Chem. Soc.*, 7534 (1965)
94. E. E. Ferguson, *J. Chem. Phys.*, **25,** 577 (1956); *J. Chem. Phys.*, **26,** 1357 (1957); *Spectrochim. Acta*, **10,** 123 (1958)
95. E. E. Ferguson and I. Y. Chang, *J. Chem. Phys.*, **34,** 628 (1961)
96. R. M. Keefer and L. J. Andrews, *J. Amer. Chem. Soc.*, **77,** 2164 (1955)
97. S. K. Garg, H. Kilp and C. P. Smyth, *J. Chem. Phys.*, **43,** 2341 (1965)
98. J. P. Kettle and A. H. Price, *J. Chem. Soc.*, *Faraday Trans.*, *II*, **68,** 1306 (1972)
99. A. G. DeRocco, T. H. Spurling and T. S. Storvick, *J. Chem. Phys.*, **46,** 599 (1967)
100. A. D. Buckingham, R. L. Disch and D. A. Dunmur, *J. Amer. Chem. Soc.*, **90,** 3104 (1968)
101. G. W. Brownson and J. Yarwood, *Spectrosc. Lett.*, **5,** 185 (1972)
102. S. Matsuoko, A. Mori, S. Hattori, *Sci. Rep. Kanazaw University*, **8,** 45 (1962)
103. S. Matsuoko and S. Hattori, *J. Phys. Soc.*, *Jap.*, **17,** 1073 (1962)
104. W. B. Person, R. E. Humphrey, W. A. Deskin and A. I. Popov, *J. Amer. Chem. Soc.*, **80,** 2049 (1958)
105. E. Augdahl and P. Klaeboe, *Chem. Acta Scand.*, **19,** 807 (1965)
106. J. Collin, L. D'Or and R. Alewaeters, *J. Chem. Phys.*, **23,** 397 (1955); *Rec. Trav. Chim. Pays-Bas*, **75,** 862 (1956)
107. G. W. Brownson and J. Yarwood, *J. Mol. Struct.*, **10,** 147 (1971)
108. G. K. Vemulapalli, *J. Amer. Chem. Soc.*, **92,** 7589 (1970)
109. H. Yada, J. Tanaka and S. Nagakura, *J. Mol. Spectrosc.*, **9,** 461 (1962)
110. W. B. Person, G. R. Anderson, J. N. Fordemwalt, H. Stammreich and R. Forneris, *J. Chem. Phys.*, **35,** 908 (1961)
111. A. G. Maki and R. Forneris, *Spectrochim. Acta*, **23A,** 867 (1967)
112. I. M. Mills (Ed., M. M. Davies), Infrared Spectroscopy and Molecular Structure, (Elsevier Ltd., Amsterdam and London, 1963), chapter 5
113. I. M. Mills, *Spectrochim. Acta*, **19,** 1585 (1963)
114. C. A. Coulson, J. Duchesne and C. Manneback, Victor Henri Memorial Volume (Maison Desoer, 1949), p. 264
115. J. Ph. Poley, *J. Appl. Sci.*, **4B,** 337 (1955)
116. N. E. Hill, *Proc. Phys. Soc.*, **82,** 723 (1963); *Chem. Phys. Lett.*, **2,** 5 (1968)
117. R. G. Gordon, *J. Chem. Phys.*, **38,** 1724 (1963)
118. Thanks are due to Dr. H. G. M. Edwards of the University of Bradford who kindly obtained the Raman spectra
119. M. M. Davies, Some Electrical and Optical Aspects of Molecular Behaviour, (Pergamon Press, London, 1965)
120. R. A. Walton, *Quart. Rev.*, **19,** 126 (1965)
121. Specialist reports on the spectroscopic properties of organometallic compounds (The Chemical Society, 1968–1972), Vols 1–4
122. C. C. Addison, D. W. Amos and D. Sutton, *J. Chem. Soc.*, **A,** 2285 (1968)
123. J. C. Evans and G. Y-S. Lo, *Spectrochim. Acta*, **21,** 1033 (1965)
124. B. J. Hathaway, D. G. Holah and A. E. Underhill, *J. Chem. Soc.*, 2444 (1962)
125. K. F. Purcell and R. S. Drago, *J. Am. Chem. Soc.*, **88,** 919 (1966)
126. M. Webster and H. E. Blayden, *J. Chem. Soc.*, **A,** 2443 (1969)

127. J. Reedijk and W. L. Groeneveld, *Rec. Trav. Chim. Pays-Bas.*, **86,** 1103 (1967)
128. J. C. Evans, *J. Chem. Soc.*, **A,** 1849 (1969)
129. E. Augdahl and P. Klaeboe, *Spectrochim. Acta*, **19,** 1665 (1963)
130. (a) H. W. Thompson, *Trans. Faraday Soc.*, **48,** 502 (1952); (b) G. A. Crowder and B. R. Cook, *J. Phys. Chem.*, **71,** 914 (1967)
131. A. Terenin, B. Filimonov and D. Bystrow, *Z. Electrochem.*, **62,** 180 (1958)
132. J. L. Hoard, S. Geller and T. B. Owen, *Acta Crystallogr.*, **4,** 405 (1951)
133. D. M. Barnhart, C. N. Caughlan and M. Ul-Haque, *Inorg. Chem.*, **7,** 1135 (1968)
134. T. L. Brown and M. Kubota, *J. Amer. Chem. Soc.*, **83,** 331, 4175 (1961)
135. H. J. Coerver and C. Curran, *J. Amer. Chem. Soc.*, **80,** 3522 (1958)
136. W. Gerrard, M. F. Lappert, H. Pyszora and J. W. Wallis, *J. Chem. Soc.*, 2182 (1960)
137. M. F. Farona and G. R. Tompkin, *Spectrochim. Acta*, **24A,** 788 (1968)
138. K. Kawai and I. Kanesaka, *Spectrochim. Acta*, **25A,** 1265 (1969)
139. K. Kawai and I. Kanesaka, *Spectrochim. Acta*, **25A,** 263 (1969)
140. G. W. A. Fowles and K. F. Gadd, *J. Chem. Soc.*, **A,** 2232 (1970)
141. A. Misono, Y. Uchida, M. Hidai and T. Kuse, *J. Chem. Soc.*, *D*, 208 (1969)
142. (a) R. E. Clarke and P. C. Ford, *Inorg. Chem.*, **9,** 227, 495 (1970); (b) R. E. Clarke, P. C. Ford and R. D. Foust, *Inorg. Chem.*, **9,** 1933 (1970)
143. S. C. Jain and R. Rivest, *Inorg. Chem.*, **6,** 467 (1967); S. C. Jain and R. Rivest, *Inorg. Chim. Acta*, **3,** 249 (1969)
144. (a) M. F. Farona and N. J. Bremer, *J. Amer. Chem. Soc.*, **88,** 3735 (1966); (b) M. F. Farona and K. F. Kraus, *Inorg. Chem.*, **9,** 1700 (1970)
145. I. R. Beattie and L. Rule, *J. Chem. Soc.*, 2995 (1965)
146. I. R. Beattie, T. R. Gilson and G. A. Ozin, *J. Chem. Soc.*, **A,** 2772 (1968)
147. M. F. Farona and J. G. Grasselli, *Inorg. Chem.*, **6,** 1675 (1967)
148. R. A. Walton, *Can. J. Chem.*, **46,** 2347 (1968)
149. (a) B. Swanson and D. F. Shriver, *Inorg. Chem.*, **9,** 1406 (1970); (b) B. Swanson and D. F. Shriver, *Inorg. Chem.*, **10,** 1354 (1971)
150. M. F. Farona, J. G. Grasselli and B. L. Ross, *Spectrochim. Acta*, **23A,** 1875 (1967)
151. G. A. Ozin and R. A. Walton, *J. Chem. Soc.*, **A,** 2236 (1970)
152. V. Devarajan and S. J. Cyvin, *Z. Naturforsch.*, **26,** 1346 (1971)
153. J. N. Murrell, *J. Chem. Soc.*, **A,** 297 (1969)
154. J. Aldous and I. M. Mills, *Spectrochim. Acta*, **18,** 1073 (1962)
155. I. W. Levin and S. Abramowitz, *J. Chem. Phys.*, **43,** 4213 (1965)
156. M. Taillandier and E. Taillandier, *Spectrochim. Acta*, **25A,** 1807 (1969)
157. F. Watari and S. Kinumaki, Sci. Reports Research Institutes, Tohuku University, **14,** 129 (1962); *ibid.*, **16,** 285 (1964)
158. N. S. Gill, R. H. Nuttall, D. E. Scaife and D. W. A. Sharp, *J. Inorg. Nucl. Chem.*, **18,** 79 (1961)
159. M. Goldstein, E. F. Mooney, A. Anderson and H. A. Gebbie, *Spectrochim. Acta*, **21,** 105 (1965)
160. J. R. Durig, B. R. Mitchell, D. W. Sink, J. N. Willis and A. S. Wilson, *Spectrochim. Acta*, **23A,** 1121 (1967)
161. F. E. Dickson, R. A. Hayden and W. G. Fateley, *Spectrochim. Acta*, **25A,** 1875 (1969)
162. G. S. Rao, *Z. Anorg. Allg. Chem.*, **304,** 176 (1960)
163. J. D. Donaldson, D. G. Nicholson and B. J. Senior, *J. Chem. Soc.*, **A,** 2928 (1968)
164. L. Corrsin, B. J. Fax and R. C. Lord, *J. Chem. Phys.*, **21,** 1170 (1953)
165. J. K. Wilmshurst and H. J. Bernstein, *Can. J. Chem.*, **35,** 1183 (1957)
166. J. H. S. Green, W. Kynaston and H. M. Paisley, *Spectrochim. Acta*, **19,** 549 (1963)

167. N. N. Greenwood and K. Wade, *J. Chem. Soc.*, 1130 (1960)
167a. R. Killean, quoted by Beattie *et al.*, reference 146
168. I. R. Beattie, M. Webster and G. W. Chantry, *J. Chem. Soc.*, 6172 (1964)
169. C. W. Frank and L. B. Rogers, *Inorg. Chem.*, **5**, 615 (1966)
170. J. Burgess, *Spectrochim. Acta*, **24A**, 277, 1645 (1968)
171. N. S. Gill and H. J. Kingdon, *Aust. J. Chem.*, **19**, 2197 (1966)
172. N. Ohkaku and K. Nakamoto, *Inorganic Chem.*, **10**, 798 (1971) and references therein
173. C. Perchard and A. Novak, *Spectrochim. Acta*, **26A**, 871 (1970)
174. D. G. Brewer, P. T. T. Wong and M. C. Sears, *Can. J. Chem.*, **46**, 131, 139, 3137 (1968); *Can. J. Chem.*, **47**, 4589 (1969)
175. J. Hiraishi, I. Nakagawa and T. Shimanouchi, *Spectrochim. Acta*, **24A**, 819 (1968)
176. J. Bradbury, K. P. Forest, R. H. Nuttall and D. W. A. Sharp, *Spectrochim. Acta*, **23A**, 2701 (1967)
177. W. Sawodny and J. Goubeau, *Z. Physik Chem.*, **44**, 227 (1965)
178. A. G. Desai and M. B. Kabadi, *J. Inorg. Nucl. Chem.*, **28**, 1279 (1966)
179. L. E. Orgel, Introduction to Transition Metal Chemistry, (Methuen, London, 1960)
180. H. Irving and R. J. P. Williams, *Nature*, **162**, 746 (1948)
181. R. C. Taylor, H. S. Gabelnick, K. Aida and R. L. Amster, *Inorg. Chem.*, **8**, 605 (1969)
182. I. R. Beattie and G. A. Ozin, *J. Chem. Soc.*, **A**, 370 (1970)
183. I. R. Beattie and G. A. Ozin, *J. Chem. Soc.*, **A**, 2373 (1968)
184. M.-T. Forel, M. Fouassier and M. Tranquille, *Spectrochim. Acta*, **26A**, 1761 (1970)
185. M. Taillandier, J. Liquier and E. Taillandier, *J. Mol. Struct.*, **2**, 437 (1968)
186. D. E. H. Jones and J. L. Wood, *J. Chem. Soc.*, **A**, 1448 (1966)
187. R. L. Amster and R. C. Taylor, *Spectrochim. Acta*, **20**, 1487 (1964)
188. S. J. Cyvin, B. N. Cyvin, J. Brunvoll and L. Schäfer, *Acta Chem. Scand.*, **24**, 3420 (1970)
189. J. Brunvoll, S. J. Cyvin and L. Schäfer, *J. Organometal. Chem.*, **27**, 69 (1971)
190. S. J. Cyvin, J. Brunvoll and L. Schäfer, *J. Chem. Phys.*, **54**, 1517 (1971)
191. L. H. Ngai, F. E. Stafford and L. Schäfer, *J. Amer. Chem. Soc.*, **91**, 48 (1969)
192. R. G. Snyder, *Spectrochim, Acta*, **15**, 807 (1959)
193. S. F. A. Kettle and I. Paul, *Adv. Organometallic Chem.*, **10**, 199 (1972)
194. R. Larsson, *Rec. Chem. Prog.*, **31**, 171 (1970)
195. K. Nakamoto, Infrared spectra of Inorganic and Coordination Compounds, (J. Wiley and Sons, New York, 1963)
196. K. Noack, *Helv. Chim. Acta*, **45**, 1847 (1962)
197. R. M. Wing and D. C. Crocker, *Inorg. Chem.*, **6**, 289 (1967)
198. J. Fahrenfort (Ed., M. M. Davies), Infrared Spectroscopy and Molecular Structure, (Elsevier Ltd., Amsterdam and London, 1963), chapter 11, p. 397
199. W. Beck, A. Melnikoff and R. Stahl, *Angew. Chem.*, **77**, 719 (1965)
200. T. L. Brown and D. J. Darensbourg, *Inorg. Chem.*, **6**, 971 (1967)
201. G. W. Bethke and M. K. Wilson, *J. Chem. Phys.*, **26**, 1118 (1957)
202. D. A. Seanor and C. H. Amberg, *J. Chem. Phys.*, **42**, 2967 (1965)
203. W. Beck and R. E. Nitzschmann, *Z. Naturforsch.*, **17B**, 577 (1962)
204. E. W. Abel and I. S. Butler, *Trans. Faraday Soc.*, **63**, 45 (1967)
205. D. J. Darensbourg and T. L. Brown, *Inorg. Chem.*, **7**, 959 (1968)
206. F. A. Cotton, *Inorg. Chem.*, **3**, 702 (1964)
207. F. A. Cotton and C. S. Kraihanzel, *J. Amer. Chem. Soc.*, **84**, 4432 (1962)
208. A. R. Manning and J. R. Miller, *J. Chem. Soc.*, **A**, 1521 (1966)
209. P. S. Braterman, R. Bau and H. D. Kaesz, *Inorg. Chem.*, **6**, 2097 (1967)
210. H. D. Kaesz, R. Bau, D. Hendrickson and J. M. Smith, *J. Amer. Chem. Soc.*, **89**, 2844 (1967)

211. B. F. G. Johnson, J. Lewis, J. R. Miller, B. H. Robinson, P. W. Robinson and A. Wojcicki, *J. Chem. Soc.*, **A,** 522 (1968)
212. A. Poletti, A. Foffani and R. Cataliotti, *Spectrochim. Acta*, **26A,** 1063 (1970)
213. D. J. Darensbourg, *Inorg. Chem.* (in press)
214. D. J. Darensbourg, *Inorg. Chem.*, **10,** 2399 (1971)
215. D. J. Darensbourg and C. L. Hyde, *Inorg. Chem.*, **10,** 431 (1971)
216. J. Chatt, D. P. Melville and R. L. Richards, *J. Chem. Soc.*, **A,** 2841 (1969)
217. Y. G. Borod'ko, S. M. Vinogradova, Y. P. Myagkov and D. D. Mozzhukhin, *Zhur. Strukt. Khim.*, **11,** 251 (1970)
218. W. J. Chambers and N. J. Fitzpatrick, *Proc. Roy. Irish Acad.*, **71,** 97 (1971)
219. K. G. Caulton, R. L. DeKock and R. F. Fenske, *J. Amer. Chem. Soc.*, **92,** 515 (1970)
220. S. Fronaeus and R. Larsson, *Acta Chem. Scand.*, **16,** 1447 (1962)
221. G. C. Pimentel and A. L. McClellan, The Hydrogen Bond, (W. H. Freeman and Co., San Francisco and London, 1960)
222. G. C. Pimentel and A. L. McClellan, *Ann. Rev. Phys. Chem.*, **22,** 347 (1971)
223. D. Hadži (Ed.), Hydrogen Bonding, (Pergamon, New York, 1959); *Pure Appl. Chem.*, **11,** 435 (1965)
224. A. S. N. Murthy and C. N. R. Rao, *Appl. Spectrosc. Rev.*, **2,** 69 (1968)
225. S. Bratož, *Adv. Quant. Chem.*, **3,** 209 (1967)
226. N. D. Solokov, *Ann. Chem. (Paris)*, **10,** 487 (1964)
227. S. N. Vinogradov and R. H. Linnell, Hydrogen Bonding, (Van Nostrand Reinhold Co., New York and London, 1971)
228. S. H. Lin, (Eds, H. Eyring, D. Henderson and W. Jost), Physical Chemistry, Vol. 5, (Academic Press, New York and London, 1970), chapter 8
229. C. A. Coulson and U. Danielsson, *Arkiv. Fysik.*, **8,** 239, 245 (1954); C. A. Coulson, *Research*, **10,** 149 (1957)
230. P. G. Puranik and V. Kumar, *Proc. Indian Acad. Sci.*, **A58,** 29 (1963)
231. K. Szczepaniak and A. Tramer, *J. Phys. Chem.*, **71,** 3035 (1967)
232. R. S. Mulliken, *J. Chim. Phys.*, **61,** 20 (1964)
233. M. R. Basila, E. L. Saier and L. R. Cousins, *J. Amer. Chem. Soc.*, **87,** 1665 (1965)
234. Z. Yoshida and E. Osawa, *J. Amer. Chem. Soc.*, **87,** 1467 (1965)
235. J. P. Hawranek and L. Sobczyk, *Acta Phys. Polon.*, **39A,** 639, 651 (1971)
236. C. M. Huggins and G. C. Pimentel, *J. Phys. Chem.*, **69,** 1615 (1956)
237. C. M. Huggins and G. C. Pimentel, *J. Chem. Phys.*, **23,** 896 (1955)
238. R. E. Kagarise, *Spectrochim. Acta*, **19,** 629 (1963)
239. H. Tsubomura, *J. Chem. Phys.*, **24,** 927 (1956)
240. R. C. Lord, B. Nolin and H. D. Stidham, *J. Amer. Chem. Soc.*, **77,** 1365 (1955)
241. N. Fuson, P. Pineau and M. L. Josien, *J. Chim. Phys.* **55,** 454 (1958)
242. R. F. Blanks and J. M. Prausnitz, *J. Chem. Phys.*, **38,** 1500 (1963)
243. M. C. Sousa Loupes and H. W. Thompson, *Spectrochim. Acta*, **24A,** 1367 (1968)
244. I. S. Perelygin and T. F. Akhunov, *Optics and Spectroscopy*, **30,** 367 (1971)
245 E. D. Becker, *Spectrochim. Acta*, **17,** 436 (1961)
246. H. H. Perkampus and F. M. A. Kerin, *Spectrochim. Acta*, **24A,** 2071 (1968)
247. J. Brandmüller and K. Seevogel, *Spectrochim. Acta*, **20,** 453 (1964)
248. A. K. Khairetdinova and I. S. Perelygin, *Optics and Spectroscopy*, **26,** 32 (1969)
249. G. J. Boobyer and W. J. Orville-Thomas, *Spectrochim. Acta*, **22,** 147 (1966)
250. E. Osawa and Z. Yoshida, *Spectrochim. Acta*, **23A,** 2029 (1967)
251. L. J. Bellamy, K. J. Morgan and R. J. Pace, *Spectrochim. Acta*, **22,** 535 (1966)
252. L. J. Bellamy and R. J. Pace, *Spectrochim. Acta*, **27A,** 705 (1971)
253. A. Allerhand and P. von R. Schleyer, *J. Amer. Chem. Soc.*, **85,** 371 (1963)
254. S. S. Mitra, *J. Chem. Phys.*, **36,** 3286 (1962)
255. M. Horak, J. Polakova, M. Jakoubkova, J. Moravec and J. Pliva, *Collect. Czech. Chem. Comm.*, **31,** 622 (1966)

256. M. Horak and J. Moravec, *Collect. Czech. Chem. Comm.*, **36,** 2757 (1971)
257. S. C. White and H. W. Thompson, *Proc. Roy. Soc.*, **A291,** 460 (1966)
258. W. H. Keesom, *Z. Phys.*, **22,** 129 (1921)
259. L. Onsager, *J. Amer. Chem. Soc.*, **58,** 1486 (1936)
260. G. W. Brownson and J. Yarwood, *Adv. Mol. Relaxation Processes*, **6,** 1 (1973)
261. R. Foster and T. J. Thompson, *Trans. Faraday Soc.*, **58,** 860 (1962)
262. J. W. Eastman, G. M. Androes and M. Calvin, *J. Chem. Phys.*, **36,** 1197 (1962)
263. I. Isenberg and S. L. Baird, *J. Amer. Chem. Soc.*, **84,** 3803 (1962)
264. Y. Matsunaga, *J. Chem. Phys.*, **41,** 1609 (1964); *Helv. Phys. Acta*, **36,** 800 (1963)
265. B. G. Anex and E. B. Hill, *J. Amer. Chem. Soc.*, **88,** 3648 (1966)
266. H. Kuroda, I. Ikemoto and H. Akamatu, *Bull. Chem. Soc. Jap.*, **39,** 547 (1966)
267. T. T. Harding and S. C. Wallwork, *Acta crystallogr.*, **8,** 787 (1955); *ibid.*, **15,** 810 (1962)
268. H. Kainer and W. Otting, *Chem. Ber.*, **88,** 1921 (1955)
269. M. A. Slifkin and R. H. Walmsley, *Spectrochim. Acta*, **26A,** 1237 (1970)
270. J. C. Moore, D. Smith, Y. Youhne and J. P. Devlin, *J. Phys. Chem.*, **75,** 325 (1971)
271. J. Stanley, D. Smith, B. Latimer and J. P. Devlin, *J. Phys. Chem.*, **70,** 2011 (1966)
272. B. Hall and J. P. Devlin, *J. Phys. Chem.*, **71,** 465 (1967)
273. B. Moszynska and A. Tramer, *J. Chem. Phys.*, **46,** 820 (1967)
274. B. Moszynska, *Acta Phys. Polon.*, **33,** 959 (1968)
275. R. N. Merrifield and W. D. Phillips, *J. Amer. Chem. Soc.*, **80,** 2778 (1958)
276. W. R. Carper and R. M. Hedges, *J. Phys. Chem.*, **70,** 3046 (1966)
277. A. R. Cooper, C. W. P. Crowne and P. G. Farrell, *Trans. Faraday Soc.*, **62,** 18 (1966)
278. E. M. Kosower (Eds, S. G. Cohen, A. Streitwieser and R. W. Taft), Progress in Physical Organic Chemistry, (Wiley-Interscience, New York, 1965), Volume 3, p. 81
279. W. D. Phillips, J. C. Rowell and S. I. Weissman, *J. Chem. Phys.*, **33,** 626 (1960)
280. B. Chakrabarti and S. Basu, *J. Chim. Physique*, **63,** 1044 (1966)
281. S. Saha, A. G. Ghosh and S. Basu, *J. Chim. Physique*, **65,** 673 (1968)
282. F. E. Prichard, *Spectrochim. Acta*, **20,** 127 (1964)
283. J. P. Larkindale and D. J. Simkin, *J. Chem. Phys.*, **56,** 3730 (1972)
283a. F. A. Miller, O. Sala, P. Devlin, J. Overend, E. Lippert, W. Lüder, H. Moser and J. Varchim, *Spectrochim. Acta*, **20,** 1233 (1964)
284. S. Bratos, J. Rios and Y. Guissani, *J. Chem. Phys.*, **52,** 439 (1970)
285. S. Bratos, J. Rios, J. C. Leicknam and Y. Guissani, *C. R. Sci. Acad.*, **B269,** 90, 137 (1969)
286. S. Bratos and E. Marechal, *Phys. Rev.*, **4A,** 1078 (1971)
287. H. Morawitz and K. B. Eisenthal, *J. Chem. Phys.*, **55,** 887 (1971)
288. R. G. Gordon and co-workers, (a) *J. Chem. Phys.*, **39,** 2788 (1963); (b) *ibid.*, **40,** 1973 (1964); (c) *ibid.*, **41,** 1819 (1964); *ibid.*, **47,** 1600 (1967)
289. R. G. Gordon, *J. Chem. Phys.*, **42,** 3658 (1965)
290. B. L. Crawford, A. C. Gilby, A. A. Clifford and T. Fujiyama, *Pure Appl. Chem.*, **18,** 373 (1969)
291. R. N. Jones, K. S. Seshadri, N. B. W. Jonathan and J. W. Hopkins, *Can. J. Chem.*, **41,** 750 (1963)
292. W. G. Rothschild and K. D. Möller, Far-infrared Spectroscopy (Wiley-Interscience, London and New York, 1971), chapter 11
293. S. Bratos, Paper presented at Wroclaw International Conference, Sept. 1972
294. M. Chalaye, E. Dayan and G. Levi, *Chem. Phys. Lett.*, **8,** 337 (1971)
295. I. Rossi-Sonnichsen, J. P. Bouanich and N-V-Thanh, *C.R. Sci. Acad.*, **273C,** 19 (1971)
296. J. Vincent-Geisse, J. Soussen-Jacob, N-T Tai and D. Descout, *Can. J. Chem.*, **48,** 3918 (1970)

297. J. Vincent-Geisse, J. Soussen-Jacob, D. Pelerin and A. M. Bize, *Ber. Bunsenges. Phys. Chem.*, **75**, 348 (1971)
298. S. Higuchi, S. Tanaka and H. Kamada, *Spectrochim. Acta*, **28A**, 1721 (1972)
299. A. K. Atakhodzhaev and E. N. Shermatov, *Spektrosk. Tr. Sib. Soveshch.*, 207 (1969) (C.A., **74**, 17523y)
300. J. Soussen-Jacob, J. Vincent-Geisse, D. Beaulieu and J. Tsakiris, *J. Chim. Physiqiue*, **67**, 1118 (1970)
301. J. Jacob, J. Leclerc and J. Vincent-Geisse, *J. Chim. Physiqiue*, **66**, 970 (1969)
302. P. V. Huong, J. Jacob and J. Vincent-Geisse, *C.R. Acad. Sci.*, **266B**, 1117 (1968)
303. C. E. Favelukes, A. A. Clifford and B. L. Crawford, *J. Phys. Chem.*, **72**, 962 (1968)
304. J. Yarwood, *Spectrosc. Lett.* **5**, 193 (1972)
305. T. Fujiyama and B. L. Crawford, *J. Phys. Chem.*, **73**, 4040 (1969)
306. T. Fujiyama and B. L. Crawford, *J. Phys. Chem.*, **72**, 2174 (1968)
306a. R. L. Fulton, *J. Chem. Phys.*, **55**, 1386 (1971)
307. W. G. Rothschild, *J. Chem. Phys.*, **55**, 1402 (1971)
308. W. G. Rothschild, *Chem. Phys. Lett.*, **9**, 149 (1971)
309. H. W. Kroto and J. Teixeira-Dias, *Mol. Phys.*, **18**, 773 (1970)
310. M. Scotto, *J. Chem. Phys.*, **49**, 5362 (1968)
311. C. H. Wang and P. A. Fleury, *J. Chem. Phys.*, **53**, 2243 (1970)
312. J. H. R. Clarke and S. Miller, *Chem. Phys. Lett.*, **13**, 97 (1972)
313. R. W. Lauver, *Diss. Abstr., Int. B*, **31**, 5239 (1971)
314. T. T. Wall, *J. Chem. Phys.*, **51**, 113 (1969)
315. J. G. David, *Spectrochim. Acta*, **28A**, 977 (1972)

Chapter 3

Complexes of *n* and π Donors with Halogens and Related σ Acceptors

M. Tamres and J. Yarwood

Chemistry Department, University of Michigan, Ann Arbor, Michigan, U.S.A., and
Chemistry Department, University of Durham, Durham, England

3.1. INTRODUCTION

3.1.1. GENERAL COMMENTS

In this chapter we consider the interaction of *n* and $b\pi$ donors with a class of $a\sigma$ acceptors, namely, halogens and related compounds. Such interaction, notably with iodine, has an extended history dating to the turn of the century. Kleinberg and Davidson[1] have reviewed the observations of iodine to 1948. Major impetus for the rapid development of the field of electron donor–acceptor interactions arose from the ultraviolet spectrophotometric study on iodine complexes by Benesi and Hildebrand in 1949[2], and the theoretical interpretation of the spectra by Mulliken in 1952[3]. Since then, in addition to continued study of electronic spectra, a wide range of techniques has been applied to the study of halogen complexes. There have been extensive vibrational studies (section 3.4) while X-ray crystallography, magnetic resonance spectroscopy (section 3.5) and dielectric absorption studies (chapter 7) have all made significant contributions. As a result the literature on halogen complexes is vast, and several recent reviews[4a,5,6] are of considerable importance.

As discussed in chapters 1 and 2, when a donor and acceptor interact to form a complex, whether of 1:1 or other stoichiometry, the physical properties of the components are perturbed and new properties arise which are attributable to the complex as a whole. The interaction may range from extremely weak, where the "association" may be considered to be of the "contact" type[7,8] (section 3.3.4), to quite strong, where the linkage is comparable with that of a normal chemical bond. The stronger the interaction, the larger is the alteration in bond distances and bond angles at the bonding sites, and also in other parts of the molecules.

The bonding in the complexes considered in this chapter is "weak" compared with that in "normal" chemical compounds. The heats of interaction range from less than 4 kJ mol^{-1} to $\sim$55 kJ mol^{-1}. These interactions are sufficient to effect changes that can be observed by a wide variety of physical methods, both spectroscopic and non-spectroscopic[4b,9].

Halogen complex studies have been made predominantly in *solution*. It is not surprising, therefore, that molecular iodine has been the most studied of the halogens, because of its ease of handling in solution. Recently, vapour phase electronic spectral studies have become prominant[6]. Much of the work has again been on iodine complexes, and this affords a direct measure of solvent effects by comparison with data in solution[6].

Besides molecular halogens (I_2, Br_2, Cl_2)[4,5,10–12] the list of *aσ* acceptors studied includes interhalogens and related compounds (*e.g.* ICl, IBr, ICN)[4,5,12], halogen atoms (I, Br, Cl)[6] and halomethanes (CCl_4, CBr_4, CF_3I)[4c,5,7]. Interactions of halides and oxyhalides of elements in groups IV–VII of the periodic table have also been investigated.[13,13a] Fluorine is much too reactive to be studied experimentally, but a theoretical treatment of NH_3–F_2 has recently been carried out[14] (see section 1.3.3).

3.1.2. SCOPE OF CHAPTER 3

In this chapter we have placed major emphasis on the results of electronic and vibrational spectroscopic studies although a short section (3.5) has been devoted to describing the most important "other" techniques which include nuclear magnetic resonance[4f,9] and nuclear quadrupole resonance[5,14a] spectroscopy. The treatment is is again illustrative rather than exhaustive and we have tried to elucidate current ideas in the light of theoretical expectations expressed in chapter 1. We have also attempted to assess the reliability of the measurements; this being important, especially for halogen complexes, since they are often used as "model" systems for the description of molecular interactions (chapters 1 and 2). Comparison of experiment and theory is made primarily from studies on iodine complexes for which the data are most complete. For more thorough compilations on complexes of n and $b\pi$ donors with halogen and halogen containing acceptors, the reader is referred to the books by Briegleb[12], Foster[4] and Andrews and Keefer[11], and to the reviews by Rao, Bhat and Dwivedi[5] and Tamres[6].

3.2. THERMODYNAMIC ASPECTS

In this section, we shall consider aspects associated with obtaining reliable thermodynamic data on molecular interactions, and the solvent influence on these data. We shall limit ourselves primarily to complexes of 1:1 stoichiometry, the most widely studied case, but many of the comments are applicable to complexes of other stoichiometry as well:

3.2.1. THE EQUILIBRIUM CONSTANT

The equilibrium constant for the interaction of a donor, D, with an acceptor, A, to form a complex, C, is

$$K_a = \frac{a_C}{a_D a_A} = K_c K_\gamma \qquad 3.1$$

where

$$K_c = \frac{C_C}{C_D C_A} \qquad 3.2$$

and

$$K_\gamma = \frac{\gamma_C}{\gamma_D \gamma_A} \qquad 3.3$$

The symbols a, C and γ represent activity, concentration and activity coefficient, respectively. Until recently, complexes had been studied almost entirely in solution. It generally had been assumed that K_γ is unity[2]; so that the calculations of equilibrium constants were based on equation 3.2, often written as,

$$K_c = \frac{C_C}{(C_D^\circ - C_C)(C_A^\circ - C_C)} \qquad 3.4$$

where C_D° and C_A° are the initial concentrations of donor and acceptor, respectively, and C_C is the concentration of complex at equilibrium. The K_c in the above equations represents the equilibrium constant in units of $dm^3\ mol^{-1}$. Another concentration unit, widely used in the earlier literature, is that of mole fraction (K_X). The concentration units of molarity, mole fraction and molality are not related in a simple way, except for very dilute solutions. Therefore, care must be taken in comparing and interpreting results based on different units.[15] A discussion of this point may be found in the book by Foster.[4g]

The reciprocal of equation 3.4 is:

$$\frac{1}{K_c} = \left[\frac{C_D^\circ C_A^\circ}{C_C} - C_D^\circ\right] + [-C_A^\circ + C_C] \qquad 3.5$$

which is a quadratic equation in C_C. It must be used in its entirety for strong complexes. Assuming the concentration of the acceptor to be smaller than that of the donor (if the reverse is the case, the C_D° and C_A° terms in the brackets of equations 3.5 can be interchanged), then $C_C \simeq C_A^\circ$ as $K_c \rightarrow \infty$. Thus, each of the quantities $[(C_D^\circ C_A^\circ/C_C) - C_D]$ and $[-C_A^\circ + C_C]$ is very small, and dropping any term within the brackets leads to appreciable error[16,17]. In the case of the fairly strong triethylamine–iodine complex, approximately a 10% error in K_c was introduced[18] by neglecting the term $[-C_A^\circ + C_C]$.[17] LaBudde and Tamres[19] have proposed a means of estimating the errors introduced by dropping terms.

For weak complexes, the extent of complexation is low, so that C_D° and $C_A^\circ \gg C_C$. Hence the last term in equation 3.5 can be neglected, reducing the equation to a linear form in C_C:

$$\frac{1}{K_c} = \frac{C_D^\circ C_A^\circ}{C_C} - C_D^\circ - C_A^\circ. \qquad 3.6$$

If, in addition, $C_D^\circ \gg C_A^\circ$, equation 3.5 reduces further to:

$$\frac{1}{K_c} = \frac{C_D^\circ C_A^\circ}{C_C} - C_D^\circ \qquad 3.7$$

the form used originally by Benesi and Hildebrand[2].

Various physical properties can be measured which, if related in a simple way to the concentration of one of the species, either free or complexed, allow determination of K_c. In the vapour phase, a manometric study (benzene–I_2[20]) and an isopiestic study (section 3.2.3; diethyl ether–I_2 and n-hexane–I_2[21]) allow C_C (and therefore K_c) to be determined from measurement of a single donor-acceptor mixture; the former calculation is based on Dalton's law of partial pressures, and the latter on a weight loss[21]. In many other methods (*e.g.* spectroscopic and calorimetric[4b]) the measured quantity is proportional to the concentration of the complex, and two mixtures must be studied to allow simultaneous determination of K_c and the proportionality factor. Generally, many measurements are made in order to establish statistical error limits for the results.

3.2.2. SPECTROPHOTOMETRIC STUDIES

Most thermodynamic data on halogen complexes have come from measurement of electronic spectra, either the charge-transfer (CT) band or a perturbed band of the acceptor. Using the Beer-Lambert law, $A = \varepsilon lC$ [where A is the absorbance, ε is the (molar) extinction coefficient and l is the path length], substitution for C_C in equations 3.5 to 3.7 gives the spectrophotochemically equivalent equations:

$$\frac{C_D^\circ l C_A^\circ}{A} = \frac{1}{\varepsilon}\left[\frac{l}{K_c} + (C_D^\circ + C_A^\circ) - \frac{A}{l\varepsilon}\right], \qquad 3.8$$

$$\frac{C_D^\circ l C_A^\circ}{A} = \frac{1}{\varepsilon}\left[\frac{1}{K_c} + C_D^\circ + C_A^\circ\right], \qquad 3.9$$

$$\frac{C_D^\circ l C_A^\circ}{A} = \frac{1}{\varepsilon}\left[\frac{1}{K_c} + C_D^\circ\right]. \qquad 3.10$$

Any other physical measurement proportional to C_C would give a similar set of equations[4b].

Equation 3.9 is experimentally more useful than 3.10 because there is no restriction in the former on the relative concentrations of C_D° and C_A°, as there is in the latter. Both equations are of linear form. For equation 3.9 (and similarly for 3.10), a plot of $Y = C_D^\circ l C_A^\circ/A$ *vs* $X = (C_D^\circ + C_A^\circ)$ gives ε from the reciprocal of the slope, and K_c from the slope/intercept ratio.

The quantitative simultaneous determination of K_c and ε is not possible for either extremely "strong" nor extremely "weak" complexes. For a "strong" complex with $C_D^\circ > C_A^\circ$, $(1/K_c) \rightarrow 0$ and $A/l\varepsilon = C_C \simeq C_A^\circ$. In effect, the component of lower concentration is converted quantitatively into product, allowing determination of ε only by measuring A. At the other extreme of a "weak" complex, $(1/K_c) \rightarrow \infty$. This term dominates all other terms in brackets in equation 3.9, allowing only the $K_c\varepsilon$ product to be determined, which is equal to $A/C_D^\circ l C_A^\circ$. Only when $C_D^\circ + C_A^\circ$ is sufficiently large with respect to $1/K_c$, so that it is reasonably greater than the experimental error limits in determining $C_D^\circ l C_A^\circ/A$, can separation of K_c and ε be achieved. Criteria for meaningful separation have been discussed by Person[22], by Deranleau[23], and by LaBudde and Tamres[19].

For charge-transfer bands, ε usually is quite large, being of the order of 10^6 dm^3 mol^{-1} m^{-1}. Therefore, very low concentrations of C_C can be measured. The advantage of this is that, except for quite "weak" complexes, relatively dilute solutions can be studied. This minimises the problems at higher concentrations with regard to changing activity and to the formation of 2:1 complexes—which often escape detection. It has been shown, for example, that aromatic hydrocarbon-tetracyanoethylene complexes in dichloromethane, previously considered to form only in 1:1 ratio, contain 2:1 species as well[24]. Evidence[25] also has been obtained quite recently that, at the high donor concentrations required for the spectrophotometric study of the weak benzene-iodine system, there is a likely possibility of the existence of a 2:1 complex in addition to that of 1:1 stiochiometry. Changes in activity and/or complex ratio could be contributing reasons why thermodynamic results reported for 1:1 complexes using methods requiring different concentration ranges sometimes are not in agreement[9].

For iodine complexes, spectral measurements can be made on the perturbed iodine band.[6] Since ε for this band is generally much smaller than that for the charge-transfer band, studies of both bands usually require a separate set of concentrations or different cell paths. Studying both regions offers a means of checking the thermodynamic results.

The halogens obey the Beer-Lambert law in the visible region, but not in one region of continuous absorption in the near ultraviolet, where analysis shows that a diatomic and a dimer band (I_4[6,26–27a] and Br_4[6,28–30a]) are superimposed. The high intensity of the dimer band, which is so much greater than that of the diatomic iodine, indicates it is due to a charge-transfer transition and not to a perturbation of molecular iodine. Also, the absorption has a frequency near that predicted for an I_4 charge-transfer transition[30b]. In the dimer, the halogen is both donor and acceptor, which changes the denominator in equation 3.4 to $(C_0 - 2C_C)^2$, where C_0 is the total iodine concentration as monomer. The spectrophotometric equation applicable in this case is a polynomial of the form[26,27]:

$$\frac{A}{lC_0} = \varepsilon_2 + (\varepsilon_4 - 2\varepsilon_2)K_4C_0(1 - 4KC_0 + 20K_4^2C_0^2 - 112K_4^3C_0^3 + \cdots) \tag{3.11}$$

where ε_2 is the monomer extinction coefficient, K_4 is the formation constant of the halogen dimer, and ε_4 is the dimer extinction coefficient. Equation 3.11 holds for the condition that $8K_4C_0 > 1$. The separation of K_4 and $(\varepsilon_4 - 2\varepsilon_2)$ depends on the contributions of the higher order terms to the curvature in equation 3.11. Separation has been reported for I_4[27] (data in tables 3.1 and 3.4), but has not been successful for Br_4[29,30].

3.2.3. CONSTANT ACTIVITY METHOD

Some years ago Kortüm *et al.*[31] calculated the equilibrium constants of iodine complexes in solution by measuring the increased solubility of iodine in the solvent upon addition of a donor; the increased solubility being attributed to 1:1 complex formation. An extension of the solubility method, one which is much more useful for spectrophotometric study because of the range in iodine concentration permitted, is based on maintaining a constant activity of iodine through an equilibrium between tetramethylammonium polyiodide solids[32] *e.g*:

$$(CH_3)_4NI_5(\text{solid}) \rightleftharpoons (CH_3)_4NI_3(\text{solid}) + I_2 \quad 3.12$$

and

$$\tfrac{1}{2}(CH_3)_4NI_9(\text{solid}) \rightleftharpoons \tfrac{1}{2}(CH_3)_4NI_5(\text{solid}) + I_2 \quad 3.13$$

Addition of a donor establishes a second equilibrium, and equation 3.12 or 3.13 is displaced to the right to replace the complexed iodine. Equilibria in the vapour phase at a single temperature have been reported for diethyl ether–iodine[21,33] and *n*-hexane–iodine[21] based on an isopiestic technique which utilizes a sensitive microbalance to determine the loss in weight of iodine from the solid polyiodide mixture when the donor is added. The constant activity method has been applied also to the vapour-phase spectrophotometric study of $(C_2H_5)_2O$–I_2, $(C_2H_5)_2S$–I_2 and $(CH_3)_2S$–I_2.[34] In solution, where the complex and free iodine have a well-defined isosbestic point, the increase in absorbance at the isosbestic point upon addition of the donor gives directly the concentration of the complex, since ε is known. In this method, the problem inherent in the conventional procedure (to separate the terms in the $K_c\varepsilon$ product for "weak" complexes) is, in principle circumvented and K_c can be calculated from the absorbance of just a single composition of donor and polyiodide mixture. The study of the classical system of iodine complexes with aromatic hydrocarbons in *n*-heptane solution[2] has been repeated using this technique[25]. The results are given in table 3.1 (also table 3.4), and will be discussed in later sections.

3.2.4. ENTHALPY DETERMINATION

Enthalpy changes ($\Delta H^\ominus$) are obtained from a study of the temperature dependence of the equilibrium constant. If K_c units are used, such a study in the vapour phase gives the change in internal energy ($\Delta E^\ominus$). For 1:1 complex formation in the vapour phase, $\Delta H^\ominus = \Delta E^\ominus - RT$; in solution $\Delta H^\ominus = \Delta E^\ominus$.

As discussed, separation of the terms in the $K_c\varepsilon$ product for quite weak complexes is not feasible, *i.e.* very large error limits are associated with K_c and with ε, often resulting in large and random variation of both with temperature. For this case, therefore, the usual procedure is to determine $\Delta H^\ominus$ from a plot of either log $K_c\varepsilon_{max}$ *vs.* $1/T$ or log K_c(ave) *vs.* $1/T$. The quantity $K_c\varepsilon_{max}$ at each temperature is obtained from equation 3.9 using the absorbances at the band maxima. This procedure will give a reliable $\Delta H^\ominus$ but, because ε_{max} is unknown, the entropy change ($\Delta S^\ominus$) is not obtained. If the variation in ε_{max} with temperature is random, but within tolerable limits, the average ε can be taken. For this case $\Delta S^\ominus$ can be obtained and K_c(ave) is calculated from the relation $K_c(\text{ave})\varepsilon(\text{ave}) = K_c\varepsilon_{max}$.

TABLE 3.1

Thermodynamic characteristics[a] of some complexes of iodine

Donor	Phase[b] (method)	$K_c^c(t°C)/$ $dm^3\ mol^{-1}$	$10^{-5}\ K_c\varepsilon^d(t°C)/$ $dm^6\ mol^{-2}\ m^{-1}$	$-\Delta E^{\ominus}/$ $kJ\ mol^{-1}$	$-\Delta H^{\ominus}/$ $kJ\ mol^{-1}$	$-\Delta S_c^{\ominus}/$ $J\ mol^{-1}\ K^{-1}$	Reference
iodine	vapour	0.4 ± 0.1 (332°) [1.7][e,f]	1.2 [5.1][e,f]		12.1 ± 1.7	27.6 ± 0.4	27
	vapour		2.67 ± 0.25[g] (110°)				26
	n-heptane		2.80[h] (22°)		4.85 ± 1.26		43
	CCl_4	[0.13 (25.5°)][i,f]	2.04[j]		4.2 ± 0.8		44
benzene	vapour		2.00 ± 0.11[k] (100°) 3.64[e,k]	7.41 ± 0.67	10.42 ± 0.67		45
	vapour	[4.5 ± 0.6][e,f]	7.43	8.4 ± 0.4	10.9 ± 0.4	[23.8 ± 1.3][f]	38
	vapour (PVT)	3.4 (27°)			10.2	23.8	46
	[l]{*n*-heptane (*ca*)[m]	0.356 ± 0.008	2.88 ± 0.09				25, 42
	n-heptane	0.211 ± 0.012	2.95 ± 0.27				25, 42
	n-heptane	0.246 (24°)	2.95		6.78	34.5[n]	47
	CCl_4	0.184 (24°)	2.48		5.90	34.1	47
toluene	[l]{*n*-heptane (*ca*)[m]	0.500 ± 0.006	3.45 ± 0.07				25
	n-heptane	0.315 ± 0.005	3.37 ± 0.08				25
	n-hexane	0.293[o]	3.46		7.5	35.5	12a, 50
	CCl_4	0.16	2.67				12a, 48
o-xylene	[l]{*n*-heptane (*ca*)[m]	0.641 ± 0.009	4.19 ± 0.10				25
	n-heptane	0.423 ± 0.007	4.10 ± 0.11				25
	n-hexane[p]	0.386	4.09			36.0	12a, 50
	CCl_4	0.27	3.38				12a, 48
m-xylene	[l]{*n*-heptane (*ca*)[m]	0.698 ± 0.006	4.57 ± 0.07				25
	n-heptane	0.539 ± 0.020	4.85 ± 0.03				25
	CCl_4	0.31	3.88				12a, 48
p-xylene	vapour		4.08 ± 0.10[q] (107°)	10.63 ± 0.59	13.77 ± 0.59		45
	[l]{*n*-heptane (*ca*)[m]	0.642 ± 0.006	3.66 ± 0.07				25
	n-heptane	0.411 ± 0.006	3.64 ± 0.08				25
	CCl_4	0.31	3.13		9.12	40.2	12a, 48, 49

mesitylene	vapour		11.0 ± 0.4(104°) 33.4[e]	13.14 ± 0.75	16.32 ± 0.75		45
	[1] *n*-heptane (*ca*)[m]	0.967 ± 0.005	7.48 ± 0.09				25
	n-heptane	0.748 ± 0.018	7.56 ± 0.30				25
	n-hexane	0.70[o]	6.90				50
	CCl_4	0.577[o]	5.88				51
	CCl_4	0.579[o]			11.97	44.7	49
diethyl ether	vapour		4.10 ± 0.11(80°) 13.4[e]	18.8 ± 0.8	21.8 ± 0.8		37
	vapour	6.4 ± 1.2	13.4	13.4 ± 0.4	15.9 ± 0.4	38.1 ± 1.3	38
	vapour (iso)	4.3 ± 0.3 (35°)					21, 33
	vapour (sat)	5.5 (35°)			13.8 ± 1.3	30.5	52
	[1] *n*-heptane (*ca*)[m]	1.44 ± 0.006(15°)	8.23 ± 0.52				25, 42
	n-heptane	1.23 ± 0.005(15°)	6.84 ± 0.39				25, 42
	n-heptane	1.3 (15°)	7.12		17.6 ± 0.8	63.1 ± 4.9	53, 54
	CCl_4	0.97 (20°)	4.6		18.0 ± 0.4	61.5	55
dimethyl sulphide	vapour	7.5 ± 1.3 (100°) 84 ± 15[e]	102 ± 3 1150 ± 31[e]	29.7 ± 1.7	32.6 ± 1.7	71.5 ± 4.7	40
	vapour	23 ± 12 (100°) 220 ± 140[e]	115 ± 60 1100	28.0 ± 2.1	31.0 ± 2.1	56.9 ± 10.0	39
	vapour (*ca*)			30.5 ± 2.5			34
	CCl_4 (*bs*)	71 ± 2					56
	CCl_4 (n.m.r.)	71 ± 5					57
diethyl sulphide	vapour	11.5 ± 1.1 (100°) 191 ± 20[e]	193.6 ± 3.0 3210 ± 55[e]	34.7 ± 0.8	37.7 ± 0.8	80.8 ± 2.3	40
	vapour	16.5 (100°) 226[e]	184.8 ± 9.2 2530[e]	32.2 ± 1.7	35.1 ± 1.7	62.8 ± 0.5	58
	vapour	58 ± 20 (100°) 750 ± 250[e]	203 ± 23 2620[e]	31.8 ± 2.1	34.7 ± 2.1	59.4 ± 7.9	39
	n-heptane	180 ± 7	4750		37.24 ± 2.51	81.2 ± 8.4	59
	n-heptane (*bs*)	180 ± 3			34.76 ± 0.63	73.6 ± 2.1	59
	CCl_4	233° (20°)	6850		33.68	68.2	60

TABLE 3.1 (continued)

Thermodynamic characteristics[a] of some complexes of iodine

Donor	Phase[b] (method)	$K_c^c(t°C)$/ $dm^3\ mol^{-1}$	$10^{-5}\ K_c\varepsilon^d(t°C)$/ $dm^6\ mol^{-2}\ m^{-1}$	$-\Delta E^{\ominus}$ $kJ\ mol^{-1}$	$-\Delta H^{\ominus}$/ $kJ\ mol^{-1}$	$-\Delta S_c^{\ominus}$/ $J\ mol^{-1}\ K^{-1}$	Reference
pyridine	*n*-heptane (*ca*)[m]	163 ± 2	(6700 ± 100)				25, 42
	n-heptane	137 ± 5	7100				61
	n-heptane	185	9300		32.6 ± 0.8	66.1	62
	n-heptane (*bs*)	157 ± 1			34.14 ± 0.92	72.4 ± 3.4	63
	CCl_4	105 ± 1			31.25 ± 0.25	66.1 ± 0.8	63

a. standard state of unit molarity.
b. method is conventional spectrophotometric study of the charge-transfer band unless specified otherwise in parentheses: (*ca*) spectrophotometry utilizing constant activity source; (PVT) manometry; (*iso*) isopiestic method; (*sat*) saturated vapour; (*bs*) blue-shifted iodine band; (n.m.r.) nuclear magnetic resonance.
c. formation constant; temperature at 25°C unless given otherwise in parentheses.
d. temperature corresponds to that of K_c; temperature given in parentheses when the $K_c\varepsilon$ product is not resolved; ε is ε_{max} unless noted otherwise.
e. extrapolated to 25°C, assuming constancy in $\Delta E^{\ominus}$.
f. value is uncertain.
g. at $\lambda = 250$ nm.
h. at $\lambda = 295$ nm.
i. estimated value.
j. at $\lambda = 288$ nm.
k. at $\lambda = 282$ nm (not max).
l. results from the same laboratory.
m. measurement at isosbestic point; ε for charge-transfer band determined by absorbance ratio (section 3.3.3).
n. changed from 38.7 $J\ mole^{-1}\ K^{-1}$ in ref. 47 for consistency with enthalpy and free energy data in that reference.
o. K_c calculated from literature K_x data using $K_c = K_xV_0$ where V_0 is the molar volume of the solvent.
p. mixture of xylene isomers.
q. at $\lambda = 294$ nm (not max).

The above procedure to obtain $\Delta H^{\ominus}$ assumes a constancy in ε over the temperature range. Actually ε is temperature dependent because of band broadening, but the variation generally is less than the experimental error in determining ε. Nevertheless, it should be noted that, as a consequence of band broadening, calculation based on ε_{max}, *i.e.* absorbance data at the band maxima, gives a value which is an upper limit for $\Delta H^{\ominus}$. It has been suggested that the change in half-width of the band with change in temperature be taken as a measure of the change in ε_{max}[35], thereby allowing correction in $\Delta H^{\ominus}$.

Equation 3.4 can be rewritten as,

$$K_c(C_D^\circ - C_C)(C_A^\circ - C_C) = C_C = A/l\varepsilon \qquad 3.14$$

and taking logarithms gives,

$$\log K_c\varepsilon + \log l + \log (C_D^\circ - C_C) + \log (C_A^\circ - C_C) = \log A \qquad 3.15$$

For "weak" complexes, and even for moderately "strong" complexes, assuming that $\log (C_D^\circ - C_C)$ and $\log (C_A^\circ - C_C)$ do not vary appreciably over the temperature range measured, a plot of log A *vs.* 1/T is equivalent to a plot of $\log K_c\varepsilon$ *vs.* 1/T. This permits $\Delta H^{\ominus}$ to be determined from the temperature dependence of only a single donor-acceptor mixture[6]. Density corrections for the concentrations must be made for solution studies, but are not needed for the vapour phase[6].

In the constant activity method [based on equation 3.12 or 3.13], the concentration C_A° is fixed at a given temperature. Letting $(C_A^\circ)_t$ be the saturation concentration of the acceptor at temperature t, then:

$$K_c = \frac{C_C}{(C_D^\circ - C_C)(C_A^\circ)_t} \qquad 3.16$$

and as long as $C_D^\circ \gg C_C$, we have:

$$K_c C_D^\circ (C_A^\circ)_t = C_C = A/l\varepsilon \qquad 3.17$$

Hence:

$$\log K_c\varepsilon + \log l + \log C_D^\circ + \log (C_A^\circ)_t = \log A \qquad 3.18$$

Differentiating with respect to 1/T gives:

$$\frac{d \log K_c\varepsilon}{d(1/T)} + \frac{d \log (C_A^\circ)_t}{d(1/T)} = \frac{d \log A}{d(1/T)} \qquad 3.19$$

The middle term can be evaluated separately by measuring the saturation concentration of the acceptor for the particular constant activity source. Therefore, equation 3.19 offers a means of determining enthalpies from a plot of log A *vs.* 1/T, and this method should be applicable to relatively "strong" complexes as well as "weak" ones.

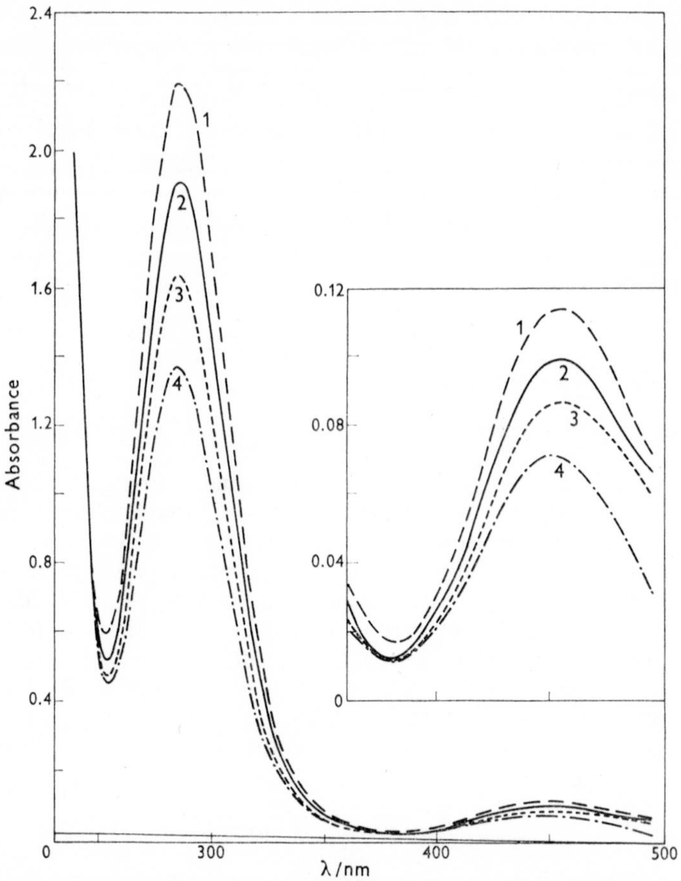

FIG 3.1 Dimethyl sulphide–iodine complex in the vapour phase using $(CH_3)_4NI_9$–$(CH_3)_4NI_5$ as a constant activity source for iodine in both the sample and the reference cell: path length = 1.000 m; $[(CH_3)_2S] = 1.44 \times 10^{-3}$ mol dm^{-3}; (1) 70.8°C, $[I_2] = 6.39 \times 10^{-5}$ mol dm^{-3}; (2) 65.5°C, $[I_2] = 4.58 \times 10^{-5}$ mol dm^{-3}; (3) 60.2°C, $[I_2] = 3.38 \times 10^{-5}$ mol dm^{-3}; (4) 55.1°C, $[I_2] = 2.44 \times 10^{-5}$ mol dm^{-3}; insert recorded with 0–0.2 absorbance scale (Reproduced from *J. Amer. Chem. Soc.*, **95**, 2516 (1973), by permission of the American Chemical Society).

Results of a vapour phase study[34] of $(CH_3)_2S$–I_2 utilizing this method are shown in figure 3.1. The system $(CH_3)_4NI_9/(CH_3)_4NI_5$ was used as the constant activity source for iodine. The value for $\Delta E^{\ominus}$ obtained in this way agrees quite well with that reported using the conventional method (data in table 3.1).

It should be noted in figure 3.1 that the absorbance *increases* with increasing temperature. This is because the increase in the second term of equation 3.19 more than compensates for the decrease in the first term. In the sequence of steps from equations 3.16 to 3.19, the absorbance refers to that of the complex only. Corrections must be made if the components also absorb. If only one component absorbs, it can be compensated for by adding it to the reference cell; the compensation being particularly precise when the component concentration is maintained by a constant activity source.

3.2.5. GENERAL SOLVENT EFFECTS

The theory of molecular interactions applies principally to the vapour phase[36a], *i.e.* it involves properties of the *isolated* donor, acceptor and adduct. Yet most of the data come from studies in solution. There has been a growing realization that neglect of activity can have a pronounced effect on the thermodynamic (and spectral, section 3.3) properties of complexes[36b]. This realization stems from studies since 1965 on charge-transfer complexes in the vapour phase[6,37–40] and, more

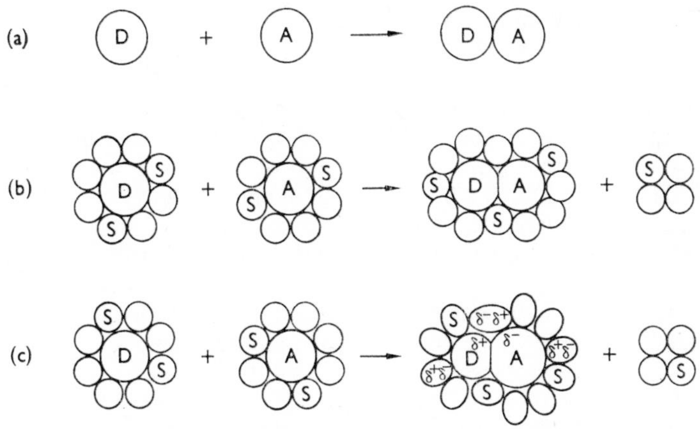

FIG 3.2. Electron donor-acceptor interactions: (a) weak, in the vapour phase; (b) weak, in solution; (c) strong, in solution.

recently, from studies in solution where activity is taken into account[6,25,41,41a,42,42a]. The effects on the thermodynamic properties of complexes are evident from the data in table 3.1. Many of the data are from recent vapour and solution studies, but several selected earlier results in relatively "inert" solvents are included for comparison. A much more complete list of vapour and solution data is to be found in a recent review[6].

The interaction of donor and acceptor in the vapour phase and in solution is illustrated in figure 3.2—which is explained in detail below (see also section 3.3.3). The relation between the thermodynamic properties in the two phases is given by the cycle,

$$\begin{array}{ccccc} \mathrm{D(g)} & + & \mathrm{A(g)} & \xrightarrow{\Delta \mathrm{X(g)}} & \mathrm{DA(g)} \\ \Delta \mathrm{X}_D \Big\downarrow \mathrm{(solv.)} & & \Delta \mathrm{X}_A \Big\downarrow \mathrm{(solv.)} & & \Delta \mathrm{X}_{DA} \Big\downarrow \mathrm{(solv.)} \\ \mathrm{D(soln)} & + & \mathrm{A(soln)} & \xrightarrow{\Delta \mathrm{X(soln)}} & \mathrm{DA(soln)} \end{array} \qquad 3.20$$

from which

$$\Delta \mathrm{X(g)} = \Delta \mathrm{X(soln)} + [\Delta \mathrm{X}_{DA}\mathrm{(solv)} - \Delta \mathrm{X}_D\mathrm{(solv)} - \Delta \mathrm{X}_A\mathrm{(solv)}] \qquad 3.21$$

where X is any thermodynamic function.

Since results from different laboratories are not always in agreement, nor are error limits always specified, quantitative comparison must be made with caution. Further the choice of units[4g] and standard states[64] sometimes varies. The thermodynamic data in table 3.1 are for a standard state of unit molarity. In spite of the complications, several trends seem to emerge from the available data[6].

(1) Parallel trends are observed for K_c and $\Delta E^\ominus$. Both are larger in the vapour phase than in solution for the weaker iodine complexes, *e.g.* I_4 and aromatic hydrocarbon–I_2. For diethyl ether–I_2, K_c in the vapour phase is larger, but the $\Delta E^\ominus$ values in the two phases are comparable. With a further increase in strength of the complex, *e.g.* dialkyl sulphide–I_2, both K_c and $\Delta E^\ominus$ are similar in magnitude in the two phases. Unfortunately, attempts to study iodine complexes in the vapour phase with still stronger donors (pyridine[65], amines[38,66], dimethyl selenide[67]) have led to reaction. However, the complete thermodynamic cycle (equation 3.20) has been obtained for trimethylamine–sulfur dioxide[35], a complex almost as "strong" as triethylamine–I_2[18]. In this case, K_c and $\Delta E^\ominus$ are larger in *n*-heptane solution (2550 ± 50 dm^3 mol^{-1} at 25°C and -46.0 ± 1.3 kJ mol^{-1}), than in the vapour phase (340 ± 35 dm^3 mol^{-1} at 298K and -38.1 ± 1.7 kJ mol^{-1})[35].

(2) The trend in the $K_c\varepsilon$ product is similar. The value is larger in the vapour phase for the weaker iodine complexes by a sizeable factor. (It should be noted that the vapour-phase results in table 3.1 for $K_c\varepsilon$ of benzene–I_2 and of *p*-xylene–I_2 which come from reference 45 are low because the ε's are *not* at λ_{max}.) For the dialkyl sulphide–I_2 complexes, $K_c\varepsilon$ is smaller for the vapour phase. For the still stronger $(CH_3)_3N$–SO_2 complex, the $K_c\varepsilon$ product is appreciably smaller for the vapour phase compared to solution (2.070×10^5 and 13.80×10^5 dm^6 mol^{-2} cm^{-1}, respectively, at 298 K)[6,35].

(3) There is a trend also in $\Delta S^\ominus$ for the iodine complexes. The vapour-phase values are noticeably smaller than those for solution for the weaker complexes. They are virtually the same for the dialkyl sulphide–I_2 complexes in the two phases, and also for $(CH_3)_3N$–SO_2

(-87.9 ± 4.2 and -89.5 ± 3.8 J mol^{-1} K^{-1} in the vapour phase and solution, respectively)[35].

(4) Results obtained for K_c (and ε) in solution studies by the constant activity (solubility) method are *not* the same as those obtained by the conventional method, at least for the weaker complexes. The latter method gives *lower* values of K_c for the weak complexes. The difference between the two methods becomes smaller as the strength of complexation increases, and for the quite strong pyridine–I_2 system K_c is the same using either method. The difference could be attributed perhaps to the different concentration ranges required for weak complexes in the two studies. Relatively dilute solutions can be used to determine K_c directly from the increase in absorbance at the isosbestic point in the constant activity method[25,42]. On the other hand, the conventional method requires higher concentrations in order to have sufficient complexation to effect a meaningful separation of terms in the $K_c\varepsilon$ product. It has been observed that, for the aromatic hydrocarbon-I_2 systems, the high concentrations lead to deviation from expected behaviour for 1:1 complexation, and this has been attributed to activity effects or to the formation of higher complexes. A reasonable fit to the data for benzene–I_2 is obtained if the formation of a 2:1 as well as a 1:1 complex is assumed[25,42].

Another consideration is the activity coefficient of iodine in solution. The equilibrium constants listed in table 3.1 for the aromatic hydrocarbon–I_2 complexes which were studied by the constant activity method were determined by *assuming* the activity coefficient to be unity. If correction based on solubility parameter theory is made for the change in the activity coefficient of iodine as a function of donor concentration, the corrected equilibrium constants come out to be remarkably close to those obtained by the conventional (Benesi-Hilderbrand) procedure[42a].

Some of the thermodynamic observations can be rationalized by the kinds of interactions shown in figure 3.2. For weak complexation in the vapour phase (model a), there should be a decrease in internal energy as a result of the van der Waals attraction between D and A. The rather loose association permits one molecule to move essentially freely over the surface of the other, so that many D–A orientations are possible; and all contribute to K_c[46,48]. For the same weak complex in solution, the van der Waals forces between D and A are not much different from those between D and S, A and S, and S and S (model b), so that there are cancelling effects which reduce the energy change[36b]. (For a further discussion of this point see section 1.2.) The solvent can be an effective competitor for either D or A, at least for the less favourable D–A orientations, thereby diminishing K_c[68]. Also, since the number of solvent molecules associated with the free donor and free acceptor is greater than that associated with the complex, it is possible that overall effects would be such as to diminish K_c and $\Delta H°$ in polar solvents, as has been cited for weak complexes of the $b\pi$–$a\sigma$ and $b\pi$–$a\pi$ type[69].

Strong D–A interaction of n–$a\sigma$ type produce complexes which are highly polar, and which are more likely to have a unique geometry. These complexes can be stabilized to a greater extent through (mutual) induced polarization with the solvent compared to the solvent interaction with the free components (model a), resulting in a larger K_c and $\Delta H^{\ominus}$. Polar solvents might be expected to further enhance the stability of the strong complexes. This has been observed for triphenylarsine–I_2[70], pyridine-I_2[71], pyridine–Br_2[69], and also $(CH_3)_3N$–SO_2[72].

These trends have been noted by Christian *et al.*[35,73] who found, for complexes in a solvent such as *n*-heptane whose interactions are non-specific, (*i.e.* the absence of strong local dipoles as in *n*-heptane), that the thermodynamic functions are affected in proportion to the strength of complexation (models b and c). In general, exact cancellation of the solvation steps in the thermodynamic cycle [equation 3.20] cannot be expected, so that only in limited cases will $\Delta X^0(g) \simeq \Delta X^0(soln)$. Drago and Wayland[74] proposed a double scale equation to calculate enthalpies from electrostatic and covalent parameters (equation 1.26). The above discussion would imply that the parameters would be slightly altered in the vapour phase from those in solution.

Carter *et al.*[75] have included the solvent displacement in their equilibrium constant expression. Their treatment leads to the result that, while the expressions for K_c and ε depend upon the inclusion or omission of the solvent (models b and a in figure 3.2), the $K_c\varepsilon$ product does not. This is not what is observed experimentally. The $K_c\varepsilon$ value generally is different in the vapour phase from that in solution, and it varies with the polarity of the solvent. This would imply a non-cancellation of solvation effects on the donor, acceptor, and the complex. For the same solvent, the same $K_c\varepsilon$ product is obtained whether the conventional or constant activity method is used.

The solvation sphere shown in model c of figure 3.2 involves only solvent molecules. If the donor or acceptor is itself an appreciably polar species, the complex may be preferentially solvated by D or A molecules, at least in part, even when the solvent is present in large excess. Such "aggregation" would undoubtedly affect the thermodynamic functions.

Contributions from the various types of interaction energies for a D–A pair in the vapour phase (model a, figure 3.2) have been estimated (see table 1.1, p. 16). Recently, Amidon[76] has made similar calculations based on model b for the interaction of chloranil and of fluoranil with benzene and a series of its methyl and fluoro derivatives in the solvent carbon tetrachloride. The calculated net energy change falls within the same range as that determined experimentally. For example, for the system, benzene/fluoranil/carbontetrachloride (*i.e.*D/A/S) with a selected (most stable) orientation, the calculated energies[76] in kJ mol^{-1} are; D–A (−18.37), D–S (−16.99), A–S (−18.47), and for the S–S term (−24.85). The calculated result is −7.76 kJ mol^{-1} compared to the experimental value of −8.4 kJ mol^{-1}.

3.2.6. SPECIFIC SOLVENT EFFECTS

Specific interactions can occur between the solvent and either the donor, or acceptor, or both, when strong local dipoles are present as in the case with hydrogen-bonding solvents. Here, the data often can be treated in terms of competing equilibria. If, say, the solvent and acceptor form a complex whose equilibrium constant is K_{SA}, then[17]:

$$K_{DA} = (1 + K_{SA}S_0)K_c \qquad 3.22$$

where K_{DA} is the (presumably) correct D–A equilibrium constant, S_0 is the concentration of the solvent, and K_c is the D–A equilibrium constant defined by equation 3.5. In the absence of solvent competition (*i.e.* $K_{SA} = 0$), we find that $K_{DA} = K_c$. The correction for solvent competition can be significant even if K_{SA} is small because S_0 for common solvents is $\sim$10 mol dm^{-3}. If the solvent complexes with the donor (equilibrium constant K_{DS}) as well as with the acceptor, then[4g]:

$$K_{DA} = (1 + K_{DS}S_0)(1 + K_{SA}S_0)K_c \qquad 3.23$$

While correction for competition with a complexing solvent may give, at times, an equilibrium constant close to that measured for the complex in a non-competing solvent[77], caution should be exercised in using this procedure. This can be illustrated with the data of McKinney and Popov[71] and of Childs[25] on the pyridine–I_2 system (table 3.2). The K_c for this complex is smaller in the aromatic hydrocarbon solvents than in n-heptane, as would be expected because these hydrocarbon solvents compete with pyridine for the iodine. Correction for this competition should lead to a K_{DA} of $\sim$157 dm^3 mol^{-1}, *i.e.* it should give the value determined directly for pyridine–I_2 in n-heptane. There is a question, however, as to the choice of the constant for the S–A interaction. That based on the constant activity method, $K_{SA}(ca)$ (uncorrected for the activity coefficient of iodine[42a]), results in a "corrected" K_{DA} which is quite high. Based on K_{SA} (*conv*), the "corrected" K_{DA} is lower, but is still high compared to the expected value. Since the conventional spectrophotometric method requires higher donor concentrations for weak complexes, there might be appreciable 2:1 complexation. Thus $K_{SA}(conv)$, as usually determined, may actually be a composite constant, intermediate in magnitude for 1:1 and 2:1 complexes. At still higher concentrations, the apparent equilibrium constant most likely approaches that of the 2:1 complex. The most concentrated solution is one where the donor is the solvent, and where the largest formation of higher complexes occurs. As shown in table 3.2, the correction based on K_{S_2A} gives values for K_{DA} which are in the range of that determined directly. Such qualitative agreement would be equally plausible in terms of changing activity rather than changing stoichiometry. In either case, use of equation 3.24 or 3.25 must be made with some care.

TABLE 3.2
Formation constants for the pyridine–iodine complex in various solvents at 298 K

	Solvent			
	benzene	toluene	*p*-xylene	*n*-heptane
K_c[a]($dm^3\ mol^{-1}$)	82.9 ± 0.5	87.8 ± 1.8	91.6 ± 0.9	157 ± 1.2[b]
S_0($mol\ dm^{-3}$)	11.242	9.403	8.107	
K_{SA}[c](*ca*)($dm^3\ mol^{-1}$)	0.356 ± 0.008	0.500 ± 0.006	0.642 ± 0.006	
K_{SA}[d](*conv*)($dm^3\ mol^{-1}$)	0.211 ± 0.012	0.315 ± 0.005	0.411 ± 0.006	
K_{S_2A}[c](*ca*)($dm^6\ mol^{-2}$)	0.063 ± 0.003	0.066 ± 0.003	0.092 ± 0.003	
K_{DA}[e](*ca*)($dm^3\ mol^{-1}$)	415	501	568	
K_{DA}[f](*conv*)($dm^3\ mol^{-1}$)	281	348	396	
K_{DA}[g](*ca*)($dm^3\ mol^{-1}$)	142	142	160	

a. ref. 71.
b. value of 163 ± 2 $dm^3\ mol^{-1}$ in ref. 25 and ref. 42.
c. constant activity method, ref. 25.
d. conventional method, ref. 25.
e. corrected constant based on K_{SA} from constant activity method.
f. corrected constant based on K_{SA} from conventional method.
g. corrected constant based on K_{S_2A}.

3.3. ULTRAVIOLET AND VISIBLE STUDIES

On complexation, electronic and vibrational bands of the components are perturbed and new spectral bands appear. In some cases, particularly when a homologous series is compared, their properties (*e.g.* band position) can be related to strength of complexation. Correlation between spectral and thermodynamic properties has often been cited in the literature.

In this section, electronic spectra of halogen complexes will be considered, again notably for iodine, with emphasis on some recent observations. For an authoritative account of spectral properties and their relation to theory, the reader is directed to the book by Mulliken and Person[36], especially chapter 10.

3.3.1. CHARGE-TRANSFER SPECTRA

Benesi and Hildebrand[2] focussed attention on the new bands which appear on complexation, although such a band apparently had been recorded earlier ($\lambda_{max} \simeq 250$ nm for ether–I_2[78]). These bands are attributed to an *intermolecular* electronic transition, *i.e.* they are charge-transfer (CT) bands[3], (figure 3.3). The transition of lowest energy (1) is that from the highest occupied molecular orbital (*HOMO*) of the donor to the lowest empty molecular orbital (*LEMO*) of the acceptor. Multiple bands arise if electronic excitation also occurs from a penultimate donor level, or if electrons are transferred to a higher empty orbital of the acceptor (2 and 3). Distinct multiple bands have been observed with some $a\pi$ acceptors[79,80]. They occur also with iodine as indicated by broad bands which suggest overlap of the CT bands[45,81].

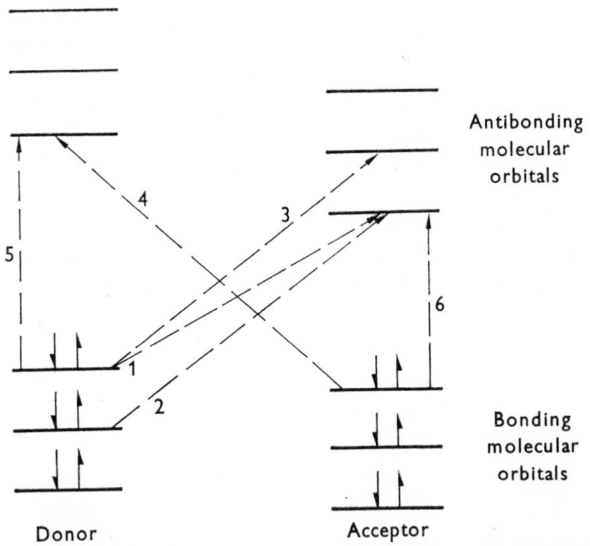

FIG 3.3 Electron excitations: (1) lowest energy CT band, (2) excitation from penultimate donor level, (3) excitation to higher empty acceptor level, (4) higher energy acceptor to donor CT band, (5) intramolecular donor excitation, and (6) intramolecular acceptor excitation (following the scheme of M. J. S. Dewar and A. R. Lepley, *J. Amer. Chem. Soc.*, 1961, **83**, 4560).

Transition (4) from an occupied level of the acceptor to an empty orbital of the donor is possible, but such transitions for weak complexes generally have not been observed because of the higher energies involved. The molecular levels are identical when the same species is both donor and acceptor, *e.g.* Br_2–Br_2 or I_2–I_2. Such species are a simple example of what may be termed "two-way" electron donor-acceptor complexes[36c].

Following the notation in figure 1.1, the CT transition of lowest energy is given by:

$$\mathbf{h}\nu_{CT}(\mathrm{g}) = \mathrm{I}_D^{\mathrm{v}} - \mathrm{E}_A^{\mathrm{v}} + \mathrm{G}_1 - \mathrm{G}_0 + \mathrm{X}_1 - \mathrm{X}_0 \qquad 3.24$$

(see equation 1.16, p. 9) where the G's are the resultant classical energies and the X's are the resonance energies. The contributions of the G and X terms are different for complexes of widely different strength, and approximations to equation 3.24 must be considered accordingly[82].

In the weakest D–A interactions, the resonance and classical energies can be neglected, except for the important coulombic energy contribution to G_1. Therefore, equation 3.24 reduces to the approximation:

$$\mathbf{h}\nu_{CT} \simeq \mathrm{I}_D^{\mathrm{v}} - \mathrm{E}_A^{\mathrm{v}} - \frac{\mathbf{e}^2}{\mathrm{d}_{12}} \qquad 3.25$$

where **e** is the electronic charge and d_{12} is the intermolecular distance between the charged moieties D^+ and A^-. This equation, already oversimplified, contains the further assumption that no interaction occurs between the ground (W_N) or excited (W_V) state of the contact pair with the excited states of any of the species[36e].

For a large number of complexes which are "weak", but whose thermodynamic properties are measurable, classical attraction and repulsion (including steric effects), reorganization energies and resonance energies are small. An equation which has been applied to these complexes for the case where they contain a common acceptor is[36d,82]:

$$\mathbf{h}\nu_{CT} \simeq I_D^V - C_1 + \frac{C_2}{I_D^V - C_1} \qquad 3.26$$

where C_1 and C_2 are parameters for the acceptor. These parameters have been determined for iodine ($C_1 = 5.2$ eV; $C_2 = 1.5$ (eV)^2) and for other acceptors[36d] from a best fit of the spectral data to equation 3.26. Actually, the parameters are not strictly constant because the terms that comprise them will differ somewhat for complexes of varying strength and type.

Equation 3.26 is quadratic in I_D^V. However, for a large portion of the data on weak complexes, the variation in the last term is small, so that $\mathbf{h}\nu_{CT}$ remains essentially a linear function of I_D^V (as indicated also by equation 3.25). Such linear relations have been reported[79,82], and hold particularly well for a *homologous* series of donors. One of the earliest series reported is that of iodine with the methylbenzenes.[48,51] It should not be expected, however, that the same linear relation is applicable to donors of widely different type because of the possibility of differences in orientation and in intermolecular distance. Mulliken and Person[36d], and also Voigt and Reid,[80] discuss in some detail aspects which must be considered in such correlations.

Equation 3.26 is not applicable to relatively strong complexes, *e.g.* amine–I_2. A different relation must be used in this case[82].

Complexes with halogen atoms are observed as transient species following intense irradiation, *e.g.* flash photolysis, pulse radiolysis and gamma-ray radiolysis[6]. Qualitative consideration of the preceding equations would indicate that the CT band maxima of complexes with atomic halogens should be shifted to longer wavelengths relative to those with molecular halogens with the same donor because the electron affinities of halogen atoms are greater than those of halogen molecules. Data for halogen acceptors, atomic and molecular, are given in table 3.3, where comparison of the spectral properties can be made for approximately similar experimental conditions. In spite of the variation in some results from different laboratories, the expected trend is observed.

The spectra of the atomic and molecular halogen complexes are

TABLE 3.3

Charge-transfer bands of some atomic and molecular halogen acceptors

Donor	I_D^a/eV	Acceptor	E_A/eV	Phase	λ_{max}/nm	$10^{-2}\ \varepsilon_{max}$/ $dm^3\ mol^{-1}\ m^{-1}$	Reference
A. Alkanes:							
1. cyclohexane	9.88	I atom	3.063[b]	vapour	310		87
		I_2	2.58[c]	vapour	<220		88
				vapour	(230)[d]		89
		I atom		cyclohexane	330	2025[e]	90
				cyclohexane	increased UV absorbance, but no maximum		91
		I_2		cyclohexane	232		92
				n-hexane	(242)[d]	(19,000)[d]	93
2. methylcyclohexane	9.85	I_2		vapour	<220		88
				vapour	(235)[d]		89
3. *n*-heptane	10.08	I_2		vapour	<220		88
				n-heptane	227		92
4. *n*-hexane	10.18	I atom		*n*-hexane	330		94
		I_2		vapour	<220		88
B. Alkyl halides:							
1. methyl chloride	11.28	I atom	3.063[b]	vapour	268		87
		Br atom	3.363[b]	vapour	300		87
2. ethyl chloride	10.98	I atom		vapour	279		87
		Br atom		vapour	310		87

TABLE 3.3 (*continued*)
Charge-transfer bands of some atomic and molecular halogen acceptors

Donor	I_D^a/eV	Acceptor	E_A/eV	Phase	λ_{max}/nm	$10^{-2}\,\varepsilon_{max}$/ $dm^3\,mol^{-1}\,m^{-1}$	Reference
3. methyl bromide	10.53	I atom		vapour	313		87
		Br atom		vapour	335		87
4. ethyl bromide	10.29	I atom		vapour	323		87
				ethyl bromide	395	1800[e]	91
		Br atom		vapour	345		87
5. methyl iodide	9.54	I atom		vapour	416, 350		87
				vapour	420–430		91
				methyl iodide	~400	9000 ± 2000[e]	95
6. ethyl iodide	9.33	I atom		vapour	425, 360		87
				vapour	no absorption detected		91
				ethyl iodide	[490][f]	13,000[e]	91
				ethyl iodide	~400	9000 ± 2000[e]	95
7. 1-chloropropane	10.82	I_2	2.58[c]	vapour	(245–270)[g]		96
				n-heptane	224		93
8. 1-bromopropane	10.18	I_2		vapour	(250–260)[h]		96
				n-heptane	244		93
9. 1-iodopropane	9.26	I_2		vapour	(– – – – –)[i]		96
10. 2-iodopropane	9.17	I_2		vapour	(– – – – –)[i]		96
				n-heptane	280		93
11. *n*-heptylbromide	≤10.10[j]	I_2		vapour	(250)[e]		89

9

C. Aromatic hydrocarbons[k]							
1. benzene	9.245	I atom	3.063[b]	vapour	430		87
				vapour	no absorption detected		91
				benzene	500		91
				benzene	465		97
				benzene	495	3500[e]	98
				benzene	500	1600[e]	91
		I_2	2.58[c]	vapour	~268		38
				n-heptane	(288–297)	(12 000–18 000)[l]	6
				benzene	(287–310)		6
		Br atom	3.363[b]	benzene	555		97
				benzene	550		99
		Br_2	2.51[c]	CCl_4	292	13 400[l]	100
		Cl atom	3.613[b]	CCl_4	490		101, 102
		Cl_2	2.38[c]	CCl_4	278	9090[l]	103
2. toluene	8 82	I atom		toluene	515	3400[e]	98
				toluene	520	1500[e]	91
		I_2		CCl_4	302	16 700[l]	48
		Br atom		toluene	575		97
		Br_2		CCl_4	301	10 500[l]	100
		Cl atom		CCl_4	475		101, 102
3. *o*-xylene	8.56	I atom		*o*-xylene	570	3600[e]	98
		I_2		CCl_4	316	12 500[l]	48
		Br_2		CCl_4	313	8200[l]	100
4. *m*-xylene	8.56	I_2		CCl_4	318	12 500[l]	48
		Br_2		CCl_4	312	10 000[l]	104
		Cl_2		CCl_4	290	6340[l]	103

TABLE 3.3 (*continued*)
Charge-transfer bands of some atomic and molecular halogen acceptors

Donor	I_D^a/eV	Acceptor	E_A/eV	Phase	λ_{max}/nm	$10^{-2}\ \varepsilon_{max}$/ $dm^3\ mol^{-1}\ m^{-1}$	Reference
5. *p*-xylene	8.445	I atom		p-xylene	520[m]	2950[e]	98
		I_2		vapour	294		89
				CCl_4	304	10 100[l]	48
		Br_2		CCl_4	306	7300[l]	100
6. mesitylene	8.40	I atom		mesitylene	590	5800[e]	98
				mesitylene	590	2600[e]	91
		I_2		vapour	301		89
				CCl_4	332	10 200[l]	51
7. hexamethylbenzene	7.95[n]	I atom		CCl_4	~770		105
		I_2		CCl_4	371	6700[l]	51
8. bromobenzene	8.98	I_2		CCl_4	290	10 400[l]	48
		Br atom		bromobenzene	560		99
		Br_2		CCl_4	288	7600[l]	100

a. ref. 83; error limit ±0.01 to ±0.03 eV.
b. ref. 84; error limit ±0.003 eV.
c. ref. 85; error limit ±0.1 eV.
d. doubtful result.
e. estimated, based on assumption of diffusion-controlled recombination in rate study.
f. band may be due to presence of molecular iodine[95].
g. low, rather flat band over this region[96].
h. possible inflection point in this region[96].
i. strong absorption by donors in the region 235–290 nm[96].
j. value for 1-bromopentane is 10.10 eV[83].
k. see table 3.4 for additional data on iodine complexes.
l. conventional spectrophotometric method.
m. broad band, may be resultant of CT transitions from different donor orbitals[98].
n. ref. 86.

similar in that both have broad and structureless bands. Further, a linear relation is found between $\mathbf{h}\nu_{CT}$ and I_D^V for the atomic halogen acceptors[87,99,101,102], just as occurs with the molecular halogen acceptors.

The positions of the band maxima of molecular halogen complexes follow, for the most part, the trend in electron affinity of the halogens, *i.e.* molecular chlorine has the lowest electron affinity and its complexes have band maxima at the shortest wavelength. But the band maxima of complexes with atomic chlorine also are at the shortest wavelength, in spite of the fact that atomic chlorine has the highest electron affinity. As mentioned previously, other factors must be considered[97].

Most of the data on halogen atom complexes pertain to their spectral properties, but some thermodynamic data also are available. Strong and Perano[106] have found that complexation with an iodine atom is stronger than with an iodine molecule. Thus, with *o*-xylene as the donor, the thermodynamic functions of the I-atom[106] and I_2[48] complexes in CCl_4 are, respectively, (in mole fraction units): $K_x(298\ K) = 7.4$ and 2.96; $\Delta H_x^\ominus = -18.4$ and -8.4 kJ mol^{-1}; and $\Delta S_x^\ominus = -45.6$ and -20.5 J mol^{-1} K^{-1}.

3.3.2. SOLVENT EFFECTS ON CHARGE-TRANSFER BANDS

The effects of the solvent and the experimental method (conventional or constant activity) on the thermodynamic properties of iodine complexes were discussed earlier with regard to the data in table 3.1. We now consider similar effects on the spectral properties of the CT bands of these complexes, the data for which are given in table 3.4.

The solvent apparently has little effect on the shape of CT bands. The half-widths ($\Delta\bar{\nu}_{1/2}$) are quite similar in the vapour phase and in solution, being of the order 5000–6000 cm^{-1}. The solvent does have an effect, however, on the CT band position. The transition energies in the two phases are related by the equation[6],

$$\mathbf{h}\nu_{CT}(\text{soln}) = \mathbf{h}\nu_{CT}(\text{g}) + [\Delta X_{DA}^{V}(\text{solv}) - \Delta X_{DA}^{N}(\text{solv})] \qquad 3.27$$

where ΔX_{DA} is the solvation energy of the complex in its ground (N) and its excited (V) states. The pattern that is observed is a large red shift from vapour to solution for the weak iodine complexes. The red shift is smaller for the stronger iodine complexes and it has been reported that a blue shift occurs with very strong complexes[35]. The solvent effect on the CT band position seems more complex for acceptors other than iodine and for solvents of varying polarity[35a]. A more detailed compilation of data and a discussion of these trends are given in the review by Tamres[6].

Since, in general, $\mathbf{h}\nu_{CT}(\text{soln})$ is different from $\mathbf{h}\nu_{CT}(\text{g})$ (although only by about a few tenths of an electron volt), there must be a modification in the parameters relating $\mathbf{h}\nu_{CT}$ and I_D^V (or E_A^V) in the two phases. Thus,

TABLE 3.4
Spectral characteristics of charge-transfer bands of some iodine complexes

Donor	Phase (method)[a]	$10^{-2}\,\varepsilon^{b}_{max}$/ $dm^3\,mole^{-1}\,m^{-1}$	λ_{max}/nm	$\Delta\bar{\nu}^{c}_{max}$/ cm^{-1}	$\Delta\bar{\nu}^{d}_{1/2}$/ cm^{-1}	f[e]	$10^{30}\,\vec{\mu}^{f}$/C m	Reference
iodine	vapour	~3000 (150–420°)	(230–255)[g]					27
	vapour		~245					26
	n-heptane		~293.5					107
	CCl_4	[16 000][h,i]	288					44
benzene	vapour		<282					45
	vapour	(1650 ± 100)[i]	~268		~5000	(0.036)[i]	(4.74)[i]	38
	[j] {*n*-heptane (*ca*)	8080 ± 80	288	2590				25, 42
	{*n*-heptane	14 200 ± 300	289	2710				25
	n-heptane	12 000	288	2590	5400	0.280	13.7	47
	CCl_4	13 460	291	2950	5500	0.319	14.7	47
	CCl_4	16 400	292	3070				12b, 48
toluene	[j] {*n*-heptane (*ca*)	6900 ± 60	300					25
	{*n*-heptane	10 700 ± 90	300					25
	n-hexane	11 800	302					12b, 50
	CCl_4	16 700	302					12b, 48
o-xylene	[j] {*n*-heptane (*ca*)	6520 ± 60	313					25
	{*n*-heptane	9680 ± 90	314					25
	n-hexane[k]	10 600	318					12b, 50
	CCl_4	12 500	316					12b, 48

m-xylene	[j] *n*-heptane (*ca*)	6550 ± 50	314					25
	n-heptane	8990 ± 160	314					25
	CCl_4	12 500	318					12b, 48
p-xylene	vapour		<320[l]					45
	[j] *n*-heptane (*ca*)	5710 ± 50	303					25
	n-heptane	8860 ± 70	302					25
	CCl_4	10 100	304					12b, 48
mesitylene	vapour		301		5500			45
	vapour		~305					39
	[j] *n*-heptane (*ca*)	7730 ± 50	327	2640				25
	n-heptane	10 110 ± 140	327	2640				25
	n-hexane	9900	333	3190				50
	CCl_4	10 200	332	3100				51
diethyl ether	vapour		234		6300[m]			37
	vapour	(2100 ± 400)[i]	234		6600[n]	(0.053)[i]	(5.76)[i]	38
	[j] *n*-heptane (*ca*)	5720 ± 70	250	2740				25, 42
	n-heptane	5560 ± 90	250	2740				25
	n-heptane	5480	252	3050	6900	0.168	9.86	54, 55
	CCl_4	4700	249	2570	6900	0.170	9.12	55
dimethyl sulphide	vapour	13 750 ± 2310	286		5,600	0.33 ± 0.06	15.0 ± 1.30	40
	vapour	5000 ± 2500	286		5,600	0.13 ± 0.07	9.0 ± 3.3	39
	n-Heptane		299	1490	5300			39

TABLE 3.4 (*continued*)
Spectral characteristics of charge-transfer bands of some iodine complexes

Donor	Phase (method)[a]	$10^{-2}\,\varepsilon^{b}_{max}$/ $dm^3\,mol^{-1}\,m^{-1}$	λ_{max}/nm	$\Delta\bar{\nu}^{c}_{max}$/ cm^{-1}	$\Delta\bar{\nu}^{d}_{1/2}$/ cm^{-1}	f[e]	$10^{30}\,\vec{\bar{\mu}}^{f}$/C m	Reference
diethyl sulphide	vapour	16 800 ± 2230	290		5800	0.41 ± 0.06	16.9 ± 1.13	40
	vapour	11 200 ± 1900	290		~5700	0.28 ± 0.05	13.6 ± 1.17	58
	vapour	3500 ± 1000	290		5800	0.09 ± 0.03	7.73 ± 1.13	39
	n-heptane	26 400 ± 1050	303	1480	5400	0.6	20	39, 59
	n-heptane	29 800 ± 600	302	1370	5400	0.695	22.3 ± 0.23	55
	CCl_4	29 400	303	1480	~(5300)[n]	~0.67	22	60
pyridine	*n*-heptane (*ca*)	41 400 ± 1000	235					25, 42
	n-heptane	50 000	235		5400	1.12	25.3	61, 62
	n-heptane	51 730	236		5250	1.40[o]	29.2[o]	61

a. conventional spectrophotometric method, except where indicated by (*ca*) for constant activity method.
b. at ~298 K unless otherwise specified.
c. taken as $\Delta\bar{\nu}_{max}$ (vapour) − $\Delta\bar{\nu}_{max}$ (soln).
d. half-width: difference in wave number at half the maximum intensity on the high and low energy side of the CT band; or twice the difference between the band peak and half the maximum intensity on the low energy side.
e. oscillator strength[36f], calculated from $f \simeq 4.32 \times 10^{-9}\,\varepsilon_{max}\,\Delta\bar{\nu}_{1/2}$.
f. transition moment[36f], calculated from $\bar{\mu} \simeq 0.0958 \left[\frac{\varepsilon_{max}\,\Delta\bar{\nu}_{1/2}}{\bar{\nu}_{max}}\right]^{1/2}$. 1 Debye = 3.335×10^{-30} C m.
g. little variation in ε_{max} over this region of λ.
h. estimated value.
i. value is uncertain.
j. results from the same laboratory.
k. mixture of xylene isomers.
l. broad spectrum; overlapping unresolved bands.
m. temperature broadening, average in range 60–98°C.
n. estimated from figure in the reference.
o. obtained by integration.

for example, it has been reported that the values in the vapour phase for the parameters C_1 and C_2 of atomic[87] and molecular[39,89] iodine obtained from equation 3.26 differ from those in solution.

It is of interest to note in the data of table 3.4 that the CT band intensity depends not only on the presence of a solvent but also, at least for the weaker iodine complexes, on the spectrophotometric method employed, *i.e.* conventional *vs* constant activity. It has long been known that, in the conventional method, increasing methyl substitution in benzene generally produced a decrease in ε_{max}. This result led to the proposal of "contact" charge-transfer to account for the trend[8]. For the constant activity method, the same $K_c\varepsilon_{max}$ product is obtained, but K_c is higher (section 3.2.5) and ε_{max} is correspondingly lower. Presumably, this method might give a better measure of the total effects (random and specific) of donor-acceptor interactions on the spectral properties of the complexes[42a]. Thus the trend in ε_{max} on methyl substitution seems much less pronounced than previously believed[25,42].

There is a reversal in the magnitude of both the $K_c\varepsilon_{max}$ product and K_c in the two phases with increasing strength of complexation, as can be seen from table 3.1. For "weak" complexes, $K_c\varepsilon_{max}$ and K_c are larger in the vapour phase, whereas the reverse is the case in solution. Such reversal is not found for ε_{max}, although there is a decided trend as the strength of complexation increases. The difficulty in determining ε_{max} of weak complexes, particularly in the vapour phase where concentrations are low, has been stressed[6]. However, the $K_c\varepsilon_{max}$ product is evaluated reasonably well, and in conjunction with the K_c obtained by non-spectrophotometric methods there is the possibility of getting a fairly reliable value for ε_{max}. Using such data for the weak benzene–I_2 and diethyl ether–I_2 complexes, (table 3.1), ε_{max} is found to be appreciably smaller in the vapour phase than in solution. With the stronger sulphide–I_2 complexes, ε_{max} (vapour) is $\sim\frac{2}{3}\varepsilon_{max}$(soln). No data for still stronger iodine complexes in the two phases are available, but for the relatively strong $(CH_3)_3N–SO_2$ system ε_{max} is virtually the same in both phases[35].

Several factors have been cited for the effect of solvent on band intensities[6,108,109]. The geometries of weak complexes, although rather flexible, correspond, on the average, to a compromise between orientations favoured by London dispersion forces and by charge-transfer forces (section 1.1.4). In the latter orientation, ε_{CT} would be greater. The large difference in ε_{CT} for weak complexes in the vapour phase and in solution could be due to a redistribution of the relative orientations in solution because of solvent competition[68]. Also, weak complexes are more compressible, and pressure from the "solvent cage" would increase orbital overlap and enhance the transition[110,111]. Strong complexes are more likely to have a unique geometry in both phases, and solvent pressure would be less effective.

3.3.3. PERTURBED INTRAMOLECULAR TRANSITIONS (BLUE-SHIFTED BANDS)

The halogen molecules absorb in the visible region. The transition, $^3\Pi_0 \leftarrow {}^1\Sigma_g$ (*i.e.* $\sigma_u \leftarrow \pi_g$), is forbidden and, therefore, is weak (reference 36, p. 137). The intensity of this transition increases with increasing molecular weight of the halogens, which is attributed to the increase in spin-orbit coupling in the heavier atoms.[36g]

It has long been known that solvents affect the visible spectrum of iodine[1]. At one time, solvents were classified according to the colour of their iodine solutions, *i.e.* "purple" or "brown". It was shown subsequently that purple solvents do not form complexes with iodine to an appreciable extent, while brown solvents do[1]. In the latter case, the iodine visible spectrum undergoes a pronounced shift to shorter wavelengths (blue shift).

Since complexation perturbs both the donor and acceptor, changes in the spectra of donors also should occur. It has been observed, for example in the iodine complexes with thioacetamide and dimethylthioacetamide, that the intense $\pi^* \leftarrow \pi$ and the weak $\pi^* \leftarrow n$ transitions of the donors in the ultraviolet spectrum undergo blue shifts[112]. The intensity of the latter transition is appreciably enhanced on complexation, perhaps due to the transition becoming more allowed as a result of the perturbation or to acquiring of intensity from the mixing of the weak intramolecular $\pi^* \leftarrow n$ band with the highly intense intermolecular CT band.

Mulliken[8] has attributed the blue shift in the visible spectrum of iodine as follows. Excitation of an electron from the occupied π_g molecular orbital in the ground state of iodine to the σ_u antibonding molecular orbital produces a larger effective size of the iodine molecule. If the molecule is complexed, the excitation results in an increase in repulsion energy between the iodine and the adjacent donor. It might be expected that this repulsion would be greater for stronger complexes, which should result in larger blue shifts. This general trend has been cited in the literature[36g,113–115].

Spectral characteristics of the blue-shifted iodine band for complexes covering a range of donor strengths are given in table 3.5. This table is not exhaustive. Thermodynamic properties of many iodine complexes have been obtained from studies in the visible region but, often, the blue-shifted iodine band has not been well-characterized, *i.e.* correction for overlap with the free iodine band or with the tail of the CT band has not been made. This may be the case for a few of the entries in table 3.5. Other data of interest can be found in a study by Voigt[121] of blue-shifted band maxima of iodine dissolved in a variety of σ-, π- and n-type donor solvents having a wide range of ionization potentials.

Most of the data in table 3.5 are for studies *in solution*. Again, there is the problem of some variation in results from different laboratories, but even this does not prevent questioning the general validity of the correlation of blue shift with donor strengths. Complexes with several donors, *e.g.* the alcohols, have larger blue shifts than would be expected from their $\Delta H^{\ominus}$ values. Further, the selenides form thermodynamically stronger complexes with iodine than the sulphides, yet the blue-shifted bands have similar maxima. This occurs also for the iodine complexes with tri-*n*-octylphosphine sulphide and tri-*n*-octylphosphine oxide. The former is a much stronger donor, yet both form complexes in *n*-heptane having a blue-shifted iodine band maximum at ~440 nm[122].

One aspect that needs to be considered in the correlation is the extent to which the solvent, which is known to affect the CT band (section 3.3.2), affects the blue-shifted iodine band. Initial attempts to establish the presence of the blue-shifted iodine band in the vapour phase were not successful, leading to some speculation about solvent influence on the spectral properties of complexed iodine.[6] Recent studies have established the existence of this band for three complexes, and the data are included in table 3.5. Some features of the vapour and solution spectra are shown in figure 3.4.

Iodine vapour has a band structure above ~500 nm and a continuum at wavelengths shorter than this (see figure 3.5A). In the presence of ~1 atm. pressure† of an "inert" gas (*e.g.* He, N_2, O_2, or saturated hydrocarbons) the lines in the band structure region are broadened towards a smooth continuum (in the limit), but there is no effect on the continuum region[88]. The change in band shape and intensity is towards that found for iodine dissolved in an "inert" solvent.

There is a pronounced enhancement in absorbance below ~500 nm with the donors $(C_2H_5)_2S$[116] (figure 3.5A) and $(CH_3)_2S$[40], and a much smaller one with $(C_2H_5)_2O$[37]. In each case, for a fixed concentration of iodine, the enhancement increases with increasing donor concentration and decreases with increasing temperature, as would be expected of complex formation.

Characterization of the blue-shifted iodine band depends on subtracting the contribution of free iodine from the total absorbance. This requires a reliable value for K_c because the determination of band position and intensity is quite sensitive to the magnitude of K_c. For the "stronger" sulphide–iodine complexes, K_c is reasonably well-determined by the conventional spectrophotometric method, giving the results for the blue-shifted bands shown in figure 3.4. But for the weaker diethyl ether-iodine complex, separation of K_c and ε from vapour-phase data is subject to very large error.

This problem for weak complexes is overcome in the constant activity method (section 3.2.3) which, by employing a differential absorbance measurement (*i.e.* having a constant activity source in the reference cell), permits direct characterization of band position and shape of

† *i.e.* a pressure of 1.01325×10^5 N m^{-2}.

TABLE 3.5
Spectral characteristics of the visible iodine band

Donor	Phase (method)[a]	$\Delta H^{\ominus}$/ kJ mol^{-1}	$10^{-2}\,\varepsilon_{max}$/ dm^3 mole^{-1} m^{-1}	λ_{max}/ nm	$\Delta\bar{\nu}^{b}_{bs}$/ cm^{-1}	λ^{c}_{iso}/ nm	$\Delta\bar{\nu}^{d}_{1/2}$/ cm^{-1}	f[e]	$10^{30}\,\vec{\mu}^{f}$/C m	Reference
—	vapour	—	~570	~498[g]						116
			350	480[h]						116
n-heptane	*n*-heptane	—	918	520			3200	0.0124	3.93	25
	n-heptane	—	897	522			3200	0.0124	3.89	36h, 55
benzene	[i] *n*-heptane (*ca*)		984	504		520	3320	0.0142	4.13	25
	n-heptane	6.78[j]	1150	498			3260	0.0162	4.33	25
	benzene		1050	500						117
toluene	[i] *n*-heptane (*ca*)		1020	502		518	3240	0.0143	4.13	25
	n-heptane	7.5[k]	1120	496			3170	0.0154	4.23	25
	toluene		1020	497						117
o-xylene	[i] *n*-heptane (*ca*)		1030	501		519	3450	0.0153	4.26	25
	n-heptane		1110	496			3320	0.0162	4.36	25
	o-xylene		1060	497						117
p-xylene	[i] *n*-heptane (*ca*)		1030	501		518	3460	0.0153	4.26	25
	n-heptane		1210	496			3350	0.0175	4.52	25
	p-xylene	9.12[l]	1080	495						117
m-xylene	[i] *n*-heptane (*ca*)		1060	499		518	3430	0.0158	4.30	25
	n-heptane		1160	496			3380	0.0169	4.46	25
mesitylene	*n*-heptane (*ca*)[m]		1110	495		520	3470	0.0169	4.38	25
	mesitylene	12.0[l]	1200	490						117
methanol	CCl_4	14.6	973	440			4250	0.0179	4.29	36h, 55

ethanol	CCl_4	18.8	1080	443			4500	0.0210	4.69	36h, 55
tri-*n*-butyl phosphate	*n*-heptane	12.1	1280	456			4200	0.0232	4.98	36h, 55
N,N-dimethyl-formamide	CH_2Cl_2	16.7	1190	441			4300	0.0221	4.79	36h, 55
diethyl ether	vapour (*ca*)	21.8[n]	870	509			~4500[o]	0.0169	4.48	34
	[i] ⎡*n*-heptane (*ca*)		852	470	1630	494	3910	0.0144	3.99	25
	⎣*n*-heptane		890	469			3710	0.0143	3.96	25
	n-heptane	17.6	950	462			4100	0.0168	4.28	36h, 55
	CCl_4	18.0	873	468			4100	0.0155	4.13	55
diethyl disulphide	*n*-heptane	19.2	1370	460			4100	0.0242	5.13	36h, 55
	n-heptane	29.8	~1400	460						118
	CCl_4	24.1	~1650	450						118
dimethyl sulphide	vapour (*ca*)	33.4	725	457			4350	0.0136	3.83	34
	vapour	32.6	~800	~457			~4500[o]	0.0155	4.10	40
	n-hexane			437	1000					119
	CCl_4		~1950[p]	437		~489[p]	~4300[p]	~0.036	~6.0	56
diethyl sulphide	vapour (*ca*)			452						34
	vapour	37.7	~1100	~453			~4800[o]	0.0224	4.90	40, 59
	n-heptane	34.8[q]	~1900	437	760	493[r]				
	n-heptane	32.6	1960	435			4200	0.0352	6.00	36h, 55
	n-heptane	37.6[s]	~1900	430						118
	CCl_4	35.6[s]	2320	430						118

TABLE 3.5 (*continued*)

Donor	Phase (method)[a]	$\Delta H^{\ominus}$/ kJ°m l^{-1}	$10^{-2}\,\varepsilon_{max}$/ dm^3 mole^{-1} m^{-1}	λ_{max}/ nm	$\Delta\bar{\nu}/^{b}_{bs}$ cm^{-1}	λ^{c}_{iso}/ nm	$\Delta\bar{\nu}^{d}_{1/2}$/ cm^{-1}	f[e]	$10^{30}\,\vec{\mu}^{f}$/C m	Reference
pyridine	*n*-heptane (*ca*)		1370	422		477	4250	0.0252	5.00	25
	n-heptane	32.6	1320	422			4300	0.0245	4.90	36h, 55
dimethyl selenide[t]	CCl_4	36.0		433		~493				56, 119, 120
	CCl_4	35.6	~2600	430			~4500	~0.05	~6.6	36h, 55
triethylamine	*n*-heptane	45.0[u]	2030	414			5700	0.0501	7.00	36h, 55

a. conventional spectrophotometric study unless specified by (*ca*) for constant activity method.
b. blue shift, taken as $\bar{\nu}_{max}$ (soln) — $\bar{\nu}_{max}$ (vapour).
c. isosbestic point.
d. half-width (see footnote d in table 3.4).
e. oscillator strength (see footnote e in table 3.4).
f. transition moment (see footnote f in table 3.4).
g. vibrational structure appears at $\lambda > 498$ nm.
h. very little temperature dependence in ε at 480 nm over the range 45°–120°C.
i. results from the same laboratory.
j. ref. 47.
k. ref. 50.
l. in CCl_4, ref. 49.
m. small uncertainty in results due to light sensitivity.
n. conventional spectrophotometric method, ref. 37.
o. taken as twice the difference between the band peak and half the maximum intensity on the *high* energy side.
p. estimated from figure 1 of ref. 56 assuming no contribution from tail of CT band.
q. a value of 37.2 kJ mol^{-1} obtained in CT study[59].
r. J. M. Goodenow, Ph.D. thesis, Univ. of Michigan, Ann Arbor, Michigan, 1965.
s. a value of 33.7 kJ mol^{-1} obtained in CT study[118].
t. reacted with iodine in the vapour phase.
u. ref. 17.

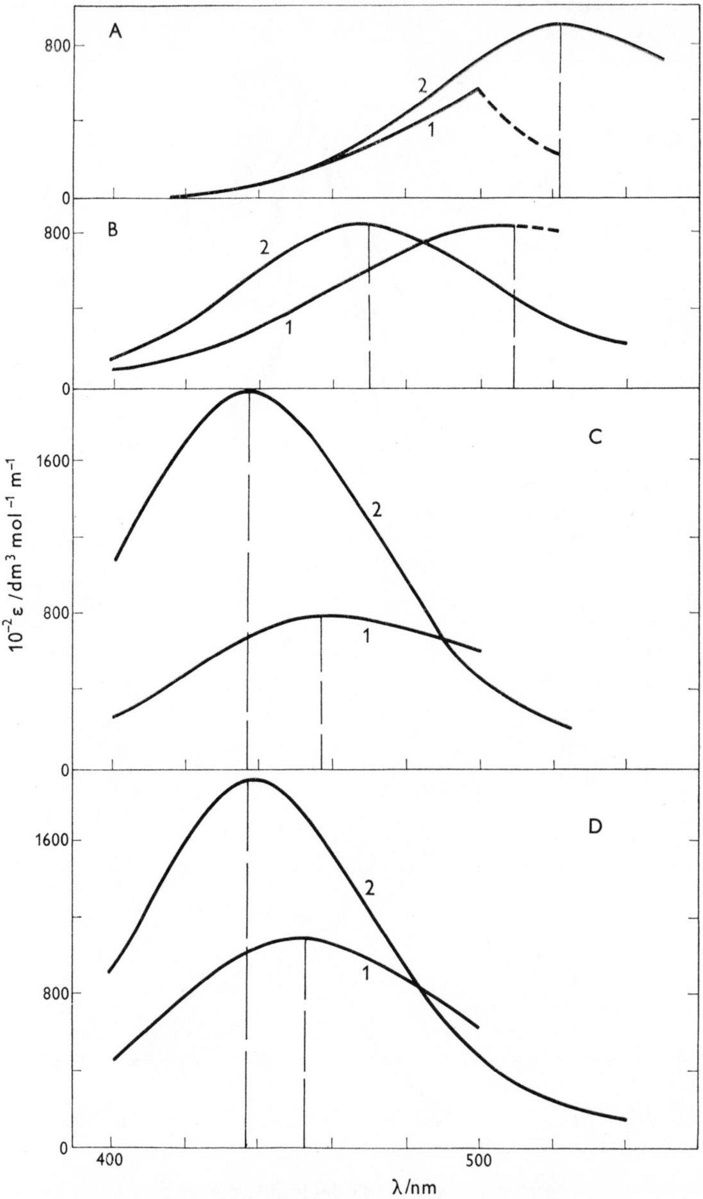

FIG 3.4 Effect of solvent on free and complexed iodine

A. (1) free iodine in vapour phase at 120°C (dashed part is the region of vibrational structure)[88], (2) iodine in *n*-heptane at 25°C[116].

B. (1) $(C_2H_5)_2O$–I_2 in vapour phase at 60°C (constant activity method[34]), (2) $(C_2H_5)_2O$–I_2 in *n*-heptane at 15°C (constant activity method[25]).

C. (1) $(CH_3)_2S$–I_2 in vapour phase at 90°C[40], (2) $(CH_3)_2S$–I_2 in CCl_4 at 25°C[56].

D. (1) $(C_2H_5)_2S$–I_2 in vapour phase at 115°C[116], (2) $(C_2H_5)_2S$–I_2 in *n*-heptane at 25°C[116].

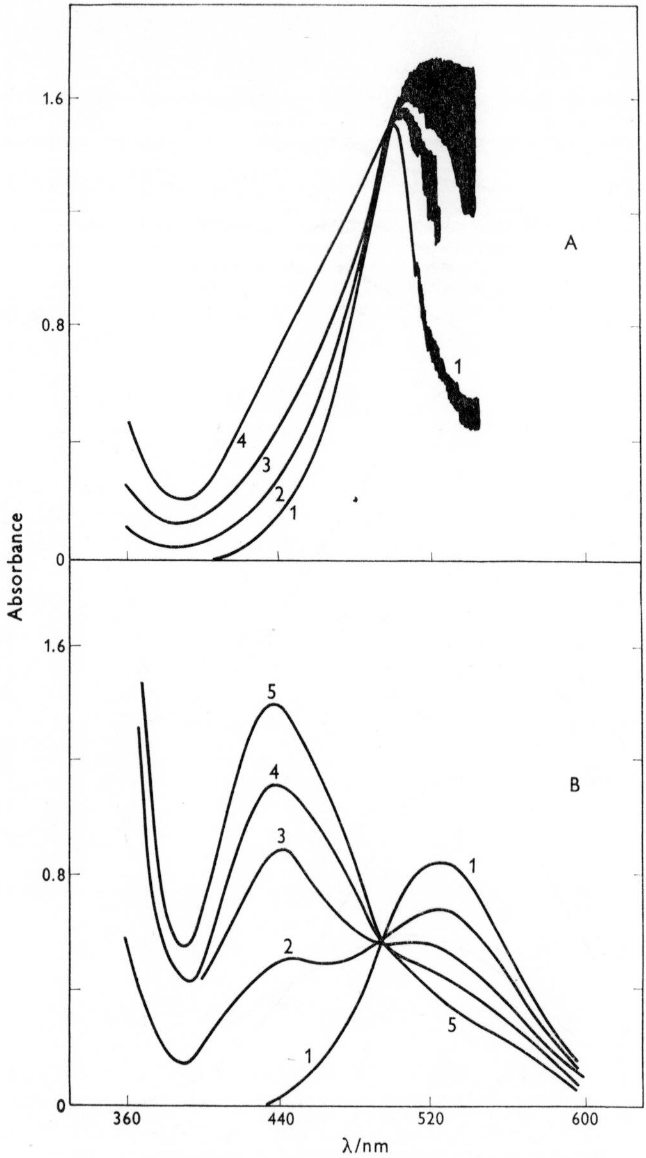

FIG 3.5 Diethyl sulfide–iodine complex

A. in vapour phase at 115°C in a 0.5 m cell[116], $[I_2] = 5.33 \times 10^{-5}$ mol dm^{-3} (1) iodine alone (2) iodine + sulphide (6.55×10^{-3} mol dm^{-3}) (3) iodine + sulphide (19.1×10^{-3} mol dm^{-3}) (4) iodine + sulphide (38.8×10^{-3} mol dm^{-3}),

B. in *n*-heptane at 25°C in a 0.01 m cell, $[I_2] = 9.2 \times 10^{-4}$ mol dm^{-3}; (1) iodine alone (2) iodine + sulphide (2.2×10^{-3} mol dm^{-3}) (3) iodine + sulphide (5.0×10^{-3} mol dm^{-3}) (4) iodine + sulphide (10.0×10^{-3} mol dm^{-3}) (5) iodine + sulphide (20.0×10^{-3} mol dm^{-3}) (from J. M. Goodenow, Ph.D. thesis, University of Michigan, Ann Arbor, Michigan, 1965).

both weak and strong complexes. The CT band can be characterized the same way, so that such a study gives directly the quantitative relation between the extinction coefficient of the blue-shifted iodine band (ε_{bs}) and ε_{CT} at their respective band maxima[25]. This is illustrated in figure 3.1. Application to the weak diethyl ether–I_2 complex[34] gave the blue-shifted band shown in figure 3.4B. The ε_{bs} was obtained by combining the K_c determined by a non-spectrophotometric study[21], the $K_c\varepsilon_{CT}$ obtained by the conventional spectrophotometric method[37], and the $\varepsilon_{CT}/\varepsilon_{bs}$ ratio found by the constant activity method[34].

A characteristic of *solution* studies is the observation of a well-defined isosbestic point, provided there is present only a 1:1 complex (figure 3.5B). This need not be a unique species, but may comprise several 1:1 isomers in equilibrium as for, say, different orientations of the complex[8]. A drift in the isosbestic "point" at higher concentrations of donor and acceptor is an indication that complexes of higher stoichiometry may be formed, although deviation from Beer's law or Henry's law also is possible. Qualitatively, the isosbestic point appears at shorter wavelengths for the stronger complexes, but quantitative correlation is not to be expected because ε for the blue-shifted bands can be appreciably different for complexes of different type.

The vapour phase blue-shifted iodine bands, unlike those in solution, show no isosbestic point (figure 3.5A). This is because ε_{bs} for the complexed iodine in the vapour phase is greater than ε_{I_2} for the free iodine over the entire continuum range. In solution, the complexed and free bands are enhanced (except, apparently, for diethyl ether–I_2) and their band maxima appear shifted in *opposite* directions, which results in their crossing. Attempts to observe an isosbestic point in the vapour phase by studying still stronger complexes to further separate the bands have not been successful because of apparent reaction with the iodine[67].

Comparison of the blue-shifted iodine bands in the vapour phase and in solution (figure 3.4) shows that the solvent effects are large. For the limited data available so far, the additional blueshift due to solvent is in the order diethyl sulphide $<$ dimethyl sulphide $<$ diethyl ether, a sequence which is opposite to their donor strengths. Overall solvent effects undoubtedly are complex. However, a qualitative explanation for the relatively large effect on a weak complex has been proposed in terms of models a and b in figure 3.2. When a complex in the vapour phase is immersed in a solvent, a cavity must be created in the liquid. Since this requires energy, the cavity will tend to be of minimum size. Therefore, the complex will be compressed by the solvent, and the intermolecular distance in model b will be smaller than that in **a**. This results in greater exchange repulsion and, hence, an additional blue shift. The effect is large for a weak complex because it is more compressible due to the shallow minimum in its potential energy curve. Compression also would increase orbital overlap, which could enhance the intensity, although this apparently is not the case for diethyl ether–I_2.

It is interesting to note that for the homologous series of iodine complexes with the methylbenzenes, ε_{bs} and K_c, as determined by the constant activity method, show a good linear relation[25].

3.3.4. "CONTACT" CHARGE-TRANSFER SPECTRA

The approach, in the vapour phase, of a donor and an acceptor molecule to van der Waals distance will always be attractive because of London dispersion forces (figure 3.2a). However, if the interaction is so weak that $-\Delta H^{\ominus}$ is less than **k**T, there will be no appreciable "sticking" of the D and A moieties. When no stable complex is formed, the D,A species is referred to as a "contact" pair. Contact pairs occur also in solution. Here, ΔH^0 may be zero or even slightly positive, depending on the nature of the D,A species and the solvent (figure 3.2b).

New spectral bands can arise as a consequence of the random "encounters"* between donor and acceptor molecules[8]. The new bands may be either vibrational (*e.g.* as found by Ketelaar[123] for gaseous mixtures with halogens) or electronic. For the latter, the band corresponds again to transition 1 in figure 3.3 and is called a "contact charge-transfer (CCT)" band. Iodine has been among the most studied acceptors in CCT spectra (another is oxygen), but other halogens and halogen-containing compounds investigated include [6]Br_2,Br atom, WF_6, MoF_6 and IF_7. The donors of the contact pair generally have been of σ-type, *e.g.* saturated hydrocarbons (although towards oxygen the donors also are of *n*- and π-type). Spectral data for some contact pairs can be found in table 3.3.

Iodine absorbs in the ultraviolet region[124] with a particularly intense band at 182 nm. In the presence of "inert" gases and saturated hydrocarbons, there is an appreciable enhancement in absorbance in this region (figure 3.6). The effect of gases such as He, N_2, O_2, and CH_4 at $\sim$1 atm pressure is very nearly the same[88,125], in spite of their high and rather different ionization potentials[83]. This is attributed to molecular collisions which broaden the rotational and vibrational structure of the iodine band at 182 nm towards that of a smooth continuum[6].

Characterization of the CCT band of alkane, iodine pairs requires subtracting the "free" iodine and alkane contributions from the total absorbance of the alkane/I_2 mixture. Based on the effect of added gases noted above, the best choice for the "free" iodine reference is the maximally pressurized iodine spectrum (curve 2 in figure 3.6).

* The term "encounter" rather than "collision", which sometimes is used, has been suggested by Mulliken[122a]. The latter term conveys an image of molecular distortion, which really is of no consequence with regard to these spectral observations.

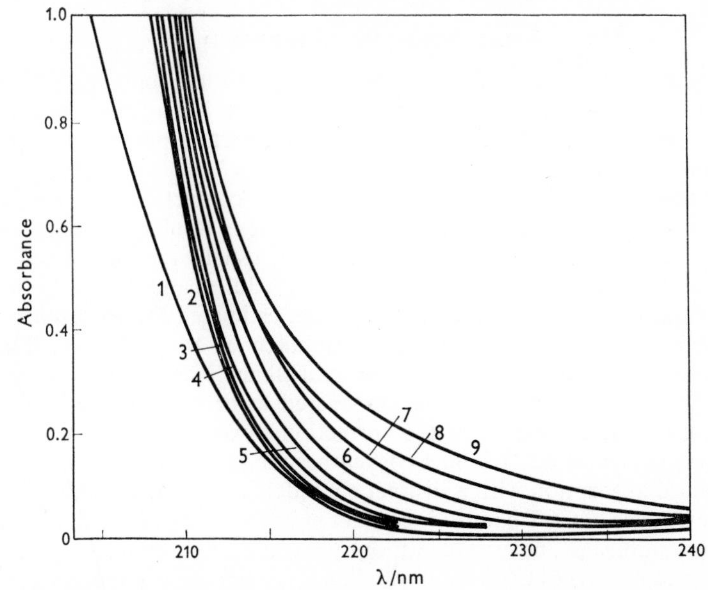

FIG 3.6 Vapour-phase ultraviolet spectra of iodine at 110°C: curves (1) through (5) for $I_2 = 9.03 \times 10^{-5}$M in a 50.0 cm cell; curves (6) through (9) for $I_2 = 6.50 \times 10^{-5}$M in a 75.0 cm cell; all gases and vapours = 3.50×10^{-2}M; (1) no added gas (2) N_2, O_2, or CH_4 (3) ethane (4) *n*-butane (5) *n*-pentane or neopentane (6) *n*-hexane (7) *n*-heptane (8) cyclohexane (9) methylcyclohexane. Curves 3-9 corrected for small alkane absorption.

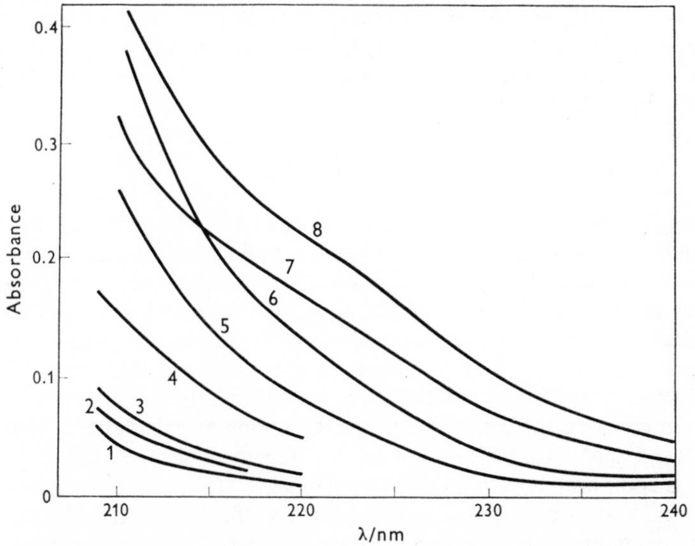

FIG 3.7 Vapour-phase CCT spectra of alkane-iodine systems at 110°C: curves (1) through (4) for $I_2 = 9.03 \times 10^{-5}$M in a 50.0 cm cell; curves (5) through (8) for $I_2 = 6.50 \times 10^{-5}$M in a 75.0 cm cell; alkanes = 3.50×10^{-2}M; (1) ethane–I_2 (2) propane–I_2 (3) *n*-butane–I_2 (4) *n*-pentane–I_2, or neopentane–I_2 (5) *n*-hexane–I_2 (6) cyclohexane–I_2 (7) (7) *n*-heptane–I_2 (8) methycycllohexane–I_2.

Subtracting this curve from all the other alkane/I_2 curves in figure 3.6 gives the CCT bands shown in figure 3.7. No distinct maxima are observed, although the presence of an inflection point is indicated in some cases.

Although CCT band maxima for alkane,I_2 'contacts' were reported in some of the early solution studies[93], these results are not valid because of the choice of a reference solvent, usually another alkane, to correct for the "free" iodine contribution[6,88]. Julien and Person[92], using the vapour spectrum of iodine as a guide, estimated the absorption of "free" iodine in solution which was taken as the reference. They reported distinct band maxima for several alkane,I_2 contacts. However, the characterization cannot be considered to be quantitative because of the qualitative nature of the correction. Evans[125a] has shown that, at low temperature, the methylcyclohexane,I_2 system shows a pronounced absorption with a distinct shoulder just below 240 nm. Thus, the existence of CCT bands for contact pairs involving molecular iodine seems established, both in the vapour phase and in solution. At present, the solvent shifts of CCT bands of alkane,I_2 pairs cannot be ascertained because no band maximum has been characterized in the vapour phase, and some uncertainty still exists about such characterization in solution. However, solvent shifts can be determined for contacts of alkanes with halogen atoms, because distinct CCT band maxima are observed in both phases (table 3.3).

The CCT absorbance has been shown to be directly proportional to both the hydrocarbon and the iodine concentrations[88]. This is expected from collision theory (see also section 3.4.2). But the same expectation holds for the condition of very weak complexation where $K_c \rightarrow 0$ [see equation 3.8]. Further, over a short temperature range, no dependence of CCT absorbance on temperature is observed, within experimental error, which again is consistent with either collision theory or a "weak" complex" model[88].

The energy of a CCT transition, $\mathbf{h}\nu_{CCT}$, is given by equation 3.25. Since I_D^v of the hydrocarbon and E_A^v of iodine are known, d_{12} can be calculated if $\mathbf{h}\nu_{CCT}$ is estimated. Tamres and Grundnes[88] assumed a CCT band maximum of $\sim$215 nm for cyclohexane,iodine. This results in a value of $\sim$5.8 Å for d_{12}. This intermolecular distance is somewhat larger than the sum obtained from van der Waals radii, as predicted by Mulliken[8,88] based on the effective large size of the strongly antibonding σ_u orbital of molecular iodine.

From collision theory, Tamres and Grundnes[88] calculated a "contact" constant, $K_c \simeq 0.13$ dm^3 mol^{-1}, for cyclohexane,iodine. This value is comparable to that estimated by Prue[126] for random contacts in solution. Values even smaller than this have been reported for K_c of weak complexes in solution. According to Scott[127], equilibrium constants obtained by the usual Benesi-Hildebrand analysis (or some modification thereof) pertain to the association of pairs in *excess* to that for random encounters. This does not mean the values obtained are precise, however, because of the inherent difficulty in separating the terms in the $K_c\varepsilon$ product for weak complexes (section 3.2.1).

The magnitude of ε can be calculated from equation 3.9 using the "contact" constant for the cyclohexane, iodine system and the absorbance data in figure 3.7 at ~215 nm. The resulting value is $\varepsilon \sim 10^6$ dm^3 mol^{-1} m^{-1}, which should be regarded only as being of a reasonable order of magnitude.

The relative positions and heights of the absorptions in figure 3.6 may be due to any one or more of the following possibilities: (1) The trend of increased absorbance of the alkane,iodine pair at longer wavelengths follows the trend of decreased ionization potential of the alkane. This would be expected from the relation between $\mathbf{h}\nu_{CCT}$ and I_D^v in equation 3.25. (2) The "contact" constant should be about the same for all the alkane,iodine systems at the same temperature. Therefore, a higher absorbance may be due to a higher ε, *i.e.* the transition probability may be related to the polarizability of the alkane. (3) In the absence of information on the CCT band maxima, their positions cannot be compared with that of the intense band (due to the excited iodine state) at 182 nm and that (due to the excited alkane state) at still lower wavelength. But the possibility that the sequence observed may arise through a mechanism whereby the CCT bands borrow intensity from a nearby intense band cannot be discounted.

Transition probabilities should vary with the geometry of the complex. For a loose donor, acceptor contact pair, different orientations should be nearly equally probable, and the observed CCT band is a composite of the contributions from all orientations. It is interesting to note that the CCT band of iodine with the open-chain *n*-pentane and with the ball-like neopentane is the same, within experimental limits. Both molecules have almost identical ionization potentials[83].

Contact charge-transfer bands with alkyl halide donors and halogen acceptors also have been observed (table 3.3). Again, distinct band maxima with halogen atoms have been found. There is some indication of CCT band maxima with molecular iodine, but characterization has been hindered by the fact that the alkyl halides also absorb in about the same region[96].

3.4. INFRARED AND RAMAN STUDIES

3.4.1. INTRODUCTION

For convenience we shall discuss changes in the spectra of the donor and halogen separately although it is clear that any interpretation made or model proposed must be consistent with all the data available. As indicated in 3.1 we have considered only a selection of the data available to illustrate the principles involved.

3.4.2. CHANGES IN THE HALOGEN MOLECULE SPECTRUM

We shall begin by summarising the changes that might be expected in the halogen molecule vibrational spectrum in the light of models which have been considered in Chapter 1. The expected ν(X-Y) band frequency and intensity perturbations have been discussed in detail in section 1.2 where it is pointed out that, in general, the magnitude of such changes depends on the nature of the donor and, therefore, on the nature and extent of the forces acting between the molecules. If the donor is such that charge transfer occurs the perturbation is expected

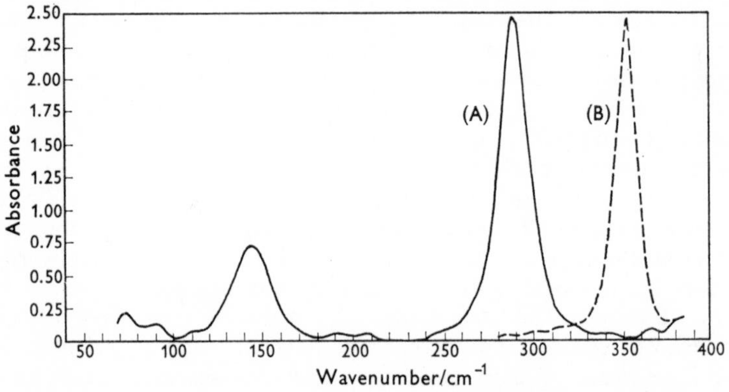

FIG 3.8 Far-infrared spectrum of (A) pyridine–ICl in benzene, and (B) "free" ICl in benzene. An illustration of the effect of complexation on the ν(I–Cl) band.

to be a large one (table 1.6, p. 42). Energy considerations show[128] (ref. 36, p. 169) that the lowest acceptor orbital (unoccupied) on the halogen is an antibonding σ_u orbital and transfer of charge to this orbital will result in a weakening of the X-Y bond (*i.e.* a decrease in the X-Y force constant). In the absence of any other effects this will result in a drop in the ν(X-Y) band frequency. This is illustrated in figure 3.8 which shows the low frequency infrared spectrum of ICl in benzene and pyridine–ICl in benzene. This mode is, however, only approximately a X-Y stretching mode since complexation results in the production of five new vibrational modes[129] (due to loss of three translational and two rotational degrees of freedom by the halogen molecule). These modes are expected to have low frequencies[129] and one or more of them may have the correct symmetry so that vibrational coupling occurs with the ν(X-Y) mode (for example in pyridine–XY the ν(D–X) and ν(X-Y) modes both belong to the species A_1). This will result in a force field change which may lead to a different force constant k_{XY}. An approximate normal co-ordinate analysis shows[130,131] that the resulting k_{XY} force constants are still considerably lower than those for the free molecules (sections 1.2.4 and 2.3.2). This supports the view that such frequency decreases are due to sacrificial action, especially since Person and Mulliken have shown[36i] (section 1.2.3) that the frequency of a diatomic molecule should increase by interaction with a donor if the k_{XY} force constant does not change as a result of the interaction. These considerations do not, of course, take into account the possibility of polarisation of the X-Y molecule by the donor which would obviously lead to a change in the ν(X-Y) frequency. Hanna and Williams[132] have attempted to calculate the contribution of electrostatic interactions to the frequency and intensity changes for halogen complexes of benzene caused by placing the halogen molecule in the electric field due to the quadrupole potential of the benzene ring. The results indicate a contribution to these spectral changes which could be up to 80%. In both the frequency and intensity case electrostatic interaction leads to changes of the same type (*i.e.* decrease in frequency and increase in intensity) as those predicted on the basis of the donor–acceptor model. There are, however, some uncertainties in the calculations and experimental data—see section 1.2.5. Clearly electrostatic interactions *are* important but they lead to the same type of change as do charge-transfer effects and so the two kinds of interaction may be very difficult to distinguish experimentally. (Note, however, possibilities through n.q.r. and infrared intensity studies—see sections 3.5.3 and 1.2.9.)

Using the Person-Mulliken model[36] the expected intensity perturbation of the ν(X-Y) band on complexation is a result of the change of electron affinity of the halogen when this vibration takes place[129] (see chapter 1 and chapter 6 of reference 36). Again, one must not ignore the effect of classical electrostatic forces which would lead to a difference in ionic character of the halogen molecule, and therefore to a perturbed ν(X-Y) band intensity; nor the possibility that the intensity changes occur by means of vibrational coupling (section 1.2.8). However, we shall see that the observed changes are in some cases very large and (as in the case of hydrogen-bonded complexes—section 2.3.4 and chapter 4) it seems unlikely that classical polarisation forces or vibrational mixing can alone account for such drastic perturbation (see section 1.2.6 for some further evidence in favour of such an opinion).

Despite the experimental difficulties associated with working in the region below 500 cm^{-1}, especially before the advent of filter/grating far-infrared spectrometers and interferometers (see section 2.2.2), considerable impetus was given to the study of the vibrational spectra of halogens when they are complexed with $b\pi$ and n donors by Person and his co-workers. A series of papers[133–137] published between 1957 and 1961 laid the foundations for future studies at the same time confirming the results of early workers[138] who had discovered that chlorine and

TABLE 3.6

Selected far-infrared data for weaker complexes of the halogens

Halogen	Donor	Solvent medium	$\bar{\nu}$(X–Y)/cm^{-1}	$-\Delta k/k_k^a$	$\Delta\bar{\nu}^d_{1/2}$/cm^{-1}	10^{-1} B_i^b	$10^{20}(\partial\vec{p}/\partial R)^c_a$/C	Reference
Cl_2		vapour	558(R)	—	—	0	0	159
		carbon tetrachloride	541(R)	—	—	0	0	159
	benzene	benzene	531	0.04	28	153	2.7	136
	benzene	benzene	525(R)	0.06	—	—	—	159
	p-dioxan	*p*-dioxan	520	0.07	—	—	—	149b
Br_2		vapour	318(R)	—	—	0	0	159
		chloroform	312(R)	—	—	0	0	159
	benzene	benzene	305	0.04	16, 22	420 ± 5	6.7	136, 137
	toluene	toluene	300(R)	0.08	—	—	—	158
	chlorobenzene	chlorobenzene	306	0.04	13	161	4.1	136
	m-dichlorobenzene	*m*-dichlorobenzene	311	0.04	—	125	3.6	155
	benzophenone	carbon tetrachloride	307.5	0.03	9	—	—	136
	iodobenzene	iodobenzene	283	0.18	—	925	9.8	155
	carbon disulphide	carbon disulphide	307	0.03	16	108 ± 20	3.3 ± 0.3	136
I_2		vapour	213(R)	—	—	0	0	136
		chloroform	207(R)	—	—	0	0	159
	benzene	benzene	200	0.06	12	160 ± 20	5.2 ± 0.3	131
	benzene	benzene	203	0.04	—	145	4.9	155
	iodobenzene	iodobenzene	195	0.11	—	440	8.5	155
	o-xylene	*o*-xylene	200	0.04	—	—	—	155
	durene	cyclohexane	200	0.06	—	640	10.3	155
	acetonitrile	*n*-heptane	205	0.02	—	—	—	131
	p-dioxan	*n*-heptane	206	0.02	6.0	680 ± 100	10.7 ± 1.6	131
	dimethyl sulphide	*n*-heptane	172	0.30	7.5	1450 ± 150	15.7 ± 1.7	131
	diethyl ether	cyclohexane	204	0.03	7.8	320	7.3	149b
	tetrahydrofuran	cyclohexane	200	0.06	7.0	490	9.1	149b
	3,3-dimethyl-1-butene	dimethyl butene	200	0.06	—	~160	5.2	149b
	dimethyl sulphide	chloroform	193	0.13	—	590	9.9	149b

ICl		vapour	381.5	—	—	—	—	133
		carbon tetrachloride	375	—	8	1050 ± 100	(8.4 ± 0.3)	133
	benzene	benzene/ carbon tetrachloride	355	0.10	15	1900 ± 300	3.3 ± 0.6	133
	toluene	toluene/carbon tetrachloride	356	0.10	14	2100 ± 400	4.0 ± 0.6	133
	p-xylene	*p*-xylene/carbon tetrachloride	350	0.13	15	1500 ± 400	2.0 ± 0.6	133
	chlorobenzene	chlorobenzene	365	0.05	20	2180 ± 400	4.2 ± 0.6	136
	benzophenone	carbon tetrachloride	353	0.11	17	2260	4.2	136
	methylene chloride	methylene chloride/ carbon tetrachloride	367	0.02	11	1350	1.7	133
	nitromethane	nitromethane/carbon tetrachloride	363	0.06	15	1000 ± 500	—	133
	acetonitrile	acetonitrile/carbon tetrachloride	352	0.12	15	2200 ± 500	4.4	133
	carbon disulphide	carbon disulphide	361	0.07	12	1400 ± 300	1.7 ± 0.6	133
	nitrobenzene	nitrobenzene/ carbon tetrachloride	358–62	0.09	15	2100 ± 500	4.0 ± 0.8	133
	p-bromoanisole	*p*-bromoanisole	352	0.12	26	2900	5.9	136
IBr		vapour	267	—	—	—	—	142
		n-heptane	262	—	—	40	(2.3)	149b
	benzene	benzene	249	0.07	—	90 ± 10	1.1 ± 0.2	149b
	diethyl ether	*n*-heptane	249	0.07	—	640	6.7	149b
	diethyl sulphide	*n*-heptane	208	0.36	—	750	7.5	149b
	p-dioxan	*n*-heptane	234	0.19	—	—	—	149b

a. Change of force constant measured from the value in a "non-complexing" solvent.
b. The total integrated intensity. Units are m mol^{-1} (*i.e.* m^3 mol^{-1} m^{-2}).
c. These transition moments are the values of the "added effective charge" in ref. 86. Values in parentheses are $(\partial\vec{p}/\partial R)_0$ values in the absence of specific interaction [*i.e.* $(\partial\vec{p}/\partial R)_T = (\partial\vec{p}/\partial R)_0 + (\partial\vec{p}/\partial R)_a$]. The values are related to those in Debye $Å^{-1}$ by 1 Debye $Å^{-1} = 3.335 \times 10^{-20}$ C.
d. The observed half-band width.

bromine exhibited infrared bands in "donor" solvents whereas they did not in the vapour nor in "non-complexing" solvents such as heptane or carbon tetrachloride. Complexes of Br_2, ICl, and IBr were studied down to the frequency limit of the instruments then available (equivalent to about 290 cm^{-1}). Many donors were used to try to produce a range of complex "strengths" and the decreasing ν(X-Y) frequency with increasing donor "strength" (measured by the first ionisation potential) for a particular halogen was interpreted in terms of the theory originally put forward by Mulliken[8,139]. This is illustrated by the selection of data given in table 3.6. Quantitative measurements of $\Delta k/k$ were obtained and semiquantitative intensity data (remarkably accurate when one considers the difficulties associated with making such measurements—even today, section 2.2.2) were used to calculate approximate values of the equilibrium constants and values for the "delocalisation moment" *i.e.* an estimate of the increase in $\partial\vec{p}/\partial R$ due to complexation (table 3.6). Work carried out on the pyridine–I_2 complex[140–141] in another laboratory helped to establish that these early results were valid.

Since 1963 most of the effort in obtaining far-infrared and Raman spectra of halogen complexes has been used to obtain spectra of "strong" complexes (mainly with amines and nitrogen heterocyclic bases) where it is also possible to observe a band usually assigned to the "stretching mode" ν(D-X) between the two molecules. Figures 3.8 and 3.9 show this band—in the 100–150 cm^{-1} region which has recently been the subject of more detailed study—see section 3.4.3. A considerable amount of data has been accumulated by Person *et al.* [130,131,142], Wood *et al.*[143–146], Yarwood[147–149] and Thompson *et al.*[150,151] Other far-infrared data have been obtained by Watari *et al.*[152,153], Nagakura *et al.*[154] and others[155–157]. Klaeboe[158] and Stammreich *et al.*[159] have published an extensive study of the Raman spectra of a wide variety of donor–halogen systems. Table 3.7 gives a selection of data for complexes in which a "ν(D–X)" band is observed.

Considering the inherent difficulties involved in studying these systems and the different instruments used in different laboratories, the agreement between data for the same systems studied more than once is remarkably good. This is especially true when one considers that the frequencies and the intensities of these bands are very much dependent on the nature of the medium (sections 1.2.9). Much of the early work[134–7,140] was carried out in mixed solvents and different concentrations of donor would give different "solvent" effects (see below). Later work, and that on complexes which can be isolated as solids, has usually been carried out in an "inert" solvent. Nevertheless there are still some discrepancies between recent results from different laboratories. For example the ν(Br-Cl) band in the solid pyridine–BrCl complex has been reported at 280 cm^{-1} [160] and 195 cm^{-1} [144]. Both authors do, however, emphasise the instability of the pyridine–BrCl complex while Wood

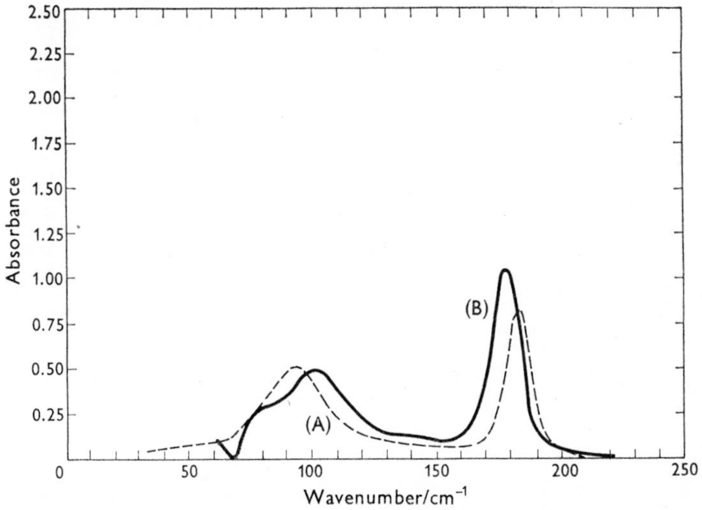

FIG 3.9 Far-infrared spectrum of pyridine–I_2 in cyclohexane (A) bands at 93 and 183 cm^{-1} for a pyridine: iodine concentration ratio of 3:1 (B) bands at 102 and 178 cm^{-1} for a corresponding ratio of 80:1. (Reproduced, by permission, from *Adv. Mol. Relaxation Processes*, **6,** 1 (1973)).

et al.[144] point out the possibility of assigning the far-infrared and Raman spectra in terms of $BrCl_2^-Pyr_2Br^+$[161]. Although they give good reasons for thinking their sample to contain 1:1 unionised complex units, it seems clear that different samples of this complex *may* contain varying amounts of ionic complexes. Some discrepancies are therefore not too surprising.

The results of these studies establish the three principal features, already mentioned, in the far infrared spectra of halogen complexes.

(a) The frequency of the halogen vibration decreases considerably on complexation and this frequency drops still further on going from a "weak" donor to a "stronger" donor. This change in halogen bond strength is indicated by an increase in $\Delta k/k$ as the complex "strength" increases (table 3.6).

(b) The intensity of the halogen stretching band ν(X-Y) increases on complexation (from zero in the case of X_2 halogens) and in most cases the intensity goes up as the donor "strength" goes up (table 3.6). There has recently been a report of some anomalous data in this respect[131] so that, until more reliable intensity data are available, one should be careful about their interpretation. (These data have led to a re-examination of the complex "model" in some cases—section 3.4.3.) Again using the diatomic model the value of $\partial\vec{p}/\partial R$ shows a general increase in most cases as $\Delta k/k$ increases.

TABLE 3.7
Far-infrared data for complexes of the halogens with amines and pyridine bases

Halogen	Donor	Solvent[a]	ν(X-Y) band		ν(D–X) band		Reference
			$\bar{\nu}_{max}/cm^{-1}$	$10^{-1} B_i^b$	$\bar{\nu}_{max}/cm^{-1}$	$10^{-1} B_i^b$	
Br_2	pyridine	benzene	231		120		151
	2-methylpyridine	benzene	226		113		151
	4-methylpyridine	benzene	225		115		151
	2,4-dimethylpyridine	benzene	223		110		151
	2,6-dimethylpyridine	benzene	230		100		151
	3-chloropyridine	benzene	243		98		151
	3-iodopyridine	benzene	239		91		156
I_2	pyridine	cyclohexane	183	2930	94	2790	150
	pyridine	cyclohexane	183	2500 ± 40	93	1950 ± 60	148
	pyridine	benzene	172		103		149b
	2-methylpyridine	cyclohexane	182	2480	87.5	2210	150
	3-methylpyridine	cyclohexane	181	2760	88.5	1660	150
	4-methylpyridine	cyclohexane	181	2850	88	1780	150
	2,4-dimethylpyridine	cyclohexane	180	3240	84.5	1710	150
	2,6-dimethylpyridine	cyclohexane	183	2980	74.5	1330	150
	3-chloropyridine	cyclohexane	188.5	2090	73	1990	150
	diethylamine	*n*-heptane	172	700[c]			131
	trimethylamine	solid (*m*)	188		148		154, 162
	trimethylamine–d_9	solid (*m*)	180		146		162
ICl	pyridine	solid (*pd*)	275, 265		179, 170		147
	pyridine	benzene	290	12000 ± 1500	140	3600 ± 600	147, 130
	pyridine	chloroform	283	12500	151	4300	149b
	pyridine	acetone	283				149b

ICl	pyridine	pyridine	277		160		143
	2-methylpyridine	benzene	290	11400 ± 800	132	2500 ± 400	149b
	3-methylpyridine	benzene	290	11900 ± 1100	133	2600 ± 300	130
	4-methylpyridine	benzene	289	11700 ± 1300	136	3400 ± 600	147
	2,4-dimethylpyridine	benzene	284	13000 ± 400	128	2900 ± 500	147
	2,3-dimethylpyridine	benzene	288	12500 ± 1000	128	2400 ± 400	147
	2,6-dimethylpyridine	benzene	287	12500 ± 1800	118	2300 ± 300	130, 147
	3,5-dimethylpyridine	benzene	289	8300 ± 1300	132	3200 ± 700	147
	trimethylamine	solid (*pd*)	249		196		152
IBr	pyridine	benzene	206	6300 ± 500	132	2800 ± 400	149b
	pyridine	pyridine	195		145		1496b, 143
	pyridine	solid (*pd*)	200, 190		159		149b
	4-methylpyridine	benzene	201		124		146
		pyridine	189		131		146
		solid (*m*)	189		146		146
	trimethylamine	solid (*pd*)	206		172		152
	trimethylamine	solid (*m*)	207		176		162
	trimethylamine–d_9	solid (*m*)	196		176		162
BrCl	pyridine	benzene	308		144		144
		pyridine	274		160		144
		solid (*m*)	295		196		144
		solid (*m*)	255(R)		171(R)		144
	4-methylpyridine	solid (*m*)	266 ± 3		173 ± 1		160
	2,6-dimethylpyridine	solid (*m*)	281		186		160

a. Solid complexes studied either as nujol mulls (*m*) or polyethylene disks (*pd*).
b. Integrated intensities in m mol^{-1} (*i.e.* m^3 mol^{-1} m^{-1}).
c. Minimum value (see ref. 131).

(c) A band at very low frequency of considerable intensity arises from some complexes (table 3.7). The interpretation of this band is by no means certain at present (section 3.4.3) but for "strong" complexes the idea of a band due to the vibration of the bond between the two molecules has been widely accepted.

For "strong" complexes notably those of substituted pyridines and amines it is, necessary of course, to consider the whole molecule[162] when making force constant calculations since in these cases there is presumably a partial covalent bond between the molecules. A number of attempts[130,131,142,148,154] have been made to calculate a force field for a linear triatomic system (*i.e.* D–X-Y) where the donor D is approximated to a point mass equal to its molecular weight. Using this treatment one allows mixing of the two low frequency a_1 modes but not mixing of these modes with those of the pyridine moiety, (section 1.2.3 and 2.3.2). The calculations on pyridine–I-X molecules[149b], using a *full* complex molecule treatment, have shown that this is a reasonable, although not completely valid, model. There is some mixing between the low frequency modes and the a_1 modes of the pyridine molecule although this primarily occurs with the lower frequency pyridine vibrations (as expected). For example, shift of the lowest pyridine, a_1 mode from 603 cm^{-1} to 631 cm^{-1} on complexation appears to be largely due to vibrational coupling with the "ν(D–I)" mode. Since there are few other major frequency shifts (table 3.13) this means that the force constants k_{IX} and k_{DI} calculated using the triatomic model are probably reasonable estimates (assuming that the interaction constant is known—see section 2.3.2). On combining the normal coordinate matrix, **L**, with the observed intensity data one is able to estimate, for the simple model, the value of the transmission moment $\partial\vec{p}/\partial R$ for both the low frequency modes. The values obtained for possible values of the interaction constant are given in table 3.8. It may be seen that the values are quite

TABLE 3.8

Transition moments[a] for the low frequency vibrations of pyridine–IX complexes

Complex	k_{13}/Nm^{-1}	$10^{20}\ \partial\vec{p}/\partial R_{IX}/C$	$10^{20}\ \partial\vec{p}/\partial R_{DI}/C$
pyridine–ICl	10	16.0	−22.7
	60	22.0	−13.3
pyridine–IBr	10	12.7	−25.4
	60	37.0	−5.5
pyridine–I_2	10	10.0	−18.7
	60	22.8	−6.95

a. Using triatomic molecule approximation; see reference 148 and section 2.3.2. Values for pyridine–ICl and pyridine–IBr have been corrected for the transition moment of the unperturbed halogen. (1 Debye $Å^{-1}$ = 3.335 × 10^{-20} C.)

TABLE 3.9

Concentration dependence of the two low frequency bands of the pyridine–I_2 complex in cyclohexane (data accurate to $\pm 2\ cm^{-1}$) (Reproduced, by permission, from *Adv. Mol. Relaxation Processes*, **6**, 1 (1973)).

Pyridine conc./ mol dm^{-3}	Iodine conc./ mol dm^{-3}	ν(I-I) band			$\tilde{\nu}$(D–I) band (composite)[a]		
		$\tilde{\nu}/cm^{-1}$	$\Delta\tilde{\nu}_{1/2}/cm^{-1}$	$10^{-1}\ B_i$[b]	$\tilde{\nu}/cm^{-1}$	$\Delta\tilde{\nu}_{1/2}/cm^{-1}$	$10^{-1}\ B_i$[b]
2.502	0.0158	176	12	3680	—[c]	—	—
1.925	0.0232	177	12	2720	—[c]	—	—
1.580	0.0216	178	11	2230	102.5	28	1950
1.206	0.0177	178	10	2930	100	29	2130
0.506	0.0216	181	10	2160	98	—	2080
0.256	0.0216	182	8	2850	94	22	1450
0.1740	0.0232	183	7	1940	92	22	1720
0.0821	0.0066	183	7	2170	94	18	1820

a. see section 3.4.3; b. intensity units are m mol^{-1} (*i.e.* $m^3\ mol^{-1}\ m^{-2}$); c. no energy at low frequency.

large; this being a reflection of the very considerable intensity of these bands. In the case of the pyridine–I_2 complex we have shown[148] that the value of $\partial\vec{p}/\partial R_{DI}$ transition moment cannot be accounted for using a simple polarisation model although it is obvious that polarisation forces must help to stabilise such a complex. It does seem likely that the intensity data are largely controlled by charge-transfer effects (and *possibly* by *repulsion* effects—section 1.2.5) for these particular complexes. The intensities of the ν(X-Y) bands for $b\pi$-$a\sigma$ complexes are, of course, much weaker (table 3.6) and so in these cases it is possible that the charge-transfer contribution is small.

Support for the charge-transfer model in the case of pyridine complexes has come from the interpretation of band energy changes which are observed on changing the solvent. Both the two low frequency bands are solvent sensitive, usually getting closer together[143,147,151] as the polarity of the solvent is increased (see tables 3.7 and 3.9). This effect has been said to be caused by an increase in the extent of charge transfer due to the increased contribution of Ψ_1 (Pyr^+–IX^-) to the ground state of the complex. If this were the case one would expect the band intensities to increase by the appropriate amount. However, it can be seen from table 3.7 that such an increase does not appear to occur for the ν(I-Cl) band. A 13 cm^{-1} shift in chloroform is, for example, a 15% increase over that shown in benzene. Assuming that the shift from the 375 cm^{-1} value in carbon tetrachloride is entirely due to charge-transfer an increase in intensity in chloroform should be, but is not, detected. It is possible that increased vibrational coupling of the low frequency modes has resulted in an intensity redistribution so more work is needed to investigate more thoroughly the normal coordinate changes. Another possibility is that part of the frequency shift from

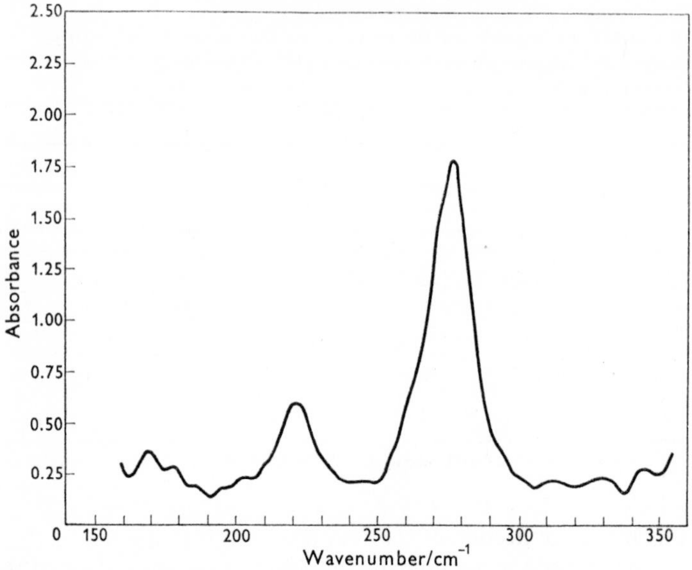

FIG 3.10 Far-infrared spectrum of pyridine–ICl in acetone. Absorption at 228 cm^{-1} shows the formation of ICl_2^-.

benzene to chloroform, acetone or pyridine is a "solvent" effect—caused by an increase in "bulk dielectric" effect due to "non-specific" interactions[155,163]. However, as is shown in section 1.2.9, the frequency change predicted[163] (see equation 1.74, p. 73) due to a change of refractive index is very small (2–3 cm^{-1} only) compared with the observed shifts. Thus, unless some unknown factor is operative, this possibility seems unlikely. The predicted "solvent" effect on the band intensities is also small in agreement with the observations, (equation 1.72 p. 71). Unfortunately one is unable to test this prediction properly since the formation of trihalide ions[142–148,150,151] prevents the measurement of accurate intensity data in the more polar solvents, (*e.g*. acetone and pyridine). The formation of ICl_2^- in this way is illustrated in figure 3.10. Another suggestion, made in section 1.2.9, is that these shifts may be caused by solvent competition in that measured frequencies may be weighted averages of those due to D–A and S–A interactions. If we consider the pyridine–ICI complex, however, (table 3.7) we see that the ν(I-Cl) frequency is higher in benzene than in chloroform. Since the benzene–ICI has a frequency (equivalent to 355 cm^{-1}—see table 2.3 and figure 3.8) which is considerably lower than that of ICI in chloroform (probably ~370 cm^{-1}) it follows that averaging of pyridine–ICI and benzene–ICI frequencies would lead to a *lower* value than for the pyridine–ICI and chloroform–ICI average (assuming that the weighting factors are the same in both cases).

One very interesting phenomenon[149] which may throw some additional light on this situation, is the spectral shift pattern which occurs for the pyridine-iodine system in cyclohexane when the amount of *excess* pyridine is increased. Table 3.9 gives details of the observed shifts and intensities for a range of pyridine concentrations. It can be seen that both the low frequency bands observed in the spectrum show distinct energy and band-width variations. The wavenumber shift is illustrated in figure 3.9. As the concentration of pyridine increases both bands get considerably broader but their energies move in opposite directions. (This effect is very similar to that mentioned above when different solvents are used.) If, however, the complex concentration is calculated assuming that both bands are dependent upon pyridine–I_2, as formed by the reaction:

$$C_5H_5N + I_2 \rightleftharpoons C_5H_5N\text{–}I_2$$

and that the equilibrium constant K_c is 157 dm^3 mol^{-1} (table 3.2), then the band areas show that there is very little intensity change of these bands (allowing for $\pm$ 10% experimental error). Since the refractive index change in this case is likely to be not more than 0.1 over the range of concentrations, the frequency changes again cannot be accounted for by the "reaction-field" effect (equation 1.74). As explained in section 2.3.6, band-shape analysis shows that the broadening, which is observed when the excess pyridine concentration is increased (or, it should be noted, when benzene is used as a solvent[149a]), is inconsistent with the idea of specific intermolecular forces between the solute (complex) and solvent. Such interaction ought to lead to a slower rate of rotational diffusion and a narrower overall half-band width. It seems likely, therefore, that vibrational relaxation (possibly collision-induced) may be important in determining the profiles of these bands, although the possibility of the formation of n:1, pyridine:I_2 species (which may also lead to band broadening) cannot be ruled out at the present time. For further discussion of this last problem see section 3.2.5 and figure 3.2.

3.4.3. VERY LOW FREQUENCY BANDS FOR HALOGEN-CONTAINING SYSTEMS

As indicated in section 3.4.2 many halogen complexes show a band in the far-infrared region which can be assigned to an intermolecular stretching mode. For thermodynamically "strong" complexes this is, of course, not unexpected, since for such a complex, vibrational analysis of the "whole" complex leads[129,136] straightforwardly to a "new" mode of the complex which may be approximately designated as "D–A stretching" (see sections 1.2.1 and 4.5.1 for further discussion of the "new modes of a complex). This mode is infrared active and is expected to be of quite high intensity on the grounds that the complex dipole moment is expected to depend quite critically on the D–A distance (section 1.2.6). For "weak" complexes such as benzene–XY it is questionable if such a "well-defined" ν(D–A) mode is to be expected since the complex lifetime is likely to be small (of the order of 10^{-12} s. see chapter 7, p. 572) and may be no more than a few vibrational periods in the far-infrared region where the band due to such a vibration is expected (see below for further discussion of this point). On the other hand, it is pointed out in section 1.2.6 that either vibronic *or* electrostatic effects are expected to produce a D–A vibration with some infrared intensity, so it is probably best to tackle the problem from an empirical point of view and to examine the observed spectra keeping *all* the possible interpretations in mind. Although absorptions in the very far-infrared (or submillimetre) spectra of simple liquids[164–181]—both polar and non-polar—have been studied since the mid-1960's it is only very recently[149,182,183] that such absorptions have been observed for donor–acceptor systems. For polar liquids such absorptions are thought[173,174,181] to be caused largely by libration (hindered rotation) of the dipole in a "cage" of surrounding molecules (either molecules of the same species or solvent). This model, known as the Poley Hill[134,135] model, is not inconsistent with the idea of a residual rotation of the molecular dipole in the liquid phase. For non-polar systems[123b,131,136] the bands are thought to be due to fluctuating induced dipoles caused by collision† between the molecules. This effect, which depends on the quadrupole (or higher) moments and polarisabilities of the colliding molecules, leads to absorption at a frequency presumably related to the frequency of collision. The absorption which is said to be "collision-induced", is obviously, at least partly, translational in nature[181].

The existence of absorptions similar to the ones mentioned above in donor–acceptor mixtures was first noticed[149,183] for systems in which halogens were dissolved in *p*-dioxan. The far-infrared spectra (after removal of the appropriate dioxan absorption) showed a very broad and quite intense band in the 80–100 cm^{-1} region. Some typical bands are shown in figure 3.11 and table 3.10 gives a summary of the data obtained so far for some donor/halogen systems. Similar absorptions have been observed[149b,182,183] for halogens and ICN in other aromatic solvents, diethyl sulphide, diethyl ether, carbon disulphide, *etc.* So for these systems, at least, the phenomenon is well established. In view of the, so far, well-accepted interpretation of the low frequency band in the pyridine–I_2 complex (section 3.4.2) it was tempting to interpret these bands in the same way (*i.e.* as arising from an intermolecular mode). There are several reasons why such an interpretation *may* be invalid.

† The word "collision" is used here since molecular motion (and distortion), rather than simply the proximity of the molecules (see footnote, p. 254), is probably more relevant here in accounting for the observed spectral effects. This is because the fluctuation of the induced dipole forces between the molecules arises from the dynamic nature of the interaction.

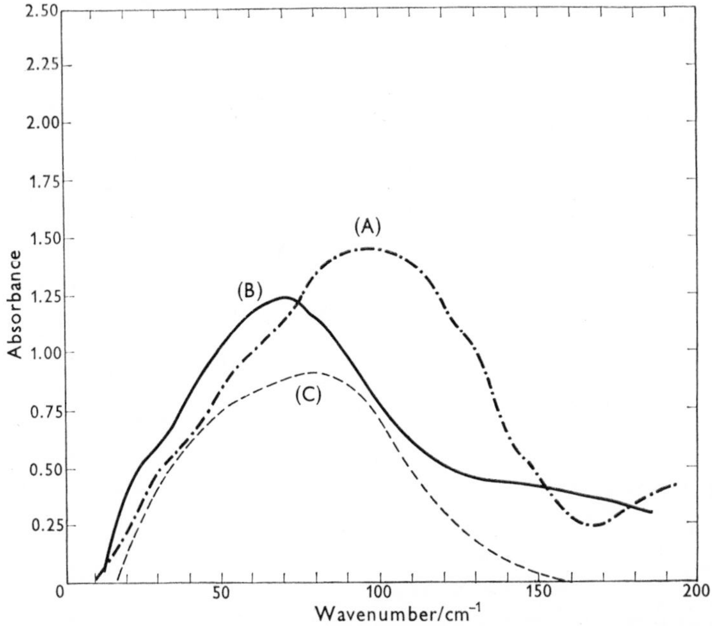

FIG 3.11 Very low frequency absorption observed for solutions of iodine in (A) benzene (0.5M) (B) carbon disulphide (0.5M) and (C) *p*-dioxan (0.2M) at 20°C. The pathlength is 0.35 cm. (Reproduced, by permission, from *Adv. Mol. Relaxation Processes*, **6,** 1 (1973)).

Firstly, the dioxan, aromatic hydrocarbon and carbon disulphide "complexes" are, thermodynamically, very weak (section 3.2) and it is debatable whether a well-defined intermolecular stretching mode is physically meaningful.

Secondly, the frequency and area of this band are dependent upon *both* the donor and halogen concentrations[149b] (see below). The computed "intensity" (B in table 3.10) depends, of course, on what one accepts as the concentration of "absorbing species". One can calculate, using the known equilibrium constant, the concentration of "complex" DA. However, if absorption is due not to a "well-defined" complex but only to the effects of sticky collisions (i.e. contacts) in solution one would not expect the band area to depend on such a computed concentration. The variations observed (table 3.11) imply that the observed band is very much "medium" dependent (since we have removed the "excess" dioxan absorption) as would be expected for a collision-induced mechanism. One would expect that the intensity of the collision-induced band would be given by[174a]:

$$\text{Total Area}^{\dagger} = K_1[\text{donor}][\text{halogen}] + K_2[\text{donor}][\text{donor}] + K_3[\text{halogen}][\text{halogen}] \quad 3.28$$

TABLE 3.10

Very low frequency infrared bands observed for mixtures of halogens with "donor" molecules at 20°C (donor "background" absorption removed except where otherwise stated).

Donor	Halogen	Halogen concentration/ mol dm^{-3}	$\tilde{\nu}_{max}$/ cm^{-1}	$\Delta\tilde{\nu}_{1/2}$/ cm^{-1}	10^{-1} B[a] (obs)/ m mol^{-1}	10^{-1} A[b] (calc)/ m mol^{-1}	Complex[d] dipole moment $10^{30}\,\vec{\mu}$/C m	Halogen polarisability[c] $10^{40}\,\alpha_{11}$/ C m^2 V^{-1}
p-dioxan[f]	I_2	0.054	80 ± 4	55 ± 5	1250	3.9	3.3	19.5
p-dioxan[f]	Br_2	0.145	85	50	800	10	4.3	11.1
p-dioxan[f]	ICl	0.132	102	51	1020	106	13.4	15.7
p-dioxan[f]	IBr	0.082	97	49	700	63	12.0	—
p-dioxan[f]	ICN	0.084	78	54	940	129	14.7	—
benzene[e,g]	I_2	0.20	82	104			2.07	19.5
benzene[e]	I_2	0.17	93	73	530		2.07	19.5
benzene[e]	I_2	0.47	102	55	320		2.48	19.5
carbon disulphide[e,g]	I_2	0.21	67	89			1.40	19.5
carbon disulphide[e,g]	Br_2	0.37	68	95			1.57	11.1

a. Observed intensity based on total halogen concentration; b. Calculated from equation 3.29; c. Values from M. W. Hanna, *J. Amer. Chem. Soc.*, **90**, 285 (1968); d. Values from chapter 7; e. Measurements in pure donor; f. Solutions in cyclohexane; dioxan concentration is 1.3 mol dm^{-3}; g. Without removal of "excess" benzene absorption.

TABLE 3.11

Effect of changing *p*-dioxan and iodine concentrations on the far-infrared band area (measurements at ~20°C in cyclohexane, pathlength is 6.5 mm)

dioxan conc./ mol dm^{-3}	iodine conc./ mol dm^{-3}	$\bar{\nu}_{max}$/ cm^{-1}	$\Delta\bar{\nu}_{1/2}/cm^{-1}$	Band area/cm^{-1}
0.23	0.058	88	54	7.0
0.38	0.056	—	—	8.8
0.68	0.055	83	52	12.0
1.28	0.054	80	—	11.7
2.81	0.054	78	44	28.0
11.38	0.054	76	—	70.0
3.44	0.032	78	54	13.9
3.44	0.067	75	52	30.0
3.44	0.123	77	56	56.0
3.44	0.136	84	50	78.0
3.44	0.169	78	55	78.0
3.44	0.209	78	—	118.0

(K values being constants for each of the individual interactions involved). Since effects due to the second term have presumably been removed and, since the third term is very small at the halogen concentrations, the observation of an integrated intensity dependent on the concentrations of *both* dioxan and iodine in the dioxan/iodine/cyclohexane system supports a collision-induced mechanism. If these bands were due to a ν(D–A) mode of a "well-defined" complex, however stabilised, one would expect an intensity dependent only on the iodine concentration (as long as [dioxan] $\gg$ [iodine]) (see section 3.2.1) with excess dioxan then showing only a (minor) "solvent" effect. A temperature variation study on these bands is now desirable since a collision-induced band is likely to be much more sensitive to temperature[173,174] than is an "internal" vibrational band.

Additional support for the idea of collision-induced absorption has been provided by the work of Kettle and Price[182] (see also section 7.3.3). They have shown that the dipole moments measured for mixtures of non-polar donors with non-polar halogens can be entirely accounted for by the mutual classical van der Waals interaction of the components[137]. Further, if the band intensities for the dioxan–IX systems (at constant dioxan concentration) are computed, based for comparision on the total halogen concentration, then it is seen (table 3.10) that the

† The equation could be extended to include contributions from both a "complex" (constant K_1) and a "sticky collision" (constant K_1').

I_2 system has the strongest band. Since I_2 has the largest polarisability of the halogens we again have support for the induced-dipole theory.

It should be emphasised that, although there appears to be good evidence in favour of the "collision-induced dipole fluctuation model", there are several aspects of these far-infrared spectra which remain incompletely explained.

(a) There are some frequency shifts which are very difficult to explain. For the dioxan/I_2 system there are variations (table 3.11) between 75 and 85 cm^{-1} which do not seem to have any obvious correlation with concentration (our error in frequency measurement for these very broad bands is about $\pm$2–3 cm^{-1}). More serious, there appear to be differences between the frequencies observed for the benzene/I_2 system by Brownson and Yarwood[183], and those of Kettle and Price[182]. This is presumably because in the first mentioned work[183] the spectrum of excess benzene was removed. The spectra shown in figure 3.11 therefore represent the absorption due only to benzene/I_2 collisions. These are at higher frequency than those for benzene–benzene collisions which is the opposite of what would be expected from the mass dependence of the frequency. (For an ideal binary collision the frequency is expected to be proportional to $[(1/M_A) + (1/M_B)]^{1/2}$ so an *increase* in mass should lead to a *decrease* in $\bar{\nu}_{max}$).

(b) It is somewhat uncertain to what extent the far-infrared absorption is due to the polar molecule "Poley-Hill" absorption which arises from libration (or "rattling") of the dipole in a cage of surrounding molecules. Although this effect is not expected to be large for the benzene/I_2 and dioxan/I_2 cases it could be important for more strongly interacting systems where the "complex" lifetime may be considerably longer (see below). Systems containing the (polar) interhalogens or a strong donor such as pyridine may well show a contribution due to this effect. One is able to calculate the intensity which would be expected for a condensed phase "rotational" band of a "rigid" complex with symmetric top symmetry whose dipole moment is $\vec{\mu}_z$. This is given by the Gordon[188] formula:

$$A = \frac{N}{3c^2}\vec{\mu}_z^2\left[\frac{1}{I_X} + \frac{1}{I_Y}\right] \qquad 3.29$$

where I_X and I_Y are the moments of inertia perpendicular to the symmetry axis. Although 3.29 gives the intensity for "pure" complex molecules one can estimate whether or not the band intensity is likely to be accounted for by such a mechanism. Table 3.10 clearly shows for homopolar halogens this is clearly not the case but for XY systems there *could* well be such a contribution due to the rotating "complex" dipole or from the dipole of the XY molecule itself. However, it should be noticed that the observed intensities do not always increase with increasing dipole moment (as expected from 3.29) so induced moments are important in these systems also.

(c) The extreme width of these bands has so far been explained only qualitatively. It should be noticed that the band width is often the same order as the frequency and to this extent the frequency variations may be of little significance[189]. Far-infrared band widths have been attributed to (i) coupling of intermolecular motion with internal vibrational modes[165–168,180] (ii) the anharmonic nature of the resonances involved[7,180] (iii) the overlapping of a series of damped resonances[190] of similar energy, the range of energies being caused by Brownian diffusion[181] of

TABLE 3.12

Effect of temperature on the far-infrared bands of the pyridine–I_2 complex in n-heptane (band positions to ± 2 cm^{-1}) (Reproduced, by permission, from Adv. Mol. Relaxation Processes, **6**, 1 (1973)).

	$\bar{\nu}$(I-I)/cm^{-1}		$\bar{\nu}$(D–I)/cm^{-1} (composite)[a]	
t°C	$\bar{\nu}_{max}$	$\Delta\bar{\nu}_{1/2}$	$\bar{\nu}_{max}$	$\Delta\bar{\nu}_{1/2}$
75	183	—	86	—
60	183	8	88	25
45	183	8	89	20
30	183	8	92	20
10	183	8	93	—
−10	183	—	97	—

a. see text.

the interacting molecule. No particular attention has been paid so far to the case of collision-induced absorption but it seems possible that uncertainty broadening may become important if the fluctuating dipoles change very rapidly. If the lifetimes for weak "complexes" are of the order of 10^{-12} s then the uncertainty broadening could be of the order of $\sim$50 cm^{-1}. A temperature variation study on the pyridine–I_2 complex in n-heptane has recently been made[149a]. The results, in table 3.12, show that the low frequency band shifts to higher energy as the temperature decreases. This is what is expected[173,174,177] if this band is due to a "Poley-Hill" hindered rotational mechanism. Since the ν(I-I) band changes only very little in frequency over this temperature range it appears that low frequency band has at least some "external" character. It is clear that this low frequency band is probably "composite" in the sense that its intensity may well arise from at least three concurrent phenomena (i) the change of dipole moment of the complex due to varying polarisation forces as the molecules vibrate against one another (ii) dipole moment fluctuation due to librational motion in the liquid (iii) a charge-transfer delocalisation moment caused by a change of overlap between the molecules. Some idea of the intensity provided by contribution (ii) can be obtained using equation 3.29. It is only about one tenth of the observed intensity (although it is not clear whether 3.29 can strictly be applied to weak solutions). It has previously been shown[148] that contribution (i) can only account for about one third of the observed dipole derivative (see section 2.3.2). Thus it appears that the charge-transfer contribution to this absorption is still considerable. It is worth mentioning here that similar results were obtained[149a] for the pyridine–IX complexes (X = Br, Cl) when the temperature is varied.

In view of the strong evidence outlined above that at least some of the spectral properties of benzene–I_2 and dioxan–I_2 are determined by classical electrostatic forces, it is important to consider the implications of this model in studying other spectral changes which occur. These, as we see (sections 3.4.2 and 3.4.4), are well-established perturbations of the ν(I-I) vibration and of certain internal modes of the donor. Not only are these perturbations severe ones but, in the case of the donor vibrations—for example in the benzene/I_2[191] case—they are quite specific since only *totally symmetric* vibrations are affected to any great extent. These changes have been shown to be consistent with the charge-transfer model[191,192], and it has been demonstrated (section 1.2) that they are probably caused by a contribution from the vibronic effect (probably a very small one for some systems). At first sight these results and the results in the far-infrared region—which can be explained quite easily for systems such as benzene/I_2 and dioxan/I_2 without invoking the idea of a "well-defined" complex—may appear to be contradictory. However, one *must* consider the effect on the interpretation of spectral data of this kind due to the relative values of the complex lifetime and the period of observation (*i.e.* the frequency of the experiment). Thus if the complex lifetime is $\sim 10^{-12}$ s then the period of observation at 1000 cm^{-1} ($\sim 10^{-13}$ s) is short compared with the lifetime and the complex may *appear* "well-defined". If, on the other hand, the observation period is 10^{-12} s or longer (in the very far-infrared) then it is not suprising that the effects of complex decomposition (*i.e.* sticky collisions between the components) predominate. It is *very important* to realise that the result to be expected very probably depends on the *frequency* of the experiment performed.

It is expected that spectral effects resulting from interactions by collision would be similar to those observed due to "solvent" effects (section 1.2.10). The observation of two bands, one due to "complexed" molecules and one due to "uncomplexed" molecules, for a given vibrational mode would therefore be regarded as good evidence for formation of a "well-defined" complex with a lifetime at least as great as the vibrational period. It is interesting therefore that for "weak" (aromatic) complexes of the halogens there are at least two cases of *two* bands having been observed, both in the Raman spectrum. Klaeboe[158] reported a "free" ν(Br-Br) stretching band to the high frequency side of the complexed Br_2 band for some aromatic hydrocarbon/Br_2 systems. However, from the spectra shown it is clear that their "interpretation" is by no means certain. Rosen *et al.*[193] have recently reported both "free" and "complexed" ν(I-I) bands in the system mesitylene/I_2 although they were unable to find the same "free" iodine band when benzene was used as the donor solvent. Since mesitylene is also non-polar these results are surprising and, if reliable, appear to indicate the formation of a "complex" with a lifetime greater than that of Ba–I_2. See section 1.2.5 for evidence of a similar phenomenon for Cl_2 in benzene/CCl_4. A careful study[149] has been made of the system benzene/IX (X = Cl, Br) in the far-infrared in an attempt to find out whether a band due to "free" IX is present. A ν(ICl) band doublet has been observed for mixtures of benzene and ICl in cyclohexane. This is shown in figure 3.12. Thus the lifetime of the C_6H_6–ICl "complex" is $>10^{-12}$ s and therefore apparently longer lived than the C_6H_6–I_2 "complex". The C_6H_6/ICl system still absorbs quite strongly in the 100 cm^{-1} region, however, so collision-induced phenomena may still be important at these frequencies. It seems reasonable to speculate that the somewhat longer lifetime of C_6H_6–ICl as compared with C_6H_6–I_2 is a result of additional dipole–induced dipole forces caused by the polar ICl molecule.

Raman studies[158] on ICl systems have also been made but without success in this respect. Studies on the donor vibration in aromatic hydrocarbon/halogen systems have also shown only one band for a particular mode. This could, of course, simply mean that the frequency shift is too small to allow detection of separate bands. For the pyridine–I_2 complex system[149b] (as well as for pyridine–phenol hydrogen-bonded complexes—section 2.3.4.) the existence of separate bands due to "complexed" and "non-complexed" species is well established. However, in these cases the presence of a stabilized (long-lived) complex is accepted.

3.4.4. CHANGES IN THE DONOR SPECTRUM

In general two types of change are expected on going from "free" donor to "complexed" donor.

(a) band frequencies are expected to change when the force field of the molecule is sufficiently perturbed by the complex formation. Large changes in force field are expected to lead to the observation of "new" bands (*i.e.* a separate "complex" bands), provided the complex lifetime is sufficiently long.

(b) band intensities are expected to be changed due to the effect of changing the electronic redistribution within the donor and due to the changes in the normal co-ordinates which result from force field changes. (see section 1.2 for a detailed discussion of the expected changes).

For halogen complexes the interaction with many donors (especially *n* donors) is quite strong and these changes are expected to be seen

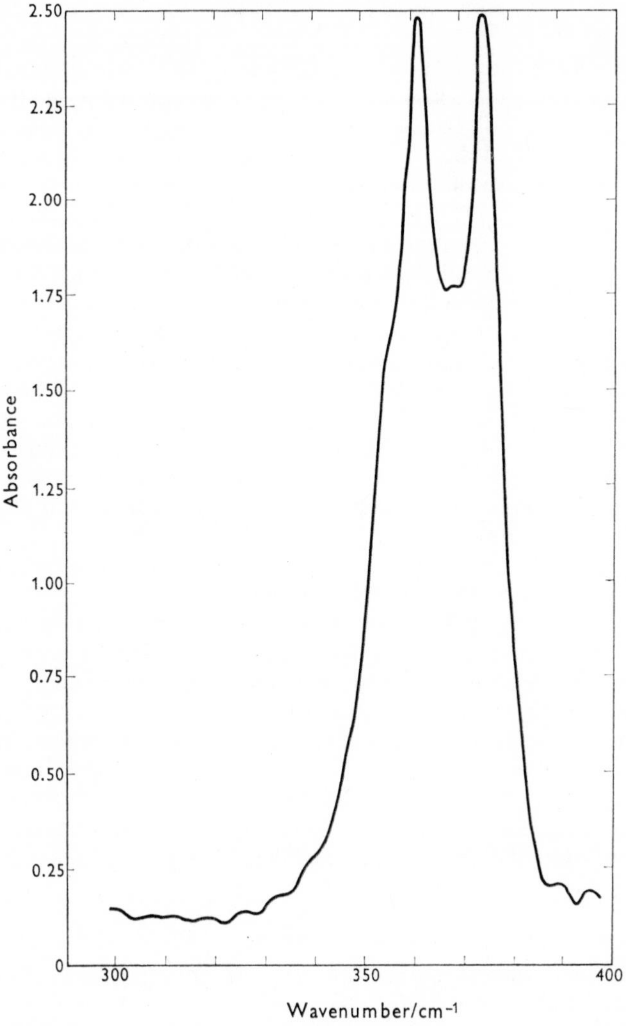

FIG 3.12 Far-infrared spectrum of a mixture of benzene and ICl in cyclohexane showing both "free" (375 cm^{-1}) and "complexed" (355 cm^{-1}) ν(I–Cl) bands. The concentrations are, benzene, 0.23M; ICl' 0.09M. The path length is 0.65 cm.

quite easily in the infrared (and Raman) spectra. In some instances a donor force field may be relatively unchanged except for those constants associated with the part of the molecule which is involved with bonding to the halogen. Examples of such donors may be molecules containing $>C{=}O$, $-C{\equiv}N$ or $>C{=}C<$ groups where lone-pair electrons or

π electrons are involved in bonding to the halogen. In such cases the multiple bond stretching force constant is invariably the largest one for the molecule and the effect of complexing may not be transmitted any further into the molecule (*i.e.* the motions of the other nuclei may be relatively unaffected). For molecules, such as benzene and pyridine, which do not contain "isolated" groups in the same sense, the changes in force-field are expected to be small (section 1.2.2) although some modes are expected to be more "sensitive" than others (see below). This is partly, of course, because coupling of modes in the donor with modes which arise as a direct result of complexation will be restricted to those belonging to the same symmetry species (section 2.3.2).

Similar considerations are relevant when considering the intensity changes expected to occur. For "strong" halogen complexes charge transfer is likely to make a considerable contribution to complex stability. According to this model[36,158,159] intensity perturbation will occur if $(\partial b/\partial Q_i)_{Q_i=0} > 0$ for a normal mode represented by Q_i (b is the mixing coefficient of the dative state—see equation 1.1). Such perturbation may, in principle, affect any mode which leads to "vibronic" coupling (*i.e.* a change of electronic wave function during vibration) but is expected[192] to occur for *symmetric* vibrations of the donor (see table 1.9, p. 63). These effects are expected to be greater, in general, than those which arise from differences in vibrational coupling (which will only occur between vibrations belonging to the same symmetry species) but it is likely that contributions from both effects will be operative as well as a contribution from purely classical electrostatic forces (section 1.2.6). The relative magnitudes *and* directions of these contributions to the total transition moment $(\partial \vec{p}/\partial Q)$ will then determine the observed band intensity. In considering which vibration will show an intensity perturbation one must, for a particular molecule, consider the form of the normal modes of vibration (not easy, because accurate force fields are rare) and how much these are likely to change in the complex. If the symmetry of the complex is different from that of the free donor then some change of the normal co-ordinates is inevitable. For the pyridine–IX complexes the symmetry is not expected to change on complexation since, in the solid at least, the N–I-X chain is linear[194]. The intensity changes[195] have therefore been interpreted in terms of the theory put forward by Ferguson[191,192]. It should be noticed, however, that small normal co-ordinate changes and possible polarisation effects should also be taken into account in any "complete" interpretation.

Halogen complexes with ketones, nitriles and other donors containing multiple bonds are expected to show intensity perturbation of the mode which approximately corresponds to stretching (or bending) of the multiple bond. It is here where electron re-distribution is likely to be greatest. Some effect may be transmitted to other parts of the molecule especially if the normal co-ordinates show strong mixing of skeletal modes, with, for example, the $\nu(C{=}O)$ or $\nu(C{\equiv}N)$ vibrations. The extent of vibrational coupling may then change on complexation. However, coupling with new modes (which arise due to complexation) is expected to be small in such cases since the complexes are relatively "weak".

The situation for complexes of benzene (and other aromatic hydrocarbons) is, as we have seen, by no means as straightforward. There is some evidence (section 3.4.2) against the formation of a "well-defined" complex, at least as far as the far-infrared spectrum is concerned. On the other hand, for a "complex" whose lifetime is $\sim 10^{-12}$ s (chapter 7) this is sufficient to allow ~ 30 vibrational periods for vibrations whose bands arise in the 1000 cm^{-1} region. Therefore perturbation of the donor vibrational modes can still be interpreted in terms of a "complex". However, the intermolecular forces which lead to such a "complex" may be different from those leading, for example, to the formation of pyridine–ICl.

There has been a large number of papers published describing frequency changes which occur on complexing *n* and $b\pi$ donors with halogens, interhalogens and pseudo-halogens (*e.g.* ICN). Some of these authors report qualitative intensity changes and a few report quantitative intensity data (usually) for one sensitive mode of the donor (*e.g.* $\nu(C{\equiv}N)$, $\nu(C{=}O)$, *etc.*—see below and section 2.3.2).

The earliest reports of changes in the donor spectrum due to complexation were made by Thompson *et al.*[196], Pimental *et al.*[197] and Ferguson *et al.*[191,192] Data presented are qualitative as far as intensity perturbations are concerned and some of the data have since been disputed[198,199].

The data for the perturbation of simple aromatic hydrocarbons has been summarised by Ferguson[192]. For benzene only two bands appear to be significantly perturbed (table 2.2 p. 124). The a_{1g} mode, assigned a band at 992 cm^{-1}, appears in the spectrum of I_2 in benzene (pure donor) as does the 850 cm^{-1} band (e_{1g}). Both these modes are forbidden by the infrared selection rules and their appearance in the complex spectrum has been attributed to changes in ionisation potential $((\partial I_D/\partial Q)_{Q=0} \neq 0)$ and overlap integral $(\partial S_{01}/\partial Q)_{Q=0} \neq 0)$ respectively. It is emphasised[129] that information on complex geometry is very difficult to obtain using infrared spectra. Perturbation of any of the modes of the donor which have the character of non-planar rotations are likely to be perturbed regardless of the complex geometry (*i.e.* even if the I_2 molecule is symmetrically placed as, for example, in the resting model (reference 36, p169–72). This is because such vibrations will lead to a value of $(\partial b/\partial Q)_{Q=0}$ which is different from zero. This point of view, originally put forward by Ferguson and Matsen[192], has been substantiated by Person and Mulliken[361]. Ferguson[192] has shown that the approximate selection rules deduced from the perturbation of the benzene spectrum apply well to the spectra of polymethylbenzenes[191,192], even though the symmetries are now much lower. In all the cases examined the modes perturbed are the ones which are expected to lead to either a change of ionisation potential (*i.e.* symmetric ring stretching modes) or a change in overlap between the molecules (*i.e.* out of plane modes which produce a "rocking" motion against the halogen) during vibration. The fact that these arguments are able to predict exactly the correct observed perturbations must lend considerable weight to the Mulliken description of such complexes[36] even though classical polarisation effects are not explicitly taken into account. It should be noted that, although classical forces have been shown to account reasonably well[132] for the perturbation of the iodine molecule (both frequency and intensity perturbation of the ν(I-I) mode), no explanation of the very specific perturbations of the benzene molecule has been achieved using the classical or "collision" complex model. On the other hand, it seems clear from the low frequency data that such a model can explain *some* of the spectral features (section 3.4.3). None of the perturbed bands in the aromatic hydrocarbon complexes are significantly different from those in the "free" hydrocarbon. This means that no great perturbation of the force constants (or normal coordinates) is caused by the interaction (as expected, section 1.2.2). In view of the relatively small changes which occur for much "stronger" complexes (see below) this is hardly surprising, and *may* be the reason why no "doublets" are observed in the donor spectrum. The perturbed bands are either Raman active only (in the benzene case) or are shifted so little in frequency that only an intensity change is detected[191].

The spectrum of pyridine changes considerably in a number of respects when the molecule is complexed to a halogen, XY[144,145,149,195,200,202] At least five bands show quite large frequency shifts. These are the bands which arise in pyridine[203–205] at 1439 cm^{-1}, 991 cm^{-1}, 1069 cm^{-1}, 603 cm^{-1} and 405 cm^{-1}. The shifts observed depend, as might be expected, on the physical state of the system under investigation, but it is now generally agreed that these five modes are the most "sensitive" to complexation, and that the shifts increase on going from I_2 to IBr to ICl. According to the data of Zingaro and Witmer[199] the shift of the 991, 1027 and 1070 cm^{-1} bands for the Br_2 complex lies between those of the I_2 and IBr complexes. Comparison is, however, difficult because some data are quoted for the solid and some for chloroform solutions. The more recent and complete data of Wood[143–146] and Yarwood[149,195] show that in benzene or chloroform solution (where the complexes are stable) the frequency changes are indeed in the order ICl $>$ IBr $>$ I_2 and that the data for Br_2 and BrCl complexes appear to parallel those of the IX series. A summary of the frequency data for these sensitive modes is given in table 3.13.

Very little quantitative intensity data have been obtained for the vibrational modes in pyridine. The only attempt at a complete intensity study is that of Yarwood[135] (for pyridine-ICl) who has shown that the a_1 vibrations of pyridine are perturbed much more than vibrations

TABLE 3.13

Wavenumber shifts (cm^{-1}) which occur for the "sensitive" modes of pyridine on complexation with halogens

Acceptor	Physical state							Reference
	pyridine (pure liquid)	1069	1439	1030	992	603	405	
I_2	solution in pyridine	1060	—	1031	1005	—	—	199
	solution in pyridine	1067	1443	1032	1007	—	—	201
	solution in pyridine	1060	1453	—	1005	—	—	146b
	pyridine/carbon tetrachloride	1064	—	—	1004	—	—	146b
	solid film	1057	1442	1029	1008	630	—	200
	benzene solution	1067	1445	1030	1006	623	420	145
	n-heptane solution	1068	—	1031	1005	617	—	149b
Br_2	chloroform solution	1065	—	1039	1010	—	—	199
	solid	1058	—	1035	1008	—	—	199
	solid film	1058	1445	1028	1010	633	—	200
	benzene solution	1062	1447	1030[a]	1009	627	418	144
IBr	solid	1057	—	1031	1011	—	—	198
	solution in pyridine	1066	1443	1032	1012	—	—	201
	solid	1057	1449	1032	1013	—	—	201
	benzene solution	1067	1449	1031	1010	627	420	145
	chloroform solution	1069	1449	1030	1010	627	420	149b
ICl	solid	1054	—	1033	1012	—	—	198
	solution in pyridine	1067	1442	1032	1012	—	—	201
	solid	1055	1452	1033	1012	—	—	201
	benzene solution	1067	1449	1030	1011	630	421	145
	chloroform solution	1067	1451	1034	1011	631	413	195
BrCl	chloroform solution	1065	—	1032	1013	—	—	199
	benzene solution	1065	1449	1033[a]	1011	630	422	144

a. solution in dichloromethane.

belonging to other symmetry classes. It is interesting to note that the quantitative data are not always in agreement with the qualitative changes which have been reported. For example, several authors[198,199] have claimed that the a_1 band at 1030 cm^{-1} (in the free molecule) is considerably decreased in going to the complex. The quantitative data show that there is no change in this band at all in going to the pyridine–ICl complex (most of the others being *increased* in intensity). This clearly illustrates the danger of making statements about apparent intensity changes without actually making the measurements. The recent intensity measurements[149] on other complexes of pyridine (section 2.3) have shown that those for pyridine–IBr are little different from those of pyridine–ICl while, for the iodine complex, those which have been measured are considerably lower. This is consistent with the frequency data. A summary of the intensity data obtained so far for pyridine complexes is given in table 2.17, p. 152.

Further discussion of these perturbations is given in section 2.3.2., but it is relevant here to consider whether the data provide any evidence for a particular complex "model". At first sight the drastic and specific perturbation of a_1 pyridine modes appears to give unequivocal support to the charge-transfer model since the selection rules deduced by Ferguson[192] appear to be obeyed. However, it should be remembered that "vibrational coupling" of the internal a_1 modes of the donor with "new" a_1 vibrations of the complex could lead to at least part of the perturbation. A whole-molecule normal co-ordinate treatment[149b] of pyridine–IX has shown that a frequency perturbation of the correct order and direction is caused without any change in the force field (section 2.3.2). These changes are caused entirely by kinetic energy effects through the **G** matrix and the resulting potential energy distribution shows considerable mixing between the ν(D–A) mode and internal a_1 modes of the pyridine. Whether or not such coupling could cause the drastic intensity perturbations is open to some doubt but it is an approach worth pursuing. It seems clear that the data may not support exclusively one model. (See section 2.3.2 for a more thorough discussion of this point). The n.q.r. data for the pyridine–ICl complex[206,207] support the idea that polarisation of the halogen by the polar pyridine molecule also plays a considerable part (see section 3.5.3). Thus the intensity changes observed for the "strong" pyridine complexes are probably a result of at least three contributory factors.

Apart from aromatic hydrocarbons and pyridine bases a number of other donors have been studied. Glusker and Thompson[146] for example studied solutions containing iodine, iodine cyanide, ketones, esters, nitriles and ether in addition to aromatic hydrocarbons and pyridine. They found that the ν(C═O) band in ketones shifted to lower frequency (by the equivalent of about 20 cm^{-1}) and that some weak new bands arose in the 1200–1300 cm^{-1} region. Since some irreversible reactions in solution occurred very weak new bands may not have been significant. For *p*-dioxan a new band arose at 1098 cm^{-1} which was assumed to be associated with the ν(C—O) band at 1119 cm^{-1}. Another new band was observed at 830 cm^{-1} this band being shifted from 874 cm^{-1}. For some esters it was found that the ν(C═O) band shifted in the same way as for ketones but that no shift of bands associated with the ether linkage occurred. However, other molecules containing both types of group did show more profound changes, indicating co-ordination of the I_2 to both sites. Surprisingly, no changes were observed in the spectrum of acetonitrile (although this would be expected to be a complexing solvent) when I_2 was added. This observation has been confirmed by Klaeboe[209] who did, however, find a reasonable shift of the ν(C≡N) band when ICl was used as acceptor.

Klaeboe and his co-workers have carried out extensive studies on various donors complexed with halogens and interhalogens. They have studied the infrared spectra of nitriles[209], dimethyl cyanamide[209], aldehydes[210], sulphoxides[211,212] and organophosphorus compounds[213,214] and in some cases determined equilibrium constants from the intensity variation of the shifted multiple bond "stretching" band. With the exception of the ν(C≡N) band in nitriles the shifts are to lower frequency

TABLE 3.14

Frequency and intensity changes for complexes of *n* donors with halogen and interhalogen molecules

Donor	Acceptor	Solvent	$\bar{\nu}_i$/cm^{-1}	$\bar{\nu}_{1/2}{}^{a}$/cm^{-1}	10^{-1} $B_i{}^{b}$	Assignment	Reference
dimethylacetamide	—	CCl_4	1662		$\varepsilon_{max} = 6740^{c}$	ν(C=O)	216
	I_2	CCl_4	1619		$\varepsilon_{max} = 780^{c}$	ν(C=O)	216
propionaldehyde	—	CS_2	1738	16	11000	ν(C=O)	210
	I_2	CS_2	1718	25	26000	ν(C=O)	210
	ICl	CS_2	1699	20	21000	ν(C=O)	210
	IBr	CS_2	1704	20	—	ν(C=O)	210
benzaldehyde	—	CS_2	1707	7	20000	ν(C=O)	210
	I_2	CS_2	1689	13	26000	ν(C=O)	210
	ICl	CS_2	1678	13	32000	ν(C=O)	210
	IBr	CS_2	1680	10	30000	ν(C=O)	210
acetone	—	CS_2	1718	—	—	ν(C=O)	196
	I_2	CS_2	1698	—	—	ν(C=O)	196
dimethylsulphoxide	—	CS_2	1072	13	16000	ν(S=O)	211
	ICN	CS_2	1036	30	21000	ν(S=O)	211
	I_2	CS_2	1021	—	—	ν(S=O)	211
	ICl	CS_2	1011	—	—	ν(S=O)	211
dimethylsulphoxide	—	CCl_4	1067	—	—	ν(S=O)	221
	I_2	CCl_4	1030	—	—	ν(S=O)	221
dimethylsulphoxide	—	CH_3NO_2	1045	—	—	ν(S=O)	223
	I_2	CH_3NO_2	1025	—	—	ν(S=O)	223
triphenylphosphine	—	CCl_4	1202	—	—	ν(P=O)	213
oxide	—	C_6H_6	1203	—	—	ν(P=O)	212
	I_2	CCl_4	1162	40	—	ν(P=O)	213
	ICN	C_6H_6	1167	—	—	ν(P=O)	212
acetonitrile	—	CCl_4	2256	7	390	ν(C≡N)	209
	ICl	CCl_4	2268	13	2500	ν(C≡N)	209
chloroacetonitrile	—	CCl_4	2263	13	20	ν(C≡N)	209
	ICl	CCl_4	2275	16	1000	ν(C≡N)	209
trimethylacetonitrile	—	CCl_4	2238	14	800	ν(C≡N)	209
	ICl	CCl_4	2252	15	4900	ν(C≡N)	209
benzonitrile	—	CCl_4	2232	10	1900	ν(C≡N)	209
	ICl	CCl_4	2240	20	11100	ν(C≡N)	209

a. the measured half-band width.
b. intensity data in m mol^{-1} (*i.e.*, m^3 mol^{-1} m^{-2}) unless otherwise stated.
c. maximum extinction coefficient at 40°C, units are dm^3 mol^{-1} m^{-1}, $K_c = 6.1 \pm 0.3$ dm^3 mol^{-1}.

(section 2.3.2). Typical magnitudes of these shifts are $\Delta\bar{\nu}$(C≡N) +7–14 cm^{-1}; $\Delta\bar{\nu}$(C=O), −15–50 cm^{-1}; $\Delta\bar{\nu}$(S=O), −30–60 cm^{-1}; $\Delta\bar{\nu}$(P=O), −30–70 cm^{-1}. These results are in general agreement with other work on carbonyl[215–219] and sulphoxide[220,221] containing compounds. In addition a number of intensity studies have been made for the most sensitive donor vibrational band. The results are summarised for some typical cases in table 3.14.

It is interesting to note that in all cases where data are reported the band width *increases* on complexation. (An exception is the ν(P=O) band which does not change appreciably in shape). Broadening of the bands is, in most cases, by about a factor of two and the intensification is also (with the exception of nitrile complexes) of this order. This *may* mean that the intensity perturbation arises simply from an increase in band width. This broadening of the infrared bands in the complex is the opposite of that expected if rotational diffusion is reduced by complexation. Broadening mechanisms which can be invoked include the possibility of vibrational relaxation during complex decomposition (see section 2.3.6).

Changes in some other donor modes occur for these complexes. For example, in acetonitrile complexes the ν(C—C) (a′) stretching band, which arises at 918 cm^{-1}, shifts to 923 cm^{-1} [209] and several combination bands show small frequency shifts. However, the two degenerate (e) stretching modes show no perturbation on complexation and this is interpreted to mean that the molecule retains its C_{3v} symmetry in the complex (presumably since destruction of the symmetry would lead to loss of degeneracy and changes in normal co-ordinates). A linear C≡N—I—X geometry would be expected anyway since the nitrogen can be thought of as approximately sp hybridised and other halogen complexes with nitrogen donors (amines) have been shown to contain linear N—I—X groups (in the solid)[222]. The carbonyl compounds studied[210,215–219] also showed frequency changes other than those related to perturbation of the ν(C—O) band. For example, in the aromatic aldehydes studied by Augdahl and Klaeboe[210] six bands between 800 and 1400 cm^{-1} show small changes in frequency but no intensity measurements were made. These vibrational modes involve b(C—H) and ν(C—C=O) skeletal modes and their perturbation supports the theory that halogens are bonded to oxygen in these complexes. In dimethyl acetamide–I_2[216] the ν(CN) band is shifted from 1460 cm^{-1} to 1530 cm^{-1} (and a band at 1000 cm^{-1} is increased in intensity). Again this is interpreted as evidence for the molecule being an "oxygen" donor even though here there is a possible nitrogen complexing site (no mention is made, however, of a detailed examination of the region where ν(C—H) changes might be expected if co-ordination occurred at the nitrogen atom). Yamada and Kozima[215] have found that several of the modes in acetone are perturbed on complexing with I_2. The b(C—C=O) band at 529 cm^{-1} is shifted to 534 cm^{-1} and the ν(C—C—C) band position is increased from 1216 to 1223 cm^{-1} (in addition to the red shift of the ν(C=O) band). The perturbed bands are again broadened slightly and the intensities increased considerably.

To summarise the situation, it is clear that vibrational studies have been made on the donor in halogen complexes with one (or more) of the following three objectives in mind.

(a) Band frequency and intensity variations of specific "group" vibration have been used to investigate qualitatively the site of complexation and the extent of the interaction. The complex symmetry has been inferred in some cases.

(b) Quantitative studies on the intensity variation of the shifted vibrational band (with concentration) have yielded thermodynamic parameters (K_c, $\Delta H°$, $\Delta S°$) for complex formation.

(c) Quantitative studies on all the vibrations of the donor (or as many as possible) have yielded information on the validity of the "models" used to describe complex formation.

3.5. OTHER TECHNIQUES

3.5.1. INTRODUCTION

A number of techniques other than electronic and vibrational spectroscopy has been used to study halogen complexes. The most important of these are X-ray crystallography, magnetic resonance spectroscopy and dielectric (microwave) absorption (including dipole moment measurements). Dielectric techniques are described in detail in chapter 7. In this section we shall consider how the other two methods are able to provide some information about the structures of halogen complexes.

3.5.2. DETERMINATION OF MOLECULAR GEOMETRY

Almost all the information on the geometries of halogen complexes has been obtained on the solid complexes by Hassel[194,222–232] and his co-workers using X-ray crystallography. These data have twice recently been critically reviewed[4,36] and it is only necessary here to emphasise the points of major importance to the foregoing discussion on halogen complexes. It is also a convenient place to mention the recent first determination[233] of the geometrical configuration of a halogen complex in solution. The data have been important for the elucidation of the complex structures in the two following ways.

1. The results have established (unambiguously in most cases) the relative orientations of the donor and acceptor in the *solid* phase. This information may be used to assess the nature of the donor–acceptor bonding orbitals through a knowledge of the shape and stoichiometry of the basic crystal unit. Thus the benzene–Br_2 complex was shown[227]

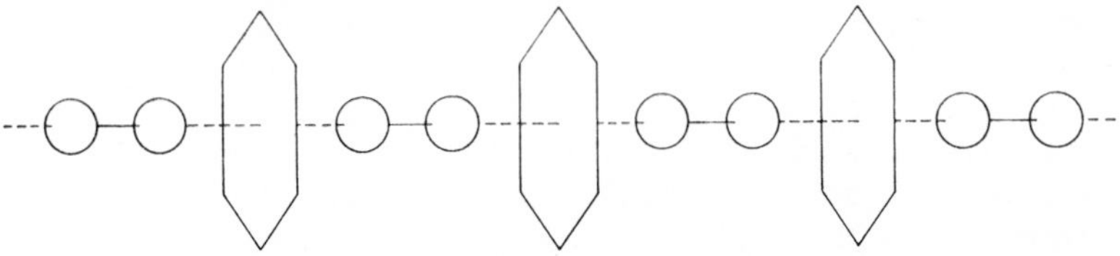

FIG 3.13 Diagrammatic representation of the structure of the solid benzene–Br_2 complex (Reproduced from *Quart. Rev.* **16,** 1 (1962), by permission of the Chemical Society).

to consist of linear chains with the bromine molecules orientated along the C_6 axis of the benzene in an *axial* configuration[139,222] (see figure 3.13). Person and Mulliken have shown (see reference 36, pp. 53–54) that if such a configuration is valid also for the "isolated" complex then the benzene donor orbital is *not* the one of highest energy (this, of course, assumes that the "isolated" complex ground state is stabilised by charge-transfer forces). For the p-dioxan and acetone complexes it has been found that the halogen molecule is not bonded in the configuration which would be expected from a consideration of the direction of the molecular orbital of lowest energy. The configuration of the dioxan–Br_2 complex is shown in figure 3.14a where, it should be noted,

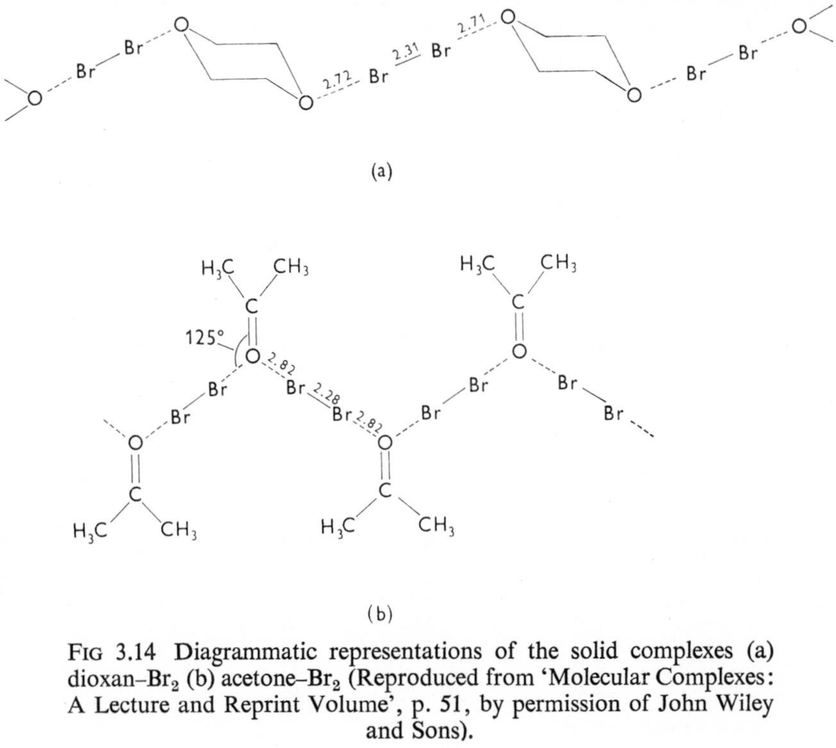

FIG 3.14 Diagrammatic representations of the solid complexes (a) dioxan–Br_2 (b) acetone–Br_2 (Reproduced from 'Molecular Complexes: A Lecture and Reprint Volume', p. 51, by permission of John Wiley and Sons).

the angle between the COC plane and the O—Br bond is 143°. If the donor orbital were the 2p oxygen "lone pair" orbital (essentially an atomic orbital) then one would expect this angle to be 90°. This means that the *n*-donor orbital, is, in fact, a hybrid orbital composed of two molecular orbitals only one of which is an essentially oxygen atomic orbital. When the oxygen in an ether is replaced by sulphur (for example, in dithiane–I_2) the angle is much smaller (about 106°) and here we see that the *n* donor orbital has almost entirely sulphur 3p "lone-pair" character. In the acetone case it is seen (figure 3.14b) that each acetone molecule is bonded to two bromine molecules and that the C=O—Br angle is 125° rather than the 90° expected if *n* donor orbital were entirely composed of the oxygen 2p orbital. As pointed out by Hassel,[222] however, these are the geometries expected if the oxygen "lone-pair" *MO*'s point in the directions expected from a simple hybridised orbital approach (*i.e.* approximately sp^3 and sp^2 hybrid orbitals in the dioxan and acetone cases respectively).

2. The crystallographic data may be used to distinguish "strong" complexes, in which sacrificial action by the halogen acceptor is reflected in the increased bond length in the complex, and "weak" ones, where no such bond extension is observed. Notice, however, that experimental error can be as high as ±0.03 Å (see reference 36, p. 46). The results, for a range of complexes with different donors, shown in table 3.15 illustrate this and also show that, in general, the donor-halogen distance (measured usually to within ±0.1 Å) is considerably smaller than the sum of the van der Waals radii of the D and X (but greater than the corresponding covalent bond length). The results for complexes such as *p*-dioxan–Br_2, methanol–Br_2 and benzene–Br_2 are usually interpreted as evidence that charge transfer is unimportant in such "weak" complexes, although the donor–Br distance is significantly less than the van der Waals radii sum. However, it may well be invalid

TABLE 3.15

X-ray crystallographic data for a selection of donor–halogen complexes (all data in Å units[a])

Complex	r(X-Y) (obs)	r(X-Y) (free)	$r(D–X)_{obs}$	Sum of van der Waals radii, D, X	Sum of covalent radii, D–X	Reference
trimethylamine–I_2	2.83	2.67	2.27	3.65	2.03	228
4-picoline–I_2	2.83	2.67	2.31	3.65	2.03	229
pyridine–ICl	2.51	2.32	2.26	3.65	2.03	194
pyridine–IBr	2.66	2.44	2.26	3.65	2.03	224
dioxan–ICl	2.33	2.32	2.57	3.55	1.99	226a
dioxan–Br_2	2.31	2.28	2.71	3.35	1.80	226b
acetonitrile–Br_2	2.33	2.28	2.84	3.45	1.84	230
acetone–Br_2	2.28	2.28	2.82	3.35	1.80	231
methanol–Br_2	2.28	2.28	2.80	3.35	1.80	232
benzene–Br_2	2.28	2.28	3.36	3.65	—	227

a. 1 Å is 0.1 nm.

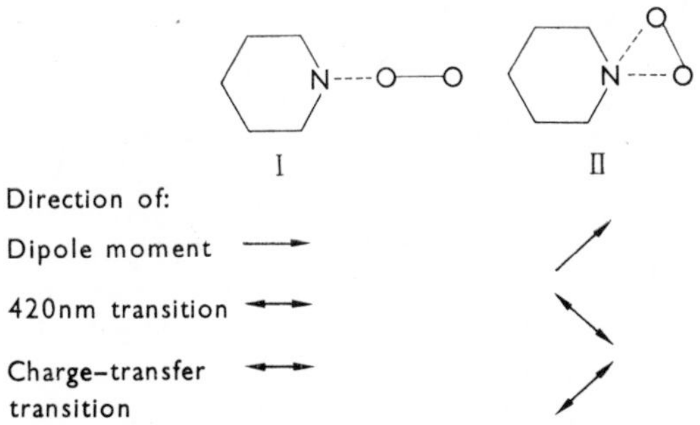

FIG 3.15 Two of the possible geometrical configurations for the pyridine–I_2 complex and the relative directions of the dipole moment and transition moments in each case (Reproduced from *J. Amer. Chem. Soc.*, **92,** 7589 (1970), by permission of the American Chemical Society).

to extrapolate from the solid phase to the geometries expected for complexes in solution where most studies have been made.

For this reason it is interesting that an attempt has recently been made[233] to determine the geometry of the pyridine–I_2 complex in solution using a method based on the effect of an electrical field on the perturbed $I_2\ {}^3\Pi_0 \leftarrow {}^1\Sigma_g^+$ transition, which in solutions containing pyridine gives rise to a band near 420 nm. Figure 3.15 shows two possible configurations of the complex and the relative directions of the dipole moment and the transition moments for each orientation. It is clear that if I is correct the component of the absorption (at 420 nm) polarised parallel to the external field should be greater than the component perpendicular to this field. If II is correct then the opposite order of components is expected. If a structure intermediate between I and II were correct then a quantitative analysis of the polarised absorption spectra for the different transitions should suffice to determine the geometry uniquely. It is shown that, at 420 nm, the component of the absorbed radiation polarised parallel to the electrical field is greater than that polarised at 90° to it, and structure I is indicated in agreement with the deduction of Mulliken[234] made from a consideration of the charge-transfer band shape. A quantitative analysis (as yet unpublished) establishes that the 420 nm band is polarised very nearly parallel to the dipole moment of the complex. This confirms the correctness of structure I.

3.5.3. MAGNETIC RESONANCE AND MÖSSBAUER STUDIES

Although relatively little work has been published on the n.m.r. and n.q.r. spectra of halogen molecular complexes, the studies which have been made provide some important information about the nature and extent of electronic redistribution due to complex formation. Only proton n.m.r. and halogen n.q.r. studies have so far been made and only relatively "strong" complexes have shown frequency shifts large enough to be properly interpreted. Thus, attempts[235,236] to study the changes in ^{1}H resonance of aromatic hydrocarbons due to dissolved halogens were unsuccessful (except for the mesitylene/I_2 case which did show a small shift) possibly due to the small proportion of hydrocarbon "complexed". Notice, here, however, that the observation time is very long (the frequency is low) compared with that used in the infrared region so that an extremely "short-lived" complex (section 3.4.3) may show only the effects of collision between the components. Such effects are expected to be small. Even systems such as pyridine–I_2[237] and ethanol–I_2[236] show only very small ^{1}H shifts of about 6–12 Hz and here the extent of interaction is known to be (thermodynamically) relatively large.

Only one systematic n.m.r. study has been made on a series of halogen complexes that being for a number of substituted pyridine–ICl complexes[147]. The proton chemical shifts approximately follow the pK_a values of the bases and, as indicated in table 3.16, the proton chemical shifts are not affected by α-methyl substituents in the same way as the infrared intensity and frequency data[130,147]. Thus the electronic redistribution appears to be uninfluenced by steric effects and, therefore, by the extent of overlap between the molecules. Indeed, the chemical shift change for the 2,6-dimethylpyridine is greater than that for the 2,4-dimethyl substituent (table 3.16) and it is speculated[147,237] that specific interaction between the -CH_3 group and an ICl lone-pair orbital is responsible for this effect. As may be seen from table 3.16, all the shifts observed are down-field and represent deshielding, presumably due to electron withdrawal from the *n*-orbital approximately located on the N atom. The effects of such donor action are clearly "felt" at all positions in the ring by inductive and (or) mesomeric effects. Magnetic anisotropic effects are presumably present but are clearly overwhelmed by the effects of donor action since a change in magnetic anisotropy of the nitrogen atom is predicted[238] to cause an *upfield* shift of the α-proton. It should be noticed that, for all the complexes with halogen molecules which have been studied[57,147,235–237,239], only "average" signals have been observed (although separate signals have been observed[239b] for ionic complexes of the type $D_2I^+I_9^-$). The observed signals are thus a weighted average of those due to "free" and "complexed" donor, despite the fact that separate "free" and "complexed" signals have been observed for nitrile–BF_3[241] complexes of comparable (or even lower) thermodynamic stability. This means that, at the temperatures used, the halogen complex lifetimes are lower than that required for observation of two signals. In view of the very high equilibrium constants obtained (for example, pyridine–ICI has a K_c† of the order of 10^5 dm^3

† At a temperature of 298 K.

TABLE 3.16

^1H chemical shift variations for halogen complexes of substituted pyridines (data in p.p.m. measured relative to a TMS internal standard in chloroform-d solution at 40°C[a])

Compound	pK^a	Ring proton shifts α	Ring proton shifts β	Ring proton shifts γ	Methyl group shifts α	Methyl group shifts β	Methyl group shifts γ	Reference
C_5H_5N	5.22							253
$C_5H_5N–I_2$		0.0	−0.07	−0.08				236
$C_5H_5N–ICl$		−0.18	−0.32	−0.50				147
$C_5H_5NH^+CF_3COO^-$		−0.25	−1.07	−1.21				254
2,3-$C_5H_3(CH_3)_2N$	6.57							253
2,3-$C_5H_3(CH_3)_2N–ICl$		−0.32	−0.18	−0.32	−0.20	−0.13		147, 237
2,3-$C_5H_3(CH_3)_2N–I_2$					−0.07	−0.01		237
2,4-$C_5H_3(CH_3)_2N$	6.63							253
2,4-$C_5H_3(CH_3)_2N–ICl$		−0.25	−0.31, −0.18		−0.20		−0.14	147, 237
2,4-$C_5H_3(CH_3)_2N–I_2$					−0.12		−0.10	237
2,6-$C_5H_3(CH_3)_2N$	6.72							253
2,6-$C_5H_3(CH_3)_2N–ICl$			−0.22	−0.21	−0.35			147, 237
2,6-$C_5H_3(CH_3)_2N–I_2$					−0.08			237
2,6-$C_5H_3(CH_3)_2NH^+CF_3COO^-$			−0.97	−1.21				252
3,5-$C_5H_3(CH_3)_2N$	6.15							253
3,5-$C_5H_3(CH_3)_2N–ICl$		−0.09		−0.40				147
3,5-$C_5H_3(CH_3)_2NH^+CF_3COO^-$		−0.93		−1.81				252
4-$C_5H_4(CH_3)N$	5.98							253
4-$C_5H_4(CH_3)N–ICl$		−0.02	−0.27					147
4-$C_5H_4(CH_3)NH^+CF_3COO^-$		−0.35	−1.07					252

a. data for iodine complexes in carbon tetrachloride.

TABLE 3.17

^{13}C chemical shift variation for halogen complexes of pyridine (data at 55°C in liquid pyridine)

Compound	Position	Chemical shift[a]/p.p.m.	Chemical shift change/p.p.m.
C_5H_5N	α	−14.4	—
	β	11.8	—
	γ	0.0	—
$C_5H_5N–I_2$	α	−13.1	+1.3
	β	8.7	−3.1
	γ	−5.4	−5.4
$C_5H_5NH^+Cl^-$	α	−6.5	+7.9
	β	7.0	−4.8
	γ	−12.4	−12.4

a. Measured relative to the pyridine γ-^{13}C signal.

mol^{-1} [240]) this is rather surprising. However for pyridine complexes the signals are broadened by quadrupolar effects and, of course, the proton chemical shifts are very much lower than those of the ^{19}F nucleus. Separate signals may therefore be difficult to detect under some circumstances.

It is interesting to compare the proton magnetic resonance data for pyridine–IX complexes with the ^{13}C data obtained recently by Larkindale and Simkin[242]. The data for C_5H_5N,$C_5H_5N–I_2$ (iodine in pure pyridine) and $C_5H_5NH^+Cl^-$ are summarised in table 3.17, from which it may be seen that the shifts are measured relative to the $^{13}C_4$ signal (*para* to the nitrogen atom). The changes in chemical shift are observed to be such that the 2- and 6-carbon atoms show an upfield shift and the 3,4 and 5 carbon signals move downfield. Generally speaking, for the corresponding proton spectra all three signals show a downfield shift on complexation (see above and table 3.16) although for pyridine–I_2[236] there appears to be very little shift of the 2- and 6-proton signals (table 3.16). This difference between the two sets of data indicates that magnetic anisotropic effects (probably at the carbon atom) are more important in the determination of ^{13}C shifts than they are for ^1H shifts.

Although the n.m.r. data described above indicate that quite severe electron migration (towards the nitrogen atom) occurs due to complexation, it is not possible to obtain an quantitative estimate of the charge transferred to the halogen molecule from such data. (Notice, however, that the ^{13}C data in table 3.17 show that $C_5H_5N–I_2$ is perturbed in the same way as, but to a lesser degree than, the $C_5H_5NH^+Cl^-$ ionic species.) Such information *is*, however, available by studying the n.q.r.

TABLE 3.18

Summary of halogen n.q.r. frequency data for halogen complexes

Compound	Signal frequency/MHz			Reference
	^{127}I	^{35}Cl	^{81}Br (or ^{79}Br)	
Br_2			319.46 (382.04)	243
I_2	332.4			255
ICl (α)	452.42	37.184		206
ICl (β)		37.202		255
C_5H_5N–ICl	464.28	21.136		206
	464.0			208
4-$C_5H_4(CH_3)N$–ICl	462.76	21.175		206
C_5H_5N–IBr	445.7			208
$(CH_3)_3N$–I_2	401.8[a,b]			208
3,5-$C_5H_3(Br)_2N$–Br_2			430.9[a] 263.3[b]	208
3,5-$C_5H_3(Br)_2N$–IBr	444.5[c]			208
3,5-$C_5H_3(Br)_2N$–ICl	463.7[c]			208
p-dioxan–Br_2			314.77[c]	244
acetone–Br_2			318.336[c]	244
diethyl ether–Br_2			320.499[c]	244
benzene–Br_2			321.83[c]	243

a. middle halogen frequency.
b. end halogen frequency.
c. only one frequency reported.

and Mössbauer spectra of the halogen atom(s) in the complex. A number of such studies has now[206–8,243–4] been made on both *n-aσ* and *bπ-aσ* complexes and a summary of the available frequency data is given in table 3.18. The values of the nuclear quadrupole coupling constant and asymmetry parameter can be used to calculate the charge distribution for the D–X–Y system. The method used is that of Townes and Dailey[245] and two basic assumptions are made. These are that the double bond (π-bond) character and the s orbital hybridisation of the halogen molecule are both zero. For the pyridine–ICl[207] complex, with these assumptions, the charge distributions† for the ICl and C_5H_5N–ICl are as shown below. These results are in good agreement the values obtained[208] for 3,5-dibromopyridine–XY complexes (obtained partly

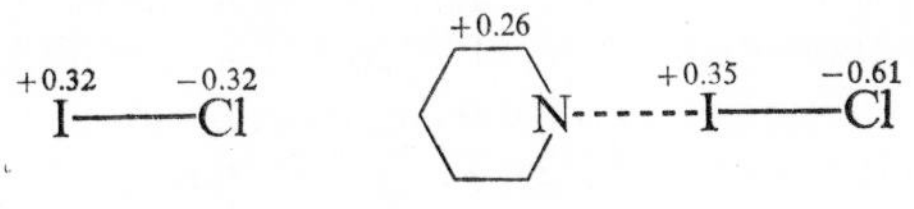

by extrapolation using a known nitrogen atom charge/substituent

† in multiples of the electronic charge e.

^{79}Br n.q.r. frequency correlation) which are,

	N charge	X charge	Y charge
3,5-dibromopyridine–Br_2	+0.2	+0.1	−0.3
3,5-dibromopyridine–IBr	+0.25	+0.3	−0.55
3,5-dibromopyridine–ICl	+0.3	+0.35	−0.65

The charge on the nitrogen (which may be partly delocalised) in each case is relative to a "zero" charge on this atom and it therefore represents the extent of charge transfer to the halogen. However, since it is known[207] that the antibonding orbital on the IX molecule (into which charge is transferred) is localised predominantly on the I atom, it follows that the results indicate a considerable polarisation of the halogen molecule by the base as might be expected. Thus the complex ground state is probably best represented by an electrostatic polarisation/charge-transfer valence bond wavefunction Ψ_{ES-CT} given by,

$$\Psi_{ES-CT} = c\phi_A + d\phi_B$$

where ϕ_A and ϕ_B represent the structures $[C_5H_5N]^+[ICl]^-$ and $[C_5H_5N][I^{\delta+} - Cl^{\delta-}]$ respectively. The n.q.r. data thus lead to a value of $c^2 = 0.26$ assuming s hybridisation is zero. If s hybridisation is estimated[207] to be 5% then the charge transferred is about 0.16e, the I and Cl charges then being +0.40 and −0.56e. The general conclusion is that between 16 and 26% of charge transfer occurs for these complexes—an important, but not overwhelming, contribution to the electronic redistribution.

This general conclusion is supported by the work on the ^{129}I Mössbauer spectra of I_2 and IX complexes of pyridine[246] and related molecules[247]. Again the charge distributions have been calculated from the quadrupole coupling constants and asymmetry parameters. The results for pyridine–IX (and related) complexes showed a consistent decrease in charge on the I atom while for some similar iodine complexes (where the signals from chemically different ^{129}I atoms were quite easily detected and analysed) it was found that the terminal iodine atom carries an increased charge. Thus the results confirm that polarisation of the IX molecule occurs and the extent of charge transfer (obtained by a total p electron count for the two iodine atoms) is, for phenazine–I_2 and hexamethylenetetramine–I_2 (HMTA–I_2), about 0.08–0.09e. This technique also gives an estimate of the π-bonding of the halogen molecule. For pyridine–ICl the ICl π-bond character percentage drops from 10% to 4% on complexation. A small amount of π-bonding between the ICl and the pyridine may be indicated by this result (see section 2.3.2). Both Mössbauer and n.q.r. spectroscopic techniques are useful in being able to detect chemically different halogen molecules in the solid phase. Thus the phenazine–I_2 complex is easily shown[247] to have only one type of I atom and therefore to be composed of units with an iodine attached to two phenazine molecules.

On the other hand, both acridine–I_2 and HMTA–I_2 crystals contain a 1:1 unit and therefore possess two chemically different iodine atoms.

It should be noted that the extent of charge transfer from donor to halogen can only be estimated in this way for relatively "strong" complexes. For crystals containing the benzene–Br_2 complex the n.q.r. ^{81}Br frequency shift is small and positive[243]. (table 3.18). Since the frequency normally drops[248] for an increase in electrical field gradient at a particular atom it is concluded that crystal field effects are important[242,243,249] and effectively mask any drop in frequency due to charge transfer.

Finally, it is important to note that n.m.r. spectroscopy provides a method of estimating complex lifetimes either from the detection of separate "free" and "complexed" signals or from a study of band widths (reference 250, pp. 222–24). However, the results which have been obtained so far for halogen complexes differ by large factors from the values estimated from dielectric relaxation data (see section 7.3). Thus for *p*-methoxyphenylmethyl sulphide–I_2 complexes the lifetime is estimated[239a] to be about 2×10^{-4} s. This, it is pointed out, is the upper limit so that a lower lifetime *could* be envisaged. The association constant is $\sim$20 dm^3 mol^{-1} for this complex as compared with a value of $\sim$1 dm^3 mol^{-1} for *p*-dioxan–I_2[131] for which the estimated dielectric relaxation time is of the order of 10^{-12} s[251]. Clearly such a large difference between complexes of this kind leads one to suspect that one (or possibly both) of the values is in error. The value of 10^{-12} s is supported by the far-infrared absorption which can be explained by a collision-induced mechanism (see sections 3.4.3 and 7.3). It is possible that the n.m.r. linewidths are partly determined by quadrupolar broadening effects or by the presence of paramagnetic species[239]. It would, however, seem desirable to carry out further n.m.r., infrared and dielectric studies on halogen complexes (as a function of temperature) in an attempt to get a more reliable estimate of their lifetime and its variation from one complex to another.

ACKNOWLEDGEMENT

It is a pleasure to acknowledge valuable discussions with Prof. R. S. Mulliken, Prof. W. B. Person, Prof. S. D. Christian, Prof. M. M. Davies and Dr. A. H. Price during the preparation of this chapter.

REFERENCES

1. J. Kleinberg and A. W. Davidson, *Chem. Rev.*, **42**, 601 (1948)
2. H. A. Benesi and J. H. Hildebrand, *J. Amer. Chem. Soc.*, **71**, 2703 (1949)
3. R. S. Mulliken, *J. Amer. Chem. Soc.*, **74**, 811 (1952)
4. R. Foster, Organic Charge-Transfer Complexes, (Academic Press, London and New York, 1969); (a) p. 17, (b) ch. 6, (c) p. 278, (d) ch. 3, (e) ch. 4, (f) ch. 5, (g) p. 127

5. C. N. R. Rao, S. N. Bhat and P. C. Dwivedi, *Applied Spectrosc. Rev.*, **5**, 1 (1971)
6. M. Tamres, (Ed. R. Foster), Molecular Complexes, Vol 1, (Paul Elek, Ltd., London, 1973), chapter 2
7. A. Mishra and A. D. E. Pullin, *Aust. J. Chem.*, **24**, 2493 (1971)
8(a). R. S. Mulliken, *Rec. trav. chim. Pays-Bas*, **75**, 845 (1956); (b) L. E. Orgel and R. S. Mulliken, *J. Amer. Chem. Soc.*, **79**, 4839 (1957)
9. R. Foster and C. A. Fyfe, (Editors, J. W. Emsley, J. Feeney and L. H. Sutcliffe) Progress in Nuclear Magnetic Resonance Spectroscopy (Pergamon Press, Oxford, 1969), Vol 4
10. L. J. Andrews and R. M. Keefer, *Adv. Inorg. Chem. Radiochem.*, **3**, 91 (1961)
11. L. J. Andrews and R. M. Keefer, Molecular Complexes in Organic Chemistry, (Holden-Day, San Francisco, 1964)
12. G. Briegleb, Elektronen-Donator-Acceptor Komplexe, (Springer-Verlag, Berlin, 1961); (a) p. 124, (b) p. 32
13. P. R. Hammond, L. A. Burkardt, R. H. Knipe, R. R. Lake, *J. Chem. Soc.*, A, 3789–3832 (1971)
13a. M. Hatano and O. Ito, *Bull. Chem. Soc. Jap.*, **44**, 916 (1971)
14. L. A. Carreira and W. B. Person, *J. Am. Chem. Soc.*, **94**, 1485 (1972)
14a. H. C. Fleming and M. W. Hanna, *J. Amer. Chem. Soc.*, **93**, 5030 (1971)
15. P. J. Trotter and M. W. Hanna, *J. Amer. Chem. Soc.*, **88**, 3724 (1966)
16. N. J. Rose and R. S. Drago, *J. Amer. Chem. Soc.*, **81**, 6138 (1959)
17. M. Tamres, *J. Phys. Chem.*, **65**, 654 (1961)
18. S. Nagakura, *J. Amer. Chem. Soc.*, **80**, 520 (1958)
19. R. A. Labudde and M. Tamres, *J. Phys. Chem.*, **74**, 4009 (1970)
20. D. Atack and O. K. Rice, *J. Phys. Chem.*, **58**, 1017 (1954)
21. S. D. Christian and J. Grundnes, *J. Amer. Chem. Soc.*, **93**, 6363 (1971)
22. W. B. Person, *J. Amer. Chem. Soc.*, **87**, 167 (1965)
23. D. A. Deranleau, *J. Amer. Chem. Soc.*, **91**, 4044 (1969)
24. P. J. Trotter and D. A. Yphantis, *J. Phys. Chem.*, **74**, 1399 (1970)
25. J. D. Childs, Ph.D. thesis, University of Oklahoma (Norman, Oklahoma, 1971)
26. M. Tamres, W. K. Duerksen and J. M. Goodenow, *J. Phys. Chem.*, **72**, 966 (1968)
27. A. A. Passchier and N. W. Gregory, *J. Phys. Chem.*, **72**, 2697 (1968)
27a. P. A. D. DeMaine, *J. Chem. Phys.*, **24**, 1091 (1956)
28. E. A. Ogryzlo and B. C. Sanctuary, *J. Phys. Chem.*, **69**, 4422 (1965)
29. A. A. Passchier, J. D. Christian and N. W. Gregory, *J. Phys. Chem.*, **71**, 937 (1967)
30. W. Y. Wen and R. M. Noyes, *J. Phys. Chem.*, **76**, 1017 (1972)
30a. D. F. Evans, *J. Chem. Phys.*, **23**, 1426 (1955)
30b. H. McConnell, *J. Chem. Phys.*, **22**, 760 (1954)
31. (a) G. Kortüm and W. M. Vogel, *Z. Elektrochem.*, **59**, 16 (1955); (b) G. Kortüm and M. Kortüm-Seiler, *Z. Naturforsch.*, **5a**, 544 (1950)
32. J. D. Childs, S. D. Christian, J. Grundnes and S. R. Roach, *Acta Chem. Scand.*, **25**, 1679 (1971)
33. J. Grundnes, S. D. Christian and V. Cheam, *Acta Chem. Scand.*, **24**, 1836 (1970)
34. M. Tamres and S. N. Bhat, *J. Amer. Chem. Soc.*, **96**, 2516 (1973)
35. J. Grundnes and S. D. Christian, *J. Am. Chem. Soc.*, **90**, 2239 (1968)
35a. O. B. Nagy, J. B. Nagy and A. Bruylants, *J. Chem. Soc.*, Perkin Trans., II, 968 (1972)
36. R. S. Mulliken and W. B. Person, Molecular Complexes: A Lecture and Reprint Volume, (J. Wiley and Sons, New York, 1969); (a) p. 2, (b) pp. 92–100; (c) ch. 17; (d) ch. 9; (e) ch. 14; (f) p. 26; (g) ch. 10; (h) p. 160; (i) p. 66
37. J. Grundnes, M. Tamres and S. N. Bhat, *J. Phys. Chem.*, **75**, 3682 (1971)
38. F. T. Lang and R. L. Strong, *J. Amer. Chem. Soc.*, **87**, 2345 (1965)
39. M. Kroll, *J. Amer. Chem. Soc.*, **90**, 1097 (1968)

40. M. Tamres and S. N. Bhat, *J. Amer. Chem. Soc.*, **94,** 2577 (1972)
41. M. W. Hanna and D. G. Rose, *J. Amer. Chem. Soc.*, **94,** 2601 (1972)
41a. F. L. Slejko and R. S. Drago, *J. Amer. Chem. Soc.*, **94,** 6546 (1972)
42. J. D. Childs, S. D. Christian and J. Grundnes, *J. Amer. Chem. Soc.*, **94,** 5657 (1972)
42a. S. D. Christian, J. D. Childs and E. H. Lane, *J. Amer. Chem. Soc.*, **94,** 6861 (1972)
43. M. M. DeMaine, P. A. D. DeMaine and G. E. McAlonie, *J. Mol. Spectrosc.*, **4,** 271 (1960)
44. R. M. Keefer and T. L. Allen, *J. Chem. Phys.*, **25,** 1059 (1956)
45. W. K. Duerksen and M. Tamres, *J. Amer. Chem. Soc.*, **90,** 1379 (1968)
46. D. Atack and O. K. Rice, *J. Phys. Chem.*, **58,** 1017 (1954)
47. B. B. Bhowmik, *Spectrochim. Acta*, **27A,** 321 (1971)
48. L. J. Andrews and R. M. Keefer, *J. Amer. Chem. Soc.*, **74,** 4500 (1952)
49. R. M. Keefer and L. J. Andrews, *J. Amer. Chem. Soc.*, **77,** 2164 (1955)
50. J. A. A. Ketelaar, *J. Phys. Rad.* (*Paris*), **15,** 197 (1954)
51. M. Tamres, D. R. Virzi and S. Searles, *J. Amer. Chem. Soc.*, **75,** 4358 (1953)
52. W. Brull and W. Ellerbrock, *Z. Anorg. Allgem. Chem.*, **216,** 353 (1934)
53. M. Brandon, M. Tamres and S. Searles, *J. Amer. Chem. Soc.*, **82,** 2129 (1960)
54. M. Tamres and M. Brandon, *J. Amer. Chem. Soc.*, **82,** 2134 (1960)
55. H. Tsubomura and R. P. Lang, *J. Amer. Chem. Soc.*, **83,** 2085 (1961)
56. N. W. Tideswell and J. D. McCullough, *J. Amer. Chem. Soc.*, **79,** 1031 (1957)
57. E. T. Strom, W. L. Orr, B. S. Snowden and D. E. Woessner, *J. Phys. Chem.*, **71,** 4017 (1967)
58. M. Tamres and J. M. Goodenow, *J. Phys. Chem.*, **71,** 1982 (1967)
59. M. Tamres and S. Searles, *J. Phys. Chem.*, **66,** 1099 (1962)
60. M. Good, A. Major, J. Nag-Chaudhuri and S. P. McGlynn, *J. Amer. Chem. Soc.*, **83,** 4329 (1961)
61. H. D. Bist and W. B. Person, *J. Phys. Chem.*, **71,** 2750 (1967)
62. C. Reid and R. S. Mulliken, *J. Amer. Chem. Soc.*, **76,** 3869 (1954)
63. W. J. McKinney and A. I. Popov, *J. Amer. Chem. Soc.*, **91,** 5215 (1969)
64. D. M. Golden, *J. Chem. Ed.*, **48,** 235 (1971)
65. I. Hanazaki, (personal communication)
66. J. M. Goodenow, Ph.D. Dissertation, University of Michigan, (Ann Arbor, Michigan, 1965)
67. M. Tamres, (unpublished results)
68. O. K. Rice, *Int. J. Quant. Chem.*, Symp. No. 2, 219 (1968)
69. J. D'Hondt, C. Dorval and Th. Zeegers-Huyskens, *J. Chim. Phys.*, **69,** 516 (1972)
70. E. Augdahl, J. Grundnes and P. Klaeboe, *Inorg. Chem.*, **4,** 1475 (1965)
71. W. J. McKinney and A. I. Popov, *J. Amer. Chem. Soc.*, **91,** 5215 (1969)
72. J. Grundnes and S. D. Christian, *Acta Chem. Scand.*, **23,** 3583 (1969)
73. (a) S. D. Christian, *J. Amer. Chem. Soc.*, **91,** 3669 (1969); (b) S. D. Christian, *J. Amer. Chem. Soc.*, **91,** 6514 (1969); (c) S. D. Christian, J. R. Johnson, H. E. Affsprung and P. J. Kilpatrick, *J. Phys. Chem.*, **70,** 3376 (1966); (d) J. R. Johnson, P. J. Kilpatrick, S. D. Christian and H. E. Affsprung, *J. Phys. Chem.*, **72,** 3223 (1968); (e) S. D. Christian and J. Grundnes, *Acta Chem. Scand.*, **22,** 1702 (1968)
74. R. S. Drago and B. B. Wayland, *J. Amer. Chem. Soc.*, **87,** 3571 (1965)
75. S. Carter, J. N. Murrell and E. J. Rosch, *J. Chem. Soc.*, 2048 (1965)
76. G. L. Amidon, Ph.D. Dissertation, University of Michigan, (Ann Arbor, Michigan, 1971)
77. R. S. Drago, T. F. Bolles and R. J. Niedzielski, *J. Amer. Chem. Soc.*, **88,** 2717 (1966)
78. G. Kortüm and G. Friedheim, *Z. Naturforsch.*, **2a,** 20 (1947)
79. E. M. Voigt, *J. Amer. Chem. Soc.*, **86,** 3611 (1964)
80. E. M. Voigt and C. Reid, *J. Amer. Chem. Soc.*, **86,** 3930 (1964)
81. M. Tamres, *J. Phys. Chem.*, **68,** 2621 (1964)

82. R. S. Mulliken and W. B. Person, *Ann. Rev. Phys. Chem.*, **13,** 107 (1962)
83. K. Watanabe, T. Nakayama and J. Mottl, *J. Quant. Spectrosc. Radiat. Transfer*, **2,** 369 (1962)
84. R. S. Berry and C. W. Reimann, *J. Chem. Phys.*, **38,** 1540 (1963)
85. W. A. Chupka, J. Berkowitz and D. Gutman, *J. Chem. Phys.*, **55,** 2724 (1971)
86. G. Briegleb and J. Czekalla, *Z. Elektrochem.*, **63,** 6 (1959)
87. V. A. Brosseau, J. R. Basila, J. F. Smalley and R. L. Strong, *J. Amer. Chem. Soc.*, **94,** 716 (1972)
88. M. Tamres and J. Grundnes, *J. Amer. Chem. Soc.*, **93,** 801 (1971)
89. C. N. R. Rao, G. C. Chaturvedi and S. N. Bhat, *J. Mol. Spectrosc.*, **33,** 554 (1970)
90. M. Ebert, J. P. Keene, E. J. Land and A. J. Swallow, *Proc. Roy. Soc.*, **287A,** 1 (1965)
91. T. A. Gover and G. Porter, *Proc. Roy. Soc.*, **262A,** 476 (1961)
92. L. M. Julien and W. B. Person, *J. Phys. Chem.*, **72,** 3059 (1968)
93. S. H. Hastings, J. L. Franklin, J. C. Schiller and F. A. Matsen, *J. Amer. Chem. Soc.*, **75,** 2900 (1953)
94. N. K. Bridge, *J. Chem. Phys.*, **32,** 945 (1960)
95. C. Capellos and A. J. Swallow, *J. Phys. Chem.*, **73,** 1077 (1969)
96. M. Tamres and J. Groff, (unpublished data)
97. N. Yamamoto, T. Kajikawa, H. Sato and H. Tsubomura, *J. Amer. Chem. Soc.*, **91,** 265 (1969)
98. R. L. Strong, *J. Phys. Chem.*, **66,** 2423 (1962)
99. J. M. Bossy, R. E. Bühler and M. Ebert, *J. Amer. Chem. Soc.*, **92,** 1099 (1970)
100. R. M. Keefer and L. J. Andrews, *J. Amer. Chem. Soc.*, **72,** 4677 (1950)
101. R. E. Bühler, *Helv. Chim. Acta*, **51,** 1558 (1968)
102. R. E. Bühler and M. Ebert, *Nature*, **214,** 1220 (1967)
103. L. J. Andrews and R. M. Keefer, *J. Amer. Chem. Soc.*, **73,** 462 (1951)
104. N. S. Bayliss, *Nature*, **163,** 764 (1949)
105. R. L. Strong and J. Pérano, *J. Amer. Chem. Soc.*, **83,** 2843 (1961)
106. R. L. Strong and J. Pérano, *J. Amer. Chem. Soc.*, **89,** 2535 (1967)
107. P. A. D. DeMaine, *J. Chem. Phys.*, **24,** 1091 (1956)
108. T. Abe, *Bull. Chem. Soc., Jap.*, **43,** 625 (1970)
109. W. Liptay, (Editor, O. Sinanoglu) Modern Quantum Chemistry; Part II, (Academic Press, N.Y., 1965), chapter 5
110. P. J. Trotter, *J. Amer. Chem. Soc.*, **88,** 5721 (1966)
111. J. Prochorow and A. Tramer, *J. Chem. Phys.*, **44,** 4545 (1966)
112. A. F. Grand and M. Tamres, *J. Phys. Chem.*, **74,** 208 (1970)
113. J. Ham, *J. Amer. Chem. Soc.*, **76,** 3875 (1954)
114. M. Brandon, M. Tamres and S. Searles, *J. Amer. Chem. Soc.*, **82,** 2129 (1960)
115. R. P. Lang, *J. Amer. Chem. Soc.*, **84,** 1185 (1962)
116. M. Tamres and S. N. Bhat, *J. Phys. Chem.*, **75,** 1057 (1971)
117. C. Van de Stolpe, Ph.D. thesis (University of Amsterdam, The Netherlands, 1953)
118. M. Good, A. Major, J. Nag-Chaudhuri and S. P. McGlynn, *J. Amer. Chem. Soc.*, **83,** 4329 (1961)
119. J. D. McCullough and I. C. Zimmermann, *J. Phys. Chem.*, **64,** 1084 (1960)
120. J. D. McCullough and D. Mulvey, *J. Phys. Chem.*, **64,** 264 (1960)
121. E. M. Voigt, *J. Phys. Chem.*, **72,** 3300 (1968)
122. R. P. Lang (personal communication)
122a. R. S. Mulliken (personal communication)
123. J. A. A. Ketelaar, *Recl. trav. chim. Pays-Bas*, **75,** 857 (1956)
124. R. S. Mulliken, *J. Chem. Phys.*, **55,** 288 (1971)
125. S. N. Bhat, M. Tamres and M. S. Rahaman (submitted for publication)
125a. D. F. Evans, *J. Chem. Soc.*, 4229 (1957)
126. J. E. Prue, *J. Chem. Soc.*, 7534 (1965)
127. R. L. Scott, *J. Phys. Chem.*, **75,** 3843 (1971)

128. R. S. Mulliken and W. B. Person, Physical Chemistry, Vol. 3 (Eds., Eyring, Jost and Henderson), (Academic Press, New York 1969), chapter 10
129. H. B. Friedrich and W. B. Person, *J. Chem. Phys.*, **44,** 2161 (1966)
130. J. Yarwood and W. B. Person, *J. Amer. Chem. Soc.*, **90,** 3930 (1968)
131. J. Yarwood and W. B. Person, *J. Amer. Chem. Soc.*, **90,** 594 (1968)
132. M. W. Hanna and D. E. Williams, *J. Amer. Chem. Soc.*, **90,** 5358 (1968)
133. W. B. Person, R. E. Humphrey, W. A. Deskin and A. I. Popov, *J. Amer. Chem. Soc.*, **80,** 2049 (1958)
134. A. I. Popov, R. E. Humphrey and W. B. Person, *J. Amer. Chem. Soc.*, **82,** 1850 (1960)
135. W. B. Person, R. E. Humphrey and A. I. Popov, *J. Amer. Chem. Soc.*, **81,** 273 (1959)
136. W. B. Person, R. E. Erickson and R. E. Buckles, *J. Amer. Chem. Soc.*, **82,** 29 (1960)
137. W. B. Person, R. E. Erickson and R. E. Buckles, *J. Chem. Phys.*, **27,** 1211 (1957)
138. J. Collin, L. D'Or and R. Alewaeters, *J. Chem. Phys.*, **23,** 397 (1955); *Rec. Trav. Chim. Pays-Bas*, **75,** 862 (1956)
139. R. S. Mulliken, *J. Amer. Chem. Soc.*, **74,** 811 (1952); *J. Phys. Chem.*, **56,** 801 (1952); *J. Amer. Chem. Soc.*, **72,** 600 (1950)
140. E. K. Plyler and R. S. Mulliken, *J. Amer. Chem. Soc.*, **81,** 823 (1959)
141. A. G. Maki and E. K. Plyler, *J. Phys. Chem.*, **66,** 766 (1962)
142. Y. Yagi, W. B. Person and A. I. Popov, *J. Phys. Chem.*, **71,** 2439 (1967)
143. S. G. W. Ginn and J. L. Wood, *Trans. Faraday Soc.*, **62,** 777 (1966)
144. S. G. W. Ginn, I. Haque and J. L. Wood, *Spectrochim. Acta*, **24A,** 1531 (1968)
145. I. Haque and J. L. Wood, *Spectrochim. Acta*, **23A,** 959 (1967)
146. I. Haque and J. L. Wood, *Spectrochim. Acta*, **23A,** 2523 (1967)
147. J. Yarwood, *Spectrochim. Acta*, **26A,** 2099 (1970)
148. G. W. Brownson and J. Yarwood, *J. Mol. Struct.*, **10,** 147 (1971)
149. (a) G. W. Brownson and J. Yarwood, *Adv. Mol. Relaxation Processes* **6,** 1 (1973); (b) (unpublished data)
150. H. W. Thompson and R. F. Lake, *Proc. Roy. Soc.*, **A297,** 440 (1967)
151. R. F. Lake and H. W. Thompson, *Spectrochim. Acta*, **24A,** 1321 (1968)
152. K. Yokobayashi, F. Watari and K. Aida, *Spectrochim. Acta*, **24A,** 1651 (1968)
153. F. Watari, *Spectrochim. Acta*, **23A,** 1917 (1967)
154. H. Yada, J. Tanaka and S. Nagakura, *J. Mol. Spectrosc.*, **9,** 461 (1962)
155. R. Cahay, *Bull. Classe. Sci. Acad. Roy. Belg.*, **56,** 818 (1970)
156. J. D'Hondt and Th. Zeegers-Huyskens, *J. Mol. Struct.*, **10,** 135 (1971)
157. V. Lorenzelli, *Gazz. Chim. Ital.*, **95,** 218 (1965); *C.R. Acad. Sci.*, **258,** 5386 (1964)
158. P. Klaeboe, *J. Amer. Chem. Soc.*, **89,** 3667 (1967)
159. H. Stammreich, R. Forneris and Y. Tavares, *Spectrochim. Acta*, **17,** 775, 1173 (1961)
160. T. Surles and A. I. Popov, (private communication)
161. J. C. Evans and G. Y-S Lo, *J. Chem. Phys.*, **44,** 4356 (1966)
162. J. N. Gayles, *J. Chem. Phys.*, **49,** 1840 (1968)
163. A. D. Buckingham, *Proc. Roy. Soc.*, **248A,** 169 (1958)
164. R. J. Jakobsen and J. W. Brasch, *J. Amer. Chem. Soc.*, **86,** 3571 (1964)
165. H. R. Wyss, R. D. Werder and Hs. H. Günthard, *Spectrochim. Acta*, **20,** 573 (1964)
166. P. Delorme, *J. Chim. Phys.*, **41,** 1437 (1964)
167. E. Decamps, A. Hadni and J. M. Munier, *Spectrochim. Acta*, **20,** 373 (1964)
168. G. W. Chantry, H. A. Gebbie, B. Lassier and G. Wyllie, *Nature*, **214,** 163 (1967)
169. H. A. Gebbie, N. W. B. Stone, F. D. Findlay and E. C. Pyatt, *Nature*, **205,** 377 (1965)
170. G. W. Chantry and H. A. Gebbie, *Nature*, **208,** 378 (1965)

171. C. C. Bradley, H. A. Gebbie, A. C. Gilby, V. V. Kechin and J. H. King, *Nature*, **211,** 839 (1966)
172. I. Darmon, A. Gerschel and C. Brot, *Chem. Phys. Lett.*, **8,** 454 (1971)
173. M. Davies, G. W. F. Pardoe, J. E. Chamberlain and H. A. Gebbie (a) *Trans. Faraday Soc.*, **64,** 847 (1968); (b) *ibid.*, **66,** 273 (1970)
174. G. W. F. Pardoe (a) *Trans. Faraday Soc.*, **66,** 2699 (1970); (b) *Spectrochim. Acta*, **27A,** 203 (1971)
175. J. E. Chamberlain, E. B. C. Werner, H. A. Gebbie and W. Slough, *Trans. Faraday Soc.*, **63,** 2605 (1967)
176. Y. Leroy and E. Constant, *C.R. Acad. Sci.*, **262B,** 1391 (1966)
177. S. G. Kroon and J. van der Elsken, *Chem. Phys. Lett.*, **1,** 285 (1967)
178. H. S. Gabelnick and H. L. Strauss, *J. Chem. Phys.*, **46,** 396 (1967)
179. S. R. Jain and S. Walker, *J. Phys. Chem.*, **75,** 2942 (1971)
180. K. D. Moller and W. G. Rothschild, Far Infrared Spectroscopy, (J. Wiley, New York and London, 1971), chapter 11
181. G. W. Chantry, Submillimetre Spectroscopy (Academic Press, New York and London, 1972), section 5.6, p. 177–187
182. J. P. Kettle and A. H. Price, *J. Chem. Soc., Faraday Trans.*, II, 1306 (1972)
183. G. W. Brownson and J. Yarwood, *Spectroscopy Letters*, **5,** 185 (1972)
184. N. E. Hill, *Proc. Phys. Soc.*, **82,** 723 (1963); *Chem. Phys. Lett.*, **2,** 5 (1968)
185. J. Ph. Poley, *J. Appl. Sci.*, **4B,** 337 (1955)
186. S. K. Garg, J. E. Bertie, H. Kilp and C. P. Smyth, *J. Chem. Phys.*, **49,** 2551 (1968)
187. S. K. Garg, H. Kilp and C. P. Smyth, *J. Chem. Phys.*, **43,** 2341 (1965)
188. R. G. Gordon, *J. Chem. Phys.*, **38,** 1724 (1963)
189. G. H. Wegdam, J. B. TeBeek, H. van der Linden and J. van der Elsken, *J Chem. Phys.*, **55,** 5207 (1971)
190. J. Chamberlain, *Chem. Phys. Lett.*, **2,** 464 (1968)
191. E. E. Ferguson and I. Y. Chang, *J. Chem. Phys.*, **34,** 628 (1961)
192. (a) E. E. Ferguson, *J. Chim. Phys.*, **61,** 257 (1964); (b) E. E. Ferguson and F. A. Matsen, *J. Chem. Phys.*, **29,** 105 (1958); *J. Amer. Chem. Soc.*, **82,** 3268 (1960)
193. H. Rosen, Y. R. Shen and F. Stenman, *Mol. Phys.*, **22,** 33 (1971)
194. O. Hassel and C. Rømming, *Acta Chem. Scand.*, **10,** 696 (1956)
195. J. Yarwood, *Trans. Faraday Soc.*, **65,** 934 (1969)
196. (a) D. L. Glusker, H. W. Thompson and R. S. Mulliken, *J. Chem. Phys.*, **21,** 1407 (1953); (b) D. L. Glusker and H. W. Thompson, *J. Chem. Soc.*, 471 (1955)
197. W. Haller, G. Jura, G. C. Pimentel and L. Grotz, *J. Chem. Phys.*, **22,** 720 (1954); *J. Chem. Phys.*, **19,** 513 (1951)
198. W. E. Tolberg and R. A. Zingaro, *J. Amer. Chem. Soc.*, **81,** 1353 (1959)
199. R. A. Zingaro and W. B. Witmer, *J. Phys. Chem.*, **64,** 1705 (1960)
200. J. N. Gayles and W. B. Person, (private communication, unpublished data, on pyridine–Br_2)
201. F. Watari and S. Kinumaki, *Sci. Rep. Res. Inst. Tohoku University*, **14A,** 64 (1962)
202. A. I. Popov, J. C. Marshall, F. B. Stute and W. B. Person, *J. Amer. Chem. Soc.*, **83,** 3586 (1961)
203. J. K. Wilmshurst and H. J. Bernstein, *Can. J. Chem.*, **35,** 1183 (1957)
204. G. Zerbi, B. L. Crawford and J. Overend, *J. Chem. Phys.*, **38,** 1383 (1966)
205. D. A. Long, F. S. Murfin and E. L. Thomas, *Trans. Faraday Soc.*, **59,** 12 (1963)
206. H. Creswell-Fleming and M. W. Hanna, *J. Amer. Chem. Soc.*, **93,** 5030 (1971)
207. G. A. Bowmaker and S. Hacobian, *Aust. J. Chem.*, **21,** 551 (1968)
208. G. A. Bowmaker and S. Hacobian, *Aust. J. Chem.*, **22,** 2047 (1969)
209. E. Augdahl and P. Klaeboe, *Spectrochim. Acta*, **19,** 1665 (1963); *Acta Chem. Scand.*, **19,** 807 (1965)
210. E. Augdahl and P. Klaeboe, *Acta Chem. Scand.*, **16,** 1637; 1647, 1655 (1962)
211. E. Augdahl and P. Klaeboe, *Acta Chem. Scand.*, **18,** 18, 27 (1964)
212. J. Grundnes and P. Klaeboe, *Acta Chem. Scand.*, **18,** 2022 (1964)

213. T. Gramstad and S. I. Snaprud, *Acta Chem. Scand.*, **16,** 999 (1962)
214. R. Dahl, P. Klaeboe and T. Gramstad, *Spectrochim. Acta*, **25A,** 207 (1969)
215. H. Yamada and K. Kozima, *J. Amer. Chem. Soc.*, **82,** 1543 (1960)
216. R. S. Drago and C. D. Schmulbach, *J. Amer. Chem. Soc.*, **82,** 4484 (1960)
217. J. Morcillo and J. Herranz, *Anales real soc. espãn fís. y quim.*, **50B,** 117 (1954)
218. M. Hasimoto, *J. Chem. Soc., Jap.*, **78,** 181 (1957)
219. J. Foster and M. Goldstein, *Spectrochim. Acta*, **24A,** 807 (1968)
220. R. S. Drago, B. Wayland and R. L. Carlson, *J. Amer. Chem. Soc.*, **85,** 3125 (1963)
221. M. C. Giordano, J. C. Bazan and A. J. Arvia, *J. Inorg. Nucl. Chem.*, **28,** 1209 (1966)
222. O. Hassel and C. Rømming, *Quart. Rev.*, **16,** 1 (1962)
223. O. Hassel and H. Hope, *Acta Chem. Scand.*, **14,** 391 (1960)
224. T. Dahl, O. Hassel and K. Sky, *Acta Chem. Scand.*, **21,** 592 (1967)
225. O. Hassel, *Acta Chem. Scand.*, **19,** 2259 (1965)
226. O. Hassel and J. Hvoslef, (a) *Acta Chem. Scand.*, **10,** 138 (1956); (b) *ibid.*, **8,** 873 (1954)
227. O. Hassel and K. O. Stromme, *Acta Chem. Scand.*, **12,** 1146 (1958); *ibid.*, **13,** 1781
228. O. Hassel, *Mol. Phys.*, **1,** 241 (1958)
229. O. Hassel, C. Rømming and T. Tufte, *Acta Chem. Scand.*, **15,** 967 (1961)
230. K-M. Marstokk and K. O. Stromme, *Acta Crystallogr.*, **B24,** 713 (1968)
231. O. Hassel and K. O. Stromme, *Acta Chem. Scand.*, **13,** 275 (1959)
232. P. Groth and O. Hassel, *Mol. Phys.*, **6,** 543 (1963); *Acta Chem. Scand.*, **18,** 402 (1964)
233. G. K. Vemulapalli, *J. Amer. Chem. Soc.*, **92,** 7589 (1970)
234. R. S. Mulliken, *J. Amer. Chem. Soc.*, **91,** 1237 (1969)
235. S. Matsuoka and S. Hattori, *J. Phys. Soc. Jap.*, **17,** 1073 (1962)
236. A. Fratiello, *J. Chem. Phys.*, **41,** 2204 (1964)
237. J. Yarwood, *Chem. Comm.*, 809 (1967)
238. J. N. Murrell and V. M. S. Gil, *Trans. Faraday Soc.*, **61,** 402 (1965)
239. D. W. Larsen and A. L. Allred, (a) *J. Amer. Chem. Soc.*, **87,** 1216 (1965); (b) *ibid.*, **87,** 1219 (1965)
240. A. I. Popov and R. H. Rygg, *J. Amer. Chem. Soc.*, **79,** 4622 (1957)
241. R. W. Taft and J. W. Carten, *J. Amer. Chem. Soc.*, **86,** 4199 (1964)
242. J. P. Larkindale and D. J. Simkin, *J. Chem. Phys.*, **55,** 5048 (1971)
243. H. O. Hooper, *J. Chem. Phys.*, **41,** 599 (1964)
244. P. Cornil, M. Read, J. Duchesne and R. Cahay, *Bull. Classe. Sci. Acad. Roy. Belg.*, **50,** 235 (1964)
245. C. H. Townes and B. P. Dailey, *J. Chem. Phys.*, **17,** 782 (1949)
246. C. I. Wynter, J. Hill, W. Bledsoe, G. K. Shenoy and S. L. Ruby, *J. Chem. Phys.*, **50,** 3872 (1969)
247. S. Ichiba, H. Sakai, H. Negita and Y. Maeda, *J. Chem. Phys.*, **54,** 1627 (1971)
248. D. C. Douglass, *J. Chem. Phys.*, **32,** 1882 (1960)
249. The author is grateful for valuable communication on this point with Dr. G. Bowmaker
250. J. A. Pople, W. G. Schneider and H. J. Bernstein, High Resolution Nuclear Magnetic Resonance (McGraw-Hill, London, 1959), chapter 10
251. V. L. Brownsell and A. H. Price, Chemical Society Special Publication, No. 20, p. 83 (1966)
252. A. R. Katritzky, F. J. Swinbourne and B. Ternai, *J. Chem. Soc.*, **B,** 235 (1966)
253. R. J. L. Andon, J. D. Cox and E. F. G. Herington, *Trans. Faraday Soc.*, **50,** 918 (1954)
254. I. C. Smith and W. G. Schneider, *Can J. Chem.*, **39,** 1158 (1961)
255. V. J. Goldansky and R. H. Harker (Editors), Chemical Applications of Mössbauer Spectroscopy, Vol. 2 (Academic Press, London, 1968)

Chapter 4

The Vibrational Spectra of Hydrogen-Bonded Complexes

J. L. Wood

Chemistry Department, Imperial College of Science and Technology, London SW72AY, England

4.1. INTRODUCTION

Hydrogen bonding is involved in an immense variety of molecular interactions, and, as Pimentel and McClellan, in their excellent monograph[1] point out, any definition risks criticism as being too restrictive, or too elastic. Furthermore, with the passage of time, it may appear that some shift of the boundaries becomes desirable. It therefore seems best to propose an *ad-hoc* definition—"Hydrogen bonding is an attractive interaction between two molecules, or between two parts of the same molecule, which requires the specific presence of a hydrogen atom in the vicinity." Here we are concerned with intermolecular complexes. Hydrogen-bonded complexes are often classified by the nature of the proton donor AH, or acid, and that of the acceptor B, or base. Since B can be expected to lose electrons to AH in forming the complex, B can also be regarded as the electron donor, and AH the electron acceptor[2]. The nature of the complex is, however, independent of which partner actually provides the proton; pyridine and phenol produce the same complex as pyridinium ion and phenate ion, so this does not serve to distinguish the moiety to be chosen as the donor. It is preferable to regard the proton donor as the moiety which retains the proton on dissociation. Often the moiety to which the proton is more closely attached can be distinguished on spectroscopic grounds, *e.g.* from the frequencies of internal modes of A and B, or from electronic transitions. In such cases the unequal sharing of the proton is well represented by the symbolisation AH—B. However, in many interesting complexes the location of the proton turns out to be surprisingly elusive, and even through a series of similar complexes, such as those formed between phenol and related bases, there may be indication of the proton switching from the phenol to the base when the relative pK's are favourable[3,4]. Since we are concerned with the nature of the complex, rather than the manner in which it was formed, it seems better to avoid the use of the terms proton donor and acceptor when the location of the proton is uncertain.

4.2. TYPES OF SYSTEMS FORMING HYDROGEN BONDED COMPLEXES

4.2.1. CHEMICAL FACTORS

It has been the purpose of many investigations to discover which chemical parameters, *e.g.* acidity, electron affinity, *etc.*, in the molecules concerned determine the tendency to hydrogen bond formation. The extent of these efforts is an indication that no clear relation between hydrogen bond formation and simple chemical parameters has emerged, and it is now becoming apparent that many factors play a part in determining whether or not a hydrogen-bonded complex will be formed. The variety of molecular species is extremely great, as is the range of functional groups involved. A particular functional group, *e.g.* C—H, may only enter into hydrogen bonding if incorporated in a suitable molecule, *e.g.* in CCl_3H. The effectiveness of a particular group naturally also depends on the nature of the other prospective partner in the complex. Thus any account of the types of system which form complexes is, of necessity, detailed and extensive—in the present context we shall rather choose examples to illustrate the variety and range of complexes that are formed, using the chemical nature of the atoms bridged by the hydrogen to order the discussion. Of the many factors involved in hydrogen bonding, one of the most important appears to be the electronegativity of the bridged atoms. Thus the strongest hydrogen bond known is in $(FHF)^-$, with a ΔH, relative to FH and F^- of 155 kJ mol^{-1} [5]. Strong hydrogen bonds are also formed between O and F, *e.g.* in $(CH_3)_2O$–HF[6,7], and between O and O, *e.g.* between β-naphthol and $(CH_3)_3NO$ (for which K is given as 6560 dm^3 mol^{-1} [8].) O—O hydrogen bonds are particularly numerous, and the range of stabilities found for these bonds show how important is the influence which the functional grouping also has on the bond strength, as does the presence of a formal charge on either of the moieties involved. As an instance of weak association in an O—O bond, the association constant of the 2,6-dimethylphenol complex with $(C_2H_5)_2O$ is only 0.67 dm^3 mol^{-1} [9]—probably a reflection of steric hindrance.

O—N hydrogen bonds are also numerous, with the range of stabilities appearing to lie rather lower than that for O—O bonds. One of the strongest O—N bonds, between *p*-chlorophenol and triethylamine has a K value of 130 dm^3 mol^{-1} [10]. Rather less numerously reported, but still common, are N—N bonds. Hydrogen bonding involving various less electronegative atoms has also been convincingly established. However, one has the impression that as the electronegativity of the atoms decreases, the variety of molecular species displaying hydrogen bonding is reduced. Hydrogen bonds involving carbon are represented by HCN self-association[11,12] or by the $CHCl_3$–$POCl_3$ complex,[13] while chlorine is involved in $(ClHCl)^-$ [14] and pyridine hydrochloride[15]. Bonds involving sulphur occur in the association of thiols[16,17] while bonds to bromine occur *inter alia*, in *e.g.* $(BrHBr)^-$ [18] and pyridine hydrobromide[15]. Hydrogen bonds linking metal atoms are rare (presumably because these atoms are insufficiently electronegative), but the complex ions $(LiHLi)^+$ [19,20] and $(BeHBe)^+$ [21] have been discussed in terms of hydrogen bonding. Association involving hydrogen bonding to π-orbitals has also been established, *e.g.* in 2-ferrocenylalcohols[22], or the interaction of $CHCl_3$ with benzene[23]. Hydrogen bonding is not limited to molecular or ionic species, but has also been demonstrated in the radical ClHCl[24] formed by passing HCl/Cl_2/argon mixtures through a glow discharge and condensing the products at 14 K.

It is pertinent to point out in any discussion of the strength or occurrence of hydrogen bonding that there is no close correspondence between the association constant and enthalpies of formation, in other words, entropies and enthalpies of formation do not fall in the same order. This is clearly shown in figure 4.1.

This discussion suggests that it is more useful to view hydrogen bonding as an effect, which, like so many others in chemistry, can play a role of varying importance in molecular association. Even in complexes which would be recognised as typical of hydrogen bonding, theoretical considerations (discussed in sections 4.7 and 1.3) suggest that *several* factors contribute to the stabilisation, and consequently

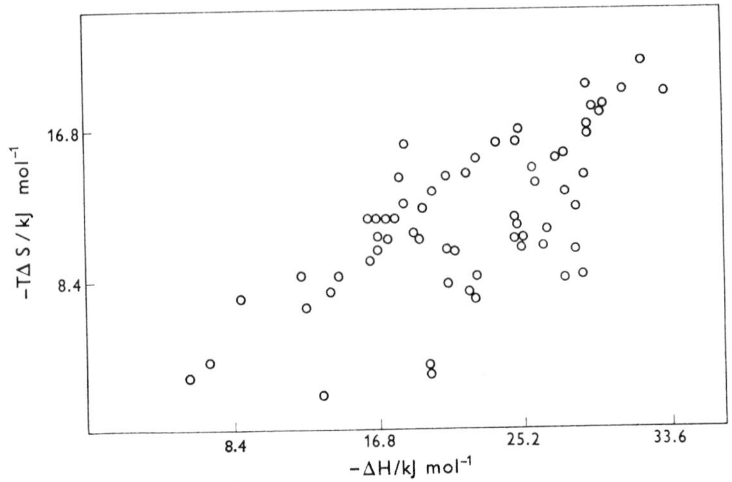

FIG 4.1 The relationship between T ΔS and ΔH, the data refers to phenol–base systems (redrawn from reference 2).

any attempt at a precise definition invites difficulties. This point of view is supported by the arguments put forward in chapter 1.

4.2.2. SYMMETRIC AND ASYMMETRIC COMPLEXES

As there is some confusion in the use of the terms symmetric or asymmetric hydrogen bond and "symmetric or asymmetric" complex, clarification is needed.

The term symmetric complex will be used for one in which the two groups joined by the hydrogen bond are formally identical, so that the complex as a whole can be represented as AHA. In symmetry terms, this means that the complex (*i.e.* its Hamiltonian) is unchanged by interchanging the A groups. If the term symmetric is used in this way, it does not involve having any knowledge of the position of the proton—in a symmetric complex the time averaged proton position must be symmetric, though it may avoid the centre of the bond.

In intermolecular hydrogen bonded systems, symmetric hydrogen bonds can only be present in systems with odd formal charge, if we exclude radicals (odd-electron systems). This follows from a simple electron count. Examples of anions, formed by the association of an acid and its conjugate base are the bihalide ions $(XHX)^-$, X = F. Cl or Br[14,25] and the ions $(ROHOR)^-$ where R = aryl[26] or acyl[27]. Symmetric complexes involving OHO bonds are also formed by pyridine oxide[28], and γ-pyridones[29]. Correspondingly, cations formed from a base and its conjugate acid *e.g.* involving NH_3,[30] various aliphatic amines[31], pyridine[31] or quinoline[32] have been identified as definite hydrogen-bonded complexes. With ammonia and other polyfunctional bases, complexes involving two or more hydrogen bonds may also occur.[30]

The requirement of an odd formal charge disappears in symmetric, intramolecular internally hydrogen-bonded systems. The systems 2′2′-bipyridyl)$^{+}$ [33] (figure 4.2a), (Hmaleate)$^{-}$ [27,34] and (Hphthalate)$^{-}$ [27,34]

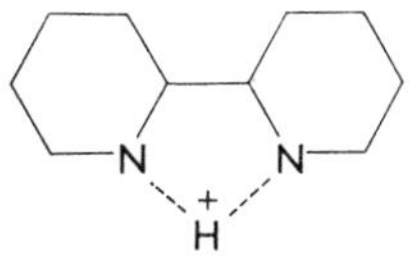

(a) 2;2′ – bypyridyl ion

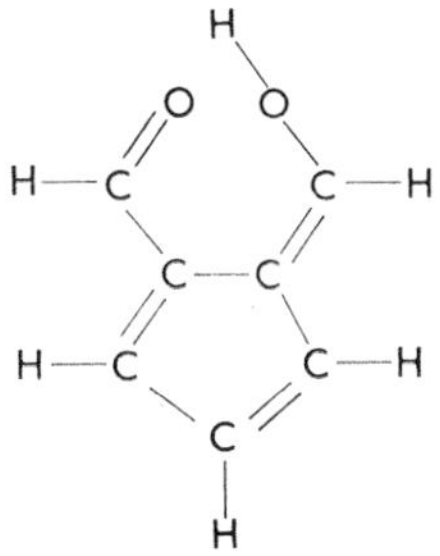

(b) 6–OH, I – (C=O)–fulvene

FIG 4.2

bear a formal charge, but neutral symmetric systems are also possible, *e.g.* 6-OH, 1(CHO)-fulvene)[35] (figure 4.2b).

In asymmetric complexes the moieties connected by the hydrogen bond are dissimilar. Both neutral and ionic complexes (with total number of electrons even) are possible. The majority of the complexes which have been examined are neutral asymmetric complexes, *e.g.* phenol–pyridine. All complexes formed by self-association are of this type if only a single hydrogen bond is present. Thus the water dimer, which appears to be open[36], is asymmetric. Asymmetric ionic complexes are also possible, *e.g.* (BrHCl)$^{-}$ [37].

In the above discussion only systems involving a single hydrogen bond have been considered. Interesting new possibilities arise if the complex is multiply hydrogen bonded. In particular, self-association can then lead to symmetric complexes. The simplest illustrations are provided by the carboxylic acid dimers[38] or by the 2-thiopyridone dimer[39].

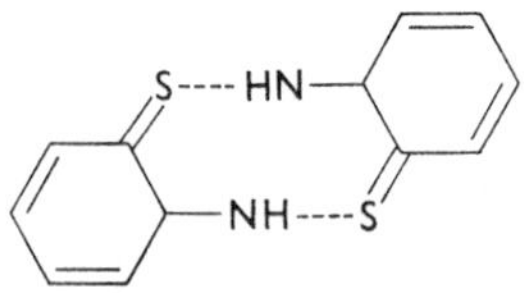

The 2-thiopyridone dimer
FIG 4.3

4.3. ROLE OF VIBRATIONAL SPECTROSCOPY IN HYDROGEN BONDING STUDIES

Infrared spectroscopy has, throughout, been the most popular method of studying hydrogen bonding, and shows no sign of diminishing in importance. The influence of a number of favourable factors account for this dominance of infrared studies. The measurements are relatively easy, and for the most part only involve commonly available commercial equipment. All phases can be readily examined, and there are no severe limitations on temperature variation. Hydrogen bonding produces substantial, and often spectacular, changes in the infrared spectra, which theory indicates can have a profound significance. Moreover, even a cursory examination shows that the infrared spectra contain a great deal of informative detail, and vibrational bands connected with proton movements are particularly strongly influenced. No other physical method can match these advantages of range and significance, and consequently infrared studies have played a relatively more important role in examining hydrogen bonding than they have for some other complexes (for example $b\pi$-$a\pi$ complexes—see section 2.3.5).

Two alternative methods of examining molecular vibrations—Raman spectroscopy and inelastic neutron scattering (I.N.S.) have so far been of much less importance in hydrogen bond studies. The paucity of Raman studies is not entirely accounted for by experimental difficulty. It is frequently the case that the only bands observed are modes little affected by hydrogen bonding—in AH–B systems the AH stretching band, frequently intense in the infrared spectrum, is often so weak in the Raman spectrum as to be unobservable. Indeed, this disadvantage is turned to account in the use of water, and other hydrogen-bonded media as solvents for Raman work. Little improvement is produced by using He/Ne or other red laser line sources, as the detector photocell sensitivity falls severely where the Raman shift of the AH stretching mode occurs. Initial results with blue or green laser excitation, using for instance an argon/krypton mixed gas laser, are also disappointing. The I.N.S. technique is relatively recent, and still rapidly evolving. A monochromatic beam of neutrons is scattered by the sample, which is most often a solid, though liquids[40], and a few vapour state samples have also been successfully examined. The energy spectrum of the scattered neutrons is observed, either to higher energy than the incident beam (energy gain spectra) or to the lower energy side (energy loss spectra), corresponding to loss or gain of vibrational energy by the sample. Two features make I.N.S. of particular interest in hydrogen-bonding studies. Firstly the symmetry restrictions that apply in infra-red or Raman spectra are absent; secondly the amplitude of the hydrogen atom motion is the predominating influence on the band intensity. Unfortunately, at present, the resolution attainable in the middle frequency range (say above 400 cm^{-1}) is much lower than in infrared or Raman spectra. As the frequency decreases, the resolution improves, and bands separated by a few cm^{-1} have been resolved in the far infrared range. A characteristic spectrum[41] of a sample containing hydrogen bonds is shown in figure 4.4. The strong band at 652 cm^{-1} is attributed to the fundamental of the transverse vibration of the H atom in the $(ClHCl)^-$ ion. (see p. 343).

4.4. OTHER PHYSICAL METHODS

Although we are concerned here mainly with vibrational studies, a brief survey of the contribution made by other physical methods to

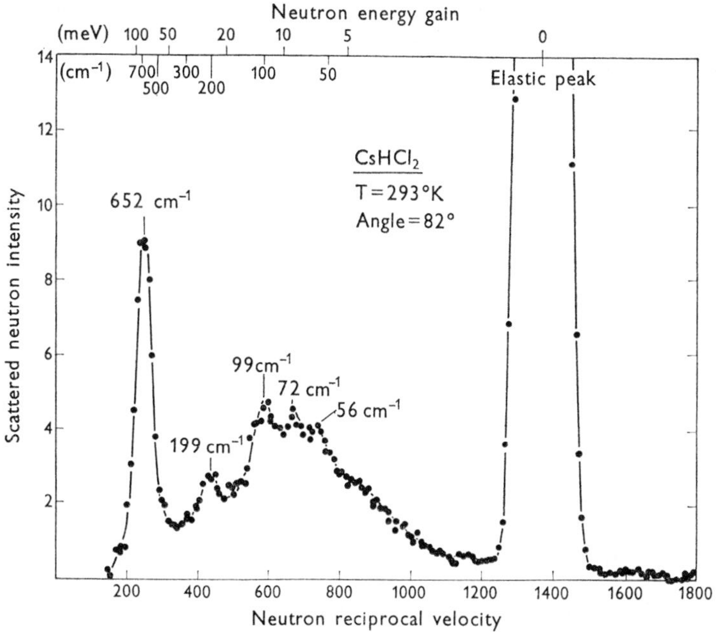

FIG 4.4 The cold-neutron spectrum of $CsHCl_2$ (in 5-mm tubes). Velocity is in microseconds per meter. (Reproduced from *J. Chem. Phys.*, **52,** 2730 (1970), by permission of the American Institute of Physics.)

current understanding of the hydrogen bond is necessary, since the models involved in the interpretation of vibrational spectra draw extensively on the results of complimentary techniques.

4.4.1. DIFFRACTION METHODS

In this respect diffraction studies on crystals incorporating hydrogen bonds have been particularly important. Both X-ray and neutron diffraction are extensively discussed in an excellent recent book by Hamilton and Ibers[42]. Since X-ray scattering is largely caused by electrons, the position of hydrogen atoms is often difficult to establish in X-ray studies. In the hydrogen bond the hydrogen atom appears to be denuded of electron density relative to the free atom[43], and this may be responsible for increasing the difficulty of location. Nevertheless, X-ray measurements have been valuable in providing the AB distances between hydrogen-bonded atoms. In stronger hydrogen bonds this distance is often significantly less than the sum of the A and B van der Waals radii. To determine the location of the hydrogen atoms, neutron diffraction is preferable. Since scattering is largely by the atomic nuclei, and increases in proportion to $M^{1/3}$ (where M is the atomic mass) the hydrogen atom gives scattering comparable with that by the heavier atoms present. Deuterium offers a further advantage. Location of the H atom has established that it is asymmetrically disposed in AB systems. The AH distance, which is greater than the normal covalent bond length, increases with the strength of the hydrogen bond, while the HB distance, which is less than the sum of the van der Waals radii, diminishes. It should always be borne in mind that diffraction methods give a distribution averaged in both space and time. Table 4.1 reproduced from Hamilton and Ibers' book, shows some representative distances. Structural studies have also shown, either directly, or implicitly from the mutual orientation of the bonded moieties, that the hydrogen bond has some tendency to be linear, *i.e.* the atoms CHD are collinear, where C and D are the atoms directly bonded. However, many exceptions to linearity are known (see the appendix of reference 42). Since the configuration adopted by a complex in the crystal is the resultant of all the forces acting on it including the non-bonded interactions between units as well as the totality of hydrogen-bonded interactions, some uncertainty arises concerning the extent to which the structural features should be ascribed to the presence of the hydrogen bond. The trends shown in table 4.1, however, are widely representative and undoubtedly reflect an intrinsic feature of the hydrogen bond.

4.4.2. NUCLEAR MAGNETIC RESONANCE STUDIES

These have mainly been concerned with the hydrogen-bond proton, though a few studies have also been made of other nuclei, *e.g.* ^{19}F[44], which may be present in the complexed molecules. A critical review has been made by Hofacker and Hofacker[45]. Most measurements relate to

TABLE 4.1

van der Waals contact distances and observed hydrogen bond distances (in Å) for some common types of hydrogen bonds (Reproduced from 'Hydrogen Bonding in Solids', W. C. Hamilton and J. A. Ibers, with permission of W. A. Benjamin Inc., Advanced Book Program, Reading, Massachusetts, U.S.A.)

Bond type	A—B (calc.)	A—B (obs.)	H---B (calc.)	H---B (obs.)
F–H–F	2.7	2.4	2.6	1.2
O–H---O	2.8	2.7	2.6	1.7
O–H---F	2.8	2.7	2.6	1.7
O–H---N	2.9	2.8	2.7	1.9
O–H---Cl	3.2	3.1	3.0	2.2
N–H---O	2.9	2.9	2.6	2.0
N–H---F	2.9	2.8	2.6	1.9
N–H---Cl	3.3	3.3	3.0	2.4
N–H---N	3.0	3.1	2.7	2.2
N–H---S	3.4	3.4	3.1	2.4
C–H---O	3.0	3.2	2.6	2.3

The values shown in the calculated columns are the sums of the van der Waals covalent radii.

the liquid phase, though an interesting examination of the vapour-phase complexation of HCl with dimethyl ether has been made by Clague, Govil and Bernstein[46]. Formation of a complex normally leads to a shift of the bonding proton resonance to lower field—the exception is when a π-orbital is the proton acceptor, when the shift is in the high field direction. The lifetimes of hydrogen bonded complexes in solution vary over a wide range[45], but at room temperature are usually too short ($<10^{-3}$ s) for the chemical shifts of "free" and complexed donor to be separately resolved. Only a single line, at the time-weighted mean chemical shift experienced by the donor is observed. From the variation of the position of this line with concentration and temperature, the form of association (dimer, trimer, *etc.*) and the appropriate association constants have been derived for numerous systems, together with the enthalpy and entropy of formation[47–50]. Solution of the simultaneous equations, either by curve fitting or computer methods, also yields the chemical shift of the complex[45]. The relation between the observables and molecular parameters is similar to that in electronic spectra, where association constants and extinction coefficients are obtained from absorption data, as discussed in chapter 3, with a corresponding need for care in the interpretation of the data. Several factors have been suggested to account for the downfield shift in 'normal' complexes[45]. Although the relative importance of these factors is still uncertain quantitatively, it appears that a reduction in the electron density around the bridging proton is one of the principal causes, and this may be compared with the results of theoretical calculations discussed in section 4.7.

4.4.3. ELECTRONIC SPECTRAL STUDIES

Electronic spectroscopy has played a far less important role in the study of hydrogen bonding, compared to its dominant position in examining complexes in which charge-transfer bands are observed—a reflection of this is that the term hydrogen bonding does not appear in the index of probably the foremost text on electronic spectra[51]. This may be because the changes produced in electronic spectra by hydrogen bonding are comparatively small compared to the vibrational effects. Indeed, difficulties have arisen in distinguishing the effects of complexing from general solvent effects on electronic spectra. Bands in the complex spectrum are usually similar to bands due to the equivalent transition in the parent moieties, in position, shape and intensity; consequently their assignment is usually unambiguous. Transitions of both the proton acceptor and the proton donor have been examined[52–54]. It is usually observed that $\pi^* \leftarrow n$ transitions in the proton acceptor shift to higher frequency, and $\pi^* \leftarrow \pi$ transitions to lower frequency. The proton donors which have been examined are predominantly phenols or naphthols, in which the transitions are $\pi^* \leftarrow \pi$ type, which also shift to lower frequency. It is obvious that a blue shift implies that hydrogen bonding has lowered the energy of the ground state more than that of the excited state, while a red shift implies the reverse. Pimentel[55] has emphasised the importance of Franck-Condon considerations in these transitions. If the excited state is more strongly hydrogen bonded than the ground state, the equilibrium length of the A—B bond will be less. As the electronic transition is vertical, the situation will be represented as in figure 4.5a (blue shift) or 4.5b (red shift). In both cases the frequency shift is $(W_0 - W_1) + \omega_1 - \omega_0$, where W_0 is the hydrogen-bond energy in the ground state, and W_1 that in the excited state, ω_0 is the zero-point energy in the ground state, and ω_1 is a correction, always in the blue direction, arising from the change in bond length. This correction will not be negligible, so the shifts do not directly give the change in stabilisation energy, but this may be obtained by the addition of data from the fluorescence spectrum.

The systems in which electronic spectra have been obtained have usually been conjugated and so lend themselves to qualitative discussion of the effects of hydrogen bonding in terms of simple *MO* descriptions. Thus in a $\pi^* \leftarrow n$ transition, the non-bonding electron is in an orbital of B directed towards the bridging proton. Delocalisation of this electron in the π^* orbital reduces the basicity of B, and results in a weaker hydrogen bond. In a similar manner, the $\pi^* \leftarrow \pi$ transition leads to an increase in the electron density in the periphery of the molecule, enhancing the electronegativity of the regions preferentially concerned in forming the hydrogen bond.

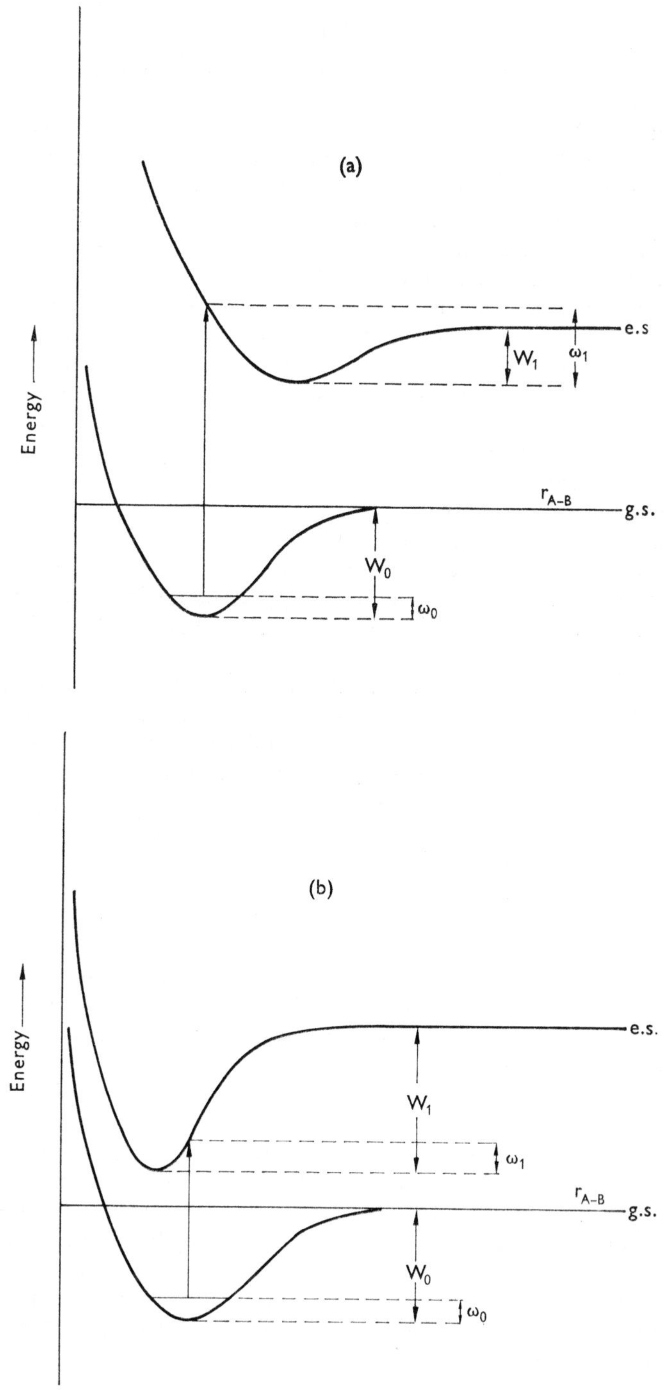

FIG 4.5 The origin of (a) a blue shift and (b) a red shift in an electronic transition, and the effect of the change in bond length.

Analogies have frequently been advanced between hydrogen bonding and complexation which is thought to primarily involve charge transfer (see section 1.2.10). Electronic spectra, however, display an important distinction. In hydrogen bonded systems new bands corresponding to charge-transfer transitions are hardly ever observed. Indeed the writer is aware of only one reference to such a transition[56], the assignment of the band at 211 nm in the hydrogen maleate ion. However, as pointed out in section 1.1.2, it is entirely conceivable, on estimating the parameters involved in determining the "charge-transfer" band position, that such a band *may* occur at higher frequencies than those due to locally-excited transitions of AH or B and may, therefore, not be seen. However, as is well-recognised (section 1.1.2) the observation of a "charge-transfer" band does *not* mean that charge-transfer interaction is responsible for the ground-state stability of any particular complex, hydrogen bonded or otherwise.

With the advent of large, high speed computers and reliable *MO* calculations (see section 4.7) the electronic spectra of hydrogen bonded systems take on increased importance. In particular they provide a gateway to an understanding of hydrogen bonding in electronically excited states, about which vibrational spectra are extremely uniformative.

A further and particularly interesting use of electronic spectroscopy has been to investigate the occurrence of proton-transferred complexes. Baba Matsuyama and Kokubun[57,58] examined mixtures of *p*-nitrophenol and triethylamine in aprotic solvents. In *iso*-octane, the *p*-nitrophenol transition giving a band at ~310 nm indicates a covalent complex OH—N, but in 1,2-dichloroethane two bands arise at ~330 nm and 390 nm, characteristic, respectively, of complexed phenol, and phenate ion. Vinogradov and co-workers[59,60] have also examined *p*-nitrophenol/amine mixtures in solution by ultraviolet spectroscopy, and confirm the presence of phenate ion. However, as these workers point out, it has not so far been possible to determine whether this is present in a hydrogen bond proton transferred complex, *i.e.* O^-—HN^+, or in non-hydrogen-bonded ion pairs. Only in the former case is a proton double-minimum potential present (see below).

Intensity measurements have been used to obtain association constants, and heats and entropies of formation from ultraviolet spectra[61]; reference to the extensive tabulation of thermodynamic parameters given by Murthy and Rao[2], however shows that ultraviolet measurements have been used much less frequently than infrared for the determination of these parameters. This is probably because the bands of the complex and of the unassociated species widely overlap in the electronic spectra which, although not an obstacle in principle, makes the solution of the simultaneous equations for the determination of the association parameters less well-conditioned.

4.5. VIBRATIONAL EFFECTS OF HYDROGEN BONDING

Following this brief outline of the parts played by some of the more important physical methods in the study of the hydrogen bond, we now turn to a more detailed account of the contribution of vibrational spectroscopy, freely drawing on results derived from these other techniques whenever appropriate.

4.5.1. FORM OF THE VIBRATIONS OF A HYDROGEN-BONDED COMPLEX

Although these vibrations were discussed by Pimentel and McClellan, it has now become desirable to extend and refine this classification. We consider a 1:1 intermolecular complex, in which the moieties bonded together are both polyatomic molecules. The classification, in general terms, is independent of whether the moieties are connected by a single hydrogen bond—*e.g.* phenol–pyridine, or multiple bonds, *e.g.* formic acid dimer. The formation of the complex results in the conversion of three translational and three rotational degrees of freedom into six intermolecular vibration modes. The nature of these will be determined by the form of the intermolecular force-field. At present very little is known about such fields, but one expects, intuitively, the following modes.

(i) Intermolecular Stretching

This is characterised by variation of the distance between the centres of mass of the partners in the complex, or of the atoms most closely associated with the hydrogen bond. This is probably the highest frequency intermolecular mode, and the only one observed so far in most of the systems examined. Following Pimentel and McClellan, it is termed ν_σ and it is illustrated in figure 4.6.

(ii) Intermolecular Bending

There are four modes of this type. The form of these is, as yet, very uncertain. However, one expects two vibrational motions in opposed

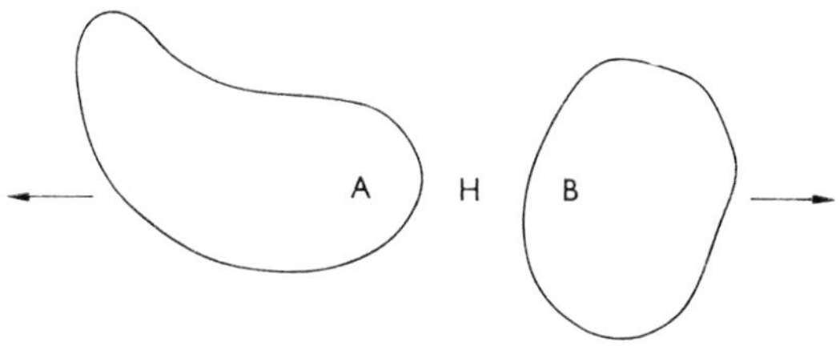

FIG 4.6 Illustration of the ν_σ intermolecular stretching mode.

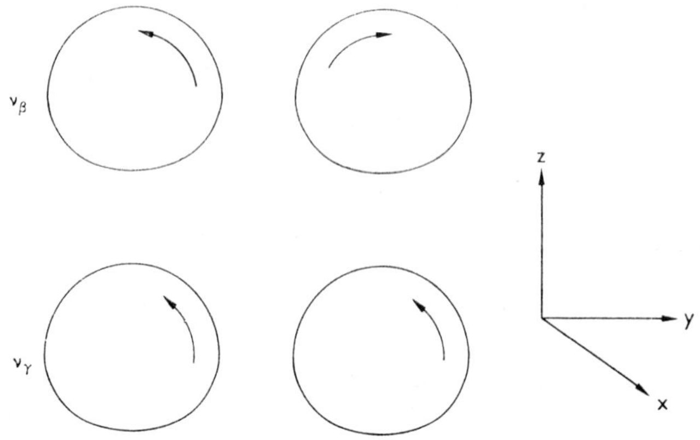

FIG 4.7 Illustration of the ν_β and ν_γ intermolecular bending modes.

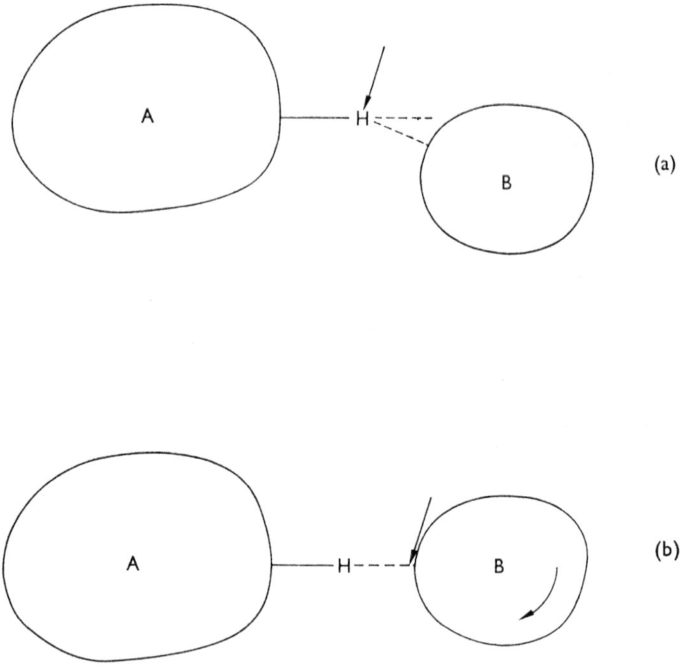

FIG 4.8 (a and b). The two ends of the hydrogen bond at which bending can occur. The arrows mark the fulcrum, which distinguishes the two bending motions.

senses ν_β and ν_δ, and two in the same sense ν_γ and ν_ϵ. Figure 4.7 illustrates the two modes ν_β and ν_γ arising from librations about the x axis. ν_δ and ν_ϵ are their counterparts arising from librations about the the z axis. The term "hydrogen bond bending" is often taken to mean bending at the H atom (figure 4.8a). Since bending at B also occurs (figure 4.8b) both deformations will be involved in the intermolecular bending modes. In the case of the HCN–HF and CH_3CN–HF complexes, Thomas[62] has recently evaluated all the intermolecular modes

Approximate normal coordinates for the bending vibrations of the complex HCN–H-F. For illustration the amplitudes of the atomic displacements have been scaled up.

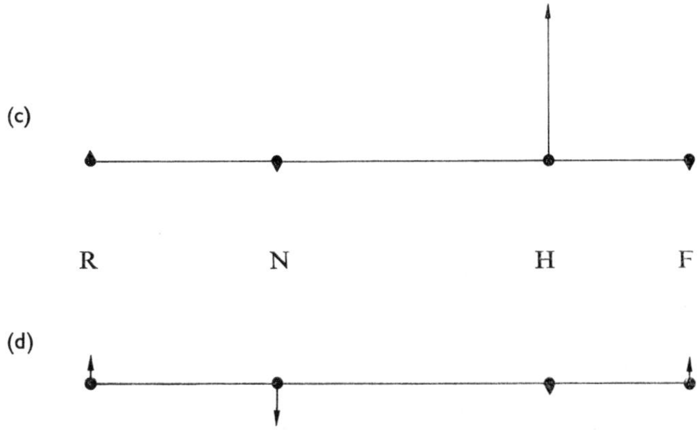

FIG 4.8 (c and d).

(see section 4.5.8). These complexes are linear, so the in-plane and out-of-plane bending modes are degenerate. Because of the small moment of inertia of the HF molecule, one of the intermolecular bending modes is similar to a restricted rotation of the HF molecule, and so has a frequency equivalent to about 500 cm^{-1}. The form of the two bending modes is shown in figures 4.8c and 4.8d. The form, and frequency, of the bending modes would however, not be representative of a typical complex in which both moieties contained more than one heavy atom.

(iii) Intermolecular Torsion ν_α

This is represented by the torsional oscillation about the y axis of one molecule, relative to the other.

When one Partner is Monatomic, e.g. in the C_5H_5N–HCl complex,† the number of intermolecular modes reduces to three, illustrated in figure 4.9. ν'_β and ν'_δ now necessarily involve bending at the H atom.

† This involves a H—Cl, not a N—H hydrogen bond, as occurs in CH_3CN—HF complex.

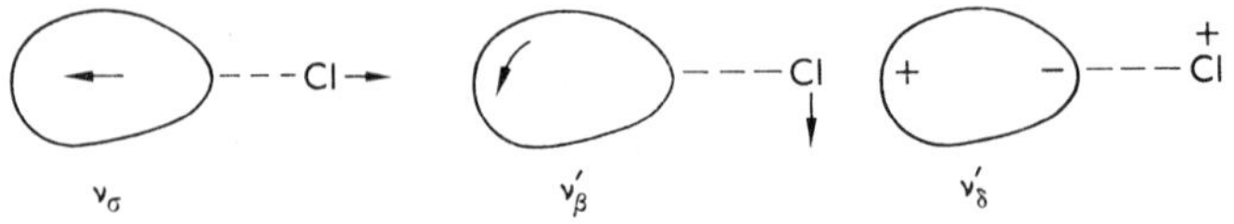

FIG 4.9 Illustration of the form of the three intermolecular vibrations when the base is monatomic, *e.g.* Cl^-.

When both Partners are Monatomic, e.g. in $(ClHCl)^-$ it seems most convenient to use the notation appropriate to a triatomic molecule.

(iv) Internal Vibrations.

Complexing usually produces substantial, and sometimes extreme changes in the internal vibrations of the associated molecules, both as a result of changes in the internal force field and configuration, and also as a result of vibrational coupling with the intermolecular modes. The modes most involved are:

The AH Stretching Mode, ν_s (figure 4.10). This is sometimes termed the hydrogen bond stretching mode.

The AH Bending Mode, ν_b (figure 4.10). This notation is applied to the in-plane bending mode when the partner A has a plane of symmetry, as is often the case, *e.g.* CH_3OH, C_6H_5OH, or $C_5H_5N^+H$. There is indirect evidence that in proton donors such as phenol the symmetry plane is often preserved on hydrogen bonding[63].

The AH Bending Mode, ν_d. This refers to the out-of-plane equivalent of ν_b. In alcohols or phenols, this is the torsion mode, which in the free molecule gives rise to a band in the far-infrared region and is comparatively weakly mixed with other internal modes. In other complexes, *e.g.* involving $C_5H_5N^+H$, several modes involve this motion and their bands lie in the mid-infrared region. The AH in-plane bending motion is similarly often involved in several modes whose bands arise in the mid-infrared region.

(v) Transitions Arising from Anharmonic Effects

In a number of instances, strong anharmonicity has been shown to be present in the vibrations of hydrogen-bonded complexes. In particular, the possibility of a double minimum in the bonding H atom potential has frequently been suggested (see pp. 335–340). This can lead to energy levels quite different from those of a harmonic oscillator, and result in new transitions, most notably inversion. This transition will be termed ν_τ, to indicate its connection with (proton) tunnelling,

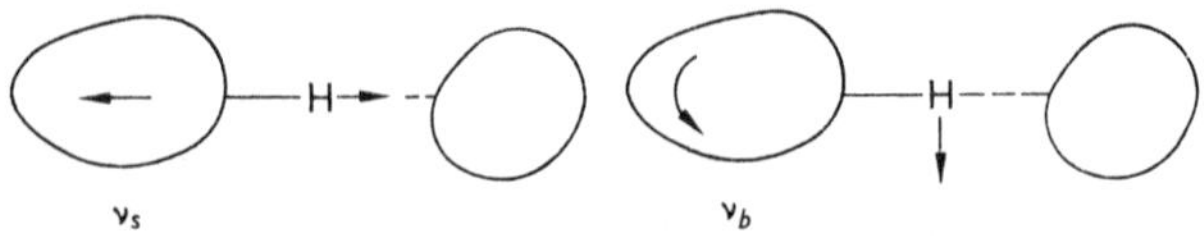

FIG 4.10. Illustration of the forms of the internal vibrations ν_s and ν_b.

the Greek alphabet denoting the intermolecular, rather than intramolecular origin. One should bear in mind that the harmonic approximation is the basis for the concept of normal modes and co-ordinates.

4.5.2 RELATION OF INTERMOLECULAR AND LATTICE MODES IN CRYSTALS

Suppose one is dealing with a molecular crystal—*i.e.* a system in which the intramolecular forces are recognisably stronger than the intermolecular forces, with m n-atomic molecules per unit cell. The 3 nm co-ordinates specifying the atomic positions in the unit cell represent 3 co-ordinates determining the cell centre of mass, leaving 3 nm − 3 representing the internal configuration. Motion in the first three co-ordinates gives rise to acoustic modes, while the remainder represent optical modes. Although each optical mode is represented by a sequence of bands rather than by discrete energy levels, infrared transitions are only strong between those parts of the band corresponding to vibrations in which the oscillations in all the unit cells are in phase[64]. Consequently when dealing with infrared spectra one can consider a single unit cell.

The 3 nm − 3 cell modes break down into m(3 n − 6) intramolecular modes, and 6 m − 3 lattice or intermolecular modes, of which 3 m derive from the rocking motion of the molecules, and 3 m − 3 from relative translational motion. When the intermolecular forces are negligible (the "oriented gas" model), the intramolecular modes are m-fold degenerate, *i.e.* they give rise to 3 n − 6 internal vibration bands, just as in the gas. As the intermolecular forces increase, this degeneracy is progressively removed. When the intermolecular forces become comparable with the internal forces, the distinction between the intramolecular modes and the lattice modes disappears. Hydrogen bonding can be a major contributor to the intermolecular force field, and will be reflected in the frequencies of the lattice vibrations, and in the splitting of the intramolecular modes.

4.5.3. THE ν_s STRETCHING MODE

The most spectacular and readily observable effects of hydrogen bonding occur in the ν_s band. The frequency of this band decreases, often by an amount corresponding to several hundred cm^{-1}, and occasionally, for very strong complexes, the decrease may exceed 1000 cm^{-1} (*e.g.* collidine-pentachlorophenol,[3] $\Delta\bar{\nu} = 1098\ cm^{-1}$). The infrared intensity increases, often several fold, in the case of phenol-hexamethylenetetramine by as much as 124-fold. The infrared band characteristically becomes very broad—often with a width of several hundred cm^{-1}, and has a very marked substructure (*e.g.* figure 4.11) with no obvious regularities. Observations such as these have usually been made on solutions, but similar behaviour is also found in the vapour and crystal phases.

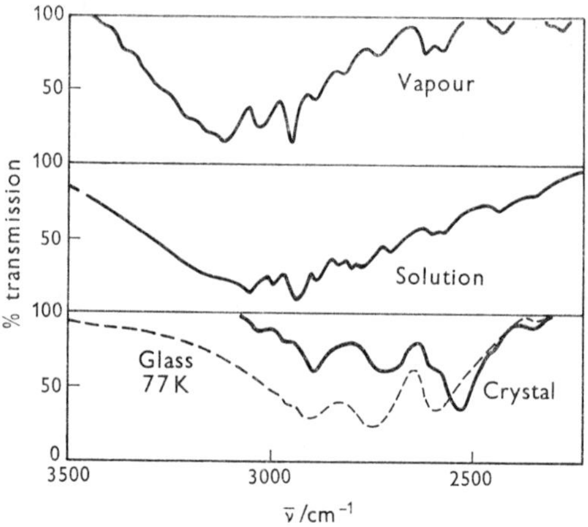

FIG 4.11 The infrared spectra of formic acid gaseous dimer, dimer in CCl_4 solution, crystal and glass. (Reproduced, by permission, from 'The Hydrogen Bond' by G. C. Pimentel and A. L. McClellan, p. 104.)

As the ν_s band has been the most important source of information about many aspects of hydrogen bonding, it will be convenient to include topics, such as multimer formation, or proton tunnelling, in the discussions of this section.

The Shifts of the ν_s Band

Quantitative measurements of $\Delta\bar{\nu}_s$ encounter two difficulties, (a) in view of the width and structure of the band corresponding to ν_s, the choice of the band centre is somewhat arbitrary, (b) the shift is solvent dependent, even if the solvent does not enter directly into the complex (see section 2.3.4). Intercomparison should therefore relate to the same solvent, but solubility limitations may prevent this in some instances. Nevertheless, quite reliable values for $\Delta\bar{\nu}_s$ have been measured for numerous systems, and correlations tested with (a) the AB distance obtained from crystal studies[65] (see figure 4.12); (b) the heat of formation, $\Delta H_f^\ominus$, of the complex[66–70]; (c) the association constant K_c[68,71]; (d) the pK_a of the proton donor[10,72–76]; (e) the pK_b of the acceptor[77]; (f) Taft and Hammett parameters.[75,78,79]

These correlations have been discussed very frequently,[2] so we shall confine ourselves to a few observations. Lippincott and Schroeder[80–82] have developed a semi-empirical expression which relates the potential, as a function of the AH and AB distances, to the Morse-like parameters of AH and BH^+. The Lippincott-Schroeder function leads to a correlation between $\Delta\bar{\nu}_s$, the bond length AB, and a heat of formation, in fair agreement with observations. These relationships depend on the bond

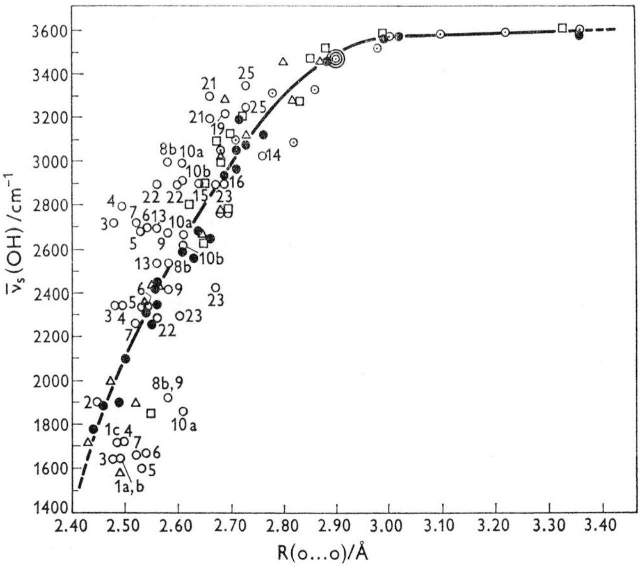

FIG 4.12 Relation between ν_s(OH) and R(O—H · · · O) (Reproduced, by permission, from *J. Mol. Struct.*, **1**, 450 (1968)).

type, *e.g.* OH—O, OH—N, *etc.*, but not on the particular chemical class of the donor or acceptor. The usual experience is that quite good correlation is observed, *e.g.* between $\Delta H_f^{\ominus}$ and $\Delta\bar{\nu}_s$ provided comparison is restricted to a particular chemical type of donor and acceptor. However, when more extensive variation in the chemical type of the donor or acceptor is tested, the correlations become less satisfactory (see section 2.3.4). This is to be expected. For instance the pK_a or $\Delta H^{\ominus}$ will depend on the relative stability of the anion A^-, but this may not make a contribution of much significance to the structure of the complex. Furthermore, the correlation cannot be expected to be significant if the spectra and thermodynamic parameters are measured in different solvents.

In very strongly hydrogen bonded systems—mainly OHO bonds in the solid phase, the ν_s band does not appear in the higher frequency region, but is replaced by a strong band, often many hundred wavenumbers broad, which may extend to as low as 600 cm^{-1}[183,84]. The origin of this absorption is not well understood.

The Structure of the ν_s Band

Not only is the ν_s band notably broadened by hydrogen bonding, but also often has a very complex structure, with numerous subsidiary peaks, the entire "massif" extending over hundreds of cm^{-1}. The phenomenon is not confined to a particular phase. Figure 4.11 shows the band in formic acid dimer in the vapour, solution, glassy and crystal states. Possible origins for the structure have received a great deal of attention, and an excellent review of the various theories has been given by Sheppard[85]. All the proposals discussed by Sheppard remain valid, and in the meanwhile, evidence which strongly supports particular mechanisms has been produced.

It now seems clear that,

1. No single universal explanation applies to all systems.
2. In certain systems, one particular mechanism may predominate and provide a satisfactory basis for the interpretation of the band structure.
3. The importance of the various contributory factors will depend on the phase. Here the main distinction is between complexes weakly coupled to the surrounding (*e.g.* in the vapor phase, or in "inert" solvents), and complexes strongly interacting with the environment. The most fully developed strong coupling arises in hydrogen-bonded crystalline networks (*e.g.* in ice[86], ferroelectrics[87,88], or crystalline imidazole[89,90], or in pure liquids (*e.g.* water[86] or alcohols). These obviously require each partner to act both as donor and acceptor.

The various suggestions that have been advanced to explain the ν_s band structure will be considered in turn, commencing with complexes which can be regarded as isolated from the surroundings.

The simplest profile that the band due to ν_s can display is a rounded profile, with ancillary peaks weak or absent (see figure 4.13). However, even in these cases, the width of the ν_s band is greatly increased by hydrogen bonding, often being $\sim$100 cm^{-1}. It is not necessary for multimeric species to be present for this broadening to occur. In pyrH$^+$—Cl$^-$ (figure 4.13), there is no reason to believe that more than a single species is present, but the ν_s band is several 100 cm^{-1} broad. In such cases three effects may contribute to the breadth of ν_s:—

a. *A Change in the Frequency of the ν_s Band in Hot Bands of other Low Lying Vibrations*

Upper levels of the intermolecular modes ν_σ, ν_β and ν_δ, *etc.* will be well populated at room temperature. The normal modes are defined by the quadratic terms in the force field of the complex. Higher order terms are certainly present, and these lead to coupling between ν_s and the intermolecular modes, with the consequence that the spacing of the $v = 1 \leftarrow 0$ transition in ν_s varies with the intermolecular level occupied. In classical language one would say that the geometry of the complex is loose, at room temperature, and that the ν_s band frequency differs in, for example, a bent complex from its value in a straight one.

b. *Sum and Difference Combinations with Intermolecular Vibrations*

Good evidence that combinations of ν_s with ν_σ can give rise to *discrete* peaks in the ν_s band "massif" has been presented, and will be discussed on p. 333. The other intermolecular modes have bands lying at

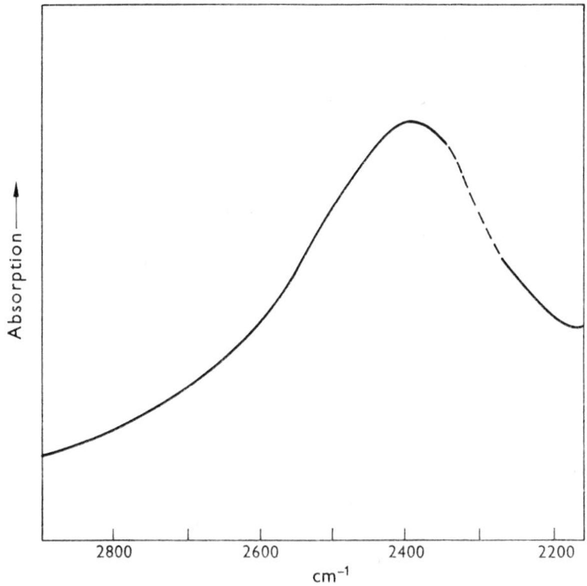

FIG 4.13. The main ν_s band of $PyrH^+Cl^-$ in CH_3CN. (Two ancilliary peaks occur at ~2000 and 2100 cm^{-1}.)

lower frequency than ν_σ and numerous combination bands could arise, and overlap, contributing to a rounded contour of the ν_s band. As Sheppard[85] has noted, sum and difference combination bands may draw intensity from a nearby fundamental by Fermi resonance if the symmetry is the same, and this applies to hot bands also. In this connection it is notable that in the HF_2^- ion, the lowest internal mode is at ~600 cm^{-1}, so that contributions from hot bands or low-lying modes are necessarily small at low temperatures. When isolated in a matrix which removes resonance broadening due to neighbour interactions, the ν_s band is extremely sharp[91]. Incidentally, this also shows that the width of the ν_s band is not related to the strength of the hydrogen bond.

c. *Relaxation Line Broadening*

The first excited state of ν_s is typically some 2500 cm^{-1} above the ground state. The dissociation energies of hydrogen-bonded complexes are usually smaller than this, frequently corresponding to 1000–2000 cm^{-1}. Most dissociation energies have been measured in solution but the few available for the vapour phase lie in the same range[7,46,92,93]. The ν_s excited state is thus metastable, in the sense that the complex would dissociate if the energy were converted into the intermolecular stretching mode ν_σ. Since coupling between ν_s and ν_σ appears to be strong, this is, at least apparently feasible. A band width of 100 cm^{-1}

would require a lifetime as short as 10^{-12} to 10^{-13} s. However Thomas and Thompson have resolved sub-bands spaced at less than 1 cm^{-1} in the ν_s band contour of the CH_3CN–HCl complex in the vapour phase[94]. These sub-bands are accounted quantitatively as arising from P-branch band heads of a series of ν_s hot bands. The point of relevance here is that the lifetime of the complex is sufficiently long to permit this detail to be observed. Similar fine structure has also been observed by Thomas in the vapour phase infrared spectra of the complexes of HF with HCN and CH_3CN[62]. The structure implies a minimum lifetime for the HCN–HF complex of 10^{-10} s, and the spectrum was unchanged when hydrogen gas was added to the mixture to a pressure of 100 kN m^{-2} (1 atm.). Under these conditions, the complex undergoes two to three collisions in 10^{-10} s. A somewhat less closely spaced structure, which appears to have a similar origin, has been observed by Jones *et al.*[93] for the H_3N–HCN complex.

The rotational line width in the microwave spectrum[95] of the CF_3COOH–HCOOH complex also indicates a substantial lifetime in the gas phase, in this case $\geqslant 10^{-7}$ s.

It therefore seems that dissociation in the vapour phase is quite unable to account for the observed broadening of the ν_s band, though it may blur out some detail arising from (a) or (b). Estimates of the lifetime of the ν_s excited state in solution do not seem to be available but, when coupling is weak, it is again unlikely that they would account for the ν_s band broadening.

In addition to a general broadening, the ν_s band frequently displays numerous subsidiary maxima. Several contributory causes have now been identified:

(*i*) *The Presence of Multimers*

Many common hydrogen-bond donors (*e.g.* water, alcohols, phenols or carboxylic acids) can also function as acceptors, so that self-association to form dimers, trimers or polymers is possible, and several species may co-exist in solution. Further possibilities arise since both open chain and cyclic multimers may occur. Variation of the ν_s band structure with concentration, in the case of self-association, or with composition, in the case of mixtures, is a good indication of more than a single form of complex. Variation with the solvent is also indicative (see section 2.3.4 for further discussion of this point).

The association of phenol in 'inert' solvents illustrates these effects[96]. In carbon tetrachloride the spectrum of monomeric phenol has a sharp band at 3611 cm^{-1}. At concentrations up to 0.1 M a second, somewhat wider, band at 3481 cm^{-1} develops, ascribed to dimer[97–99]. At increasing concentrations a much broader band $\Delta\bar{\nu}_{1/2} \sim 200$ cm^{-1} arises, centred at 3350 cm^{-1}, which has been assigned to either trimer or higher polymer. The greater breadth in comparison with the 3481 cm^{-1} band in itself suggests that several multimeric species contribute.

Attempts to determine the molecularity of the associated species have been based on n.m.r.[45], ultraviolet[97] or infrared measurements[96]. Each species is characterised by a chemical shift or extinction coefficient, which is combined with equilibrium constants to give expressions for the observed chemical shift or extinction coefficient of the solution as a function of concentration. Curve fitting methods are then used to determine which equilibria best represent the observations. However, decisive answers have been difficult to obtain. Several factors seem likely to have caused these uncertainties.

(a) The difficulty of knowing which species, or whether more than one species, is contributing to a particular absorption band. (b) The possibility that the measured values are not sufficiently sensitive to the form of association, so that varying proportions of complexed species could result in the same observations. For a particular associated species, the parameter itself may be solvent and temperature sensitive, so that the changes cannot be entirely ascribed to an alteration in the proportions present.

However, more detailed spectroscopic considerations may also throw light on the form of association. Thus Bellamy and Pace[99], by careful compensation, were able to observe a further band at 3599 cm^{-1} in solutions of phenol in CCl_4. They argue convincingly that this represents the ν_s band of an OH group which is acting as a hydrogen-bond acceptor, but not donor. The 3481 cm^{-1} band correspondingly represents the ν_s of an OH group which is donor but not acceptor, and 3350 cm^{-1} that of OH acting both as donor and acceptor[100]. The decrease in frequency and force constant for the OH stretching motion when the oxygen atom acts as an hydrogen-bond acceptor indicates an increase in acidity, so that the H atom will more readily hydrogen bond. Similarly the donor function of the OH group can be expected to enhance the tendency for the oxygen atom to accept hydrogen bonds. Both these effects will favour the formation of chains or large rings. If this interpretation is substantiated, and the absorption arising from the linking and end hydroxyl groups can be distinguished and measured, the ratio of open to closed polymer can be obtained, provided the polymer molecular weight can be independently measured.

Alcohols can also give complex multimers[101–104], with consequent influence on the ν_s band. Thus for monomeric CH_3OH in carbon tetrachloride this mode is assigned a band at 3642 cm^{-1}, a band ascribed to dimer is at 3542 cm^{-1}, and one ascribed to polymer at 3346 cm^{-1} [99]. Bellamy and Pace have also been able to distinguish a band at 3637 cm^{-1}, due to an OH group acting as acceptor, but not donor.

In addition to uncertainties about the molecularity of association in phenol and alcohol solutions (and in alcohol in the vapour phase[102]), variations in structure are possible, and each may have a different ν_s frequency. For both phenol and methanol dimers in solution, Bellamy and Pace favour an open, rather than a ring dimer (see figure 4.14).

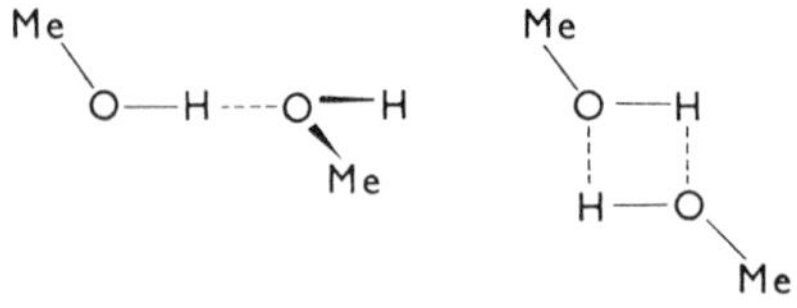

(a) (b)

FIG 4.14 Illustrating open (a) and ring (b) conformations for the methanol dimer.

This conflicts with the conclusion of van Thiel *et al.*[105] based on study of the ν_s band of CH_3OH multimers isolated in a solid nitrogen matrix.

Tursi and Nixon[36] have re-examined H_2O, D_2O and HDO in nitrogen matrices (which were also believed to be present as cyclic dimers[106]) and decided in favour of the open structure. It is interesting that recent molecular orbital calculations also favour the open dimer structure, rather than the cyclic one, for both $(H_2O)_2$[107–111] and $(CH_3OH)_2$[112], and indicate stronger hydrogen bonding in the trimer than in the dimer[109,112].

Among the many other systems in which multimer formation contributes to the ν_s band contour, HF vapour[113,114] and in solution[115], and HCl in matrices[116–118] are notable. Several forms of complexes may also occur in binary systems—indeed the solvent may itself function as acceptor in solutions, *e.g.* HF in "inert" solvents, leading to complexes of the form[115].

B----HF----HF

In a similar way, pentachlorophenol forms 2:1 and 3:1 complexes with amines, in addition to the 1:1 complex, which have been assigned the structures such as[3],

$$\mathrm{Bu_3NH\cdots\underset{\displaystyle R}{\underset{|}{O}}\cdots H{-}\underset{\displaystyle R}{\underset{|}{O}}}$$

Complications due to multimer formation can also be expected, of course, whenever a polyfunctional donor or acceptor is present.

(ii) *Fermi Resonance*

Even when only a single, well defined complex is present, the ν_s band frequently displays numerous subsidiary peaks. One possible origin for these is Fermi resonance of ν_s with overtones or combinations of other internal vibrations. When two vibrational levels of the same symmetry are accidentally degenerate, or lie close together, the presence of cubic or higher order terms in the potential energy function results in interaction between the levels, which are displaced. The intensities of transitions involving these levels may also be greatly altered[119]. This effect was first recognised by Fermi[120] in 1931.

If Ψ_1° and Ψ_2° are the wavefunctions of the unperturbed levels (*i.e.* the levels which would be present if only quadratic terms were present in the potential energy), and the interaction is:

$$W_{12} = \langle \Psi_1^\circ | H' | \Psi_2^\circ \rangle \tag{4.1}$$

where H′ represents the additional terms, then the new energy levels are:

$$E_1 \text{ or } E_2 = \tfrac{1}{2}(E_1^\circ + E_2^\circ) \pm \tfrac{1}{2}\Delta \tag{4.2}$$

where E_1° and E_2° are the energies of the unperturbed levels. Δ, the separation of the new levels is:

$$\Delta = (\delta^2 + 4W_{12}^2)^{1/2} \tag{4.3}$$

and $\delta = E_1^\circ - E_2^\circ$[119,121,122].

The corresponding wavefunctions for the perturbed levels are:

$$\begin{aligned} \Psi_1 &= a\Psi_1^\circ - b\Psi_2^\circ \\ \Psi_2 &= b\Psi_1^\circ + a\Psi_2^\circ \end{aligned} \tag{4.4}$$

the coefficients a and b being given by:

$$a = \left(\frac{\Delta + \delta}{2\Delta}\right)^{1/2}, \quad b = \left(\frac{\Delta - \delta}{2\Delta}\right)^{1/2} \tag{4.5}$$

If the intensity expected for a transition to the state Ψ_1° of the unperturbed system is I_1°, and I_2° is that of a transition to Ψ_2°, originating from a common level, then the effect of Fermi resonance is to change the intensities to I_1 and I_2 where:

$$\begin{aligned} I_1 &= [a(I_1^\circ)^{1/2} - b(I_2^\circ)^{1/2}]^2 \\ I_2 &= [b(I_1^\circ)^{1/2} + a(I_2^\circ)^{1/2}]^2 \end{aligned} \tag{4.6}$$

In applying this expression, the sign of $(I_1^\circ)^{1/2}$ is the same as that of $\langle \Psi_{\text{initial}} | \mu | \Psi_1^\circ \rangle$[121].

The consequences of Fermi resonance are,

(a) The interacting levels are pushed apart, the perturbation increasing with W_{12}, and decreasing with δ (figure 4.15). If one of the levels concerned is an overtone or combination band, its approximate unperturbed frequency follows from that of the fundamentals. Combination with the observed value of Δ permits δ and hence W_{12}, a and b to be obtained.

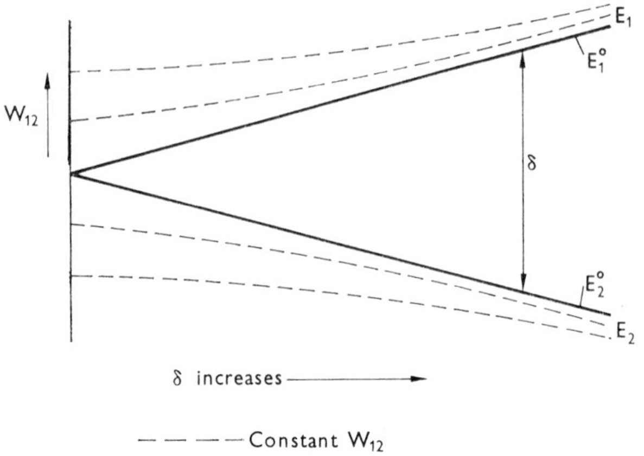

FIG 4.15 The general dependence of the interaction between two levels on (a) the interaction constant W_{12} and (b) their unperturbed separation δ.

(b) The intensity is shared between the two transitions. If $I_1^\circ \gg I_2^\circ$ then:

$$I_2/I_1 = b^2/a^2 \tag{4.7}$$

b/a ranges from zero when $W_{12}/\delta \to 0$ to one when $\delta/W_{12} \to 0$. Thus the intensity of an extremely weak combination band may become comparable to that of the band with which it is interacting (for example, the ν_s band) when W_{12} is large compared to δ.

Figure 4.16 shows how the intensity varies with b/a and I_1°/I_2°. The right hand side of the figure represents no interaction, while the left hand side represents complete mixing. From this figure it can be seen that Fermi resonance never reverses the order of intensity—at the most the intensities are equalised. However, it may, for weak interactions when the bands have comparable initial intensities, lead to a disproportionation rather than a sharing of intensity.

There is strong evidence that Fermi resonance makes a major contribution to the ν_s band structure in some complexes. Hall and the writer[123] found that in a series of complexes of *p*-substituted phenols with *p*-substituted pyridines, the ν_s band structure was characteristic of the phenol, but almost independent of the base (see figure 4.17). This would not be so if combination with ν_σ were involved (section iii) for ν_σ varies appreciably with both donor and acceptor. The possibilities of proton tunnelling (section iv) and multimer formation can equally be eliminated. The Fermi resonance origin of the structure was checked by finding the internal modes of the donor whose frequency was affected by complexing. Only combinations or overtones of these

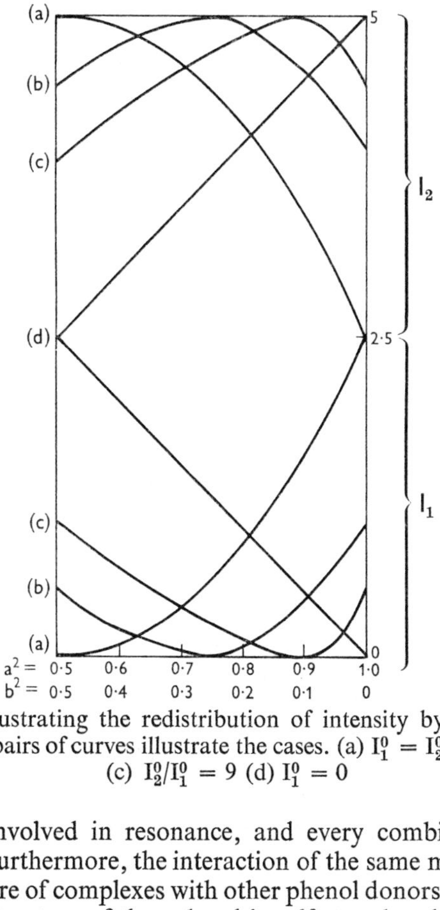

FIG 4.16 Illustrating the redistribution of intensity by Fermi resonance. The pairs of curves illustrate the cases. (a) $I_1^0 = I_2^0$ (b) $I_2^0/I_1^0 = 4$ (c) $I_2^0/I_1^0 = 9$ (d) $I_1^0 = 0$

modes were involved in resonance, and every combination was accounted for. Furthermore, the interaction of the same modes allows the ν_s band structure of complexes with other phenol donors to be predicted, and also the structure of the ν_s band in self-complexed phenols[100]. The effect of Fermi resonance may be particularly pronounced in the complexes formed by chloroform with various acceptors such as triethylamine[124] where ν_s can interact with the first overtone of the CH bending mode ν_4 of the chloroform. As the ν_4 fundamental is observable the shift caused by the resonance interaction can be estimated as about 60 cm^{-1}, yielding $b^2/a^2 \sim 1/4$, so that Fermi resonance alone can account for the observed enhancement of the intensity of $2\nu_4$, and it is not necessary to infer that the intensity of $2\nu_4$ itself has been enhanced by complexing, as Thompson and Pimentel suggest. In fact, it is probable that both effects are present. If the original intensity of $2\nu_4$ is appreciable, the full expression for I_2/I_1 must be used.

Fermi resonance can also produce a transmission window in what would otherwise be a broad absorption band or continuum. The requirement is for a sharp band of low intrinsic intensity to interact with

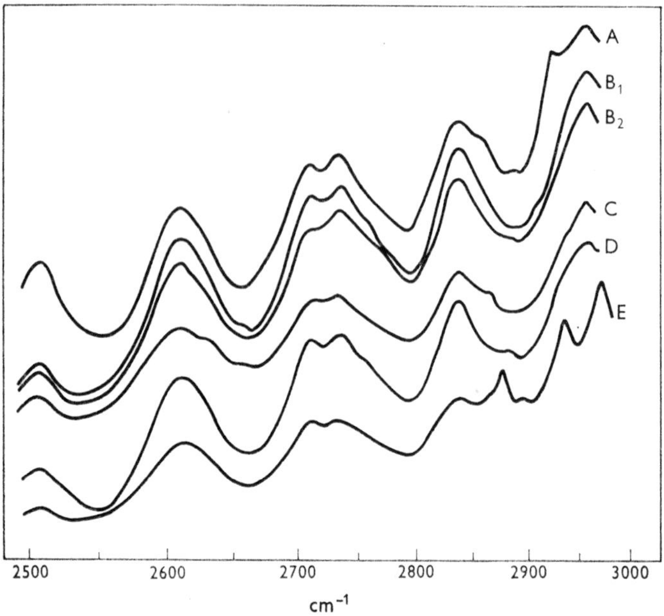

Phenol + pyridines. *A*—γ-picoline, B_1—pyridine, B_2—pyridine d_5, *C*—*p*-chloropyridine, *D*—*p*-phenylpyridine, *E*—*p*-ethylpyridine.

FIG 4.17 Sub-structure of the ν_s band in phenol-amine complexes. (The structure is independent of the amine, but varies with the phenol.) (Reproduced, by permission, from *Spectrochim. Acta*, **23A**, 1259 (1966).)

all the individual levels which combine to form the continuum[121,125–7]. The window results because the continuum levels are pushed away from the frequency of the sharp band, as shown in figure 4.18.

Claydon and Sheppard[128] have pointed out that transmission windows may be formed in this way in the broad ν_s band of hydrogen-bonded complexes by interaction with the weak, but sharp $2\nu_b$ or $2\nu_d$ bands (p. 318). Figure 4.19 reproduced from reference (128) shows how the appearance of the 'window' changes to the normal resonance enhancement as the overtone moves away from the centre of the continuum. The occurrence of such windows provides the most satisfactory explanation[128] of the remarkable ν_s bands found in a wide range of strong O–H—O complexes, such as self-associated dimethyl-arsinic acid or diphenylphosphinic acid, examined by Hadzi[84]. The ν_s band in such systems spreads over a range from ~1600 to 2700 cm^{-1}, and typically has three broad maxima, which have been termed the 'A', 'B' and 'C' bands. Claydon and Sheppard[128] have shown that the minima between these correspond to the $2\nu_b$ and $2\nu_d$ frequencies respectively, and this explanation fits the spectra of the deuterated analogues.

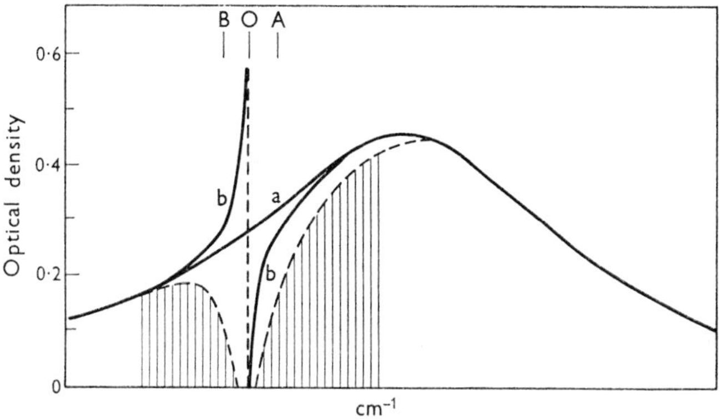

FIG 4.18 The formation of a Fermi resonance window. The observed absorption spectrum is given by the line b. Line a is the undistorted Lorentz curve. (Reproduced, by permission, from *Spectrochim. Acta*, **16,** 996 (1966).)

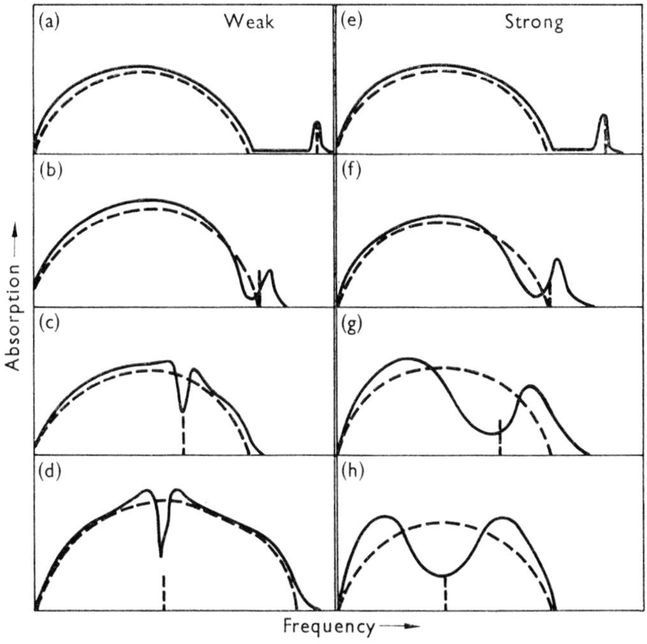

FIG 4.19 Schematic diagram showing the possible effect of increasing Fermi resonance as the frequency of a sharp overtone approaches that of a very broad fundamental. (a)–(d) moderate-to-weak Fermi resonance; (e)–(h) strong Fermi resonance. – – – – spectrum in absence of Fermi resonance; ——— resultant vibrational spectrum when Fermi resonance occurs. (Reproduced from *Chem. Comm.*, 1432 (1969), by permission of the Chemical Society.)

(iii) *Sum and Difference Bands*

The possibility that sum and difference bands arising from combinations of ν_s with low frequency intermolecular modes, particularly ν_σ, can account for some of the ν_s band features has been recognised for some time[66,129–131,85]. The proposal was first expressed in wave-mechanical terms by Stepanov[129,130], and it seems preferable to call this the *Stepanov mechanism*, rather than by the rather misleading term of *predissociation theory*, which could designate what is here called *relaxation line broadening*.

The observations that the equilibrium AH length is increased by complexing, and that the AH stretching frequency falls can be represented on a diagram[132] which shows the potential energy as a function of the AH and AB distances (figure 4.20). Although the potential shown is based on the Lippincott-Schroeder potential, it probably provides a satisfactory model of the general potential surface. The vibrational energy levels implied by this figure can often be obtained quite satisfactorily by a Born-Oppenheimer type separation of the two vibrational modes. Keeping the AB distance fixed, the energy levels for the AH stretching motion are found. This is repeated over the range of AB distances, giving curves for the potential energy as a function of the distance AB for the ν_s ground state and the ν_s excited states (figure 4.21). These potential curves are then used to obtain the overall energy levels of the system, as indicated by the horizontal lines. It is convenient to designate these levels by (n, m), n indicating the quantum number in

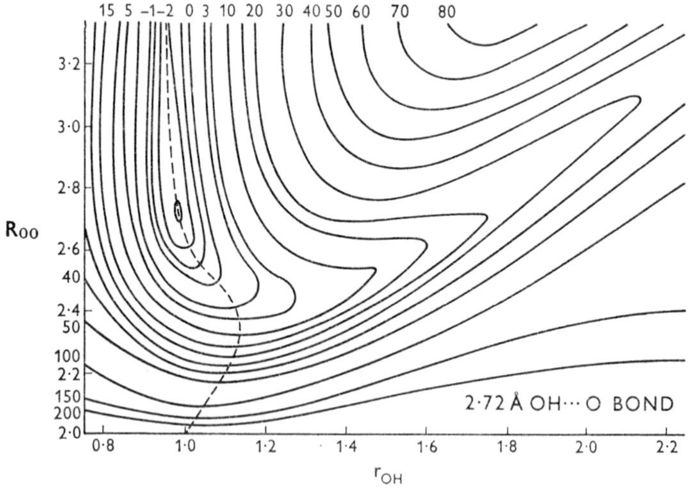

FIG 4.20 Potential surface for the H-bonded proton in a bond of equilibrium length 2.74 Å, allowing both changing r_{OH} and changing O–O distance. (Reproduced from *J. Chem. Phys.*, **30**, 184 (1959), by permission of the American Institute of Physics.)

ν_s and m in ν_σ. Three features of figure 4.21 merit attention. First, the equilibrium AB length will as a rule differ with the ν_s level. Second, the spacing of the ν_σ levels varies somewhat, even with fixed n. Third, the ν_σ spacing in general changes more appreciably for different ν_s levels. The manifold of levels is thus similar to those of a vibronic transition, ν_s taking the place of the electronic excitation and ν_σ the vibrational excitation. Figure 4.21 shows the situation when the hydrogen bond is shorter, and stronger, in the ν_s excited state. To obtain the ν_s band contour, transition moments are required—as the frequency of the ν_s mode is much greater than that of the ν_σ mode, the approximation can be made that there is little change in AB during the transition—*i.e.* excitation is vertical, and the Franck-Condon method can be used. Thus only exceptionally will a series of well-spaced lines result, for at room temperature the (0, 1) and (0, 2) levels are appreciably populated, and transitions starting from these levels may not closely overlie those

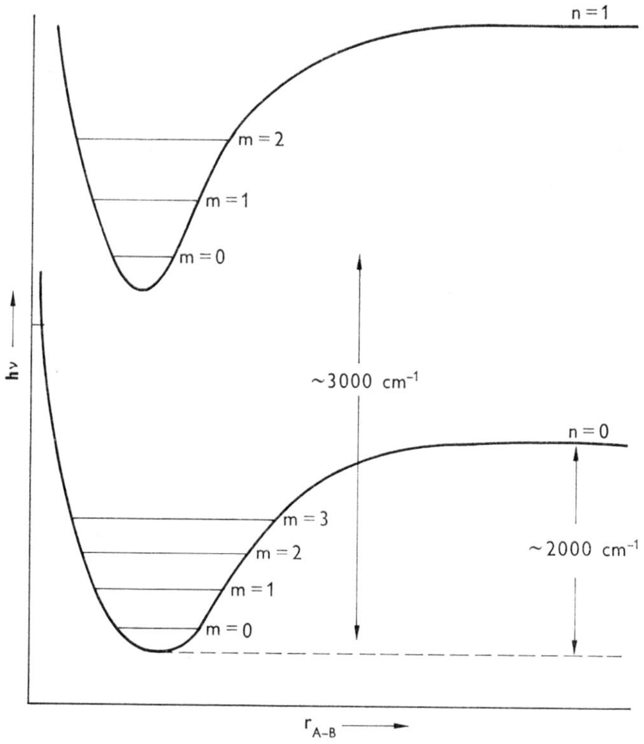

FIG 4.21 Illustration of the Stephanov separation of the AH and AB motion. To improve the legibility, the spacing of the sublevels is not shown to scale. It is $\sim$150 cm^{-1} in the lower state, and a little more in the upper state.

from the (0, 0) level. Attempts to match the spacing of the ν_s band structure with the ν_σ frequency are then unlikely to be successful. Reference to figure 4.21 shows that even when the complications due to hot-band transitions are unimportant, the sum bands will show the spacing of ν_σ^*, the intermolecular stretching frequency in the ν_s *excited* state. This is *not* the experimentally observable frequency ν_σ.

There are, however, several cases in which sum and difference contributions to the ν_s band contour are well established. Millen and his co-workers have interpreted the structure of the ν_s bands of a number of complexes examined in the vapour phase in this way. These include ether–HCl[133–4], ether–HF[6] and ketone–HF[135]. Figure 4.22 shows a typical band, the shoulder at ~100 cm^{-1} on the high frequency side of the central peak being ascribed to the sum modes ($\nu_s + \nu_\sigma^*$) *i.e.* transitions in which $\Delta m = +1$. The low frequency shoulder is similarly assigned to difference modes ($\nu_s - \nu_\sigma$) *i.e.* $\Delta m = -1$ and ν_σ in this case refers to the ν_s ground level. Couzi[7] and his co-workers, however, consider that in the case of the complexes involving HF, multimeric association of the type F—H---F—H---B also contributes to the ν_s structure. Confirmation of Millen's interpretation can be sought in the direct observation of ν_σ in the far-infrared system. In $(CH_3)_2O$/nitric acid mixtures, a band is observed at 175 cm^{-1} in the vapour[136], in agreement with shoulders displaced ~185 cm^{-1} from the main ν_s band[137]. Thomas[138] has similarly observed bands at ~180 cm^{-1} in the far-infrared spectra of HF and DF complexes with dimethyl and

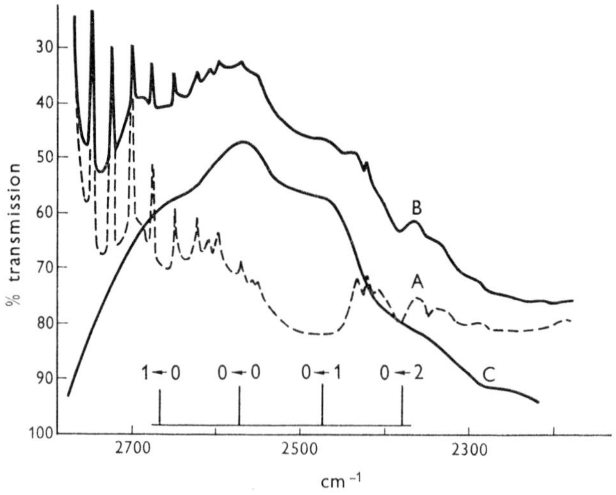

FIG 4.22 Hydrogen chloride and dimethyl ether. (A) Unmixed gases. (B) Mixture: HCl, 250 mm; ether, 125 mm; 10-cm path-length. (C) Subtraction spectrum. (Reproduced from *J. Chem. Soc.*, 498 (1965), by permission of the Chemical Society.)

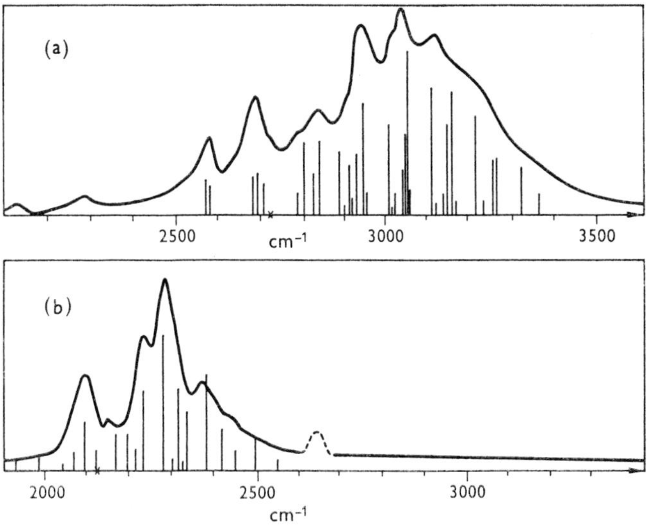

FIG 4.23 Experimental infrared spectra (room temperature) and theoretical reconstitution of (A) CD_3COOH dimer (gas), (B) CD_3COOD dimer (gas). (Reproduced from *J. Chem. Phys.*, **48,** 3701 (1968), by permission of the American Institute of Physics.)

diethyl ether, in support of Arnold and Millen's[6] interpretation of the ν_s band structure. The complex structure of the ν_s band of acetic acid dimers, also in the vapour phase, has also been explained in terms of the Stepanov mechanism by Marechal and Witkowski[139]. The situation is complicated by a further factor in this case—there are two equivalent hydrogen bonds, with strong interaction between them. However, the observed band contour in both the CD_3COOH dimer and the CD_3COOD dimer is very well accounted for (see figure 4.23).

(iv) *Double Minima in the Proton Potential Function*

The possibility that in some hydrogen bonds the proton potential function may have two minima, corresponding to the arrangements AH—B and A—HB, has received a great deal of attention, particularly in view of the role that proton tunnelling between the two minima may play in ferroelectrics[88], mutations[140] and reaction kinetics[141]. The possibility that double minima arise is strongly suggested by the relation found between the AH and AB distance. Figure 4.24 shows the data for O—O bonds, the most extensively examined. The lower curve relates to systems in which A and B differ, and it is noted that the proton is asymmetrically placed and that the same r/R relationship in symmetric bonds would imply a double minimum. Unfortunately, as Ibers has pointed out, it is not easy to determine directly by neutron diffraction the proton position in symmetric (*i.e.* A—H—A) systems,

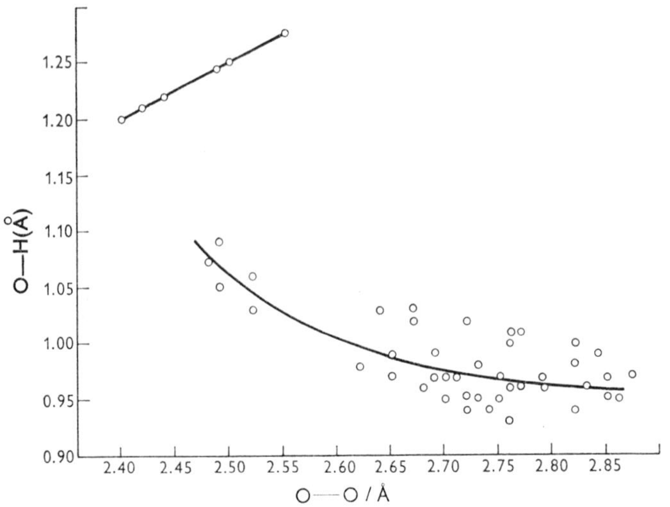

FIG 4.24 O–H distance as a function of O–O distance as determined by neutron diffraction for a number of compounds containing O–H–O hydrogen bonds. The curved line represents the best least squares fit to the points, but the deviations of the points from the line are significant in many cases. In particular, the O–O distance characteristic of the shift from an asymmetric hydrogen bond is not well defined. (Reproduced from 'Hydrogen Bonding in Solids', W. C. Hamilton and J. A. Ibers, p. 53, by permission of W. A. Benjamin Inc., Advanced Book Program, Reading, Massachusetts, U.S.A.)

since when thermal motion is taken into account the superposition of two symmetric proton distributions is difficult to distinguish from a single minimum distribution[142] (figure 4.25). The upper curve of figure 4.24 thus represents the mean proton position in symmetric systems.

For this reason, the spectroscopic examination of possible double minimum systems is particularly valuable. Ultraviolet measurements, discussed in section 4.4.3 suggest that both a polar and a non-polar complex are present in *p*-nitrophenol–amine systems. This could imply a double minimum potential for the proton motion. Such a potential would be expected to produce notable effects in the infrared spectrum.

In a model which only considers the motion of the H atom between fixed atoms, the potential in symmetric systems can be represented most simply by:

$$V = ax^2 + bx^4. \qquad 4.8$$

The wave equation incorporating this potential can be reduced to the simple form,[143]

$$d^2\Psi/dz^2 - (z^4 - Bz^2 - \lambda)\Psi = 0, \qquad 4.9$$

by substituting $z = (2\mu/\hbar^2)^{1/6}b^{1/6}x$. $b = (2\mu a^3/\hbar^2 b^2)^{1/3}$ characterises the barrier shape, a single minimum quartic potential corresponding to B = 0, while increasing B represents a deepening double minimum (figure 4.26). The eigenvalues of the double minimum oscillator, related to the energy E by $\lambda = (2\mu/\hbar^2)^{2/3}b^{-1/3}E$ have been obtained in a variety of forms[143–146]. Figure 4.27 is reproduced from reference 143. The eigenfunctions are of alternatively even and odd parity, and are conveniently labelled $0_+, 0_-, 1_+, 1_-$, etc., so that the integer indicates the

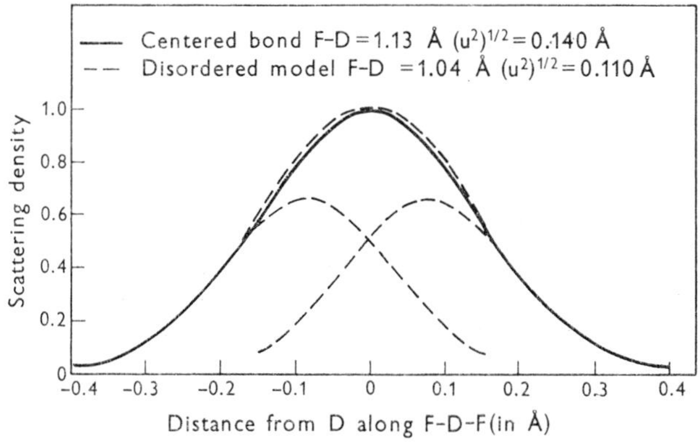

FIG 4.25 An illustration of the equivalence in scattering density of FDF bonds. The dotted line that nearly superimposes on the solid line is the sum of the two half-deuterium peaks. (Reproduced from 'Hydrogen Bonding in Solids', W. C. Hamilton and J. A. Ibers, p. 110, by permission of W. A. Benjamin Inc., Advanced Book Program, Reading, Massachusetts, U.S.A.)

limiting harmonic oscillator level as the barrier becomes large. The four lowest lying levels are displayed on figure 4.26. The infrared selection rule requires a change of parity, and figure 4.26 also illustrates the principal transitions (a) with no barrier, (b) a low barrier and (c) a high barrier. With the high barrier ($B > \sim 8$) the two transitions, $0_+ \rightarrow 1_-$ and $0_- \rightarrow 1_+$, starting from the ground state will closely coincide, so that a single band results. As the Raman transition is also at the same frequency, the spectra are the same as those given by a truly asymmetric system. As the barrier falls, splitting occurs first in the upper levels, and two bands of comparable intensity arise in the infrared spectrum from these transitions. The splitting depends critically on the barrier height, and is greatly reduced when D replaces H. As the barrier falls further, the relative intensities of the two bands vary in a complex manner, as a result of the influence of both the Boltzmann and transition moment factors. Figure 4.28, calculated from the results of Bevan[146a] shows the variation of the transition factor $(\langle\Psi_i| z |\Psi_j\rangle)^2$ with B. These curves apply to any quadratic quartic oscillator, characterised by a given

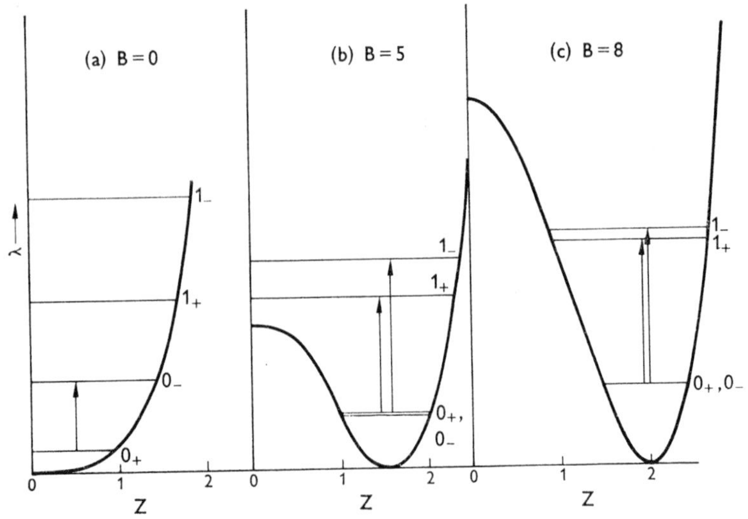

FIG 4.26 Energy levels, and principal infrared transitions in (a) a quartic well, (b) a shallow double minimum, (c) a deep double minimum. To save space, only the right hand half of each potential profile is shown.

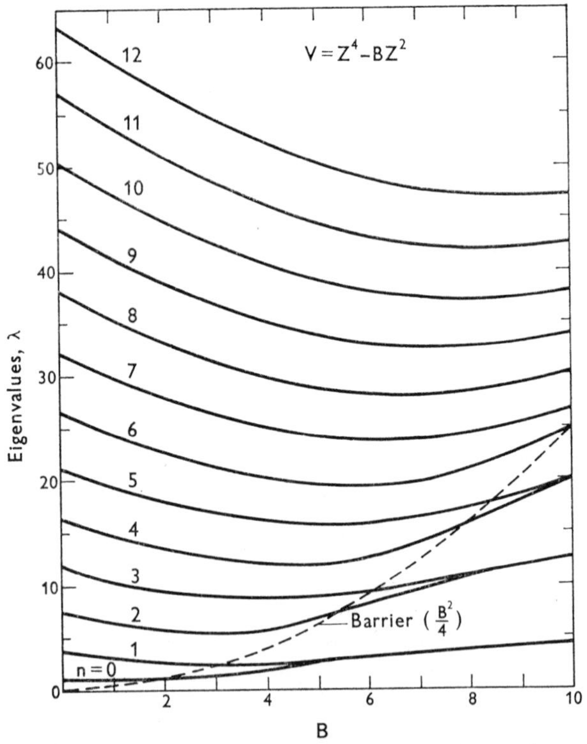

FIG 4.27 Eigenvalues for the potential $V = z^4 - Bz^2$. (Reproduced from *J. Chem. Phys.*, **47**, 4943 (1967), by permission of the American Institute of Physics.)

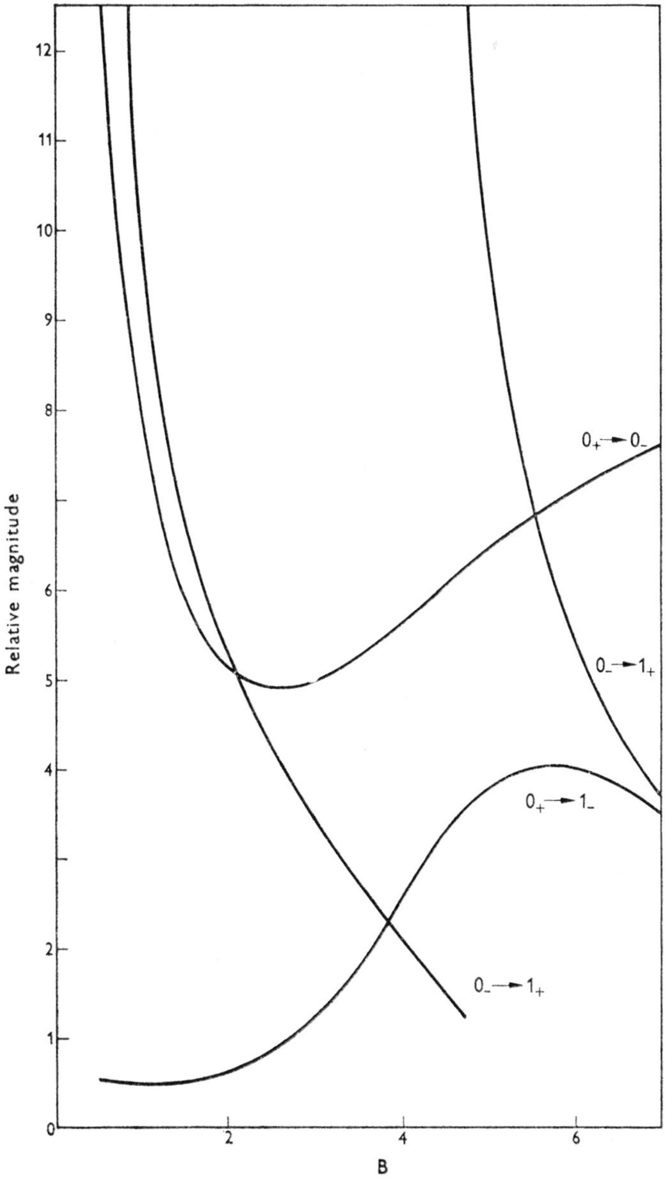

FIG 4.28 Values of the transition factor $(\langle\Psi_i|\, z\, |\Psi_j\rangle)^2$ for the quartic oscillator. The intensity scale for the $0_+ \rightarrow 1_-$ transition, and the right hand portion of the $0_- \rightarrow 1_+$ transition is enlarged ten-fold compared to the $0_+ \rightarrow 0_-$ and left hand portion of the $0_- \rightarrow 1_+$ transition, which are on the same scale.

value of B. To obtain relative intensities the transition factor must first be divided by $2\mu a/\hbar^2)^{1/3}$ to give $\langle\Psi_i| x |\Psi_j\rangle^2$ and then multiplied by the appropriate frequency and Boltzmann factor. Finally, as the barrier diminishes yet further, the splitting of the two lowest levels becomes appreciable, and the population factor ensures that only transitions from the 0_+ level need be considered. The transition $(0_+ \rightarrow 1_-)$ now represents, in effect, an overtone, with a very small transition moment. Thus with a single minimum potential, neither of the transitions $(0_+ \rightarrow 1_-$ or $0_- \rightarrow 1_+)$ has appreciable infrared intensity. However, in this case the transition between the 0_+ and 0_- levels, which may be thought of as the inversion or proton tunnelling frequency, becomes important. The frequency, ν_τ, of this transition is equal to 1/(twice the lingering time of the proton in a potential well). With a higher barrier, the frequency of this transition rapidly diminishes and the intensity, as a result, tends to zero.

The proton potential must be strongly dependent on the A—A distance, which will be reflected by a strong interaction between the transitions just considered and the ν_σ mode in intermolecularly hydrogen-bonded systems. In linear symmetric systems ν_σ itself is dipole inactive, whatever the barrier height. However, with a high barrier, two components corresponding to $(\nu_\sigma + \nu_\tau)$ and $(\nu_\sigma - \nu_\tau)$ are active, and the splitting $2\nu_\tau$ is very slight, so that one will observe a single band corresponding to ν_σ, as though the symmetry was, in effect, reduced[147]. As the barrier falls, the ν_σ band will first split into components $2\nu_\tau$ apart, which on further separation will decrease in intensity. Thus, examination of the low frequency region, where the ν_σ band is expected, may be indicative of the form of the barrier potential. The transitions arising from the various potentials in symmetric systems are summarised in table 4.2 which refers to the simplest triatomic prototype.

As only two parameters, a and b, are needed to provide an indication of the barrier shape, the measurements of two frequencies is sufficient.† As one might hope to observe the three transitions $0_+ \rightarrow 0_-$, $0_+ \rightarrow 1_-$, $0_- \rightarrow 1_+$, it would be expected that it would be relatively easy to determine the barrier shape in symmetric systems. With the availability of additional measurements on the deuterated counterparts, a check on the model becomes possible. However, in the systems examined so far, several complicating factors have emerged, and the interpretation is not as straightforward as the above considerations suggested.

Probably the best known symmetric intermolecular system is the linear[142,148] $(FHF)^-$ ion. The infrared spectra show two strong bands, at ~1470 cm^{-1} and ~1240 cm^{-1} (the exact frequency depending on the salt, and medium) and polarization measurements on single crystal spectra show the lower frequency band to be the bending mode ν_2[149]

† It is not sufficient to determine B only, as this gives the barrier shape only in terms of the reduced distance z.

TABLE 4.2
Transitions of a linear triatomic system AHA

Type of transition	Single minimum		Low barrier with appreciable tunnelling		High barrier with no appreciable tunnelling	
	IR	R	IR	R	IR	R
Tunnelling ν_T	Strong, replaces ν_s	Forbidden	Less strong, at low frequency	Forbidden	Very weak Very low ν	Forbidden
AH stretching $\nu_s(\nu_3)$*	Weak, corresponds to overtone or hot transitions		Strong, split $0_+ \to 1_-$ $0_- \to 1_+$	Allowed, split $0_+ \to 1_+$ $0_- \to 1_-$	Splitting negligible	
					Permitted	Permitted
					"normal ν_s"	
A–A stretch $\nu_\sigma(\nu_1)$*	Forbidden	Permitted	Permitted as $\nu_\sigma \pm \nu_T$ (split)	Permitted	Permitted as $\nu_\sigma \pm \nu_T$ (splitting negligible)	Permitted
A–H–A Bending (ν_2)*	Permitted	Forbidden	Permitted	Weakly permitted	Permitted	Permitted
Effective Symmetry	$D_{\infty h}$				$C_{\infty v}$	

* Parentheses indicate usual notation for modes in rigid triatomic molecules.

(see table 4.3). The absence of further infrared bands indicates a single minimum potential, (*c.f.* table 4.2.) The F—F stretching mode, *i.e.* the ν_σ transition, is Raman active, and occurs near 600 cm^{-1} [150,151], the high frequency reflecting the strength of this hydrogen bond. The spectra of the $(FDF)^-$ salts completely confirm the single minimum potential.

TABLE 4.3
Vibrational bands of acid bifluoride anions
(values in cm^{-1})

KHF_2	KDF_2	$\bar{\nu}_H/\bar{\nu}_D$	Assignment[a]
600 (R)	600	1.0	ν_1 (ν_σ)
1222 (IR)	888	1.37	ν_2 bending
1450 (IR)	1023	1.415	(ν_s)

a. notation as in rigid triatomic $D_{\infty h}$ molecule.

The spectra and structure of the $(ClHCl)^-$ salts have been extensively investigated, but continue to give rise to much discussion. Evans and Lo[152] showed that the spectra were strongly dependent on the nature of the cation, and divided the salts into two types; I, including the Cs^+, and $N(CH_3)_4^+$ salts, and II, which includes the $N(C_2H_5)_4^+$ and $N(C_3H_7)_4^+$ salts. Further work has produced a better understanding of the spectra of type I salts. Table 4.4 shows the principal features of the vibrational spectra, the infrared frequencies referring to Nibler and Pimentel's[14] data, taken at 20 K, while the Raman and neutron scattering results were obtained by Stirling, Ludman and Waddington[41]. Nibler and Pimentel suggested, on the basis of comparison with the $(FHF)^-$ spectra, that the pair of bands at 602 and 660 cm^{-1} arose from the fundamental of the bending motion, the splitting appearing to arise from crystal splitting. This contrasts with Evans and Lo's assignment of absorption around 1200 cm^{-1} to this mode, which Nibler and Pimentel explain as an overtone of the bending mode, enhanced by interaction with ν_s. Support for this assignment comes from neutron scattering measurements[41] showing a frequency at ~650 cm^{-1} involving appreciable proton movement. There is general agreement that the low frequency band is ν_σ– the smallness of the deuteration shift shows it is not ν_T.

Of the possible straight structures, the infrared activity† of ν_σ eliminates $D_{\infty h}$. Nibler and Pimentel's assignment would also argue against $D_{\infty h}$, for $2\nu_2$ would be infrared inactive, and ν_3 Raman inactive. A low barrier with tunnelling would imply a ν_s band doublet in the H salt, with a much reduced splitting in the D salt and an abnormal $\bar{\nu}_H/\bar{\nu}_D$ ratio. The correspondence of the HCl_2^- and DCl_2^- spectra shown in table 4.4 seems reasonable, so that assignment of the ~1200 cm^{-1} and ~1670 cm^{-1} features of the HCl_2^- spectrum to the split ν_s band does not seem likely. The 1790 cm^{-1} and 1875 cm^{-1} bands, ascribed by Nibler

† Since this is a critical observation, the data would be worth checking.

TABLE 4.4
Main features of the vibrational spectra of Type I, $(HCl_2)^-$ + $(DCl_2)^-$ salts

	CsHCl$_2$				CsDCl$_2$				
			Neutron scattering				Neutron scattering		
	Infrared	Raman	Energy gain	Energy loss	Infrared	Raman	Energy gain	Energy loss	ν_H/ν_D
Lattice modes			55				56		
			73				74		
			105				105		
ν_1			199				195		1.02
ν_2	602.0 (vs)	612			429.7 (vs)	450			1.41
	660.5 (s)		652	652	468.0 (s)		480	438	1.41
	1168.5 (vs)	1165		1198	851.5 (vs)	850			1.37
ν_2	1208 (m)								
(2 ← 0)	1294 (m)				925 (s)				1.40
	1340 (w)								
ν_3	1670 (vs)	1630			1320 (sh)				1.26
polymer	1790 (vvs)				1370 (vvs)				
	1875 (vvs)				1420 (sh)				

and Pimentel to higher polymers, may, however, contain the second member of the doublet, with 1370 cm^{-1} the single corresponding band in the DCl_2^- salt. The potential implied by this assignment would involve a tunnelling transition ν_τ of ~8 cm^{-1}, but no evidence for this is found in the I.N.S. spectrum. However, there is no other conflict with the infrared and Raman spectra, and a double minimum potential cannot be excluded on present evidence. The $C_{\infty v}$ structure, (*i.e.* with high barrier), is consistent with all the spectroscopic observations, and must also be included as a possibility.

Of the bent structures (figure 4.29), that with C_{2v} symmetry (single minimum) would imply no enhancement of $2\nu_2$ by ν_3, and so is unlikely on Nibler and Pimentel's assignment. The bent structures with a low or high barrier, however, are possible.

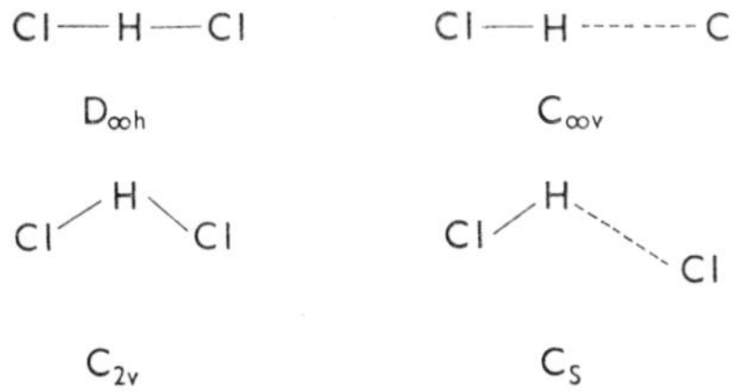

FIG 4.29 Illustration of some of the possible structures for the (ClHCl)$^-$ ion.

To summarise, it seems clear that the (ClHCl)$^-$ ion in type I salts does not have the centrosymmetric structure ($D_{\infty h}$), present in the (FHF)$^-$ ion. This could well be a consequence of the weaker hydrogen bond in (ClHCl)$^-$, which would be more easily distorted to fit the lattice.

The spectra of the type II salts have been interpreted in terms of a linear centrosymmetric model[152], but the evidence available as yet is far from conclusive, as is also the data on these ions in solution. The (BrHBr)$^-$ salts also have spectra that can be grouped into two types, each type being similar, after allowance for the change in mass, to the corresponding (ClHCl)$^-$ spectra[18,153]. Similar structures are therefore anticipated.

Solutions of ammonium salts in liquid ammonia have been examined by Corset, Huong and Lascombe[30]. The NH_4^+ ions appear to interact with the neighbouring NH_3 molecules by hydrogen bonding, and also with the anion (Cl^-, Br^-, I^-, NO_3^- or ClO_4^-). Although a broad and complex absorption band occurs in the NH stretching region, it has not been possible to analyse its origin, or determine whether manifestation of proton tunnelling is present in the spectra.

The vibrational spectra of the $(H_5O_2)^+$ ion, present in perchloric acid dihydrate, has also been studied recently[154], and the spectra interpreted on the basis of a centrosymmetric $H_2O—H—OH_2$ structure of trans configuration (C_{2h}). A neutron diffraction study of $HAuCl_4.4H_2O$[154a]

which contains this ion, reveals a structure with the two water molecules in the *trans* configuration, and an O—O length of 2.57 ± 0.01 Å. The bridging proton is found to be symmetrically disordered and also off the O—O axis. The distance between the two "half-proton" sites ~0.62 Å. It is now apparent that the detailed structure of the $H_5O_2^+$ ion is very sensitive to its crystal environment, but at least this one example of a symmetric double minimum occurs. Unfortunately no decisive spectroscopic demonstration of proton tunnelling has yet been furnished by symmetric intermolecular complexes.

(a) *Systems with Strong Interaction Between Hydrogen Bonds*

In these systems containing linked hydrogen bonds the proton potential function depends on the position of the proton in an adjacent bond.

A prototype which displays the essential features in a clear cut manner is the series of carboxylic acid dimers. Karle and Brockway established the inequality in the length of the C=O and C—O bonds in an early electron diffraction study[155], *i.e.* the structure is I of C_{2h} symmetry

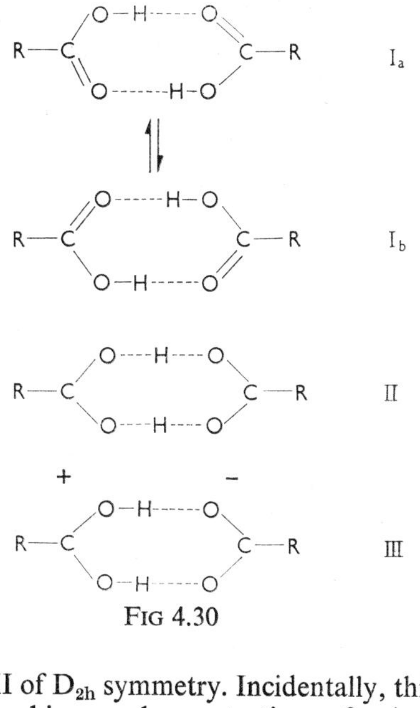

FIG 4.30

(figure 4.30) and not II of D_{2h} symmetry. Incidentally, this is one of the most decisive and unambiguous demonstrations of a double minimum potential I_a and I_b representing the two configurations. Undoubtedly the polar form (III) is of higher energy, *i.e.* the potential of the proton in bond (a) depends on the position of the proton in bond (b). Confirmation of the C_{2h} structure comes from infrared spectra. Only one vibrational fundamental differs in activity in the C_{2h} and D_{2h} structures;

this is the skeletal deformation shown in figure 4.31 which is of a_u symmetry. It is infrared active in C_{2h}, and inactive in D_{2h}. The band observed at 68 cm^{-1} in formic acid dimer has been assigned[156,157] to this mode. Further support for a double minimum is found in microwave spectra although the need for a permanent dipole moment

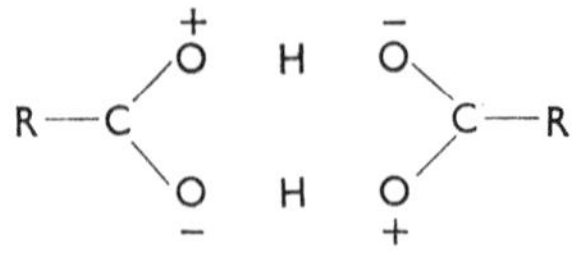

FIG 4.31 The a_u intermolecular mode in a carboxylic acid dimer.

requires the choice of somewhat different complexes. The two complexes CF_3COOD–HCOOH and CF_3COOH–HCOOD have distinguishable microwave spectra. Hence the proton and deuteron are not in a central minimum, and since the tunnelling rate must be slow compared to the rotation of the complex, the barrier is estimated $>6000\ cm^{-1}$[95]. Incidentally this is in interesting contrast to the CF_3COOH–CH_3COOD complex, which has a microwave spectrum indistinguishable from CF_3COOD–CH_3COOH and therefore a barrier to the combined proton/deuteron tunnelling estimated as $<5000\ cm^{-1}$. In principle one might hope to determine the barrier height from splittings in the vibrational fundamental bands of carboxylic acid dimers, similar to the effect of inversion on the NH_3 spectrum[158]. However, only one band has a doublet structure, and this is not caused by tunnelling.

A preliminary *SCF–MO* calculation of the simultaneous two proton tunnelling in formic acid dimer has been carried out by Clementi and co-workers[159]. This supports a double minimum potential.

Ice

Nine crystalline forms of ice are known of which ordinary ice is known as Ice Ih[42].

The positions of the D atoms in D_2O ice can be located more precisely than their counterparts in H_2O ice, and Peterson and Levy[160] showed by neutron diffraction that the D atoms are asymmetrically distributed in ordinary D_2O ice. Each D atom is displaced about 0.37 Å from the midpoint of the O—O bond. This is a clear indication that the structure is built out of discrete molecules, *i.e.*, each oxygen atom has, at any instant, two D atoms uniquely associated with it. The oxygen atoms form a regular structure based on tetrahedral co-ordination, so that around each O atom two of the D atoms are covalently linked, and two hydrogen bonded. These requirements are, however, not sufficient to determine a regular arrangement throughout the crystal of the D atoms —the crystal is orientationally disordered[161]. H_2O ice is expected to have a similar structure.

The vibrational spectra of the various forms of ice have been extensively examined, (see figure 4.32) particularly by Whalley and his

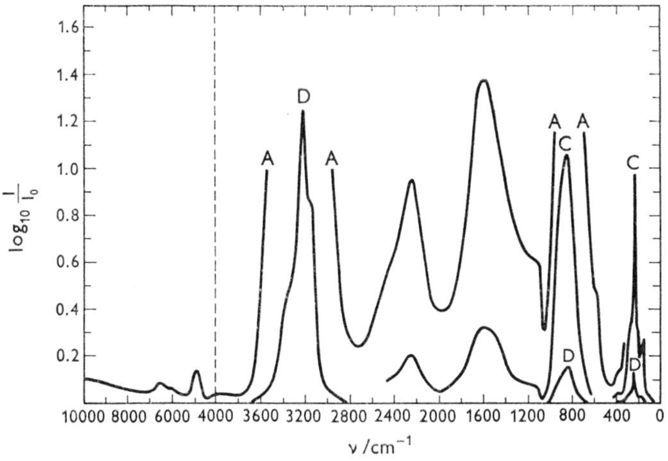

FIG 4.32 Infrared spectrum of ice in the range of 10 000 to 30 cm^{-1}. (Reproduced from *Developments in Applied Spectroscopy*, **6**, 278 (1968), by permission of Plenum Pub. Corp. N.Y.)

co-workers, who has recently critically reviewed the position[162]. Although many details of the spectra have yet to be understood, the interpretation in general is in terms of discrete, though strongly interacting, water molecules, in corroboration of the neutron diffraction studies. The occurrence of a structure based on discrete molecules implies a proton double minimum potential.

The potential function for a proton within a single bond depends on the positions of the adjacent protons, and will have an asymmetric single minimum if these are "frozen" in their positions. However, for each possible configuration there will be a corresponding one, with the same energy, in which all the protons have moved to the "mirror" positions on the other side of each O—O bond. Thus there is a symmetric, double minimum potential as a result of linked proton motions. The situation is similar to that in the carboxylic acid dimers, the coupled proton motions now extending over a network of hydrogen bonds. The ν_s band in ice is broad (see figure 4.32) and has considerable structure. Several factors have been recognised as possibly contributing to the observed contour: intramolecular coupling between the two OH stretching motions; correlation coupling with the vibrations in neighbouring molecules; variations in the environment, reflected in the static field at a molecule and intensity variations through a band, as well as Fermi resonance and hot band effects[162]. As Whalley concludes, any detailed interpretation will require information obtained by other techniques.

Water

Liquid water is undoubtedly the most important multiple hydrogen-bonded system, and the general similarity of its vibrational spectrum to that of Ice Ih[162] suggest that its structure is similarly based on discrete, though strongly interacting water molecules—a view which is usually taken for granted without regard to the need for evidence. An X-ray diffraction study of H_2O water by Danford and Levy[163] shows a small maximum in the electron density radial distribution function at 1.10 Å, less than half the O—O distance (2.88 Å), supporting an asymmetric proton distribution. Further support for the molecular model comes from entropy calculations[164,165]. The ν_s band in water is broader than in ice, and as expected, shows a smoother contour[166] (figure 4.33). After a critical review, Whalley summarises the situation

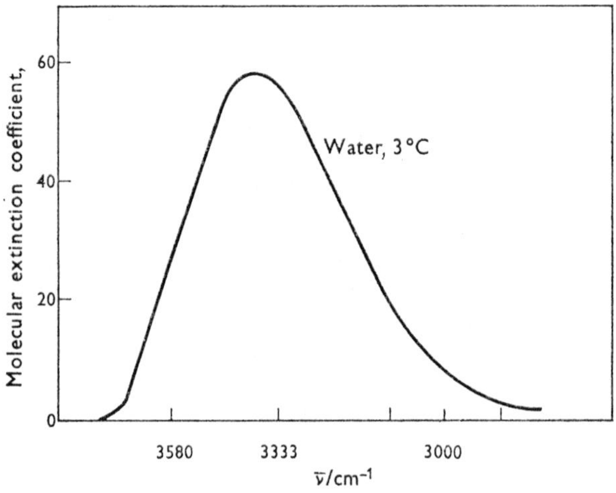

FIG 4.33 O–H stretching band of liquid H_2O [after J. J. Fox and A. E. Martin, Proc. Roy. Soc. London. **A174**:234 1940] (Reproduced from *Developments in Applied Spectroscopy*, **6**, 293 (1968), by permission of Plenum Pub. Corp. N.Y.)

"Since the OH stretching vibrations of ice are not understood in detail, those of the liquid are even less well understood. There is therefore no useful structural information at present in the band". However, in both ice and water the spectroscopic data does indicate the presence of molecules, in accordance with other evidence, with a co-operative double minimum potential as a result[167].

Some other complexes in which a co-operative double minimum potential is indicated are the phenol cyclic trimer and thiopyridone[168] dimer (figure 4.34). It is notable that the only systems in which a double minimum potential has been established with reasonable certainty are those in which concerted proton movements due to strong interaction

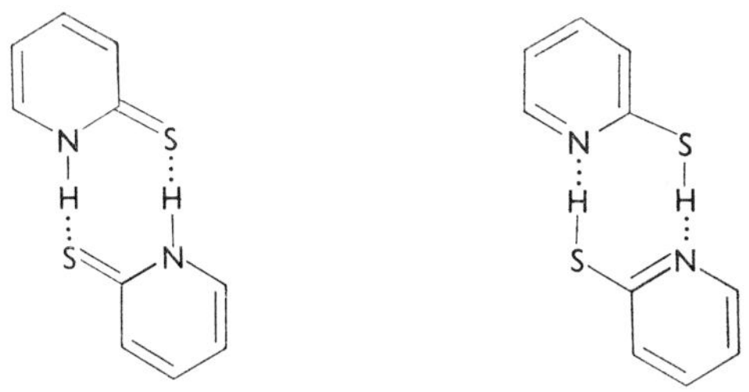

I II

FIG 4.34 Illustration of synchronous proton tunnelling in 2-thiopyridone. (Reproduced from *Proc. Roy. Soc.*, **A257**, 98 (1960), by permission of the Royal Society.)

are involved. A very simple model suggests why barriers have proved so much easier to detect in these cases. If we consider a system with two interacting protons, each in a double minimum potential, then this could be represented by the Hamiltonian:

$$H = [p_1^2/2m + (k_2/2)q_1^2 + (k_4/2)q_1^4] + [p_2^2/2m + (k_2^2/2)q_2^2 + (k_4/2)q_2^4] + kq_1q_2 \quad 4.10$$

where the first term in parentheses represents the kinetic energy $p_1^2/2m$ and the potential energy $(k_2/2)q_1^2 + (k_4/2)q_1^4$ of the first proton in its double minimum potential, the second term in parentheses refers to the second proton in an equal potential and kq_1q_2 is the interaction term. The potential energy surface is as shown in figure 4.35. Transforming to the co-ordinates 4.11:

$$\left.\begin{aligned} Q_+ &= \frac{1}{\sqrt{2}}(q_1 + q_2) \\ Q_- &= \frac{1}{\sqrt{2}}(q_1 - q_2) \end{aligned}\right\} \quad 4.11$$

we get:

$$H = (p+)^2/2m + (p-)^2/2m + (k_2/2)Q_+^2 + (k_2/2)Q_-^2 + \frac{k_4}{4}[Q_+^4 + 6Q_+^2Q_-^2 + Q_-^4] + \frac{k}{2}(Q_+^2 - Q_-^2) \quad 4.12$$

The barrier experienced by a single proton, before interaction is $k_2^2/8k_4$, and the width, the distance between the minima, is $2(k_2/2k_4)^{1/2}$. The coupling leaves the effective mass unaltered, but tends to switch the motion along the co-ordinate Q_+ (if k is negative). (Compare figures 4.35a and 4.35b.) The barrier is now $(k_2 + k)^2/4k_4$ and the width $2[(k_2 + k)/k_4]^{1/2}$. Not only does k directly increase the effective height and width of the barrier, but the concerted motion imposed by

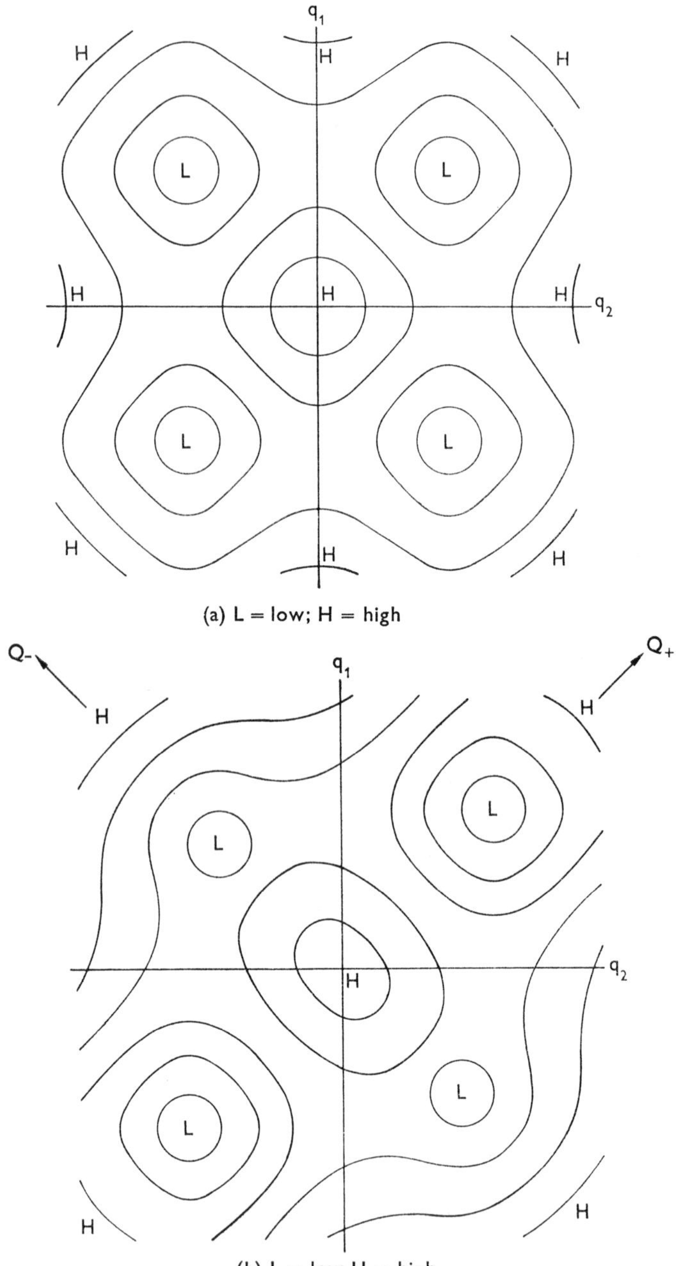

FIG 4.35 (a) Double minimum potentials in the co-ordinates q_1 and q_2 with no interaction. (b) With interaction represented by an additional term kq_1q_2 (k is taken to be negative).

the interaction itself doubles the height of the barrier, and increases the width by a factor of $(2)^{1/2}$. Although the model is only for a pair of protons tunnelling in concert, it is apparent that the argument can be extended to systems involving larger number of interacting protons, such as the phenol trimer, or liquid water. Thus it will be easier to observe barriers to tunnelling if strong interaction is present.

(b) *Systems with Less Strong Interactions between Hydrogen Bonds—Ferroelectric Crystals*

Ferroelectric crystals polarise spontaneously, *i.e.* in the absence of an electric field, below a characteristic critical temperature—the Curie point, T_c. An important class of ferroelectrics, of which Rochelle salt, (NaK tartrate.$4H_2O$), KH_2PO_4, and KH_2AsO_4 are examples, contain hydrogen bonds, and the marked change in the Curie point on deuterium substitution[169], *e.g.*

	KH_2PO_4	KD_2PO_4	KH_2AsO_4	KD_2AsO_4
T_c(K)	123	213	96	162

shows the intimate connection between the ferroelectric property and the proton motion. It has been suggested that the proton potential along the O—O bonds has a double minimum, with the depths of the two wells depending on the positions of the protons in neighbouring hydrogen bonds[170]—this represents less strong coupling than in water, where the interaction is sufficient to suppress the second minimum for the motion of an individual proton. It is supposed that below the Curie temperature, the co-operative effect leads to a regularity in the proton positions in the double minimum potentials—a condensation in double minimum phase space. Above the Curie temperature the protons are randomly distributed between the two positions in a bond. As a result the interactions average out to make each individual double minimum potential symmetrical. This provides a natural explanation for the deuteration effect, and would lead to the expectation of an abrupt change in the vibrational spectra at the Curie temperature. For this reason, the spectra of ferroelectrics such as KH_2PO_4 have received a great deal of attention. The ν_s region in KH_2PO_4 displays a very broad band extending from 1500 to over 3000 cm^{-1}[87], with several distinct sub-maxima (figure 4.36). Surprisingly, the band shows no abrupt changes on passing through the Curie point, even the sub-structure being unaltered. This argues against earlier suggestions[171] that the maxima arise from the 0_+ to 1_- and 0_- to 1_+ transitions in a double minimum potential, as do also the normal deuteration shifts[87]. Indeed it would appear that Claydon and Sheppard's[128] model of transmission windows due to Fermi resonance fit the ν_s band structure rather well. We have $2 \times \gamma(OH) = 2076$ cm^{-1} and $2 \times \delta(OH)$† $= 2600, 2632$ in

† Notation as in reference 87, γ(OH) and δ(OH) are described as out-of-plane and in-plane transverse hydrogen vibrations.

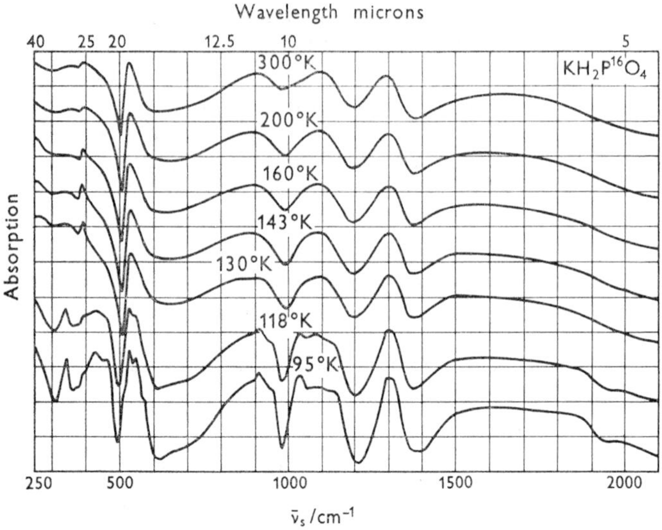

FIG 4.36 Infrared absorption spectra (250–2000 cm^{-1}) of KH_2PO_4 at different temperatures (T_c = 123 K). (Reproduced from *J. Chem. Phys.*, **52,** 2883 (1970), by permission of the American Institute of Physics.)

KH_2PO_4; minima occur at both positions. KD_2PO_4 similarly has minima in the ν_s band continuum near $2 \times \gamma(OD) = 1420$ cm^{-1} and $2 \times \delta(OD) = 1920$ cm^{-1}. This does not necessarily account for all the ν_s band structure in these salts, but it certainly offers an explanation of the major features in the ν_s band structure. Although the ν_s band gives no definite indication of a double minimum potential, this is indicated by evidence from the lower frequency range. It seems probable that the co-operative effect is sufficient to make the potential minima unequal, but weak enough for a barrier to remain between the two positions (figure 4.37).

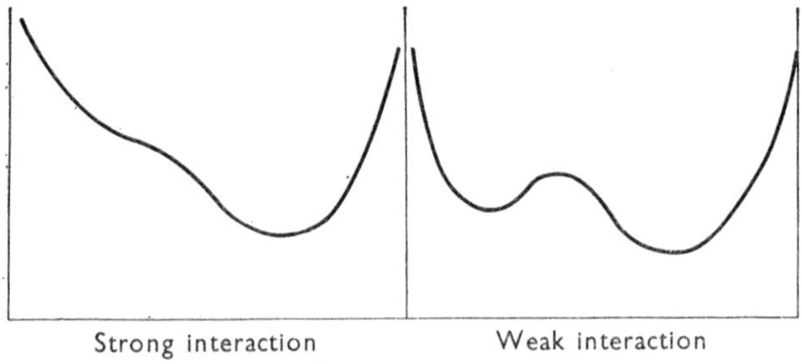

FIG 4.37 The development of a double minimum potential. Neighbouring proton positions are "frozen".

Temperature Effects on the ν_s Band

The ν_s band "massif" is unusual in that it often moves towards higher frequency on raising the temperature. This contrasts with the usual behaviour of unresolved vibrational bands, in which the increasing proportionate contribution of difference bands leads to a fall in the band centroid as the temperature rises. Examples of the effect occur in the spectra of solid and liquid imidazole polymer[90], and in associated alcohols Asselin and Sandorfy[172] have made quantitative measurements on ν_s and also its first overtone, for both *iso*-propanol and *t*-butanol. The temperature dependence was almost linear, with $\partial\bar{\nu}_{01}/\partial T = +0.59$, $\partial\bar{\nu}_{02}/\partial T = +0.59$ for propanol, and $\partial\bar{\nu}_{01}/\partial T = +0.44$, $\partial\bar{\nu}_{02}/\partial T = +0.47$ for butanol, all in cm^{-1}/degree The coefficient were some 30–40% smaller in the OD analogs. These findings are completely consistent with those of an extensive earlier examination of the effect in a range of self-complexed alcohols and phenols by Finch and Lippincott[173].

Several explanations for the effect have been offered. (a) In self-complexed systems, the shift of the ν_s band reflects a change in the proportions of the associated species. Certainly $\bar{\nu}_s$ falls with increase in multimer size (*e.g.* in phenol), so that the decrease in association as the temperature rises would give a shift in the observed direction. Finch and Lippincott argue against this mechanism in alcohols, and the temperature effect occurs in systems where the association does not alter, showing the need for alternative suggestions. (b) As the temperature falls, the intermolecular hydrogen bond contracts. Shorter hydrogen bonds are stronger, with a larger $\Delta\bar{\nu}_s$—as a result $\bar{\nu}_s$ falls. This general argument can be interpreted in two ways. One may regard the contraction as intramolecular, the result of anharmonic coupling between the ν_s and ν_σ modes. If this is represented by a Stepanov type approach, in the $v = 1$ level of ν_s the equilibrium bond length is shorter, and the force constant greater than in the $v = 0$ level. Increasing the temperature leads to a change in the population of the ν_σ levels, and by evaluating, via Franck-Condon type factors the position of the ν_s fundamental centroid, the temperature effect can be calculated. Unless very large amounts of anharmonicity are introduced into the H-bond stretching potential, effects of the observed magnitude are not obtained. Thus an intramolecular origin seems unlikely. Direct support for this view comes from measurements on the $CF_3CH_2OH–N(CH_3)_3$ complex in the gas phase[174], where the ν_s band shows hardly any shift over a temperature range of 60°C. There remains the possibility that the hydrogen bond is 'squeezed' by pressure exerted by neighbouring molecules. To this one would expect the intermolecular hydrogen bond to be very susceptible, as it has a low force constant (section 4.6). However, the marked deuteration effect on $\partial\bar{\nu}_s/\partial T$ noted by Asselin and Sandorfy is surely significant, and indicates that a satisfactory explanation remains to be found†.

† *Note added in proof.* This will be dealt with in a forthcoming publication by the author.

ν_s Overtone Bands and Deuteration Shifts; Mechanical Anharmonicity.

Most of the effects discussed above, *e.g.* Fermi resonance, proton tunnelling, sum and difference bands, depend on the presence of anharmonicity. Even if these particular effects are small, it is certain that the potential function for the proton motion in the hydrogen bond departs markedly from a harmonic one. This is reflected in the numerous potential functions which have been proposed at various times[80–82, 129,130,132,175,176]. Mechanical anharmonicity will be manifest in the positions of the ν_s overtone bands, and also in the shifts on deuteration. Sandorfy and his colleagues[177–180] have extensively examined the frequencies of the ν_s overtone bands. Although, as expected, hydrogen bonding is clearly observed to affect the anharmonicity, the situation turns out to be far from simple. In general complexing increases the anharmonicity, *i.e.* $\Delta = 2\nu(\mathrm{v} = 1 \leftarrow 0) - \nu(\mathrm{v} = 2 \leftarrow 0)$ is positive and increases, but in some weakly hydrogen bonded systems, Δ decreases on complexing. The frequently used expression:

$$\nu(\mathrm{n} \leftarrow 0) = n\omega_0 - \mathrm{n}(\mathrm{n}+1)\mathrm{x}\omega_0 \qquad 4.13$$

fails to fit the frequency of the $\mathrm{v} = 3 \leftarrow 0$ band when this is observable[180]. The shifts of both the ν_s fundamental band and the overtone bands on deuteration also reflect the anharmonicity. The overtones are weak, and rarely observed, but the deuteration shifts of the ν_s fundamental band have been recorded for a very large variety of complexes. It is frequently found that the ratio $(\nu_s)_H/(\nu_s)_D$ decreases on complexing, compared to that for the free AH[177,178], values of this ratio as low as 1.256 having been observed[181]. In strong hydrogen bonds the complexity of the ν_s band may prevent an unambiguous determination of the shift. High values of the ratio are also occasionally observed, *e.g.* 1.415 in HF_2^-[175], 1.47 in crystals of type II, HCl_2^- salts[152] and even 1.50 in the HCl_2 radical[24]. In the few cases where both overtone frequencies and deuteration shifts are known, the relationship, not unexpectedly, does not fit expressions:

$$\frac{\omega_D}{\omega_H} = \left(\frac{\mu_H}{\mu_D}\right)^{1/2}$$
$$\frac{(\mathrm{x}\omega)_D}{(\mathrm{x}\omega)_H} = \frac{\mu_H}{\mu_D} \qquad 4.14$$

obtained by second order perturbation theory for a one-dimensional anharmonic oscillator[180].

A useful way of considering mechanical anharmonicity is to start with only the harmonic (quadratic) terms in the potential function, and to establish normal co-ordinates based on these in the usual way. The additional anharmonic terms are of two kinds, (a) terms such as $K_{3i}Q_i^3$ and $K_{4i}Q_i^4$, which do not link normal modes, and (b) terms such as $k_{ij}Q_iQ_j^2$, which do[134]. Two modes which can be expected to have strong anharmonicity of type (a) are ν_σ, since this leads to dissociation of the complex, and ν_s. These two modes will also strongly interact, involving type (b) terms. The Fermi resonance, which is such a notable feature of the spectra of hydrogen bond complexes, also requires the presence of type 'b' anharmonic terms. Suppose all 'b' type coupling terms are negligible, *i.e.* one is dealing with a one-dimensional oscillator. For a harmonic potential ω_H/ω_D will vary from 1.37 to 1.40, depending on the reduced mass. The addition of a quartic term K_4Q^4 with positive coefficient will make the potential more box-like, and increase ω_H/ω_D. For a proton in a square well potential $\omega_H/\omega_D = 2.0$. This explains the high ratios in HCl_2 and HF_2^-. The addition of a cubic anharmonic term widens out the potential as v increases, and reduces the ω_H/ω_D ratio, providing an explanation of the low ratios. Both terms may be present, so the observation of a 'normal' ratio does not prove that anharmonicity is absent.

Laane[181a] has recently calculated the ω_H/ω_D ratio for barriers of various shapes, including $V = K_2O^2 + K_4Q^4$. The ratio of the reduced masses is not given, but is presumably 1:2. For a high barrier, ω_H/ω_D approaches 1.4. The ratio falls as the barrier decreases reaching a smallest value of ~1.2 when the top of the barrier is close to the upper level. The ratio then increases with further fall of the barrier, passing through 1.4 and reaching a maximum of 1.5–1.6 with a flat-bottomed well, finally falling to 1.4 again for a single minimum harmonic oscillator. Earlier results[145] giving isotope ratios as low as 1.0 appear to be in error. An important result of these calculations is that the general features of the isotopic ratio are rather insensitive to the precise form of the barrier potential. Consequently the isotopic ratio can be a good indication of the barrier shape. However, many factors, as has been seen, can also influence the appearance of the ν_s band, and it would be unwise to ascribe an apparently low ω_H/ω_D ratio to a double minimum unless these other factors have also been taken into account.

If type (b) anharmonic terms are present, their influence on the energy level may be quite as important as that of the type (a) terms, and this may account, at least in part, for the difficulty in establishing a simple rationale of the behaviour of the ν_s overtone bands and the deuteration shifts using a one-dimensional model.

Provided there is no accidental co-incidence between the spacing of the levels in a low frequency mode and a splitting (*e.g.* by tunnelling) in a higher frequency mode, the interaction between them can be quantitatively represented by the Stepanov scheme[147,182,183]. The low frequency mode need not be confined to the intermolecular stretching—other low frequency modes can be treated similarly.

No account of the deuteration effect on the ν_s band can omit the remarkable observations of Ibers and his co-workers[184,185] on crystals of chromous and cobaltous acids. Figure 4.38 shows the infrared spectra[42] of $HCrO_2$ and $DCrO_2$. The remarkable differences are (1) the ν_s band in $HCrO_2$ is broad, and centred at 1640 cm^{-1}. In $DCrO_2$, this is replaced by two bands, with maxima at 1594 and 1916 cm^{-1}. The centroid of the $DCrO_2$ band has in fact moved *up* in frequency to 1750 cm^{-1}. (2) The ratio of the frequencies of the bending modes, *i.e.* $\nu_d(OHO)/\nu_d(ODO)$ is 1.44 at 24°C, compared to a harmonic value of 1.386. In other hydrogen bonded systems this ratio is usually low. (3) Only two bands are distinguishable in the lattice mode region of $HCrO_2$ even at −196°C, whereas six bands are present in this region in $DCrO_2$.

The spectra of the crystalline cobaltous acids $HCoO_2$ and $DCoO_2$ also differ in a very similar manner. Ibers has suggested that the spectral differences result from a change in the effective one-dimensional potential curves, the H acids having a barrier, if any, below the ground

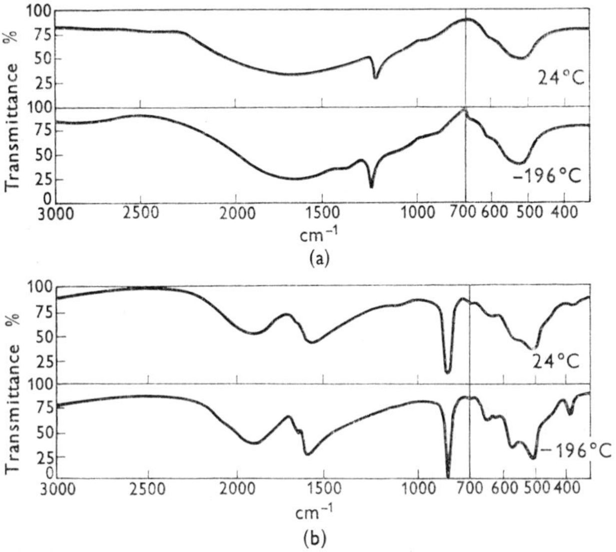

FIG 4.38 (a) Infrared spectrum of $HCrO_2$ at 24° and at −196°C (in 1 inch KBr disks). (b) Infrared spectrum of $DCrO_2$ at 24° and at −196°C (in 1 inch KBr disks). (Reproduced from 'Hydrogen Bonding in Solids', W. C. Hamilton and J. B. Ibers, p. 121, by permission of W. A. Benjamin Inc., Advanced Book Program, Reading, Massachusetts, U.S.A.)

vibrational level, while that in the D acids is much higher (figure 4.39). Such potentials can be chosen to fit the observed ν_s bands. The tunnelling of the D, (~10 cm^{-1}) would be sufficiently low frequency to give an effectively asymmetric ODO bond, with the consequence that all nine lattice modes (the unit cell contains one formula weight) are infrared allowed. The H would be symmetrically located, when averaged over the period of a lattice vibration, so that only four lattice modes are allowed. Thus the change in potential function satisfactorily explains

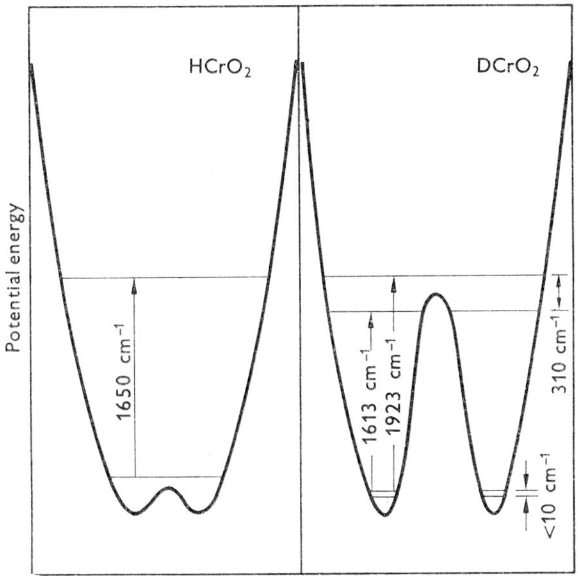

FIG 4.39 Potential energy curves for the hydrogen bond of $HCrO_2$ and $DCrO_2$. (Reproduced from 'Hydrogen Bonding in Solids', W. C. Hamilton and J. B. Ibers, p 122, by permission of W. A. Benjamin Inc., Advanced Book Program, Reading, Massachusetts, U.S.A.)

the increased number of lattice modes seen in $DCrO_2$. The smaller ν_s shift for the D acid implied by this interpretation corresponds to a weaker bond; as the OHO bending mode often increases with the bond strength, this can also explain the anomalously high ν_d(OHO)/ν_d(ODO) ratio. The change in potential would suggest a longer O—O distance in the deuterated acid in order to fit in the barrier. This is observed as shown in table 4.5. Neutron diffraction measurements confirm the asymmetric D position in $DCrO_2$ but do not permit the H position to be determined. However, despite this supporting evidence, several aspects of the problem remain to be understood. As Ibers has pointed out, the one-dimensional potential shown in figure 4.39 is only intended as an effective potential. Obviously coupling between the O—O and O—H displacements is large, and a two-dimensional model incorporating this interaction should be examined to determine whether different potential functions are still required for the H and D systems when the second dimension is included. Our estimates suggest that a common potential function still does not suffice with this model. A breakdown in the Born-Oppenheimer separation of electronic and nuclear motion† leading to different potentials for the H and D systems seems unlikely.

† *N.B.* Not the Stepanov separation.

TABLE 4.5
Hydrogen bond lengths in chromous and cobaltous acids (Å units)

Parameter	Chromous acid	Cobaltous acid
OHO	2.49	2.50
ODO	2.55	2.57

A more probable origin for the unusual spectral changes lies in the interaction between neighbouring hydrogen-bonds, as suggested by Hofacker[186,187] and co-workers. This interaction is represented by coupling via the lattice vibrations. It is shown that with suitable conditions, a double minimum potential can give rise to a single ν_s band maximum, and further, that the absorption spectrum can be highly sensitive to a change of mass. It seems certain that this solid state theory approach to the interpretation of the vibration spectrum of systems of strongly interacting hydrogen bonds will prove to be capable of extensive application. In particular a quantitative treatment of the chromous and cobaltous acids problem would be of great interest, and also of the "Ubbelohde effects"[188,189] the change in the lattice parameters of hydrogen-bonded crystals on deuteration.

Intensity of the ν_s Band

The integrated intensity of the ν_s band almost always increases on hydrogen bonding, often by a many-fold factor. This is one of the most characteristic and reliable indications of hydrogen bond formation, and was early recognised as such.[190] Numerical values of the intensity enhancement are tabulated in Pimentel and McClellan[1] (see table 4.6 and section 2.3.4), but in contrast to the strong interest in the ν_s band structure, relatively few measurements of the intensity enhancement have been made in the past ten years. This is probably as a consequence of the great breadth of the band, and the uncertainty of the contribution from the wings.

Any account of the intensity enhancement would naturally be in terms of the change, as the proton moves, in the electric moment of the complex as a whole. Some increase in intensity results from the greater amplitude of the proton oscillation in the complex, but this is not sufficient to account for the magnitude of the intensity change.

Friedrich and Person[191] have suggested that there is a similarity in this respect between hydrogen-bonded complexes and donor-halogen complexes (chapters 2 and 3). Thus the amount of charge transfer changes considerably with the proton position, with a consequently large dipole moment derivative (a more detailed discussion of this mechanism and its effect on the vibrational spectrum is given in section 1.2). It is notable that some of the most recent whole complex MO

TABLE 4.6
Factors of intensification of ν_s on hydrogen-bond formation

Acid	Base	10^{-1} B(complex)/ m mol^{-1}†	B(complex)/B(monomer)
Phenol	Benzene	16	2.1
Phenol	Ethyl acetate	84	7.2
Phenol	Acetonitrile	50	4.3
Phenol	Diethyl ether	80.3	6.9
Phenol	Hexamethylene tetramine	124	10.7
n-Butyl alcohol	Diethyl ether	38	5.3
sec-Butyl alcohol	Diethyl ether	32	5.5
t-Butyl alcohol	Diethyl ether	26	6.2
γ-Butyrolactam dimer		41	14
Acetic acid dimer		74	37
Methanol polymer		24	5.0
H_2O polymer		—	12
N-Ethylacetamide polymer		44	19
$CDCl_3$	Benzene	0.31	6.2
$CDCl_3$	Ethyl acetate	—	9.5
$CDCl_3$	Acetone	0.54	10.8
$CDCl_3$	*N*-Ethylacetamide	1.70	34
$CDCl_3$	Pyridine	—	20.5
$CDCl_3$	Triethylamine	1.80	36

† 1 $dm^3 mol^{-1} cm^{-2} = 10 m^3 mol^{-1} m^{-2} \equiv 10 m mol^{-1}$. (B is the integrated intensity.) (Reproduced from 'The Hydrogen Bond' by G. C. Pimentel and A. L. McClellan, by permission of W. H. Freeman and Co. Copyright © (1960).)

treatments (discussed in sections 4.7 and 1.3) also lead naturally to a substantial ν_s band intensity enhancement, but the charge-transfer mechanism usually contributes only a small part of the effect. (See section 1.3 for some detailed estimates of the size of this contribution to the intensity of the ν_s band.)

A qualitative model which plausibly accounts for a contribution to the intensity enhancement has been proposed by Boobyer and Orville-Thomas[192,193]. As the AH bond extends, an induced polarisation occurs in B, particularly in the lone pair electrons. The sign of the induced dipole enhances the dipole moment *change* in AH, with a resulting increase in intensity.

4.5.4. THE IN-PLANE BENDING MODE, ν_b

The study of the effect of hydrogen-bonding on the bending mode ν_b has been much less fruitful than the study of the mode ν_s. This is

TABLE 4.7

The ν_s (stretching) and ν_b (bending) vibrational bands in hydrogen-bonded pyridinium salts, and the free cations $PyrH^+X^-$ (values in cm^{-1})

X^-	Cl^-	Br^-	I^-	$SnBr_6^-$	BF_4^-	$SbCl_6^-$
ν_s	{2439 2375	2591	2833	3236	3290	3300
ν_b	1244	1238	1235	1235	1255	1245

primarily because the bending motion is much less concentrated in a single mode than the stretching motion—frequently several modes in the 1000–1500 cm^{-1} range involve bending—*e.g.* in phenol both the b_2 modes at 1180 cm^{-1} and 1342 cm^{-1} [123].

In the free molecule this mixing in manifested by small deuteration shifts occurring in several modes. The same modes are sensitive to complexing. In general ν_b-type modes increase in frequency on hydrogen bond formation, (see reference 1, p 150), but a recent review[165] of O—H—O systems shows that there is considerable scatter, and even a tendency to decrease in very strongly bonded complexes. It might be hoped that the complications introduced by varying structural factors could be avoided by studying a series such as the pyridinium salts $PyrH^+ X^-$ with a series of anions of varying hydrogen bonding power. The assignment of a band at 1244 cm^{-1} (in the chloride) to ν_b by Cook[194] has been confirmed by Foglizzo and Novak[15]. Although the deuteration shift shows this is a fairly unmixed NH bending mode, the frequency shows no clear correlation with the H-bond strength, as indicated by ν_s (see table 4.7).

The intensity of the ν_b band certainly does not show the spectacular changes that the ν_s band displays. The ν_b band is in any case often not very strong, and although it may at times seem to disappear on complexing, it has probably been lost by shifting under a stronger band.

4.5.5. THE OUT-OF-PLANE BENDING MODE ν_d

When the proton donor molecule is an alcohol, phenol, or is a structurally similar molecule, this is the torsion mode lying in the far infrared range. For proton donors such as the pyridinium ion, or imidazole, the out-of-plane bending has no torsional character, and lies in the middle frequency range. We shall first discuss the non-torsional type of bending.

As with ν_b, deuteration shifts indicate mixing of the out-of-plane bending with other modes, the extent varying from case to case. This complicates any clear cut correlation between ν_d band frequency shift and the strength of hydrogen bonding. An increase in the ν_d band frequency is generally observed. In imidazole, this mode does not couple strongly with other fundamentals, and the relative shift $\Delta\bar{\nu}_d/\bar{\nu}_d$ of this mode on complexing is greater than $\Delta\bar{\nu}_s/\bar{\nu}_s$[89]. Large shifts of both the in-plane and out-of-plane bending modes are also observed in the hydrogen-bonded pyrimidinium halides[195]. In the pyridinium salts the out-of-plane motion is less concentrated in a single mode, and $\Delta\bar{\nu}_d/\bar{\nu}_d$ is much less[15].

Again the intensity change of ν_d is not particularly pronounced.

The additional constraint imposed on the H atom on forming the A—H—B bond would itself lead to an increase in the frequency of modes involving H bending. It is not possible to say whether this alone can account for the usually observed rise, without a much fuller knowledge of the entire force field of the complex.

4.5.6. THE TORSIONAL MODE

In proton donors such as alcohols or phenols, the out-of-plane motion of the bridging proton corresponds to the torsional mode in the free donor. If, as is probable, the hydrogen bond is linear, then a change in the torsional angle by π radians will increase the length of the bridge by such an extent that it will energetically be completely broken

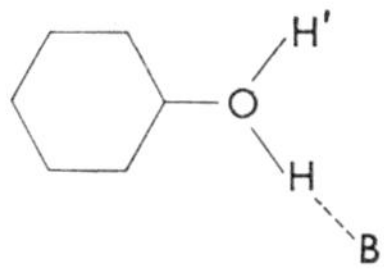

FIG 4.40 Effect of bonding on a torsional oscillation. H' is the position of the H atom when the torsional angle changes by π radians.

(see figure 4.40). Referring, for example, to phenol, the torsional frequency of the monomer, in either the vapour or in inert solvents, is equivalent to ~310 cm^{-1} corresponding to a twofold barrier of ~14.4 kJ/mol^{-1} [196] (1215 cm^{-1}). We might take 20 kJ mol^{-1} as typical of the energy for breaking the hydrogen bond[197]. Thus in the complex, the energy increases by this amount when the torsional angle increases by π (see figure 4.41). The simplest assumption is to represent this additional energy by a single cosine potential, as shown. The resulting increase in the torsion frequency would be about 17%. It is therefore surprising that in self-complexed phenol, this band appears to rise to over 600 cm^{-1} [198,199]. Concentration and deuteration studies serve to check the assignment.

Huong, Couzi and Lascombe[200] examined the torsion frequency of phenol in various solvents, which serve as proton acceptor. Their results are shown in figure 4.42, together with the torsion frequencies of intramolecularly hydrogen-bonded phenols, observed by Nyquist[201]. It is seen that there is a good correlation between the torsional frequency and $\Delta\bar{\nu}_s$ which holds for both internal and intermolecular hydrogen

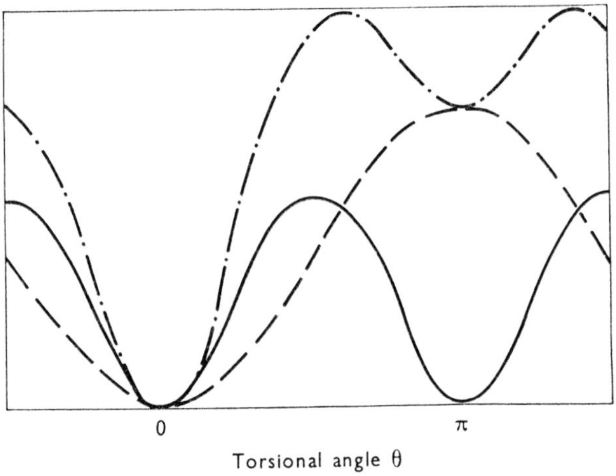

——— torsional barrier before hydrogen-bonding. - - - - hydrogen-bonding energy. – . – . – total energy.

FIG 4.41 The effect of hydrogen bonding on a torsional potential.

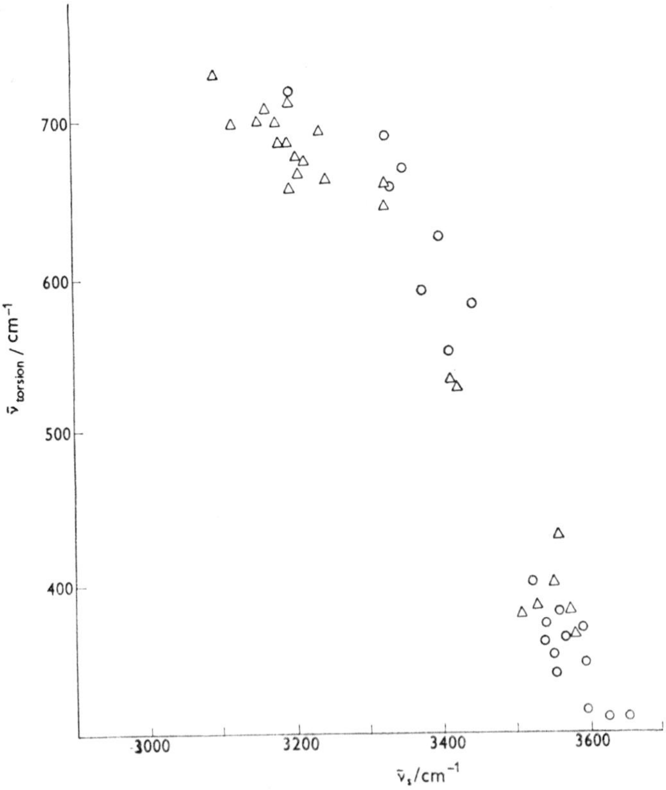

FIG 4.42 The relation between $\bar{\nu}_s$ and $\bar{\nu}_{torsion}$ for phenol complexes. ⊙ data of Couzi, Huong and Lascombe[200] △ data of Nyquist[201]

bonding. The observations cluster into two ranges—a lower one with torsion frequencies up to about 400 cm^{-1}, about the increase expected for a bond energy of ~30 kJ mol^{-1}, and a region around 600–700 cm^{-1}. No satisfactory account of the cause of an increase of this magnitude has been put forward, but it seems possible that the clustering of the observations is not fortuitous, but reflects the occurrence of two differing types of vibration.

The effect of hydrogen bonding on the torsional mode in alcohols has also been examined[202]. Bands observed in the vapour at ~200 cm^{-1} were assigned to the torsion of the monomeric forms and bands in the liquid at ~300 cm^{-1} to the torsion of self-associated forms.

However, the far-infrared spectra of alcohols are complicated by interaction of the torsion with the overall rotation (*e.g.* in methanol[203]), and by torsional isomerism in the higher alcohols[204]. In these respects the complexes will differ, so that in order to make any exact inferences, a more extensive investigation is required.

4.5.7. OTHER VIBRATIONAL TRANSITIONS

The tunnelling transition ν_r has already been mentioned in section 4.5.3. In symmetric systems this is infrared allowed, and might be expected to be observed anywhere in a broad lower frequency region. Identification would be easy because of the extreme effect of deuteration on the frequency. Nevertheless, although this transition has been much sought for, there appears to be no well substantiated instance of its observation. In view of the interest in double minimum potentials, and the key importance of ν_r type transitions in substantiating their occurrence, the absence of observations is most remarkable. Certainly with low barriers neither the band frequency or intensity should cause

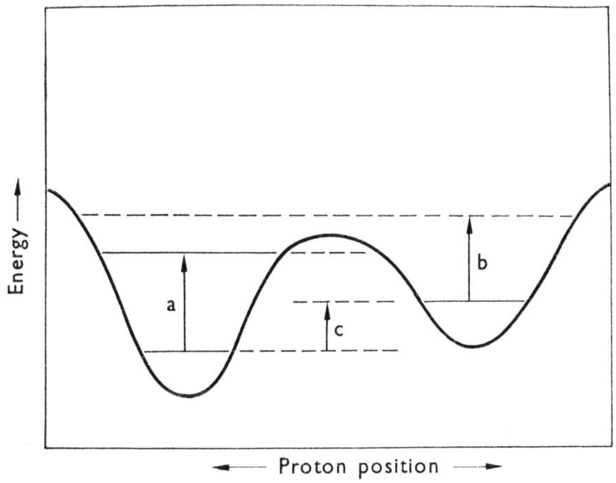

FIG 4.43 Transitions in an asymmetric double minimum potential.

undue difficulties. Inelastic neutron scattering offers a promising alternative method for checking on the presence of these transitions.

Related transitions are possible in systems with an asymmetric double minimum potential (figure 4.43) and these will be Raman as well as infrared allowed. Somorjai and Hornig have calculated the energy levels and infrared transition intensities for some representative asymmetric double minimum potentials[145]. As asymmetry is introduced, the wavefunctions rapidly localise in the two potential wells, and tunnelling correspondingly decreases. The intensity of well-to-well transitions (*e.g.* a and b, in figure 4.43), is much greater than that of tunnelling transitions, (*e.g.* c in figure 4.43). A broad infrared absorption about 400 cm^{-1} in KH_2PO_4 above the Curie point has been ascribed to such a transition[170] but further support is required for the interpretation to become convincing.

4.5.8. ν_σ AND OTHER INTERMOLECULAR MODES

Although the bending and symmetric stretching vibrations of bihalide ions (HX_2^-) can be compared to intermolecular modes, it was more appropriate to discuss the vibrations of these ions as a whole, in section 4.5.3 and the present section deals with intermolecular modes as usually understood, *i.e.* low frequency modes with no close counterparts in the separate moieties.

Because of the strength of association, intermolecular modes were first observed in the carboxylic acid dimers. As long ago as 1940 Bonner and Kirby-Smith[205] reported a Raman band at 232 cm^{-1} in formic acid vapour, which is predominantly the cyclic dimer. Far-infrared observations followed in 1958, Millikan and Pitzer[206] using a home-built spectrometer to observe bands, at ~240 and 160 cm^{-1}, again for formic acid vapour. Assuming that the effective point group of the dimer is C_{2h}—which corresponds to a high tunnelling barrier, three infrared and three Raman active intermolecular modes are expected. Carlson *et al.*[207] have more recently confirmed the presence of infrared bands at 248 and 164 cm^{-1}, and located the third infrared active fundamental band at 68 cm^{-1}. In $(HCOOD)_2$ they find the corresponding bands at 240, 158 and 68 cm^{-1}. In $(CH_3COOH)_2$ vapour three infrared bands are again found, at 186, 164 and 50 cm^{-1} [157,207] but it has been shown[156,158] that the two higher frequency maxima arise from the structure of a single fundamental. A summary of these data is given in table 4.8. A similar situation probably arises in $(CD_3COOH)_2$ and $(CD_3COOD)_2$. The third infrared active intermolecular mode of the acetic acid dimer is expected to be near 80 cm^{-1}. Unfortunately knowledge of the Raman active intermolecular modes is far less complete.

The intermolecular vibrations have been studied in several other self-complexed $(AH)_n$ type systems. Such systems appear to offer the advantage of simplicity—usually only the single substance is examined.

TABLE 4.8

Infrared active intermolecular bands of carboxylic acid dimers (values in cm^{-1})

Mode	Symmetry	$(HCOOH)_2$	$(HCOOD)_2$	$(CH_3COOH)_2$
ν(OH—O)	b_u	248	240	{186, 164}*
γ(OH—O)	a_u	164	158	—
γ twist	a_u	68	68	50

* Reference 208; all other data from reference 207.

However, it is always desirable to show that a band ascribed to an intermolecular mode has an intensity proportional to the amount of complex, and in some cases this has not been considered. This criticism does not apply to the work of Jakobsen and Brasch, who used an interesting method to support their assignment of the ν_σ mode of self-complexed phenols[199]. By monitoring the ν_s band region, they showed that phenols on absorption into polythene film at room temperature are predominantly polymerised—spectrum A of figure 4.44. On annealing, the phenol disperses in the film as monomer (B). On cooling, partial repolymerisation takes place (C). The intensity of the band ascribed to ν_σ ($\sim$165 cm^{-1} in the example shown) follows that of the polymer ν_s band, and disappears entirely when only monomer is present. Thus the assignment of the intermolecular mode is firmly established. Since the OH group of phenols can act both as donor and acceptor, numerous polymeric forms are possible[96]. It seems probable that the observed low

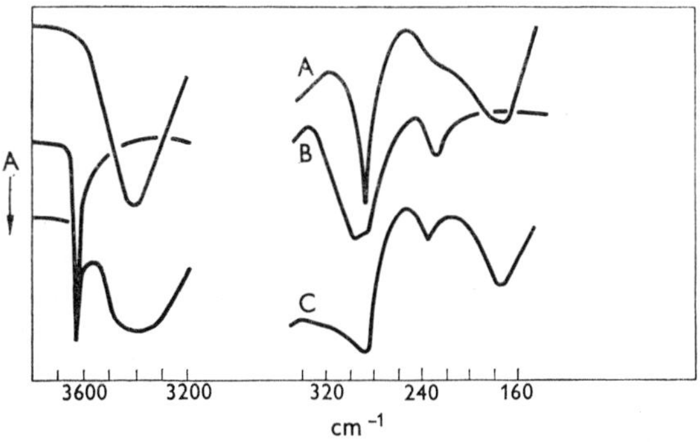

FIG 4.44 Infrared spectra of 2,6-dimethylphenol. The shifted ν_s and the ν_σ bands appear and disappear together. A, initial spectrum, B, after annealing, C, after cooling to -190°C. (Reproduced, by permission, from *Spectrochim. Acta*, **21,** 1756 (1956).)

frequency band represents the superposition of the ν_σ bands of several associated species.

The association between phenol and pyridine (in excess) in CCl_4 solution has been well characterised and yields only a 1:1 complex, with the proton on the phenol side. A quite sharp band at 134 cm^{-1} is absent in the separate components, and relates to the amount of complex present, so that its assignment to the ν_σ mode is well established[208]. The narrowness of the band $\Delta\bar{\nu}_{1/2} \sim 15$ cm^{-1} probably reflects the fact that only a single complex species is present. In the corresponding phenol D complex, $\bar{\nu}_\sigma$ falls to 130 cm^{-1}. The shift is small, but well established experimentally. Since phenols and pyridines form a well characterised series of 1:1 complexes, Hall and Wood extended this work to examine the influence of chemical factors on $\bar{\nu}_\sigma$[209]. This mode was observed in 24 complexes (see table 4.9). Although it is apparent that $\bar{\nu}_\sigma$ is influenced by changes in mass, confirming, if proof were still

TABLE 4.9

The ν_σ band positions (in cm^{-1}) for phenol–pyridine complexes

	Acceptors			
Donors	Pyridine	4-methyl-pyridine	4-ethyl-pyridine	4-phenyl-pyridine
Phenol	130.5	128.0	125.0	115.5
4-methylphenol	122.3	121.0	117.5	109.7
4-fluorophenol	125.7	123.7	121.4	112.7
4-chlorophenol	123.0	118.6	116.5	108.0
4-bromophenol	117.5	115.0	110.6	103.0
4-iodophenol	115.3	111.8	108.7	101.0

needed, that $\bar{\nu}_\sigma$ is not localised in the O–H—N unit, the 4-methylphenol complexes fall outside the mass order, showing that chemical factors are also influential.

Lake and Thompson[202] have assigned a band which is present in liquid alcohol (*i.e.* self-complexed) to ν_σ. Unexpectedly, the frequency of the band rises with the molecular weight of the alcohol, and in methanol was not found in the frequency range scanned (80 cm^{-1} upward). The nature of the relaxation process giving rise to this absorption demands further examination, and is obviously of importance in connection with the properties of liquid alcohols.

Low frequency bands in a number of crystalline systems have been ascribed to ν_σ-type vibrational modes. The interaction represented by the hydrogen bond is now one among the components of the lattice force-field. Foglizzo and Novak have made a careful study of the lattice modes of pyridinium chloride, bromide and iodide[210]. Fortunately

TABLE 4.10
Vibrational bands of pyridinium halides ascribed to lattice modes involving hydrogen bond deformation (data from reference 210)

Complex	ν_σ	ν_δ	ν_γ
$C_5H_5NH^+Cl^-$	192	116	92
$C_5H_5NH^+Br^-$	133	97	88
$C_5H_5NH^+I^-$	110	87	75

the crystal structure of the chloride is known, the unit cell containing two complexes and having a centre of inversion. By obtaining both Raman and infrared spectra, and also examining the D_1 and D_5 salts, it was possible to assign the three lattice modes deemed to have most hydrogen bond deformation character. These are shown in table 4.10. The fall in $\bar{\nu}_\sigma$ does not appear to be entirely due to the increase in the anion mass, and in part reflects the decreasing strength of the hydrogen bond, clearly shown by the lessening $\bar{\nu}_s$ shifts. The investigation has also been extended to the pyrimidinium and pyrazinium halides[211]. The ν_σ mode is assigned bands at 198 cm^{-1}, 146 cm^{-1} and (115 and 110) cm^{-1} in the pyrimidinium Cl^-, Br^- and I^- respectively, and at 186 cm^{-1}, 136 cm^{-1} and 107 cm^{-1} in the corresponding pyrazinium halides. Using a simple diatomic model for the intermolecular force constant, k_σ, a fairly linear correlation is found with the strength of the hydrogen bond measured by $\Delta\bar{\nu}_s/\bar{\nu}_s$. The results for the pyridinium salts also fit this plot. It is interesting that when the linear relation is extrapolated to $\Delta\bar{\nu}_s/\bar{\nu}_s = 0$, the value of k_σ is not zero, but 32 N m^{-1}. This agrees well with the interionic force constants for $C_5H_5NCH_3\ Cl^-$ (33 N m^{-1}) and $C_5H_5NCH_3\ I^-$ (30 N m^{-1}) in which there is no hydrogen bonding, obtained using the same model. This value then represents the residual Coulomb attraction, arising from the formal charges.

Similar studies have been made of the lattice modes of crystalline imidazole, which is composed of chains of hydrogen-bonded molecules[90]. Six of the nine infrared active lattice modes were observed,

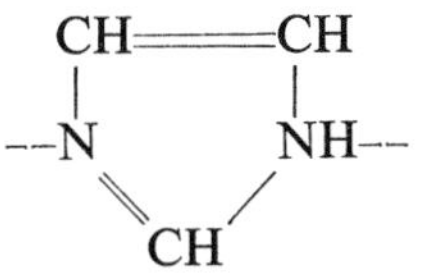

and tentative assignments to intermolecular hydrogen bond modes made.

Although the ν_σ mode was not observed directly, mention should be made here of a most notable investigation by Thomas of the HCN–HF and —CH_3CN–HF complexes in the vapour phase.[62] Both complexes are linear, so that there are only five intermolecular modes, ν_σ, and two bending modes, each of which is degenerate. ν_s and one of these bending

TABLE 4.11

Fundamental vibrational bands (in cm^{-1}) for cyanide–HF complexes (data from reference 62)

	A_1 species		E species	
Complex	$\tilde{\nu}_s$	$\tilde{\nu}_\sigma{}^e$	see figure 4.8c	see figure 4.8d[d,e]
HCN–HF	3710	155[a]	555	70
DCN–DF	2720	—	416	62
CH_3CN–HF	3627	168[b]	620	40
CD_3CN–HF	3627	165[b]	620	37
CH_3CN–DF	2667	160[c]	463	39
CD_3CN–DF	2668	160[c]	461	36

a. $\pm 10\ cm^{-1}$.
b. $\pm 3\ cm^{-1}$.
c. $\pm 20\ cm^{-1}$.
d. $\pm 35\%$.
e. inferred values.

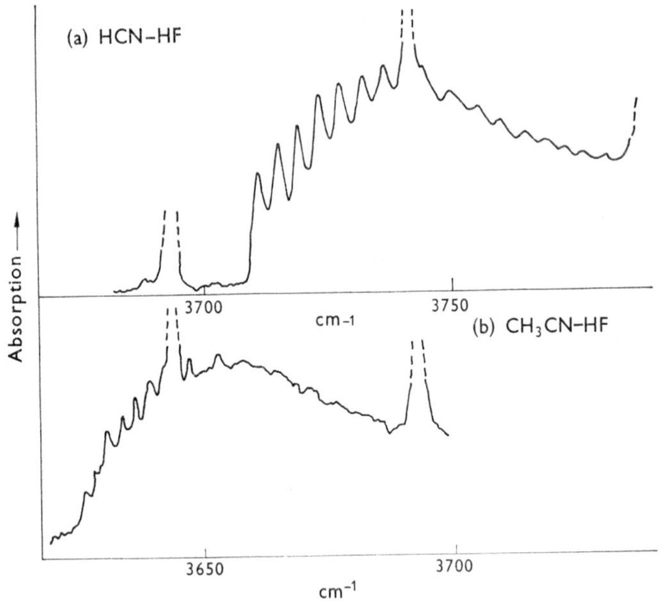

FIG 4.45 The fine structure associated with ν_1 of the cyanide-HF complexes (a) HCN–HF (b) CH_3CN–HF (The dotted lines indicate vibration-rotation lines of HF). (Reproduced from *Proc. Roy. Soc.*, **325A,** 137 (1971), by permission of the Royal Society.)

modes gave rise to bands within the range observed (>200 cm^{-1}), as shown in table 4.11. Both bands show extensive fine structure (see figure 4.45). This structure has been shown to arise from hot bands arising from excited levels of the other intermolecular bending vibration, in a similar manner to the CH_3CN–HCl complex (see section 4.5.3). The frequency of the bending mode is obtained from the sub-band intensities. Shoulders on the ν_s band due to the combinations $\nu_s + \nu_\sigma$, similar to those observed by Millen and co-workers in other gas phase complexes, gives a value of $\bar{\nu}_\sigma$. Thus, the frequencies of all the intermolecular modes are obtained, at least approximately (see table 4.11). This work gives us the most complete picture at present of the vibrational behaviour of a hydrogen-bonded complex. As well as the frequencies of the intermolecular modes, the lifetime of the complex is estimated (see section 4.5.3) and direct confirmation of the hot band mechanism in broadening the ν_s band is obtained.

4.6. THE EFFECT OF HYDROGEN BONDING ON MOLECULAR FORCE-FIELDS

4.6.1. INTERMOLECULAR FORCE-FIELDS

For a number of reasons, the carboxylic acid dimers have provided the most favourable system for examining intermolecular hydrogen-bonding force-fields. The advantages are (1) the structure of the complexes are known, (2) the spectral data refers to the vapour phase[38], (3) data has been obtained on complexes deuterated both in the H or methyl group and in the hydrogen bond, (4) symmetry simplifies the force field, (5) bending as well as stretching intermolecular modes have been observed, (6) the *intra*molecular vibrations of the complexes have been documented[38]. Although the calculation was carried out before more recent data became available, the treatment of Kishida and Nakamoto[212] is of great interest, because all the 17 in-plane modes of the $(HCOOH)_2$ complex were taken into account. Using a modified Urey-Bradley force field, the intermolecular O—H stretching force constant was obtained as 36 N m^{-1}, the OH—O bending constant was 1.5 N m^{-1}, and C═O—H bending constant was 1.0 N m^{-1}. Thus the intermolecular stretching force constant for this, quite strong hydrogen bond, is about one tenth of that typical of a covalent single bond. Although there is probably more uncertainty in the magnitude of the bending force constant, it appears to be considerably less than the typical ratio of one tenth of the stretching force constant.

It is convenient to include here mention of the changes in the internal force constants, which were also evaluated in these calculations. These are shown in table 4.12. As expected, the K(O—H) constant falls and H(C—O—H) rises on complexing. Interestingly, K(C═O) falls, and K(C—O) rises, showing the C═O and C—O bond orders are less distinct in the complex.

TABLE 4.12

Urey-Bradley force constants of the formic acid monomer and dimer (all $\times 10^{-2}$ N m^{-1})

	Monomer	Dimer
Stretching		
K(O—H)	6.9	4.7
K(C—O)	4.6	5.5
K(C=O)	11.2	10.0
K(C—H)	4.0	4.0
Bending		
H(O—C=O)	0.50	0.45
H(H—C—O)	0.19	0.19
H(H—C=O)	0.25	0.25
H(C—O—H)	0.40	0.45
Repulsive		
F(C · · · H)	0.55	0.60
F(H · · · O)	0.60	0.60
F(H · · · O_2)	0.80	0.70
F(O_1 · · · O_2)	1.00	0.80

Jacobsen *et al.*[156] extended the treatment to the out-of-plane vibrations, leading to the assignment of the infrared active intermolecular modes given in table 4.8 (p. 364). Unfortunately they do not report the force constants obtained.

A normal co-ordinate treatment of acetic acid monomer and dimer has been carried out, for both the in-plane and out-of-plane modes, by Fukushima and Zwolinski[213], also using a modified Urey-Bradley force field. Again all the degrees of freedom were taken into account. The intermolecular O—H stretching force constant obtained was 39 N m^{-1}, the OH—O bending constant was 6.6 N m^{-1} and the C=O—H bending constant was 0.5 N m^{-1}, indicative of slightly stronger hydrogen bonding than in the formic acid dimer. However the uncertainty in the values of several of the intermolecular frequencies prevents the validity of these constants being fully tested. The changes in the internal Urey-Bradley parameters were also obtained in this treatment, and some of the values are shown in table 4.13. The changes in the stretching force constants are similar to those in the formic acid dimer. Remarkable however, is the very large increase in the C—C stretching constant. The bending force constant changes are also quite different from those in the formic acid system. Although the treatments of the formic and acetic acid dimers are at present the most complete available, taking into account both internal and intermolecular modes, and also data from the deuterium substituted species, caution should be exercised before placing too fine an interpretation on the results. Not only are some of the frequencies lacking, and anharmonicity not

TABLE 4.13 Urey-Bradley force constants of acetic acid monomer and dimer (all $\times 10^{-2}$ N m^{-1})

	Monomer	Dimer
Stretching		
K(O—H)	7.05	4.89
K(C—O)	4.43	5.22
K(C=O)	10.05	9.00
K(C—H)	4.54	4.68
K(C—C)	1.90	2.90
Bending		
H(O—C=O)	0.20	0.34
H(C—C—O)	0.63	0.31
H(C—C=O)	0.93	0.63
H(C—O—H)	0.44	0.30

allowed for, but the effects of the simplification involved in using a Urey-Bradley force field in these systems has not been tested.

The situation is far less favourable for the calculation of intermolecular force constants for complexes in solution, since often only one frequency, that of ν_σ, is observed. It is, however, of great interest to try to form some estimate of this direct measure of the attractive force holding the complex together. Two models have frequently been used: (a) to treat the complex as a diatomic molecule, using the entire masses of the linked moieties; (b) as (a), but using the masses of only the directly linked atoms. Although (a) may be preferable to (b), neither is satisfactory, for model (a) implies that the internal force constants of the moieties are infinite, and (b) that they are zero. Any realistic treatment must include all the internal flexibility of the linked moieties. The value obtained for the hydrogen bond stretching force constant, *i.e.* of the intermolecular bond, using a simple valence force field, is critically dependent on the model taken. Some figures shown in table 4.14 relating to the phenol–pyridine complex[208] suffice to show this.

Turning to crystalline systems, it would seem that quantitative treatments of the lattice vibrations offer encouraging prospects. In several cases the structures are well established, and many of the

TABLE 4.14
Hydrogen bond stretching force constant for phenol–pyridine complex

Model	10^{-2} K/N m^{-1}
Simplified normal co-ordinate	0.23
Diatomic, linked atom masses	0.08
Diatomic, linked molecule masses	0.45

lattice modes observed. By extending the data with single crystal measurements, and N.I.S. spectra, a fairly complete assignment of the lattice modes could be expected. Although so far there is rather little experience of the appropriate choice of intermolecular force field in molecular crystals, it is an area where much progress can be expected.

4.6.2. THE EFFECT OF HYDROGEN BONDING ON "INTERNAL VIBRATIONS" OF THE MOIETIES

Hydrogen bonding affects some, but not all, of the internal vibrations of the complexed moieties. Frequency shifts are often of the order of the equivalent of 5 cm^{-1}, but exceptionally, shifts of up to 20 cm^{-1} are found[15,214,215]. The objective is to determine the changes in the internal force field, and to interpret these in the light of theories of hydrogen bonding. As the vibrations of the H atom ν_s, ν_b and ν_d are not completely localised (particularly for ν_b and ν_d modes), any discussion of the changes in the internal force-field must include the totality of the vibrations of the complex. The difficulties are very great. Firstly, data from vibrational frequencies alone are far from adequate to determine uniquely the general harmonic force field (*i.e.* with $n(n + 1)/2$ parameters) for a molecule with n modes. Usually the force-field of the uncomplexed molecules is itself uncertain. A reliable evaluation of the force-field in the complex is at first sight an order of magnitude more difficult. Secondly, anharmonicity is certain to be large in the complex, and to differ from, or have no counterpart in the free molecules, so that cancellation cannot be relied on to compensate for its neglect. A third, but lesser problem is that complexing causes changes in the molecular shapes, which can result, via the kinetic energy (**G**) matrix in appreciable frequency changes. Thus any examination of force-fields must be made in the spirit of exploring the possibilities, rather than in the expectation of decisive answers. Three types of system will be discussed, each highlighting particular aspects of the problem.

Fifer and Schiffer[216] have examined the force-field of H_2O in the hydrates $MnCl_2.2H_2O$, $CoCl_2.2H_2O$, $CuCl_2.2H_2O$ and $FeCl_2.2H_2O$ and of their D_2O analogues. Although there are no internal vibrations in the usual sense, of the water molecule, the simplicity of the system permits a more penetrating treatment of the force-field. A general quadratic force field for water has four parameters, and Fifer and Schiffer represent the potential by:

$$2V = f_r(\Delta r_1^2 + \Delta r_2^2) + f_\alpha \Delta\alpha^2 + 2f_{rr} \Delta r_1 \Delta r_2 + 2f_{r\alpha}(\Delta r_1 + \Delta r_2) \Delta\alpha, \tag{4.15}$$

where Δr_1, Δr_2 and $\Delta\alpha$ are the displacements of the bonds and apex angle from their equilibrium positions. Even in the free molecules, the effect of anharmonicity is such that a common set of quadratic force constants will not simultaneously fit the frequencies of H_2O, HOD and D_2O. In the present work it was found that this could be allowed for

TABLE 4.15
Intramolecular force constants of water in the vapour and in condensed systems (all values are $\times 10^{-2}$ N m^{-1})

	$(f_r)_D$	$(f_r)_H$	f_α/r^2	$f_{r\alpha}/r$	f_{rr}
H_2O[a]	8.45	8.45	0.76	0.23	−0.10
H_2O	7.89	7.68	0.72	0.24	−0.08
H_2O in CCl_4	7.69	7.49	0.71	0.13	−0.07
$MnCl_2.2H_2O$	6.94	6.71	0.73	0.22	0.02
$CoCl_2.2H_2O$	6.85	6.60	0.75	0.30	0.00
$CuCl_2.2H_2O$	6.65	6.38	0.74	0.30	0.06
$FeCl_2.2H_2O$	6.79	6.55	0.75	0.30	0.01

a. This set is based on frequencies corrected for anharmonicity. The remaining sets of parameters use the observed (uncorrected) frequencies.

on an *ad hoc* basis by choosing f_r individually for each isotopic species, retaining the other parameters throughout. In the hydrates, both OH bonds are hydrogen bonded to Cl^-, so that the site retains the C_{2v} symmetry of the free molecule. In the experiments, isotopic dilution methods were used to eliminate intermolecular coupling. For the calculations, the force field was perturbed to match the frequencies, with a reduced weighting factor for the match of the bending frequencies. Table 4.15 gives a summary of the results. The reduction in f_r stands out clearly, and the force-field used leads to an indication that the reduction is proportionately greater for H_2O than D_2O suggesting that the hydrogen bond is stronger than the deuterium bond. f_α shows a small, but consistent increase, in line with expectation. Both interaction constants show a tendency to increase on complexing, but particularly f_{rr}. It would be interesting to know whether an *SCF MO* treatment would check with this finding.

A disadvantage of the hydrates is that they are in the condensed phase, so that forces other than those of the hydrogen bond may be acting on the molecules. To eliminate this influence one must turn to gas-phase complexes. Formic and acetic acid dimers have been the most intensively investigated systems, and an account of the changes in their intramolecular force constants was included for convenience in section 4.6.1., along with a discussion of the intermolecular force field. Very few force-field calculations on other vapour phase complexes have been carried out. There are various difficulties—for H_2O dimer the spectrum is not well known[217,218], while for HF several multimers are present[114]. Although numerous complexes have in recent years been shown to exist in the gas phase, usually by observation of the ν_s band[6,7,92,93,133–135,137,219–226], association is usually very incomplete, so that many parts of the spectrum are strongly overlaid with bands from the uncomplexed molecules, making the determination of the frequencies of the internal modes in the complex particularly difficult.

A typical hydrogen-bonded system, involving polyatomic molecules, is pyridine-phenol, which has been examined in solution by the author and colleagues[227]. Association is strong, so that nearly all the internal modes of the interacting molecules can be followed in the complex. There has been a substantial amount of work on the assignments of the modes in the free molecules, and also on their force fields[198,228,196,229,230]. The principal difficulty is that the vibrational frequencies provide insufficient data to uniquely fix the general harmonic force-field, even if only "reasonable" values of the parameters are accepted. This problem carries over into the complex. However, if the derivatives $\partial \nu_i / \partial f_j$ are calculated (where f_j is a force constant), it is found that there is a substantial measure of consistency in these, even though one commences with different initial force fields. This reflects the fact that the assignment of a vibration, expressed in the usual way, has a measure of meaning. Consequently, a set of changes in force constants can be proposed which would reproduce the observed frequency shifts, even though the original force-field was in some doubt. In this way, for the phenol–pyridine complex, the effect of hydrogen bonding on the internal modes can be reproduced by varying only the CO and OH stretching and the COH and CCO bending diagonal force constants of phenol, and the NC and CC stretching diagonal force constants in pyridine, together with the influence of the intermolecular force constants.

Thus it can be seen that the calculation of the changes in the internal force field produced by hydrogen bonding is not so unpromising as seems at first sight. In particular, simple vapour phase complexes invite attention, as does comparison with *MO* theoretical approaches.

4.7. ELECTRONIC THEORIES OF THE HYDROGEN BOND

Efforts to explain hydrogen bonding in electronic terms followed closely on the discovery of the effect, and the progress made up to about 1966 has been excellently reviewed by Bratoz[231]. Since hydrogen-bonded systems necessarily involve a considerable number of electrons, work carried out before the advent of high speed computing methods required the use of massive simplifications, either in the formulation of the calculation, and/or by the introduction of semi-empirical parameters. Although the inferences drawn from these earlier treatments of the hydrogen bond were not quantitatively reliable they have been valuable in two respects: not only as pointing the direction of advance, and particularly in showing the extreme sensitivity of the calculations to the underlying assumptions, but also by instilling a sense of the temporary nature of any current assessment. This caveat made, calculations done in recent years which do not involve semi-empirical parameters show promise of convergence on a common result, and the discussion will be limited to these so-called ab-initio

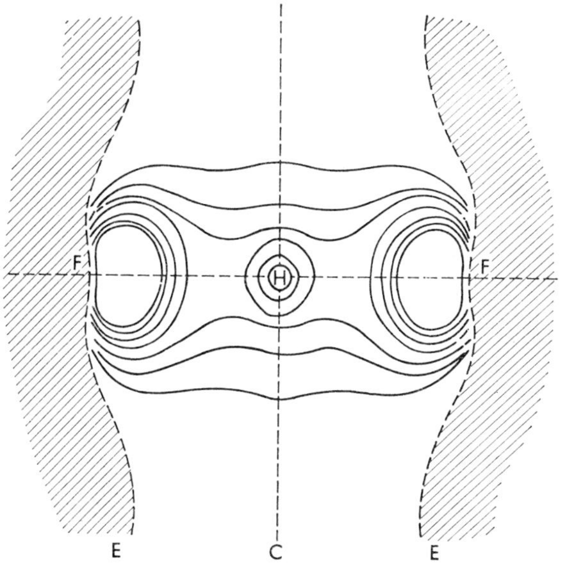

FIG 4.46 δ function for FHF^-. (Reproduced from *Advances in Quant. Chem.*, **3**, 226 (1967), by permission of Academic Press Inc.)

treatments (see section 1.3 for a discussion of the results obtained using *CNDO/INDO* methods).

Because of its relative simplicity, the first system to receive attention was the $(FHF)^-$ ion[232,233], though the strength of the bond makes this far from a typical system. The *SCF LCAO MO* method is applied to the entire system, and in this early work, the dimensions were fixed at the experimental values. The results depend considerably on the number and type of AOs used in the basis set, but agree in two important respects: (a) the ion has a stabilisation energy, with respect to $F^- + HF$ of the right magnitude, (b) the H atom is nearly neutral, while the negative charge is pushed away from the centre of the ion. Figure 4.46 shows the calculated[231] change in electron density $\delta(r) = \rho(\text{ion}) - \rho(\text{free atoms})$. In the shaded region the electron density has increased, while the contours indicate the extent of the decrease.

It also appeared from this work that the calculated energy of the complex is substantially above the SCF limit (depending on the basis set), which itself is above the true energy. However, some compensation takes place in obtaining the binding energy, and δ probably gives a fair qualitative indication of the changes in electron distribution on forming the complex.

The calculations have been repeated using a different basis set by Kollman and Allen[234], and give good general agreement with the earlier results—in particular the electron density map is quite similar. In these latest calculations the geometry was allowed to vary; the linear complex, with r(F—F) = 2.285 Å (experimental r(F—F) = 2.26 to 2.28 Å, depending on the salt) was most stable. A single minimum for the proton potential function was also found. The good measure of agreement of the various *ab-initio* treatments of the bifluoride ion encourage confidence in this method.

Numerous calculations have also been carried out on the water dimer[107–111,235]. These are particularly interesting for several reasons. First they allow a comparison of the influence of the various computational approximations—it should be made clear that these do not involve empirical parameters, but reflect the truncation in the representation of the wave-function. Secondly, at the time of the calculations the structure of the dimer was in doubt, and various geometries were explored to determine the preferred configuration. Thirdly, in several cases the calculations were extended to examine polymer formation.

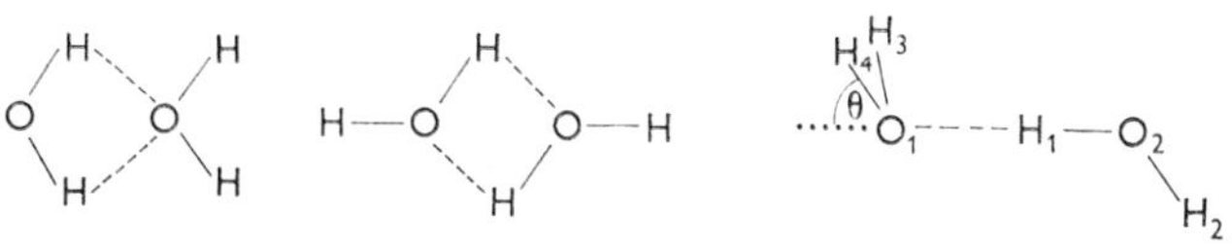

FIG 4.47 Possible structures for the water dimer.

Despite differences in the choice of basis functions, the results show a very encouraging measure of agreement. All concur in finding the bifurcated and ring dimers less stable than the open dimer, (see figure 4.47) even though at the time the ring dimer was preferred on spectroscopic evidence. The preferred open structure has a linear hydrogen bond $O_1H_1O_2$, and the energy minimises for a reasonable O_1O_2 length (near 3.0 Å). The most stable configuration has a plane of symmetry ($H_1O_2H_2$), and the plane $H_3O_1H_4$ is inclined about 50° (θ) to the $O_1H_1O_2$ axis. The binding energy is comparable to the experimental estimate of ~21 kJ mol^{-1} [236,237]. Although there is agreement on the magnitude of θ, the sign can correspond to the *trans*[108,235] (as shown) or *cis*[111] configuration. In fact the energies may be very close. When it is appreciated that the structural variations only result in changes in the total calculated energy of the order of 1 part in 10^6, the agreement is impressive. The earlier spectroscopic evidence[106], indicating a ring structure in conflict with the MO results has recently been re-examined by Tursi and Nixon[36] who conclude on the basis of new measurements, that the data is consistent with the open structure. MO calculations[111,235] also reproduce the extension of the OH bond length on complexing, and the lowering of the stretching force constant.

Thus the results with water dimer are very encouraging, especially in view of the fact that the discrepancy between the calculated (total) energy and the *SCF* limit is estimated[108] to be 100–1000-fold greater than the energy variations following structural changes. Obviously there is much cancellation of errors.

The water dimer calculations have also been extended to trimers[109,235]. These concur in favouring the open polymer,

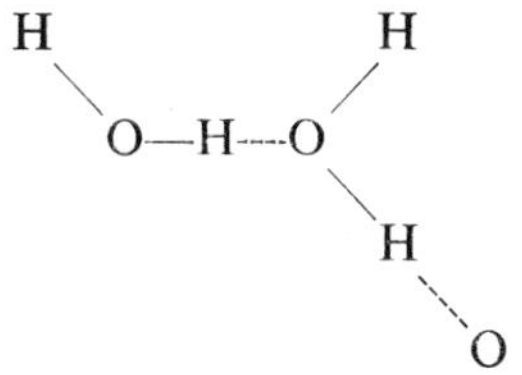

and indicate that the binding energy per bond is greater than in the dimer—*i.e.* the proton acceptor function of the central H_2O molecule has enhanced its donor power. Infrared spectroscopic evidence drawn from other systems supports this conclusion[7,99]. (See section 1.3 for a further discussion of calculations on the water dimer.)

The symmetric $(H_5O_2)^+$ ion has also been treated by the *ab-initio* method[234]. One of the most interesting findings is that at the most stable O—O distance, (2.3 Å), the proton potential turned out to be a flat-bottomed single minimum or 'U' shaped well. On stretching the O—O distance to 2.49 Å a double minimum developed.

Recently the *LCAO–SCF* method has been applied to larger systems, particularly by Pullman, and by Clementi, using advanced computing methods—thus Clementi has carried out a calculation on the guanine-cytosine base pair, involving the computation of $>10^9$ integrals[159]. Pullman and co-workers[238,239] have examined the formamide dimer, as a model for interactions in proteins. The results are particularly illuminating. As well as the SCF energy (E_{SCF}), Dreyfus and Pullman[239] calculated the coulomb energy (E_C) from the simple product wave function for the dimer, $\Psi_{dimer} = \Psi^{\circ}_{mono\,1}\,\Psi^{\circ}_{mono\,2}$ where Ψ°_{mono} is the wave function for the isolated monomer. The addition of antisymmetrisation, *i.e*:

$$\Psi_{dimer} = \mathscr{A}[\Psi^{\circ}_{mono\,1}\Psi^{\circ}_{mono\,2}]$$

gives the energy $E_1 = E_C + E_E$ where E_E is the exchange repulsion energy. The difference between E_{SCF} and E_1 is due to polarisation and charge-transfer effects. Figure 4.48 shows one of the configurations examined and figure 4.49 shows these energies as a function of the O—N distance in the open dimer configuration. The coulomb energy is appreciable at long range, while the exchange repulsion cuts on much more sharply. The polarisation and charge-transfer contributions are small, but not negligible at long range, and though larger at the equilibrium distance, is still smaller than either E_E or E_C individually. The latter alone would lead to a stable complex, and explains the relative success of early electrostatic models of the hydrogen bond. Figure 4.50

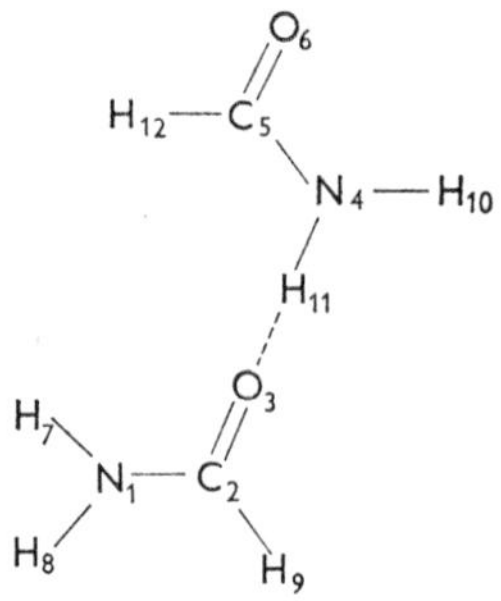

FIG 4.48 One of the configurations of the formamide dimer on which *ab-initio MO* calculations have been made. (Reproduced from *Theor. Chem. Acta*, **19**, 20 (1970), by permission of Springer Verlag Inc.)

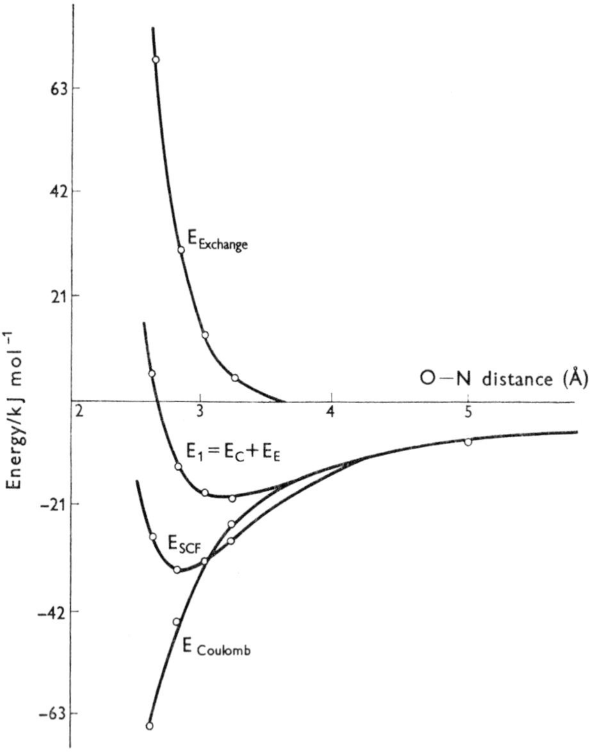

FIG 4.49 Dependence of the contributions to the hydrogen bonding energy on the O—N distance. (Reproduced from *Theor. Chem. Acta*, **19**, 20 (1970), by permission of Springer Verlag Inc.)

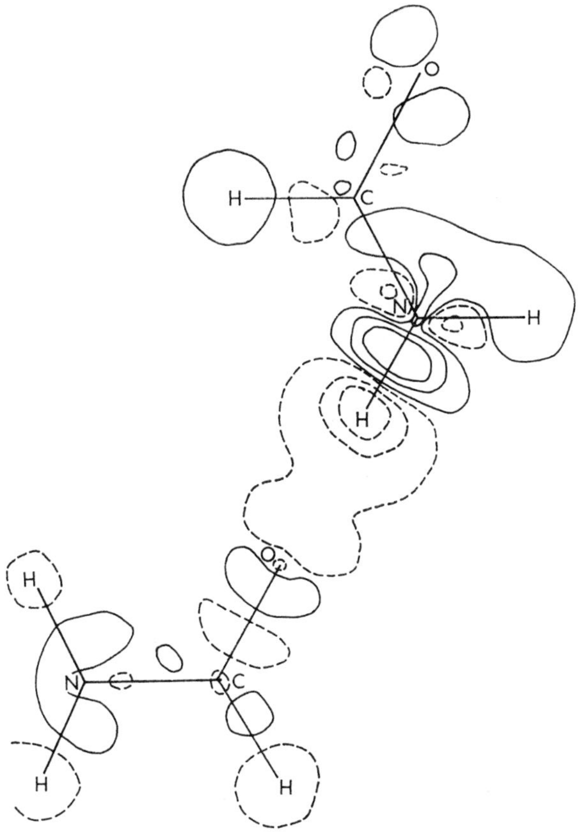

FIG 4.50 Calculated electron density change on forming the O—N hydrogen bond in the formamide dimer – – – indicates decrease and ——— increase in electron density. (Reproduced from *Theor. Chem. Acta*, **19,** 20 (1970), by permission of Springer Verlag. Inc.)

shows the electron density changes on forming the equilibrium configuration of the complex. The dotted contours indicate reduced density. It is notable that the electron density between the H atom and the oxygen has actually decreased, *i.e.* there is no indication of covalent character in the hydrogen bond. Electrons have been pushed back from the vicinity of the H atom, the most important effect being the polarisation of the NH bond under the influence of the approaching polar carbonyl group. The electron density changes in the body of the molecules away from the vicinity of the H bond are relatively small, and probably reflect the nature of the individual molecules, without particular significance in hydrogen-bonding. The extent of charge-transfer in this complex is also small, about 0.025 ε at the equilibrium distance.

It has been remarked that one of the most notable spectroscopic consequences of hydrogen bonding is the great increase in the intensity of the AH stretching band. Dreyfus and Pullman have calculated the total complex dipole moment as a function of r_{A-H}, and evaluate an eleven-fold intensity increase. Less than one third of this is due to the fluctuation of the amount of charge-transfer as the AH bond vibrates.

An interesting experimental indication of the pushing of the electrons back into the AH bond is shown by electron density contours obtained from X-ray measurements on a complex pharmaceutical containing the N—H—Br bond[240].

From these investigations a much clearer picture of the hydrogen bond in its electronic ground state is emerging. The unique function of hydrogen arises because it has no inner shell electrons—as a result the (repulsive) exchange energy is low. The Li bond is consequently much weaker, and does not give rise to a parallel range of complexes. In weak or medium strength hydrogen bonds the largest attractive term is the coulomb energy. This is only weakly directional, so that the hydrogen bond can be easily bent. The exchange repulsion is also large, and if the molecules joined by the bond do not have symmetry about the bond, as is usually the case, the bond may bend to reduce the repulsive interactions between the bodies of the molecules. Charge-transfer and polarisation contributions only appear to become important in short, *i.e.* strong, bonds. However, although of smaller magnitude than the coulomb and exchange energies, they are sufficient to significantly influence the equilibrium geometry. These conclusions now seem to have been established by sufficiently different approaches (see, for example, reference 241) to be likely to remain valid even though, for instance, the attractive dispersion energy is omitted by an *SCF* treatment, and the *LCAO–MO* energies obtained are probably a good way off the *SCF* limit.

Nothing in this discussion relates to electronically excited states of the complex. Theoretical treatments are as yet not advanced far enough for any consistent picture to emerge, but undoubtedly this will shortly become an active area.

REFERENCES

1. G. C. Pimentel and A. L. McClellan, The Hydrogen Bond, (W. H. Freeman and Co., San Francisco and London, 1960)
2. A. S. N. Murthy and C. N. R. Rao, *Appl. Spect. Rev.*, **2,** 69 (1968)
3. T. Gramstad, *Acta Chem. Scand.*, **16,** 807 (1962)
4. T. Zeegers-Huyskens, *Spectrochim. Acta*, **21,** 221 (1965)
5. S. A. Harrell and D. H. McDaniel, *J. Amer. Chem. Soc.*, **86**, 4497 (1964)
6. J. Arnold and D. J. Millen, *J. Chem. Soc.*, 503 (1965)
7. M. Couzi, J. Le Calve, P. V. Huong and J. Lascombe, *J. Mol. Struct.*, **5,** 363 (1970)
8. T. Kubota, *J. Amer. Chem. Soc.*, **88,** 211 (1966)
9. L. J. Bellamy, G. Eglinton and J. F. Morman, *J. Chem. Soc.*, 4762 (1961)
10. S. Singh, A. S. N. Murthy and C. N. R. Rao, *Trans. Faraday Soc.*, **62,** 1056 (1966)

11. G. E. Hyde and D. F. Hornig, *J. Chem. Phys.*, **20,** 647 (1952)
12. C. M. King and E. R. Nixon, *J. Chem. Phys.*, **48,** 1685 (1968)
13. N. P. Razaev and K. Shchepanyak, *Optics and Spectroscopy*, **16,** 283 (1964)
14. J. W. Nibler and G. C. Pimentel, *J. Chem. Phys.*, **47,** 710 (1967)
15. R. Foglizzo and A. Novak, *J. Chim. Phys.*, **66,** 1539 (1969)
16. L. J. Bellamy, Advances in Infrared Group Frequencies, (Methuen, London, 1968), p. 288
17. J. G. David and H. E. Hallam, *Spectrochim. Acta*, **21,** 841 (1965)
18. J. C. Evans and G. Y-S. Lo, *J. Phys. Chem.*, **73,** 448 (1969)
19. G. H. F. Diercksen, H. Preuss, *Int. J. Quantum Chem.*, **1,** 637 (1967)
20. G. Shaw, *Int. J. Quantum Chem.*, **3,** 219 (1969)
21. R. Janoschek, G. Diercksen and H. Preuss, *Int. J. Quantum Chem.*, **2,** 159 (1968)
22. A. W. Baker and D. E. Bublitz, *Spectrochim. Acta.*, **22,** 1787 (1966)
23. O. F. Bezrukov, M. F. Vuks and V. A. Suchkov, *Optics and Spectroscopy*, **2,** 234 (1963)
24. P. N. Noble and G. C. Pimentel, *J. Chem. Phys.*, **49,** 3165 (1968)
25. T. C. Waddington, *J. Chem. Soc.*, 1708 (1958)
26. D. Hadži, A. Novak and J. E. Gordon, *J. Phys. Chem.*, **67,** 1118 (1963)
27. D. Hadži and A. Novak, *Spectrochim. Acta.*, **18,** 1059 (1962)
28. D. Cook, *Chem. and Industry*, 607 (1963)
29. D. Cook, *Can. J. Chem.*, **41,** 2575 (1963)
30. J. Corset, P. V. Huong and J. Lascombe, *Spectrochim. Acta.*, **24A,** 2045 (1968)
31. J. F. Coetzee, G. R. Padmanabhan and G. P. Cunningham, *Talanta*, **11,** 93 (1964)
32. (a) R. Clements, R. L. Dean, T. R. Singh and J. L. Wood, *J. Chem. Soc., D*, 1125 (1971); (b) R. Clements, R. L. Dean and J. L. Wood, *J. Chem. Soc., D*, 1127 (1971)
33. Z. Dega-Szafran, *Bull. Acad. Polon. Sci., Ser. Sci. Chim.*, **15,** 393 (1967)
34. R. Blinc and D. Hadži, *Spectrochim. Acta.*, **16,** 853 (1960)
35. E. B. Wilson, Centenary Lecture of the Chemical Society, London (1971)
36. A. J. Tursi and E. R. Nixon, *J. Chem. Phys.*, **52,** 1521 (1970)
37. J. C. Evans and G. Y-S. Lo, *J. Phys. Chem.*, **70,** 20; 543 (1966)
38. M. Haurie and A. Novak, *J. Chim. Phys.*, **62,** 146 (1965)
39. L. J. Bellamy and P. E. Rogasch, *Proc. Roy. Soc.*, **A257,** 98 (1960)
40. H. Boutin and S. Yip, Molecular Spectroscopy with Neutrons, (M.I.T. Press, Boston, 1968)
41. G. C. Stirling, C. J. Ludman and T. C. Waddington, *J. Chem. Phys.*, **52,** 2730 (1970)
42. W. C. Hamilton and J. A. Ibers, Hydrogen Bonding in Solids, (W. A. Benjamin Inc., New York, 1968)
43. D. W. Rogers, private communication
44. H. B. Yang and R. W. Taft, *J. Amer. Chem. Soc.*, **93,** 1310 (1971)
45. G. L. Hofacker and U. A. H. Hofacker, 14th Colloque Ampere, (North-Holland Publications Co., 1967), p. 502
46. A. D. H. Clague, G. Govil and H. J. Bernstein, *Can. J. Chem.*, **47,** 625 (1969)
47. P. J. Berkeley and M. W. Hanna, *J. Phys. Chem.*, **67,** 846 (1963)
48. C. J. Creswell and A. L. Allred, *J. Amer. Chem. Soc.*, **85,** 1723 (1963)
49. F. Takahashi and N. C. Li, *J. Phys. Chem.*, **68,** 2136 (1964)
50. S. N. Vinogradov and R. H. Linnell, Hydrogen Bonding, (Van Nostrand, New York, 1971)
51. H. H. Jaffe and M. Orchin, Theory and Applications of Ultraviolet Spectroscopy, (J. Wiley and Sons, London and New York, 1962)
52. A bibliography is given in reference 2
53. E. Lippert (Ed., D. Hadži), Hydrogen Bonding, (Pergamon Press, London, 1959)

54. H. Baba and S. Suzuki, *J. Chem. Phys.*, **35,** 1118 (1961)
55. G. C. Pimentel, *J. Amer. Chem. Soc.*, **79,** 3323 (1957)
56. S. Nagakura, *J. Chim. Phys.*, **61,** 217 (1964)
57. H. Baba, A. Matsuyama and H. Kokubun, *J. Chem. Phys.*, **41,** 895 (1964)
58. H. Baba, A. Matsuyama and H. Kokubun, *Spectrochim. Acta*, **25A,** 1709 (1969)
59. R. M. Scott, D. de Palma and S. N. Vinogradov, *J. Phys. Chem.*, **72,** 3192 (1968)
60. R. A. Hudson, R. M. Scott and S. N. Vinogradov, *Spectrochim. Acta*, **26A,** 337 (1970)
61. See reference 50, pp. 106–108
62. R. K. Thomas, *Proc. Roy. Soc.*, **325A,** 133 (1971)
63. A. Hall and J. L. Wood, unpublished data
64. W. Vedder and D. F. Hornig, *Advances in Spectroscopy*, **2,** 189
65. H. Ratajczak and W. J. Orville-Thomas, *J. Mol. Struct.*, **1,** 449 (1968)
66. R. M. Badger and S. H. Bauer, *J. Chem. Phys.*, **5,** 839 (1937)
67. T. D. Epley and R. S. Drago, *J. Amer. Chem. Soc.*, **89,** 5770 (1967)
68. M. D. Joesten and R. S. Drago, *J. Amer. Chem. Soc.*, **84,** 3817 (1962)
69. M. S. Nozari and R. S. Drago, *J. Amer. Chem. Soc.*, **92,** 7086 (1970)
70. R. West, D. L. Powell, L. S. Whatley, M. K. T. Lee and P. von R. Schleyer, *J. Amer. Chem. Soc.*, **84,** 3221 (1962)
71. T. Gramstad, *Spectrochim. Acta*, **19,** 497 (1963)
72. A. M. Dierckx, P. Huyskens and T. Zeegers-Huyskens, *J. Chim. Phys.*, **62,** 336 (1965)
73. T. Gramstad, *Acta Chem. Scand.*, **15,** 1337 (1961)
74. T. Gramstad and W. J. Fuglevik, *Acta Chem. Scand.*, **16,** 1369 (1962)
75. D. Neerinck and L. Lamberts, *Bull. Soc. Chem. Belges*, **75,** 473; 483 (1966)
76. T. Kitao and C. H. Jarboe, *J. Org. Chem.*, **32,** 407 (1967)
77. A. Hall, Ph.D. Thesis, University of London, 1969
78. B. B. Wayland and R. S. Drago, *J. Amer. Chem. Soc.*, **86,** 5240 (1964)
79. E. Ōsawa, T. Katō and Z. Yoshida, *J. Org. Chem.*, **32,** 2803 (1967)
80. E. R. Lippincott and R. Schroeder, *J. Chem. Phys.*, **23,** 1099 (1955)
81. R. Schroeder and E. R. Lippincott, *J. Phys. Chem.*, **61,** 921 (1957)
82. E. R. Lippincott, J. N. Finch and R. Schroeder, (Ed. A. Hadži) Hydrogen Bonding, (Pergamon Press, London, 1959), p. 362
83. D. Hadži, *J. Chem. Soc.*, 5128 (1962)
84. D. Hadži, *Pure Appl. Chem.*, **11,** 435 (1965)
85. N. Sheppard (Ed., A. Hadži), Hydrogen Bonding, (Pergamon Press, London, 1959), p. 85
86. E. Whalley, *Developments in Applied Spectroscopy*, **6,** 277 (1968)
87. E. Wiener, S. Levin and I. Pelah, *J. Chem. Phys.*, **52,** 2881, 2891 (1970)
88. R. Blinc, *Adv. in Mag. Resonance*, **3,** 141 (1968)
89. C. Perchard, A-M. Bellocq and A. Novak, *J. Chim. Phys.*, **62,** 1344 (1965)
90. C. Perchard and A. Novak, *J. Chem. Phys.*, **48,** 3079 (1968)
91. J. A. Salthouse and T. C. Waddington, *J. Chem. Phys.*, **48,** 5274 (1968)
92. L. J. Sacks, R. S. Drago and D. P. Eyman, *Inorg. Chem.*, **7,** 1484 (1968)
93. W. J. Jones, R. M. Seel and N. Sheppard, *Spectrochim. Acta*, **25A,** 385 (1969)
94. R. K. Thomas and H. W. Thompson, *Proc. Roy. Soc.*, **A316,** 303 (1970)
95. C. C. Costain and G. P. Srivastava, *J. Chem. Phys.*, **41,** 1620 (1964)
96. K. B. Whetsel and J. H. Lady, Spectroscopy of Fuels, (Plenum Press, New York, 1970), Chapter 20
97. R. Mecke, *Disc. Faraday Soc.*, **9,** 161 (1950)
98. M. M. Maguire and R. West, *Spectrochim. Acta*, **17,** 369 (1960)
99. L. J. Bellamy and R. J. Pace, *Spectrochim. Acta*, **22,** 525 (1966)
100. A. Hall and J. L. Wood, *Spectrochim. Acta*, **23A,** 2657 (1967)
101. U. Liddel and E. D. Becker, *J. Chem. Phys.*, **25,** 173 (1956)

102. R. G. Inskeep, J. M. Kelliher, P. E. McMahon and B. G. Somers, *J. Chem. Phys.*, **28,** 1033 (1958)
103. L. J. Bellamy, Advances in Infrared Group Frequencies, (Methuen, London 1968), p. 255 et seq
104. A. N. Fletcher and C. A. Heller, *J. Phys. Chem.*, **71,** 3742 (1967); **72,** 1839 (1968)
105. M. Van Thiel, E. D. Becker and G. C. Pimentel, *J. Chem. Phys.*, **27,** 95 (1957)
106. M. Van Thiel, E. D. Becker and G. C. Pimentel, *J. Chem. Phys.*, **27,** 486 (1957)
107. K. Morokuma and L. Pedersen, *J. Chem. Phys.*, **48,** 3275 (1968)
108. G. H. F. Diercksen, *Chem. Phys. Lett.*, **4,** 373 (1969)
109. D. Hankins, J. W. Moskowitz and F. H. Stillinger, *Chem. Phys. Lett.*, **4,** 527 (1969)
110. P. A. Kollman and L. C. Allen, *J. Chem. Phys.*, **51,** 3286 (1969)
110a. P. A. Kollman and L. C. Allen, *J. Chem. Phys.*, **52,** 5085 (1970)
111. K. Morokuma and J. R. Winick, *J. Chem. Phys.*, **52,** 1301 (1970)
112. A. S. N. Murthy, R. E. Davis and C. N. R. Rao, *Theor. Chim. Acta*, **13,** 81 (1969)
113. D. F. Smith, *J. Chem. Phys.*, **48,** 1429 (1968)
114. P. V. Huong and M. Couzi, *J. Chim. Phys.*, **66,** 1309 (1969)
115. P. V. Huong and M. Couzi, *J. Chim. Phys.*, **67,** 1994 (1970)
116. M. T. Bowers and W. H. Flygare, *J. Chem. Phys.*, **44,** 1389 (1966)
117. L. F. Keyser and G. W. Robinson, *J. Chem. Phys.*, **45,** 1694 (1966)
118. A. J. Barnes, H. E. Hallam and G. F. Scrimshaw, *Trans. Faraday Soc.*, **65,** 3150 (1969)
119. G. Herzberg, Infrared and Raman Spectra of Polyatomic Molecules, (Van Nostrand, Ltd., New York and London, 1945)
120. E. Fermi, *Z. Phys.*, **71,** 250 (1931)
121. J. C. Evans, *Spectrochim. Acta*, **16,** 994 (1960)
122. R. N. Dixon, *J. Chem. Phys.*, **31,** 258 (1959)
123. A. Hall and J. L. Wood, *Spectrochim. Acta*, **23A,** 1257 (1967)
124. W. E. Thompson and G. C. Pimentel, *Z. Elektrochem.*, **64,** 748 (1960)
125. J. C. Evans, *Spectrochim. Acta*, **17,** 129 (1961)
126. J. C. Evans, *Spectrochim. Acta*, **18,** 507 (1962)
127. J. C. Evans and N. Wright, *Spectrochim. Acta*, **16,** 352 (1960)
128. M. F. Claydon and N. Sheppard, *Chem. Comm.*, 1431 (1969)
129. B. I. Stepanov, *Zh. Fiz. Khim.*, **19,** 507 (1945)
130. B. I. Stepanov, *Zh. Fiz. Khim.*, **20,** 408 (1948)
131. M. I. Batuev, *Zh. Fiz. Khim.*, **23,** 1399 (1949)
132. C. Reid, *J. Chem. Phys.*, **30,** 182 (1959)
133. J. E. Bertie and D. J. Millen, *J. Chem. Soc.*, 497 (1965)
134. J. E. Bertie and D. J. Millen, *J. Chem. Soc.*, 514 (1965)
135. J. Arnold and D. J. Millen, *J. Chem. Soc.*, 510 (1965)
136. G. L. Carlson, R. E. Witkowski and W. G. Fateley, *Nature*, **211,** 1289 (1966)
137. L. Al-Adhami and D. J. Millen, *Nature*, **211,** 1291 (1966)
138. R. K. Thomas, *Proc. Roy. Soc.*, **A322,** 137 (1971)
139. Y. Marechal and A. Witkowski, *J. Chem. Phys.*, **48,** 3697 (1968)
140. P.-O. Löwdin, *Rev. Mod. Physics*, **35,** 724 (1963); *Biopolymers Symposia*, **1,** 161, 293 (1964)
141. R. P. Bell, J. A. Fendley and J. R. Hulett, *Proc., Roy. Soc.*, **A235,** 453 (1956)
142. J. A. Ibers, *J. Chem. Phys.*, **40,** 402 (1964)
143. J. Laane and R. C. Lord, *J. Chem. Phys.*, **47,** 4941 (1967)
144. S. I. Chan, T. R. Borgers, J. W. Russell, H. L. Strauss and W. D. Gwinn, *J. Chem. Phys.*, **44,** 1103 (1966)
145. R. L. Somorjai and D. F. Hornig, *J. Chem. Phys.*, **36,** 1980 (1962)
146. J. Laane, *J. Chem. Phys.*, **55,** 2514 (1971)
146a. J. W. Bevan, M.Sc. Thesis, University of Surrey, 1969

147. T. R. Singh and J. L. Wood, *J. Chem. Phys.*, **48,** 4567 (1968)
148. B. L. McGraw and J. A. Ibers, *J. Chem. Phys.*, **39,** 2677 (1963)
149. J. A. A. Ketelaar and W. Vedder, *J. Chem. Phys.*, **19,** 654 (1951)
150. L. Couture and J-P. Mathieu, *C.R. Acad. Sci.*, **228,** 555 (1949)
151. L. Couture and J. P. Mathieu, *C.R. Acad. Sci.*, **230,** 1054 (1950)
152. J. C. Evans and G. Y-S Lo, *J. Phys. Chem.*, **70,** 11 (1966)
153. J. C. Evans and G. Y.-S. Lo, *J. Phys. Chem.*, **71,** 3942 (1967)
154. A. C. Pavia and P. A. Giguère, *J. Chem. Phys.*, **52,** 3551 (1970)
155. J. Karle and L. O. Brockway, *J. Amer. Chem. Soc.*, **66,** 574 (1944)
156. R. J. Jakobsen, Y. Mikawa and J. W. Brasch, *Spectrochim. Acta*, **23A,** 2199 (1967)
157. D. Clague and A. Novak, *J. Mol. Struct.*, **5,** 149 (1970)
158. S. G. W. Ginn and J. L. Wood, *J. Chem. Phys.*, **46,** 2735 (1967)
159. E. Clementi, J. Mehl and W. von Niessen, I.B.M. Research Report (1969)
160. S. W. Peterson and H. A. Levy, *Acta Crystallogr.*, **10,** 70 (1957)
161. P. G. Owston, *Quart Rev. Chem. Soc.*, **5,** 344 (1951)
162. E. Whalley, *Developments in Applied Spectroscopy*, **6,** 277 (1968)
163. M. D. Danford and H. A. Levy, *J. Amer. Chem. Soc.*, **84,** 3965 (1962)
164. L. Pauling, *J. Amer. Chem. Soc.*, **57,** 2680 (1935)
165. J. L. Lebowitz, *Ann. Rev. Phys. Chem.*, **19,** 407 (1968)
166. J. J. Fox and A. E. Martin, *Proc. Roy. Soc.*, **A174,** 234 (1940)
167. J. L. Kavanau, Water and Water-solute Interactions (Holden Day, San Francisco, 1964)
168. L. J. Bellamy and P. E. Rogasch, *Proc. Roy. Soc.*, **A257,** 98 (1960)
169. C. Kittell, Introduction to solid state physics, 2nd ed. (J. Wiley and Sons, New York, 1956), p. 182
170. Y. Imry, I. Pelah and E. Wiener, *J. Chem. Phys.*, **43,** 2332 (1965)
171. R. Blinc and D. Hadži, *Mol. Phys.*, **1,** 391 (1958)
172. M. Asselin and C. Sandorfy (to be published)
173. J. N. Finch and E. R. Lippincott, *J. Phys. Chem.*, **61,** 894 (1957)
174. S. A. Rice and J. L. Wood (to be published)
175. J. A. Ibers, *J. Chem. Phys.*, **41,** 25 (1964)
176. R. Chidambaram, R. Balasubramanian and G. Ramachandran, *Biochim. Biophys. Acta*, **221,** 182, 196 (1970)
177. C. Berthomieu and C. Sandorfy, *J. Mol. Spectrosc.*, **15,** 15 (1965)
178. G. Durocher and C. Sandorfy, *J. Mol. Spectrosc.*, **15,** 22 (1965)
179. A. Foldes and C. Sandorfy, *J. Mol. Spectrosc.*, **20,** 262 (1966)
180. M. Asselin, G. Bélanger and C. Sandorfy, *J. Mol. Spectrosc.*, **30,** 96 (1969)
181. L. J. Bellamy, Advances in Infrared Group Frequencies, (Methuen, London, (1968)), p. 244
181a. J. Laane, *J. Chem. Phys.*, **55,** 2514 (1971)
182. T. R. Singh and J. L. Wood, *J. Chem. Phys.*, **50,** 3572 (1969)
183. A. Witkowski (private communication)
184. R. G. Snyder and J. A. Ibers, *J. Chem. Phys.*, **36,** 1356 (1962)
185. R. G. Delaplane, J. A. Ibers, J. R. Ferraro and J. J. Rush, *J. Chem. Phys.*, **50,** 1920 (1969)
186. N. Riehl, B. Bullemer and H. Engelhardt, (Eds.). Physics of Ice, (Plenum Press, London and New York, 1969), p. 1867
187. S. F. Fischer, G. L. Hofacker and M. A. Ratner, *J. Chem. Phys.*, **52,** 1934 (1970)
188. J. M. Robertson and A. R. Ubbelohde, *Proc. Roy. Soc.*, **A170,** 222 (1939)
189. K. J. Gallagher (Ed., D. Hadži), Hydrogen Bonding, (Pergamon Press, London, 1959), p. 45
190. D. Williams and E. K. Plyler, *J. Chem. Phys.*, **4,** 154 (1936)
191. H. B. Friedrich and W. B. Person, *J. Chem. Phys.*, **44,** 2161 (1966)
192. G. J. Boobyer and W. J. Orville-Thomas, *Spectrochim. Acta*, **22,** 147 (1966)

193. C. M. Huggins and G. C. Pimentel, *J. Phys. Chem.*, **60,** 1615 (1956)
194. D. Cook, *Canad. J. Chem.*, **39,** 2009 (1961)
195. R. Foglizzo and A. Novak, *Spectrochim. Acta*, **26A,** 2281 (1970)
196. H. D. Bist, J. C. D. Brand and D. R. Williams, *J. Mol. Spectrosc.*, **24,** 402 (1967)
197. S. Singh and C. N. R. Rao, *J. Phys. Chem.*, **71,** 1074 (1967)
198. J. C. Evans, *Spectrochim. Acta*, **16,** 1382 (1960)
199. R. J. Jakobsen and J. W. Brasch, *Spectrochim. Acta*, **21,** 1753 (1965)
200. P. V. Huong, M. Couzi and J. Lascombe, *J. Chim. Phys.*, **64,** 1056 (1967)
201. R. A. Nyquist, *Spectrochim. Acta*, **19,** 1655 (1963)
202. R. F. Lake and H. W. Thompson, *Proc. Roy. Soc.*, **A291,** 469 (1966)
203. J. S. Koehler and D. M. Dennison, *Phys. Rev.*, **57,** 1006 (1940)
204. J. Michielsen-Effinger, *J. Mol. Spectrosc.*, **29,** 489 (1969)
205. L. G. Bonner and J. S. Kirby-Smith, *Phys. Rev.*, **57,** 1078 (1940)
206. R. C. Millikan and K. S. Pitzer, *J. Amer. Chem. Soc.*, **80,** 3515 (1958)
207. G. L. Carlson, R. E. Witkowski and W. G. Fately, *Spectrochim. Acta*, **22,** 1117 (1966)
208. S. G. W. Ginn and J. L. Wood, *Spectrochim. Acta*, **23A,** 611 (1967)
209. A. Hall and J. L. Wood, (to be published)
210. R. Foglizzo and A. Novak, *J. Chem. Phys.*, **50,** 5366 (1969)
211. R. Foglizzo and A. Novak, *J. Mol. Struct.*, **7,** 205 (1971)
212. S. Kishida and K. Nakamoto, *J. Chem. Phys.*, **41,** 1558 (1964)
213. K. Fukushima and B. J. Zwolinski, *J. Chem. Phys.*, **50,** 737 (1969)
214. H. Takahashi, K. Mamola and E. K. Plyler, *J. Mol. Spectrosc.*, **21,** 217 (1966)
215. E. Spinner, *J. Chem. Soc.*, 3860 (1963)
216. R. A. Fifer and J. Schiffer, *J. Chem. Phys.*, **52,** 2664 (1970)
217. R. M. MacQueen, J. A. Eddy and P. J. Léna, *Nature*, **220,** 1112 (1968)
218. H. A. Gebbie, W. J. Burroughs, J. Chamberlain, J. E. Harries and R. G. Jones, *Nature*, **221,** 143 (1969)
219. J. Arnold, J. E. Bertie and D. J. Millen, *Proc. Chem. Soc.*, **121,** (1961)
220. D. J. Millen and O. A. Samsonov, *J. Chem. Soc.*, 3085 (1965)
221. D. J. Millen and J. Zabicky, *J. Chem. Soc.*, 3080 (1965)
222. I. Rossi, A. Levy and C. Haeusler, *C.R. Acad. Sci.*, **264C,** 133 (1967)
223. M. Couzi and P. V. Huong, *C.R. Acad. Sci. Ser. B*, **270B,** 832 (1970)
224. H. E. Hallam, *J. Mol. Struct.*, **3,** 43 (1969)
225. M. Haurie and A. Novak, *J. Chim. Phys.*, **64,** 679 (1967)
226. M. A. Hussein, D. J. Millen and G. W. Mines, *J. Chem. Soc.*, **D,** 178 (1970)
227. D. Cummings, A. Hall and J. L. Wood, (to be published)
228. J. H. S. Green, *J. Chem. Soc.*, 2236 (1961)
229. E. Castellucci, G. Sbrana and F. D. Verderame, *J. Chem. Phys.*, **51,** 3762 (1969)
230. D. A. Long, F. S. Murfin and E. L. Thomas, *Trans. Faraday Soc.*, **59,** 12 (1963)
231. S. Bratož, *Advances in Quant. Chem.*, **3,** 209 (1967)
232. G. Bessis and S. Bratoz, *J. Chim. Phys.*, **57,** 769 (1960)
233. E. Clementi and A. D. McLean, *J. Chem. Phys.*, **36,** 745 (1962)
234. P. A. Kollman and L. C. Allen, *J. Amer. Chem. Soc.*, **92,** 6101 (1970)
235. J. Del Bene and J. A. Pople, *Chem. Phys. Lett.*, **4,** 426 (1969)
236. J. D. Lambert, *Disc. Faraday Soc.*, **15,** 226 (1953)
237. J. S. Rowlinson, *Trans. Faraday Soc.*, **47,** 120 (1951)
238. M. Dreyfus, B. Maigret and A. Pullman, *Theor. Chim. Acta*, **17,** 109 (1970)
239. M. Dreyfus and A. Pullman, *Theor. Chem. Acta*, **19,** 20 (1970)
240. F. H. Allen, D. Rogers and P. G. Troughton, *Acta Crystallogr.* (in press)
241. J. N. Murrell, *Chem. Brit.*, **5,** 107 (1969)

Chapter 5

Complexes of *n* and π Donors with Transition Metal Acceptors

D. A. Duddell

Chemistry Department, City of Leeds and Carnegie College, Beckett Park, Leeds, LS6 3QS

5.1. INTRODUCTION

5.1.1. GENERAL COMMENTS AND SCOPE OF CHAPTER

The primary objective of this chapter is to illustrate the many applications of vibrational spectroscopy to the study of transition metal complexes. At the same time, the results obtained from other investigations are discussed when they have particular bearing on the vibrational studies. It is hoped that none of the work discussed has been subsequently disproven, except when this is pointed out in the text, but in several cases interpretations are presented which can only be considered as tentative suggestions from which it is hoped a better understanding will emerge. An attempt has been made to emphasise the relation between vibrational spectra and the nature of the bonding. In particular, little mention is made of purely stereochemical investigations (this aspect being discussed to some extent in section 2.3.3). This is partly a matter of personal choice but also partly justified for two reasons. First, the complexes can usually be isolated in crystalline form and therefore their structures can be determined by X-ray crystallography. Secondly, the determination of the stereochemistry of a complex of the common type D_n–MX_m by vibrational spectroscopy is usually best approached by considering the bands arising from the MX stretching modes (or XY stretching modes in complexes of the type D_n–$M(XY)_m$) and is therefore a general method which does not require special discussion in the present context. The stereochemical investigations which are discussed are either those relating to non-rigid structures or those where the stereochemistry changes from solid to solution. The donors that have been chosen for discussion are considered to be a representative selection illustrating the different possible kinds of donor–acceptor interaction but the length at which they are discussed naturally reflects the volume of research published in each case. On the other hand, no attempt has been made to treat the transition metal acceptors systematically. The investigations so far reported have concentrated very heavily on the relatively inert complexes of metal ions with d^6, d^8 or d^{10} electronic configurations and this is reflected in the content of the chapter.

In section 5.1.2 the various types of possible donor–acceptor interaction are briefly summarised and this is followed in section 5.1.3 by a discussion of certain general aspects of the vibrational spectroscopic approach which, it is felt, are particularly important in the present context. The remainder of the chapter is divided into four parts. In the first two, the donors are considered individually and studies of their interaction with different metals are discussed. In the next, comparisons of different donors with particular acceptors are reported. Finally, certain specialised areas of research are considered.

5.1.2. DONOR-TRANSITION METAL INTERACTION

The simplest interaction that can occur is one of *n-v* type involving a lone pair on the donor and a vacant orbital in the coordination sphere of the transition metal ion or atom. However, in very few cases will this σ-bond be the only possibility because of the d orbitals of π and δ symmetry located on the metal which are not involved in the σ-bonding. As an illustration we can consider the formation of the complex D–ML_n with the D–M bond direction lying along the z axis. There will be two metal d*π*-orbitals, d_{xz} and d_{yz}, and two d*δ*-orbitals, d_{xy} and $d_{x^2-y^2}$. (In particular cases certain of these may, of course, be involved in *σ*-bonding.) The d*δ*-orbitals can probably be neglected but when the d*π*-orbitals are empty they may also contribute to the *v* acceptor properties of the metal in complexes with suitable donors. Although such multiple interaction is rarely claimed to be significant in the formation of molecular complexes it could contribute to the bonding to a metal in a high oxidation state when the high positive charge would stabilise all of the possible acceptor orbitals.

More commonly, complexes are formed between metals with occupied d*π*-orbitals and donors with vacant *π*-orbitals. There is then the possibility that the *σ*-bonding will be accompanied by a *π* interaction with the metal acting as the donor. The relative importance of the *σ*- and *π*-bonding will depend on the metal and its oxidation state. For a high oxidation state the high positive charge will simultaneously stabilise the *σ*-acceptor orbital and bind the d*π*-electrons strongly. At the other extreme, when the metal has a filled d-shell and is in a low oxidation state, for example as in $Ni(CO)_4$, the *π* interaction may well be dominant and it will no longer be really meaningful to talk of the metal as the "acceptor."

There are two other possible interactions that may be significant in complexes of *n* donors. The first is repulsion between lone pairs on the bound donor and on the metal. This may affect both the donor–metal bond strength and the conformation of the complex (see section 5.2.7). The second is hyperconjugation, *i.e.* a *π* donor–metal interaction between *σ*-bonding orbitals on a donor and vacant metal d*π* orbitals, which could occur, particularly when the donor atom were a first-row element. Wayland and Rice[1] suggested that spin delocalisation in hexammine complexes of Co(II) and Mn(II), as determined by proton contact shift studies, involved hyperconjugative overlap but they emphasised that there was no evidence that such interaction contributed to the donor–metal bond strength.

Donors that are capable of *bπ–v* complex formation can be divided into two categories. In the one, donation is from a localised *π*-bond, in the other, from a delocalised ring system. In almost every case the metal that can act as an acceptor to such donors will have filled d*π* orbitals so that there will be a possibility of "back bonding" into the *aπ* (*i.e.* *π**) donor orbitals to stabilise the complex. With cyclic C_nH_n type *bπ* donors there is the further possibility of "back donation" from filled d*δ* metal orbitals to be considered.

Although *π*-bonding has been discussed at some length it must be emphasised that in many cases its importance is still a matter of debate. For certain donors such as CO, C_2H_4 or PF_3, which do not form strong complexes with acceptors like BX_3 and H^+ but are known to stabilise low metal oxidation states, there is no doubt that the *π* interaction significantly contributes to the donor–metal bond strength. However, for donors such as PR_3, C_5H_5N(pyr) or RCN in complexes with metals in normal oxidation states the situation is unclear. It is true that the complexes of these donors with metals having filled dπ-orbitals are particularly stable but this may simply be because such metal ions are capable of forming particularly strong covalent σ-bonds. This problem is discussed further in section 5.2.6.

5.1.3. APPLICATION OF VIBRATIONAL SPECTROSCOPY

A correct assignment of a spectrum is essential if it is to yield reliable information. It may, therefore, be of value to review the methods used to confirm assignments in the light of recent developments. The first step in an assignment should be the determination of the symmetry of the vibrations giving rise to the spectral bands. For this, the most detailed information can be obtained from the infrared and Raman spectra of oriented single crystals. Although the interpretation of a crystal spectrum and the relationship between the crystal vibrations and molecular vibrations is not always straightforward this approach is being increasingly used and a few spectra from molecular complexes have been reported. (see, for example, section 5.2.5.) A technique which is free from these problems but retains the advantage of orientation is the use of an aligned nematic liquid crystal as a solvent. These solvents have found extensive use in n.m.r. but, so far, only limited application in vibrational spectroscopy[2]. The identification of totally symmetric vibrations by Raman polarisation studies in fluids is as widely used and as valuable as ever. The attribution of bands to specific modes of vibration has always been a subjective process for all but the simplest molecules. However, the study of isotopically substituted molecules gives valuable information which can lead to more definite assignments. The scope of such investigations has recently been extended by the availability of a wider range of isotopes, particularly of metals. (Examples are given in sections 5.2.3 and 5.2.6.)

The question as to how much information concerning structure and bonding can be reliably obtained from a vibrational spectrum is discussed at some length in other chapters of this book. There is, however, one particular aspect that needs emphasising here. More often than not the complexes will involve two-way bonding between certain pairs of atoms as discussed in the previous section but there will only be one bond stretching force constant describing their vibration. It is not known *a priori* what contribution the individual components of the bonding will make to the force constant. For the $\nu(C{\equiv}O)$ vibration in metal carbonyl complexes Kettle[3] has attempted to estimate the contributions of the σ- and π-bonding to the force constant by means of a simple calculation. Using Slater atomic orbitals for carbon and oxygen 2p orbitals he has calculated the σ- and π-overlap as a function of internuclear distance. The results are shown in figure 5.1 where the solid vertical line corresponds to a typical equilibrium bond length and

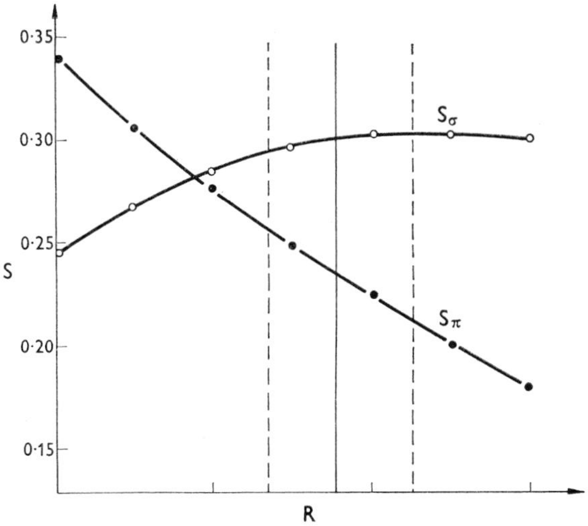

FIG 5.1 The overlap, S, as a function of internuclear distance, R, for 2p (σ and π) orbitals (Reproduced, by permission, from *Spectrochim. Acta*, **22A**, 1388 (1966)).

the two dotted lines to a reasonable vibrational amplitude. Although the calculation can only be considered as a first order approximation it does lend support to the theory that comparisons of CO stretching force constants, and therefore frequencies, will reflect changes in π-bonding. It is clear, therefore, that care has to be taken in comparing force constants or frequencies for two-way bonded interactions with data from experiments that relate to total bond strength.

5.2. COMPLEXES WITH n DONORS

5.2.1. CARBON MONOXIDE COMPLEXES

Of the many donors that form transition metal complexes, carbon monoxide is the most common and it would be impossible to describe more than a few of the spectroscopic investigations of these compounds that have been reported. Since the earlier literature has been reviewed in considerable detail by Haines and Stiddard[4] the present discussion will be restricted to the more recent developments. A discussion of infrared intensity studies on metal–carbon monoxide complexes is given in section 2.3.3.

Force Constant Analyses

Carbon monoxide is a sufficiently simple donor that complete harmonic force fields can be calculated for certain symmetric complexes. General quadratic valence force field treatments have been reported for $Ni(CO)_4$ by Bouquet and Bigorgne[5] and for $Ni(CO)_4$ and for $Cr(CO)_6$, $Mo(CO)_6$ and $W(CO)_6$ by Jones, McDowell and Goldblatt[6,7].

For the tetracarbonylnickel calculations, Jones and coworkers used data from the infrared spectra of gaseous $Ni(CO)_4$, $Ni(^{13}CO)_4$ and $Ni(C^{18}O)_4$, the infrared and Raman spectra of solutions in carbon tetrachloride and the Raman spectrum of $Ni(CO)_4$ vapour. Because the Raman data were incomplete many of the frequencies for the Raman active fundamentals were obtained from infrared combination bands. Harmonic frequencies were calculated for the $\nu(C{\equiv}O)$ vibrations but for all others the observed spectral frequencies were used because combination and overtone frequencies showed that the anharmonicity corrections would have been less than the equivalent of 2 cm^{-1}. There should have been sufficient data for a calculation of all of the force constants but in practice it was not possible to obtain a converging solution for the F_2 symmetry species without constraints for some of the off-diagonal elements. Bouquet and Bigorgne tackled the problem of the equations for the F_2 species by calculating certain force constants by extrapolation from values for $Ni(CO_3)PF_3$ and $Ni(CO)_2(PF_3)_2$ for which these authors had previously found linear correlations between frequencies and degree of substitution. They claimed that their results, which were in good agreement with those of Jones *et al.*,[7] were a little more accurate. The vibrational data used in the calculations and the values of the bond stretching force constants calculated for the gaseous molecule are given in table 5.1. Jones and coworkers[7] calculated a potential energy distribution for the F_2 species and surprisingly found very little mixing of the modes, in particular the 458.5 cm^{-1} band arose mainly from NiCO bending and the 423.3 cm^{-1} band from NiC stretching.

The calculations for the three hexacarbonyls were based on infrared spectra for gases and solutions of $M(CO)_6$, $M(^{13}CO)_6$ and $M(C^{18}O)_6$ and Raman spectra of solutions and solids. Again, frequencies for Raman active modes of the vapours were taken from combination bands in the infrared and CO stretching frequencies were corrected for anharmonicity. For the A_g, E_g, F_{1g}, F_{2g} and F_{2u} symmetry species the calculations were straightforward although the agreement between observed and calculated frequencies was not as good as expected. However, for the F_{1u} species there were, in effect, only ten frequencies with which to calculate ten force constants because the isotope shifts of the CMC deformation frequencies were insensitive to the force constant solution. As a result a converging solution could only be obtained if certain constraints were applied. Table 5.2 lists the vibrational data for the gas phase normal species and the calculated valence stretching force constants. The potential energy distribution showed, in contrast to the results for $Ni(CO)_4$, that there was considerable mixing of MCO and CMC bending in all of the low frequency modes and that for the f_{1u} vibrations this also involved MC stretching motion.

TABLE 5.1

Vibrational data (cm^{-1}) and bond stretching force constants ($N\,m^{-1}$) for gaseous $Ni(CO)_4$ and $Ni(C^{18}O)_4$

		Bouquet and Bigorgne[5]		Jones *et al.*[6]	
Symmetry	Mode	$Ni(CO)_4$	$Ni(C^{18}O)_4$	$Ni(CO)_4$	$Ni(C^{18}O)_4$
a_1	ν_1	2131	2084	2132.4	2085.5
	ω_1	2151.3	2103.3	2154.1	2106.3
	ν_2	367.5	354.4	370.8	359.9
e	ν_3	380	376	380	—
	ν_4	64	61	62	—
f_2	ν_5	2057.6	2010.6	2057.8	2010.4
	ω_5	2090.8	2042.2	2092.2	2043.5
	ν_6	458.8	454.2	458.9	453.8
	ν_7	421	417.4	423.1	416.6
	ν_8	80	76	79	76
f_1	ν_9	300	296	—	—
f_{CO}		1782 ± 5		1785 ± 9	
f_{NiC}		206 ± 5		208 ± 10	
$f_{CO,C'O'}$		10 ± 3		12 ± 4	
$f_{NiC,NiC'}$		8 ± 2		10 ± 4	
$f_{NiC,CO}$		51 ± 6		52 ± 10	
$f_{NiC,C'O'}$		-11 ± 4		-10 ± 4	

Because π-"back-bonding" involves CO antibonding orbitals the CO stretching force constants suggested stronger π-bonding for these complexes than for $Ni(CO)_4$ with the strongest in $W(CO)_6$. The differences in the MC stretching force constants were greater than those in the CO stretching force constants and indicated that the MC σ-bonding was significantly stronger in $W(CO)_6$ than in $Mo(CO)_6$ and $Cr(CO)_6$. The stretching-stretching interaction force constants between the CO groups were compared with those predicted by simple models. The π-bond interaction theory of Cotton and Kraihanzel[8] predicted $f^t_{CO,C'O'} \approx 2f^c_{CO,C'O'}$ whereas Haas and Sheline[9] had predicted that on the basis of interacting oscillating dipoles $f^c_{CO,C'O'}$ should be greater than $f^t_{CO,C'O'}$. The calculations showed that the interaction constants followed the oscillating dipole behaviour and that the apparent agreement between the π-bonding interaction theory and experiment previously reported had been due to the use of uncorrected anharmonic frequencies for the calculations.

TABLE 5.2

Vibrational data (cm^{-1}) and bond stretching force constants[7] ($N\ m^{-1}$) for gaseous $M(CO)_6$

Symmetry	Mode	$Cr(CO)_6$	$Mo(CO)_6$	$W(CO)_6$
a_{1g}	ν_1	2118.7	2120.7	2126.2
	ω_1	2139.2	2144.2	2153.2
	ν_2	379.2	391.2	426
e_g	ν_3	2026.7	2024.8	2021.1
	ω_3	2045.2	2043.3	2037.6
	ν_4	390.6	381	410
f_{1g}	ν_5	364.1	341.6	361.6
f_{1u}	ν_6	2000.4 ± 0.1	2003.0 ± 0.1	1997.6 ± 0.1
	ω_6	2043.7	2043.1	2037.6
	ν_7	668.1 ± 0.3	595.6 ± 0.4	586.6 ± 0.2
	ν_8	440.5 ± 0.5	367.2 ± 0.5	374.4 ± 0.2
	ν_9	97.2 ± 0.5	81.6 ± 0.5	82.0 ± 0.5
f_{2g}	ν_{10}	532.1	477.4	482.0
	ν_{11}	89.7	79.2	81.4
f_{2u}	ν_{12}	510.9	507.2	521.3
	ν_{13}	67.9	60	61.4
f_{CO}		17.24 ± 0.07	17.33 ± 0.06	17.22 ± 0.04
f_{MC}		2.08 ± 0.08	1.96 ± 0.06	2.36 ± 0.04
$f^c_{CO,C'O'}$		0.21 ± 0.03	0.22 ± 0.05	0.22 ± 0.02
$f^t_{CO,C'O'}$		0.02 ± 0.07	−0.06 ± 0.06	0.00 ± 0.04
$f^c_{MC,MC'}$		−0.019 ± 0.003	0.031 ± 0.009	0.049 ± 0.002
$f^t_{MC,MC'}$		0.44 ± 0.08	0.53 ± 0.06	0.56 ± 0.04
$f_{MC,CO}$		0.68 ± 0.07	0.73 ± 0.06	0.79 ± 0.04
$f^c_{MC,C'O'}$		−0.05 ± 0.03	−0.05 ± 0.04	−0.08 ± 0.02
$f^t_{MC,C'O'}$		−0.10 ± 0.07	−0.15 ± 0.06	−0.12 ± 0.04

Molecular Orbital Calculations

There have been several semi-empirical molecular orbital calculations of the bonding in carbonyl complexes but the results obtained have been conflicting. However, Hillier and Saunders[10] published the results of an *ab initio* all electron *SCF–MO* treatment of the electronic structures of $Ni(CO)_4$ and $Cr(CO)_6$ which were in agreement with experimental data from valence band[11] and X-ray photoelectron spectroscopy[12]. These calculations were amongst the largest ever reported and came near to the limits of the capabilities of present day computers. The computing time on ATLAS computer for the $Cr(CO)_6$ calculation was more than 100 hours. The results, as presented, bear little relationship to the chemist's intuitive ideas of bonding but they *do* show a decrease in 2s and an increase in 2pπ population for the carbon on complex formation which is consistent with the expected two-way bonding interaction. They also showed that for both molecules the π-bonding is stronger than the σ-bonding with the result that the metals bear a net positive charge. The authors pointed out that the calculated CO bond overlap populations were 1.06, 1.19 and 1.22 electrons for CO, $Ni(CO)_4$ and $Cr(CO)_6$ respectively—the major differences arising from σ-bonding changes—and did not correlate with the respective force constants of 1850, 1785 and 1724 N m^{-1}. However, it is interesting to note that if the CO pπ-pπ overlap alone is considered there is a correlation between its magnitude and the CO stretching force constants. For CO the pσ-pσ and pπ-pπ overlaps are 0.290 and 0.848 respectively compared to 0.290 and 0.811 for $Cr(CO)_6$. In $Ni(CO)_4$ the σ and π overlaps have the same symmetry and are not determined independently but if the pσ-pσ overlap were again 0.290 the pπ-pπ overlap would be 0.837 and the CO π-bonding order would be $Cr(CO)_6 < Ni(CO)_4 <$ CO which is the same as the order of the force constants. This lends support to the conclusion drawn from Kettle's simple calculation discussed in section 5.1.3.

Raman Intensities

There have been two recent studies of the intensities of the bands in the Raman spectra of simple carbonyl complexes. In both cases the validity of the interpretation has yet to be proven but together they illustrate how bonding information could be obtained from Raman intensity data.

Terzis and Spiro[13] measured the intensities of the a_{1g} modes of the group VI hexacarbonyls in carbon tetrachloride solution using the 459 cm^{-1} $(\nu_1)CCl_4$ band as an internal standard for conversion to absolute values. They calculated normal mode mean molecular polarisability derivatives $\alpha'_{Q_i} = \partial\alpha/\partial Q_i$, where α is the mean molecular polarisability, from the band intensities using the standard equation for the intensity I_i of a band corresponding to a totally symmetric normal mode Q_i at a Raman shift of $\Delta\bar{\nu}$ from the exciting line at $\bar{\nu}_0$; *i.e.*:

$$I_i = \frac{KM(\bar{\nu}_0 - \Delta\bar{\nu})}{\Delta\bar{\nu}[1 - \exp(-\mathbf{h}\Delta\bar{\nu}/\mathbf{k}T)]}\left[45\left(\frac{6}{6 - 7\rho}\right)\alpha'^2_{Q_i}\right] \qquad 5.1$$

where M is the molar concentration, ρ the depolarisation ratio and K is a correction factor for the instrumental response. In order to interpret the results a bond polarisability model was used, for which the assumption was made that the normal mode mean molecular polarisability derivative can be treated as a suitable sum of internal coordinate, or bond, mean molecular polarisability derivatives. The relationship between normal mode and bond polarisability derivatives is:

$$\alpha'_{Q_i} = \sum_j (N_j)^{1/2} L_{ji} \alpha'_{uj} \qquad 5.2$$

where $\alpha'_{uj} = \partial\alpha/\partial_{uj}$ is the bond polarisability derivative for internal coordinate u_j of which there are N_j in symmetry coordinate S_j. L_{ji} is the eigenvector element relating S_j to the normal coordinate Q_i. The eigenvectors were calculated from the force constant data discussed earlier in this section. Unfortunately α'_{Q_i} is only determined in magnitude from the intensities which, in this case, with two normal modes of a_{1g} symmetry, meant two sets of values for the bond polarisability derivatives.

Since the change in polarisability with internuclear separation of a purely ionic bond is zero it has always been considered that a bond polarisability derivative should be a measure of covalency. The results were, therefore, further interpreted using an equation proposed by Long and Plane[14], following a suggestion by Lippincott and Stutman[15], which relates bond polarisability derivative to bond order:

$$n/2 = \frac{3}{2} \frac{Za_0}{g\sigma'} \frac{\alpha'_u}{r^3} \qquad 5.3$$

where n is the number of electrons in the bond, a_0 the Bohr radius, r the bond length, σ' the Pauling covalent bond character, Z the effective nuclear charge and g the δ function "strength".

Equation 5.3 is derived from an empirical model which considers the bonding electrons between any two nuclei to be subject to a potential which is constant except at the nuclei where it becomes minus infinity as described by a δ function. The validity of this equation must be in some doubt considering its approximate derivation, particularly as it effectively treats the derived polarisability as a scalar not as a tensor quantity, so the numbers obtained from it must be treated with caution. However, for the MC stretching vibrations the indicated bond order W > Cr > Mo was consistent with the force constant data discussed above. The bond polarisability derivatives did not show this order because of their dependence on bond length as shown by the $1/r^3$ factor in the Long and Plane equation. The results for the CO stretching modes were unsatisfactory, the one set of bond orders varied from 1.5 to −0.6, the other from 4.2 to 4.4. The latter set was preferred in that the similarity of the values compared well with the similarity of the CO stretching force constants. However, the bond orders were very high even allowing for the approximations involved. Terzis and Spiro concluded that the bond polarisability approximation was breaking down because of the significant percentage of MC stretching involved in the CO stretching modes as shown by the eigenvector elements. They pointed out that this was not evident in the potential energy distribution because of the low force constant for MC stretching compared with that for CO stretching. However, it could lead to significant transfer of π-overlap between the CO and MC groups during a "$\nu(C\equiv O)$" vibration and thus to unexpectedly high α'_{CO} values. The results of these calculations are given in table 5.3.

Kettle, Paul and Stamper[16] considered the interpretation of Raman intensities from a different viewpoint. Their investigations were prompted

TABLE 5.3

Raman intensity data[13] and calculated parameters[a] for $M(CO)_6$

	$Cr(CO)_6$	$Mo(CO)_6$	$W(CO)_6$
a_{1g}, 'MC stretching mode.'			
$\bar{\nu}/cm^{-1}$	381	402	427
Molar intensity relative to ν_1 band in CCl_4	3.06	3.76	5.04
$10^{30} \, \alpha'_Q/C \, m \, V^{-1}$ a.m.u.$^{-1/2}$	1.17	1.35	1.72
Contribution of S_{CO} to the Q_{MC} eigenvector	0.013	0.014	0.016
Contribution of S_{MC} to the Q_{MC} eigenvector	0.181	0.181	0.180
$10^{30} \, \alpha'_u/C \, m \, V^{-1}$			
Set 1	2.40	2.82	3.62
Set 2	2.66	3.04	3.80
Bond Order			
Set 1	1.62	1.27	1.80
Set 2	1.82	1.37	1.89
a_{1g}, 'CO stretching mode.'			
$\bar{\nu}/cm^{-1}$			
Molar intensity relative to ν_1 band in CCl_4	0.77	0.53	0.22
$10^{30} \, \alpha'_Q/C \, m \, V^{-1}$ a.m.u.$^{-1/2}$	1.94	1.60	1.03
Contribution of S_{CO} to the Q_{CO} eigenvector	0.382	0.382	0.381
Contribution of S_{MC} to the Q_{CO} eigenvector	-0.224	-0.224	-0.226
$10^{30} \, \alpha'_u/C \, m \, V^{-1}$			
Set 1	3.49	3.39	3.24
Set 2	-0.50	0.07	1.16
Bond Order			
Set 1	4.36	4.58	4.38
Set 2	-0.63	0.09	1.57

a. The polarisability derivatives α' are given in terms of polarisability per unit length (*i.e.* in $C \, m^2 \, V^{-1}/m$ or $C \, m \, V^{-1}$; $1 \, Å^2 \equiv 1.115 \times 10^{-30} \, C \, m \, V^{-1}$).

by the observation that, contrary to expectations, the totally symmetric CO stretching modes gave very weak Raman bands compared with those arising from non-totally symmetric vibrations. They analysed this phenomenon in terms of a bond polarisability model considering just the CO stretching vibrations, arguing that, although this approach had not in the past yielded transferable parameters, it could be applicable to the interpretation of relative intensities for a particular molecule. It was assumed that the bond polarisability tensor derivative α'_{CO} had a diagonal form when related to the bond axes, *i.e.*:

$$\alpha'_{CO} = \begin{bmatrix} a & 0 & 0 \\ 0 & b & 0 \\ 0 & 0 & c \end{bmatrix} \qquad 5.4$$

where b is the longitudinal tensor element†. The change in molecular polarisability with respect to normal coordinate displacement $Q_\beta(t)$, the tth normal coordinate of symmetry species Γ_β, is then given by:

$$\frac{\partial \alpha_m}{\partial Q_\beta}(t) = \sum_i C_{i\beta}(t) \frac{\partial \alpha_m}{\partial r_i} \qquad 5.5$$

where $C_{i\beta}(t)$ is symmetry determined, r_i is the ith internal coordinate and

$$\frac{\partial \alpha_m}{\partial r_i} = T_i(\alpha'_{CO})_i T_i^t \qquad 5.6$$

where T is the transformation matrix relating bond and molecular axes and T^t is its transpose.

For the octahedral hexacarbonyls the CO bond is cylindrically symmetrical thus a = c in equation 5.4. The following expression can then be derived for the relative intensities of the e_g and a_{1g} CO stretching modes:

$$\frac{I(e_g)}{I(a_{1g})}(\perp) = \frac{7k_{e_g}(b - a)^2 \nu_{a_{1g}}}{5k_{a_{1g}}(b + 2a)^2 \nu_{e_g}} \qquad 5.7$$

where $\nu_{a_{1g}}$ and ν_{e_g} are the frequencies of the a_{1g} and e_g CO stretching vibrations and k_{e_g} and $k_{a_{1g}}$ are defined by the scattering equations for polarised excitation. To a good approximation, $\nu_{a_{1g}} = \nu_{e_g}$ and $k_{a_{1g}} = k_{e_g}$ and so there are three possible situations. When it is found experimentally that $I(a_{1g}) > 2.86\ I(e_g)$, b and a must have the same sign; when it is found that $I(a_{1g}) < 0.71\ I(e_g)$, b and a must have different signs, and when the ratio of the intensities has a value between these two limits the situation is indeterminate. Kettle *et al.*[16] quoted relative intensities from their own studies and from previous work, the results

† It is more usual, however, to use the bond as the Z axis.

TABLE 5.4
Relative intensities of the a_{1g} and e_g CO stretching bands in the Raman spectra of $M(CO)_6$ species

Complex	Solvent	$I(a_{1g})/I(e_g)$	Reference
$Re(CO)_6^+$	CH_3CN	0.44	17
$Cr(CO)_6$	CH_2Cl_2	0.14	16
$Mo(CO)_6$	CH_2Cl_2	0.14	16
$W(CO)_6$	CH_2Cl_2	0.16	16
$V(CO)_6^-$	CH_3CN	0.08	17

are given in table 5.4. They concluded that the changes in polarisability along and perpendicular to the CO axis during the vibration were of opposite sign and noted that the intensity ratio varied with the formal oxidation state of the metal. Their interpretation was that the Co σ- and π-electron densities responded differently to bond length changes and this caused the unusual carbonyl behaviour.

The expressions for the intensities of the bands tricarbonyl complexes (with C_{3v} symmetry) are more complicated in that there is no longer cylindrical CO symmetry. Since the totally symmetric band is now not expected to have a depolarisation ratio of zero there are two equations to consider:

$$\frac{I(e)}{I(a_1)}(\perp) \approx \frac{[3.5(a^2 + c^2) + 4.7(b^2 - ab - bc) - 2.3ac]}{[4.5(a^2 + c^2) + 6.7(ab + bc) + 4.3ac + 3.3b^2]} \quad 5.8$$

$$\frac{I(e)}{I(a_1)}(\|) \approx \frac{[3.0(a^2 + c^2) + 4.0(b^2 - ab - bc) - 2.0ac]}{(a - c)^2} \quad 5.9$$

(these assume a CMC angle of 90° which is close to that observed in structural studies). Experimental results were obtained for π-$C_6H_6Cr(CO)_3$ and π-$C_5H_5Mn(CO)_3$ and are given in table 5.5. The results for the manganese compound clearly indicated that a $\approx$ c, in other words, the CO bond was cylindrically symmetrical whereas for the chromium

TABLE 5.5
Relative intensities of the a_1 and e CO stretching bands in the Raman spectra of tricarbonyl complexes[16]

Complex	Solvent	$I(e)/I(a_1)(\perp)$	$I(e)/I(a_1)(\|)$
π-$C_6H_6Cr(CO)_3$	CH_2Cl_2	8.3	11.4
π-$C_5H_5Mn(CO)_3$	CH_2Cl_2	3.7	67.0
π-$C_5H_5Mn(CO)_3$	C_6H_6	3.7	∞

complex there was considerable asymmetry, the calculated ratios for the tensor elements being b/a = 9.1, b/c = −1.28 or *vice-versa*. The conclusion was that the difference between the behaviour of the two complexes had to be electronic in origin because there was little difference in the geometries of the tricarbonyl groups.

Solid State Studies

Most of the spectra of carbonyl complexes have been obtained for solutions. However, in some cases it is not possible to obtain solution spectra, particularly a Raman spectrum, because the complex is insoluble, is unstable in solution or because of isomerisation. On the other hand the solid can usually be studied without difficulty. For this reason alone it is valuable to have guidelines to the interpretation of solid state spectra. Hopefully, such studies may also provide further information which can be related to the bonding and structure of the complexes. Certain vibrations of a molecular crystal lattice, the 'internal modes', can be related to the vibrations of the free molecule and the modification of the vibrations from molecule to crystal can be considered in three stages. Firstly, the local molecular symmetry may be lower in the crystal; secondly, there may be interaction between the vibrations of symmetry related molecules in a primitive unit cell; and finally, there may be interaction between the vibrations of the unit cells. Chemists usually analyse the spectra on the basis of an approximate model. In the 'site group' approach the vibrational couplings between molecules in a unit cell and between unit cells are neglected and the spectrum is analysed in terms of the molecular site symmetry. The 'factor group' or 'unit cell group' method allows for vibrational coupling within a primitive unit cell but disregards the interaction between unit cells. Since the wavelength of the light used in vibrational spectroscopy is very great compared with a normal unit cell dimension the only crystal vibrations that are detected spectroscopically are those in which primitive cells vibrate essentially in phase and so the factor group approach should be a good qualitative approximation. There are cases where it may break down but this problem will not be considered here.

Kettle and coworkers[18,19] studied the spectra of several crystalline carbonyl complexes. The first was benzenetricarbonylchromium[18] which crystallises with two molecules per primitive unit cell in a lattice of symmetry $P2_1/m(C_{2h}^2)$. A 'site group' analysis predicts, on the basis of C_s symmetry, three CO stretching bands which would be both infrared and Raman active. The 'factor group' prediction is three infrared and three Raman bands but at non-coincident frequencies because of the centre of symmetry which relates the two molecules in a unit cell. The infrared spectrum showed absorptions at 1966, 1879 and 1858 cm^{-1}, the Raman spectrum had bands at 1945, 1887 and 1865 cm^{-1}. It was clear that vibrational coupling in the unit cell was significant and so a 'factor group' approach was needed. However, the molecular identity was clearly still evident in the spectra, in that the high frequency band and the lower frequency doublet in each spectrum could be related to the solution a_1 and e vibrational bands respectively.

In a later paper Kettle and coworkers[19] outlined a method of qualitatively correlating molecule and crystal spectra on the basis that the molecular identity of the bands was retained in the latter. Their treatment was based on the relationships between crystal and molecular dipole moment and polarisability derivatives. The equation for the dipole moment derivatives $\vec{\mu}'$ is,

$$\vec{\mu}'_{cryst} = N \sum_{m} T_m \cdot \vec{\mu}'_m \qquad 5.10$$

where N is a normalising term, T_m is the transformation matrix relating crystal and molecular axes for the mth molecule in the unit cell and the summation is a symmetry-allowed linear combination of the molecular dipole derivatives $\vec{\mu}'_m$. The corresponding expression for the crystal polarisability tensor derivatives is:

$$\alpha'_{cryst} = N \sum_{m} T_m \cdot \alpha'_m \cdot T^t_m \qquad 5.11$$

They tested the validity of this approach by considering the spectra of *cis*-diethylenetriaminetricarbonyl-chromium, -molybdenum and -tungsten. These crystallise with four molecules per unit cell in a lattice of $P2_12_12_1(D_2^4)$ space group symmetry. The correlation table for the factor group prediction is given in table 5.6 and it indicates that twelve Raman

TABLE 5.6
Correlation between molecular, site and unit cell group symmetry for the carbonyl groups of *cis*-dienM(CO)$_3$

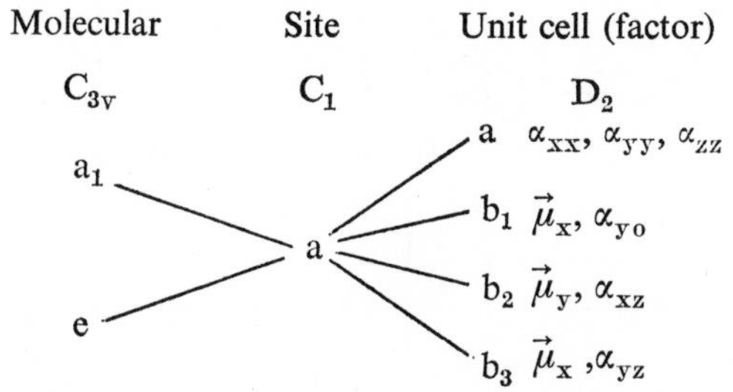
Molecular
Site
Unit cell (factor)
C_{3v}
C_1
D_2
a_1
e
a
a $\alpha_{xx}, \alpha_{yy}, \alpha_{zz}$
b_1 $\vec{\mu}_x, \alpha_{yo}$
b_2 $\vec{\mu}_y, \alpha_{xz}$
b_3 $\vec{\mu}_x, \alpha_{yz}$

active CO stretching vibrations are expected with nine of them also infrared active. However, this is modified in the treatment outlined above. An examination of the crystal structure shows that to a very good approximation the transformation matrices T in this particular instance have the forms,

$$\begin{bmatrix} 0 & \frac{1}{\sqrt{2}} & \frac{1}{\sqrt{2}} \\ 0 & -\frac{1}{\sqrt{2}} & \frac{1}{\sqrt{2}} \\ 1 & 0 & 0 \end{bmatrix}, \quad \begin{bmatrix} 0 & -\frac{1}{\sqrt{2}} & -\frac{1}{\sqrt{2}} \\ 0 & -\frac{1}{\sqrt{2}} & -\frac{1}{\sqrt{2}} \\ -1 & 0 & 0 \end{bmatrix},$$

$$\begin{bmatrix} 0 & -\frac{1}{\sqrt{2}} & -\frac{1}{\sqrt{2}} \\ 0 & \frac{1}{\sqrt{2}} & -\frac{1}{\sqrt{2}} \\ 1 & 0 & 0 \end{bmatrix} \quad \text{and} \quad \begin{bmatrix} 0 & \frac{1}{\sqrt{2}} & \frac{1}{\sqrt{2}} \\ 0 & \frac{1}{\sqrt{2}} & -\frac{1}{\sqrt{2}} \\ -1 & 0 & 0 \end{bmatrix}$$

This is because instead of the expected general orientation the local molecular $(CO)_3$ groups are so aligned that their C_3 axes lie in the crystal xy plane making angles of very nearly 45° with the axes. The molecular dipole derivatives are,

$$\vec{\mu}'_m\,(a_1) = \begin{bmatrix} 0 \\ 0 \\ \vec{\mu}'_z \end{bmatrix}, \quad \vec{\mu}'_m(e_x) = \begin{bmatrix} \vec{\mu}'_x \\ 0 \\ 0 \end{bmatrix} \quad \text{and} \quad \vec{\mu}'_m(e_y) = \begin{bmatrix} 0 \\ \vec{\mu}'_y \\ 0 \end{bmatrix} \qquad 5.12$$

but, since local C_{3v} symmetry is very nearly preserved and the molecular y and z axes are practically equivalent in the crystal, then $\vec{\mu}'_x = \vec{\mu}'_y = \vec{\mu}'_z = \vec{\mu}'$ to a good approximation. The crystal dipole derivatives as given by 5.10 are, therefore:

$$\vec{\mu}'_{cryst}(a_1) = \begin{bmatrix} 0 \\ \sqrt{2}\,\vec{\mu}' \\ 0 \end{bmatrix} \quad \text{and} \quad \begin{bmatrix} \sqrt{2}\,\vec{\mu}' \\ 0 \\ 0 \end{bmatrix}$$

$$\vec{\mu}'_{cryst}(e_x) = \begin{bmatrix} 0 \\ 0 \\ \vec{\mu}2' \end{bmatrix} \qquad 5.13$$

$$\vec{\mu}'_{cryst}(e_y) = \begin{bmatrix} 0 \\ \sqrt{2}\,\vec{\mu}' \\ 0 \end{bmatrix} \quad \text{and} \quad \begin{bmatrix} \sqrt{2}\,\vec{\mu}' \\ 0 \\ 0 \end{bmatrix}$$

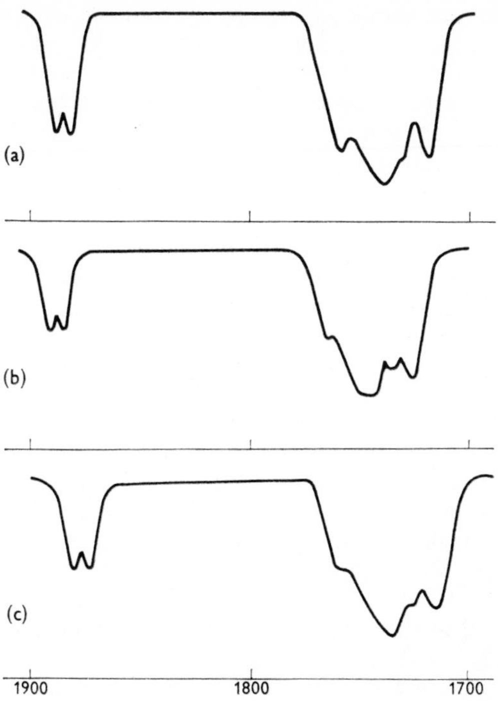

FIG 5.2 Infrared spectra of crystalline (a) *cis*-dien $Cr(CO)_3$, (b) *cis*-dien $Mo(CO)_3$ and (c) *cis*-dien $W(CO)_3$. (Reproduced from *J. Chem. Soc.*, **A**, 3148 (1971), by permission of the Chemical Society.)

There are two important results of this analysis. The first is that because of the molecular orientation in the crystal, the b_3 vibration originating from the molecular a_1 mode, the e_x derived b_2 and b_3 modes and the e_y derived b_1 crystal vibration are all predicted to have zero intensity so that five not nine infrared bands are expected. The second is that the two molecular a_1 derived vibrational bands should have equal intensity whereas the region where e vibrational bands are expected should show two equally intense bands and one of double this intensity. The spectra that were obtained are illustrated in figure 5.2; they show a striking resemblance to the expected pattern the only difference being the presence of an additional weak feature in the 1700 cm^{-1} region. These authors also carried through the corresponding treatment for the Raman spectra and again found good agreement with the experimental results.

5.2.2. AMMONIA COMPLEXES

If the possibility of hyperconjugative interaction is disregarded ammonia is the simplest example of a purely *n*-donor. Studies of ammine complexes should, therefore, give direct information about the effect of σ-bonding on a donor and an acceptor and of the strength of such interaction. Unfortunately, in many of the complexes so far investigated there has been considerable hydrogen bonding involving the ammonia protons and this has made the interpretation of the spectra of the complexes less straightforward than might have been expected.

Assignments

The characteristic frequencies for the internal donor vibrations and for the skeletal modes of ammines are now established beyond doubt. However, at the time of investigation the problem was a subject of some controversy. The papers discussed here have stood the test of time and together illustrate how such a problem can be tackled.

Mizushima and coworkers[20] established the ranges for the bands due to the deformation and rocking modes. They calculated the frequencies using a Urey–Bradley force-field for a simple $H_3N–M$ model system. The values of the force constants which were used were H(HNH) 54, H(HNM) 18, F(HH) 6 and F(HM) 10 N m^{-1} and the band positions predicted for the degenerate deformation, symmetric deformation and rocking modes were, respectively, 1579, 1340 and 854 cm^{-1} for $Co(NH_3)_6^{3+}$ and 1156, 1021 and 628 cm^{-1} for $Co(ND_3)_6^{3+}$. These correspond very well to the regions of absorption observed in the infrared spectra. In a later publication, Mizushima *et al.*[21] pointed out that this model also predicted which bands would be metal sensitive. The degenerate deformation, involving only H(HNH) and F(HH), should be metal independent whereas the rocking mode depended on H(HNM) and F(HM) and should be strongly metal sensitive and the symmetric deformation, involving all four constants, should show some metal dependence. They measured the band positions for these vibrations in a series of complexes with different metals and confirmed the predictions.

Powell and Sheppard[22] studied the infrared spectra of a series of ammines and deuteroammines and confirmed Mizushima's assignments of the ammonia vibrations. They also observed weak bands around 500 cm^{-1} in some of the spectra which were only slightly isotope dependent and could be assigned to the metal–nitrogen stretching vibration. In support of this conclusion they quoted the observation by Mathieu[23] of a polarised line at 538 cm^{-1} in the Raman spectrum of an aqueous solution of $[Pt(NH_3)_4]Cl_2$ pointing out that there would be no totally symmetric skeletal deformation mode for such a complex.

Force Constant Studies

Although ammonia is a simple donor a full harmonic force-field determination for, say, a hexammine complex would be a formidable undertaking even if a set of vibrational data relating to the complex in the same state and free from hydrogen-bonding effects were available. In these circumstances it is not surprising that the calculations so far attempted have involved approximate force-fields. The most extensive series was that reported by Hiraishi, Nakagawa and Shimanouchi[24]. They used data from the infrared spectra of solid chloride salts of ammine complex cations for calculations using a modified Urey-Bradley force-field. Because there is considerable hydrogen bonding in the solid chlorides the results cannot be used to give a direct comparison of free and complexed ammonia. The calculated values of the F matrix diagonal elements for the MN stretching, NH_3 symmetric deformation and NH_3 rocking coordinates are given in table 5.7. The authors

TABLE 5.7
Force constants ($N\,m^{-1}$) calculated for ammine complexes from infrared spectra using a modified Urey-Bradley force-field

	F_{dia} (MN stretch)	F_{dia} (NH_3 symmetric deformation)	F_{dia} (NH_3 rock)
$Pt(NH_3)_6^{4+}$	239	49	56
$Hg(NH_3)_2^{2+}$	230	44	33
$Pt(NH_3)_4^{2+}$	222	45	44
$Pd(NH_3)_4^{2+}$	197	43	35
$Co(NH_3)_6^{3+}$	175	44	39
$Cr(NH_3)_6^{3+}$	155	42	33
$Cu(NH_3)_4^{2+}$	124	40	30
$Ni(NH_3)_6^{2+}$	86	37	27
$Co(NH_3)_6^{2+}$	82	35	24

claimed that the order of the metal–nitrogen stretching force constants was probably the order of covalency in the complexes. The variation in the deformation and rocking force constants clearly supports the ideas discussed earlier in this section.

Stereochemical Studies

Because the bands arising from the metal–nitrogen stretching vibrations occur in an otherwise transparent region of the spectrum they can be confidently correlated with the stereochemistry of complexes when they have a reasonable intensity. Durig and Mitchell[25] took advantage of this in studying the kinetics of the *cis-trans* isomerisation of $(NH_3)_2$-PdX_2 in the solid state. As the isomerisations take five days and two days (to reach completion) for the chloride and bromide respectively, infrared spectroscopy is ideally suited for following the reaction. The bands due to the ν(PdN) vibrations are observed at 496 cm^{-1} (antisymmetric) and 476 cm^{-1} (symmetric) for the *cis* chloride and at 496 cm^{-1} for the *trans* chloride. For the corresponding bromides these bands are at 480, 460 and 490 cm^{-1} respectively. The band intensity for the Pd N symmetric stretching mode of the *cis* complex was therefore used to follow the reaction in each case. One problem encountered was that the reaction was catalysed by impurity and so the samples had to be prepared with care. Both sets of experimental data gave a reasonable fit to a first-order rate plot; the bromide "fit" was not as good as that of the chloride but this was attributed to the faster reaction time in this case.

Vibronic Spectra

The determination of the frequencies of fundamental vibrations which are both infrared and Raman inactive is always something of a problem. One solution, the interpretation of overtone and combination

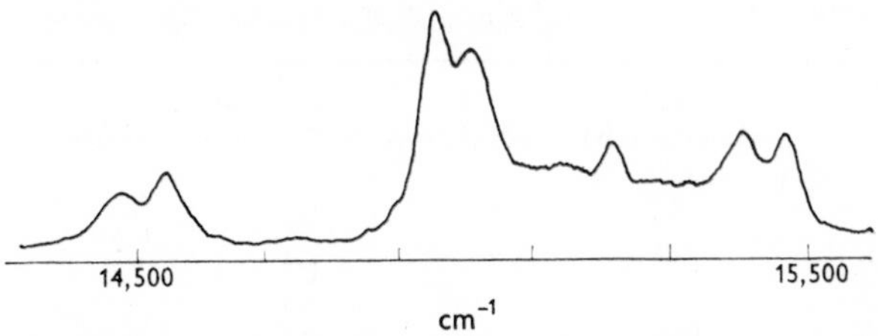

FIG 5.3 He–Ne (632.8 nm) excited emission spectrum of crystalline $Cr(NH_3)_6(ClO_4)_3$. (Reproduced from *J. Amer. Chem. Soc.*, **93**, 634 (1971), by permission of the American Chemical Society.)

bands, was discussed in the carbonyl section (section 5.2.1) but is a possibility for any complex. A different approach which has found application in the study of certain ammine complexes is the analysis of vibronic spectra. For an octahedral d^3 complex the transition between the ground, $^4A_{2g}$, state and the lowest doublet state, ^2E_g, is both spin[26] and Laporte[27] forbidden but becomes allowed through spin-orbit coupling of the $^4T_{2g}$ and E_g states and vibronic mixing of T_{2u} or T_{1u} vibrational and $^4T_{2g}$ electronic states. There have been several studies of this transition in $Cr(NH_3)_6^{3+}$ salts and these have in certain cases shown some vibrational structure but recently Long and Penrose[27] studied the transition in a well-resolved resonance-phosphorescence spectrum using He-Ne laser excitation and made reliable assignments for the vibrational levels. The observed spectrum (figure 5.3) showed the emission bands and Raman lines superimposed but they had no assignment problems because the emission spectrum had a symmetric Stokes-Anti Stokes pattern around the 0–0 band position and showed a marked anion dependence in the spectra of a series of different salts of the complex cation. The band positions could only be measured to ± 10 cm^{-1} but there was good agreement with the infrared spectra for the t_{1u} vibrations so the frequencies obtained for the $(t_{2u})NH_3$ and the $(t_{2u})CrN_6$ skeletal deformation can be confidently accepted. The assignment for the spectrum shown in figure 5.3 is given in table 5.8.

The spectrum of the corresponding deuteroammine complex ion in solution was also obtained and the isotopic shift of the band positions was in agreement with the proposed assignments.

Internal Rotation

When the vibrational spectrum of an ammine complex is analysed on the basis of a high point group symmetry it is assumed that there is a low barrier to internal rotation of the ammonia molecules. This

TABLE 5.8
Assignment of the emission spectrum of crystalline $Cr(NH_3)_6(ClO_4)_3$. The $[^4A_{2g}\text{-}^2E_g]$ system[28]

$\bar{\nu}$/cm^{-1}	$\bar{\nu} - \bar{\nu}_{00}$/cm^{-1}	Assignment
15,962	739	(0-0) + $(t_{1u})NH_3$ rock*
15,894	671	(0-0) + $(t_{2u})NH_3$ rock*
15,478	255	(0-0) + $(t_{1u})CrN_6$ def.*
15,414	191	(0-0) + $(t_{2u})CrN_6$ def.*
15,289	66	(0-0) + $(t_{2g})CrN_6$ def.* – $(t_{2u})CrN_6$ def.*
15,223	0	(0-0)
15,153	70	(0-0) – $(t_{2g})CrN_6$ def. + $(t_{2u})CrN_6$ def.
15,016	207	(0-0) – $(t_{2u})CrN_6$ def.
14,963	260	(0-0) – $(t_{1u})CrN_6$ def.
14,757	466	(0-0) – $(t_{1u})CrN_6$ stretch
14,555	668	(0-0) – $(t_{2u})NH_3$ rock
14,483	740	(0-0) – $(t_{1u})NH_3$ rock

* excited-state fundamental.

has been established in certain cases where crystallographic studies are consistent only with high site symmetries for the complexes[29]. However, in favourable circumstances it is possible to obtain quantitative estimates of the barriers to internal rotation using vibrational spectroscopy. For a symmetric rotor like ammonia internal rotation can only be detected by its effect on the perpendicular vibrations, *i.e.* the degenerate $\nu_a(NH_3)$, $\delta_{as}(NH_3)$ and $r(NH_3)$ modes. In the absence of internal rotation the infrared bands from these vibrations have normal vibrational contours and little temperature dependence. For molecules where there is partially hindered rotation the sharp vibrational contour may be superimposed on rotational 'wings' and the band shape will show a marked temperature dependence. This situation can be analysed by a method proposed by Bulanin and Orlova[30] where it is assumed that the intensity of the central peak is due to molecules whose energies are too low for the ammonia to rotate and that of the wings is due to molecules with energies of rotation greater than the barrier height.

Assuming a Boltzmann energy distribution the intensities are given by the expression:

$$\frac{I_{wings}}{I_{total}} = \frac{\int_{K^*}^{\infty} \exp(-K^2\mathbf{h}^2/8\pi^2 I_B \mathbf{k}T)\, dK}{\int_0^{\infty} \exp(-K^2\mathbf{h}^2/8\pi^2 I_B \mathbf{k}T)\, dK} \qquad 5.14$$

where K^* is the quantum number for the rotational level which is at the energy of the barrier and I_B is the moment of inertia. The barrier height U is then calculated from the usual quantum-mechanical expression for the energy of a rotor at a high K value (with an additional factor of two, appropriate to a sinusoidal barrier):

$$U = 2B\mathbf{h}K^{*2} \qquad 5.15$$

where B is the rotational constant $= \mathbf{h}/8\pi^2 I_B$.

In other cases the internal rotation does not give rise to rotational wings as such but simply to a temperature dependent broadening of the vibrational bands. In such circumstances the broadening can be analysed using a theoretical treatment due to Rakov.[31] He argued that there were three mechanisms for broadening: dissipation of molecular vibrational energy as thermal energy (γ_1), interaction of molecular vibrations with the molecular environment (γ_2), and rotational re-orientation which would involve crossing the rotational barrier (γ_3). Of these, only the last would depend on temperature as given by the expression:

$$\gamma_3 = \text{constant.}\exp(-U/\mathbf{k}T) \qquad 5.16$$

thus a plot of log γ_3 against 1/T should yield a straight line of slope U. (For further discussion of band broadening mechanisms and band shape analysis see section 2.3.6.)

Leech, Powell and Sheppard[32] investigated the infrared band contours of the perpendicular vibrations of ammonia in the *trans*-complexes $(NH_3)_2PdCl_2$ and $(NH_3)_2PdI_2$ in the solid state. From the temperature dependence of the spectra, the three perpendicular bands were easily identified and, in particular, that due to the degenerate deformation which showed the rotational wings and the central peaks clearly differentiated. They analysed the intensities to give the following barrier heights; $(NH_3)_2PdCl_2$, 430 cm^{-1} at 470 K and $(NH_3)_2PdI_2$, 210, 225 and 210 cm^{-1} at 90, 298 and 470 K respectively. Janik *et al.*[33] studied the corresponding vibrations in solid $[Co(NH_3)_6]I_2$ and $[Ni(NH_3)_6]I_2$. In these spectra the bands did not have rotational wings so they applied the Rakov method. They measured the broadening of the bands due to the $\delta_{as}(NH_3)$ and $r(NH_3)$ modes several times in each case and found average values for the barrier height of 225 and 209 cm^{-1} for $[Co(NH_3)_6]I_2$ and $[Ni(NH_3)_6]I_2$ respectively. No great accuracy was claimed for either sets of results but, where comparisons are possible, they appear to be reasonable estimates and therefore the theoretical approach to the interpretation of the spectra seems justified.

5.2.3. NITRILE COMPLEXES

A nitrile can act as both an *n* donor and an *aπ* acceptor using orbitals of the same type as are involved in a carbonyl complex. Studies of complexes of nitriles with simple *v* acceptors have shown that the *ν*(CN) frequency in the complex is always higher than that in the free nitrile and this has been attributed to a stronger CN *σ*-bond in the complex. (This point is discussed in more detail in section 2.3.3 and also in section 6.3.3). There is, therefore, a possibility of obtaining unequivocal evidence for *π* "back-bonding" in nitrile complexes from the *ν*(CN) frequency since such interaction involving CN *aπ*-orbitals will necessarily weaken the CN bond. However, a low *ν*(CN) stretching frequency will, of course, also be observed if a nitrile acts as a *π* donor so it is important that the structure is established before the spectra are so interpreted.

Ford and coworkers[34,35] studied the infrared spectra of several cationic complexes of the type $RCN–M(NH_3)_5$. In particular they measured the differences in the *ν*(CN) frequencies of the complexes and of the free donors. Because there was uncertainty as to the effect of a change of anion or of medium on the *ν*(CN) frequency they investigated a wide range of salts under different conditions. The spectra showed that in every case the formation of a complex with $Ru(NH_3)_5^{2+}$ resulted in a *ν*(CN) frequency lower than that in the free nitrile whereas the corresponding complexes with $Ru(NH_3)_5^{3+}$ and $Rh(NH_3)_5^{3+}$ always showed higher *ν*(CN) frequencies. Some typical results are given in table 5.9. The authors claimed that the spectra clearly indicated that *π* "back bonding" from the $Ru(NH_3)_5^{2+}$ cation was exceptionally strong

TABLE 5.9
Positions (cm^{-1}) of the *ν*(CN) bands in free and complexed organonitriles[34,35]

Compound (cation)	Anion	$\tilde{\nu}_{max}(CN)/cm^{-1}$	Reference
CH_3CN		2254	34
$[CH_3CN–Ru(NH_3)_5]^{2+}$	$2[BF_4]^-$	2239	34
$[CH_3CN–Ru(NH_3)_5]^{3+}$	$3[ClO_4]^-$	2286	34
$[CH_3CN–Rh(NH_3)_5]^{3+}$	$3[ClO_4]^-$	2323	35
C_6H_5CN		2231	34
$[C_6H_5CN–Ru(NH_3)_5]^{2+}$	$2[ClO_4]^-$	2188	34
$[C_6H_5CN–Ru(NH_3)_5]^{3+}$	$3[ClO_4]^-$	2267	34
$[C_6H_5CN–Rh(NH_3)_5]^{3+}$	$3[ClO_4]^-$	2287	35

and argued that this was an important factor in the stability of the closely related N_2 complexes.

5.2.4. PYRIDINE COMPLEXES

Although pyridines, like nitriles, are potential n donors and π acceptors they are different in two important respects. The first is that the position of the π-orbitals on pyridine is determined by the plane of the ring whereas there is no such restriction in the axially symmetric CN group (except possibly for a donor such as C_6H_5CN where there is conjugation of the CN π-orbitals with the aromatic ring π-orbitals). Consequently, the π "back bonding" properties of pyridine will be dependent on steric effects. The second is that the energies of the pyridine σ- and π-orbitals, which can be considered to be mutually independent, can be perturbed in well-understood ways by substitution of the pyridine ring.

Brewer and coworkers[36,37] undertook an extensive study of the bonding of substituted pyridines to first row divalent metal ions. They first measured the heats of formation of the complexes with cupric ion, in aqueous solution, of pyridine, 4-methylpyridine, 4-cyanopyridine, 4-acetylpyridine, 4-carbinylpyridine, 4-pyridine-carboxamide and 4-carbomethoxypyridine[36] and found that the changes followed the same order as did those of the acid dissociation constants of the corresponding pyridinium ions. The results are given in table 5.10. They correlated the two sets of values in terms of two quantities σ_H and δ defined as:

$$\sigma_H \equiv \log K - \log K^\circ$$

and

$$\delta \equiv \Delta H_c - \Delta H_c^\ominus \qquad 5.17$$

where K, K°, ΔH_c and $\Delta H_c^\ominus$ are the acid dissociation constants of the pyridine ions and the heats of coordination of the cupric complexes for substituted and unsubstituted pyridines respectively. A plot of δ against σ_H was approximately linear with the greatest deviation for 4-cyanopyridine. Since there can only be σ-bonding between a pyridine and a proton the nearly linear relationship indicated that this was the important interaction in the copper complexes. In a subsequent investigation[37] they measured the infrared frequencies of the bands assigned to vibrations which could be approximately described as ν(CuN) modes in the complexes $2(Pyr)_2$–$CuCl_2$ with the same set of pyridines. They also calculated CuN stretching force constants using a modified Urey-Bradley force field treatment of an approximate 1:1 complex model treating the substituent as a single mass. The positions of these bands and the force constants, which are also given in table 5.10, followed the same trend as did the heats of formation.

In order to determine the relative magnitudes of possible π "back-bonding" interactions in such complexes it is essential to measure some quantity which is independent of the σ-bond strength. Wong and

TABLE 5.10

Heats of formation of pyridine complexes with Cu^{2+}, association constants of pyridinium ions, spectral data and force constants for Cu^{2+} complexes[36,37]

R in R– 	$\Delta H_c{}^a$/ kJ mol^{-1}	$\Delta H_a{}^b$/ kJ mol^{-1}	$K_A{}^c$/ dm^3 mol^{-1}	$\bar{\nu}^d$/ cm^{-1}	K(CuN)^e N m^{-1}
CH_3	67.4	19.7	1.06×10^6	285	186
CH_2OH	56.5	18.5	2.55×10^5	276	174
H	54.8	18.3	1.84×10^5	268	158
$COCH_3$	29.3		3.13×10^3	265	150
$CONH_2$	28.4	12.9	2.66×10^3	264	149
$COOCH_3$	27.6	12.7	1.76×10^3	264	149
CN	13.8	10.4	80.0	243	105

a. ΔH_c is the heat of reaction for n pyr + $Cu^{2+} \rightarrow Cu(pyr)_n^{2+}$.
b. ΔH_a is the average heat of coordination per mole of donor complexed to the metal.
c. K_A is the equilibrium constant for the reaction $pyr_{aq} + H^+_{aq} \rightleftharpoons pyr\,H^+_{aq}$.
d. $\bar{\nu}$ refers to the band assigned to the "metal–nitrogen stretching" vibration.
e. K(CuN) is the Urey-Bradley force constant relating to the CuN coordinate.

Brewer[36] measured the frequencies of the ν_{12} pyridine vibration. They claimed that this had been shown in previous studies to be a ring vibration which was independent of the mass of the substituent and did not involve any substituent-ring or hydrogen-ring bending. The simple force constant analysis mentioned above indicated that the vibration involved mainly the CN stretching motion. The frequencies for the free pyridines could apparently be directly correlated with the electron density in the ring and the virtually unchanged values for the corresponding pyridinium ions was taken to be evidence that this was little affected by coordination via the nitrogen lone pair. However, for the metal complexes the bands were shifted to higher frequency, particularly so in the complexes of the pyridines substituted by electron withdrawing groups, with the result that the trend in the frequencies for the different pyridines when complexes was the reverse of that for the free molecules. This was interpreted as being due to "back-donation" from the metal increasing the electron density near the nitrogen atom. The band positions are given in table 5.11. It must be emphasised here that the actual form of the normal vibrations in pyridine is open to discussion—Yarwood[38] has discussed these problems recently—and so there must be some doubt as to the validity of these conclusions. One aspect of their interpretation which is particularly open to question is that the *π* "back-bonding" is said to strengthen the CN bond in pyridine whereas this would be expected to be a sacrificial interaction

TABLE 5.11

$\bar{\nu}_{12}$ vibrational data for substituted pyridines and their complexes with Cu^{2+}, Zn^{2+} and H^+

R in R— 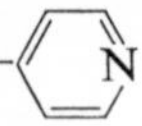 N	$(pyr)_2$–$CuCl_2$ $\bar{\nu}_{max}/cm^{-1}$	$(pyr)_2$–$ZnCl_2$ $\bar{\nu}_{max}/cm^{-1}$	pyr H^+ $\bar{\nu}_{max}/cm^{-1}$	pyridine $\bar{\nu}_{max}/cm^{-1}$
CH_3	1042	1033	1033	1040
CH_2OH	1049	1035	1037	1038
H	1043	1043	1028	1029
$COCH_3$	1055	1058	1018	1020
$CONH_3$	1055	1057	1023	1025
$COOCH_3$	1056	1055	1023	1024
CN	1063	1063	1003	1006

involving the pyridine $a\pi$-orbitals. The same methods were used for a comparison of the bonding of 4-methylpyridine in the Mn(II), Co(II), Ni(II), Cu(II) and Zn(II) complexes of the type $(pyr)_2$–MCl_2[39]. The frequencies of the "metal-nitrogen stretching" modes and the corresponding force constants indicated that the bond strengths decreased in the order Cu(II) > Ni(II) > Zn(II) > Co(II) > Mn(II) which is the expected Irving-Williams order[40] based on ligand field stabilisation effects. The bands due to the ν_{12} internal modes showed little change in the series of complexes and little change from that in the free molecule which was taken to be evidence that the π "back-bonding" interaction was weak as expected.

Since strong π "back-bonding" to a pyridine donor can only occur if the plane of the pyridine molecule is favourably oriented with respect to the metal π-orbitals there is particular interest in complexes of bidentate donors such as 2,2′-bipyridine and 1,10-phenanthroline where stereochemical rigidity is guaranteed. Unfortunately the spectra of complexes of these donors are even more complicated, particularly in the low frequency region, so it is not at all easy to identify the bands due to the "metal–nitrogen stretching" vibrations. However, two recent studies have shown that information *can* be obtained from the vibrational spectra.

Hutchinson, Takemoto and Nakamoto[41] investigated the effect of metal isotope substitution on the low frequency infrared spectra of salts containing the complex cations $Fe(bipyr)_3^{2+}$, $Ni(bipyr)_3^{2+}$, $Ni(phen)_3^{2+}$, $Zn(bipyr)_3^{2+}$ and $Zn(phen)_3^{2+}$. The isotopes used were ^{58}Ni, ^{62}Ni, ^{54}Fe, ^{57}Fe, ^{64}Zn and ^{68}Zn and were all of at least 90% purity. The complexes containing these pure isotopes were prepared on a milligram scale. Since the complex ions can only have D_3 symmetry, three ν(MN) modes are expected to be infrared active. In some of the spectra only two bands showed isotope shifts; nevertheless the frequency ranges for "metal-nitrogen stretching" in the different complexes were established beyond doubt. These indicated that the order of bond strength in these complexes was Fe(II) ≫ Ni(II) > Zn(II). The $\bar{\nu}$(MN) data are given in table 5.12.

Strukl and Walter[42] investigated the infrared spectra of a series of complexes of formula 2,2′-bipyridyl–MX_2 containing Mn, Fe, Co, Zn, Pd and Pt and with X = Cl and Br. The high frequency bands were assigned by comparison with the spectra of the free donor and hydrogen-dependent modes identified by a comparison of the spectra of the normal and perdeutero palladium complexes. In the region below

TABLE 5.12
Vibrational data (cm^{-1}) for the ν(MN) bands in 2,2′-bipyridyl and 1,10-phenanthroline complexes of isotopically pure metal ions[41]

$[Fe(bipyr)_3][ClO_3]_2$	^{54}Fe	386.0	376.2	
	^{57}Fe	380.0	371.0	
$[Ni(bipyr)_3][ClO_4]_2$	^{58}Ni	293.1	282.9	267.3
	^{62}Ni	286.0	277.5	260.0
$[Ni(phen)_3][ClO_4]_2$	^{58}Ni	301.5	260.2	246.8
	^{62}Ni	299.2	256.3	244.0
$[Zn(phen)_3][ClO_4]_2$	^{64}Zn	203.5	175.0	
	^{68}Zn	199.5	172.0	

400 cm^{-1} it was possible to obtain reasonable assignments for some of the bands on the basis of the following arguments. The ν(MX) bands could be identified by their strong dependence on halogen substitution. The ν(MN) bands were evident from their insensitivity to deuteration, their slight dependence on halogen substitution and their variation with change of metal. Certain internal bipyridyl vibrations could be assigned as their frequencies did not change throughout the series of complexes studied. In order to obtain some information about the relative bond strengths, Urey–Bradley force constants were calculated for the complexes. The spectral data for the donor-metal dependent bands and the related force constants are given in table 5.13. The authors pointed out that the data showed that the complexes with Pt, Pd and Fe were much stronger than those with Mn, Co and Zn, the difference being due to strong π "back-bonding" in the former. It is interesting to note that the values of the nitrogen–carbon stretching force constant indicate that the "back-bonding" results in a weakening of the NC bond. This is the opposite to the trend observed by Brewer, as mentioned above, and would be consistent with electron transfer into antibonding ring orbitals.

TABLE 5.13

Spectral data (cm^{-1}) for the donor–metal dependent bands and related force constants ($N\ m^{-1}$) for 2,2′-bipyridyl–MCl_2 complexes[42]

Mode	Pt	Pd	Fe	Co	Zn	Mn
a_1, $\nu(MN) + \nu(MCl)$ + chelate ring stretch	385	409	423	—	369	357
b_1, $\nu(MN)$	295	280	318	270	243	239
a_1, $\delta(MN_2) + \delta(MCl_2)$ + chelate ring bend	201	206	219	194	183	182
K(MN)	297	264	219	62	54	37

5.2.5. NITRIC OXIDE COMPLEXES

The simplest possible donor-acceptor interaction in nitrosyl complex formation would involve *n*-donation from the lone pair of electrons on the less electronegative atom nitrogen. However, the resulting complex would then retain the paramagnetism arising from the unpaired electron in the NO $a\pi$-orbital. Since the evidence from magnetic studies is that such paramagnetism does not occur it is necessary to consider other forms of interaction. Disregarding those cases where there is a drastic modification of the nitric oxide in the complex it is convenient to consider the bonding in terms of NO^+ or NO^- and a metal in the particular oxidation state that this would imply. It must be emphasised that this is a formalism; in particular a complex of "NO^+" could in reality contain a net negative charge on the nitric oxide if there were sufficiently strong π "back-bonding" from the metal. Unfortunately, in some of the earlier vibrational work this was ignored and many attempts were made simply to deduce directly from the $\nu(NO)$ frequencies in the complexes whether they should be described as involving NO^+, NO or NO^-. The important difference between the two types is that NO^+ is isoelectronic with CO and would be expected to form complexes with a linear MNO arrangement whereas NO^- is isoelectronic with O_2 and would be expected to coordinate with a bent MNO structure, the non-linearity arising from a Jahn-Teller[43] distortion relieving the degeneracy of the $(\pi)^2$ electronic configuration of NO^-. Alternatively, since NO^- is isoelectronic with O_2, it might be anticipated that it would likewise act as a π donor.

Crystallographic Studies

Since gross structural differences are expected between NO^+ and NO^- complexes it is extremely valuable to have crystallographic data in cases where there could be some doubt as to the nature of the bonding. Ibers and coworkers[44,45] determined the structures of a number of complexes and found that they could be clearly classified.

A typical example was $IrCl_2(NO)(P(C_6H_5)_3)_2$[44] which was found to have an essentially square pyramidal coordination of the iridium with an IrNO angle of 123°. Since five-coordinate d^6 metal complexes usually have such a stereochemistry there was no doubt that it could be described as an NO^- complex of iridium(III). On the other hand, the closely related complex cation $[IrH(NO)(P(C_6H_5)_3)_2]^+$ [45], in which the hydrogen was not located, was essentially trigonal bipyramidal with an approximately linear IrNO arrangement. Again, the metal stereochemistry and the IrNO angle were consistent and indicated a formulation as an NO^+ complex of iridium(I). In discussing the bonding in these complexes they recalled the argument presented by Griffith[46] that although O_2 and NO^- were isoelectronic they bonded differently because their orbital energies did not have the same order. For O_2 the π-orbitals were of higher energy than the σ_{2p} orbital whereas in NO^- the order was reversed and therefore a π-complex would not be expected. They also produced the following arguments, based on a simple qualitative bonding picture, to explain why two such apparently similar complexes should have such different structures. In the absence of π "back-bonding" there are two possibilities; either the NO $a\pi$-orbitals are of higher energy than the metal $d_{x^2-y^2}$ and d_{z^2} orbitals, which is to be expected for a high metal oxidation state, and they remain empty giving an NO^+ complex or they are of lower energy in which case they are occupied and the result is an NO^- complex. However, in the latter situation π "back-bonding" from the other metal d-orbitals may raise the energy of the NO $a\pi$-orbitals to above that of the metal $d_{x^2-y^2}$ and d_{z^2} orbitals so that an NO^+ complex is again favoured. They concluded that in general NO^+ complexes could be expected for high and low oxidation states and NO^- complexes for intermediate states. In the particular cases of the two iridium complexes studied the electron density on the iridium atom in $[IrH(NO)(P(C_6H_5)_3)_2]^+$ would be relatively high, due to the low electronegativity of the ligands, and so π "back-bonding" to NO would be strong and consequently the energies of the NO $a\pi$-orbitals would be raised above those of the metal d-orbitals. For $IrCl_2(NO)(P(C_6H_5)_3)_2$ the electron density at the metal would be lower, the NO $a\pi$-orbitals would remain at lower energy than the highest energy metal d-orbitals and so an NO^- complex would be formed.

Structural Isomerism

Mercer, McAllister and Durig[47] provided a good illustration as to how careful examination of vibrational spectra can lead to correct structural interpretations. They investigated the infrared spectra of the two series of isomeric salts of empirical formula $[Co(NO)(NH_3)_5]X_2$ whose structures had been subject to controversy for many years. At the time of the study, crystal structure determinations[48] of the black isomer of the chloride salt had shown that it was monomeric, with structure I in figure 5.4, and conductivity measurements[49] had indicated that the red isomers were dimeric. In the spectrum of the black chloride they identified the ν(NO) band at 1610 cm^{-1} and the ν(CoN) band at 644 cm^{-1} (since they shifted on substitution by ^{15}NO). The NH_3 vibrational bands were identified from their sensitivity to deuteration. An absorption at 1172 cm^{-1}, previously assigned to the ν(NO) vibration, was shown to arise from an impurity because its intensity varied with extent of recrystallisation. They pointed out that the 630 cm^{-1} feature varied in the same way and suggested that the impurity was hexammine cobalt(II) chloride which they were unable to completely remove and was likely to have been the cause of variable magnetic moments which had been obtained in different investigations[50].

They studied the acid decomposition of the red salts and found that the only gaseous product was nitrous oxide. This suggested a hyponitrite structure. The infrared spectrum of the nitrate salt of this series was unchanged above 1200 cm^{-1} by ^{15}NO substitution but the bands at 1136, 1046 and 932 cm^{-1} were all shifted. The shift for the 1136 cm^{-1} band was too high for it to be due to a ν(NO) mode so it was assigned to a ν(NN) vibration. By comparison with studies of the $N_2O_2^=$ ion, the 1046 and 932 cm^{-1} bands were attributed to the antisymmetric and symmetric ν(NO) modes. However, they emphasised that these descriptions were highly idealised and in reality the vibrations were considerably mixed. Lower frequency absorptions were tentatively assigned by comparison with the spectra of the $N_2O_2^=$ ion and of the deuterated compound. In considering the possible stereochemistry of the dimer they argued that a *cis*-hyponitrite configuration would be consistent with the infrared activity of the ν(NN) mode but would not be expected because of steric interactions. Alternatively, a *trans* structure, which would be expected intuitively, would not be expected to show a strong infrared ν(NN) band. However, a structure with one cobalt bonded to oxygen and the other bonded to nitrogen, as shown in figure 5.4, II, although of an uncommon type, fitted the spectra best. A subsequent X-ray diffraction study proved that this was indeed the structure[51].

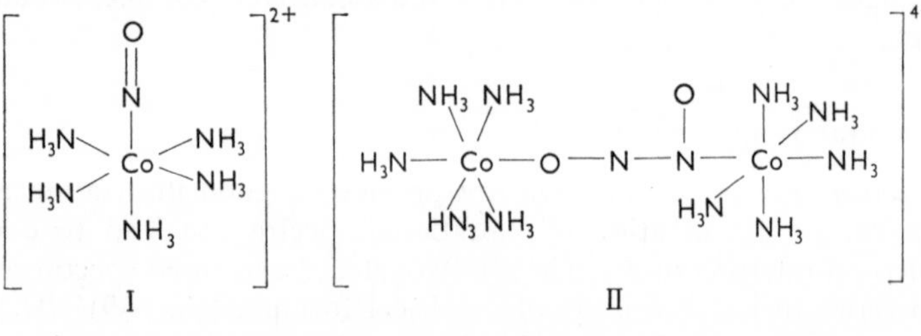

FIG 5.4 Structures of the black (I) and red (II) cations in $[Co(NO)(NH_3)_5]X_2$ salts.

Single Crystal Spectroscopy

The investigation discussed above showed how the ν(MN) and the δ(MNO) bands could be differentiated by ^{15}N isotopic substitution. A different method was used by Sabatini[52] to assign the bands in the infrared spectrum of $Na_2[Fe(NO)(CN)_5].2H_2O$. This crystallises with space group symmetry Pnnm (D_{2h}^{15}) and site symmetry C_s. The orientation in the crystal is such that the z axes of the approximately C_{4v} molecules and the z axis of the crystal are mutually perpendicular so that the molecular ν(MN) vibration gives rise to infrared active crystal vibrations with x and y polarisations only. However, the δ(MNO) mode is responsible for crystal vibrations polarised along x, y and z. Sabatini prepared single crystals of the salt which were then ground and polished to thicknesses of about 0.03 mm with sections perpendicular to either the crystal x or y axes. Three spectra were obtained; the first with z polarisation and x propagation, the second with y polarisation and x propagation and the third with x polarisation and y propagation. A band at 662 cm^{-1} was only seen in the last two. The latter was, therefore, assigned unambiguously to the ν(MN) mode.

5.2.6. PHOSPHINE COMPLEXES

In discussing the donor properties of phosphines it is inevitable that they would be compared with those of the corresponding amines. On the basis of thermodynamic data, Ahrland, Chatt and Davies[53] proposed that acceptors could be classified as being of either (a) or (b) type depending on whether they formed more stable complexes with donors where the lone pair was located on a first row or second row atom respectively. In the transition metal series the class (a) acceptors are found in the early groups and the class (b) acceptors in the later groups. There is also a tendency for class (b) character in any one group to increase with atomic number. Ahrland, Chatt and Davies considered that class (b) character was exhibited by those metals that could π "back-bond" into the empty d-orbitals on a second row donor atom. In a related classification, Pearson[54] proposed that acceptors and donors could be considered as either "hard" or "soft" acids and bases. A soft base has easily polarised valence electrons but a hard base does not. A hard acid has a small acceptor atom or ion, high positive charge and no polarisable valence electrons whereas a soft acid has a large acceptor atom, low charge or several valence electrons which are easily polarisable or removable. Hard acids prefer hard bases and soft acids prefer soft bases. A phosphine is a soft base and prefers a soft acid which, as far as the transition metals are concerned, corresponds to a class (b) acceptor. Although this classification also allows the difference between a phosphine and an amine donor to be attributed to the empty dπ-orbitals of the former, it does allow for the possibility that stronger σ-bonding arising from the higher polarisability of the phosphine lone pair contributes to the high stability of certain phosphine complexes.

Complexes with Acceptors in Low Oxidation States

Although phosphine complexes are known with metals throughout the transition series in a wide variety of oxidation states there are two areas where there have been particularly thorough investigations. One of these comprises the complexes of metals of the later groups in low oxidation states, particularly of Ni(0).

Edwards and Woodward[55] obtained the infrared spectra of gaseous $Ni(PF_3)_4$, $Pd(PF_3)_4$ and $Pt(PF_3)_4$ and the Raman spectra of the liquids. The Raman spectrum of $Pd(PF_3)_4$ was very complicated and could only be tentatively interpreted but those of the nickel and platinum complexes showed just three polarised bands which could be assigned to the ν(PF), ν(MP) and $\delta(PF_3)$ modes of a_1 symmetry by comparison with the spectrum of free PF_3. With the spectral ranges for these vibrations thus established they were able to assign the infrared bands with confidence. They calculated simple valence force constants using the a_1 vibrational frequencies and found that the values of the metal–phosphorus stretching force constants were in the range normally associated with single bonds. They pointed out that this was somewhat surprising since the π "back-bonding" would be expected to be significant in these complexes but that a similar situation was encountered in $Ni(CO)_4$. Another surprising result was that the metal–phosphorus stretching force constant increased on going from Ni to Pd to Pt whereas the nickel complex had the highest thermal stability and the palladium complex was particularly unstable. Some of the spectral and force constant data are given in table 5.14.

Benazeth, Loutellier and Bigorgne[56] investigated the effect of ionic charge on the bonding in trifluorophosphine complexes by comparing the infrared and Raman spectra of the compounds $HCo(PF_3)_4$ and $K[Co(PF_3)_4]$ with those of $Ni(PF_3)_4$. They found that the increase in frequency of the ν(MP) band and the decrease of that of the ν(PF) band from $Ni(PF_3)_4$ to the isoelectronic $Co(PF_3)_4^-$ were clearly consistent with increased π "back-bonding" from metal to phosphorus in the latter leading to a decrease in PF π-bonding. (In this particular case the metal to phosphorus π-bonding could be considered to be partly sacrificial in nature.) However, the percentage change in the ν(MP) stretching frequency was considerably more than that in the ν(PF) stretching frequency. The results for $HCo(PF_3)_4$ were anomalous in that this compound had the highest ν(PF) stretching frequency of the three whereas its ν(CoP) stretching band fell at an intermediate position. However, as the authors pointed out, a straightforward comparison would not be expected because of its different structure. The vibrational data for the ν(MP) and ν(PF) stretching modes are given in table 5.15.

The other phosphorus donor for which unambiguous assignments of metal–phosphorus stretching vibrations can be made is phosphine itself. Unfortunately very few complexes with this donor are known but

TABLE 5.14
Spectral (cm^{-1}) and force constant ($N\ m^{-1}$) data for $M(PF_3)_4$ complexes[55]

Symmetry	Mode	Description	$Ni(PF_3)_4$ Raman (liquid)	$Ni(PF_3)_4$ I.R. (gas)	$Pd(PF_3)_4$ Raman (liquid)	$Pd(PF_3)_4$ I.R. (gas)	$Pt(PF_3)_4$ Raman (liquid)	$Pt(PF_3)_4$ I.R. (gas)
a_1	ν_1	PF str.	954		958		958	
	ν_2	PF_3 def.	534		545		546	
	ν_3	MP str.	195		204		213	
a_2	ν_4	PF_3 tors.						
e	ν_5	PF str.	851		852		855	
	ν_6	PF_3 def.	332		327		334	
	ν_7	PF_3 rock					283	
	ν_8	MP_4 def.	54				47	
f_1	ν_9	PF str.						
	ν_{10}	PF_3 def.						
	ν_{11}	PF_3 rock						
	ν_{12}	PF_3 tors.						
f_2	ν_{13}	PF str.	883	898	914	904	905	903
	ν_{14}	PF str.	883	860	880	865	867	867
	ν_{15}	PF_3 def.	505	508	495 or 518	508	514	513
	ν_{16}	PF_3 def.	385	390	390	389	382	384
	ν_{17}	PF_3 rock		287	284		283	281
	ν_{18}	MP str.	219	217	222	223	219	220
	ν_{19}	MP_4 def.	54	52			47	53
		$k_{PF\ str.}$	773		773		766	
		$k_{PF_3\ def.}$	53		53		48	
		$k_{MP\ str.}$	271		317		382	

the vibrational spectra of a few have been studied by Bigorgne *et al.*[57] Again, the ν(MP) bands could be identified by polarisation studies on the Raman spectra of solutions of the complexes. These were located (in cm^{-1}) as follows, $Ni(CO)_3PH_3$, 295; $Fe(CO)_4PH_3$, 334; $Mo(CO)_5$-PH_3, 276; *cis*-$Mn(CO)_4PH_3I$, 307; *cis*-$Mn(CO)_3(PH_3)_2I$, 302; $Fe(CO)_3$-PH_3I_2, 320 and $Fe(CO)_2(PH_3)_2I_2$, 293.

Loutellier and Bigorgne[58] also undertook an extensive study of the set of the complexes $Ni(CO)_{4-n}(PR_3)_n$ (with $PR_3 = PF_3$, $P(CH_3)_3$ or $P(OCH_3)_3$). They obtained spectra for the complete series with trifluorophosphine and were able to assign them on the basis of the

TABLE 5.15
Vibrational data (cm^{-1}) for ν(MP) and ν(PF) stretching modes in complexes of trifluorophosphine[56]

	$Ni(PF_3)_4$	$Co(PF_3)_4^-$	$HCo(PF_3)_4$
ν(PF)	892	820	897
	850	800	856
ν(MP)	218	245	230 or 238

characteristic frequencies established in the spectra of $Ni(PF_3)_4$ and $Ni(CO)_4$. Only the mono- and di-substituted complexes could be prepared in the trimethylphosphine series. For this donor the identification of the PC_3 group symmetric deformation bands in the spectra of the complexes was not immediately obvious from a comparison with the spectrum of the free donor. For $Ni(CO)_3P(CH_3)_3$ either of the polarised bands observed at 350 and 221 cm^{-1} could conceivably be related to the 305 cm^{-1} band established for trimethylphosphine itself. (The corresponding band positions for $Ni(CO)_3PF_3$ were 510 and 262 as compared with 487 cm^{-1} in PF_3.) However, they presented reasons for expecting an increase in frequency for the deformation when the donor was complexed and therefore the higher band was assigned to this mode. Their arguments were based on an assumed geometry for the complex and the hypothesis that the force constant for the CPC internal coordinate would be the same in the complex and in the free donor. Nevertheless, the assignment was consistent with the behaviour of the corresponding vibration in PF_3 and certainly the ν(NiP) stretching band would not be expected above the 295 cm^{-1} observed for $Ni(CO_3)$-PH_3. The full series of complexes was again studied for trimethylphosphite. The vibrational spectrum of this donor is far more complicated than the others and therefore, although they presented full assignments for the spectra, they can only be considered as tentative.

In order to interpret these spectra in terms of the differences in the donor properties of the various phosphines they calculated force constants for all of the complexes treating the donor as PR_3 in each case. In the later publication[57] they reported a similar calculation for $Ni(CO_3)PH_3$. A simple valence force-field treatment was used except that interaction constants for the pairs of internal coordinates (r(NiC), $Ni\hat{C}O$) (r(NiP), $R\hat{P}R$) were included. The nickel–phosphorus stretching force constants showed a strong dependence on the group R, the values being 124, 132, 167 and 192 N m^{-1} for $Ni(CO)_3P(CH_3)_3$, $Ni(CO)_3PH_3$, $Ni(CO)_3P(OCH_3)_3$ and $Ni(CO)_3PF_3$ respectively. The simplest interpretation of these force constants would be that they reflected an increasing degree of π "back-bonding" from the $P(CH_3)_3$ to the PF_3 complex. However, the frequencies of the ν(CO) and ν(MC) bands for the four complexes indicated that the donor and acceptor properties of PH_3 were virtually the same as those of $P(OCH_3)_3$ in agreement with the results of an earlier investigation by Barlow and Holywell[59]. They concluded that when there was a possibility of internal π-bonding in PR_3; for instance, when R was OCH_3 or F, the nickel–phosphorus force constant would no longer depend solely on the electrons constituting the nickel–phosphorus bond but could reflect changes in the PR π-bonding during a vibration. It is possible, however, that the anomalous NiP stretching force constants for the $P(OCH_3)_3$ complexes result from the approximation of treating OCH_3 as a rigid group.

Complexes of Ni(II), Pd(II), Pt(II) and Au(I)

A wide variety of simple phosphine complexes of these acceptors can be obtained, consequently they present an excellent opportunity for systematic study. Since neither PH_3 or PF_3 complexes are available the simplest donor that can be investigated is $P(CH_3)_3$. Goggin *et al.*[60] obtained the infrared and Raman spectra of as complete as possible a series of trimethylphosphine complexes with palladium(II), platinum(II) and gold(I) halides. As before, the identification of the metal–phosphorus stretching bands and the symmetric phosphine deformation bands was not immediately obvious. For example, polarised bands were observed at 383, 326 and 210 cm^{-1} in the Raman spectrum of a solution of $(CH_3)_3PAuCl$. The 326 cm^{-1} feature could be attributed to the ν(AuCl) mode as it was not present in the spectra of the corresponding bromide and iodide. One of the reasons for assigning the 383 cm^{-1} band to the metal–phosphorus stretching mode was that in the complex $(CH_3)_2SAuCl$ the corresponding bands were found at 345 and 279 cm^{-1} [61]. All of the available evidence shows that phosphines are stronger donors than sulphides with these acceptors so the assignment of the 210 cm^{-1} band to the ν(MP) vibration would have been inconsistent with this. Another reason for preferring the assignment of the higher frequency bands to metal-phosphorus stretching mode was that they showed exactly the behaviour expected of skeletal modes when the stereochemistry and composition of a series of complexes were changed. This was particularly evident in the trimethylphosphine platinum(II) halide series[62] for which the spectral data are given in table 5.16.

TABLE 5.16

$\bar{\nu}$(PtP) data (in cm^{-1}) for trimethylphosphine–platinum(II) halide complexes[62]

Complex	Skeletal symmetry			
$(P(CH_3)_3)_4Pt^{2+}$	D_{4h}	393 (a_{1g})	357 (e_u)	327 (b_{2g})
$(P(CH_3)_3)_3PtCl^+$	C_{2v}	403 (a_1)	370 (a_1)	365 (b_1)
$(P(CH_3)_3)_2PtCl_2$(*cis*)	C_{2v}	403 (a_1)	378 (b_1)	
$(P(CH_3)_3)_2PtCl_2$(*trans*)	D_{2h}	371 (a_g)	346 (b_{2u})	
$P(CH_3)_3PtCl_3^-$	C_{2v}	390 (a_1)		
$(P(CH_3)_3)_3PtBr^+$	C_{2v}	402 (a_1)	370 (a_1)	365 (b_1)
$(P(CH_3)_3)_2PtBr_2$(*cis*)	C_{2v}	399 (a_1)	376 (b_1)	
$P(CH_3)_3PtBr_3^-$	C_{2v}	386 (a_1)		
$(P(CH_3)_3)_3PtI^+$	C_{2v}	402 (a_1)	366 (a_1)	365 (b_1)
$(P(CH_3)_3)_2PtI_2$(*cis*)	C_{2v}	395 (a_1)	371 (b_1)	
$(P(CH_3)_3)_2PtI_2$(*trans*)	D_{2h}	370 (a_g)	347 (b_{2u})	
$P(CH_3)_3PtI_3^-$	C_{2v}	384 (a_1)		

Shobatake and Nakamoto[63] used metal isotope substitution to identify the metal–phosphorus stretching vibrations in nickel and palladium complexes of triethyl and triphenylphosphine. Just two bands in each case were found to be significantly sensitive to a change in metal isotope in the spectra of the *trans* complexes $((C_2H_5)_3P)_2NiCl_2$, $((C_2H_5)_3P)_2NiBr_2$ and $((C_2H_5)_3P)_2PdCl_2$, one of which could be assigned to the metal-halogen stretching mode. The bands at 273, 265 and 234 cm^{-1} respectively, had, therefore, to be attributed to the metal–phosphorus stretching modes. They pointed out that these results disagreed with certain earlier assignments of these vibrations to bands around 400 cm^{-1} [64]. They also claimed that their results invalidated the assignments of the corresponding trimethylphosphine complexes discussed above. However, since trimethylphosphine is lighter than triethylphosphine and there is no evidence to suggest that their donor properties are significantly different the results do, in fact, support the proposed assignments.

It is interesting to compare the assignments proposed for the trimethylphosphine complexes of Ni(0) and of Pt(II) and Pd(II). For a complex with Ni(0) a ν(NiP) band at about 220 cm^{-1} was proposed, together with an increase from 305 to 350 cm^{-1} for the symmetric deformation band of the free donor on going to the complex. However, in a Pt(II) complex a ν(PtP) band at about 390 cm^{-1} and a decrease from 305 to 220–240 cm^{-1} for the deformation band were postulated. Three reasons for these differences seem possible. The first is simply that the assignments are in error in either or both cases. However, if they are accepted, the most obvious explanation for the different behaviour is that the nature of the bonding is significantly different in the two cases and that this affects both the metal–phosphorus stretching and the deformation modes. Alternatively, it could be that when the frequency of the MP stretching band is lower than that of the deformation band the latter is shifted to higher frequency whereas when the frequency of the metal–phosphorus stretching band is higher the deformation is shifted to lower frequency. Whatever the explanation, it is clear that extreme care has to be taken in interpreting the spectra of complexes of these donors, simple as they may appear.

Nuclear Magnetic Resonance Studies

Since phosphorus is isotopically pure, with a nucleus of spin $\frac{1}{2}$, ^{31}P n.m.r. spectroscopy is a powerful technique for studying phosphine complexes. There are many different possible applications of the technique—the comprehensive review by Nixon and Pidcock[65] is recommended to interested readers—but the only one that will be discussed here is the determination and interpretation of metal-phosphorus coupling constants. Pidcock, Richards and Venanzi[66] measured the ^{195}Pt–^{31}P coupling constants for a series of complexes of

Pt(II) and Pt(IV). They interpreted the results in terms of the approximate equation 5.18 derived by Pople and Santry[67] for the Fermi contact term which is considered to be the dominant factor governing the coupling constants in these complexes:

$$J(Pt\text{–}P) \propto \gamma_{Pt}\,\gamma_P\,(\Delta E)^{-1}\,\alpha_{Pt}^2\,\alpha_P^2\;|\Psi_{Pt(6s)}(0)|^2\,|\Psi_{P(3s)}(0)|^2 \qquad 5.18$$

where γ_{Pt} and γ_P are the magnetogyric ratios for the nuclei with spin $\frac{1}{2}$, ΔE is an average excitation energy, α_{Pt}^2 and α_P^2 are the s characters of the orbitals used by Pt and P to form the bond and the $|\Psi(0)|^2$ factors are the electron densities for the particular orbitals evaluated at the nuclei. They argued that ΔE could be considered a constant for a series of related complexes and that $\alpha_P^2\,|\Psi_{P(3s)}(0)|^2$ would be constant for complexes of one particular phosphine with one particular metal. They further claimed that the coupling constants measured for the complex *cis*-$((C_2H_5)_3P)_2PtCl(CH_3)$ were so different for the two phosphines, which had to share a common $|\Psi_{Pt(6s)}(0)|^2$, that measured changes in coupling constants could be correlated with the α_{Pt}^2 term. In other words, the PtP coupling constants would be a measure of the s character of the platinum–phosphorus bonds.

The coupling constant for *cis*-$((C_4H_9)_3P)_2PtCl_2$ was found to be 1.47 times that in the *trans* complex and a *cis trans* ratio of 1.41 was found for the corresponding Pt(IV) complexes *cis*-$((C_4H_9)_3P)_2PtCl_4$. They pointed out that there was considerable experimental evidence that platinum(II)–phosphorus bonds were stronger in *cis* complexes than in *trans* complexes but that this had been interpreted in terms of different strengths for the π-bonding in the isomeric complexes. The argument was that in a *trans* complex the phosphines had to share two platinum dπ-orbitals but that in a *cis* complex there were three available. They considered that their results were inconsistent with this hypothesis because the π "back-bonding" interaction in a Pt(IV) complex would be expected to be small because of the high positive charge on the metal. Therefore, the similar results found for Pt(II) and Pt(IV) strongly suggested that the differences between the platinum phosphorus bonds in *cis* and *trans* isomers had its origin in σ-bond energy differences.

Grim and Wheatland[68] measured the $^{183}W\text{–}^{31}P$ coupling constants in a series of complexes $(R_3P)_2W(CO)_4$ of *cis* and *trans* stereochemistries. Tributyl, phenyldibutyl and diphenylbutyl phosphines were used and the values obtained for J_{WP} were 225, 220 and 230 Hz for the *cis* complexes and 265, 270 and 275 Hz for the *trans*, respectively. Two trends were apparent from these results. The first was that the coupling constants increased with increasing phenyl substitution of the phosphine, and the second that the *trans* complexes had higher J values than did the *cis* complexes. They argued that since alkylphosphines are expected to be better σ donors but poorer π acceptors than arylphosphines the first trend, which was in agreement with that observed in an earlier study[69], indicated that the changes in the coupling constants reflected changes in π "back-bonding". The comparison between the *cis* and *trans* complexes supported this conclusion since in a *cis* complex a phosphine is competing for metal dπ-orbitals with the *trans* carbonyl group which is known to be a stronger π acceptor. They pointed out that the π-bonding from the tungsten 5d orbitals would lead to de-shielding of the tungsten 6s electrons but that π-bonding to the phosphorus 3d orbital would not affect the shielding of the phosphorus 3s electrons because these had the same principal quantum number. Therefore, the observed trends in the coupling constants were consistent with the π-bonding hypothesis.

It must be emphasised here that the σ- versus π-bonding dependences of spin-spin coupling constants are still the subject of controversy and both sets of results presented above have led to alternative interpretations. What is important is that spin-spin coupling constants do appear to correlate with bond strengths and therefore can be used in conjunction with vibrational spectroscopic data to provide insight into the nature of the donor–acceptor bonding.

5.2.7. ALKYL SULPHIDE COMPLEXES

It might be expected that the donor properties of alkyl sulphides would be similar to those of alkyl phosphines but some studies of complexes with the platinum group metals have shown that the additional lone pair of electrons on the sulphur atom can lead to significantly different properties.

X-ray and Infrared Studies

Woodward *et al.*[70] determined the crystal structures of $((CH_3)_2S)_2$-Pd_2Br_4 and $((C_2H_5)_2S)_2Pt_2Br_4$ following a suggestion by Goggin et al.[71] (who had studied the vibrational spectra of these and closely related complexes) that they had markedly different structures. The palladium complex was found to have the *trans* bridged structure I (figure 5.5) typical of complexes of platinum and palladium of this stoichiometry[72]. One aspect of the structure of particular interest was that the dimethyl-sulphide was so oriented with respect to the plane of the molecule as

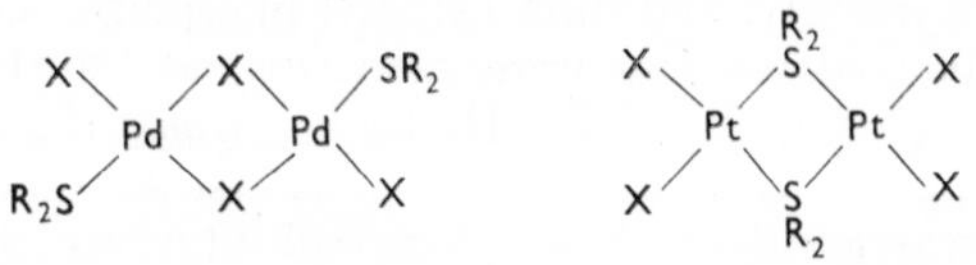

I II

FIG 5.5 Structures of bridged dinuclear complexes of palladium and platinum with dialkyl sulphides.

defined by the Pd_2Cl_4 skeleton that one of the SC bonds lay approximately in this plane. They pointed out that this conformation would be expected if there were repulsion between the lone pair on the sulphur and the filled metal dπ-orbitals.

The platinum complex, on the other hand, was shown to adopt structure II (figure 5.5) in which the sulphides acted as bidentate bridging donors. This was the structure expected on the basis of the infrared spectrum which had shown no absorptions in the region where "bridged" PdBr stretching bands are expected. The remarkable feature of the structure was that the PtS bond length of 2.21 Å was considerably shorter than the PdS bond length of 2.30 Å found in $((CH_3)_2S)_2Pd_2Br_4$. They claimed that this difference indicated that dialkyl sulphides were more strongly bound when bridging since the bond lengths in related platinum and palladium complexes were usually similar, dimethyl and diethyl sulphide had similar donor properties, and bridging bonds were usually weaker than the corresponding terminal bonds. The infrared spectra supported this conclusion; a band at about 420 cm^{-1} for every platinum complex containing alkyl sulphide bridges could only be assigned to a PtS stretching vibration whereas corresponding bands in the spectra of $((CH_3)_2S)_2Pd_2Cl_4$[72] and *cis*-$((CH_3)_2S)_2PtCl_2$[61] were observed at 340 cm^{-1} and at 349 and 338 cm^{-1} respectively. It seems clear from these studies that the platinum– or palladium–sulphur bond in terminal alkyl sulphide complexes is significantly weakened by repulsion between the sulphur lone pair and the electrons in metal dπ-orbitals.

Nuclear Magnetic Resonance Studies

The time scale for n.m.r. spectroscopy is considerably longer than that for vibrational spectroscopy and it is possible to study, by their effect on the n.m.r. signal, the rates of processes which are too fast to be followed by conventional kinetic techniques but which would be undetected by vibrational spectroscopy. Turley and Haake[73] used this technique to investigate the conformational changes in alkyl sulphide complexes of platinum. They first studied qualitatively the effect of temperature on the methylene proton n.m.r. spectrum of *cis*-$((C_6H_5CH_2)_2S)_2PtCl_2$ dissolved in a 1:1 chloroform-nitrobenzene mixture. They found that at low temperature the spectrum was that predicted for an AB system but that at higher temperatures (330 K upwards) the spectrum coalesced into one typical of an A_2 system. They considered that the low temperature pattern was consistent with pyramidal coordination at the sulphur atom and a conformation of the benzylsulphide with respect to the platinum such that the methylene groups were in different environments. It is worth noting here that this would be entirely consistent with the results of the crystallographic study of $((CH_3)_2S)_2Pd_2Cl_4$ discussed above. The coalescence at high temperature could then be attributed to an inversion at the sulphur atoms. It could not involve dissociation of the sulphide from the platinum because PtSCH coupling was still observed above the coalescence temperature and, furthermore, the value of J_{PtH} was equal to the average of the two values found at low temperature.

In order to gain more information about the inversion they studied the rate of the process. For the analysis of the low temperature spectrum they used equation 5.19 which had been proposed by Gutowsky and Holm[74] giving the rate constant, k, to a good approximation:

$$k = \frac{\pi}{\sqrt{2}}(\Delta\nu^2 - \Delta\nu_e^2)^{1/2} \qquad 5.19$$

where $\Delta\nu_e$ is the observed line separation and $\Delta\nu$ is the actual chemical shift separation (in the absence of inversion.) Above the coalescence temperature the rate constant can be determined from the line width and for this they used equation 5.20 which was also derived by Gutowsky and coworkers[75]:

$$k = \frac{\pi\{W'' + W^*[1 + 2(W^*/\Delta\nu)^2 - (W^*/\Delta\nu)^4]^{1/2}\}}{2[(W^*/\Delta\nu)^2 - (W''/\Delta\nu)^2]} \qquad 5.20$$

where W″ is the half-height band width in the absence of exchange and W* is the observed band width. This reduces to equation 5.21 when W* ≫ W″.

$$k = \frac{\pi}{2}\Delta\nu[(\Delta\nu/W^*)^2 - (W^*/\Delta\nu)^2 + 2]^{1/2} \qquad 5.21$$

They also studied the rates of inversion in *cis*- and *trans*-$((C_2H_5)_2S)_2PtCl_2$ and compared the results with those which had been previously obtained, using polarimetry, for a related compound containing pyramidalsulphur, CH_3SO-*p*-$C_6H_4CH_3$[76]. The results are given in table 5.17. They argued that the increase in the rate by a factor of 10^{18}–10^{20} from the sulphoxide to the platinum complex was inconsistent with a simple inversion at sulphur as had been proposed for the former[76]. They suggested, therefore, an inversion process which was, in fact, an

TABLE 5.17
Rates of inversion in tricoordinated sulphur compounds

Compound	Temperature/K	Rate constant k/s^{-1}	Relative rate at 298 K	Reference
CH_3SO-*p*-$C_6H_4CH_3$	523	6.7×10^{-5}	1	76
cis-$((C_6H_5CH_2)_2S)_2PtCl_2$	298	10	10	73
cis-$((C_2H_5)_2S)_2PtCl_2$	333	65	10	73
trans-$((C_2H_5)_2S)_2PtCl_2$	274	36	10^{20}	73

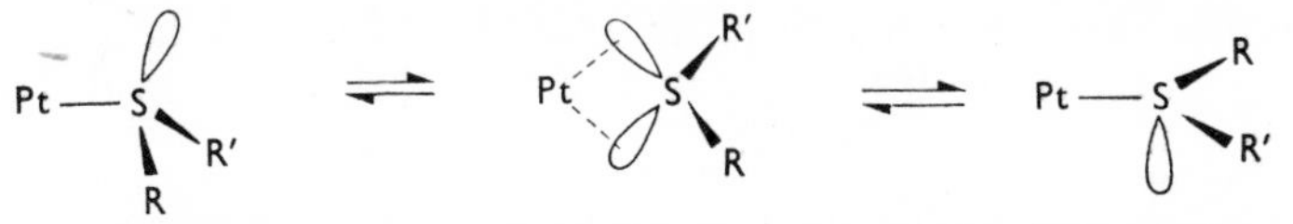

FIG 5.6 Mechanism of sulphur inversion in dialkylsulphide complexes of platinum. (Reproduced from *J. Amer. Chem. Soc.*, **89**, 4611 (1967), by permission of the American Chemical Society.)

internal displacement of one sulphide lone pair by the other. This is illustrated in figure 5.6. The important difference between this and the simple inversion mechanism was that it involved tetrahedrally coordinated sulphur in the transition state or intermediate and not a planar trigonal stereochemistry.

5.3 π DONORS

5.3.1. OLEFIN COMPLEXES

Understandably, there were considerable differences of opinion as to the nature of the bonding in olefin complexes of transition metals before the structure proposed by Dewar[77] was eventually proved by the X-ray diffraction study of Zeise's salt $K[C_2H_4PtCl_3]H_2O$ by Wunderlich and Mellor[78]. However, even after the stereochemistry had been established the important question as to how the bonding could best be described still remained unanswered. The three models illustrated in figure 5.7 represent extreme cases with which the actual bonding could be compared. In I there is a simple π-donation from the ethylene to the vacant metal σ-orbital whereas in II this is accompanied by "back-bonding" from the metal $d\pi$-orbitals into the ethylene $a\pi$-orbitals. In III the olefinic character has been destroyed and a three-membered ring has been formed involving two MC σ-bonds and a CC single bond. Vibrational spectroscopists have attempted to show which of these three describes the bonding best but the problems of interpretation of the spectra have not been easily solved. It is instructive, therefore, to follow these investigations chronologically and to note how the interpretations of the spectra changed as further information became available. There are, of course, two obvious measures of the bonding; the one being the effect of coordination on the C═C stretching

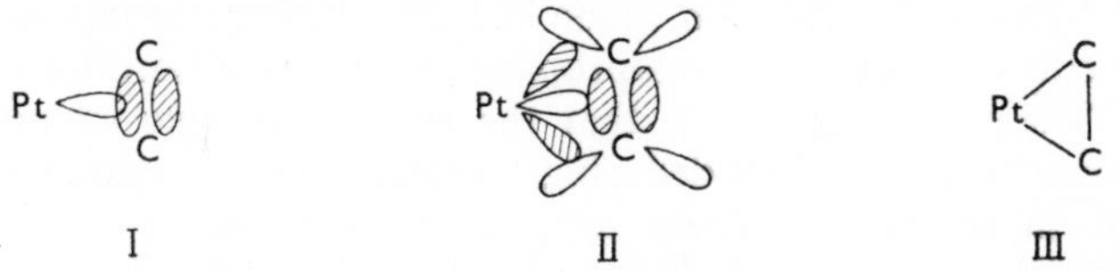

FIG 5.7 Possible models for the bonding in ethylene complexes.

vibration, the other the frequencies of the M–C stretching modes. Although certain workers have studied both they will be treated separately in the following discussion.

Carbon–Carbon Stretching Frequencies

Taufen, Murray and Cleveland[79] obtained the Raman spectra of aqueous solutions of olefin complexes of Ag^+. The carbon-carbon stretching vibrations gave rise to strong bands and were, therefore, easily identified. In each case there was a shift to lower frequency of about 60–70 cm^{-1}. The band positions for the free olefin and the complex (respectively) were 1653 and 1584 cm^{-1} for cyclohexene, 1613 and 1539 for cyclopentene, 1660 and 1598 for *cis*-2-butene and 1677 and 1612 cm^{-1} for *trans*-2-butene. An infrared study of propylene complexes of platinum by Chatt and Duncanson[80] showed that the metal–olefin bonding was stronger in these complexes since the band assigned to the ν(CC) stretching vibration shifted by 143 cm^{-1} on coordination to the metal. A similar effect was observed in the infrared spectra of several substituted ethylene complexes of platinum by Jonassen and Field[81]. Powell and Sheppard[82] attempted a fairly complete assignment of the ethylene vibrations in platinum complexes on the basis that the frequencies should have intermediate values between those in free ethylene and in the 'strongly bound' ethylene in ethylene sulphide. They followed the earlier workers in assigning the ν(CC) vibration a weak band in the complexes around 1500 cm^{-1}. The other frequencies were found to be similar to those in free ethylene, however, and supported the description of the bonding in terms of a slightly perturbed ethylene. The comparison of the band positions (in cm^{-1}) for Zeises' salt and ethylene (in parentheses) was as follows: 3085 (3106), 3020 (3019), 2990 (2990), 1516 (1623), 1428 (1444), 1402 (1342), 1241 (1236), 1022 (949), 1010 (943), 975 (1027) and 841 (810).

The assignments proposed by Chatt and Duncanson and Powell and Sheppard were disputed by Babushkin *et al.*[83] who assigned the infrared spectrum by comparison with the spectra of ethylene oxide and cyclopropane. They pointed out that the ν(CH) vibrations in ethylene oxide and cyclopropane gave rise to bands above 3000 cm^{-1} in their infrared spectra and so the observation of absorptions above 3000 cm^{-1} in the ethylene complexes did not prove that the olefinic character was retained, as had been claimed. They also argued that the band at about 1500 cm^{-1} in the spectrum of each complex could be attributed to a $\delta(CH_2)$ 'scissor' deformation mode because the corresponding vibrations in ethylene oxide and cyclopropane gave rise to absorptions in that region. Their conclusion was that the spectra strongly supported a three-membered single-bonded ring structure for the complexes *i.e.* type III in figure 5.7. These arguments were almost immediately challenged by Adams and Chatt[84] and Powell and Sheppard[85]. Independently they pointed out that bands around 1500 cm^{-1} had been observed by Jonassen and Kirsch[86] in the spectra of 1,2-disubstituted ethylenes (which do not contain CH_2 groups) and also were observed for similar complexes in their own studies. It was, therefore, inconsistent for these to be assigned to $\delta(CH_2)$ deformations in the spectra of the ethylene complexes.

Grogan and Nakamoto[87] measured the infrared spectra of Zeise's salt, deuterated Zeise's salt, and of the bridged dimers $2(C_2H_4)Pt_2Cl_4$ and $2(C_2H_4)Pd_2Cl_4$). They followed the original assignments for the carbon-carbon stretching vibrations. Although they considered that the appearance of four $\nu(CH)$ stretching bands was evidence for a lowering of the ethylene symmetry from D_{2h} to C_{2v} they carried out a modified Urey-Bradley force-field calculation for the ethylene vibrations on the basis of D_{2h} symmetry. They found that the calculated value for K(CC) of 600 N m^{-1} was some 20% lower than the 740 N m^{-1} previously calculated for ethylene itself[88]. At the same time, they pointed out that the potential energy distribution indicated that there was strong coupling between the $\nu(CC)$ stretching mode and the symmetric $\delta(CH_2)$ 'scissor' deformation mode. Therefore, the frequency of the former could not be used as a quantitative estimate of the strength of the metal–olefin bonding. However, this coupling did not occur in the deuterated complex.

Hiraishi[89], taking advantage of the improved instrumentation available, measured the Raman spectra of Zeise's salt, deuterated Zeise's salt and the bridged dimer both as solids and solutions and the results were somewhat surprising. In the Raman spectrum of the aqueous solution of Zeise's salt he found a strong polarised band at 1243 cm^{-1} which, he claimed, could only be assigned to the $\nu(CC)$ stretching vibration. Because of this, he attributed the 1515 cm^{-1} infrared absorption to the symmetric $\delta(CH_2)$ deformation. For the deuterated species, he agreed with the previous order of energies and assigned the 1353 cm^{-1} band to the $\nu(CC)$ stretching mode and the 962 cm^{-1} band to the deformation. He attributed the increase in frequency of the stretching vibration on deuteration to interactions between these two internal modes which were of the same symmetry. This resulted in an apparently low stretching frequency for the normal complex and an apparently high stretching frequency for the deuterated compound. In support of these assignments, he pointed out that where two closely spaced bands were observed in the spectra of solid samples only one was found in the corresponding spectra of solutions. This indicated that these were solid state splittings and, therefore, could not be attributed to two different molecular vibrations in each case as had been done previously. He observed this experimentally for the infrared absorptions at 1022 and 1010 cm^{-1} and the Raman bands at 1353 and 1346 cm^{-1}. He also claimed that this was true for the 1415 and 1426 cm^{-1} pair but could not prove this was so because of interference by the solvent. Finally, he undertook a force constant calculation, using the frequencies for the normal and deuterated complex, and compared the results with those from similar unpublished calculations for ethylene and ethylene oxide. The calculated carbon-carbon stretching force constants for ethylene, Zeise's salt and ethylene oxide were 858, 655 and 491 N m^{-1} respectively. He claimed that these results together with his proposed assignments of the platinum–carbon stretching vibrations suggested that the ethylene was considerably modified in these complexes and had 'almost the same structure as that of ethylene oxide'.

In a subsequent study Hiraishi *et al.*[90] considered the spectra of the 2-butene complexes which had previously been cited as evidence against assignment of the weak infrared band around 1500 cm^{-1} in the spectrum of Zeise's salt to the $\delta(CH_2)$ deformation mode. They, first of all, pointed out that in the spectra of the butylene oxides there were bands around 1490 cm^{-1} which could be attributed to δ(CH) bending vibrations. Therefore, the arguments of Adams and Chatt and Powell and Sheppard presented above were invalid. Furthermore, strong polarised bands were found at 1267 cm^{-1} for the *trans*-butene complex and at 1250 cm^{-1} for the *cis*-butene complex which could be attributed to the C=C stretching vibrations. They did admit, however, that these assignments were no longer clear-cut in that several bands of comparable intensity and polarisation were observed in the spectra in this region.

Powell, Scott and Sheppard[91] decided that the problem of assignment could be approached by a comparison between the spectra of corresponding complexes of platinum(II) and silver(I) since there was no doubt that, for the latter acceptor, the bonding was weak and so the ethylene would be little changed in the complexes. They also obtained the infrared spectrum of the complex $2(CH_3)_2CC(CH_3)_2Pt_2Cl_4$ and again found absorptions around 1500 cm^{-1} which could not be attributed to either $\delta(CH_2)$ or δ(CH) deformation modes but could reasonably be assigned to the ν(CC) stretching mode. There was a 10% lowering in frequency from tetramethylethylene itself, where the corresponding band was found at 1670 cm^{-1}. This approximated very closely to the 10.5% found if the 1515 cm^{-1} and 1353 cm^{-1} band positions were compared for C_2D_4 and the deuterated Zeise's salt. They pointed out that for these two olefins there would be no coupling between the C=C stretching mode and a $\delta(CH_2)$ symmetric deformation and thus the comparison between free and complexed donor was straightforward. For other olefins some allowance for the vibrational mixing would have to be made if the strength of the bonding were to be deduced from the frequency changes. Two bands were always observed in the Raman spectrum, one in the 1500–1700 cm^{-1} region, the other in the 1200–1350 cm^{-1} region, which together could be attributed to the C=C stretching vibration and the symmetric $\delta(CH_2)$ or δ(CH) deformation. Hence, it seemed reasonable to correlate the total percentage change for the two

TABLE 5.18
Spectral data and percentage frequency lowerings on coordination for olefin complexes of Pt(II) and Ag(I)[91]

Compound	$\tilde{\nu}_{max}$/ cm^{-1}	Percentage lowering	$\tilde{\nu}_{max}$/ cm^{-1}	Percentage lowering	Total percentage lowering
C_2H_4	1623		1342		
$(C_2H_4)PtCl_3^-$	1515	6.5	1240	7.5	14.0
$(C_2H_4)Ag^+$	1579	2.5	1320	1.5	4.0
C_2D_4	1515		981		
$(C_2D_4)PtCl_3^-$	1353	10.5	961	2.0	12.5
cis-C_4H_8	1660		1255		
(*cis*-C_4H_8)$PtCl_3^-$	1503	9.5	1242	1.0	10.5
(*cis*-C_4H_8)Ag^+	1597	4.0	1250	0.5	4.5
trans-C_4H_8	1675		1308		
(*trans*-C_4H_8)$PtCl_3^-$	1526	9.0	1263	3.5	12.5
(*trans*-C_4H_8)Ag^+	1615	3.5	1302	0.5	4.0

bands on coordination with the strength of the metal–olefin bond. These percentage changes are given in table 5.18 and two results are immediately apparent. The first is that the difference between the bond strengths for the platinum complexes and those for the silver complexes is clearly brought out by the figures. The second is that the similar total percentage lowerings found for the platinum complexes gave support to their interpretation. They concluded that the band which showed the greatest percentage lowering in each case could be considered to have the greater ν(CC) character and, furthermore, that this band would be strong in the Raman spectrum. On this basis, they agreed with Hiraishi's assignment for the ethylene complex but considered that, for substituted ethylenes, the band in the 1500–1600 cm^{-1} region of the spectra of the complexes arose mainly from ν(CC) vibration.

Powell and coworkers extended these studies by measuring the spectra of complexes of rhodium(I), palladium(II) and platinum(II)[92] and of copper(I), silver(I), gold(I) and gold(III)[93] with the bidentate donor 1,5-cyclooctadiene. They again identified the bands which together could be attributed to the C=C stretching vibrations and the symmetric δ(CH) deformation modes. The positions found for these bands are given in table 5.19 together with the percentage frequency lowerings on coordination. They concluded that, for the d^8 metal ions the observed trend in frequency lowering, Rh(I) $>$ Pt(II) $>$ Pd(II) was consistent with a decrease in π "back-bonding" along this series which would be expected from the decreasing ionic radius and increasing electronegativity. Furthermore, this was the same as the trend observed for the metal–olefin stretching frequencies (see below). However, the high value for the total percentage frequency lowering for the copper(I)

TABLE 5.19
Spectral data and percentage frequency lowerings on coordination for 1,5-cyclooctadiene[a] complexes[929,93]

Species	$\tilde{\nu}_{max}/cm^{-1}$	Percentage lowering	$\tilde{\nu}_{max}/cm^{-1}$	Percentage lowering	Total percentage lowering
COD	1658 I.R. 1644 Raman		1280 Raman		
COD–$PdCl_2$	1534, 1511 I.R. 1522 Raman	8.5	1271 Raman	0.5	9.0
COD–$PtCl_2$	1496 I.R. 1500 Raman	9.5	1267 Raman	1.0	10.5
[COD–RhCl]$_2$	1475 I.R. 1476 Raman	11.5	1241 Raman	3.0	14.5
[COD–CuCl]$_2$	1490 I.R. 1490 Raman	10.3	1265 Raman	1.2	11.5
COD–Ag^+	1602 I.R. 1605 Raman	3.8	1276 Raman	0.3	4.1
COD–Au_2Cl_2	1488 I.R.	10.5			

a. abbreviated to COD.

complex was not consistent with the comparatively weak bonding in this complex. In postulating a possible reason for this behaviour, they recalled the suggestion by Nyholm[94] that π "back-bonding" from d^{10} metal ions was weaker than it was from d^8 ions. Thus, they proposed the tentative hypothesis that the high percentage of σ-bond character in the metal–olefin bond caused a relatively greater weakening of the CC bond than occurred in the d^8 complexes although the total metal–olefin bond strengths were higher in the latter series.

Metal–Olefin Stretching Vibrations

The first attempts at complete assignments of the low frequency region of the spectra were published almost simultaneously. Grogan and Nakamoto[87] interpreted the low frequency spectrum of Zeise's salt in terms of a simple D–$PtCl_3$ model, having assigned the higher frequency region to ethylene vibrations, and therefore ignored the three vibrations which originate in the loss of rotational degrees of freedom by the ethylene. They assigned the band at 407 cm^{-1} to the DPt stretching mode. Pradilla-Sorzano and Fackler[95] also used a simple D–$PtCl_3$ model for a preliminary analysis but then extended the treatment to include the vibrations of the PtC_2 ring. Alone of all the investigators, they analysed the spectrum on the basis of strict C_{2v} symmetry and, therefore, did not assign any bands to a_2 vibrational modes. Since they were studying the spectra of solid samples where the site symmetry and factor group symmetry was known to be low there seems little justification for this approach. However, it did not affect the assignment of the metal–carbon stretching vibrations which they attributed to the absorptions at 491 and 403 cm^{-1}. The higher frequency band which they assigned to the symmetric mode was obscured by a vibration of the water molecule in Zeise's salt but was found in the spectrum of a dehydrated sample. They also found comparable features in the spectra of the corresponding bromide complex and of other olefin complexes. Hiraishi[89] agreed with Pradilla-Sorzano and Fackler that the 493 and 405 cm^{-1} bands could be attributed to the platinum–olefin stretching vibrations. However, Raman polarisation studies of an aqueous solution indicated that the lower frequency band was due to the symmetric stretching mode. He considered that the higher frequency for the asymmetric ν(PtC) stretching vibrational band was evidence for a strongly bound olefin which could be described in terms of a PtC_2 single-bonded ring. Had the bonding been of type I (figure 5.7) the asymmetric platinum–carbon stretching mode (or 'olefin tilting' vibration) would have had a much lower frequency than that of the symmetric ν(PtC) mode.

Powell and coworkers[91,92,93] accepted these assignments for the platinum complexes but found the corresponding bands around 280 cm^{-1} in the spectra of the silver complexes. They pointed out that this was consistent with the weaker bonding in the silver complexes. They also found the order of decreasing frequency Rh(I) > Pt(II) > Pd(II) for cyclooctadiene complexes which again was that expected if it were postulated that metal-to-olefin π "back-bonding" was the important stabilising interaction. For the copper complexes of cyclooctadiene they found no bands in the 350–600 cm^{-1} region but did observe features at 200 and 300 cm^{-1} which they tentatively suggested could be attributed to the copper–olefin stretching vibrations. They concluded that the bonding was not strong in these complexes. Their assignments for the platinum, palladium and rhodium complexes are given in table 5.20.

TABLE 5.20

Assignment of metal–olefin stretching vibrations in complexes of 1,5-cyclooctadiene[92]

	COD–$PdCl_2$		COD–$PtCl_2$		$[COD\text{–}RhCl]_2$	
	I.R.	Raman	I.R.	Raman	I.R.	Raman
	$\tilde{\nu}_{max}/cm^{-1}$		$\tilde{\nu}_{max}/cm^{-1}$		$\tilde{\nu}_{max}/cm^{-1}$	
'Tilt 1'	570	569	588	587	583	586
'Tilt' 2	464	464	480	482	490	480
'Stretch' 1	415	413	461	461	476	480
'Stretch' 2	350	352	378	385	388	393

'Tilting vibration' is an asymmetric ν(M—D) mode.
'Stretching vibration' is a symmetric ν(M—D) mode.
'Tilt 1' and 'Stretch 1' are out-of-phase vibrations of the bidentate system.
'Tilt 2' and 'Stretch 2' are in-phase modes.

Although the vibrational spectra of these complexes are now better understood than they were, the progress towards this understanding has been extremely erratic and even now there has not been a complete proven assignment for even the simplest complex. The difficulties involved in these investigations should not be underestimated, but it is only fair to comment that the chances of successfully interpreting spectra are much improved if both infrared and Raman data are available, if there is no uncertainty as to whether or not certain spectral features have their origin in solid state effects and if the geometry of the complexed donor is known.

Nuclear Magnetic Resonance Studies

In principle, information concerning the strength of the donor–metal interaction in olefin complexes can be obtained using n.m.r. spectroscopy in a number of ways. Two approaches will be considered here. The determination of the barriers to olefin rotation and the measurement of changes in chemical shifts and coupling constants for olefin nuclei when the complex is formed.

Powell and Sheppard[85] found that the chemical shifts of olefinic hydrogen nuclei in platinum complexes were close to those observed for the free olefins and much removed from those found for cyclopropane and ethylene sulphide. The considered this to be evidence for the 'perturbed ethylene' type of structure in the complexes. Cramer *et al.*[96] compared the coupling constants in free and coordinated ethylene and tetrafluoroethylene in rhodium(I) complexes. For the ethylene protons they found fairly small changes, for instance the geminal coupling in ethylene was 2.4 Hz and in $(C_2H_4)Rh(C_5H_5)SO_2$ was -0.07 Hz which represented a change of about 20% towards the value expected for a saturated geminal methylene group. However, the changes in fluorine coupling constants on coordination of tetrafluoroethylene were large, indicating that π "back-bonding" to this donor was very strong and that the ethylene was severely modified in the complex. Lewis *et al.*[97] measured the coupling constants in the complexes olefin–Pt(acac)X(acac = $CH_3CO{\cdot}CH{\cdot}CO{\cdot}CH_3$). Although they believed that their results were not open to a quantitative interpretation they did consider that they indicated some contribution from a PtC_2 single-bonded ring structure which could have been interpreted in terms of a carbon hybridisation halfway between sp^2 and sp^3. A comparison with the data for the rhodium(I) complexes indicated slightly weaker bonding for the platinum complexes in agreement with the vibrational spectroscopic results discussed earlier. Some of the coupling constants are given in table 5.21.

Cramer[102] studied the temperature dependence of the n.m.r. spectrum of bis-(ethylene)-π-cyclopentadienylrhodium(I). At 253 K he found a band pattern that could be attributed to non-equivalent ethylene protons, which would be expected from the lack of symmetry about the

TABLE 5.21
Coupling constants (Hz) for ethylene complexes

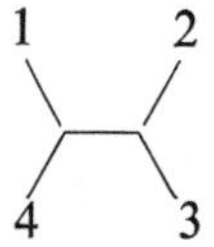

	J_{12}	J_{13}	J_{14}	Reference
ethylene	11.5	19.1	2.3	98
$(C_2H_4)Rh(C_5H_5)SO_2$	8.8	14.4	−0.07	96
propene	10.02	16.81	2.05	99
$(C_3H_6)Pt(acac)Cl$		14.0	0	97
$(C_3H_6)Pt(acac)Br$	8.0	13.0	0	
alkanes		12		100
Cyclopropane	9	7	−5	101

ethylene–rhodium bond, but at 330 K the ethylene proton bands coalesced. He considered the following possible mechanisms; proton exchange involving dissociation or tunnelling, ethylene exchange involving dissociation, ethylene exchange via a bimolecular non-dissociative mechanism and olefin rotation. Proton exchange and ethylene exchange by dissociation were shown not to occur, the tunneling mechanism was precluded by the similarity in behaviour of C_2H_4 and C_2D_4 complexes and a bimolecular mechanism was ruled out by the concentration independence of the phenomena. He concluded that at higher temperatures the olefin rotated with the olefin–rhodium σ-bond direction as the axis. In a later study Cramer *et al.*[96] obtained quantitative data for the rotation process in this complex and in the complex $(C_2F_4)(C_2H_4)Rh(C_5H_5)$. They found activation energies of about 60 J mol^{-1} for ethylene rotation but the fluorine spectrum in the latter complex showed no change up to 373K, indicating that the stronger bonding to tetrafluoroethylene inhibited the rotation. Lewis *et al.*[97] obtained thermodynamic data for the olefin rotations in the complexes olefin–Pt(acac)Cl. They argued that the barrier to rotation was the difference in energy between the configuration with the CC bond perpendicular to the platinum coordination plane and the one with the CC bond parallel and that there could be both steric and electronic factors affecting the barrier height. Accordingly, they studied the rotation in complexes containing different ethylene substituents to determine the importance of the steric effects. They found that the variations in the steric effects were too small for these alone to account for the observed barrier height. They concluded that the orientation with the CC bond perpendicular to the coordination plane had a favourable electronic energy which could well be due to hybridisation of the d_{xz}- and p_z-orbitals on platinum to form an orbital capable of stronger π "back-bonding" as had been suggested by Chatt and Duncanson[80].

Crystallographic Results

Most of the investigations which have been discussed in this section were carried out at a time when there was no crystallographic information available concerning the geometry of coordinated ethylene. However, the situation has recently changed. Hamilton *et al.*[103] determined the structure of Zeise's salt using neutron diffraction and found that the hydrogen atoms were displaced from a planar configuration. Unfortunately, it has since been shown by Owston *et al.*[104] that the structure was analysed in the wrong space group but it is unlikely that this will have led to a grossly incorrect structure. Guggenberger and Cramer[105] used X-ray diffraction to obtain the crystal structure of $(C_2H_4)(C_2F_4)Rh(C_5H_5)$ and were able to determine both the ethylene and the tetrafluoroethylene geometries. The carbon-hydrogen bonds in the ethylene were bent back such that the planes defined by each CH_2 group were at an angle of approximately 20° from a planar configuration whereas the CF_2 groups were each displaced by an angle of 37°. These results confirmed the conclusion drawn from their earlier n.m.r. work that the bonding to C_2F_4 was much stronger than was that to C_2H_2.

5.3.2. ACETYLENE COMPLEXES

Although acetylenes are simpler donors than ethylenes, from the point of view of the vibrational spectroscopist, they have been less studied because fewer simple complexes have been isolated; in particular, very few complexes of acetylene itself have been prepared. However, Iwashita and coworkers[106] studied the vibrational spectrum of the complex $(C_2H_2)Co_2(CO)_6$. They identified seven bands in the infrared spectrum of the liquid as being due to the coordinated acetylene by their sensitivity to ^2H and ^{13}C substitution. Two of these, at 605 and 551 cm^{-1}, could be attributed to cobalt–acetylene stretching vibrations leaving five that were assigned to internal acetylene modes. Of these, the 3116 and 3086 cm^{-1} bands could be immediately assigned to ν(CH) modes, the 1402.5 cm^{-1} feature was attributed to the ν(CC) mode because it was shifted by 23.5 cm^{-1} on ^{13}C substitution of the acetylene and the 894 and 768 cm^{-1} absorptions were assigned to δ(CH) modes.

It was evident from the appearance of five infrared bands that the symmetry of the coordinated donor was lower than that of free acetylene. An asymmetric structure was impossible because the n.m.r. spectrum showed that the protons were equivalent. Furthermore, the n.m.r. spectrum of the triphenylphosphine substituted complex $(C_2H_2)CO_2$-$(CO)_5P(C_6H_5)_3$ which had a virtually identical acetylene vibrational spectrum, showed only one pH spin-spin coupling constant. Hence, the structure of C_{2v} symmetry shown in figure 5.8 was proposed for the complex. This was similar to that determined crystallographically for

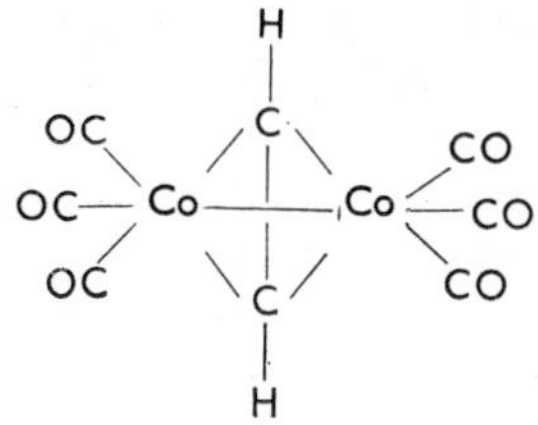

FIG 5.8 Proposed structure of $(C_2H_2)Co_2(CO)_6$ (reference 106).

the complex $(C_6H_5CCC_6H_5)Co_2(CO)_6$[107]. The frequencies of the vibrations of the coordinated acetylene were substantially different from those of the corresponding vibrations in acetylene itself indicating that there was considerable distortion of the donor in the complex. The frequencies did, in fact, compare closely with those of the first excited state of acetylene which had been determined by Ingold and King[108]. The comparison is shown in table 5.22. Although the excited state molecule had a *trans* structure they argued that this would be unlikely in a complex because of steric effects and that calculations showed that the *trans* and *cis* excited states would have similar energies[109]. In a later publication, Iwashita[110] produced further evidence for this similarity by calculating force constants for the ^1A_u excited state of acetylene and for complexed acetylene using an approximate $(CHCH)Co_2$ model. He used a Urey-Bradley force-field for one calculation and a general valence force-field (but only for the acetylene vibrations in $(CHCH)Co_2$) for the other. The calculated CC stretching force constants were 590 (UBFF) or 730 (GVFF) N m^{-1} for the ^1A_u excited state and 620 (UBFF) or 730–750 (GVFF) N m^{-1} for coordinated acetylene. There was a similar correlation for the CH stretching force constants; 469 (UBFF) or 496 (GVFF) and 496 (UBFF) and 495–502 (GVFF) N m^{-1} for the ^1A_u state and coordinated molecule respectively. These were

TABLE 5.22
Infrared data[a] of acetylene in normal, excited[108] and coordinated states[107]

State		$\bar{\nu}_1$	$\bar{\nu}_2$	$\bar{\nu}_3$	$\bar{\nu}_4$	$\bar{\nu}_5$
$^1\Sigma_g^+$ $(D_{\infty h})$	C_2H_2	3373.7	1973.8	3287.0	611.8	729.1
	C_2D_2	2700.5	1762.4	2427	505	539.1
^1A_u (C_{2h})	C_2H_2		1380	3020		1049
	C_2D_2		1310	2215		844
Coordinated (C_{2v})	C_2H_2	3116.0	1402.5	3086.0	768.0	894.0
	C_2D_2	2359.0	1346.5	2297.0	602.0	751.4

a. band positions in cm^{-1}.

very different from the 1580 and 592–599 N m^{-1} for the CC and CH stretching force constants (respectively) of ground-state acetylene.

The conclusion was this similarity existed because the nature of the bonding of the acetylene in this particular complex was such that one electron had been donated from the highest energy bonding π-orbital while one electron had been accepted into the lowest energy $a\pi$-orbital. The electronic structure of the complexed acetylene was, therefore, directly comparable with that in the ^1A_u excited state of acetylene. A similar comparison between a coordinated π donor and an excited state of the donor had previously been made by Wilkinson *et al.*[111] for CS_2 in $(CS_2)Pt(P(C_6H_5)_3)_2$ on the basis of the geometry of the molecule determined by X-ray diffraction and Iwashita and Hayata[112] produced evidence for the same behaviour by CO_2.

5.3.3. BENZENE COMPLEXES

Almost all of the π complexes of benzene that are known can be formally described as involving a six-coordinate metal atom or ion with the benzene occupying three sites as if the six benzene π-electrons were involved in three $b\pi$-v interactions. Because of this apparently stringent requirement and because an 18 valence electron 'inert gas' configuration for the metal (which must be in a low oxidation state) is required for stability, the number of transition metal acceptors that can form stable π-complexes with benzene is small. Most of the studies so far reported have concerned the two series of complexes; $(C_6H_6)_2M$, particularly $(C_6H_6)_2Cr$, and $(C_6H_6)M(CO)_3$ where M is Cr(0), Mo(0) and W(0).

Dibenzenechromium

The molecular structure of dibenzenechromium has been studied many times in an attempt to determine beyond doubt whether benzene retained its hexagonal symmetry in the complexes or whether there was some degree of localisation of the π-cloud, corresponding to the three formal metal coordination sites mentioned above, resulting in a reduction to trigonal symmetry. Jellinek[113] claimed that the X-ray data were consistent with a trigonally distorted, Kekulé-type structure for the benzene rings as had been predicted earlier by Ruch[114]. However, Cotton *et al.*[115] redetermined the crystal structure and found that the difference in the CC bond lengths could only be 0.02 Å at most and was, therefore, of the magnitude to be expected from the effect of the crystalline environment. Ibers[116], on the other hand, considered that the data of Cotton *et al.* were entirely consistent with D_{6h} symmetry for the benzene molecules. In a further repeat of the X-ray work, Keulen and Jellinek[117] agreed that D_{6h} symmetry was indicated by their results but that this could have arisen from rotational disorder of molecules of lower symmetry although they considered this to be unlikely. Haaland[118] determined the structure of the gaseous molecules by electron diffraction and again found that the results favoured a D_{6h} benzene structure although he could fit the data to a D_{3h} symmetry ring if rather unlikely values for the vibrational amplitudes were accepted. More recently, Albrecht and coworkers[119] interpreted their neutron diffraction data as categorically favouring D_{3h} symmetry.

Snyder[120] obtained the infrared spectra of crystalline $(C_6H_6)_2Cr$ and $(C_6D_6)_2Cr$ at 93 K which are shown in figure 5.9. At the time of this

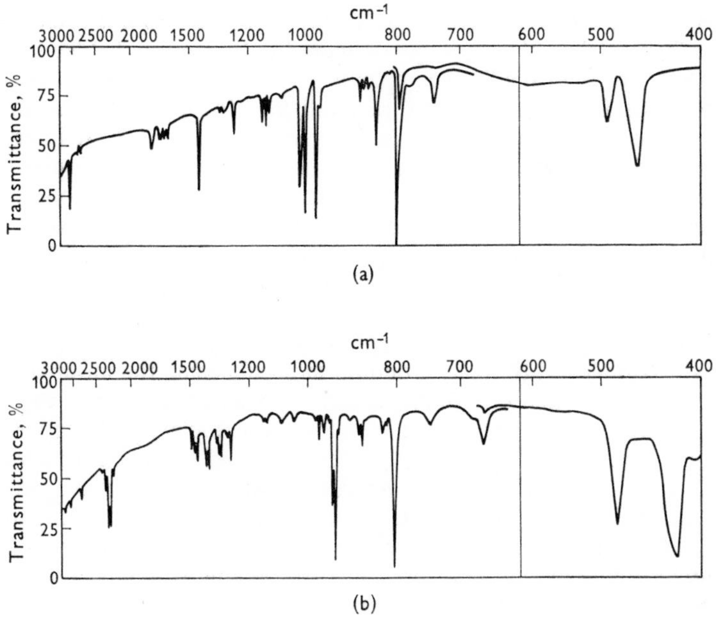

FIG 5.9 Infrared spectra of (a) $Cr(C_6H_6)_2$ and (b) $Cr(C_6D_6)_2$ at 93 K. (Reproduced, by permission, from *Spectrochim. Acta*, **15**, 807 (1959).)

study, the crystallographic results (which were to be interpreted by Jellinek as indicating trigonal symmetry for the benzene rings) had not appeared and he considered that the spectra could be readily attributed to a molecule with D_{6h} symmetry. Accordingly, he proposed that the strong absorptions at 1430, 1014, 1002, 970 and 796 cm^{-1} could be assigned to the vibrations which gave rise to the bands at 1478, 1036, 990, 854 and 687 cm^{-1}, respectively, in the spectrum of benzene vapour. The remaining spectral features were attributed to overtones and combination transitions and 'site' and 'factor group' effects. However, Fritz *et al.*[121] measured the infrared spectra of dibenzenechromium as a solid, as a gas and in solution. He concluded that there were too many bands in all three spectra for the complex to have D_{6h} symmetry. In a later publication, Fritz and Fischer[122] compared the infrared spectra of crystalline $(C_6H_6)_2Cr$ and of the two crystalline modifications of $(C_6H_6)_2V$. They found that the low frequency region could be interpreted in terms of the site symmetries of the complexes. Thus, four bands were found for the cubic crystals where the site symmetry is S_6 and six were found for the monoclinic crystals of the vanadium complex which had C_i site symmetry. However, no difference was observed between the high frequency spectra of the two vanadium modifications. They argued, therefore, that the observation of many more bands in the high frequency region of the chromium complex could not be attributed to either 'site group' or 'factor group' effects but could only arise from a trigonal symmetry as opposed to the undisputed hexagonal symmetry of the vanadium complexes.

Ngai, Stafford and Schäfer[123] reinvestigated the infrared spectrum of gas phase dibenzenechromium taking measurements at 20 K intervals between 458 and 528 K to ensure that the spectrum of the complex was differentiated from that of the free benzene present. They found the spectrum could be unambiguously interpreted in terms of D_{6h} symmetry and eight of the nine infrared active vibrations were detected in the region 400–4000 cm^{-1}. Further support for the hexagonal structure in the gas phase has come from calculations of force constants and vibrational amplitudes. Cyvin *et al.*[124] calculated valence force constants for the complete dibenzenechromium molecule using the infrared data of Fritz *et al.*[121] and the Raman data of Schäfer *et al.*[125] Their calculations were based on D_{6h} symmetry for the complex. Two important results were found. One was that the force-field showed considerable interaction between one of the internal vibrational modes and one of the skeletal vibrations which could not be accounted for in approximate calculations involving the donor vibrations alone. The other was that the vibrational amplitudes were found to be almost identical to those of benzene itself and to those observed in the electron diffraction study of Haaland[118] if hexagonal symmetry was assumed. They concluded that these calculations proved beyond doubt that, in the gas phase, there was no trigonal distortion of the benzene rings in dibenzenechromium. Similar calculations for the corresponding complexes of the metal acceptors Cr(I), V(0), Mo(0), Mo(I), W(0), W(I), Tc(I) and Re(I), (using the internal force constants calculated for the chromium complex and the experimental low frequency data of Fritz *et al.*[121]) were also reported by Cyvin *et al.*[126] They found vibrational coupling between internal and skeletal modes as before. They also found the rather surprising result that the observed differences in the low frequency spectra could be accounted for simply in terms of the different metal masses with no differences in the donor–metal force constants (this point is also discussed in section 2.3.3).

It is perhaps not surprising that there have been problems in the interpretation of the spectra of complexes of a donor such as benzene. For the gaseous dibenzenechromium molecule there seems no doubt that the molecule has hexagonal symmetry but for the crystalline complex there must remain some doubt until all of the observed bands in the infrared and Raman spectra have been accounted for.

5.3.4. OXYGEN COMPLEXES

Although oxygen can act as a monodentate[127] or bidentate[128] n donor with certain metals, most of the studies have concerned complexes in which it functions as a π donor. Dioxygen complexes are classified as 'reversible' or 'irreversible' depending on whether or not oxygen gas can be regenerated from them. Crystal structure determinations by Ibers *et al.*[129,130,131] showed that the 'irreversible' complexes were characterised by long O_2 bonds, *e.g.*, 1.509 Å in $O_2Ir(CO)I(P(C_6H_5)_3)_2$[129] and 1.625 Å in $O_2Ir((C_6H_5)_2PCH_2CH_2P(C_6H_5)_2)_2^+PF_6^-$[130] whereas the 'reversible' complexes had shorter O_2 bonds, *e.g.*, 1.30 Å in $O_2IrCl(CO)(P(C_6H_5)_3)_2$[131] and 1.418 Å in $O_2Rh((C_6H_5)_2PCH_2CH_2P(C_6H_5)_2)_2^+PF_6^-$.[130] They considered that these results were consistent with increased π "back-bonding" in the 'irreversible' complexes weakening the O_2 bond. However, Amma and coworkers[132] found an O_2 distance of 1.461 Å in the reversible complex $O_2Ir(CO)Cl(P(C_6H_5)_2C_6H_5)_2$ which was closer to that in the corresponding 'irreversible' triphenylphosphine chloride complex than to that in the 'reversible' triphenylphosphine complex. They concluded that it could be misleading to correlate O_2 bond lengths with chemical stabilities.

Infrared Studies

Nakamura *et al.*[133] investigated the infrared spectra of a number of dioxygen complexes using ^{18}O isotopic substitution to aid the assignments. The infrared spectrum of solid $O_2Pt(P(C_6H_5)_3)_2$ showed a band at 828 cm^{-1} and two at 472 and 462 cm^{-1} which split when a complex prepared from a mixture of $^{16}O_2$ and $^{18}O_2$ was examined. However, the spectrum of a solution of the 'normal' complex showed only one absorption in the lower region and it was concluded that the 472, 462 cm^{-1} doublet probably arose from a crystal splitting. Two bands, at 893 and 484 cm^{-1}, in the spectrum of $O_2Pd(\textit{tert}\text{-}C_4H_9NC)_2$ showed splittings into three components on partial isotopic substitution but, on careful examination, the 869 cm^{-1} band present in the spectrum of the $^{16}O^{18}O$ species appeared to be split into two bands. They concluded that the oxygen in this complex was not symmetrically coordinated. For the two complexes $O_2RhCl(P(C_6H_5)_3)_2(\textit{tert}\text{-}C_4H_9NC)$ and $O_2Ni(\textit{tert}\text{-}C_4H_9NC)_2$ the identification of two isotope sensitive bands in each was straightforward.

They considered two possible models for interpreting these results. These were the 'monodentate' and the 'isosceles' as illustrated in figure 5.10. They expected that the 'monodentate' structure would be characterised by an oxygen-oxygen stretching force constant close to the 1150 N m^{-1} in free O_2 and just one metal–oxygen stretching vibration,

FIG 5.10 Possible structures for dioxygen π-complexes.

whereas the 'isosceles' structure would have a much lower oxygen-oxygen stretching force constant and two metal oxygen–stretching modes. However, a simple calculation of f_{OO} from the observed frequency for the ν(OO) mode in the platinum complex (where there would be least movement of the metal in the normal vibration) gave a value of 310 N m^{-1}. This was considered to be good evidence for the bonding to be best described by the isosceles model. The only problem was that just one band had been found in the metal–oxygen stretching region. They were unable to obtain Raman spectra, which would possibly have resolved the problem, because the complexes decomposed in the laser beam. They pointed out, however, that the one ν(MO) band could have been too weak to be observed or, alternatively, that the two vibrations could have the same frequency. This latter possibility was indicated by their simple valence force constant calculations.

They compared their calculated values for simple harmonic force constants in the MO_2 ring with the chemical stabilities of the complexes. The values for f_{MO} were 210, 210, 240 and 320 N m^{-1} for the Pt(0), Pd(0), Ni(0) and Rh(I) complexes respectively. They argued that the high value for the rhodium complex was consistent with its comparatively high stability. However, the values for f_{OO} of 300, 350, 330 and 350 N m^{-1} respectively showed no correlation with chemical stability. They also studied the spectra of certain rhodium complexes with different ligands and found the following trends in the metal–oxygen stretching frequency, Cl > Br > I; $As(C_6H_5)_3 > P(C_6H_5)_3$; *tert*-$C_4H_9NC$ < *cyclo*-$C_6H_{11}NC$ < *p*-$CH_3C_6H_4NC$. They pointed out that mass effects could be significant in the halogen or phosphine/arsine comparisons but that the frequency increase with increasing electron-attracting nature of the isocyanide substituent could only arise from an electronic effect. They concluded that the strength of the rhodium–oxygen bond was due to the positive charge on the metal which strengthened the σ interaction.

5.4. COMPARATIVE STUDIES OF DONORS

5.4.1. NITROGEN AND CARBON MONOXIDE

An immediate comparison that can be made between the isoelectronic donors N_2 and CO is that complexes of the former are much less stable and so far have only been prepared for a limited range of transition metal acceptors. One consequence of this has been that their vibrational spectra have not been investigated in any detail. However, a few attempts have been made to compare corresponding dinitrogen and carbonyl complexes using vibrational spectroscopy (see section 2.3.3 for further discussion).

Two groups of workers studied the frequencies of the donor stretching vibration and interpreted essentially the same set of observations in different ways. Collman *et al.*[134] prepared two dinitrogen complexes of iridium(I) and compared their ν(NN) frequencies with those previously reported for dinitrogen complexes of Ru(II), Os(II), Co(I) and Rh(I) and carbonyl complexes of Rh(I) and Ir(I). The vibrational data are given in table 5.23. They pointed out that the lowering of the ν(NN) frequency on coordination of nitrogen was greater than the lowering of the ν(CO) frequency on coordination of carbon monoxide and concluded that N_2 was a stronger π-acceptor than CO. Chatt *et al.*[135] considered a series of Ir(I) complexes and made the comparison in terms of the percentage lowering of the stretching frequency on coordination and found that this was virtually the same in corresponding nitrogen and carbonyl complexes. Their results are also presented in table 5.23. However, they believed that this indicated that the bonding to nitrogen was weaker than that to carbon monoxide. Purcell[137] also argued that the frequency shifts on coordination of nitrogen and carbon monoxide could not be taken as direct measures of the strengths of the donor-acceptor bonding. One of his arguments was that the acceptor $a\pi$-orbital in N_2 was more strongly antibonding than was the acceptor $a\pi$-orbital in CO. Thus, equal degrees of "back-bonding" in corresponding dinitrogen and carbonyl complexes would result in a greater degree of lowering of the dinitrogen stretching frequency. This assumes, of course, that in dinitrogen complexes as in carbonyl complexes the vibrational frequencies are dominated by π-bonding effects as discussed in section 5.1.3.

TABLE 5.23
ν(NN) and ν(CO) bands in dinitrogen and carbonyl complexes

Complex	$\bar{\nu}$/cm^{-1}	$\bar{\nu}_{donor} - \bar{\nu}_{complex}$/ cm^{-1}	$\bar{\nu}_{complex}$/ $\bar{\nu}_{donor}$	Reference
$N_2Ru(NH_3)_5^{2+}$	2129	201		134
$N_2Os(NH_3)_5^{2+}$	2033	297		134
$N_2RhCl(P(C_6H_5)_3)_2$	2152	178		136
$CORhCl(P(C_6H_5)_3)_2$	1977	166		134
$N_2IrCl(P(C_6H_5)_3)_2$	2105	225	0.903	134
$COIrCl(P(C_6H_5)_3)_2$	1962	181	0.916	134
$N_2IrI(P(C_6H_5)_3)_2$	2113	217	0.906	135
$COIrI(P(C_6H_5)_3)_2$	1971	172	0.920	135
$N_2IrCl(P(CH_3)_2C_6H_5)_2$	2051	279	0.880	135
$COIrCl(P(CH_3)_2C_6H_5)_2$	1958	185	0.914	135

Darensbourg made the comparison between dinitrogen and carbonyl complexes of iridium(I)[138] and osmium(II)[139] in terms of the absolute infrared intensities of the donor stretching vibrations. For each complex the intensity was measured at a number of concentrations, corrected for decomposition each time and extrapolated to zero concentration. The data may be interpreted in terms of a local oscillating dipole model using equation 5.22, *i.e.*

$$\vec{\mu}'_{MD} = (\partial\vec{p}/\partial R) = 1.537(B_i/M_D)^{1/2} \qquad 5.22$$

where M_D is the inverse mass of a CO or N_2 group, $\vec{\mu}'_{MD}$ is the dipole moment derivative for the "perturbed" donor stretching vibration in the complex and B_i is the measured integrated intensity as defined in equation 2.2. The results, which are given in table 5.24, showed that the

TABLE 5.24

Absolute intensity data[a] for dinitrogen and carbonyl complexes[138,139]

	$\bar{\nu}/cm^{-1}$	$10^5\ B_i/m\ mol^{-1}$	$10^{20}\ \vec{\mu}'_{MD}$
$N_2IrCl(P(C_6H_5)_3)_2$	2109	3.29 ± 0.06	24.6
$COIrCl(P(C_6H_5)_3)_2$	1965	8.82 ± 0.12	39.9
cis-$N_2OsCl_2(P(C_2H_5)_2C_6H_5)_3$	2075	6.46 ± 0.07	34.5
cis-$COOsCl_2(P(C_2H_5)_2C_6H_5)_3$	1924, 1916	9.05 ± 0.18	40.4

a. Original data have been quoted here in SI units. The intensities, B, are in $m^3\ mol^{-1}\ m^{-2}$ (or $m\ mol^{-1}$) and the $\vec{\mu}'_{MD}$ values have been converted to C (*i.e.* $C\ mm^{-1}$) from their original arbitrary units. The total $(\partial\vec{p}/\partial R)$ has been calculated (in $D\ Å^{-1}$) using equation 5.22 and these parameters have been converted using $1\ D\ Å^{-1} = 3.335 \times 10^{-20}$ C.

value of $\vec{\mu}'_{MNN}$ was significantly less than that of $\vec{\mu}'_{MCO}$ in each case. They concluded that there was weaker π "back-bonding" to nitrogen since there was clear evidence that the high intensity of carbonyl stretching vibrations, in transition metal complexes could be almost wholly attributed to the π interaction[140] and the behaviour in dinitrogen complexes was expected to be similar. Infrared intensity studies on carbonyl and nitrogen complexes are described more fully in section 2.3.3.

5.4.2. *TRANS* EFFECT AND *TRANS* INFLUENCE IN d^8 SYSTEMS

It was first observed many years ago by Chernyaev[141] that the stereochemical courses of the reactions of square-planar d^8 metal complexes could be rationalised if it were postulated that the ease of substitution of a coordinated group was dependent on the nature of the substituent *trans* to it. Subsequent work showed that donors could be placed in the following series of decreasing "*trans* effect": $SC(NH_2)_2$, PR_3, SR_2 > NH_3, Pyr, RNH_2 > H_2O on the basis of their labilising effect on a *trans* group. The series was later extended by Hel'man[142] to include ethylenes, carbon monoxide and nitric oxide which were found to have very high *trans* effects. The work was given a quantitative basis by means of kinetic measurements which were reviewed by Basolo and Pearson[143] who also proposed the definition; "the *trans* effect is that effect of a coordinated group A upon the rate of substitution reactions of the group opposite to A". Not unexpectedly, there has been much speculation as to the cause of a donor's *trans* effect and, in particular, as to whether high *trans* effect donors simply weaken the *trans* metal–ligand bond or whether they lower the activation energy for a substitution process by lowering the energy of the transition state. Because of the probability that different donors operated through different mechanisms, Pidcock *et al.*[66] found it convenient to define the "*trans* influence" of a donor as "the extent to which that donor weakens the bond *trans* to itself in the equilibrium state of a substrate". Therefore, spectroscopic investigations can be interpreted in terms of *trans* influence.

Goggin *et al.*[144] recorded the infrared spectra of solutions of tetra-alkylammonium salts of the complex anions D–MCl_3^- with M = Pt and Pd and identified the bands arising from the metal chlorine stretching vibrations. The particular advantage of this study was that it allowed simultaneous observation of the vibrational frequencies for MCl bonds both *cis* and *trans* to the donor–metal bond and the comparison between the different complexes was made under constant experimental conditions. The three metal-chlorine stretching bands expected were found for the palladium complexes and, with the exception of the carbonyl complex, the strongest was observed at about 340 cm^{-1} in each case and the weakest at about 290 cm^{-1}. The relative intensity of these two absorptions allowed an immediate assignment to the asymmetric and symmetric $\nu(PdCl_2)$ modes respectively. The third band had an intermediate intensity, and a frequency dependent on the nature of the donor, and so could be attributed to the ν(PdCl) mode for the chlorine *trans* to D. Only two metal-chlorine stretching bands were observed for the platinum complexes. Since the symmetric and asymmetric $\nu(PtCl_2)$ modes would be expected to have more nearly equal frequencies for the heavier platinum, it was assumed that the two bands were not resolved under the experimental conditions. Once again, the strongest absorption occurred at about the same frequency in each spectrum with the exception of that of the carbonyl complex whereas the frequency of the other varied as the donor was changed. The vibrational data are given in table 5.25. The frequencies of the ν(MCl) mode gave the following orders of *trans* influence; CO < SMe_2 ≈ C_2H_4 ≈ SEt_2 ≪ $AsEt_3$ < PMe_3 ≈ $AsMe_3$ ≈ PEt_3 for the platinum

TABLE 5.25
$\bar{\nu}$(M–Cl) data for D–MCl_3 complexes[144]

	$\bar{\nu}_s(MCl_2)/cm^{-1}$	$\bar{\nu}_{as}(MCl_2)/cm^{-1}$	$\bar{\nu}(MCl)/cm^{-1}$
$P(CH_3)_3PdCl_3^-$	295 (5)[a]	343 (10)	265 (9)
$P(C_2H_5)_3PdCl_3^-$	293 (5)	340 (10)	265 (8)
$As(CH_3)_3PdCl_3^-$	294 (sh)[b]	338 (10)	253 (9)
$As(C_2H_5)_3PdCl_3^-$	296 (7)	338 (10)	272 (9)
$S(CH_3)_2PdCl_3^-$	293 (5)	344 (10)	307 (8)
$S(C_2H_5)_2PdCl_3^-$	288 (6)	339 (10)	311 (9)
$C_2H_4PdCl_3^-$	288 (4)	337 (10)	317 (sh)[b]
$COPdCl_3^-$	300 (3)	352 (10)	331 (6)
$P(CH_3)_3PtCl_3^-$		332 (10)	275 (9)
$P(C_2H_5)_3PtCl_3^-$		330 (10)	271 (9)
$As(CH_3)_3PtCl_3^-$		329 (10)	272 (10)
$As(C_2H_5)_3PtCl_3^-$		328 (10)	280 (9)
$S(CH_3)_2PtCl_3^-$		325 (10)	310 (9)
$S(C_2H_5)_2PtCl_3^-$		325 (10)	307 (8)
$C_2H_4PtCl_3^-$		330 (10)	309 (8)
$COPtCl_3$		343 (10)	322 (4)

a. The number in brackets is a relative peak height. b. shoulder.

series and $CO < C_2H_4 < SEt_2 \approx SMe_2 \ll AsEt_3 < PEt_3 \approx PMe_3 < AsMe_3$ for the palladium complexes. They considered that the differences in the orders were probably insignificant since no allowance had been made for possible vibrational interactions. However, there was clearly a good correlation between the order of *trans* influence and the order of σ donor strength. In particular, it was evident that the high *trans* effects of ethylene and carbon monoxide could not be attributed to bond weakening whereas those of phosphines and arsines possibly could be.

Church and Mays[145] carried out a similar investigation of a series of complexes *trans*-D–$PtX(P(C_2H_5)_3)_2^+ClO_4^-$ with X = Cl or H and again found that the order of *trans* influence as measured by the platinum-chlorine stretching frequency was the same as the order of σ donor strength. However, the platinum-hydrogen stretching frequencies did not show as good a correlation even after those for the carbonyl and isocyanide complexes had been corrected for Fermi resonance[146] using measurements of the changes in frequency of the ν(CO) and ν(CN) modes on deuteration. They attributed this to a changing solvent sensitivity of the vibration in the different complexes. On the other hand, the values of J_{PtH}, obtained from proton n.m.r. spectra of the same set of complexes, had a similar order to that of the platinum-chlorine stretching frequencies observed for the chlorides. They considered that this order could be interpreted on the assumption that the magnitude of

J_{PtH} was dominated by the Fermi contact term in the same way that Pidcock, Richards and Venanzi[66] had interpreted the values of J_{PtP} in phosphine complexes (see section 5.2.6). However, they did point out that the change in J_{PtH} did not necessarily reflect changes in $(\alpha_{Pt})^2$ arising from changes in the s character of the D–Pt bond but could be due to changes in the degree of covalency of the PtH bond. Their results are given in table 5.26.

TABLE 5.26

$\tilde{\nu}$(Pt–Cl) data and platinum-hydrogen coupling constants in *trans*-D–PtX$(P(C_2H_5)_3)_2$ complexes[145]

D	$\tilde{\nu}$(PtCl)/cm^{-1} for D–PtCl$(P(C_2H_5)_3)_2$	J_{PtH}/Hz for D–PtH$(P(C_2H_5)_3)_2$
CO	344	967
$(CH_3)_3CNC$	341	895
p-$MeOC_6H_4NC$	335	890
$P(OC_6H_5)_3$	316	872
$P(OCH_3)_3$	316	846
$P(C_2H_5)_3$	295	790

Although the terms *trans* effect and *trans* influence traditionally referred to planar complexes of d^8 metal ions similar phenomena have been reported for linear d^{10} complexes[60,147] and for several series of octahedral complexes[148]. The investigations of these have followed similar lines to those discussed above.

5.4.3. DETERMINATION OF OXIDATION STATE

Although most complexes can be formulated in terms of a particular oxidation state of the metal there are cases where there is some doubt as to which is the most appropriate. This problem was discussed in section 5.2.5 in relation to nitric oxide complexes where the geometry of the coordinated donor allowed a meaningful choice to be made. However, for certain π complexes the same uncertainty exists but the situation cannot always be clarified by a determination of the donor geometry. Since electron spectroscopy is a technique which allows the determination of atomic core-level binding energies, which are dependent on the atomic charge, it provides an opportunity for determining oxidation states experimentally.

Cook *et al.*[149] investigated the ESCA† spectra of a series of platinum complexes of general formula D–Pt$(P(C_6H_5)_3)_2$ with D = $2P(C_6H_5)_3$, C_2H_4, $C_6H_5CCC_6H_5$, CS_2, O_2 and 2Cl. For the first and last mentioned the oxidation states are 0 and 2 respectively but for the others either 0 or 2 could be claimed. The binding energies of the Pt($4f_{7/2}$) and P(2p)

† Electron Spectroscopy for Chemical Analysis, *i.e.* X-ray photoelectron spectra.

levels were determined relative to that of the triphenylphosphine C(1s) level each time since this could reasonably be expected to remain constant. The near constancy of the P(2p) binding energies indicated that any change in the Pt($4f_{7/2}$) energy could be attributed to the change in the donor D. The chemical shifts of the Pt($4f_{7/2}$) levels suggested that the state of oxidation increased with changing D along the series $P(C_6H_5)_3 < C_6H_5CCC_6H_5 \approx C_2H_4 < CS_2 < O_2 < Cl_2$. They further interpreted the data by assuming that platinum transferred no electrons to $P(C_6H_5)_3$ in $(P(C_6H_5)_3)_4Pt$ but one to each chlorine in $Cl_2Pt(P(C_6H_5)_3)_2$ and that the binding energies depended linearly on the metal donor charge transfer. In this way metal to donor charge transfers were calculated as 0, 0.7, 0.8, 1.3, 1.8 and 2.0 electrons for D = $(P(C_6H_5)_3)_2$, $C_6H_5CCC_6H_5$, C_2H_4, CS_2, O_2 and Cl_2 respectively. They pointed out that the assumed value for Cl was certainly too high since a 30–50% covalency for the PtCl bond had been proposed as a result of n.q.r. measurements[150] but that, nevertheless, the numbers were significant if due allowance were made for this overestimation.

They argued that the charge transfers determined for the acetylene and ethylene complexes were consistent with a formulation as Pt(0) complexes but with considerable π-donation from metal to donor. However, the value obtained for the carbon disulphide complex indicated that it was best described as a complex of platinum(I). Although this would be a novel description of a platinum complex, they claimed that it would be consistent with the geometry of the CS_2 as determined by X-ray diffraction[111] which had an angle close to that anticipated for the CS_2^- ion if, as expected, it had a similar geometry to that of the formally isoelectronic molecule NO_2. This interpretation is, of course, at variance with that originally proposed which is mentioned in section 5.3.2.

5.5. SPECIALISED AREAS OF RESEARCH

5.5.1. MATRIX ISOLATION STUDIES

It can be claimed that matrix isolation is the best technique for studying the vibrational spectra of any molecule. It has the advantage that the molecule is isolated in inert surroundings, is not rotating and is at sufficiently low temperature for spectra to be obtained free of any "hot" bands. However, the present discussion will only be concerned with its particular use in the isolation and study of 'unstable' species.

Carbonyl complexes of nickel have been studied using two different approaches. Rest and Turner[151] found that photolysis of $Ni(CO)_4$ in an inert gas matrix at 15 K with an iodine line source or a medium pressure arc gave two additional infrared bands in the region expected for ν(CO) modes. They assigned these to the a_1 and e modes of a pyrimidal $Ni(CO)_3$ species as their relative intensities remained constant for any given matrix as the photolysis proceeded. However, the relative intensities varied with the size of the matrix atom, the ratio of e to a_1 intensities being greatest for xenon, the largest atom used. They argued that this behaviour indicated that the pyramidal configuration was enforced by the size of the matrix cavities and as these became larger the molecules became more nearly planar. DeKock[152] obtained infrared spectra for the complete series $Ni(CO)_{1-4}$ by depositing nickel atoms from the vapour in a 500:1, Ar:CO matrix at liquid helium temperature and then gradually raising the temperature. The first band appeared at 1996 cm^{-1} and was attributed to NiCO, the next to appear was one at 1967 cm^{-1} which was assigned to $Ni(CO)_2$ followed by bands at 2017 and 2052 cm^{-1} arising from $Ni(CO)_3$ and $Ni(CO)_4$ respectively. Finally, a feature at 2035 cm^{-1} was detected which he attributed to a dimeric species. Although the band positions for the tri- and tetracarbonyl agreed with those reported by Rest and Turner he did not observe the weak band at 2065 cm^{-1} attributed to the a_1 tricarbonyl mode. The spectra are shown in figure 5.11. DeKock also obtained the corresponding spectra using a 200:1 Ar:$C^{18}O$ matrix.

Darling and Ogden[153] established the existence of palladium tetracarbonyl in an argon matrix. They condensed palladium atoms with a large excess of 95:5, Ar:$C^{16}O$ at 27 K and observed one strong infrared band at 2070.3 cm^{-1}. A similar experiment with nickel gave $Ni(CO)_4$, as evidenced by the strong infrared band at 2050 cm^{-1}, suggesting that $Pd(CO)_4$ had been prepared. This was confirmed by repeating the preparation with 95:2.7:2.3, Ar:$C^{16}O$:$C^{18}O$. The product from this showed five infrared absorptions at 2070.3, 2047.5, 2037.0, 2029.0 and 2022.0 cm^{-1}. They showed that the positions and relative intensities for the bands in this spectrum were entirely consistent with their assignment to the modes derived from the f_2 vibration of the normal species. These calculations followed the method of Bor[154] and Haas and Sheline[9] and involved the usual approximation that the $\nu(CO)$ modes could be "factored off" from the other vibrations.

5.5.2. ADSORPTION STUDIES

In this section a few examples of the use of vibrational spectroscopy to elucidate the nature of the bonding of chemisorbed molecules to transition metal surfaces will be considered. Experiments concerning hydrocarbon complexes are discussed because there is a particularly wide variety of possible structures for these and they include types that are unknown in normal stoichiometric complexes. Other donors that have been studied in some detail are CO[155] and N_2[156].

Sheppard and coworkers[157,158] investigated the adsorption of several hydrocarbons on silica-supported nickel and platinum using infrared spectroscopy. They used three criteria, first applied by Eischens and Pliskin[159], for establishing the nature of the adsorbed species. The first was the frequencies of the infrared $\nu(CH)$ bands which were compared with those established for the different possible CH_n groupings. They

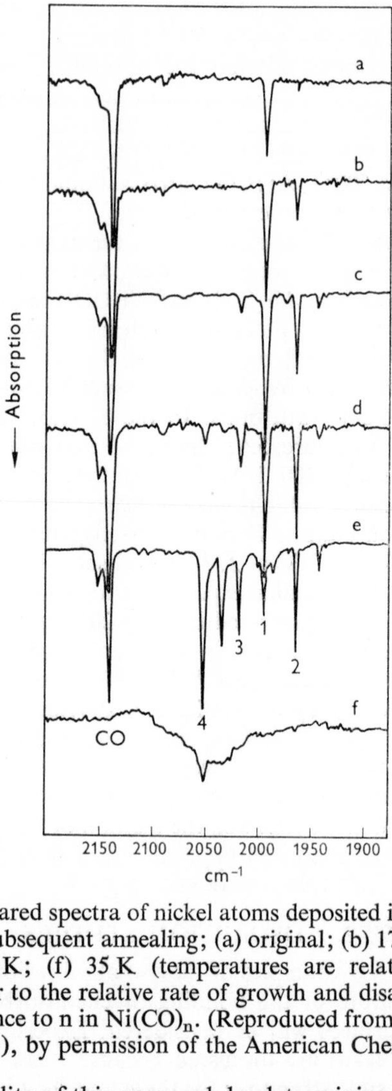

FIG 5.11 Infrared spectra of nickel atoms deposited in a 500:1 Ar:CO matrix with subsequent annealing; (a) original; (b) 17 K; (c) 18 K; (d) 19 K; (e) 26 K; (f) 35 K (temperatures are relative). The arabic numerals refer to the relative rate of growth and disappearance of the bands and hence to n in $Ni(CO)_n$. (Reproduced from *Inorg. Chem.*, **10,** 1207 (1971), by permission of the American Chemical Society.)

extended the utility of this approach by determining typical frequencies for CH_3M groups using the compounds $(CH_3)_4Sn$, $(CH_3)_4Pb$, $CH_3Fe(CO)_2(C_5H_5)$, $(CH_3)_2Pd(P(C_6H_5)_3)_2$, $CH_3PtCl(P(CH_3)_3)_2$ and $(CH_3)_2Pt(P(CH_3)_3)_2$ as models. As the CH_3M bands were observed to be at slightly higher frequencies than the corresponding CH_3C-bands it was assumed that RCH_2M, $\nu(CH)$ frequencies would be higher than those for the RCH_2C-group. The second criterion was the ratio of the intensities of the antisymmetric $\nu(CH_3)$ and $\nu(CH_2)$ absorptions, near 2955 and 2925 cm^{-1} respectively, which varies with the length of an *n*-alkyl chain; or the ratio of the intensities of the =CH— and =CH_2 absorptions, near 3025 and 2925 cm^{-1} respectively, which similarly varies with the length of a polyethylene chain. Finally, the total integrated intensities of bands in the *ν*(CH) stretching regions in the spectra of the adsorbed species before and after hydrogenation was used. This gave some indication as to how many CH bonds were present in the adsorbed species. In certain cases the spectra of both the adsorbed species and of the vapour phase following hydrogenation could be used as additional evidence to support the postulated structures for the original product.

Three distinct chemisorption processes are possible. These are 'associative', 'dissociative' and 'self-hydrogenation'. The difference between 'associative' and 'dissociative' adsorption is that, in the latter, CH bonds are broken and MH bonds formed whereas in the former no CH bonds are broken. 'Self-hydrogenation' involves disproportionation into a hydrogen rich and a hydrogen deficient surface species. In all three cases the product may be monomeric or polymeric. From the point of view of the present discussion only an associative adsorption could be described as donor–acceptor complex formation but it is, of course, impossible to study this in isolation.

Sheppard and Ward[157] obtained the infrared spectra of adsorbed acetylene. In the spectrum of acetylene on nickel some absorption at 3020 cm^{-1} was observed which could be attributed to =CH— groups in either MCH=CHM or polymeric MCH=CH—CH=CHM type species. However, stronger bands were found in the region where the sp^3 hybridised carbon *ν*(CH) bands are expected. The optical density ratio of bands at 2925 and 2955 cm^{-1} suggested that, on average, there were two CH_2 groups for every CH_3 group. They concluded that a mixture of surface ethyl and butyl groups was formed by a 'self-hydrogenation' process. The spectrum of acetylene on platinum had very weak *ν*(CH) bands which increased greatly in intensity on hydrogenation indicating that the adsorption was mainly dissociative and very little complex formation occurred.

Adsorption of ethylene on platinum and nickel was studied by Morrow and Sheppard[158]. Many spectra were obtained including a series for the adsorption on silica-supported platinum at 195, 293, 368 and 423 K, and a series where the ethylene was adsorbed on a very thick silica/platinum disc at 195 K and the sample was then heated to 473 K in stages. Spectra were also obtained for adsorption at 128 K and for adsorption of a 20:1, C_2D_4:C_2D_3H mixture at room temperature. These spectra are shown in figure 5.12. There was a clear indication that at least three different surface species were present. For temperatures of 195 K and higher the dominant species detected in the spectra had bands at 2880, 2795 cm^{-1} and contributed to the absorption in the 2920 cm^{-1} region. The simplicity of the spectrum and the observation that ethane predominated in the gas phase after hydrogenation suggested a monomeric structure. The band positions were those expected for

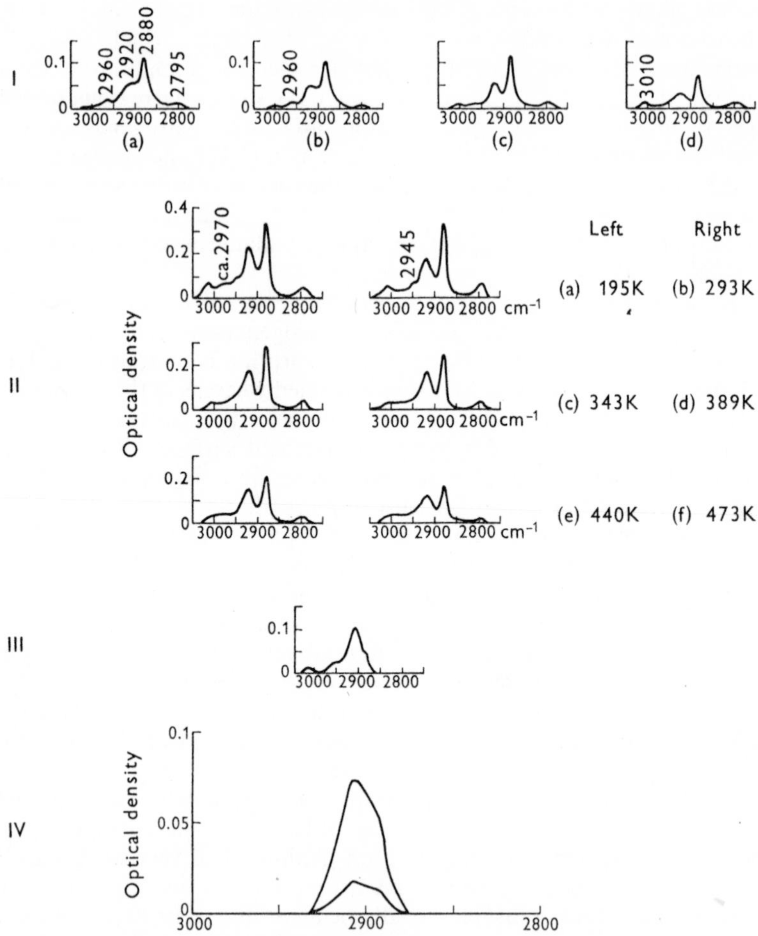

Fig 5.12 Spectra of ethylene adsorbed on platinum I (a) 423 K, (b) 368 K, (c) 293 K, (d) 195 K. II Adsorption on a thick disc at 195 K (a) then heating in stages to 473 K, (b) to (f). III Adsorption at 128 K. IV Adsorption of a 20:1, C_2D_4:C_2D_3H mixture at 293 K. (Reproduced from *Proc. Roy. Soc.*, **311A,** 397 (1969), by permission of the Royal Society.)

ν(CH) vibrations of sp^3 hybridised carbon atoms. The following structures were possible, therefore: $Pt_2CHCHPt_2$, Pt_2CHCH_2Pt, $PtCH_2CH_2Pt$, Pt_2CHCH_3 and $PtCH_2CH_3$. The last two were ruled out because a CCH_3 group always gives rise to a pair of bands at 2960 and 2870 cm^{-1} and these were not observed. However, the study of the spectra of alkyl metal compounds, which was mentioned above, had shown that a MCH_2- group always gave rise to a weak band near 2800 cm^{-1} (which had been attributed to an overtone transition, enhanced in intensity by Fermi resonance[146]). Thus, it was concluded that an 'associative' adsorption of ethylene to give $PtCH_2CH_2Pt$ complexes occurred and that the symmetric and asymmetric $\nu(CH_2)$ modes for this species gave rise to the bands at 2880 and 2920 cm^{-1} respectively. In support of this, the band at 2907 cm^{-1} in the spectrum of the adsorbed deuterated molecule was attributed to a $PtCHDCD_2Pt$ complex. The remainder of the intensity around 2920 cm^{-1} was attributed to a 'dissociatively' adsorbed species such as $Pt_2CHCHPt_2$ and the 3010 cm^{-1} feature to an alternative product of 'dissociative' adsorption, PtCHCHPt. However, the overall intensity increase by a factor of about ten on hydrogenation showed that a considerable proportion of the adsorption resulted in a fourth product, a 'surface carbide' species which of course, was undetected in the $\nu(CH)$ region of the spectra. Finally, the spectrum obtained for adsorption at 138 K could only be explained in terms of 'dissociative' formation of $Pt_2CHCHPt_2$ with very little 'associative' complex formation.

The experiments with silica-supported nickel gave time-dependent spectra indicating that a reaction was taken place after adsorption. The spectrum obtained after a short period of adsorption at 195 K resembled that obtained with platinum and bands at 2925, 2870 and 2790 cm^{-1} were attributed to the complex $NiCH_2CH_2Ni$. However, on standing, the band also observed at 2958 cm^{-1}, typical of a methyl group, became gradually stronger as did the absorption at 2885 cm^{-1}. At 293 K the spectrum, after a short period of adsorption, was similar to that obtained, after standing at 195 K. After two hours at 293 K just three bands were present—at 2958, 2923 and 2875 cm^{-1} with the last, which arose from a merging of the 2885 and 2870 cm^{-1} bands, being the strongest and after eighteen hours the same three bands were present but the 2923 cm^{-1} band was then the most intense. Since the 2923 cm^{-1} band could be attributed to CH_2- groups it was apparent that a polymerisation was occurring. The spectra of the gas phase after hydrogenation showed the presence only of n-butane indicating that C_4 species were formed. It was concluded, therefore, that $NiCH_2CH_2Ni$ was an initial product which polymerised to a C_4 species and underwent 'self-hydrogenation' to give the n-butyl group. The 2958 and 2923 cm^{-1} bands arose from the CH_3- and $-CH_2$- groups respectively and the 2875 cm^{-1} absorption could be attributed to either or both —CHNi— and —CH_2Ni groupings. The following reaction scheme between adsorbed species and hydrogen was proposed.

$$CH_2CHCHCH_2 \xrightarrow{\ H\ } CH_3CHCHCH_2$$
$$\begin{array}{cccc} | & | & | & | \\ Ni & Ni & Ni & Ni \end{array} \qquad \begin{array}{ccc} | & | & | \\ Ni & Ni & Ni \end{array}$$

$$\longrightarrow CH_3CH_2CHCH_2 \xrightarrow{\ H\ } CH_3CH_2CH_2CH_2$$
$$\begin{array}{cc} | & | \\ Ni & Ni \end{array} \qquad\qquad\qquad \begin{array}{c} | \\ Ni \end{array}$$

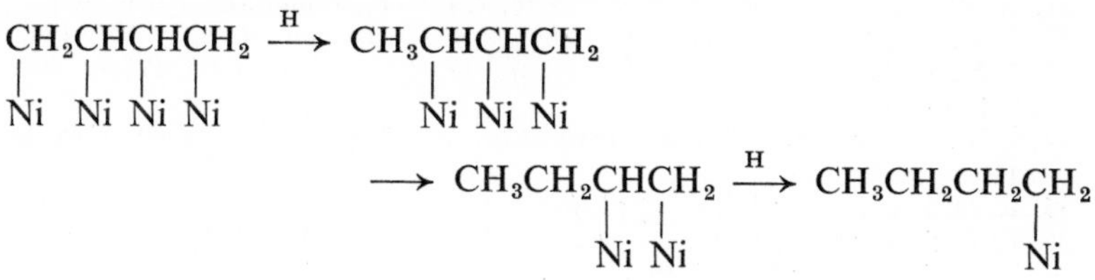

REFERENCES

1. B. B. Wayland and W. L. Rice, *Inorg. Chem.*, **6**, 2270 (1967)
2. R. A. Levenson, H. B. Gray and G. P. Ceasar, *J. Amer. Chem. Soc.*, **92**, 3653 (1970)
3. S. F. A. Kettle, *Spectrochim. Acta.*, **22A**, 1388 (1966)
4. L. M. Haines and M. H. B. Stiddard, *Advan. Inorg. Chem. Radiochem.*, **12**, 53 (1969)
5. G. Bouquet and M. Bigorgne, *Spectrochim. Acta*, **27A**, 139 (1971)
6. L. H. Jones, R. S. McDowell and M. Goldblatt, *J. Chem. Phys.*, **48**, 2663 (1968)
7. L. H. Jones, R. S. McDowell and M. Goldblatt, *Inorg. Chem.*, **8**, 2349 (1969)
8. F. A. Cotton and C. S. Kraihanzel, *J. Amer. Chem. Soc.*, **84**, 4432 (1962)
9. H. Haas and R. K. Sheline, *J. Chem. Phys.*, **47**, 2996 (1967)
10. I. H. Hillier and V. R. Saunders, *Mol. Phys.*, **22**, 1025 (1971); *J. Chem. Soc.*, **D**, 642 (1971)
11. D. R. Lloyd and E. W. Schlag, *Inorg. Chem.*, **8**, 2544 (1969)
12. M. Barber, J. A. Connor, I. H. Hillier and V. R. Saunders, *J. Chem. Soc.*, **D**, 682 (1971)
13. A. Terzis and T. G. Spiro, *Inorg. Chem.*, **10**, 643 (1971)
14. T. V. Long and R. A. Plane, *J. Chem. Phys.*, **43**, 457 (1965)
15. E. R. Lippincott and J. M. Stutman, *J. Phys. Chem.*, **68**, 2926 (1964)
16. S. F. A. Kettle, I. Paul and P. J. Stamper, *J. Chem. Soc. Dalton Trans*, 2413 (1972); *J. Chem. Soc.*, **D**, 1724 (1970) and 235 (1971)
17. E. W. Abel, R. A. N. McLean, S. P. Tyfield, P. S. Braterman, A. P. Walker and P. J. Hendra, *J. Mol. Spectrosc.*, **30**, 29 (1969)
18. H. J. Buttery, G. Keeling, S. F. A. Kettle, I. Paul and P. J. Stamper, *J. Chem. Soc.*, **A**, 2077 (1969)
19. H. J. Buttery, S. F. A. Kettle, G. Keeling, P. J. Stamper and I. Paul, *J. Chem. Soc.*, **A**, 3148 (1971)
20. S-I. Mizushima, I. Nakagawa and J. V. Quagliano, *J. Chem. Phys.*, **23**, 1367 (1955); I. Nakagawa and S-I. Mizushima, *Bull. Chem. Soc. Jap.*, **28**, 589 (1955)
21. G. F. Svatos, D. M. Sweeny, S-I. Mizushima, C. Curran and J. V. Quagliano, *J. Amer. Chem. Soc.*, **79**, 3313 (1957)
22. D. B. Powell and N. Sheppard, *J. Chem. Soc.*, 3108 (1956)
23. J. P. Mathieu, *J. Chim. Phys.*, **36**, 308 (1939)
24. I. Nakagawa and T. Shimanouchi, *Spectrochim. Acta*, **22**, 759 (1966); J. Hiraishi, I. Nakagawa and T. Shimanouchi, *Spectrochim. Acta*, **24A**, 819 (1968)
25. J. R. Durig and B. R. Mitchell, *Appl. Spectrosc.*, **21**, 221 (1967)
26. E. U. Condon and G. H. Shortley, The Theory of Atomic Spectra, (Cambridge University Press, 1935), 236
27. O. Laporte, *Z. Physik*, **23**, 135 (1924)
28. T. V. Long and D. J. B. Penrose, *J. Amer. Chem. Soc.*, **93**, 632 (1971)
29. M. T. Barnet, B. M. Craven, H. C. Freeman, N. E. Kime and J. A. Ibers, *J. Chem. Soc.*, **D**, 307 (1966)
30. M. O. Bulanin and N. D. Orlova, *Optics and Spectroscopy*, **15**, 112 (1963)
31. A. W. Rakov, *Optics and Spectroscopy*, **7**, 128 (1959) *ibid.*, **13**, 203 (1962)
32. R. C. Leech, D. B. Powell and N. Sheppard, *Spectrochim. Acta*, **21**, 559 (1965)
33. J. M. Janik, J. A. Janik, A. Migdal and G. Pytasz, *Acta. Phys. Polon.*, **40A**, 741 (1971)
34. R. E. Clarke and P. C. Ford, *Inorg. Chem.*, **9**, 227 (1970) *J. Chem. Soc.*, **D**, 1109 (1968)
35. R. D. Foust, Jr. and P. C. Ford, *Inorg. Chem.*, **11**, 899 (1972)
36. P. T. T. Wong and D. G. Brewer, *Can. J. Chem.*, **46**, 131 (1968)
37. P. T. T. Wong and D. G. Brewer, *Can. J. Chem.*, **46**, 139 (1968)

38. J. Yarwood, *Trans. Faraday Soc.*, **65,** 934 (1969)
39. D. G. Brewer, P. T. T. Wong and M. C. Sears, *Can. J. Chem.*, **46,** 3137 (1968)
40. H. Irving and R. J. P. Williams, *Nature*, **162,** 746 (1948)
41. B. Hutchinson, J. Takemoto and K. Nakamoto, *J. Amer. Chem. Soc.*, **92,** 3335 (1970)
42. J. S. Strukl and J. L. Walter, *Spectrochim. Acta*, **27A,** 223 (1971)
43. H. A. Jahn and E. Teller, *Proc. Roy. Soc.*, **161A,** 220 (1937)
44. D. M. P. Mingos and J. A. Ibers, *Inorg. Chem.*, **10,** 1035 (1971)
45. D. M. P. Mingos and J. A. Ibers, *Inorg. Chem.*, **10,** 1479 (1971)
46. J. S. Griffith, *Proc. Roy. Soc.*, **235A,** 23 (1956)
47. E. E. Mercer, W. A. McAllister and J. R. Durig, *Inorg. Chem.*, **6,** 1816 (1967)
48. D. Hall and A. A. Taggart, *J. Chem. Soc.*, 1359 (1965); D. Dale and D. C. Hodgkin, *J. Chem. Soc.*, 1364 (1965)
49. R. D. Feltham, *Inorg. Chem.*, **3,** 1038 (1964)
50. J. L. Milward, W. Wardlaw and W. J. R. Way, *J. Chem. Soc.*, 233 (1958)
51. B. F. Hoskins, F. D. Whillans, D. H. Dale and D. C. Hodgkin, *J. Chem. Soc.*, **D,** 69 (1969)
52. A. Sabatini, *Inorg. Chem.*, **6,** 1756 (1967)
53. S. Ahrland, J. Chatt and N. R. Davies, *Quart. Rev.*, 265 (1958)
54. R. G. Pearson, *J. Amer. Chem. Soc.*, **85,** 3533 (1963)
55. H. G. M. Edwards and L. A. Woodward, *Spectrochim. Acta*, **26A,** 897 (1970)
56. S. Bénazeth, A. Loutellier and M. Bigorgne, *J. Organomet. Chem.*, **24,** 479 (1970)
57. M. Bigorgne, A. Loutellier and M. Pańkowski, *J. Organometal. Chem.*, **23,** 201 (1970)
58. A. Loutellier and M. Bigorgne, *J. Chim. Phys., Physiochem. Biol.*, **67,** 78, 99 and 107 (1970)
59. C. G. Barlow and G. C. Holywell, *J. Organometal. Chem.*, **16,** 439 (1969)
60. D. A. Duddell, P. L. Goggin, R. J. Goodfellow, M. G. Norton and J. G. Smith, *J. Chem. Soc.*, **A,** 545 (1970)
61. P. L. Goggin, R. J. Goodfellow, S. R. Haddick, F. J. S. Reed, J. G. Smith and K. M. Thomas, *J. Chem. Soc., Dalton Trans.*, 1904 (1972)
62. D. A. Duddell, P. L. Goggin, R. J. Goodfellow and M. G. Norton, *J. Chem. Soc.*, **D,** 879 (1968)
63. K. Shobatake and K. Nakamoto, *J. Amer. Chem. Soc.*, **92,** 3332 (1970)
64. P. L. Goggin and R. J. Goodfellow, *J. Chem. Soc.*, **A,** 1462 (1966); G. E. Coates and C. Parkin, *J. Chem. Soc.*, 421 (1963)
65. J. F. Nixon and A. Pidcock, *Ann. Rev. Nucl. Magn. Resonance Spectrosc.*, **2,** 345 (1962)
66. A. Pidcock, R. E. Richards and L. M. Venanzi, *J. Chem. Soc.*, **A,** 1707 (1966)
67. J. A. Pople and D. P. Santry, *Mol. Phys.*, **8,** 1 (1964)
68. S. O. Grim and D. A. Wheatland, *Inorg. Chem.*, **8,** 1716 (1969)
69. S. O. Grim, D. A. Wheatland and W. McFarlane, *J. Amer. Chem. Soc.*, **89,** 5573 (1967)
70. D. L. Sales, J. Stokes and P. Woodward, *J. Chem. Soc.*, **A,** 1852 (1968)
71. P. L. Goggin, R. J. Goodfellow, D. L. Sales, J. Stokes and P. Woodward, *J. Chem. Soc.*, **D,** 31 (1968)
72. R. J. Goodfellow, P. L. Goggin and L. M. Venanzi, *J. Chem. Soc.*, **A,** 1897 (1967)
73. P. Haake and P. C. Turley, *J. Amer. Chem. Soc.*, **89,** 4611 and 4617 (1967); *Inorg. Nucl. Chem. Lett.*, **2,** 173 (1966)
74. H. S. Gutowsky and C. H. Holm, *J. Chem. Phys.*, **25,** 1228 (1956)
75. A. Allerhand, H. S. Gutowsky, J. Jonas and R. Meinzer, *J. Amer. Chem. Soc.*, **88,** 3185 (1966)
76. D. R. Rayner, E. G. Miller, P. Bickart, A. J. Gordon and K. Mislow, *J. Amer. Chem. Soc.*, **88,** 3138 (1966)

77. M. J. S. Dewar, *Bull. Soc. Chim. France*, C71 (1951)
78. J. A. Wunderlich and D. P. Mellor, *Acta. Crystallogr.*, **7**, 130 (1954) *ibid.*, **8**, 57
79. H. J. Taufen, M. J. Murray and F. F. Cleveland, *J. Amer. Chem. Soc.*, **63**, 3500 (1941)
80. J. Chatt and L. A. Duncanson, *J. Chem. Soc.*, 2939 (1953)
81. H. B. Jonassen and J. E. Field, *J. Amer. Chem. Soc.*, **79**, 1275 (1957)
82. D. B. Powell and N. Sheppard, *Spectrochim. Acta*, **13**, 69 (1958)
83. A. A. Babushkin, L. A. Gribov and A. D. Hel'man, *Russ. J. Inorg. Chem.*, **4**, 695 (1959)
84. D. M. Adams and J. Chatt, *Chem. and Ind.* (*London*), 149 (1960)
85. D. B. Powell and N. Sheppard, *J. Chem. Soc.*, 2519 (1960)
86. H. B. Jonassen and B. Kirsch, *J. Amer. Chem. Soc.*, **79**, 1279 (1957)
87. M. J. Grogan and K. Nakamoto, *J. Amer. Chem. Soc.*, **88**, 5454 (1966); *ibid.*, **90**, 918 (1968)
88. T. Shimanouchi, *J. Chem. Phys.*, **26**, 594 (1957)
89. J. Hiraishi, *Spectrochim. Acta*, **25A**, 749 (1969)
90. J. Hiraishi, D. Finseth and F. A. Miller, *Spectrochim. Acta*, **25A**, 1657 (1969)
91. D. B. Powell, J. G. V. Scott and N. Sheppard, *Spectrochim. Acta*, **28A**, 327 (1972)
92. D. B. Powell and T. J. Leedham, *Spectrochim. Acta*, **28A**, 337 (1972)
93. T. J. Leedham, D. B. Powell and J. G. V. Scott, *Spectrochim. Acta*, **28A**, 559 (1972)
94. R. Nyholm, *Proc. Chem. Soc.*, 273 (1961)
95. J. Pradilla-Sorzano and J. P. Fackler Jr., *J. Mol. Spectrosc.*, **22**, 80 (1967)
96. R. Cramer, J. B. Kline and J. D. Roberts, *J. Amer. Chem. Soc.*, **91**, 2519 (1969)
97. C. E. Holloway, G. Hulley, B. F. G. Johnson and J. Lewis, *J. Chem. Soc.*, **A** 1653 (1970)
98. D. M. Graham and C. E. Holloway, *Can. J. Chem.*, **41**, 2114 (1963)
99. A. A. Bothner-By and C. Naar-Colin, *J. Amer. Chem. Soc.*, **83**, 231 (1961)
100. H. S. Gutowsky and C. Juan, *Discuss. Faraday Soc.*, **34**, 52 (1962)
101. J. W. Emsley, J. Feeney and L. H. Sutcliffe, High Resolution Nuclear Magnetic Resonance Spectroscopy, Volume 2 (Pergamon, Oxford, 1965), p. 694
102. R. Cramer, *J. Amer. Chem. Soc.*, **86**, 217 (1964)
103. W. C. Hamilton, K. A. Klanderman and R. Spratley, *Acta Crystallogr.*, **25A**, S172 (1969)
104. J. A. J. Jarvis, B. T. Kilbourn and P. G. Owston, *Acta Crystallogr.*, **27B**, 366 (1971)
105. L. J. Guggenberger and R. Cramer, *J. Amer. Chem. Soc.*, **94**, 3779 (1972)
106. Y. Iwashita, F. Tamura and A. Nakamura, *Inorg. Chem.*, **8**, 1179 (1969)
107. W. G. Sly, *J. Amer. Chem. Soc.*, **81**, 18 (1959)
108. C. K. Ingold and G. W. King, *J. Chem. Soc.*, 2702 (1953)
109. H. W. Kroto and D. P. Santry, *J. Chem. Phys.*, **47**, 792 (1967)
110. Y. Iwashita, *Inorg. Chem.*, **9**, 1178 (1970)
111. M. Baird, G. Hartwell, R. Mason, A. I. M. Rae and G. Wilkinson, *J. Chem. Soc.*, **D**, 92 (1967)
112. Y. Iwashita and A. Hayata, *J. Amer. Chem. Soc.*, **91**, 2525 (1969)
113. F. Jellinek, *Nature*, **187**, 871 (1960)
114. E. Ruch, *Ber. Jahrestagung Chem. Ges. DDR* (*Leipzig*), 125 (1959)
115. F. A. Cotton, W. A. Dollase and J. S. Wood, *J. Amer. Chem. Soc.*, **85**, 1543 (1963)
116. J. A. Ibers, *J. Chem. Phys.*, **40**, 3129 (1964)
117. E. Keulen and F. Jellinek, *J. Organometal. Chem.*, **5**, 490 (1966)
118. A. Haaland, *Acta Chem. Scand.*, **19**, 41 (1965)
119. E. Förster, G. Albrecht, W. Dürselen and E. Kurras, *J. Organometal. Chem.*, **19**, 215 (1969)

120. R. G. Snyder, *Spectrochim. Acta*, **15,** 807 (1959)
121. H. P. Fritz, W. Lüttke, H. Stammereich and R. Forneris, *Chem. Ber.*, **92,** 3246 (1959); *Spectrochim. Acta*, **17,** 1068 (1961)
122. H. P. Fritz and E. O. Fischer, *J. Organometal. Chem.*, **7,** 121 (1967)
123. L. H. Ngai, F. E. Stafford and L. Schäfer, *J. Amer. Chem. Soc.*, **91,** 48 (1969)
124. S. J. Cyvin, J. Brunvoll and L. Schäfer, *J. Chem. Phys.*, **54,** 1517 (1971); *J. Organometal. Chem.*, **27,** 69 (1971)
125. L. Schäfer, J. F. Southern and S. J. Cyvin, *Spectrochim. Acta*, **27A,** 1083 (1971)
126. S. J. Cyvin, B. N. Cyvin, J. Brunvoll and L. Schäfer, *Acta Chem. Scand.*, **24,** 3420 (1970)
127. A. L. Crumbliss and F. Basolo, *J. Amer. Chem. Soc.*, **92,** 55 (1970)
128. W. P. Schaefer and R. E. Marsh, *Acta Crystallogr.*, **21,** 735 (1966)
129. J. A. McGinnety, R. J. Doedens and J. A. Ibers, *Inorg. Chem.*, **6,** 2243 (1967)
130. J. A. McGinnety, N. O. Payne and J. A. Ibers, *J. Amer. Chem. Soc.*, **91,** 6301 (1969)
131. S. J. La Placa and J. A. Ibers, *J. Amer. Chem. Soc.*, **87,** 2581 (1965)
132. M. S. Weininger, I. F. Taylor and E. L. Amma, *J. Chem. Soc.*, **D,** 1172 (1971)
133. A. Nakamura, Y. Tatsuno, M. Yamamoto and S. Otsuka, *J. Amer. Chem. Soc.*, **93,** 6052 (1971)
134. J. P. Collman, M. Kubota, F. D. Vastine, J. Y. Sun and J. W. Kang, *J. Amer. Chem. Soc.*, **90,** 5430 (1968)
135. J. Chatt, D. P. Melville and R. L. Richards, *J. Chem. Soc.*, **A,** 2841 (1969)
136. L. Yu. Ukhin, Yu. A. Shvetsov and M. L. Khidekel, *Izv. Akad. Nauk, SSSR, Ser. Khim.*, 957 (1967)
137. K. F. Purcell, *Inorg. Chim. Acta*, **3,** 540 (1969)
138. D. J. Darensbourg and C. L. Hyde, *Inorg. Chem.*, **10,** 431 (1971)
139. D. J. Darensbourg, *Inorg. Chem.*, **10,** 2399 (1971)
140. T. L. Brown and D. J. Darensbourg, *Inorg. Chem.*, **6,** 971 (1967)
141. I. I. Chernyaev, *Ann. Inst. Platine (USSR)*, **5,** 102 (1927)
142. A. D. Hel'man, *Compt. rend. acad. sci. U.R.S.S.*, **24,** 549 (1939)
143. F. Basolo and R. G. Pearson, *Prog. Inorg. Chem.*, **4,** 381 (1962)
144. R. J. Goodfellow, P. L. Goggin and D. A. Duddell, *J. Chem. Soc.*, **A,** 504 (1968)
145. M. J. Church and M. J. Mays, *J. Chem. Soc.*, **A,** 3074 (1968)
146. E. Fermi, *Z. Physik*, **71,** 250 (1931)
147. P. L. Goggin, R. J. Goodfellow, S. R. Haddock and J. G. Eary, *J. Chem. Soc., Dalton Trans.*, 647 (1972)
148. R. Mason and A. D. C. Towl, *J. Chem. Soc.*, **A,** 1601 (1970)
149. C. D. Cook, K. Y. Wan, U. Gelius, K. Hamrin, G. Johansson, E. Olsson, H. Siegbahn, C. Nordling and K. Siegbahn, *J. Amer. Chem. Soc.*, **93,** 1904 (1971)
150. R. Ugo, *Coord. Chem. Rev.*, **3,** 319 (1968)
151. A. J. Rest and J. J. Turner, *J. Chem. Soc.*, **D,** 1026 (1969)
152. R. L. DeKock, *Inorg. Chem.*, **10,** 1205 (1971)
153. J. H. Darling and J. S. Ogden, *Inorg. Chem.*, **11,** 666 (1972)
154. G. Bor, *Inorg. Chim. Acta*, **1,** 81 (1967)
155. R. P. Eischens, W. A. Pliskin and S. A. Francis, *J. Chem. Phys.*, **22,** 1786 (1954); *J. Phys. Chem.*, **60,** 194 (1956)
156. A. Ravi, D. A. King and N. Sheppard, *Trans. Faraday Soc.*, **64,** 3358 (1968)
157. N. Sheppard and J. W. Ward, *J. Catal.*, **15,** 50 (1969)
158. B. A. Morrow and N. Sheppard, *Proc. Roy. Soc.*, **311A,** 391 (1969)
159. R. P. Eischens and W. A. Pliskin, *Adv. Catal.*, **10,** 1 (1958)

Chapter 6

Complexes of *n* and π Donors with Main Group Metal Compounds and Other Vacant Orbital Acceptors

P. N. Gates and D. Steele

Chemistry Department, Royal Holloway College, Englefield Green, Surrey, England.

6.1. INTRODUCTION

6.1.1. ORGANISATION

The number of complexes covered by the title of this chapter is vast and the spectroscopic literature is too large to be covered comprehensively in the space of a single chapter. We have therefore attempted to outline the main approaches and reasoning in the applications of vibrational spectroscopy (section 6.1.3) and, in the rest of the chapter, to illustrate the use of these approaches with selected examples from the literature. The contents of this latter part of the chapter have been organised according to the chemical species involved rather than according to bonding type as this forms a more natural classification of the subject matter. Some subjects cross the boundaries of our headings, belonging to more specialised classifications in addition to, say, periodic group classifications. Where this occurs more detailed discussion is reserved for the special categories.

6.1.2. VACANT ORBITAL ACCEPTORS

Vacant orbital acceptors are clearly one of the most important acceptor types to be considered when discussing donor–acceptor complexes as, in conjunction with *n* donors, they form a large number of relatively "strongly" bonded adducts. A vacant orbital acceptor is a species which possesses a vacant orbital of relatively high electron affinity (denoted by *v*—see section 1.1.3). Such species can be broadly classified into two main groups although further detailed subdivisions can also be made[1]. These are (i) neutral species into which class fall large numbers of metallic and non-metallic halides which have unoccupied orbitals of suitable energy and (ii) vacant orbital cation species such as Li^+, Na^+, etc.

One of the most commonly encountered series of class (i) acceptors are the group III trihalides of which a vast number of complexes are known. There has been a good deal of interest in complexes involving these acceptors and, in particular, the boron trihalides. In view of this some discussion of the properties of the boron trihalides as acceptors will be useful.

The ground state electronic configuration of boron is $2s^2 2p^1$ and, since the BX_3 (X = halide) molecules are known to be planar, the boron atom can be considered as using the three valence electrons to form three σ-bonds. In the case of BF_3 three σ-bonds with the three fluorine atoms are formed leaving a vacant 2p orbital on boron. Each fluorine then has, besides the electron in the BF, σ-bond three lone pairs, one of which is suitably placed to donate back to the boron, forming a $p\pi$-$p\pi$ bond. In effect this can be considered as intramolecular donor-acceptor action with the fluorine helping to satisfy the electron deficiency of boron. Support for this explanation comes from bond length data, as the BF bonds in BF_3 (1.309 Å)[2] are shorter than would be expected from the sum of their covalent radii.

One aspect of the boron trihalides which has evoked much interest is the relative order of their electron-accepting powers. There is now a large body of experimental evidence (see, for example, reference 6 and Cannon and Martin, reference 7, p. 399; also further discussion in section 6.3.5) which suggests that the order decreases as follows:

$$BI_3 > BBr_3 > BCl_3 \gg BF_3$$

This is certainly a reversal of the order expected on the basis of electronegativity or steric grounds and various explanations have been put forward to explain it. When complex formation takes place the π-bonding in the trihalide would probably be partially or completely lost as the vacant p_z orbital of the boron would be "occupied" by electron density from the donor. Experimental support for this again comes from bond length data as, in general, the BF bond length in various adducts (1.334 and 1.353 Å in $CH_3CN{-}BF_3$[4] and 1.38 to 1.40 Å in some alkylamine complexes[5]) is seen to be greater than with BF_3 itself (1.309 Å)[2]. Therefore, on this basis alone, for a given donor the addition compound with the trihalide having the greatest degree of π-bonding would be the least stable as more energy has to be expended to create a "vacant" orbital. Molecular orbital calculations[3] have shown that the π-bond stabilisation energy decreases in the order $BF_3 > BCl_3 > BBr_3$, thus qualitatively accounting for the observed order of acceptor power. These calculations were also used to compute the energy required to reorganise the planar boron trihalide to the pyramidal form in the complex and these decreased in the order $BF_3 > BCl_3 > BBr_3$. Experimentally-determined heats of reaction decrease in the order $BBr_3 > BCl_3 > BF_3$ for several $D{-}BX_3$ series of complexes[6] (where D represents a standard reference donor in each series) but when the estimated reorganisation energies were combined with these (and certain assumptions made about the other terms in the thermochemical cycle) the heat of the donor–acceptor bond formation was calculated to decrease in the order[6]

$$BF_3 > BCl_3 > BBr_3$$

In other words, if the reorganisation energy term is dominant in the thermochemical cycle the order of acceptor power (as measured by the stability of the complex) decreases in the same direction as expectations based on electronegativity or steric factors. However, apart from estimates based on heats of reaction, various spectroscopic methods have also been used to infer qualitatively the order of acceptor power. For example, low-frequency carbonyl shifts on complex formation between BX_3 and the carbonyl group (see section 6.2.3) and low field shifts of the α-protons in the n.m.r. spectra of trialkylamine–BX_3

complexes all increase in the order

$$BF_3 < BCl_3 < BBr_3 < BI_3$$

Such spectroscopic measurements would not measure the effect due to reorganisation energy and yet they apparently indicate the same order as the *experimental heats of reaction*. These spectroscopic quantities cannot, at present, be related quantitatively to acceptor power. However, some normal coordinate calculations[8,9] on $CH_3CN–BX_3$ systems have indicated that the strength of the B–N bond (as measured by the force constant) may indeed follow the same order. Thus it appears that there is a discrepancy between the spectroscopic results and the estimates based on the reorganisation energy calculations. This point is discussed further in section 6.3.4.

Common examples of vacant orbital cation acceptors (type (ii)) are ions such as Li^+, Na^+, K^+, Ag^+, etc. A manifestation of their action can be seen in the effects on solvent molecules in which they are dissolved. Ions in solution are invariably co-ordinated with solvent and oppositely-charged species and, if these species are polyatomic, then their vibrational spectra will be modified by the association. Numerous investigations of the interactions have been made, both on melts and solutions, by Raman and infrared techniques. Evidence has been obtained, both for ion association where the ions are in direct contact, and where they are separated by solvent molecules. In addition, multi-ionic association can also occur and may be detected by frequency and intensity variations of the vibrational bands.

The distinction between contact ion pairs and co-ordination complexes is essentially that the bonding in the former is electrostatic with the electrons remaining primarily in the sphere of their respective ions whereas no such sharp distinction can be made for ion-neutral solvent association. In this latter case the percentage contribution of the covalency to the bond is likely to be much greater although a dipole-induced dipole contribution to the bonding will still be important (section 1.1). A brief discussion of nitrate spectra in solution is given in section 6.8.

6.1.3. APPLICATION OF VIBRATIONAL SPECTROSCOPY

Vibrational spectroscopy has been used widely to investigate *n-v* donor–acceptor complexes and the literature is large. We have made no attempt to be comprehensive in our coverage but rather have sought to demonstrate the main approaches and reasoning used to study molecular structures and intermolecular interactions by these techniques. The main applications of vibrational spectroscopy which will be considered in this chapter are,

(i) the deduction of structural information.
(ii) the study of the strengths of donor-acceptor interactions.

(iii) the use as a monitor of concentrations of the species present in order to calculate thermodynamic data.

Apart from vibrational spectroscopy, if the complex is stable in the vapour state it should be possible to measure the pure rotational spectrum in the microwave region and hence determine the bond lengths and angles. It appears that complexes of group VI donors have not been successfully studied by this latter method as yet although a few investigations have been carried out on some complexes with group V donors (see section 6.3.2).

In addition, since nuclear magnetic resonance studies have played an important role in the identification of complexed species involved in Friedel-Crafts type intermediates, these are discussed in conjunction with the vibrational spectra (section 6.5.2).

(i) Structural Studies

Extensive use of vibrational spectroscopy has been made to elicit structural information about *n-v* complexes, such studies falling into two main groups,

(a) inference of the shape of the complex by measurement of the infrared and Raman spectra and a consideration of the selection rules governing the expected pattern of the spectra for various conceivable geometries of the complex, (b) deduction of the structure from a consideration of shifts in characteristic group frequencies of donor and acceptor on complex formation. In studies of type (a), once the stoichiometry has been established there will, in general, be only a limited number of possible shapes for the complex. In principle, from a consideration of the point group to which each possible structure belongs, the nature of the vibrational spectra (that is, infrared and Raman activities, degeneracies, polarisation properties of the Raman lines, etc.) expected in each case can be predicted in the usual way, subject to the limitations of the experimental data (see, for example, references 10 and 12). Comparison of these predictions with the observed spectra may decide between the possibilities in favourable circumstances (see chapter 2 for a more detailed discussion of the difficulties and limitations involved).

The deduction of the structure from the perturbation of characteristic group frequencies of donor and acceptor has also proved fruitful, especially in complexes involving carbonyl groups. It is expected that the "carbonyl stretching" frequency, ν(C=O), will be affected by donation of electrons from the oxygen to the acceptor. The use of this observation as an indicator of the site of donation has proved fruitful, subject to the limitations discussed below.

(ii) Strength of Donor–Acceptor Interaction

Apart from structural inferences, the other main area of application of vibrational spectroscopy is in the deduction of information about the strength of the donor–acceptor interaction. The two main approaches here are (a) study of the extent of perturbation of the characteristic group frequencies of donor and acceptor and any new modes associated with the complex formed and (b) calculation of force constants, particularly for the stretching of the donor–acceptor bond (see below, and chapters 1 and 2).

(a) *Studies Involving Characteristic Group Frequencies*

In principle, the perturbation of characteristic group frequencies of the donor molecule on complex formation might yield qualitative information about the strength of the electron acceptor species involved. Probably the most widely investigated systems employing this approach are those containing the carbonyl group in organic donors and it is convenient to illustrate the arguments with reference to this group.

In general, it is found that when a Lewis acid forms a complex with a carbonyl-containing donor, the frequency characteristic of the carbonyl stretching mode falls. It is also found that, for a given donor, the decrease in frequency varies with the electron acceptor. In a carbonyl group, the oxygen atom of which is the n-donor site, the extent of polarisation of the π-bonds (and hence the force constant and vibrational frequency) will depend on the electronic nature of the substituent groups. When a complex is formed it might be expected that the polarisation would be increased by an overall flow of electrons from the double bond to the intermolecular bond, thus reducing the stretching force constant and vibrational frequency. Hence, for a given donor, the magnitude of the characteristic frequency shift, $\Delta\nu$(CO) (that is, ν(CO) in the uncomplexed donor minus the perturbed value in the complex) might give a measure of the electron-accepting power of the Lewis acid. That this is the dominant effect in carbonyl containing complexes seems almost certain but the possibility of other effects which might operate in the reverse direction must also be considered. Firstly, there is the effect of kinematic coupling between the new vibrational modes which arise as a result of complex formation and the internal modes of the donor. In particular, for the $>$C=O–M- system there will be coupling between the C=O and M–O stretching vibrations, and, if the relative acceptor powers are inferred from the value of $\Delta\nu$(CO) for a series of Lewis acids, then the assumption is made that any such coupling does not vary from complex to complex. Secondly, any change in the hybridisation at the oxygen atom could affect σ-overlap in the carbon-oxygen bond with consequent effect on the force constant and stretching frequency. A number of such studies have been made and some of these are discussed in section 6.2.3. A similar situation is not found, however, for complexes of simple alkyl cyanides with electron acceptors. In this case the characteristic ν(CN) band frequency usually increases on complex formation. This is discussed at greater length in section 6.3.5 (and in section 2.3.3).

At first sight it might be expected that the most direct evidence for conclusions about the strength of the intermolecular bond from vibrational frequencies could be inferred from a characteristic donor–acceptor bond stretching frequency. However, this ignores several difficulties and it seems likely that only in certain special circumstances could qualitative conclusions be drawn. Some examples of this approach and a discussion of the limitations are given in sections (6.2.3 and 2.3.3).

(b) *Force Constant Calculations*

A number of normal coordinate calculations have been made on *n-v* donor–acceptor systems, the primary object usually being the estimation of the intermolecular bond stretching force constant which may throw light on the nature of the bonding involved.

The general techniques by which force constant calculations are carried out are well-known and are described in a number of texts[10,11,12,13] together with the limitation and difficulties inherent in the procedure. In most of the *n-v* systems considered in this chapter there is a relatively large number (8 or more) of atoms involved. This considerably increases the complexity of the calculations. With modern computing methods this presents no great difficulty but it is important to appreciate the consequences of the assumptions and approximations which are inevitable in calculations on larger molecules, especially if conclusions about bond strengths are to be drawn from the force constants obtained.

In the case of *weak* complexes it is usual to assume that the restoring forces in the complexed entities are close to those in the constituent donor and acceptor (see sections 1.2.1 and 1.2.2). The true harmonic force field for the isolated donor and acceptor will often not be known. However, any reasonable force field which reproduces the frequencies of the parent molecules will generally suffice for the purpose of understanding the *interplay* between the deformations of the donor–acceptor bonds and the deformations of the donor and acceptor. In such weakly bound systems it is often possible, for example, to ignore interaction force constants involving the internal co-ordinates created by forming the complex. However, since the eigenvectors deriving from such a calculation will be unreliable as a result of errors in the force field great caution must be exercised in interpreting any data which are sensitive functions of these eigenvectors.

In the case of *strong* complex formation both the geometry and force constants of the constituent molecules may be seriously perturbed. Force constants may then have to be estimated by comparison with data from analogous systems or, if these are not available, crude guesses have to be made. The resulting calculated frequencies will then be subject to larger errors with consequent doubt as to the validity of any conclusions drawn. An additional problem concerns the actual geometry in the complex. While the overall shape may be inferred from vibrational studies of the type discussed in sections 6.2.2 and 6.3.2, the actual bond lengths and bond angles can often be only crudely estimated. Fortunately, small uncertainties of, say $\pm 5°$ in angles and ± 0.3 Å in bond lengths usually have little effect on computed frequencies.

In view of the above problems most calculations have been carried out on relatively simple systems. In many instances the problem has been simplified even further by treating one, or sometimes both, interacting species as a monatomic unit (sections 1.2 and 2.3). With modern computing methods[13] however, such drastic simplifications are becoming less common and less justified.

In summary, the reliance to be placed on the conclusions based on the computed force constants will depend largely on the approximations made in the particular calculation and on the relative magnitudes of the quantities being compared (see, for example, discussion of acetonitrile complexes on p 483).

Apart from the above considerations results of *approximate* normal co-ordinate calculations can also give guidance to the location of fundamentals and are often a useful aid to making detailed assignments.

A further aspect which should be considered here is the inference of the strength of the donor-acceptor bond from calculated force constants for stretching of the bond. The basis of this is the empirical observation that the greater the bond strength, the greater is the stretching force constant, in various simple systems (for example, the hydrogen halides (see reference 14, page 9)). There is little theoretical understanding of the correlation at present, although support for it is forthcoming from studies of empirical potential functions (see, for example, references 15 and 16). From a simple qualitative viewpoint, however, it is not an unreasonable assumption that the greater the depth of the potential well (dissociation energy) in which the nuclei are moving, then the greater must be the curvature at the potential minimum (the second derivative, $(d^2V/dr^2)_{r_e}$, being the force constant for the motion). It must be appreciated however, that in a polyatomic complex the stretching of the donor-acceptor bond will generally be accompanied by a readjustment of the nuclear configurations in the constituent molecules, thereby reducing the bond dissociation energy. Thus, if a diatomic empirical potential function is used as an approximation to compute the bond dissociation energy from calculated force constants, an estimate of the upper limit will usually be obtained, assuming that the function itself is reliable.

(iii) Measurement of Thermodynamic Data

In principle, if the peak height or area of an absorption or Raman band is proportional to the concentration of a particular species, then it may be possible to use this property to monitor that concentration. If a suitable band for each species present at equilibrium in a particular system can be found and, provided it is free of overlap from neighbouring bands, it may be possible to obtain an estimate of equilibrium constant and other thermodynamic data for such a system. Some examples of this type of application are given in section 6.2.4, the principles of the method being described in detail in section 3.2.

6.2. COMPLEXES BETWEEN *n* DONORS CONTAINING GROUP VI ELEMENTS AND VACANT ORBITAL ACCEPTORS

6.2.1. INTRODUCTION

Elements of group VI possess the ns^2np^4 electron configuration in their ground state. Commonly these elements form divalent compounds in which the bonding can be regarded as being based on four sp^3 hybrid orbitals on the central element, two of which form bonds by overlap with suitable orbitals on the other atoms involved, the other two each being occupied by a lone pair of electrons. This qualitatively explains the bent configuration as exemplified in the water molecule or the ethers in the case of oxygen.

Such species readily act as electron donors and many complexes involving O and S are known. The donor properties of Se and Te compounds are less well-known. This means that a large number of oxygen-containing organic molecules are potential donors, for example, aldehydes, ethers, ketones, lactones, carboxylic acids, esters, amides, *etc*. Many analogous sulphur compounds are known and these can also be regarded as potential donors. Apart from these there are large numbers of other oxygen-containing species in which oxygen is bonded to other elements, *e.g*. sulphur, selenium, phosphorus, arsenic, nitrogen, *etc*.

Among such group VI complexes are the synthetically important intermediate species involved in Grignard (section 6.2.2) and Friedel-Crafts (sections 6.4 and 6.5) reactions, both of which have been investigated by vibrational spectroscopy.

The most conclusive structural studies have been carried out by X-ray diffraction methods and a number of these involving oxygen-containing donors have been summarised by Lindquist[17]. However, this type of structural information is very sparse for ketone complexes with electron acceptors and, in this case, vibrational spectroscopy has been used very successfully to infer complex structure.

6.2.2. STRUCTURAL STUDIES

(i) Studies of Complex Shape from Vibrational Spectra and Symmetry Considerations

It is well established that the carbon atom in ketones has a planar, trigonal arrangement of bonds around it. On this basis the bonding can be considered in terms of three sp^2 hybrid orbitals forming σ-bonds with the two substituent groups and with the oxygen atom. The remaining p orbital, together with the p-orbital from the oxygen, then forms the π-bond. This formally leaves two lone-pairs of electrons on the oxygen atom which are potentially available for donation. It is therefore probable that, in 1:1 donor–acceptor complexes the CO and intermolecular bonds will not be collinear. In view of the lack of crystallographic and other data for the ketone complexes vibrational spectra might be expected to be an important source of information about the overall shape of such complexes.

Some very thorough studies of this nature have been made on the acetone–boron trifluoride by Forel *et al.*[18] using various isotopic species. Preparation of $(CH_3)_2CO{-}^{11}BF_3$, $(CH_3)_2CO{-}^{10}BF_3$, $(CD_3)_2CO{-}^{11}BF_3$ and $(CD_3)CO{-}^{10}BF_3$ enabled nearly all the fundamentals to be assigned by consideration of the isotopic shifts and comparison with the spectra of $(CH_3)_2CO$ and $(CD_3)_2CO$. The vibrational spectra were found to be consistent with the 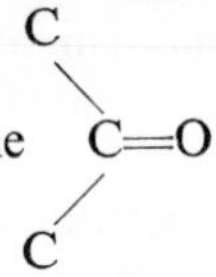 plane of the acetone remaining as a plane of symmetry in the complex with BF_3 but with non-collinear O–B and C═O bonds.

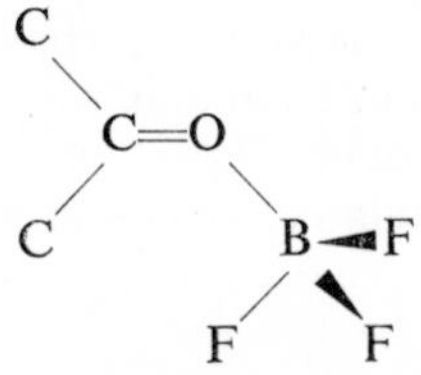

This is similar to the structure involved in the acetone–Br_2 complex[19] which is further discussed in chapter 3. For other complexes in which oxygen is the donor atom in groups such as -P═O, -S═O, -Se═O, -As═O etc. there is a good deal of independent structural information. For example the S-O–B angle in $(CH_3)_2SO{-}BF_3$ is 119.2° and the Nb–O-P bond in $Cl_3PO{-}NbCl_5$ is 148.8° [see reference 17, page 96].

Complexes of electron acceptors with ethers are well known and interest in the magnesium halide/ether systems centres chiefly on the nature of the species involved in the synthetically important Grignard reactions.

It is known from X-ray diffraction studies[20] that the solid complex $[(CH_3)_2O]_2{-}MgBr_2$ has a tetrahedral arrangement of bonds around the Mg atom and a similar arrangement of bonds around Mg is found in the C_6H_5MgBr dietherate[21]. Recently Wieser and Krueger[22] have studied the vibrational spectra of the solid $[(CH_3)_2O]{-}MgI_2$ and diethyl ether solutions of $MgBr_2$ and MgI_2. Since the MgX_2O_2 skeleton apparently has approximately C_{2v} symmetry there should be two Mg–X and two Mg–O stretching vibrations ($2a_1 + b_1 + b_2$) which would all be active in the infrared and Raman spectra. However, various polymeric structures have been postulated for ethereal solutions of MgI_2 and $MgBr_2$[23,24] which involve either 4- or 6-co-ordinate magnesium. In the 4-co-ordinate species the basic unit would be tetrahedral MgX_4, forming chains via halogen bridges with two ether molecules bonded to the last magnesium atom in a chain. The T_d symmetry of the isolated MgX_4 unit would be reduced to C_{2v} in the chain which would give four Mg-X stretching modes ($2a_1 + b_1 + b_2$) and, in addition, two Mg–O stretching modes, the intensities of which might be weak relative to those of the Mg-X modes, depending on the length of the chain. For the six-coordinate species, the basic unit of which is MgX_4O_2 the symmetry would be D_{2h} for a polymeric chain with planar halogen bridges. This would give rise to four Mg–X stretching modes, $a_g + b_{1g} + b_{2u} + b_{3u}$, the former two being Raman active and the latter two infrared active together with one infrared (b_{1u}) and one Raman active (a_g) stretching mode of the Mg–O bonds.

In a concentrated solution of MgI_2 in diethyl ether only two Raman bands (at 147 and 163 cm^{-1}) were found in the region where MgI and MgO stretching vibrational bands would be expected, these being assigned to the MgI stretching vibrations of the polymeric species. Since only the two bands were observed it seems likely that the basic unit is the 6-coordinated MgI_4O_2 (D_{2h} symmetry) as more would be expected for the other models. On the other hand, in the Raman spectra of the solid dietherate there was a weak band at 183 cm^{-1} and a strong one at 111 cm^{-1} which were assigned to the MgI stretching and bending vibrations respectively of the monomeric, tetrahedral MgI_2O_2 skeleton. In dilute solution a new band at 112 cm^{-1} occurred together with a pair of bands at 143 and 163 cm^{-1} similar to those in the concentrated solution but with reduced intensity. This suggests that some monomeric complex exists in dilute solution but that the polymeric structure is also still present.

(ii) Deduction of Structure from Characteristic Group Frequency Considerations

Although the arguments used to deduce the structure of the diethyl-ether–MgX_2 complexes in the last section partly relied on characteristic regions for bands due to particular modes the main consideration was that of symmetry. However, the concept of characteristic group frequencies and their perturbation on complex formation has proved extremely fruitful in the inference of structure, especially for complexes involving carbonyl groups. As mentioned above (section 6.1.3), it is expected that the "carbonyl stretching" frequency will be affected by donation of electrons from oxygen to the acceptor and use of this observation as an indicator of the site of donation has proved useful.

For example, a series of donor–acceptor complexes between xanthone (I) and numerous acceptors have been investigated by Cook[25]. That

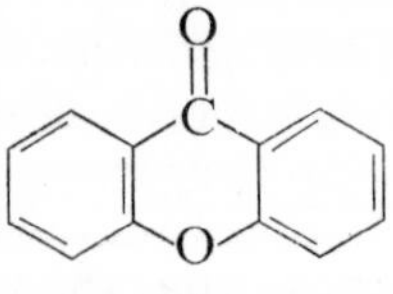

I

donation occurs from the carbonyl oxygen is certain as the shift observed from the parent carbonyl stretching band (at 1663 cm^{-1}) ranges from -34 cm^{-1} (for the $CdCl_2$ complex) to -263 cm^{-1} (for the BI_3 complex) whereas the pattern is relatively unchanged for frequencies associated with vibrations of that part of the ring containing the heterocyclic oxygen atom, thus making donation from this site unlikely.

Many amides of general type $RCONH_2$ (R = alkyl) will form 1:1 complexes with electron acceptors and again there are two potential donor atoms present. Gerrard *et al.*[26] have investigated the infrared spectra of complexes of BCl_3 and BBr_3 with various amides. The two alternatives are,

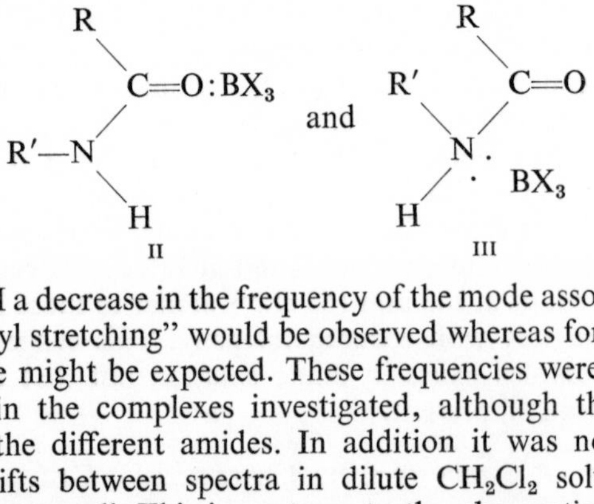

II and III

In structure II a decrease in the frequency of the mode associated mainly with "carbonyl stretching" would be observed whereas for structure III some increase might be expected. These frequencies were all observed to decrease in the complexes investigated, although the magnitude varied with the different amides. In addition it was noted that the frequency shifts between spectra in dilute CH_2Cl_2 solution and in mulls were very small. This is contrary to the observation for uncomplexed amides[27] in which absorptions associated with motions of the NH and CO bonds showed large shifts on passing from solid (as a mull) to solution. These shifts were interpreted as arising as a result of intermolecular hydrogen bonding. The absence of comparable solution-solid shifts in the complexes indicates that association does not occur in this case. This points towards the non-availability of the oxygen lone-pair, and implies strongly that the carbonyl oxygen is the donor site.

As discussed further in section 6.4.3, characteristic carbonyl stretching frequencies yield a rapid and reliable method of determining between structures of type IV and type V for $RCOX{-}MX_n$ complexes (R =

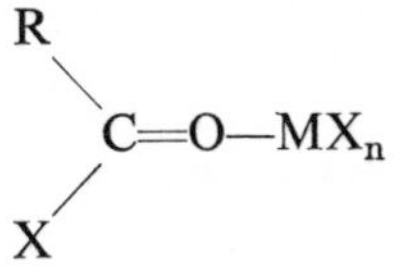

$[R{-}C{\equiv}\overset{+}{O}][MX_{n+1}]^-$

IV V

alkyl or aryl, X = halogen, MX_n = electron acceptor). For the solid 1:1 complex, *o*-$CH_3C_6H_4COCl{-}TiCl_4$[28], a downward ν(CO) frequency shift is observed in the infrared spectrum with no absorption in the 2200–2300 cm^{-1} region, a type IV structure thus being inferred. Many examples of type V are known in which the characteristic ν(CO) frequency is replaced by a strong absorption in the region 2200–2300 cm^{-1}, characteristic of the $\nu(C{\equiv}O^+)$ stretching band (for further discussion see section 6.4). Analogous arguments have been used to decide between ionic and molecular structures in complexes involving $POCl_3$. Due to low-frequency shifts in the bands assigned to the "PO stretching" frequency, it was concluded[29] that $POCl_3{-}AlCl_3$ and $POCl_3{-}GaCl_3$ were not ionic but that coordination occurred at the phosphoryl oxygen.

In the absence of an X-ray structural determination some disagreement over the vibrational spectra and their interpretation in terms of structure of the $POCl_3{-}BCl_3$ complex has arisen. It was originally proposed by Waddington and Klanberg[30] that a covalent structure of type VI was formed with coordination at the phosphoryl oxygen.

$Cl_3P{=}O{-}BCl_3$ $[POCl_2]^+[BCl_4]^-$

VI VII

They obtained infrared spectra of nujol mulls and found a strong band at 1290 cm^{-1} which was similar to the "PO stretching" band position in $POCl_3$ itself and was assigned as the ν(PO) mode of the complex. On the basis of a weak band at 1190 cm^{-1} which they assigned as a "B–O stretching" mode, and a large decrease in the ^{10}BCl and ^{11}BCl antisymmetric stretching frequencies (assigned to bands at 700 and 667 cm^{-1}) they concluded that structure VI was the correct one. Gerrard *et al.*[31], again studying the infrared spectra of nujol mull samples of the complex, also found that the band at 1290 cm^{-1} remained unchanged but observed no sign of the 1190 cm^{-1} band which had previously been assigned to a "B–O stretching" mode. Furthermore, a broad region of absorption around 700 cm^{-1} was assigned to the antisymmetric stretching of ^{10}BCl and ^{11}BCl in BCl_4^- and from this it was concluded that an ionic formulation of type VII was present. However, Peach and Waddington[32] reinvestigated $POCl_3{-}BCl_3$ together with the BCl_3 complexes of $(C_6H_5)_2POCl$ and $(C_6H_5)_3PO$ and found a progressive decrease in the "PO stretching" frequency which strongly suggested that the coordination site was oxygen. Wartenberg and Goubeau[33] obtained infrared spectra of the solid complex at low temperature (−90°C) by subliming it onto alkali halide plates, and Raman spectra at room temperature. In the low temperature infrared spectra no band was observed at 1290 cm^{-1} but a strong band at 1167 cm^{-1} appeared which was assigned as a perturbed "PO stretching" band. Strong bands at 713 and 382 cm^{-1} were assigned to the B–O stretching and B-O-P bending modes respectively of the covalent complex. The antisymmetric ^{10}BCl and ^{11}BCl stretching modes were assigned to bands in the 727–795 cm^{-1} region. The room temperature Raman spectrum gave substantially the same result and it was concluded that structure VI was the correct one. It was also suggested that the conflicting results of the other two groups of workers were primarily due to hydrolysis and dissociation of the complexes.

Very recently, some further observations[34] are in substantial agreement with the work of Goubeau for the complex sublimed onto KBr plates and the spectrum recorded at −100°C. However, when the complex was allowed to warm to room temperature and then removed by pumping, all the bands disappeared except for those due to BCl_4^- which indicates that reaction with the plates had occurred. The intensity of these bands increased with the time for which the complex was in contact with the plates at room temperature and it is possible that this may have occurred in the spectrum observed by Gerrard *et al.*[31] When the complex was examined as a nujol mull at room temperature the 1190 cm^{-1} band (observed by Waddington *et al.* but not by Gerrard *et al.*) did not appear immediately and then increased in intensity with time, together with a band at 3200 cm^{-1}. These are almost certainly due to a hydrolysis product such as boric acid. Also, in view of more recent force constant calculations on B–O containing systems, the mode which contains a large proportion of B–O stretching in its normal coordinate usually gives rise to a band in the 600–700 cm^{-1} region. This makes the assignment of the band at 1190 cm^{-1} to a ν(BO) mode rather unlikely. In the spectra recorded as nujol mulls the presence of the strong band at 1290 cm^{-1} could have been due to $POCl_3$ itself from dissociation of the complex as bands due to liquid BCl_3 itself were also observed.

Hence, the weight of the evidence points strongly towards structure VI as the correct one.

6.2.3. STRENGTH OF DONOR–ACCEPTOR INTERACTION

(i) Studies Involving Shifts in Characteristic Group Frequencies

The assumptions upon which studies of this type are based are discussed in section 6.1.3 (above). The majority of such investigations have involved complexes with carbonyl compounds in which the value for $\Delta\bar{\nu}$(CO) has been used as the criterion of acceptor power towards a particular carbonyl-containing donor.

Lappert[35] has measured the values of $\Delta\bar{\nu}(CO)$ for complexes between ethyl acetate (as reference donor) and several group III trihalide acceptors (see table 6.1). Various other carbonyl-containing species have been used as reference donors, some of which are listed in table 6.1. In the examples selected above the relative orders are the

TABLE 6.1

$\Delta\bar{\nu}(CO)$ (in cm^{-1}) for some complexes of carbonyl-containing donors with various electron acceptors

Acceptor	$\Delta\bar{\nu}(CO)$			
	Xanthone[25]	1-Ethyl-2,6-dimethyl-4-pyridone[36]	Ethyl acetate[35]	Benzophenone[38]
$CdCl_2$	34	56		
$CaCl_2$	47			
$CoCl_2$	50			
$HgCl_2$	52	73		
$HgBr_2$	54			
$ZnCl_2$	57	89		
$BiCl_3$	59			
$InCl_3$	59		113	
$SnCl_4$	133		128	
$SnBr_4$	136	96	111	
BF_3	138		119	103*
$FeCl_3$	141			145
$FeBr_3$	154			
$AlCl_3$			117	122
$AlBr_3$	163		138	142
PF_5	166			
$ZrCl_4$	167		104	
$TiCl_4$	175	102	128	144
$TiBr_4$	181		128	
BCl_3	229		176	166*
SbF_5	243			
$SbCl_3$	252	105		
BBr_3	253	96	191	181*
BI_3	263			

* Reference 37.

same for the boron trihalides towards the different reference donors. However, it can also be seen that when ethyl acetate is used as the reference donor BF_3 and $InCl_3$ show very similar shifts whereas the shift for BF_3 is over twice that for $InCl_3$ with xanthone as reference donor. Thus, vague statements about acceptor power without specification of the reference donor are meaningless. In any event, more than a qualitative order can hardly be expected from measurements of this type as, for example, the polarisation of the carbonyl bond by the different donors will vary and any effects due to coupling of vibrations are ignored. In the light of these factors it is important to realise that measurements of this nature are not (as yet) capable of being related directly to acceptor power in a quantitative manner. This is especially the case in the absence of normal co-ordinate calculations since normal coordinate changes may well contribute to the $v(CO)$ band shift.

(ii) Characteristic Frequencies of the Donor–Acceptor Bond Stretching mode

It might be expected that the most direct indication of the strength of the intermolecular interaction would be from a characteristic donor–acceptor bond stretching frequency. The chief disadvantage to such an approach is that there is likely to be coupling between the vibration of the intermolecular bond and internal vibrations of appropriate symmetry in the donor and acceptor. (See section 1.2.3 for a more detailed discussion of this point.) When normal coordinate calculations are carried out on such systems it is usually found that the stretching of this bond is considerably coupled with other vibrations in the normal mode (see, for example, references 18, 39, 40). In such circumstances it is meaningless to talk about a characteristic stretching frequency as there may be a large contribution to the potential energy of the mode from stretching and deformations of other bonds and bond angles. This means that such an approach is unlikely to yield useful information for systems of the type R_2O-MX_n or R_3N-MX_n. On the other hand, a series of ketone–boron trifluoride complexes have been examined and the $v(B-O)$ stretching mode assigned[41] bands in the 620–710 cm^{-1} region. In this instance a single reference acceptor is being used and in each case it is coordinated to a carbonyl grouping. In these circumstances it may be reasonable to suppose that the coupling effect will remain reasonably constant throughout the series and that the stretching frequency will reflect, to some extent, variations in the force constant of the bond. In effect it is being assumed that there is a vibration which is localised within a common group of bonds namely:

$$\begin{array}{c} C \\ \diagdown \\ \quad C{=}O{-}BF_3 \\ \diagup \\ C \end{array}$$

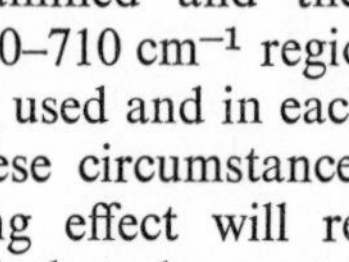

and that the observed frequency variation is a measure of the varying contribution from the O–B bond. The contribution to λ_i $(= 4\pi^2 v_i^2)$ from the stretching of this bond will be $(l_{ji})^2 f_{jj}$ where l_{ji} is the jth

component of the eigenvector of the ith vibration and f_{jj} is the force constant for the jth internal coordinate, in this case the B–O stretch. If we can assume that the change in the eigenvector is small then the variation in λ_i reflects the variations in the changing force constant. In the above it is assumed that variations in the interaction force constants may be neglected. Since in general $f_{ij} < f_{ii}, f_{jj}$ this is normally reasonable. For the ketone complexes of the type $R_2CO–BF_3$ the values assigned to the ν(B–O) mode were found to decrease in the order $CH_3(690\ cm^{-1}) > C_2H_5(672\ cm^{-1}) >$ iso-$C_3H_7(654\ cm^{-1})$ suggesting that acetone forms the strongest complex[41]. However, in the light of the more recent normal coordinate calculations on the $(CH_3)_2CO–BF_3$ complex[18] it is probable that this mode, although mixed in character, does not contain a preponderance of B–O stretching motion in its potential energy distribution. (See below and section 2.3.3 for further discussion of these normal coordinate calculations.)

(iii) Force Constant Studies

Some approximate force constant calculations have been made, chiefly on complexes involving oxygen donors with group III trihalides.

An investigation of the acetone-boron trifluoride complex[18] included some calculations on the $X–BF_3$ grouping (X = O, S, Se). This was done on the basis of an isolated C_{3v}, XBF_3 system with the X group separated from the rest of the donor molecule. Various sets of frequencies were calculated by allowing the value of the B–X stretching force constant to vary. For the A_1 species of the $O—BF_3$ group (in $(CH_3)_2CO–BF_3$) computed on this basis the best fit between calculated and observed frequencies was obtained using a force constant of 200 N m^{-1}. By using observed frequencies for the complexes $(CH_3)_2S–BF_3$ and $(CH_3)_2Se–BF_3$ and transferring the force field employed above and varying the B-X stretching force constant, the best fit for the A_1 species was obtained by using stretching force constants of 150 N m^{-1} for both the B–S and B–Se bonds. It was, however, also pointed out that the B-X vibration could not be regarded as a localised mode with a characteristic absorption.

A similar study[42] has been made on the $(CH_3)_2SO–BF_3$ complex and, using isotopic substitution and comparison with the spectra of $(CH_3)_2SO$ and $(CD_3)_2SO$ together with Raman polarisation data, a vibrational assignment made. By making similar approximations (see ref. 18) it was proposed that a B–O stretching force constant of 250 to 300 N m^{-1} gave the best fit between calculated and experimental frequencies.

Infrared spectra of complexes between $AlCl_3$ and various oxygen-containing donors such as $(CH_3)_2O$, $(C_2H_5)_2O$, CH_3NO_2 and $C_6H_5NO_2$ have been investigated by Jones and Wood[40] and assignments made.

The $AlCl_3X$ framework with all tetrahedral angles (C_{3v} symmetry) was taken as the basis for the calculations. A simplified force field was used and the mass of X was allowed to vary from 16 to 32 units to allow, to some extent, for the effect of coupling with internal donor vibrations. The calculations showed that it was not possible to consider any particular band characteristic of the intermolecular bond stretching frequency which is common experience in such calculations (see for example references 18, 39, 42). The Al–O stretching force constants which best reproduced the observed frequencies on the basis of the calculations were found to be 160, 150, 150 and 210 N m^{-1} for $(CH_3)_2O$, $(C_2H_5)_2O$, CH_3NO_2 and $C_6H_5NO_2$ respectively. This is of the same order as the B–O, B–S and B–Se values above[18].

6.2.4. MEASUREMENT OF THERMODYNAMIC DATA

Infrared and Raman methods have been used to monitor concentrations of species present at equilibrium in solutions of donors and acceptors and from these calculations equilibrium constants for the associations have been estimated.

Duyckaerts and Michel[43] have used this approach to study the donor properties of methylbenzoate and acetophenone towards $SbCl_3$ in diethyl ether. Raman spectra of the acetophenone–$SbCl_3$ system showed bands at 1692 cm^{-1} ("free" acetophenone) and 1671 cm^{-1} (complexed acetophenone) and for methyl benzoate the corresponding bands were at 1729 and 1699 cm^{-1}. Application of Job's method of continuous variations was made by measuring the relative areas of two peaks for various mixtures of the Lewis acid and the donor. On the basis of this and employing the expression:

$$K_{a:b} = \frac{[D_aA_b]}{[D_a]^a[A]^b} \qquad 6.1$$

the association constants calculated (a = b = 1) were 0.79 ± 0.01 dm^3 mol^{-1} for the $C_6H_5COOCH_3$–$SbCl_3$ and 2.70 ± 0.40 dm^3 mol^{-1} for the $C_6H_5COCH_3$–$SbCl_3$ system. An investigation of the $(C_2H_5O)_3PO$–$AsCl_3$ and $(C_4H_9O)_3PO$–$AsCl_3$ has also been made by Duyckaerts and Roland[44] using an infrared method to monitor concentrations. The complexity of the spectra made it difficult to measure directly the intensity of the "complexed" ν(PO) band but Job's method was applied by using the intensity variation with concentration for the "free" ν(PO) band. The complex formed with trialkyl phosphate was found to have 1:1 stoichiometry with an association constant $K_{1:1}$ of 14.2 ± 1.7 dm^3 mol^{-1}. Both 1:1 and 2:1 donor acceptor species were identified in the tributyl phosphate system with association constants of 15.4 ± 1.6 dm^3 mol^{-1} and 4.2 ± 0.25 dm^6 mol^{-2} respectively.

6.3. COMPLEXES BETWEEN *n* DONORS CONTAINING GROUP V ELEMENTS AND VACANT ORBITAL ACCEPTORS

6.3.1. INTRODUCTION

The elements of group V, nitrogen, phosphorus, arsenic, antimony and bismuth, possess the ns^2np^3 ground state electronic configuration. Commonly these elements form trivalent compounds in which there is a pyramidal distribution of bonds around the element together with a lone pair of electrons in the fourth sp^3 hybrid orbital. Such species can thus act as electron-pair donors and large numbers of complexes involving them are known. Nitrogen differs from the lower members of the series in that it has no d-orbitals and thus can *only* act as a simple electron-pair donor. On the other hand, phosphorus, arsenic and antimony possess empty d-orbitals of low energy. This allows the possibility of multiple bonding if the acceptor has electrons in orbitals of appropriate symmetry.

6.3.2. GENERAL STRUCTURAL STUDIES

As discussed above, in the case of a nitrogen-containing donor (with the exception of those containing -C≡N, etc.) there is usually a pyrimidal distribution of bonds around the central atom with an electron pair occupying the fourth sp^3 hybrid orbital. In view of this an approximately tetrahedral disposition of bonds is expected on complex formation and vibrational spectra of various complexes of the type $R_3N–MX_3$ have been satisfactorily explained on the basis of a staggered, ethane-like configuration (with C_{3v} symmetry)[39,45]. In view of this there has been less spectroscopic interest in structural aspects of these complexes than the corresponding group VI species. However, two types of investigation have been carried out;

(a) the inference of molecular shape from vibrational spectra.
(b) the accurate determination of molecular geometry from pure rotational spectra.

(a) Inference of Molecular Shape from Vibrational Spectra

It would be difficult (and pointless) to always separate structural investigations into those which rely solely on characteristic frequency considerations or solely on symmetry arguments. Frequently the two approaches are used in combination as in, for example, the prediction of a certain number of stretching frequencies (based on symmetry arguments) expected to lie in a particular region of the spectrum (on the basis of characteristic group frequencies).

An example of such an application is the inference of the structure of phosphine–borane in the solid state. Rudolph *et al.*[45] investigated the

^1H and ^{11}B n.m.r. spectra of the compound originally reported as $2PH_3$–B_2H_6 in the liquid state. It was concluded that the complex was monomeric PH_3–BH_3 in the liquid state but the vibrational spectra in the solid were investigated to determine whether or not the same structure persisted in this state. On the basis of the liquid state structure with a staggered C_{3v} symmetry there should be eleven infrared active and eleven Raman active modes. Two possible alternatives proposed for the solid were:

$$[PH_4]^+[H_2P(BH_3)_2]^- \qquad [(PH_3)_2H_2B]^+[BH_4]^-$$

VIII IX

If the BH_3 group in VIII and the PH_3 groups in IX are considered as point masses and all angles assumed to be tetrahedral, each structure contains ions of T_d (PH_4^+ and BH_4^-) and C_{2v} [$H_2P(BH_3)_2^-$ and $(PH_3)_2$-BH_2^-] symmetries from which the skeletal frequencies can be predicted. For T_d, two infrared and four Raman bands (two coincidences) are expected, and for C_{2v}, eight infrared and nine Raman bands (eight coincidences) should occur. Internal motions of the MH_3 groups would increase the complexity of the observed spectrum in the appropriate regions. Thus, if either VIII or IX were the correct structure the infrared and Raman spectra should differ significantly. It was found however, that the infrared and Raman spectra were similar, relatively simple and could be easily related to the C_{3v} model and it was inferred that the solid complex exists as monomeric units. Further, the known characteristic frequencies[14] for PH_4^+ or BH_4^- were not observed.

Recently the gas phase structure of $2PF_3$–B_2H_4 has been determined by electron diffraction.[46] The PF_3 groups are *trans* about the B-B axis with coplanar B and P atoms. The hydrogen positions could not be uniquely determined but the data favoured a structure with four terminal hydrogens rather than bridging H bonds. Matrix-isolated infrared and gas-phase Raman spectra showed no coincidence of any bands which is completely consistent with a planar PBBP skeleton and a centre of symmetry. To decide between the presence of a structure involving H bridges and the one involving all terminal H atoms the characteristic infrared frequencies were employed. There was no absorption between 1600 and 1800 cm^{-1} which is the region associated with the stretching of bridging boron-hydrogen bonds[47]. Since the most intense band of B_2H_6 occurs in this region it seems unlikely that $2PF_3$–B_2H_4 involves any bridging hydrogen atoms.

Various *gem*-dibasic ligands such as $(CH_3)_2PCH_2N(CH_3)_2$, $(CH_3)_2$-PCH_2SCH_3 and $(CH_3)_2NCH_2SCH_3$ all readily form 1:1 complexes with BH_3 by reaction with diborane and, in addition, a stable $(CH_3)_2$-$PCH_2N(CH_3)_2$–$2BH_3$ is formed. BH_3 bonding to P (as distinct from N or S) was inferred[48] for the first two ligands because the $P(CH_3)_2$ methyl proton resonances were shifted more downfield than the other proton resonances on complex formation with BH_3. An infrared absorption at 2790 cm^{-1} in the $(CH_3)_2PCH_2N(CH_3)_2$–BH_3 complex indicates the presence of 3 coordinate nitrogen with a methyl group attached. This conclusion is based on the characteristic frequency range of 2760–2820 cm^{-1} found for compounds containing the NCH_3 grouping[49]. Similar N-methyl compounds in which coordination of the nitrogen lone pair occurs do not show this absorption. In the case of the $(CH_3)_2PCH_2N(CH_3)_2$–$2BH_3$ complex the structure involving simple coordination (at N and P) is favoured rather than an ionic formulation involving BH_4^- since the infrared spectrum shows no sign of the strong bending absorption characteristic of BH_4^- in the 1070–1170 cm^{-1} region.[14]

Infrared and Raman spectra of $2N(CH_3)_3$–AlH_3 and $2N(CH_3)_3$–AlD_3 have been investigated[50]. A number of possible geometries could be envisaged for this complex including a trigonal bipyramid of D_{3h} symmetry, a distorted bipyramid (D_3), a square pyramid (C_{2v}) or a dimer of fused octahedra bridged by two Al-H-Al three centre links (D_{2h}). The D_{3h} structure would be expected to be the most chemically-probable but the observed spectrum was most easily assigned on the basis of a distorted trigonal bipyramid (D_3 point group) structure. The main point of the argument is the infrared activity of the NalH deformation mode which would be inconsistent with the D_{3h} structure. In either case the spectra are completely consistent with a planar arrangement of the AlH_3 part of the molecule.

(b) Molecular Geometry Determination from Pure Rotational Spectra

Accurate determinations of geometry from pure rotational spectra are relatively rare for donor–acceptor complexes as such studies are limited to the gas phase. Such a limitation is severe either because of involatility of the complex or because of a high degree of dissociation at the temperature necessary to produce a sufficient vapour pressure.

However, complete determinations of molecular geometry have now been made by microwave spectroscopy for several complexes involving P–B and N–B donor–acceptor bonds. The intermolecular bond lengths determined in this way are summarised in table 6.2.

Lide *et al.*[51] investigated the microwave spectrum of the vapour above crystalline $(CH_3)_3N$–$B(CH_3)_3$. Four rotational transitions were observed and hence rotational constants calculated, the four values being in good agreement with one another. The mean rotational constant calculated from these was 1575 ± 5 MHz. Making some assumptions about the geometry (tetrahedral CNC and CBC angles with CN and CB bond lengths the same as those in the parent donor and acceptor respectively) an estimate of the N–B bond length was made from the observed moment of inertia. Allowing for an error

TABLE 6.2
Some microwave determinations of the donor–acceptor bond length in complexes between group V donors and boron-containing acceptors

Complex	D—A bond	Bond length[a]/Å	Reference
$(CH_3)_3N\text{–}B(CH_3)_3$	N–B	1.80 ± 0.15	51
$(CH_3)_3N\text{–}BF_3$	N–B	1.636 ± 0.004	53
$(CH_3)_3N\text{–}BH_3$	N–B	1.65 ± 0.02	54
$F_3P\text{–}BH_3$	P–B	1.836 ± 0.012	55
$HF_2P\text{–}BH_3$	P–B	1.832 ± 0.009	56

a. 1 Å = 0.1 nm.

of ± 4° in the tetrahedral angles assumed, the N–B bond distance thus calculated was 1.80 ± 0.15 Å.

Prokhorov and Shipulo[52] have measured ten rotational lines in the microwave spectrum of $(CH_3)_3N\text{–}BF_3$ and calculated an average value for the rotational constant of 1750 MHz. More recently Bryan and Kuczkowski[53] have examined the $J = 9 \leftarrow 8$ and $J = 10 \leftarrow 9$ rotational transitions for several isotopic species of the same complex, namely $(CH_3)_3{}^{14}N\text{–}BF_3$, $(CH_3)_3{}^{15}N\text{–}{}^{11}BF_3$ and $(CH_3)_3{}^{14}N\text{–}{}^{10}BF_3$. The N–B bond length calculated from these results was 1.636 ± 0.004 Å but the structure of the rest of the molecule was not determined.

From a microwave study of eight isotopic species of $(CH_3)_3N\text{–}BH_3$ Schirdewahn[54] has obtained a value of 1.65 ± 0.02 Å for the B–N distance. Using several isotopic species for each complex the microwave spectra of $PF_3\text{–}BH_3$[55] and $PF_2H\text{–}BH_3$[56] have been investigated and complete structural determinations made. The P–B bond lengths were found to be essentially identical in the two compounds, 1.832 ± 0.009 Å for $PF_2H\text{–}BH_3$ and 1.836 ± 0.012 Å for $PF_3\text{–}BH_3$. For the $PF_2H\text{–}BH_3$ complex the expected staggered configuration with C_s symmetry was confirmed but two of the PBH angles indicate that the BH_3 is tilted away from the fluorine atoms. On electronegativity grounds it might be expected that PF_2H would be a better donor than PF_3 towards BH_3 and it is thus surprising that the two B–P bonds lengths are virtually identical. In accord with this, $PF_2H\text{–}BH_3$ is more stable towards dissociation[57] than $PF_3\text{–}BH_3$ and, on this basis, a shorter B–P bond in this complex might be expected. Various reasons for this stability were advanced although it is difficult to single out any particular one as likely to be dominant.

Since the possibility exists in these complexes that the P–B σ-bond might be supplemented by $d\pi\text{–}p\pi$ bonding, between the d-electrons of phosphorus and the p orbital of boron, some extended Hückel molecular orbital (*EHMO*) calculations were made to determine whether such a bonding mechanism could account for the tilt. However, although the calculated energies were improved by including d-orbitals, a minimum energy was obtained for a structure in which the BH_3 group was coaxial about the P–B bond. The calculations on this basis indicated that the experimentally-observed tilt should result in a *less* stable situation. It was suggested that non-bonded electrostatic interactions (which were not explicitly included in the *EHMO* calculations) could account for the effect. Electrostatic calculations for a symmetrical and a tilted BH_3 group made by placing plausible charges (from the *EHMO* calculations) on the terminal atoms supported this argument. It was thus considered that electrostatic interaction was the most likely explanation for the tilt.

It is evident that a normal coordinate analysis for this and similar complexes would yield valuable information.

A further point of interest is the evaluation of barriers to rotation about the B–P bond in PF_3–BH_3[55] and PF_2H–BH_3[56]. For the PF_2H–BH_3 complex the torsion mode was assigned a band at either 221 cm^{-1} or 247 cm^{-1}. Using a three-fold potential function, and assuming a BH_3 group symmetrical about the B–P bond, a range for the barrier of 15.1–18.9 kJ mol^{-1} was obtained. It was pointed out that this is high compared to the isoelectronic SiF_2HCH_3 (5.2 kJ mol^{-1}) but is consistent with the difference between PF_3–BH_3 (13.5 kJ $mole^{-1}$)[55] and SiF_3CH_3 (5.8 kJ $mole^{-1}$). As yet no detailed explanation for these high barriers is forthcoming.

6.3.3. STRENGTH OF DONOR–ACCEPTOR INTERACTION

As before, studies of this type can be divided into those involving characteristic group frequencies and force constant calculations.

(i) Studies Using Characteristic Group Frequencies

Donors containing the -C≡N group might be expected to be analogous to the carbonyl grouping and show a decrease in the characteristic ν(CN) band frequency on complex formation. However, this is not so and in many complexes[58,59] an *increase* has been observed in the stretching frequency when a complex is formed (section 2.3.3). Thus, qualitative estimates of acceptor order based on such shifts cannot be drawn in the same simple fashion as with carbonyl groups and obviously the dominant effect cannot be electron withdrawal from the CN bond. This fact has been the centre of some interest. It is discussed in more detail in sections 6.4.3 and 2.3.3.

As might be expected characteristic donor–acceptor bond stretching frequencies are not meaningful in complexes of the type R_3N–MX_n as the 'N–M stretching mode' is a considerably "mixed" mode (see above and section 2.3.3).

Taylor and Cluff[60] have investigated the NH_3–BH_3 system and the deuterated analogues and assigned bands in the 708–785 cm^{-1} region to the "B–N stretching" modes on the basis of the isotopic shifts. Based on a pseudo-diatomic model the force constant calculated for NH_3–BH_3

was 279 N m^{-1}. This is obviously a much oversimplified calculation but the work definitely established that a mode, at least largely, associated with the N–B stretching lay in this region. In further studies on NH_3–BF_3 Taylor[61] found that considerable mixing occurred between the N–B bond stretch and the appropriate BF vibrations and subsequent workers[18,39,40] emphasise that such mixing always occurs. Hence, in view of this, force constant studies (even relatively elementary ones) are usually more useful for drawing conclusions about the strength of the bonding than frequency correlations. (This is emphasised in chapter 1 and in section 2.3.3.)

(ii) Force Constant Studies

Subject to the same limitations discussed in section 6.1.3 the donor–acceptor stretching force constant should give some indication of the strength of the interaction. Several of the simple systems involving nitrogen and phosphorus donors with group III acceptors have been investigated by normal coordinate analyses of varying degrees of complexity. An approximate normal coordinate analysis[62] of the complex H_3N–$AlCl_3$ gave the best fit between calculated and observed frequencies for a Al–N bond stretching force constant of 220 N m^{-1}.

The complexes PH_3–BH_3 and PF_3–BH_3 have been studied by Sawodny and Goubeau[63] who obtained values for the PB stretching force constants of 195.8 and 229.5 N m^{-1} respectively. From these values estimates of the bond order of 0.78 and 0.92 for the P–B bond in PH_3–BH_3 and PF_3–BH_3 respectively were made. The same pair of complexes have been the subject of an *ab initio SCF–MO* calculation[64]. It was concluded that, in both cases, the intermolecular bond is essentially a σ-bond involving the 3p and 3s phosphorus orbitals and the 2p boron orbital. The calculations indicated that the PH_3–BH_3 was more stable than PF_3–BH_3 suggesting that PH_3 is a more effective donor towards BH_3 than PF_3. This correlates with a calculated larger increase in the formal phosphorus charge on formation of PH_3–BH_3 than on formation of PF_3–BH_3. However, these results appear to be contrary to the conclusions reached for the force constant calculations above[63]. In addition, some structural information is available for these complexes. A value of 1.836 $\pm$ 0.012 Å was determined for the B–P distance in gaseous PF_3–BH_3 by microwave spectroscopy[55] (see section 6.3.2). The corresponding distance in solid PH_3–BH_3 was determined by X-ray crystallography[65] to be 1.93 Å. Unfortunately a direct comparison of these distances is less valuable than if both determinations had been carried out in the same phase but it appears that the B–P bond in PF_3–BH_3 is the shorter (and presumably stronger) which is in accord with the order based on the force constant calculations and contrary to the *MO* predictions.

The infrared spectrum of PI_3–BI_3 has been investigated and some approximate force constant calculations carried out on the complete eight atom system[39]. The best fit between experimental and calculated frequencies was obtained by using a P–B stretching force constant of 100 N m^{-1}. The source of greatest uncertainty was the vibrational assignment. The inability to dissolve the material without decomposing it prevented Raman polarisation measurements being made, and the lack of a single crystal prevented the orientation of the transition moment being used as an aid to assignment. These problems are common in this area of molecular spectroscopy. From the P–B force constant a very rough estimate of the dissociation energy for the P–B bond was made by plotting bond dissociation energy against stretching force constant for several diatomic species and extrapolating the plot to the value of 100 N m^{-1} to give a value of about 67 kJ mol^{-1}. It should be emphasised that such a procedure can hardly be expected to give better than an order of magnitude for the bond dissociation energy.

6.3.4. ACETONITRILE COMPLEXES

As mentioned above (section 6.3.3) the ν(CN) stretching frequency of acetonitrile increases upon complex formation with electron acceptors. A few examples of the characteristic ν(CN) band positions are listed in table 6.3. (A more detailed selection is given in table 2.10). Factors which might be expected to influence the analogous ν(C═O) stretching frequency in complexes of carbonyl donors are discussed in section 6.1.3. In these systems the coordination at the oxygen atom weakens the bond, presumably by polarisation of the π-bond electrons due to electron withdrawal by the acceptor. If any of the other effects (kinematic coupling or rehybridisation at the oxygen atom) are present they must either reinforce the polarisation effect or, if they oppose it (as expected, see section 1.2.3), must be outweighed by it. In the case of the increase in the characteristic CN stretching frequency it seems probable that one of these "other effects" opposes the bond polarisation effect and is dominant.

Beattie and Gilson[69] have carried out a normal coordinate analysis of the $CH_3CN–BX_3$ (X = F, Cl, Br) and $CH_3CN–GaCl_3$ systems. They allowed the CN force constant to assume the values 1500, 1000 and 500 N m^{-1}. For a value of 1500 N m^{-1} (which is of the order expected for a triple bond) the ν(CN) frequency was very insensitive to variations of the halogen attached to boron and to the value given to the B–N stretching force constant. The calculated frequency shift for CH_3CN corresponded to +25 cm^{-1} whereas that observed experimentally was in the range +70 to +100 cm^{-1} (see table 6.3).

The calculations suggested that the increase is not likely to be a purely kinematic coupling effect because only by giving the appropriate interaction force constant a high negative value could the correct order of frequency shift be computed and it is not easy to justify why this should be done. It was considered that the CN force constant

TABLE 6.3
$\tilde{\nu}$(CN) for various complexes of the type CH_3CN–MX_n

MX_n	$\tilde{\nu}$(CN)/cm^{-1}	Reference
ICl	2278	16
$SnCl_4$	2302	67
$AlCl_3$	2330	68
BF_3	2376	8
BCl_3	2380	9
BBr_3	2350	9

must increase to some extent to cause the upward frequency shift. As a qualitative explanation it was suggested that, in the case of the >C=O—MX_n complexes where the C-O–M system is non-linear, the X atoms will be close to the electrons forming the CO link and close to the carbon so that repulsions from X may effectively weaken the CO link. In CH_3CN–BX_3 the CNB angle is 180° and it was suggested that the effect of X may be to repel electrons from the nitrogen into the CN link, with the possibility of an increased force constant.

Purcell and Drago[66] carried out force constant and molecular orbital calculations for the complexes of CH_3CN with ICl, $SnCl_4$ and BF_3. Their calculations, in agreement with the previous work[69], showed that the ν(CN) frequency only increased slightly as a result of kinematic coupling between ν(CN) and the ν(donor-acceptor) modes, and that the chief reason for the frequency rise was an increase in force constant on complex formation. More recently, X-ray measurements on CH_3CN–BX_3 complexes (X = F and Cl) have shown[4] definitely that there is a small, but real, decrease in the CN bond length on passing from CH_3CN (1.157 Å)[70] to CH_3CN–BF_3 (1.135 Å) and CH_3CN–BCl_3 (1.122 Å). Furthermore, recent force constant calculations [8,9] (discussed in more detail below) have shown (see table 6.4,) that the CN stretching force constant increases on complex formation with BX_3 (X = F, Cl, Br), in line with the CN bond shortening. Thus, it now seems almost certain that the increase in the ν(CN) frequency observed in these complexes is primarily due to the effect of an increase

TABLE 6.4
Valence force constants f_{BN} and f_{CN} (in N m^{-1}) for CH_3CN–BX_3 complexes (X = F, Cl, Br)[9]

	CH_3CN	CH_3CN–BF_3	CH_3CN–BCl_3	CH_3CN–BBr_3
f_{CN}	1740	1880	1870	1860
f_{BN}	—	250	340	350

in force constant consequent upon bond strengthening. An explanation for this behaviour has been sought[66] in terms of changes in the bonding which occur in the CN group on complex formation. Qualitatively, it would be expected that any weakening of the bond due to a polarisation effect of electrons in the π-bond (as in the case of the carbonyl group) must be more than outweighed by an increased stabilisation of the σ molecular orbitals. On the basis of the *MO* calculations it was shown that the major cause of the strengthening of the CN bond following coordination was the overlapping of the N_{2s} orbital with the C_{2s} and C_σ orbitals in all the σ-molecular orbitals.

A further interesting aspect arising from these force constant and *MO* calculations was an explanation of the experimentally-observed increase in intensity in the "CN stretching" band in the infrared. This had previously been ascribed[71] to an increased charge separation in the CN bond with consequent larger dipole moment on complex formation. However, such an explanation does not take into account the fact that the normal mode involved will be different for the free CH_3CN and the complex. On the basis of the MO calculations it was found that the "CN" dipole was actually *smaller* in the BF_3 complex than in CH_3CN and also the *difference* in the C and N net charges *decreased.* On the other hand the normal coordinate calculations showed that there was kinematic coupling between the N–B and CN bond stretching vibrations in the "ν(CN)" normal mode of the complex. The MO calculations indicated a charge on the boron atom of about 1 atomic unit and inclusion of motion of such a charge in the normal mode would cause an increase in the absorption intensity. (See section 2.3.3 for further discussion of this point.)

Recently, some very thorough studies have been made on the CH_3CN–BX_3 (X = F, Cl and Br) systems by Shriver *et al.*[4,8,9] X-ray structural determinations on the CH_3CN–BF_3 and CH_3CN–BCl_3 complexes[4] yielded B–N bond lengths of 1.630 Å and 1.562 Å respectively which very strongly suggests that BCl_3 forms a stronger bond with CH_3CN than does BF_3. This seems to be the most definite evidence available for the stronger B–N bond in CH_3CN–BCl_3. (See section 6.1.2 for discussion of the acceptor abilities of BX_3 molecules.)

The same order of decreasing B–N bond lengths has recently been found by X-ray investigations[72] of the series of $(CH_3)_3N$–BX_3 (X = Cl, Br, I) complexes. The B–N distances were found to be 1.610 ± 0.006 Å (Cl), 1.603 ± 0.02 Å (Br) and 1.584 ± 0.025 Å(I) and although the latter two values have relatively large uncertainties there appears to be a small but real decrease along the series.

Following the X-ray work on the CH_3CN complexes further studies on the vibrational spectra of the complexes with BF_3,[8] BCl_3 and BBr_3[9] were made. Infrared and Raman spectra were investigated in the solid state at −196°C using five different isotope species for each compound; CH_3CN–$^{11}BX_3$, CD_3CN–$^{11}BX_3$, CH_3CN–$^{10}BX_3$, $CH_3C^{15}N$–$^{10}BX_3$, and

CD_3CN–$^{10}BX_3$ (where X = F, Cl and Br). The Raman spectrum was also obtained in nitromethane solution to measure polarisation data. These data were used as the basis for the vibrational assignments. In this work, since some reasonably quantitative deductions about bond strengths were being sought, considerable care was taken over the assumptions and approximations necessarily made in the calculations and these are discussed at some length. Basically, although the vibrational problem was not seriously under-determined it was necessary to constrain a number of the force constants to zero. On the basis of the C_{3v} symmetry of the complexes, $14A_1$ and 17E class symmetry force constants were used in the calculations. The remaining force constants involved interaction between symmetry coordinates not sharing a common atom (stretch-stretch interactions) or a common bond (stretch-bend or bend-bend interactions) and were constrained to zero.

The CN and BN stretching force constants employed in the final calculations are listed in table 6.4. Variation of the value of f_{CN} along the series is hardly significant indicating that simple correlations of the value of $\bar{\nu}$(CN) are of little value in understanding the bonding in such complexes. The most interesting results in this study are the computed force constants for the B–N bond. From table 6.4 it can be seen that, while the difference between the BCl_3 and BBr_3 complexes is not significant there is an increase of 90 N m^{-1} from the BF_3 to the BCl_3 complex. This difference is greater than twice the estimated error and since the constraints used in the force field did not seriously affect the B–N stretching force constant the 90 N m^{-1} difference can be considered to be significant. These results reinforce the conclusions from the structural data that there is a real increase in the bond strength on passing from the BF_3 to the BCl_3 complex; that is, in the same direction as the heats of formation (see section 6.1.2). Such a conclusion is inconsistent with the idea that the reorganisation energy and not the B–N bond strength is the dominant factor in determining the stability of the complex. It thus appears that the donor–acceptor bond energy must more than compensate for any reorganisation energy required. In view of this some alternative description of the bonding was sought using some *CNDO/2* molecular orbital calculations (see section 1.3 for a general description of such calculations). These were carried out on BF_3 and BCl_3 for varying degrees of angular distortion from the planar case. The important result was a calculated decrease in energy (or increase in electron affinity) of the acceptor orbital in each case when such a reorganisation occurred. It also appeared that BCl_3 has the higher electron affinity of the two when it is distorted. (This originates from the smaller degree of B_p and X_p interaction in BCl_3).

The whole explanation of the order of acceptor power in those trihalides as measured by bond lengths, heats of formation, force constants etc. is as yet not certain but, apart from the arguments based on the loss of π-bonding energy (discussed above in section 6.1.2), another possibility was proposed in the light of the molecular orbital calculations[9]. It was suggested that the most easily reorganised electron acceptor (presumably BCl_3 on the basis of calculations) will be distorted to a greater extent and therefore have a greater electron affinity than the less flexible BF_3. It seems probable that both effects mentioned above may operate to yield the observed order of acceptor power.

6.4. CARBONIUM IONS AND OTHER FRIEDEL-CRAFTS INTERMEDIATES

6.4.1. INTRODUCTION

The group of reactions known as Friedel-Crafts reactions are amongst the most versatile and important known to synthetic chemists. They are defined as any reaction taking place under the catalytic effect of a Lewis acid (e.g. an acidic halide) or of a proton acid and this clearly represents an enormously wide range of systems.

Many methods have been employed to study the nature of the intermediates—notably vapour pressure studies, nuclear magnetic resonance and infrared absorption spectroscopy (see ref. 7 for detailed review). Raman spectroscopy is of great potential use, but so far the technique has been employed in only a very limited number of cases.

The earliest reaction studies were between two alkyl halides in the presence of aluminium chloride to yield, *inter alia*, higher hydrocarbons[73]. The usual concept of a Friedel-Crafts reaction is better typified, however, by the reaction between benzene and *n*-amyl chloride in the presence of aluminium chloride.

$$C_6H_6 + CH_3(CH_2)_4Cl \xrightarrow{AlCl_3} C_6H_5(CH_2)_4CH_3 + HCl \qquad 6.2$$

It is common in such reactions for the aluminium chloride to dissolve and for two layers to separate, the lower being an involatile layer which reacts violently with water. This immediately suggests strong complex formation. In addition to the aluminium halides, other strongly acidic metal halides will catalyse such reactions including those of zinc, boron and iron.

It was soon established that these Lewis acid halide—organic halide complexes are ionic or highly polar in active Friedel-Crafts reactions, but the detailed structures have only recently been elucidated by spectroscopic techniques.

Before categorising the complexes according to different groups of Lewis acids and bases it is worth enumerating the effects expected of association between the acids and bases of type

$$RX' + MX_n = RX'\text{–}MX_n \qquad 6.3$$

(a) The creation of the bond X′–M will usually involve a change in the electron density of the bond R-X′. This will have the effect of altering

the ability of the bond to resist deformation. If, as is usual, the electrons are withdrawn from a bonding orbital, there will be a decrease in the bond strength and, in accord with experience, this leads to a decrease in the stretching force constant. Of course this bond cannot be treated in isolation. The bond X′–M exists and interaction between the R-X′ and X′–M bonds occurs, as discussed in chapters 1 and 2. Nevertheless, since the bond strengths are usually very different, the vibrational interaction is often weak and, to a fair approximation, we may treat each separately.
(b) In addition to a new bond X′–M being formed there will generally be rehybridisation of the electrons around M and at the atom of R to which X′ is attached. This will lead to a change in vibrational modes and transition probabilities. Not only the vibrational spectrum but also the n.m.r. spectrum may be drastically affected. An excellent example of this afforded by the interaction of benzene and $HAlCl_4$ (see p 507)[74].
(c) The electron distribution in R may be strongly affected. This will show as changes in force constants of R, but more noticeably as changes in the chemical shifts associated with the atoms of R[75]. Changes in the ultraviolet spectra will also result, though, in general, these are difficult to interpret.

6.4.2. REACTIONS BETWEEN GROUP IV AND GROUP V HALIDES AND ALKYL HALIDES

The infrared spectra of mixtures of methyl chloride and stannic chloride and of methyl chloride and antimony pentachloride have been studied by Nelson[76]. In the case of the stannic chloride solution the band position's were, on an average, 11 cm^{-1} lower than in the vapour phase, but no new features, such as new bands or splitting of degenerate bands, appeared. In antimony pentachloride the spectrum of CH_3Cl showed an extra band in the C-Cl stretching region. This band was at 688 cm^{-1}, which is to be compared with 714 cm^{-1} for the unassociated ν(C-Cl) stretch. Assignment of this to an associated species follows from the increase in its intensity with a decrease in temperature. No splitting of the twofold degeneracies of the perpendicular modes (E species) occurs. This might have been interpreted as indicating that the threefold symmetry axis was retained, but in all likelihood it is probably because the interaction is too weak to lead to noticeable splitting. Thus a complex is believed to exist between CH_3Cl and $SbCl_5$, but it is clear from these results that it is weak.

It might be anticipated that stronger complexes could be obtained by increasing the electrophilic character of the halogen and the Lewis acid strength of the radical R. Indeed carbonium ion salts can be made from tertiary butyl fluoride and antimony pentafluoride. The n.m.r spectrum of this system is a singlet with a chemical shift, δ of 4.5 showing that the effect of the fluorine nuclear spin has been uncoupled from the protons[77]. ^{2}H and ^{13}C resonance studies of this ion were also carried out and confirmed the planar structure of the carbon skeleton. ^{13}C magnetic resonance is particularly useful in recognising charge density at carbon atoms. Thus in the above instance at −20°C a ^{13}C resonance was found[192] at −135.4 p.p.m. from $^{13}CS_2$. Such a large deshielding of the nucleus, as is implied by this shift, can only be explained by the carbon atom carrying a substantial positive charge. The weight of evidence is strongly in favour of a carbonium ion rather than fast fluorine exchange. The infrared spectra of $(CH_3)_3C^+$ and $(CD_3)_3C^+$ have been measured over the range 4000 to 8000 cm^{-1} (figure 6.1). Moisture must

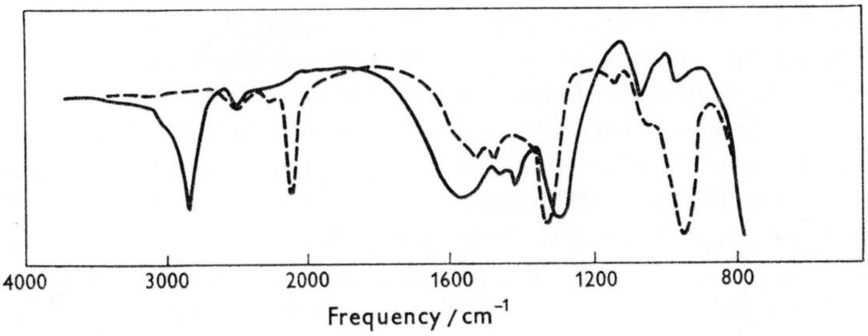

FIG 6.1

be strictly excluded from these highly reactive complexes and spectra were measured using Irtran 3 (calcium fluoride) plates. The following features were observed.

(a) The bands due to CH stretching were near 2830 cm^{-1}, which is very low. This shows that the positive charge has drawn electrons from the CH bonding orbitals and thus weakened the bonds.

(b) the degenerate, antisymmetric CC stretching mode gives rise to a band at 1290 cm^{-1} in $(CH_3)_3C^+$ and 1330 cm^{-1} in the perdeutero compound. These are to be compared with 1260 cm^{-1} for the corresponding mode in tertiary butyl chloride. It should be noted, however, that this mode designation as "antisymmetric CC stretch" is artificial. Hüttner and Zeil[78] have calculated the potential energy distribution in tertiary butyl chloride and have shown that there is considerable mixing between the CC stretching and HCC bending modes.

The Raman spectrum of the trimethyl carbonium ion, and also that of its perdeutero form, has been reported recently[79]. The similarity of the spectrum with that of boron trimethyl, which is known to be planar, was taken as evidence of a planar skeleton. The strongest band in the spectrum is at 347 cm^{-1} and is highly polarised having a depolarisation ratio of 0.25 for incident polarised light with the analyser in the emitted beam. Since the depolarisation ratio is less than 0.75 the transition involved must transform as the fully symmetric species of the group to which the molecular configuration belongs. Analogy with $B(CH_3)_3$ leads to its assignment as the out of plane umbrella mode,

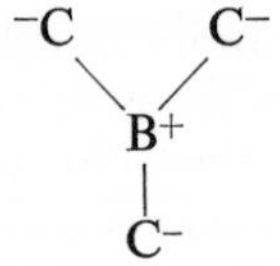

This can only belong to the fully symmetric species if the overall molecular point group is less than D_{3h} as a result of the proton disposition. The observed spectral activities are best accounted for by assuming C_{3v} symmetry. It appears rather surprising that the proton configurations can lead to such a low depolarisation ratio for what must be an almost pure skeletal deformation.

If a reliable set of force constants for the tertiary butyl cation could be derived from the vibrational spectroscopic data it would be interesting to determine the effects of the configuration change, and also of the positive charge, on the CC bond strengths. Evans[80] calculated the vibrational frequencies and the potential energy distribution amongst the various internal co-ordinate distortions of the carbonium ions transferring the force-field from a Urey-Bradley type field derived for hydrocarbons. The only change made from the hydrocarbon field was to increase the CC stretching force constant. Agreement between calculated and observed frequencies was only moderate as would be expected.

A further example of the structure of a complex being determined by the combined use of several spectroscopic methods is afforded by a study of methyl fluoride in antimony pentafluoride[193]. In a 1:1 mixture the proton n.m.r. spectrum is a singlet showing that the fluorine has either been separated from the methyl or is undergoing rapid exchange on the n.m.r. time scale. The ^{13}C resonance is only slightly displaced from the $^{13}CH_3F$ frequency (δ, 117.8 $\rightarrow$ δ, 118.9) suggesting that there is little deshielding and the ionic structure can be discounted. Excess methyl fluoride caused no shift in the proton magnetic resonance and the normal doublet of CH_3F was seen to be superimposed on the bands due to the complex. This shows no intermolecular exchange on the n.m.r. time scale. The Raman spectrum of the 1:1 complex was very similar to that expected for non-interacting CH_3F and antimony pentafluoride, thus confirming a stable co-ordination complex in which the donor and acceptor retain much of their original identity.

Olah and his co-workers have studied the norbornyl and 2-alkyl substituted norbornyl cations in some detail by a variety of spectroscopic techniques[81–83]. Using the INDOR method they measured the ^{13}C n.m.r. spectrum of the norbornyl cation itself and noted a pentuplet at 101.8 p.p.m. from $^{13}CS_2$. This indicates three equivalent carbon atoms interacting with four equivalent protons and suggests a fast

exchange of the protons indicated by,

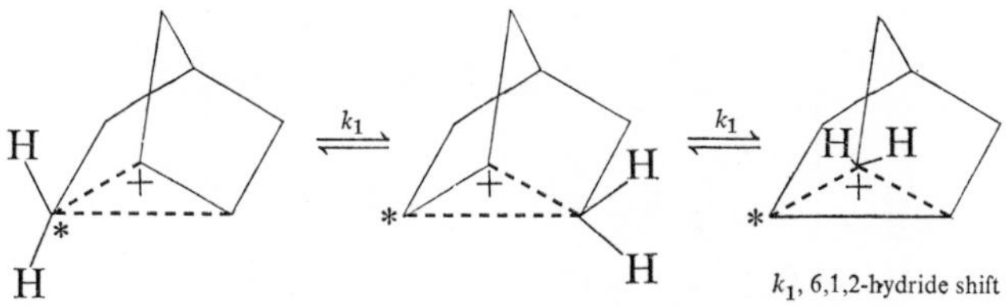

(Reproduced from *J. Amer. Chem. Soc.*, **92,** 4633 (1970), by permission of the American Chemical Society.)

The corresponding proton resonances appear as a singlet at temperatures above −120°C showing that their exchange rate is fast compared with the differences in the transition frequencies of the non-equivalent protons. On cooling to −120°C the absorption band broadens and then at below −130°C splits into two resonances, each corresponding to two protons. The observed proton and ^{13}C shifts differ considerably from those expected on the basis of rapid exchange of classical structures. This is taken to indicate a non-classical multi-centre bond system such as in X.

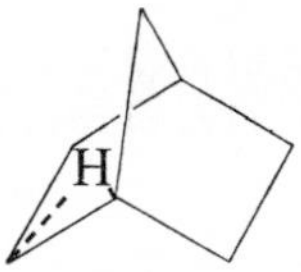

X

The laser Raman spectrum of the norbornyl cation bears a strong resemblance to that of nortricyclene, which in turn differs considerably from those of norbornane and substituted norbornanes[83] (see table 6.5). This strongly suggests a similarity between the skeletal structures and, in conjunction with the n.m.r. data, gives good support for the non-classical structure. The Raman spectra of the 2-alkyl substituted norbornyl cations differ significantly from that of norbornane but, as is also shown in table 6.5, are very similar to those of norbornane and its derivatives. However, the Raman spectrum of 1-methyl-nortricyclene is also very similar, so that one must be cautious in deducing that the skeletal structure of the alkyl substituted norbornyl cations is classical from this data.

The spectra of a number of other carbonium ions have been studied by Olah and his coworkers using antimony pentafluoride as a strong Lewis acid to remove fluorine from an alkyl fluoride. Many of these ions, such as the *n*-alkyl carbonium ions, are unstable and rearrange to form tertiary alkyl carbonium ions. For a more detailed discussion of these the reader is referred to a review of Olah and Pittman[84].

TABLE 6.5

Main Raman spectral lines (cm^{-1}) of the norbornyl and 2-methyl- and 2-ethylnorbornyl cations and vibrational data of nortricyclene and norbornane[a]

	C–H region (3100–2700 cm^{-1})					C–C region (1000–250 cm^{-1})			
Norbornyl cation	3110	3030	3010	2947	2860		972 (vs, p)	796 (w)	390
3-Br-Nortricyclene	3087	3006	2975	2955	2875		959 (vs, p)	732 (m)	300
Nortricyclene	3080	2990	2945	2915	2867		951 (vs, p)	783 (w)	
2-Me-Norbornyl cation			2984	2960	2855	911 (s)	880 (s, p)	770 (m)	350 (m, p)
2-Et-Norbornyl cation			2980	2965	2873	920 (s)	884 (s, p)	765 (m)	350 (m, p)
Norcamphor			2964	2920	2882	938 (s)	880 (m)	786 (m)	
Norbornane			2964	2936	2873	920 (s, p)	871 (s, p)	753 (m)	
2-Cl-Norbornane			2976	2933	2878	925 (s, p)	879 (s, p)	764 (m)	345
2-Br-Norbornane			2978	2933	2878	925 (s)	878 (s)	762 (m)	294
1-Me-Nortricyclene	3069	2994	2967	2930	2871	905 (s, wp)	857 (vs, wp)	794 (m)	
2-Me-Norbornane[b]			2952	Source	2869	923 (s)	873 (m)	719 (m)	
2-Et-Norbornane[b]			2947	Source	2869	923 (m)	878 (m)	696 (m)	

a. s = strong; m = medium; w = weak; p = polarized; vs = very strong.

b. V. T. Alexsanyan and Kh. E. Sterin, *Fiz. Sb. L'vov. Gos. Univ.*, **1**, 59 (1957); *cf. Chem. Abstr.*, **53**, 21158*a* (1959).

(Reproduced from *J. Amer. Chem. Soc.*, **92**, 4627 (1970), by permission of the American Chemical Society.)

The X-ray photoelectron spectra of the tertiary butyl and the trityl cations have been measured.[85] In non-ionic organic systems the ionisation energies of the carbon 1s electrons are quite insensitive to the environment as these electrons do not participate in the bonding to any significant extent. In the tertiary butyl cation (or trimethyl carbonium ion in the alternative carbonium ion nomenclature) the binding energy of the 1s electrons of the central atom was found to be increased by about 3 eV whereas the other 1s electron energies were relatively unaltered. This is in accord with the expectation that the charge is localised on the central carbon atom. In contrast, no shift was found for the trityl ion showing that the excess charge is delocalised by conjugation of the rings through the central atom:

6.4.3. REACTIONS BETWEEN MAIN GROUP METAL HALIDES AND ACYL HALIDES

The presence of a ketonic grouping might be expected to increase the stability of the carbonium ions, and indeed this is so. Many papers have been published on the infrared, and in some cases also the Raman, spectra of the products of reaction of acyl halides with group III and V halides. The reactions may be represented generally by:

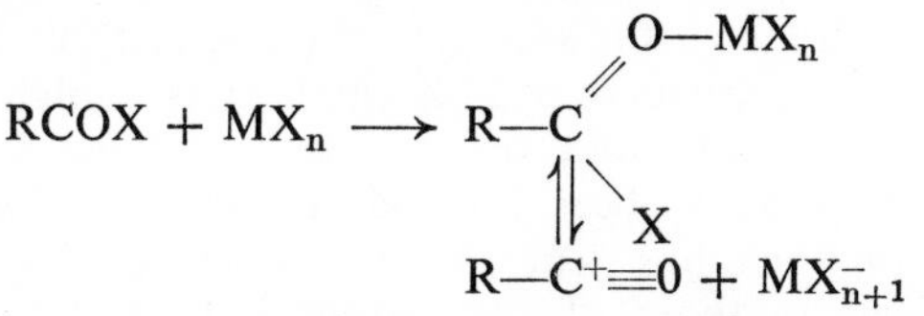

$$RCOX + MX_n \longrightarrow R-C \overset{\displaystyle O-MX_n \qquad XI}{\underset{\displaystyle R-C^+\!\equiv\!0 \, + \, MX_{n+1}^- \quad XII}{\overset{\Vert}{\diagdown X}}} \qquad 6.4$$

The extent to which the ionisation proceeds depends on the Lewis acid strength of MX_n and on the ability of the radical R to stabilise the ion. The vibrational spectrum affords a convenient means of studying these reactions. Association of the acid with lone pair electrons of the acyl halide leads to withdrawal of the electrons from the C=O bond and a consequential bond polarisation and weakening (see also section 6.1.3). This has the effect of decreasing the so-called $\nu(CO)$ stretching frequency. It should be appreciated that there is now an extra restoring force on the oxygen atom arising from the O–M bond. In general, however, the unperturbed $\nu(O-M)$ stretching frequency is far less than the $\nu(CO)$ frequency and coupling between the two oscillators is weak.

The earliest spectroscopic investigation of the acetylium ion, CH_3CO^+, was that of Susz and Wuhrmann[86] who mixed acetyl fluoride and boron trifluoride at low temperatures. The spectrum of the white crystalline solid so formed was measured as a mull at $-40°C$. The $\nu(CO)$ band at 1879 cm^{-1} in acetyl fluoride had completely disappeared and was replaced by a strong sharp band at 2300 cm^{-1}. When the acylium ion is formed by halogen transfer to the MX_n

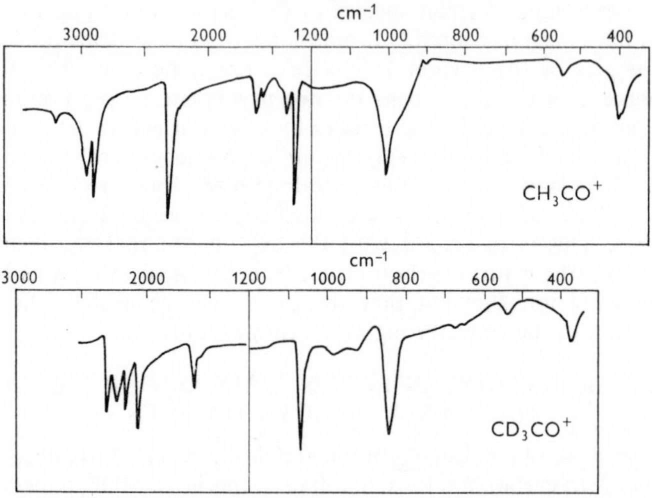

FIG 6.2 (Reproduced from *J. Mol. Struct.*, **1,** 349 (1968), by permission of the Elsevier Publishing Co.)

then a formally triple bond is created with the positive charge centred on the carbon atom. This is shown by the considerable increase in the ν(CO) band frequency to over 2200 cm^{-1}, thus implying an increase in bond strength and bond order. The acetylium and perdeutero acetylium ions have also been prepared by mixing acetyl chloride (or CD_3COCl) with $SbCl_5$ in 1:1 molar ratios. Their spectra (see figure 6.2) are very similar to those of acetonitrile and CD_3CN[87]. The ν(CO) stretching frequency is higher than the ν(CN) stretching frequency in CH_3CN (table 6.6) and this is reflected in derived force constants. (Unfortunately the dispersions on the force constants are misquoted in

TABLE 6.6
Calculated and assigned frequencies[87] of CH_3CO^+ and a comparison of these with the assigned frequencies[88,89] of CH_3CN

Species	CH_3CO^+ (solid) calc.	obs.	CH_3CN (vapour) obs.
A_1	2843	2870	2954
	2301	2295	2267
	1289	1319	1385
	944	950	920
E	2956	2940	3009
	1391	1360	1453
	1010	1001	1041
	382	390	362

ref. 87. The actual dispersions are the square roots of those given. Since all these are less than unity this increases all dispersions.) In addition the CC force constant is higher for CH_3CO^+ than for CH_3CN. An explanation of these facts is readily given in terms of molecular orbital theory as follows. A triple bond description for first row elements is $(2p\sigma)^2(2p\pi)^4$. *Any difference in the electronegativities of the atoms constituting the bond will tend to draw electrons from these bonding orbitals.* The very existence of such a high ν(CO) frequency shows that the excess positive charge is achieved by drawing electron density from the carbon non-bonding orbitals into the bond. This positive charge will tend to counter the pull of the oxygen atom for the electrons, thus allowing an increase in the availability of the electrons to the bonding orbitals. In CH_3CN there is no such factor to counter the electronegativity of the nitrogen, so the CN force constant is lower. Notice that the reduction in CN bond polarity due to donor action by the N lone pair in nitriles may be used to rationalise the *increase* in ν(CN) stretching frequency in nitrile complexes. This mechanism is described in more detail in section 2.3.3.

In addition to these investigations, where an apparently fully ionic compound was obtained, there have been other cases where the spectra appear to show the presence of associated species of type XI in addition to the ionic system. Three groups of workers have studied the interaction between acetyl chloride and aluminium chloride. The absorption spectrum of Susz and Wuhrmann[90] shows the main bands of the CH_3CO^+, but in addition to these and those of the tetrachloroaluminate ion a number of additional bands appear. Of particular interest are bands at about 2200 cm^{-1} and 1639 cm^{-1} (see below). While the compound of Susz and Wuhrmann was a white crystalline solid, Cook[91] reported that a 1:1 mixture gave a viscous liquid for which the extra bands were now much enhanced.

Various other Lewis acid-acetyl chloride systems have been studied including CH_3COF with SbF_5 and CH_3COCl with $AlCl_3$ as well as with $TiCl_4$ and $GaCl_3$. Varying amounts of the (presumably) different types of complex were found. For the CH_3COCl–$TiCl_4$ complex no bands were found in 2300–2000 cm^{-1} region[92,93] and the main ν(CO) band was at 1620 cm^{-1}. In an attempt to clarify the nature of the bonding, Cassimatis and Susz[92] calculated the vibrational frequencies treating the system as a penta-atomic grouping, *viz*:

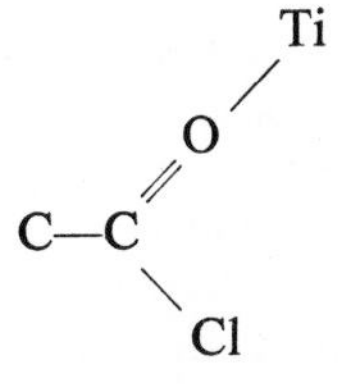

and using a simple valence force-field. By varying the f_{CO}, f_{OTi}

and f_{COTi} force constants, to obtain a good fit between observed and calculated frequencies, they derived the approximate set of force constants shown in table 6.7. The approximations in such a calculation are obviously drastic and the absolute values should not be taken too literally, but the magnitudes are probably significant.

The complex between gallium chloride and acetyl chloride as prepared by Cook[94] was a viscous liquid. The main band occurred at 2300 cm^{-1}, but a moderate band at 1616 cm^{-1} indicated that the neutral complex was also present. In nitrobenzene solution the 2300 cm^{-1} absorption almost completely disappeared and was replaced by a band at 2200 cm^{-1}. There seems little doubt that the band occurring near 1600 cm^{-1} is due

TABLE 6.7

The calculated and observed wavenumbers corresponding to the ν(CO) and ν(OTi) modes of the acetyl chloride–titanium chloride complex and the force constants used to obtain this fit (from ref. 92)

(obs.) for complex	1620, 1145 cm^{-1}
(calc.) for complex	1621, 1134 cm^{-1}
f_{CO} (CH_3COCl)	1175 N m^{-1}
f_{CO} (complex)	905 N m^{-1}
f_{OTi} (complex)	275 N m^{-1}

to the neutral complex of type XI (p 493). There is less agreement about the interpretation of the 2300 and 2200 cm^{-1} bands. It has been suggested that the lower frequency band is due to the free ion and that the band near 2300 cm^{-1} arises from ionic association. That is, it is postulated that the ionic interaction between the anion and the polar charged C≡O bond leads to an extra restoring force on the oxygen atom. The writers do not agree with this interpretation since the X-ray crystallographic study of the complex $CH_3CO^+SbF_6^-$ does not show any strong ion pair associations[95] and furthermore the higher frequency seems quite insensitive to the anion. It seems more likely that the lower frequency band is of an another complex species of some form.

It might be anticipated that the substitution of fluoroacyl or aroyl halides for acyl halides would lead to increased tendency for formation of the neutral complexes. In fact, trifluoro acetylium hexafluoroantimonate can be prepared at temperatures below −25°C and has been shown[96] to be primarily ionic below the boiling point of CF_3COF (−57°C). At low temperatures the ν(CO) frequency corresponds to 2371 cm^{-1}, but as the temperature is raised the intensity of this band decreases and bands at 1780 and 1631 cm^{-1} (cf. 1901 cm^{-1} for CF_3COF) grow in strength. At room temperature the 2371 cm^{-1} band has virtually disappeared.

$CF_3COF–AsF_5$ decomposes above −57°C whereas $CF_3COF–PF_5$ has not been isolated at all[97].

The higher ν(CO) frequency in CF_3CO^+ as compared to CH_3CO^+ may be due in part to kinematic mixing of the CO stretching and C—F stretching modes. Any two oscillators, if coupled, will lead to new modes with frequencies above and below each component. In CH_3CO^+ this implies that the CH and CO stretching motions will interact to lower ν(CO) whereas in CF_3CO^+ the analogous interaction will lead to an increase in ν(CO). In addition to this kinematic effect the high electronegativity of the fluorines will lead to an enhancement of the effect (discussed earlier) of the C^+ in countering the dipole of the CO bond. Of course, too great an electron withdrawal will lead to a reduction in the bond restoring force, but the evidence is that in CF_3CO^+ the restoring force is greater than in CH_3CO^+.

The higher perfluoro-acyl halides do not form the fully ionic complexes even with antimony pentafluoride. In accord with this the complexes do not react with benzene whereas the $CF_3COF–SbF_5$ complex and benzene react slowly at 6°C[98].

Aroyl halides also react less readily with the group III and V halides than do the acyl halides. However, ortho and para nucleophilic substituents increase the trend towards the ionic form. Benzoyl chloride and aluminium chloride associated to give only a complex of type XI[99,28]. The ν(CO) vibrational band decreases from 1773 cm^{-1} in benzoyl chloride to 1710 cm^{-1} in the complex. With 2-methylbenzoyl chloride, aluminium chloride forms some ionic and some neutral complex[100] and with mesitoyl chloride the complex appears to be entirely in the ionic form[90]. In these aroylium ions the ν(CO) band is at 2170 to 2220 cm^{-1}. The lower frequencies compared with acylium ions is presumably due to the conjugation to the aromatic ring and a consequential lowering of the bond order. By increasing the Lewis acid strength the benzoylium ion can be prepared. Thus AsF_5 and SbF_5 convert benzoyl fluoride to the ion (D. Cook, reference 7, p. 801).

Perkampus and Weiss[101] sealed benzoyl chloride and aluminium bromide in a cuvet at −196°C and studied the spectral absorption changes on raising the temperature. At +7°C a band began to appear at 2209 cm^{-1} reaching maximum intensity in the range 40 to 55°C. The spectral changes were entirely reversible showing that the following equilibrium (perhaps via an intermediate complex) exists:

$$C_6H_5COCl + AlBr_3 \rightleftharpoons C_6H_5CO^+(AlBr_3Cl)^- \qquad 6.5$$

Only one thioacyl halide complex with metal halides has been confirmed and investigated at the present time. Thiobenzoyl chloride reacts with silver hexafluoro-antimonate in liquid SO_2 at −40°C to give a 1:1 complex[102]. Conductivity measurements and infrared studies show that the compound is predominantly in the ionic form. The ν(CS) vibrational band is raised to 1332 cm^{-1} from its value in thiobenzoyl chloride of 1250 cm^{-1}† and the t_{1u} vibration of SbF_6^- gives rise to a strong band at 673 cm^{-1}.

6.4.4. REACTIONS BETWEEN METAL HALIDES AND THE OXYHALIDES OF GROUPS V AND VI

No complex between $Ag^+SbF_6^-$ and phosphoryl halides has been isolated. In all cases the fluoroantimonate acts as a fluorinating agent and PF_3O is formed. No ionic complex of the type $[RR'R''PO]^+SbF_6^-$ has been prepared. However, the infrared spectrum of the reaction product of diphenyl-phosphinyl chloride and $AgSbF_6$ in liquid SO_2 or acetonitrile shows the presence of a donor-acceptor complex[103].

The phosphoryl halides and thiophosphoryl halides react with triphenyl-phosphine or triphenylarsine in the presence of a trace of water in the following manner:

$$OPCl_3 + P(C_6H_5)_3 \rightleftharpoons PCl_3 + OP(C_6H_5)_3 \qquad 6.6$$

In the case of the phosphoryl halides this reaction proceeds via an intermediate which can be isolated. It has been studied by, amongst other techniques, infrared spectroscopy[104] and has been shown to have the structure:

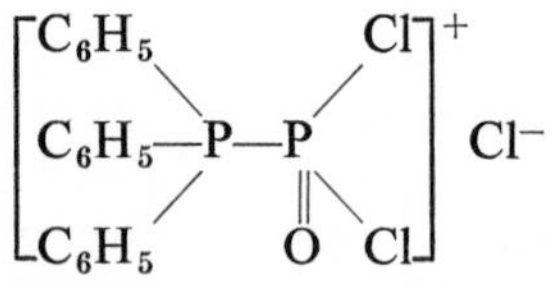

The arsenic adduct has a similar structure but since the triphenylarsine is a weaker base than the phosphine it is less stable and slowly decomposes above room temperature[104]. The vibrations arising predominantly from P-halogen and PO stretching increase in frequency in the complex as compared with the phosphorus oxyhalides themselves[97] (table 6.8). This is interpreted as an increase in the bond orders due to the positive charge on the $P(C_6H_5)_3$ entity lowering the actual bond polarities of the $POCl_2$ units. These dihalogenophosphoryltriphenylphosphonium halides are of some importance since they react readily with metal halides to form the corresponding triphenylphosphine oxide–metal halide complexes.

A group of complexes formally similar to the acyl halide–metal halide complexes are those formed from aryl sulphonyl halides and the Lewis acids, MF_n, where M = Sb, As, P or B. The aryl sulphonylium ions can only be isolated if the ion can transfer its charge to the

† These are the $\bar{\nu}$(CS) values quoted by the authors. In fact the ν(CS) vibration mixes strongly with other modes and there is almost certainly a large contribution of the CS stretching motion to other normal modes. The above quoted frequency shift is probably a low measure of the bond order change.

TABLE 6.8
The ν(PO) and ν(P-halogen) vibrational bands of phosphoryl halides and their substitution products[97] $[(C_6H_5)_3P.PX_2O]^+X^-$

Compound	Molecular point group	$\tilde{\nu}$(PO)/ cm^{-1}	$\tilde{\nu}$(PX)/ cm^{-1}
$OPCl_3$	C_{3v}	1290 (a_1)	577 (e) 486 (a_1)
$OPBr_3$	C_{3v}	1261 (a_1)	488 (e) 340 (a_1)
$[(C_6H_5)_3.P.P(Cl)_2O]Cl$	C_s	1320 (a′)	587 (a′) 564 (a″)
$[(C_6H_5)_3P.P(Br)_2O]Br$	C_s	1305 (a′)	492 (a′) 479 (a″)

organic residue by a mechanism such as the quinonoid structure shown below for the p-dimethylamino-benzene sulphonylium ion[105].

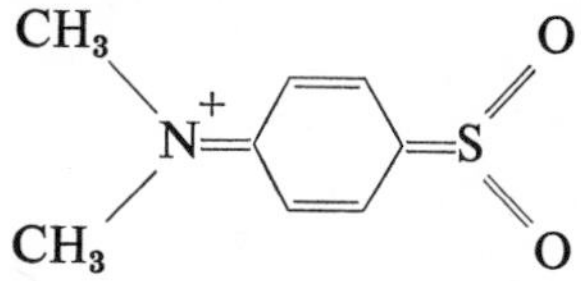

The infrared spectra of such complexes show an increase in bond order of the CS bond. An increase in both the symmetric and antisymmetric ν(SO) stretching frequencies occurs. Thus in the system $[p\text{-}N(CH_3)_2C_6H_4SO_2)^+[SbF_6]^-$ the symmetric and antisymmetric ν(SO) vibrational bands are at 1195 and 1418cm^{-1} which is to be compared with 1177 and 1376 cm^{-1} for p-$N(CH_3)_2C_6H_4SO_2Cl$[97]. There is a danger in correlating these frequency increments with increases in bond order. The configuration of the C–SO_2 entity is changed on removal of Cl^-. In the ion the kinematic coupling with the CS bond is greater as a result of the coplanarity of the bond system. This could be enough to lead to the frequency increases observed over the sulphonyl chloride.

6.4.5. THE DIMETHYL HALONIUM IONS

An interesting set of ions has been prepared by dissolving methyl halides in "super-acid" (see p 507). These are the dimethylhalonium ions such as $(CH_3BrCH_3)^+$. Solutions of these ions in liquid SO_2 have yielded good Raman spectra[106]. For a linear structure we would expect only one polarised Raman band associated with the skeleton. In fact, two polarised bands are observed at low frequencies, which is in accord with a structure of C_{2v} symmetry. There is good agreement between the band positions for the dimethylbrominium ion and the isoelectronic dimethylselenide as shown below[106] and this has been taken to suggest similar angles. Raman depolarisation ratios are given in parentheses.

$CH_3BrCH_3^+$	282 (0.46),	544 (0.31)	and	561 (dp) cm^{-1}
CH_3SeCH_3	236 (dp)	586 (p)	and	602 (dp) cm^{-1}

6.5. ARENONIUM COMPLEXES

6.5.1. INTRODUCTION

When hydrogen halides are passed through suspensions of aluminium halides in aromatic systems a two phase system is frequently formed due to the formation of a stable carbonium ion complex of the type $ArH^+Al(Hal)_4^-$. The aromatic proton complex is known as an arenonium ion. It can be thought of as an "inner" complex from the Brönsted acid species (e.g. $HAlCl_4$) and the $b\pi$ aromatic donor (section 1.1.3). With alkyl benzenes this phase separation invariably occurs, but the lower ionic phase is rarely the simple salt designated above. Rather, it it is more likely to be complexed with more solvated aromatic and aluminium halide molecules. Under certain conditions a well-defined complex can be formed. Thus at $-80°C$ aluminium chloride dissolves in toluene on addition of hydrogen chloride to form a green solution with molar ratios 1:1:1 of toluene: hydrogen chloride: $AlCl_3$.[107]

It was originally postulated that the extra proton in arenonium ions added to the π electron cloud.[108] However, ultraviolet studies by Gold and Tye[109] on aromatic hydrocarbons in sulphuric acid gave strong evidence that the interaction in these cases, at least, was much stronger and involved rehybridisation at one of the carbon atoms to give structures such as XIII. Structures such as this must lead to substantial

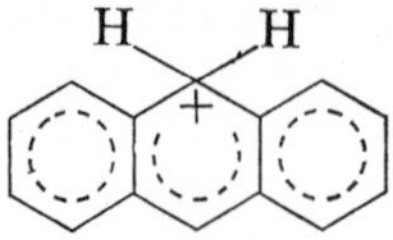

XIII

changes in the n.m.r. and the vibrational spectra from those of the parent hydrocarbons. Some elegant infrared and n.m.r. investigations have indeed confirmed the σ-bond type structures (described below).

6.5.2. NUCLEAR MAGNETIC RESONANCE SPECTRAL STUDIES

MacLean and Mackor and their colleagues have measured the n.m.r. spectra of many arenonium complexes (e.g. refs. 110–114) such as 1,3,5-trimethylbenzonium tetrafluoroborate[111] and 7,12-dimethylbenzanthracenonium tetrafluoroborate in liquid hydrogen fluoride[110]. The creation of a σ-bond as in structure XIV leads to a new absorption

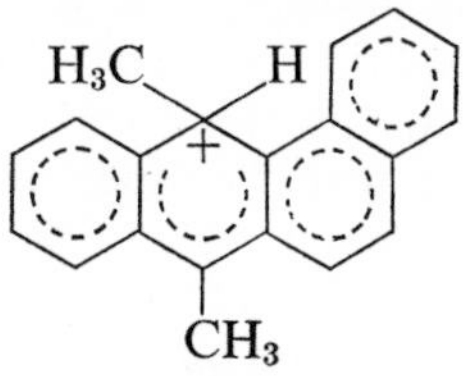

XIV

in the aliphatic proton region (figure 6.3B). The observed spectrum of the proton addition complex of 7,12-dimethylbenzanthracene is compared with that of the parent compound in figures 6.3A and 6.3B. From these we see that the absorption due to the methyl group in the 12 position (the high field signal) is displaced to high field and split into a

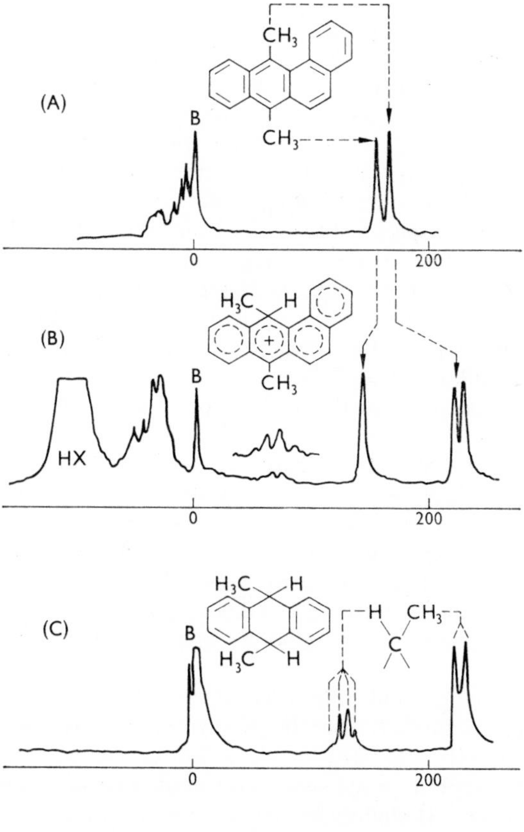

FIG 6.3 (Reproduced from *Angew. Chem., Int. Ed.*, **3**, 779 (1964), by permission of Verlag Chemie, GMBH.)

doublet. The new doublet absorption is at a frequency typical of an aliphatic methyl and the splitting is clearly due to interaction with the added proton attached to the same carbon. The interpretation is confirmed by comparison with the spectrum of 9,10-dihydro-9,10-dimethylanthracene (figure 6.3C).

In addition to the structural information afforded by the n.m.r. spectrum, information is forthcoming about exchange phenomena. Three types of proton exchange can be visualised

A. Exchange with the solvent. If we represent the aromatic proton complex by AH^+ then, for example in hydrogen fluoride, we can write:

$$AH^+ + HF_2^- \underset{}{\overset{k_1}{\rightleftharpoons}} A + 2HF \qquad 6.7$$

B. Intermolecular exchange between the same base, A:

$$AH^+ + A \underset{}{\overset{k_2}{\rightleftharpoons}} A + AH^+ \qquad 6.8$$

C. Intramolecular rearrangement in which the proton moves from one site to another site in the same molecule.

All three types of exchange can be studied by nuclear magnetic resonance spectroscopy provided that the exchange rates ($1/\tau$) are comparable to the frequency separations ($\Delta\nu$) of absorptions arising from the individual species. In the situation where the exchange rate is far too slow to satisfy the above requirement absorptions due to the different species are simply additive. In the event of much faster exchange the effective environment of the nucleus whose transition energies are being measured will be the concentration-weighted average from the different molecular species. The absorption lines in both the above situations will be sharp in the absence of effects such as quadrupolar relaxation.

In the intermediate case, where $\Delta\nu \sim 1/\tau$, the Heisenberg uncertainty condition dominates the line widths. As this condition is approached the lines broaden, and eventually coalesce. Then, as the exchange rate is increased, by increasing concentrations of the reacting species or by increasing the temperature of the mixture, the (coalesced) lines sharpen. A more thorough and detailed discussion of this phenomenon is to be found in many textbooks on n.m.r. spectroscopy (e.g. ref. 115).

Solvent exchange and intermolecular exchange effects are both demonstrated beautifully by the n.m.r. spectra of the mesitylene complex with HF and BF_3[111]. At -100°C the 60 MHz spectrum consists of the superimposed spectra of mesitylene and its proton complex (figure 6.4a). At slightly higher temperatures those signals due to the ring protons broaden and merge, finally coalescing into a single sharp line. The methyl proton resonances do likewise (figure 6.4d). Up to -30°C the absorption due to the hydrogen fluoride solvent is

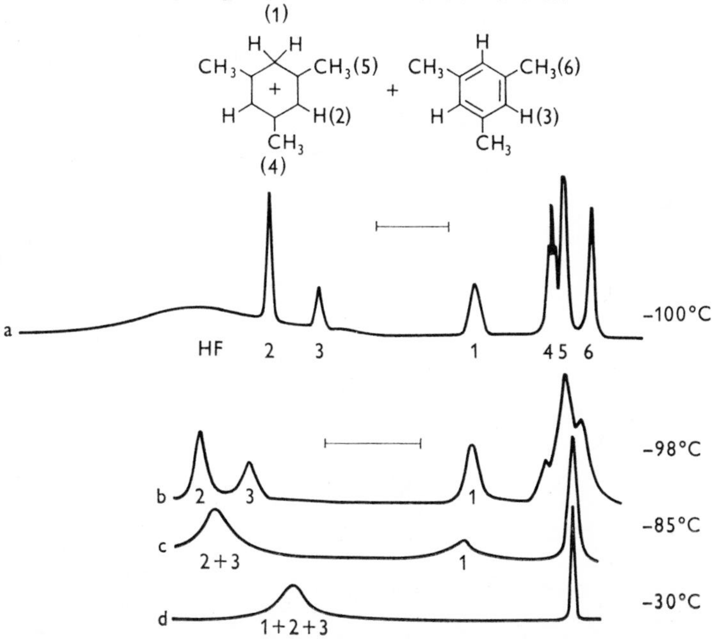

FIG 6.4 The n.m.r. spectra (60 MHz) of mesitylene and its proton complex in HF + BF_3 at various temperatures. (Two different experiments; a and b, c, d; concentrations of AH^+ and A were smaller in expt. a. In spectra b, c and d the solvent peak has been omitted; the methyl groups were recorded at different amplifications.) Field increases from left to right. Scale (50 Hz) is indicated. (Reproduced from C. Maclean and E. L. Mackor, *Disc. Faraday Soc.* **34,** 170 (1962), by permission, of the Chemical Society.)

unaffected, but above this temperature the HF and ring proton signals broaden and finally at about +50°C they merge. From these observations estimates of the rate constants k_1 and k_2 are derivable from a knowledge of the spin lattice relaxation times of the reacting species and their concentrations. The values so derived are $k_1 = 10^{10.3}$ exp (−7400/RT) $dm^3\ mol^{-1}\ s^{-1}$ and $k_2 = 10^{11.0}$ exp (−8000/RT) $dm^3\ mol^{-1}\ s^{-1}$. These rate constants appear to be typical for exchange with solvent and for intramolecular exchange[112]. It has been pointed out that the high activation energy contrasts these reactions to those such as:

$$H_3O^+ + HF_2^- \rightleftharpoons H_2O + HF \cdots HF. \qquad 6.9$$

for which the activation energy is very low. This has been explained as due to the fact that the CH_2 group is aliphatic and not likely to interact with the solvent whereas the HF_2^- ion forms strong hydrogen bonds.

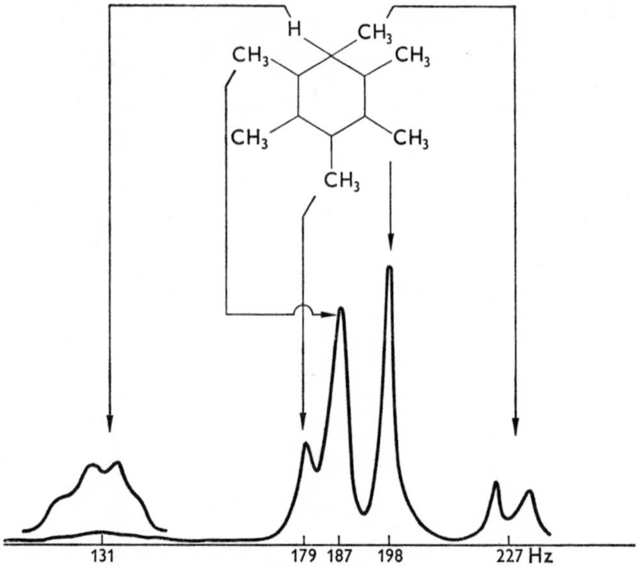

FIG 6.5 The proton resonance spectrum (40 MHz) of the hexamethylbenzenium ion in HF + BF_3 at −85°C. The internal reference is benzene.
(Redrawn from Maclean and Mackor, *Mol. Phys.* **4**, 241 (1961).)

Intramolecular exchange has been proven for several ions. The n.m.r. spectrum of the proton complex of hexamethylbenzene demonstrates this phenomenon particularly well.[111] The low temperature spectrum (−110°C figure 6.5) is as expected for the static species XV.

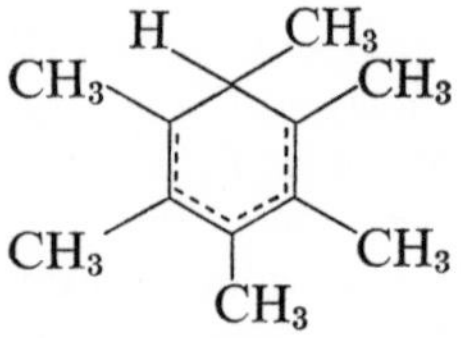

XV

The spin-spin coupling constants required to describe the observed splitting are J_a (for H and CH_3 on same carbon) = 6.8 Hz, $J_o \sim 1$ Hz, $J_p = 3.5$ Hz and $J_m \sim 0$. At higher temperatures (−30°C) the proton migrates between sites sufficiently rapidly to lead to a collapse of the methyl group signals and an averaging of the proton-methyl spin-spin coupling constants. The CH signal, in fact, consists of a series of lines of spacing 2.1 Hz. This spacing can be rationalised in terms of the previously mentioned coupling constants if it is assumed that all the coupling constants have the same sign. On averaging then:

$$J_{eff} = \tfrac{1}{6}(J_a + 2J_o + 2J_m + J_p) = 2.1 \text{ Hz} \qquad 6.10$$

The CH multiplet arises from the various possible spin states of the methyl protons. That the exchange is intramolecular is confirmed by the independence of the exchange rate to the concentration of BF_3 and hence of hexamethylbenzene and HF_2^-.

The halogeno methyl-benzenes have also been shown to form ionic proton complexes. It is to be expected that the halogen substituents will weaken the complexes due to their electron withdrawal from and hence deactivation of, the ring. Brouwer[116] has studied the structures of these ions as formed in HF–SbF_5. In 2-halomesitylenium ions it is found that the proton goes onto one of the unsubstituted carbon atoms. The halogen tends to deactivate the 4,6-positions (meta to halogen). However, substitution here is still preferred relative to substitution at the methylated carbon atoms since such substitution would destroy conjugation of the methyl group with the aromatic ring through hyperconjugative effects. Addition to the halogen-bearing carbon atom is even less favoured as a result of the strong conjugative interaction between the halogen and the aromatic ring.

Protonation of 2-, 4- and 5-bromo-*m*-xylenes was found to give identical spectra which were shown to be due to proton complexes of 5-bromo-m-xylene. Both the 2-protonated and 4-protonated xylenenium ions (structures XVI and XVII respectively) were formed in the ratio of 2.0 ($\pm$ 0.1) to 1. The 5-chloro and 5-fluoro-m-xylenes also gave similar stable ions, but in different ratios due to the varying relative basicities of the 2 and 4 carbon atoms. Similar studies have been reported for halotoluenium and halobenzenium ions[117].

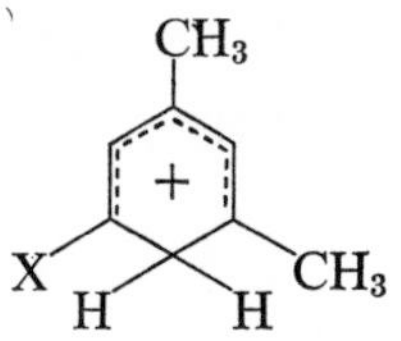

XVI

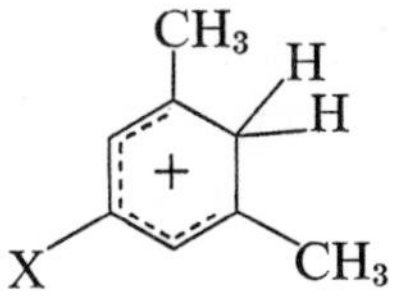

XVII

It follows from the description of the spectra resulting from proton exchange that a simple spectrum may result from intermolecular, or intramolecular, proton exchange between several species. The halobenzenium ions give examples of this[117]. Chlorobenzene in HF/SbF_5 gives a single peak at $\delta = 7.5$. On increasing the temperature from -20°C to $+10$°C the line narrows from 55 Hz to 28 Hz. The chemical shift is in agreement with what one would expect for the weighted average of the shifts of CH_2 and $-\overset{\displaystyle H}{\overset{|}{C}}-$ in the chloro-benzenium ion.

In aromatic ethers protonation could occur either on the ring or at the oxygen. The n.m.r. spectrum of anisole[118] in fluorosulphuric acid containing some SbF_5 to increase the acidity is shown in figure 6.6. The relative areas of the aromatic:methoxy:CH_2 peaks are 3.89:3.00:1.90 (approx. 4:3:2) as required for C protonation. The simple AB type pattern of the aromatic signals and the triplet splitting of the

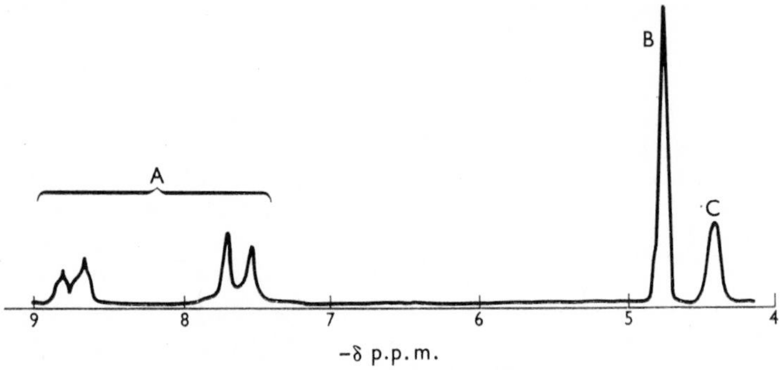

FIG 6.6 (Reproduced by permission of the National Research Council of Canada from *Can. J. Chem.*, **42,** 1435 (1964).)

low field component indicates that protonation occurs para to the methoxy substituent[118]. Previously it had been suggested that the ultraviolet spectrum was incompatible with C protonation[119].

The anisole complex has no absorption down to wavelengths of 250 nm whereas the proton complexes of 1,3,5-trihydroxybenzene, dimethoxy-benzene and the methyl benzenes all show bands near 240 and 350 nm. The difference in the spectra was explained by Birchall *et al.*[118] as due to the quinoid structure (XVIII). The ultraviolet spectrum

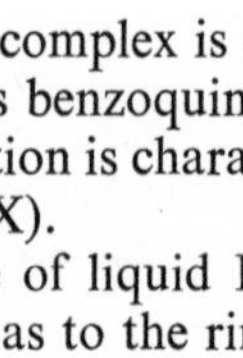

XVIII

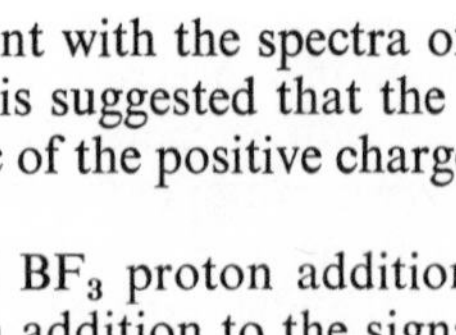

XIX

for the anisole complex is consistent with the spectra of other quinone systems such as benzoquinone. It is suggested that the presence of the 350 nm absorption is characteristic of the positive charge residing in the ring, as in (XIX).

In a mixture of liquid HF and BF_3 proton addition occurs to the oxygen as well as to the ring[120]. In addition to the signals arising from the carbonium ion further signals appear at low temperatures (−80°C). A doublet at $\delta = 5.0$ arises from the methoxy grouping split by the OH and the *o*-proton resonance at $\delta = 12.8$ is split into a quartet by the CH_3 protons. The ratio of carbonium to oxonium ion varies from 1.5 at −80°C to over 50 at 0°C. It also clearly depends on the solvent as evidenced by the failure to observe the oxonium ion in $HFSO_3$ at temperatures down to −64°C.

Finally it ought to be mentioned that the benzenium complex itself has been observed by n.m.r.[121] Benzene dissolves in the "super-acid" $HF/SbF_5/SO_2F_2$ system. At −50°C the p.m.r. spectrum is a singlet but at −134°C the intramolecular exchange is frozen out to allow individual identification of the resonances of the various non-equivalent protons.

6.5.3. VIBRATIONAL SPECTRAL STUDIES

There have been few studies of the vibrational spectra of arenonium complexes, mainly, presumably, because of the difficulty in handling solutions of these in such solvents as hydrogen fluoride and antimony pentafluoride in strictly water-free conditions. Perkampus and Baumgarten[74,122–124] have studied the infrared spectra by laying down thin films of the components on a suitable substrate at low temperatures. On allowing these to warm, reaction occurred to form complexes. As with the n.m.r. spectra marked changes occur in the infrared spectra.

Figure 6.7 compares the spectrum, presented in simple line form,

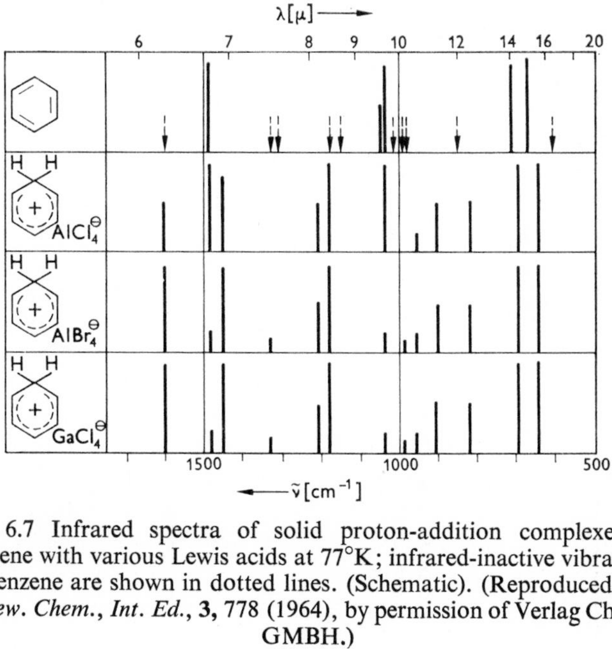

FIG 6.7 Infrared spectra of solid proton-addition complexes of benzene with various Lewis acids at 77°K; infrared-inactive vibrations of benzene are shown in dotted lines. (Schematic). (Reproduced from *Angew. Chem., Int. Ed.*, **3**, 778 (1964), by permission of Verlag Chemie, GMBH.)

TABLE 6.6

Correlation of the bands which occur in the range 670–900 cm^{-1} with the number of mutually adjacent H atoms in the ternary complex aromatic substance $-GaCl_3-HCl$

	Adjacent H-atoms 5	4	3	2	1	
Expected range	730–770 (vs) 690–715 (s)	735–770 (vs)	750–810 (vs) 660–725 (m)	800–860 (vs) 700–750 (m)	860–900 (m)	Not assigned
Aromatic substance						
benzene	690 (vs)					
toluene		742 (vs)		848 (vs) 709 (vs)		
o-xylene			797 (m) 710 (s)			898 (vs) 893 (m)
m-xylene				829 (s) 732 (s)	875 (s)	
p-xylene				857 (vs) 720 (s)	873 (m)	697 (m) 779 (s) 830 (m)
mesitylene[a] (1, 3, 5)					867 (m)	
hemimellitene[a] (1, 2, 3)				784 (vs) 729 (s)		712 (m)
pseudocumene[a] (1, 2, 4)				824 (s) 734 (s)	876 (m) 888 (m)	743 (m)
durene[a] (1, 2, 4, 5)					889 (m)	

m = medium, s = strong, vs = very strong.
a. positions of methyl groups shown in parentheses.

resulting from the system benzene/$AlCl_3$/HCl with that of benzene itself[74]. The positions of infrared-inactive modes of benzene are marked. It is apparent that interaction has resulted in a lowering of the effective molecular symmetry of the aromatic entity. The results appear to be compatible with C_{2v} symmetry as required for a benzenonium complex in which the C_6 skeleton remains planar and the CH_2 is symmetrically disposed around it. It does not appear to the authors, however, that this spectrum, on its own gives evidence, of retaining ring planarity. Perkampus and Baumgarten[74] related the bands of the complex to those of benzene itself by showing that the ratio of the frequencies of corresponding bands in C_6H_6 and C_6D_6 were close to the ratios of similar bands in the $C_6H_7^+$ and $C_6D_7^+$ ions. This is unlikely, however, to be sensitive to small degrees of non-planarity. The best evidence for a planar skeleton comes from the correlations Perkampus and Baumgarten[123] were able to draw between the very strong absorptions in the spectral range 900 to 660 cm^{-1} and the number of adjacent CH units excluding the $>CH_2$. It is well known[125–127] that there is a good correlation between the number and position of these 'out-of-plane' bending mode absorptions and the number of adjacent unsubstituted CH bonds around the ring. These bands are additive in the sense that a 1,2,4-trisubstituted benzene shows bands characteristic of a lone CH bond (the bond at C_3) and of two adjacent bonds (those at C_5 and C_6). The position of the strong bands in the proton complexes are in the expected ranges when the $>CH_2$ of the complex is treated as a substituted aromatic ring carbon. Table 6.6 gives a typical set of results[124]. These correlations are clearly of important diagnostic value in the structural analysis of these complexes. Figure 6.8 shows the spectrum of toluene and of its proton addition complex in the presence of the base $GaCl_3$[123].

A full vibrational analysis of an arenonium complex has not yet been carried out. If successful, such an analysis would be of considerable interest in that it might allow estimates of the changes in the bond and angle restoring forces arising from the proton addition. These in turn could be related to the nature of chemical bonding. The major obstacle to such an analysis at the moment is the lack of any Raman data. Using a technique such as that of Perkampus and Baumgarten, and with laser excitation, Raman data should be attainable.

The feasibility of such Raman studies is confirmed by a number of related investigations by Olah and his co-workers. Commeyras and Olah[128] have carried out a thorough investigation of mixed systems involving antimony pentafluoride, hydrogen fluoride, fluorosulphonic acid, sulphur dioxide and water using n.m.r. and Raman spectroscopy. The conclusions of earlier n.m.r. work on SbF_5–HSO_3F[129] on SO_3–HSO_3F and on HF–SbF_5[130] have been confirmed and extended. The spectrum of antimony pentafluoride itself was reinterpreted on the basis of a weakly associated pair of SbF_5 molecules. While it was easier

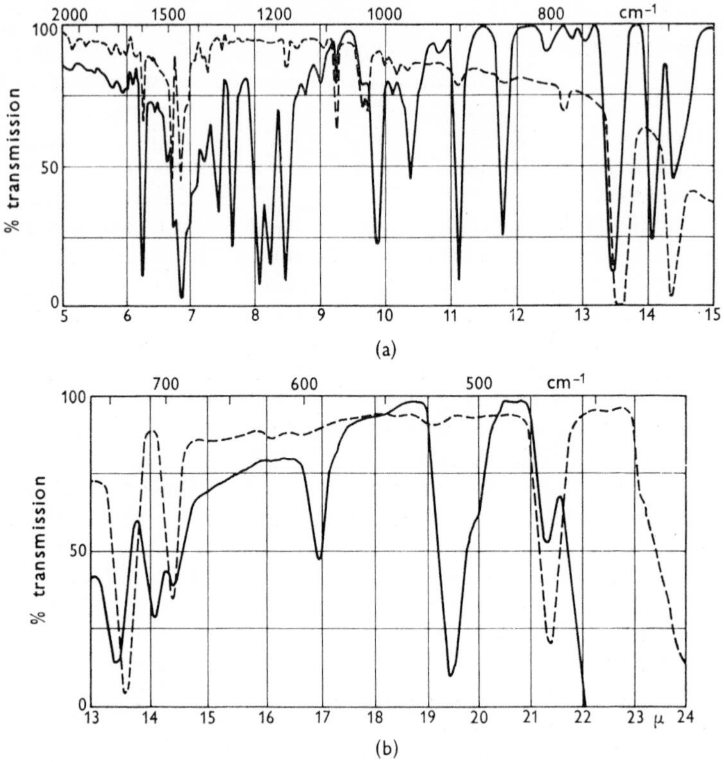

FIG 6.8 The infrared spectra of toluene $GaCl_3$ HCl (solid line) and of toluene (broken line). Measurements were made at −196°C. The upper spectrum (a) is as measured with NaCl prism, the lower (b) with KBr prism. (Reproduced from *Z. Elektrochem.*, **68,** 70 (1964), by permission of Verlag Chemie, GMBH.)

to analyse for the species present using n.m.r. spectra, bands due to complexes such as SbF_5SO_2 and $HSbF_5SO_3F$ were identified by comparison of spectra of different mixtures of known composition[131].

One of the most stable of arenonium ions is the terminal product in the Friedel-Crafts methylation of benzene, namely the heptamethylbenzenonium ion[132]. It is commerically important in that this stable ion consumes, and hence deactivates, large quantities of the catalyst. Its n.m.r. and infrared spectra have been reported and favour its structural assignment as 4-methylene-1,1,2,3,5,6-hexamethyl-cyclohexadiene-2,5 (XX). Thus an absorption doublet in the methyl "umbrella" region at 1380 and 1359 cm^{-1} is quoted as evidence for a *gem*-dimethyl group (see ref. 127, p 24) and a strong band at 855 cm^{-1} is assigned to the vinylidene (>C=CH_2) group deformation. Such a grouping normally gives rise to absorption in the range 895 to 885 cm^{-1}. In view of the conjugation and positive charge the shift to 855 cm^{-1} is not unreasonable.

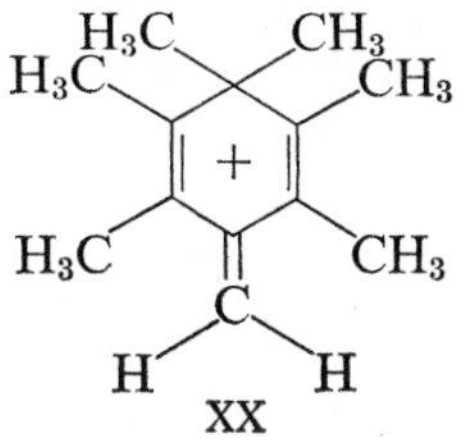

XX

6.6. CYCLOPENTADIENYL COMPLEXES OF THE NON-TRANSITION METALS

The metal cyclopentadienyls form an interesting group of molecules in which bonding between the ring and the metal ranges in type from ionic, through $b\pi$-v (with "back-bonding") to pure σ. The inclusion of these options under the title 'complexes' is, in certain cases, open to criticism, but, on the grounds that the bonding problems and the techniques used to study them are similar, they are being discussed in this text. Vibrational spectroscopy has been the most widely used technique in studying both this bonding and the molecular geometry. Despite all that has been achieved there is still controversy and uncertainty of the details of the interpretation in certain instances. Some of these troubles will be seen to arise from changes in bonding between the gaseous and the solid state.

The bond between a cyclopentadienyl group and a metal, or metalloid, can be thought of in the first instance as belonging to one of three groups—an ionic bond, a 'non-classical' π–metal bond or a σ-bond. These are depicted in figure 6.9.

In the first instance we shall consider an isolated M–Cp unit (Cp = C_5H_5). The three structures represented in figure 6.9 should give rise to very different spectra. In the purely ionic structure (a) the cyclopentadienyl anion has D_{5h} symmetry. In a system of such high symmetry the selection rules are very restrictive. The representation generated by the

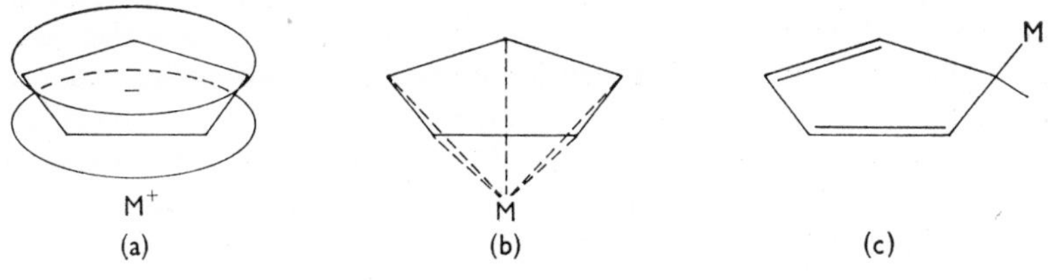

FIG 6.9

nuclear coordinates reduces to:

$$\Gamma = 2A_1' + A_2' + A_2'' + 3E_1' + E_1'' + 4E_2' + 2E_2''$$

Only the fundamental vibrations of symmetry a_2'' and e_1' are infrared active, and only those of a_1', e_1'' and e_2' are Raman active. That is, there ought to be four infrared bands and seven Raman bands belonging to the ion. It has been said in the past that since there is no bond between the metal and the ring no metal–Cp stretching vibration can be expected. Electrostatic forces are strong, however, and, for the solid, lead to lattice vibrations which may give rise to absorption bands lying in the region of interest. Even a small admixture of covalent bonding may lead to significant frequency shifts while scarcely perturbing the spectra of the ring vibrations (see below).

In the second bonding group (b) the cyclopentadienyl unit is assumed to retain its 5-fold axis—though this is something which must generally be proved. The bonding from M reduces the symmetry to C_{5v}. To a first approximation, the normal vibrations can be treated as the twenty four vibrations of the $C_5H_5^-$ ion perturbed by the unit M, plus those vibrations involving M and derived from the three new degrees of freedom arising from the addition of the new nuclear point M. With this artificial separation of the ring vibrations the normal modes now classify as Ring modes $3a_1$(p, ir) + a_2(ia, ia) + $4e_1$(dp, ir) + $6e_2$(dp, ia) Cp–M modes a_1(p, ir) + e_1(dp, ir).

In the event that bonding is through a classical type covalent bond to one carbon atom of the ring we would not expect to be able to separate the C_5H_5 group vibrations from those of the M–Cp unit. We might anticipate that the spectrum could be understood as that of a perturbed cyclopentadiene system. However, the coupling between the metal–carbon bond and the ring vibrations will generally be strong. An intensity perturbation of low frequency infrared and Raman bands can be expected, since coupling of ring and skeletal modes will be greatest here as well as the appearance of new bands associated with M. (See section 1.2.8 for a general discussion of intensity changes expected due to vibrational mixing.)

In the above classification it has been assumed that any metal cyclopentadienyl can be classified in one of the above three categories. Such artificial divisions are rarely honoured in nature. A consideration of the bonding of sodium chloride in the crystal phase gives even this classical ionic material 15% covalent character[133]. It is natural then that all mixtures of bonding between ionic and covalent bonding of types a and b will be encountered. On symmetry grounds, of course, there is a clear distinction between bonding along the C_5 axis and the classical σ-bond structure of figure 6.9c.

The group I cyclopentadienyls are likely to be the most ionic and indeed the cyclopentadienyls of potassium, and the group I elements lower in the periodic table, have been satisfactorily interpreted on this basis[134]. Sodium and lithium cyclopentadienyls have, however, been accorded a σ-bond structure, mainly on the grounds that absorption bands are found at 315 and 538 cm^{-1} respectively. These are not due to ring vibrations and must therefore be considered as M–Cp vibrational bands. The metal-cyclopentadiene stretching frequencies associated with ferrocene type structures (like figure 6.9b) are generally much higher. On the other hand it must be pointed out that the $k = 0$ lattice optic mode of sodium bromide, which has a similar reduced mass to sodium cyclopentadienyl, has a frequency corresponding to 135 cm^{-1} [136]. A small increase in the π-bonding could, no doubt, explain the observed frequency. Indeed, a wave mechanical calculation of lithium cyclopentadienyl[136] using the *SCF LCAO* method predicts a C_{5v} structure. The calculated distance between the lithium atom and the centre of the ring is 1.68 ± 0.05 Å and the $\tilde{\nu}$(Li–Cp) value is calculated to be 750 cm^{-1}. This result, while it is too high, does emphasise that the π-type bonding is reasonable.

If the bonding in these compounds does have considerable ionic character then an intense metal-cyclopentadienyl stretching band is to be expected. For a 100% ionic character the intensity is easily predicted[137] since the dipole gradient $\partial\vec{\mu}/\partial Q = 2\mathbf{e}$ where $\mathbf{e}$ is the electronic charge. This gradient is likely to drop faster than the percentage ionic character due to electron flow during the bond deformation. This model thus cannot be used in its simplest form except for a qualitative demonstration of the importance or otherwise of ionic bonding. It would be interesting to know to what extent the observed bands of $K(C_5H_5)$, $Na(C_5H_5)$ and $Li(C_5H_5)$ deviate from this prediction.

The spectra of the alkali earth metals from calcium downwards have also been interpreted as indicating primarily ionic structures. Both beryllium and magnesium dicyclopentadienyls have been extensively studied, and, rather surprisingly, their structures are very different. An electron diffraction study showed that in $Be(C_5H_5)_2$ the rings were parallel with the beryllium atom along the 5 fold axis but nearer to one ring than to the other[138]. The actual distances between the beryllium atoms and the centres of the rings were 1.48 and 1.98 Å, the inter-ring distance being of the order of that in solid benzene (3.4 Å). As was known earlier[139] this compound has a large dipole of $8.2 \pm 0.2 \times 10^{-30}$ C m in benzene and $7.5 \pm 0.3 \times 10^{-30}$ C m in cyclohexane. Such a large dipole is compatible with a fully ionic structure, and the non-centrosymmetric location of the beryllium atom would be expected to be more stable than the centrosymmetric one on electrostatic grounds. Nevertheless, other models are also compatible with these experimental observations.

The infrared spectra of solid and gaseous $Be(C_5H_5)_2$ as well as of solutions in benzene and cyclohexane have been measured[140,141]. No Raman data has been reported. Morgan and McVicker[142] suggested on the basis of the n.m.r. data that specific solvent interactions occurred.

With this in mind, it is perhaps advisable to consider what deductions can be made for the gaseous and solid state spectra first. These same two authors prepared $Be(C_5H_5)_2$, taking great care over its purification, excluding all water and oxygen. The spectra of solid samples were measured as mulls and absorption bands due to decomposition products identified by allowing short exposures to the atmosphere after the initial spectra were run[140]. The spectra of the gaseous phase were measured over a range of temperatures from 25° to 70°C and were found to be essentially unchanged. Considerable differences existed, however, between these and the solid state spectra. The latter were quite complex, there almost being two bands for every one of ferrocene. This is compatible with a symmetry lower than D_{5h} and, in particular, compatible with C_{5v} symmetry. Particular interest is centred around the bands near 1100 cm^{-1}. It has been observed that when a cyclopentadienyl unit is covalently linked to a metal then a band appears near 1100 cm^{-1}. When the bonding is ionic this band is no longer observed[143]. This empirical correlation is explained by the assignment of the band to a symmetric ring breathing mode. In the isolated ion there can be no dipole gradient for such a vibration and hence no absorption. However, under the influence of a perturbation along the C_5 axis due to metal bonding there could well be an appreciable transition moment. It is possible that the mechanism for generation of the transition moment is a "vibronic" one similar to that described for halogen complexes (see chapters 1 and 3). For crystalline dicyclopentadienyl beryllium two absorption bands of moderate strength occur at 1122 and 1102 cm^{-1} [140]. This is in accord with significant covalent bonding with both rings, the latter being non-equivalent.

The spectrum of the compound in the gaseous state was much simpler. Indeed there were only two strong bands, which were near 969 and 736 cm^{-1}. The absence of any band near 1100 cm^{-1} was taken to indicate a highly ionic character for the metal-ring bonding.

Dicyclopentadienyl magnesium has been shown by X-ray diffraction studies to have the magnesium atom centrally placed between the rings, 2.0 Å from each[144]. In accord with this it does not have a dipole moment[139]. Since magnesium has no d electrons which it can use to form a bond direct to the π-bonds of the rings it was assumed originally to be ionic[145,146] and that its central location between the rings was due to the larger cation size compared with beryllium. Its solubility in benzene was against such a model and its infrared and Raman spectra were shown to be very similar to those of ferrocene[138]. Furthermore a moderately strong absorption band at 1108 cm^{-1} is in accord with a covalent type bond. The antisymmetric metal-ring stretching mode was identified with a band at 526 cm^{-1} which is to be compared with 492 cm^{-1} for ferrocene. The intensity of this band was measured and compared with that predicted for a purely ionic system. It was found that the ratio of the observed to calculated intensity was approximately

1 to 68, again emphasising the essentially covalent character of the bonding.

Cotton and Reynolds[147] examined the ν(CH) stretching region of cyclopentadienyl thallium and dicyclopentadienyl magnesium. In both cases they observed only a single ν(CH) band which they took as evidence for ionic bonding. For D_{5h} symmetry only the e_1' mode would be infrared active. Making the assumptions (a) that a perturbation through covalent bonding leads to an angle α between the CH bond and C_5 ring plane, and (b) that on stretching a bond the only dipole change was along that bond, then they pointed out that the intensities, Γ, (see p 52) of the a_1 and e_1 modes (C_{5v} symmetry) will be in the ratio $\frac{1}{2}\tan^2\alpha$. For $\alpha = 18°$, this ratio $= 0.05$.

It is doubtful if covalent bonding would produce anything like an angle $\alpha = 18°$. In the electron diffraction studies[138] of $Be(C_5H_5)_2$ the best fit with the experimental data are obtained with $\alpha \sim 0°$. For $In(C_5H_5)$ electron diffraction[148] gave $\alpha = 4.5 \pm 2°$ whereas recent microwave studies[149] give a value of only 20′ with an uncertainty of the same order. For $Tl(C_5H_5)$, $\alpha = 46' \pm 20'$[149]. The greater value of the dihedral angle in $In(C_5H_5)$ as given by electron diffraction is probably due to vibrational distortions. The metal-ring stretching frequency has been derived as equivalent to 160 ± 20 cm^{-1} from the intensities of rotational lines arising in the microwave spectrum as a result of transitions from excited states[150]. In addition, a considerable amount of vibration-rotation interaction data, such as centrifugal distortion and coriolis coupling constants were determined. This shows that microwave spectroscopy can give data for quite complex molecules which is of great value in determining general quadratic force-fields.

Cotton and Reynolds[147] calculated the metal-ring bonding energy for both $Mg(C_5H_5)_2$ and $In(C_5H_5)$ using a simple LCAO model. For the magnesium compound significant covalent bonding was deduced. An error was made in calculating the overlap integrals for $In(C_5H_5)$ which was subsequently corrected by Shibata, Bartell and Gavin[148], who again deduced that the bonding was essentially covalent. The essence of the calculation is as follows. The wavefunctions of the 5s and 5p electrons of indium can be classified in the C_{5v} subgroup as $a_1(5s)$, $a_1(5p_z)$ and $e_1(5p_x, 5p_y)$. If we allow complete s, p mixing of the a_1 wavefunctions then we can describe two orthogonal wavefunctions:

$$a_1' = \frac{1}{(2)^{1/2}}(p_z + s)$$

and

$$a_1'' = \frac{1}{(2)^{1/2}}(p_z - s).$$

The a_1'' orbital is taken to have the lower energy and to be filled whereas the a_1' orbital is vacant. Bonding can occur through overlap of a_1'

with the C_5H_5 a_1 orbital ($b\pi$-v) and through overlap of the C_5H_5 e_1 orbitals of the ring with the occupied indium orbital. The energy is now minimised with the ring atom distance as a variable. The magnitude of the overlap integrals gives a measure of the covalency. This simple model was refined a little by the later investigators[148].

One of the most interesting of the group III cyclopentadienyls is the thallium compound. It is remarkably stable towards hydrolysis and oxidation—two properties which render it particularly attractive for spectroscopic investigations. By analogy with cyclopentadienyl indium it would be expected to be essentially covalent. This is in accord with the stability of the compound towards hydrolysis. A recent investigation[151] on the solid phase has shown that the vibrational spectra contain few bands, and the ring mode near 1100 cm^{-1} absorbs very weakly (see table 6.7). This strongly suggests that there is a considerable ionic contribution to the metal–ring bond. However, these results refer to the crystal phase. X-ray[152] and microwave[153] results show that the metal-ring distance in the solid is much greater (3.19 Å) than in the vapour (2.41 Å). This strongly suggests that for the isolated $Tl(C_5H_5)$

TABLE 6.10

The vibrational bands (in cm^{-1}) of crystalline cyclopentadienyl thallium

Raman	Infrared	Assignment
3092 (s)		a_1'
	3070 w	e_1'
3065 (m)		e_2'
	2920 w	
	1425 w	e_1'
1370 (m, sh)		} e_2'
1342 (s)		
	1300 w	e_2'
1208 (s)		
	1158 w	
1118 (vs)	1120 vw	a_1'
1058 (m)		e_1''
	1000 s	e_1'
841 (m)		e_2''
762 (s)		} e_2'
751 (s)	755 m	
	727 s	a_2''
445 (w)		} e_2'
418 (w)		

w = weak, m = medium, sh = shoulder, s = strong, vs = very strong.

species there will be considerable covalent bonding. These results indicate that the structural changes accompanying phase changes parallel those in dicyclopentadienyl beryllium and that it is unwise to attempt to correlate calculated properties for the isolated metal cyclopentadienyls with those existing in the condensed state.

Both dicyclopentadienyl tin and lead have considerable dipole moments. As measured in cyclohexane[145] they are $3.2 \pm 0.3 \times 10^{-30}$ C m and $4.30 \pm 0.13 \times 10^{-30}$ C m respectively. In accord with these moments electron diffraction of the gaseous compounds show that the rings are non parallel, the angle between the planes being $45 \pm 15°$ and about 55° respectively for lead and the tin cyclopentadienyls[154]. The absorption spectra of these compounds strongly resemble that of ferrocene, showing strong covalent bonding[155]. On the other hand the n.m.r. spectra are both singlets at room temperature[156]. This must mean that the rings rotate freely about the metal ring bond. There are other cases known where the n.m.r. spectrum shows that the rings are rotating on a time scale similar to, or much less than the n.m.r. one as defined by the inverse of the differences in chemical shifts. One interesting case is afforded by dicyclopentadienyl mercury whose physical and chemical properties are not in accord with the metal to ring bond being primarily ionic. The infrared spectrum[157] would appear to support a diene structure by virtue of absorption near to 1700 cm^{-1}. The n.m.r. spectrum of this compound[158] is a sharp singlet ($\delta = 5.8$) at room temperature. This indicates that even with σ-bonding the point of attachment of the metal to the ring changes rapidly. The line remains sharp and is a singlet down to −70°C for the compound dissolved in tetrahydrofuran or triethylamine. In carbon disulphide the peak is broadened at −70°C and in sulphur dioxide at this temperature three features are apparent at $\delta = 2.6$, 6.0 and 6.4. On the assumption that there are indeed solvent effects on the intrinsic chemical shifts of the protons, and not on the nature of the chemical bond, deductions can be made from the SO_2 solvent data concerning the changes occurring in the bonding site to the ring. The breadth of the $\delta = 6.4$ peak, which was assigned as due to the protons α to the metal ring bond, indicates that the mercury atom rotates around the ring by a series of 1,2 shifts[158]. Further examples of proton averaging in what appears to be δ-bonded cyclopentadienyls are afforded by cyclopentadienyl triphenylphosphine gold(I)[159] and cyclopentadienyl triethylphosphine copper (I)[160].

Numerous examples exist of substituted cyclopentadienyls whose structures have been studied by vibrational spectroscopy. Since the symmetries of such systems are generally low deductions are generally possible only on the basis of empirical correlations. The correlation between a band near 1100 cm^{-1} and the presence of an unsubstituted covalently π-bonded cyclopentadiene ring is easily the most useful in this field at the present time. As an example of its use we may quote the reaction between a 1-argenta-1′-bromoferrocene and bismuth tribromide to yield *tris* (1′-bromoferrocenyl) bismuthine:

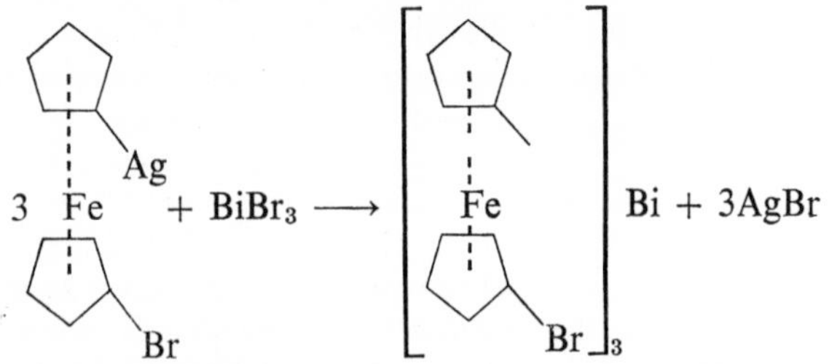

$$3 \; Fe\substack{Ag \\ Br} + BiBr_3 \longrightarrow \left[Fe\substack{ \\ Br} \right]_3 Bi + 3AgBr \qquad 6.11$$

The absence of an absorption near 1100 cm^{-1} was taken as adequate evidence that bonding is through a sigma bond to the ring other than that containing the bromine[161].

6.7. COMPLEXES OF π DONORS AND POTENTIAL π DONORS WITH NON-TRANSITION METALS

In general the π-electron system of an aromatic ring can act as donor or acceptor. With group III acceptors the role of the π-electron system is unambiguous. Nevertheless, complexing between aromatics and group III compounds may occur through other than the π-electron system. The spectroscopic evidence is quite strong for σ-bonding in many instances.

The boron trihalides are planar systems with D$_{3h}$ symmetry in their unperturbed ground state configurations (see section 6.1.2). It was observed[162] that dissolution of BBr$_3$ or BI$_3$ in benzene resulted in the appearance of an absorption band at the frequency of the symmetric stretching mode. The absorption coefficient was concentration dependent and could be explained on the basis of a 1:1 equilibrium of the form:

$$BBr_3 + C_6H_6 \rightleftharpoons C_6H_6\text{-}BBr_3 \qquad 6.12$$

Such an interaction would result in a symmetry of, at the most, C$_{3v}$ and therefore permit activity of the symmetric stretch. It should be noted that this observation does not require that the boron trihalide is non-planar in this complex. The intensity may arise from a vibronic charge-transfer mechanism as postulated by Friedrich and Person for halogen complexes[163] (see chapters 1 and 2). The dipole moment of BBr$_3$ in benzene has been measured and is small (0.63 × 10^{-30} C m), though significantly different from zero, thus suggesting a small deviation from planarity in the average configuration.[164] It should be noted that an excellent study[195] of the dielectric constants and refractive indices of the hexafluorobenzene–benzene system showed that small dipole moments of complexes derived from dielectric data alone could be seriously in error. This is due to temperature dependence of the concentration of complex and due to non additivity of polarisabilities.

The changes in chemical shifts on dissolving compounds in benzene (the 'aromatic solvent induced shift' or ASIS) has been a popular method of studying aromatic complexes. The original observations were that if hydrogen halides or chloroform were dissolved in benzene then the proton resonances of the solutes, after correction for magnetic susceptibilities and breaking of hydrogen bonds in the original compound, moved to higher field[165]. This was explained on the basis of diamagnetic susceptibilities arising from ring currents in the π-electron system—that is, in the magnetic field the electrons in the π-electron system circulated in a fashion such as to generate an opposing (diamagnetic) field[166]. This effect seems to be remarkably general and has been widely used to show molecular complexing. Boron tribromide in benzene proved to be no exception. The ^{11}B resonance moved to high field in benzene[162]. However, it has recently been elegantly demonstrated that such shifts may not imply any specific chemical bond[167]. It was shown that the ASIS for the various protons of camphor could be expressed as the product of two terms—one characteristic of the solvent and the other dependent on the site of the proton in the molecule. Thus:

$$\text{ASIS} = (\text{site factor}) \times (\text{solvent parameter}). \qquad 6.13$$

Following from this, plots of the ASIS of one proton in various solvents against the ASIS of a second, non-equivalent, proton yielded a very good linear plot (see figure 6.10). The site factor is given to a good approximation by $(3 \cos^2 \theta - 1)/r^3$ where r is the distance of the spin site from the end of the oxygen atom and θ is the inclination between the vector $\vec{r}$ and the CO bond. This crude model implies that the oxygen atom is located on an average near to the site of the effective dipole from the diamagnetic circulation current and that no specific chemical bonds exist which would perturb the chemical shift.

It should be emphasised that while this work explains the common occurrence of the diamagnetic ring current effect it would be quite wrong to assume that all ASIS can be explained without invoking chemical bonding.

The normal molecular state of the aluminium halides is a halogen-bridged dimer. Clearly, we have a very different situation from that of the boron halides. Aluminium chloride, *if* it interacts with benzene, does so very weakly. Aluminium bromide dissolves readily and the solute-solvent interaction responsible for this has been frequently investigated. An X-ray analysis of the crystalline solid shows benzene and Al_2Br_6 interleaved throughout the crystal[168]. Dipole moment studies[169] show that the complex has a quite large dipole of 8.67×10^{-30} C m. The interaction is clearly strong.

Of the interaction between aromatic systems and the group V halides the most thoroughly studied has been that between antimony trichloride and benzene. Two groups of workers have recorded the

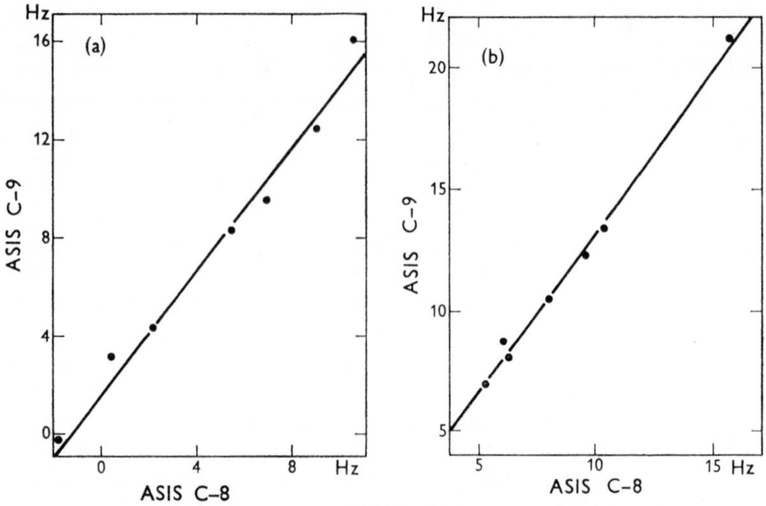

FIG 6.10 Plot of the C-9 methyl ASIS of camphor *vs.* the C-8 methyl ASIS of camphor for a single series of substituted benzene solvents: (a) for solvent shifts referenced to camphane, correlation coefficient = 0.990; (b) for solvent shifts referenced to C-10, correlation coefficient = 0.995. (Reproduced from *J. Amer. Chem. Soc.*, **93**, 1317 (1971), by permission of the American Chemical Society.)

Raman spectra and reported shifts of between 3 and 30 cm^{-1} in the bands associated with vibrations of the halide molecule[170–172]. One of the groups[170] reported two new bands at 477 and 1236 cm^{-1} which they assigned to the bending and stretching vibrations of the bond between antimony and benzene. This, we feel, would require a much stronger bond than there is any evidence for. Daasch[173] and Perkampus and Baumgarten[174] have studied the infrared spectra. Shifts in the out-of-plane γ(CH) vibrational bands were noted as well as intensity enhancements of some of the bands arising from transitions which are forbidden for D_{6h} symmetry. An analysis of the bands showed that, in the solution, the effective complex symmetry is C_{3v} while in the solid state C_{2v} symmetry is indicated[173]. No evidence for interaction between benzene and $SbCl_5$ was found[174].

Many studies have been made of the interaction between metallic halides and non-aromatic systems containing π-bonds which are potential π donors. Terenin and his co-workers have observed vibrational frequency changes accompanying interaction in many such systems[175]. Thus the absorption of nitric oxide at 1876 cm^{-1} shifts to 2141 and 2200 cm^{-1} on formation of complexes with aluminium chloride and tin tetrachloride respectively. It is clear that an electron has been largely or wholly removed from a p orbital of the oxygen atom thereby allowing participation of the remaining electron in the degenerate π-bonding orbital. Complexing between acetonitrile and aluminium chloride leads to an increase in the ν(CN) stretching frequency corresponding to 77 cm^{-1} and the ν(CC) frequency also rises[175] (see sections 6.3.4 and 2.3.3). This makes it clear that interaction is not through the π electrons themselves, but through electrons localised on one of the atoms. (Some nitriles *do* complex through the CN π electrons—section 2.3.3.) With ethyl acetate the characteristic acetate bands at 1745 and 1250 cm^{-1} are considerably displaced, again in accord with strong participation of oxygen lone-pair electrons[175].

Boron tribromide dissolved in acetonitrile to yield a conducting solution. The spectra of the complex in the solid phase has already been discussed (section 6.3.5) and the results indicate coordination to the nitrogen. In solution the bands in the 650–700 cm^{-1} region sharpen and a new band at 590 cm^{-1} appears which is characteristic of the BBr_4^- ion[176]. Based on this and additional evidence the structure $(CH_3CN)_2BBr_2^+BBr_4^-$ is proposed for the complex in solution.

An interesting Raman study has been made of the complex between aluminium chloride and cyanogen chloride[177]. The conductivity of cyanogen chloride increases dramatically on addition of aluminium chloride. For example the conductivity of a 0.02 M solution of aluminium chloride is a thousand times greater than that of cyanogen chloride itself[178]. The explanation proposed is that the following reaction sequences occur.

$$Al_2Cl_6 \rightleftharpoons 2AlCl_3 \qquad 6.14(a)$$

$$AlCl_3 + ClCN \rightleftharpoons AlCl_3\text{–}NCCl \qquad 6.14(b)$$

$$AlCl_3\text{–}NCCl \rightleftharpoons AlCl_4^- + CN^+ \qquad 6.14(c)$$

$$CN^+ + ClCN \rightleftharpoons Cl\text{—}C{\equiv}N\text{—}C{\equiv}N \qquad 6.14(d)$$

The reaction (d), or something like it, would seem unavoidable in view of very high Lewis acid strength of the nitrile cation. The Raman spectra confirm the main features of these postulated reactions. Two bands were found in the nitrile region. The first at 2286 cm^{-1} compares with 2203 cm^{-1} in pure ClCN and, by analogy with the effects of aluminium chloride on methyl cyanide (*vide supra*), this band must be identified with the ν(CN) mode of the complexed species Cl—C≡N–$AlCl_3$. The only other new band in the ν(CN) stretching region was at 2327 cm^{-1}. The CN^+ ion in the gaseous phase and in the $^1\Sigma^+$ state, which is probably the electronic ground state, has a vibrational frequency equivalent to 2107 cm^{-1} [179]. The high frequency observed in the complex cannot be due to the isolated ion but is most likely due to the central C≡N group of the complex Cl—C≡N—C≡N. No band was observed which could be assigned to the terminal nitrile stretching mode of the complex but reasons were given for believing that it was close to the Cl—C≡N–$AlCl_3$ absorption.

A corresponding infrared study of complexes of aluminium, antimony, and titanium halides with cyanogen halides has been carried out in solid state and in solutions in thionyl and phosphoryl chlorides[180]. The displacements of the bands resulting from the formation of the coordination bond, and the identification of a metal–nitrogen stretching band indicates that bonding is through the lone pair electrons of the nitrogen atom of the cyanogen halides. For the aluminium halide complexes the spectral activities and assignments are compatible with C_{3v} symmetry in the complex, that is, a linear Al—C≡N—Cl bond system. Similar results are obtained for the other complexes. In solution dissociation of the complex is apparent through the appearance of ν(CN) band near that for the parent cyanogen halide, but in addition, bands occur which are displaced less than 10 cm^{-1} from those for the metal halide-cyanogen halide complex. This is interpreted as due to complexes of the type Cl—C≡N–$OPCl_3$. A force constant analysis based on a Urey–Bradley field shows a significant increase in the CN force constant for the complex.

6.8. THE SPECTRA OF IONIC MELTS AND SOLUTIONS

As indicated in 6.1.2 cations and anions are powerful vacant orbital acceptors and donors respectively. For polyatomic ions association of ions, or complexing with solvent, may show as perturbations to the intra-ionic vibrations. For monoatomic species no intra-ionic vibrations exist and association can show only through modification of the ligand vibrations or through the appearance of new vibrational modes associated with the new bond. The more polar solvents absorb extremely strongly in the infrared and therefore studies of ion-ion and ion-solvent interactions are difficult by infrared spectroscopy. The extremely strong absorption of water is well known. Nevertheless, a number of investigations have been made in aqueous media. An example of this is the study of the equilibria between nickel cyanide anions $Ni(CN)_n^{-(n-2)}$ and CN^- in aqueous solution of $Na_2Ni(CN)_4$ and KCN[181]. This was possible because water has a transmission window in the 2100 cm^{-1} region. The intensities of absorptions at 2124 and 2103 cm^{-1} could be satisfactorily interpreted as due to $n = 4$ and $n = 5$ respectively. Earlier references to other similar examples are given in reference 182.

Whereas increased polarity tends to lead to larger dipole derivatives in vibrational transitions and hence greater absorption strength, in Raman spectroscopy increased polarity leads to a decrease in polarisability derivatives $(\partial\alpha/\partial Q)$ and a consequent decrease in Raman emission intensities. In simple terms the more polar the bonds then the more tightly the electrons are held in the locality of the nuclei. As a polar bond is deformed the polarisability is little altered and hence the Raman scattering cross section (proportional to the square of $(\partial\alpha/\partial Q)$) is small. Water is not only a very polar compound but has a low electron density in the molecule as a whole. It follows that it should be an excellent solvent for Raman studies and this is found to be so. A further corollary of the above considerations is that Raman scattering from ion pairs will be weak. George, Rolfe and Woodward[183] computed the Raman intensities for ion pairs using various effective values of the dielectric constant and approximate wavefunctions derived for incompressible spheres held under coulombic forces. Comparison of the results with known polarisability gradients associated with vibrations in methane and carbon tetrachloride indicated that the scattering intensities would be lower for the ions by a factor of between 10^{-2} and 10^{-5}.

While the interionic and ion-polarised dipole forces produce no new observable vibrations they may perturb the frequencies and intensities of the intra-ionic and intramolecular vibrations of the ions and solvent. A considerable amount of information can be obtained in this manner which is conveniently demonstrated by reference to the literature on the nitrate ion.

The isolated nitrate ion has D_{3h} symmetry and therefore its vibrations transform as the irreducible representations $A_1'(R) + A_2''(ir) + 2E'(R, ir)$. Coordination with a metal ion could be envisaged as monodentate or bidentate:

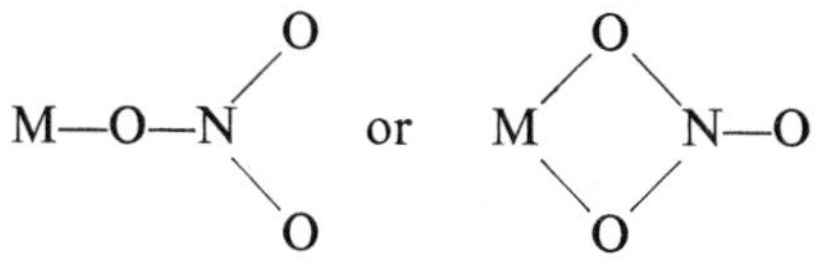

In either case the symmetry of the ion is reduced to C_{2v} and the degeneracies of the e′ vibrations are lifted. The experimental observation[184] is that even in very dilute solutions the band arising from excitation of the degenerate antisymmetric stretching mode (ν_3) centred at 1373 cm^{-1} is split by about 56 cm^{-1}. This splitting is independent of the cation. These facts strongly suggest that the splitting arises from hydrogen bonding to the nitrate. This is supported by the observation of a band at 689 cm^{-1} attributable to the anion–water librational mode. As the concentration of salt is increased the splitting of the ν_3 band becomes concentration and cation sensitive and the band due to the other degenerate mode (ν_4), at about 720 cm^{-1}, also splits. For $LiNO_3$ and $Zn(NO_3)_2$ the splitting of ν_4 does not occur until there are less than six molecules of water per nitrate ion[185] which strongly indicates that the splitting is a criterion of inner-sphere interaction. By contrast even in dilute solutions the nitrates of calcium, cadmium, mercury (II), copper (II), indium (III), cerium (IV) and bismuth (III) nitrates show evidence for inner sphere ion pairing[185].

Having established association of the ions the next natural step is to consider whether the evidence favours monodentate or bidentate structures. Qualitative predictions and calculations show that for the monodentate structure the lower frequency component of ν_3 should belong to the fully symmetric species whereas for the bidentate structure the reverse applies[186]. Polarisation measurements of the Raman bands show that Ca^{2+}, In^{3+}, Zn^{2+}, Hg^{2+}, CH_3Hg^+, and Cd^{2+} nitrates are monodentate whereas the bidentate coordination is preferred in Sc^{3+}, La^{3+}, Ce^{4+}, In^{4+} and Bi^{3+} (see reference 181).

Many infrared and Raman studies have been made on molten nitrates (see, for example, the review by Hester[187]). In addition to the types of problems exemplified by the work described above on aqueous solutions there is much interest in the existence of order in the melt of the crystalline lattice type. Low frequency bands in the Raman and infrared spectra have been interpreted as lattice modes [e.g. reference 187, 188]. The second frequency moment

$$\int \log\left(\frac{I_0}{I}\right)\left(\frac{\omega - \omega_0}{\omega}\right)^2 d\omega$$

where ω_0 is the band maximum ($\equiv 2\pi\nu_0$) of the out-of-plane NO_3^- deformation is far greater than expected on the basis of a free ion with a classical rotational energy[189]. This has explained as due to sums and differences bands with lattice phonons. A more detailed discussion of these phenomenon would take us beyond the attempted scope of this text and the interested reader is referred to the original literature[189–191]. (A brief discussion of the usefulness of band-shape analysis for studying molecular interactions is given in section 2.3.6.)

REFERENCES

1. R. S. Mulliken and W. B. Person, Molecular Complexes, (Wiley-Interscience, New York, 1969)
2. S. G. W. Ginn, J. K. Kenney and J. Overend, *J. Chem. Phys.*, **48**, 1571 (1968)
3. F. A. Cotton and J. R. Leto, *J. Chem. Phys.*, **30**, 993 (1959)
4. B. Swanson, D. F. Shriver and J. A. Ibers, *Inorg. Chem.*, **8**, 2182 (1969)
5. J. L. Hoard and S. Geller, *Acta Crystallogr*, **4**, 399 (1951)
6. C. T. Mortimer, Reaction Heats and Bond Strengths, (Pergamon Press, London, 1962)
7. G. A. Olah, Friedel Crafts and Related Reactions, (Interscience, New York, 1963)
8. B. Swanson and D. F. Shriver, *Inorg. Chem.*, **9**, 1406 (1970)
9. D. F. Shriver and B. Swanson, *Inorg. Chem.*, **10**, 1354 (1971)
10. E. B. Wilson, J. C. Decius and P. C. Cross, Molecular Vibrations, (McGraw-Hill, New York, 1955)
11. I. M. Mills (Ed. M. M. Davies) Infrared Spectroscopy and Molecular Structure, (Elsevier, Amsterdam, 1963), p 166
12. M. J. Ware (Eds. H. A. O. Hill and P. Day) Physical Methods in Advanced Inorganic Chemistry, (Interscience, 1968), p 214
13. D. Steele, Theory of Vibrational Spectroscopy, (W. B. Saunders and Co., Philadelphia, 1971)
14. K. Nakamoto, Infrared Spectra of Inorganic and Coordination Compounds, (Wiley-Interscience, New York, 1970), p 9

15. E. R. Lippincott, R. Schroeder and D. Steele, *J. Chem. Phys.*, **34,** 1448 (1961)
16. D. Steele and E. R. Lippincott, *J. Chem. Phys.*, **35,** 2065 (1961)
17. I. Lindquist, Inorganic Adduct Molecules of Oxo-Compounds, (Springer-Verlag, Berlin, 1963)
18. M.-T. Forel, M. Fouassier and M. Tranquille, *Spectrochim. Acta*, **26A,** 1761 (1971)
19. O. Hassel and K. O. Strømme, *Acta. Chem. Scand.*, **13,** 275 (1959)
20. H. Schibilla and M.-T. LeBihan, *Acta. Crystallogr*, **23,** 332 (1967)
21. G. Stucky and R. E. Rundle, *J. Amer. Chem. Soc.*, **86,** 4825 (1964)
22. H. Wieser and P. J. Krueger, *Spectrochim. Acta*, **26A,** 1349 (1970)
23. S. Hayes, *Bull. Soc. Chim. France.*, 1963, 1404.
24. A. D. Vreugdenhil and C. Blomberg, *Rec. Trav. Chim. Pays-Bas*, **82,** 453 (1963)
25. D. Cook, *Can. J. Chem.*, **41,** 522 (1963)
26. W. Gerrard, M. F. Lappert, H. Pyszora and J. W. Wallis, *J. Chem. Soc.*, 2144 (1960)
27. W. Klemperer, M. W. Cronyn, A. G. Maki and G. C. Pimentel, *J. Amer. Chem. Soc.*, **76,** 5846 (1954)
28. B.-P. Susz and D. Cassimatis, *Helv. Chim. Acta*, **44,** 395 (1961)
29. H. Gerding, J. A. Koningstein and E. R. van der Worm, *Spectrochim. Acta*, **16,** 881 (1960)
30. T. C. Waddington and F. Klanberg, *J. Chem. Soc.*, 2339 (1960)
31. W. Gerrard, E. F. Mooney and H. A. Willis, *J. Chem. Soc.*, 4255 (1961)
32. M. E. Peach and T. C. Waddington, *J. Chem. Soc.*, 3450 (1962)
33. E. W. Wartenberg and J. Goubeau, *Z. Anorg. Allg. Chem.*, **329,** 269 (1964)
34. M. E. Anthoney, Ph.D. Thesis, University of London (1970)
35. M. F. Lappert, *J. Chem. Soc.*, 542 (1962)
36. D. Cook, *Can. J. Chem.*, **41,** 515 (1963)
37. P. N. Gates, E. J. McLauchlan and E. F. Mooney, *Spectrochim. Acta.*, **21,** 1445 (1965)
38. B. P. Susz and P. Chalandon, *Helv. Chim. Acta*, **41,** 1332 (1958)
39. G. W. Chantry, A. Finch, P. N. Gates and D. Steele, *J. Chem. Soc.*, **A,** 896 (1966)
40. D. E. H. Jones and J. L. Wood, *J. Chem. Soc.*, **A,** 1448 (1966)
41. P. N. Gates and E. F. Mooney, *J. Inorg. Nucl. Chem.*, **30,** 839 (1968)
42. M.-T. Forel, M. Tranquille and M. Fouassier, *Spectrochim. Acta*, **26A,** 1777 (1970)
43. G. Michel and G. Duyckaerts, *Spectrochim. Acta.*, **21,** 279 (1965)
44. G. Roland and G. Duyckaerts, *Spectrochim. Acta.*, **27A,** 975 (1971)
45. R. W. Rudolph, R. W. Parry and C. F. Farran, *Inorg. Chem.*, **5,** 723 (1966)
46. E. R. Lory, R. F. Porter and S. H. Bauer, *Inorg. Chem.*, **10,** 1072 (1971)
47. R. M. Adams, (Editor), Boron, Metallo-Boron Compounds and Boranes, (Interscience, New York, 1964), page 527
48. K. L. Lundberg, R. J. Rowatt and N. E. Miller, *Inorg. Chem.*, **8,** 1336 (1969)
49. J. T. Braunholtz, E. A. V. Ebsworth, F. G. Mann and N. Sheppard, *J. Chem. Soc.*, 2780 (1958)
50. C. W. Heitsch and R. N. Kniseley, *Spectrochim. Acta*, **19,** 1385 (1963)
51. D. R. Lide, R. W. Taft and P. Love, *J. Chem. Phys.*, **31,** 561 (1959)
52. A. M. Prokhorov and G. P. Shipulo, *Opt. Specktrosk.*, **8,** 218 (1960)
53. P. S. Bryan and R. L. Kuczkowski, *Inorg. Chem.*, **10,** 200 (1971)
54. H. G. Schirdewahn, Doctoral Thesis, University of Freiburg (1965)
55. R. L. Kuczkowski and D. R. Lide, *J. Chem. Phys.*, **46,** 357 (1967)
56. J. P. Pasinski and R. L. Kuczkowski, *J. Chem. Phys.*, **54,** 1903 (1971)
57. R. W. Rudolph and R. W. Parry, *J. Amer. Chem. Soc.*, **89,** 1621 (1967)
58. H. J. Coerver and C. Curran, *J. Amer. Chem. Soc.*, **80,** 3522 (1958)
59. W. Gerrard, M. F. Lappert, H. Pyszora and J. W. Wallis, *J. Chem. Soc.*, 2182 (1960)

60. R. C. Taylor and C. L. Cluff, *Nature*, **182,** 390 (1958)
61. R. C. Taylor, Advances in Chemistry Series, No. 42 (American Chem. Soc., Washington D.C. 1964), page 59
62. H. Gerding and H. Houtgraaf, *Rec. Trav. Chim. Pays-Bas*, **74,** 15 (1955)
63. Von W. Sawodny and J. Goubeau, *Z. Anorg. Allg. Chem.*, **356,** 289 (1968)
64. I. H. Hillier and V. R. Saunders, *J. Chem. Soc.*, **A,** 664 (1971)
65. E. L. McGandy, Doctoral Thesis, Boston University, Boston (1961)
66. K. F. Purcell and R. S. Drago, *J. Amer. Chem. Soc.*, **88,** 919 (1966)
67. E. Augdahl and P. Klaboe, *Spectrochim. Acta*, **19,** 1665 (1963)
68. A. Terenin, B. Filiminov and D. Bystrow, *Z. Elektrochem.*, **62,** 180 (1958)
69. I. R. Beattie and T. Gilson, *J. Chem. Soc.*, 2292 (1964)
70. C. C. Costain, *J. Chem. Phys.*, **29,** 864 (1958)
71. R. A. Walton, *Quart. Rev.*, **19,** 126 (1965)
72. P. H. Clippard, R. C. Taylor and J. C. Hanson, *J. Cryst. Mol. Struct.*, **1,** 363 (1971)
73. C. Friedel and J.-M. Crafts, *Bull. Soc. Chim. France*, **27,** 530 (1877); *C.R. Acad. Sci.*, **84,** 1450 (1877)
74. H. H. Perkampus and E. Baumgarten, *Angew. Chem.*, **76,** 965 (1964); *Int. Ed.*, **3,** 776
75. G. A. Olah, *J. Amer. Chem. Soc.*, **87,** 1103 (1965)
76. H. M. Nelson, *J. Phys. Chem.*, **66,** 1380 (1962)
77. G. A. Olah, E. B. Baker, J. C. Evans, W. S. Tolgyesi, J. S. McIntyre and I. J. Bastien, *J. Amer. Chem. Soc.*, **86,** 1360 (1964)
78. W. Hüttner and W. Zeil, *Spectrochim. Acta*, **22,** 1007 (1966)
79. G. A. Olah, J. R. de Member, A. Commeyras and J. L. Bribes, *J. Amer. Chem. Soc.*, **93,** 459 (1971)
80. J. C. Evans (see reference 77)
81. G. A. Olah and A. M. White, *J. Amer. Chem. Soc.*, **91,** 3954 (1969)
82. G. A. Olah and A. M. White, *J. Amer. Chem. Soc.*, **91,** 3956 (1969)
83. G. A. Olah, A. M. White, A. Commeyras, J. R. de Member and C. Y. Lui, *J. Amer. Chem. Soc.*, **92,** 4627 (1970)
84. G. A. Olah and C. U. Pittman, (Ed., V. Gold) Advances in Physical Organic Chemistry, Vol. 4 (Academic Press, New York and London, 1966), p 305
85. G. R. Olah, G. D. Mateescu, L. A. Wilson and M. H. Gross, *J. Amer. Chem. Soc.*, **92,** 7231 (1970)
86. B. P. Susz and J.-J. Wuhrmann, *Helv. Chim. Acta.*, **40,** 722 (1957)
87. P. N. Gates and D. Steele, *J. Mol. Struct.*, **1,** 349 (1968)
88. I. Nakagawa and T. Shimanouchi, *Spectrochim. Acta*, **18,** 513 (1962)
89. J. L. Duncan, *Spectrochim. Acta*, **20,** 1197 (1964)
90. B. P. Susz and J.-J. Wuhrmann, *Helv. Chim. Acta*, **40,** 971 (1957)
91. D. Cook, *Can. J. Chem.*, **37,** 48 (1959)
92. D. Cassimatis and B. P. Susz, *Helv. Chim. Acta*, **44,** 943 (1961)
93. D. Cassimatis, P. Gagnaux and B. P. Susz, *Helv. Chim. Acta.*, **43,** 424 (1960)
94. D. Cook, *Can. J. Chem.*, **40,** 480 (1962)
95. F. P. Boer, *J. Amer. Chem. Soc.*, **88,** 1572 (1966)
96. E. Lindner and H. Kranz, *Z. Naturforsch.*, **20b,** 1305 (1965)
97. E. Lindner, *Angew Chem. Int. Ed.*, **9,** 114 (1970)
98. E. Lindner and H. Kranz, *Chem. Ber.*, **99,** 3800 (1966)
99. I. Cooke, B. P. Susz and C. Herschmann, *Helv. Chim. Acta*, **37,** 1280 (1954)
100. B. P. Susz and D. Cassimatis, *Helv. Chim. Acta*, **44,** 395 (1961)
101. H.-H. Perkampus and W. Weiss, *Angew. Chem., Int. Ed.*, **7,** 70 (1968)
102. E. Lindner and H.-G. Karmann, *Angew. Chem.*, **80,** 567 (1968) or *Angew. Chem. Int. Ed.*, **7,** 548 (1968)
103. E. Lindner and K. M. Matejcek, *Z. Anorg. Allg. Chem.*, in press
104. E. Lindner and H. Schless, *Chem. Ber.*, **99,** 3331 (1966)
105. E. Lindner and H. Weber, *Chem. Ber.*, **101,** 2832 (1968)

106. G. A. Olah and J. R. de Member, *J. Amer. Chem. Soc.*, **92,** 718 (1970)
107. H. C. Brown and H. W. Pearsall, *J. Amer. Chem. Soc.*, **74,** 191 (1952)
108. M. J. S. Dewar, The Electronic Theory of Organic Chemistry, (Oxford University Press, 1949)
109. V. Gold and F. L. Tye, *J. Chem. Soc.*, 2173, 2181, 2184 (1952)
110. C. MacLean, J. H. van der Waals and E. L. Mackor, *Mol. Phys.*, **1,** 247 (1958)
111. C. MacLean and E. L. Mackor, *Mol. Phys.*, **4,** 241 (1961)
112. C. MacLean and E. L. Mackor, *Disc. Faraday Soc.*, **34,** 165 (1962)
113. E. L. Mackor and C. MacLean, *Pure Appl. Chem.*, **8,** 393 (1964)
114. D. M. Brouwer, E. L. Mackor and C. MacLean, *Rec. Trav. Chim. Pays-Bas*, **84,** 1564 (1965)
115. C. S. Johnson, Jr., (Ed. J. S. Waugh) Advances in Magnetic Resonance, Volume 1, (Academic Press, New York and London 1965)
116. D. M. Brouwer, *Rec. Trav. Chim. Pays-Bas*, **87,** 335 (1968)
117. D. M. Brouwer, *Rec. Trav. Chim. Pays-Bas*, **87,** 342 (1968)
118. T. Birchall, A. N. Bourns, R. J. Gillespie and P. J. Smith, *Can. J. Chem.*, **42,** 1433 (1964)
119. E. M. Arnett and C. Y. Wu, *J. Amer. Chem. Soc.*, **82,** 5660 (1960)
120. D. M. Brouwer, E. L. Mackor and C. MacLean, *Rec. Trav. Chim., Pay-Bas*, **85,** 109 (1966)
121. G. A. Olah, R. H. Schlosberg, D. P. Kelly and Gh. D. Mateescu, *J. Amer. Chem. Soc.*, **92,** 2546 (1970)
122. H. H. Perkampus and E. Baumgarten, *Z. Elektrochem.*, **67,** 16 (1963)
123. H. H. Perkampus and E. Baumgarten, *Z. Elektrochem.*, **68,** 70 (1964)
124. H. H. Perkampus, (Ed. V. Gold) Advances in Physical Organic Chemistry, Vol. 4, (Academic Press, New York and London, 1966), p 195
125. D. H. Whiffen, P. Torkington and H. W. Thompson, *Trans. Faraday Soc.*, **41,** 200 (1945)
126. R. R. Randle and D. H. Whiffen, Proceedings of a Conference on Molecular Spectroscopy, (Institute of Petroleum, London, 1955), p 111
127. L. J. Bellamy, The Infrared Spectra of Complex Molecules, 2nd edition, (Methuen & Co. Ltd., London, 1968), p 75
128. A. Commeyras and G. A. Olah, *J. Amer. Chem. Soc.*, **91,** 2929 (1969)
129. R. C. Thompson, J. Barr, R. J. Gillespie, J. B. Milne and R. A. Rothenbury, *Inorg. Chem.*, **4,** 1641 (1965)
130. R. J. Gillespie and E. A. Robinson, *Can. J. Chem.*, **40,** 675 (1962)
131. R. J. Gillespie and K. C. Moss, *J. Chem. Soc.*, **A,** 1170 (1966)
132. W. v. E. Doering, M. Saunders, H. G. Boyton, H. W. Earhart, E. F. Wadley, W. R. Edwards and G. Laber *Tetrahedron*, **4,** 178 (1958)
133. C. A. Coulson, Valence, 2nd edition, (Oxford University Press, 1961), page 313
134. H. P. Fritz and L. Schäfer, *Chem. Ber.*, **97,** 1829 (1964)
135. G. O. Jones, D. H. Martin, P. A. Mawer and C. H. Perry, *Proc. Roy. Soc.*, **261A,** 10 (1961)
136. R. Janoschek, G. Diercksen, H. Preuss, *Int. J. Quantum Chem., Symp.* (No. 1) 205 (1967)
137. E. R. Lippincott, J. Xavier and D. Steele, *J. Amer. Chem. Soc.*, **83,** 2262 (1961)
138. A. Almenningen, O. Bastiansen and A. Haaland, *J. Chem. Phys.*, **40,** 3434 (1964)
139. E. O. Fischer and S. Schreiner, *Chem. Ber.*, **92,** 938 (1959)
140. G. B. McVicker and G. L. Morgan, *Spectrochim. Acta*, **26A,** 23 (1970)
141. H. P. Fritz and D. Sellmann, *J. Organometal. Chem.*, **5,** 501 (1966)
142. G. L. Morgan and G. B. McVicker, *J. Amer. Chem. Soc.*, **90,** 2789 (1968)
143. H. P. Fritz, (Eds. F. G. A. Stone and R. West) Advances in Organometallic Chemistry, Vol. 1 (Academic Press, New York, 1964), p. 239ff
144. E. Weiss and E. O. Fischer, *Z. Anorg. Allg. Chem.*, **278,** 219 (1955)

145. L. Friedman, A. P. Irsa and G. Wilkinson, *J. Amer. Chem. Soc.*, **77,** 3689 (1955)
146. G. Wilkinson, F. A. Cotton and J. M. Birmingham, *J. Inorg. Nucl. Chem.*, **2,** 95 (1956)
147. F. A. Cotton and L. T. Reynolds, *J. Amer. Chem. Soc.*, **80,** 269 (1958)
148. S. Shibata, L. S. Bartell and R. M. Gavin Jr., *J. Chem. Phys.*, **41,** 717 (1964)
149. A. P. Cox and C. Roberts—to be published; private communication to authors
150. C. Roberts, A. P. Cox and M. J. Whittle, *J. Mol. Spectrosc.*, **35,** 476 (1970)
151. R. T. Bailey and R. H. Curran, *J. Mol. Struct.*, **6,** 391 (1970)
152. E. Frasson, F. Menegus and C. Panattoni, *Nature*, **199,** 1087 (1963)
153. J. K. Tyler, A. P. Cox and J. Sheridan, *Nature*, **183,** 1182 (1959)
154. A. Almenningen, A. Haaland and T. Motzfeldt., *J. Organometal. Chem.*, **7,** 97 (1967)
155. E. O. Fischer and H. Grubert, *Z. Anorg. Allg. Chem.*, **286,** 237 (1956)
156. H. H. Dearman, W. W. Porterfield and H. M. McConnell, *J. Chem. Phys.*, **34,** 696 (1961)
157. G. Wilkinson and T. S. Piper, *J. Inorg. Nucl. Chem.*, **2,** 32 (1956)
158. E. Maslowsky and K. Nakamoto, *Chem. Comm.*, 257 (1968)
159. R. Hüttel, U. Raffay and H. Reinheimer, *Angew. Chem., Int. Ed.*, **6,** 862 (1967)
160. G. M. Whitesides and J. S. Fleming, *J. Amer. Chem. Soc.*, **89,** 2855 (1967)
161. A. N. Nesmeyanov, N. S. Sazonova, V. A. Sazonova and L. M. Meskhi, *Izvest. Akad. Nauk. S.S.S.R., Ser. Khim.*, 1827 (1969)
162. A. Finch, P. N. Gates and D. Steele, *Trans. Faraday Soc.*, **61,** 2623 (1965)
163. H. B. Friedrich and W. B. Person, *J. Chem. Phys.*, **44,** 2161 (1966)
164. M. A. Rollier, *Gazz Chim. Ital.*, **77,** 372 (1947)
165. L. W. Reeves and W. G. Schneider, *Can. J. Chem.*, **35,** 251 (1957)
166. J. Pople, *J. Chem. Phys.*, **24,** 1111 (1956)
167. E. M. Engler and P. Laszlo, *J. Amer. Chem. Soc.*, **93,** 1317 (1971)
168. D. D. Eley, J. H. Taylor and S. C. Wallwork, *J. Chem. Soc.*, 3867 (1961)
169. H. H. Perkampus and G. Orth, *Z. Phys. Chem.*, **58,** 327 (1968)
170. M. S. Ashkinazi, P. V. Kurnosova and V. S. Finkel'stein, *J. Phys. Chem. (USSR)*, **7,** 438 (1936)
171. Sh. Sh. Raskin, *Dokl. Acad. Nauk., S.S.S.R.*, **100,** 485 (1955)
172. Sh. Sh. Raskin, *Opt. Specktrosk.*, **1,** 516 (1956)
173. L. W. Daasch, *Spectrochim. Acta*, **15,** 726 (1959)
174. H.-H. Perkampus and E. Baumgarten, *Z. Phys. Chem.*, **39,** 1 (1963)
175. A. N. Terenin, V. N. Filimonov and D. S. Bystrov, *Isvest. Akad. Nauk. SSSR., Ser Fiz.*, **22,** 1100 (1958) (*Chem. Abs.*, **53,** 860 (1959))
176. C. D. Schmulbach and I. Y. Ahmed, *Inorg. Chem.*, **8,** 1414 (1969)
177. (a) D. A. Long and R. T. Bailey, Unpublished material, private communication; (b) R. T. Bailey, Ph.D. thesis, University College, Swansea (1962)
178. A. A. Woolf, *J. Chem. Soc.*, 252 (1954)
179. A. E. Douglas and P. M. Routby, *J. Astrophys.*, **119,** 303 (1954)
180. K. Kawai and I. Kanesaka, *Spectrochim. Acta*, **25A,** 263 (1969)
181. J. S. Coleman, H. Petersen and R. A. Penneman, *Inorg. Chem.*, **4,** 135 (1965)
182. D. N. Waters, (Eds. A. J. Downs, D. A. Long and L. A. K. Staveley) Essays in Structural Chemistry, (Macmillan and Co. Ltd, London, 1971), Chapter 13
183. J. H. B. George, J. A. Rolfe and L. A. Woodward, *Trans. Faraday Soc.*, **49,** 375 (1953)
184. D. E. Irish and A. R. Davis, *Can. J. Chem.*, **46,** 943 (1968)
185. D. E. Irish, A. R. Davis and R. A. Plane, *J. Chem. Phys.*, **50,** 2262 (1969)
186. R. E. Hester and W. E. L. Grossman, *Inorg. Chem.*, **5,** 1308 (1966)
187. R. E. Hester, *Ann. Rep. Chem. Soc.*, **66A,** 79 (1969)
188. D. W. James and W. H. Leong, *J. Chem. Phys.*, **51,** 640 (1969)
189. R. Bonn, G. H. Wegdam and J. Van Der Elsken, *J. Chem. Phys.*, **50,** 1901 (1969)

190. R. G. Gordon, *J. Chem. Phys.*, **42,** 3658 (1965); *ibid.*, **43,** 1307 (1965)
191. R. G. Gordon, (Ed., J. S. Waugh) Advances in Magnetic Resonance, Vol. 3 (Academic Press, New York and London, 1968)
192. G. A. Olah and A. M. White, *J. Amer. Chem. Soc.*, **91,** 5801 (1969)
193. G. A. Olah, J. R. de Member, R. H. Schlosberg and Y. Halpern, *J. Amer. Chem. Soc.*, **94,** 156 (1972)
194. G. A. Olah, R. H. Schlosberg, R. D. Porter, Y. K. Mo, D. P. Kelly and G. D. Mateescu, *J. Amer. Chem. Soc.*, **94,** 2034 (1972)
195. M. E. Baur, D. A. Horsma, C. M. Knobler and P. Perez, *J. Phys. Chem.*, **73,** 641 (1969)

Chapter 7

The Dielectric Properties of Molecular Complexes

A. H. Price

Edward Davies Chemical Laboratories, University College of Wales, Aberystwyth, Wales.

7.1. INTRODUCTION

This chapter describes some results obtained from the study of the dipole moments and the dielectric relaxation behaviour of molecular complexes in the electronic ground state.

Matter is polarized by external electric fields. The molar polarization (P) is related to the relative permittivity (ε) of the material by:

$$P = \frac{\varepsilon - 1}{\varepsilon + 2} \frac{M}{\rho} \qquad 7.1$$

where M is the molar mass and ρ the density. Three contributions to the molar polarization are distinguished: (i) a dipolar molecule tends to become aligned in an external electric field and reduces the effective electric charge at the electrodes. This effect is the orientation polarization (P_0) and is proportional to the square of the molecular dipole moment and to the dipolar concentration. (ii) interaction between the applied field and the nuclear charges produces a charge displacement and a polarization contribution which is dispersed in the infrared spectral region. This is the atomic polarization (P_A); (iii) the applied field displaces the electronic distribution within a molecule. This is the electronic polarization (P_E) and is dispersed in the visible and ultra-violet spectral regions. The molar polarization is thus $P = P_O + P_A + P_E = P_O + P_D$ where P_D is the distortion polarization.

Dielectric studies provide information on the molecular charge redistribution on complex formation and also information about the lifetime of the complex. Comparison of the dipole moment of the complex with the dipole moment of the component molecules provides information on charge redistribution on complex formation. Dispersion of the orientation polarization in liquids usually occurs in the frequency range 10^8 to 10^{12} Hz. A rigid dipole tends to align itself in the direction of the applied field. Removal of this field allows the dipole to return to random orientation and the orientation polarization relaxes to zero. This is termed "dielectric relaxation" and the relaxation rate coefficient for rigid dipolar molecules is of the order of 10^{11} s^{-1}. In molecular complexes two mechanisms contribute to the relaxation of the orientation polarization. One is by rotation of the dipolar species (as in rigid molecules), the other is by the decomposition of the dipolar complex into its (sometimes non-polar) components. Dielectric relaxation measurements often distinguish between these two processes and allow of an estimation of the lifetime of the complex.

7.2. THE DIPOLE MOMENTS OF MOLECULAR COMPLEXES

7.2.1. EVALUATION OF THE DIPOLE MOMENT

Clausius and Mossotti[1] related the molar polarization to the polarizability (α) of non-polar molecules:

$$P = \frac{\varepsilon - 1}{\varepsilon + 2}\frac{M}{\rho} = \frac{N}{3\epsilon}\alpha \qquad 7.2$$

(where N is the Avogadro constant and ϵ is the permittivity of vacuum). This equation converts to the Lorenz-Lorentz[2] equation for the molar refraction (R):

$$R = \frac{n^2 - 1}{n^2 + 2}\frac{M}{\rho} = \frac{N}{3\epsilon}\alpha \qquad 7.3$$

(where n is the refractive index) by Maxwell's relation $\varepsilon = n^2$ (ε and n measured at the same frequency). Debye[3] extended the Clausius-Mossotti equation to dipolar systems and deduced:

$$P = \frac{\varepsilon - 1}{\varepsilon + 2}\frac{M}{\rho} = \frac{N}{3\epsilon}\left(\alpha + \frac{\mu^2}{3kT}\right) \qquad 7.4$$

The orientation polarization P_O equals $N\mu^2/9\epsilon kT$ (μ is the magnitude of the dipole moment vector) and the distortion polarization $P_D = P_E + P_A = N(\alpha_E + \alpha_A)/3\epsilon$ where α_E and α_A are the electronic and atomic polarizabilities respectively. The Debye equation forms the basis of most evaluations of molecular dipole moments. It is not a completely satisfactory relation as intermolecular interactions in the medium are ignored. For pure liquids the Onsager[5] equation is more appropriate. However in dilute solutions and in the absence of molecular interactions (*e.g.* hydrogen bonding) the Debye and Onsager equations yield dipole moments in reasonable agreement with each other.

In dilute solution the molar polarization is written as:

$$P_{12} = P_1x_1 + P_2x_2 = \left(\frac{\varepsilon_{12} - 1}{\varepsilon_{12} + 2}\right)\frac{M_1x_1 + M_2x_2}{\rho_{12}} \qquad 7.5$$

where x is the mole fraction, and the subscripts 12, 1 and 2 refer to the solution, solvent and solute respectively. The solute molar polarization (P_2) should be independent of concentration. In practice molecular interactions produce deviations from a constant P_2 and extrapolation to infinite dilution is usually necessary to obtain a reasonable value for P_2. Even then solute-solvent interactions are still present and P_2 is solvent dependent (see p 543). The most generally used extrapolation procedures are due to Hedestrand[5], Halverstadt and Kumler[6] and Guggenheim[7]. Hedestrand shows that if $\varepsilon_{12} = \varepsilon_1 + ax_2$ and $\rho_{12} = \rho_1 + bx_2$ then:

$${}_{\infty}P_2 = \frac{3a\varepsilon_1M_1}{(\varepsilon_1 + 2)^2\rho_1} + \left(\frac{\varepsilon_1 - 1}{\varepsilon_1 + 2}\right)\frac{M_2 - M_1b}{\rho_1} \qquad 7.6$$

where ${}_{\infty}P_2$ is the solute molar polarization at infinite dilution. Halverstadt and Kumler express the concentration in weight fraction units (w) and assume $\varepsilon_{12} = \varepsilon_1 + a'w_2$ and $v_{12} = v_1 + b'w_2$ (v is the specific volume). They deduce:

$${}_{\infty}P_2 = \frac{3a'v_1M_2}{(\varepsilon_1 + 2)^2} + \left(\frac{\varepsilon_1 - 1}{\varepsilon_1 + 2}\right)(v_1 + b')M_2 \qquad 7.7$$

Guggenheim rearranged the Debye equation (equation 7.4) into the form:

$$\mu^2 = \frac{9kT\epsilon}{N}\frac{3}{(\varepsilon_1 + 2)(n_1^2 + 2)}\left(\frac{\Delta}{c}\right)_{c\to 0} \qquad 7.8$$

where $\Delta = (\varepsilon_{12} - \varepsilon_1) - (n_{12}^2 - n_1^2)$ and the concentration c is expressed in units of mole per unit volume. $(\Delta/c)_{c\to 0}$ is the slope (as $c \to 0$) of the graph of Δ against c. Density does not enter this expression explicitly and the Guggenheim method is often preferred experimentally. Frequently the graph of Δ against c is linear.

Within experimental error these extrapolation methods yield the same effective dipole moment. However, they all suffer from the disadvantage that reliable determinations are required at low concentrations where the solution permittivity is only slightly different from the solvent permittivity and the experimental discrimination is correspondingly low.

Permittivity determinations involve the measurement of the capacitance of a standard capacitor when empty and when filled with sample. The ratio of the two capacitances gives the relative permittivity (ε). Bridge, resonance and heterodyne methods are the ones most commonly used for accurate capacitance determinations. These are adequately described in the general literature on dielectrics[8–14] and will not be discussed here. Some typical experimental results taken from measurements by Briegleb *et al.*[15] are quoted in table 7.1. These results were analysed using the Hedestrand equation and gave a dipole moment (μ) of $(3.0 \pm 1.3) \times 10^{-30}$ C m for the complex. A discrimination ($\Delta\varepsilon/\varepsilon$) of about 5×10^{-5} is usually required in the permittivity measurements on solutions containing molecules with low dipole moments. In the above results the complex is the only dipolar species in solution and its concentration must be known before the dipole moment of the complex can be calculated. The polar complex contribution to the solution polarization is proportional to $c\mu^2$ (c is the complex concentration). A 10% error in the concentration produces a 5% error in the calculated dipole moment. The situation is more complicated in solutions where the donor or the acceptor molecule is dipolar. The polarization contribution from the uncomplexed dipolar component must be considered before the dipole moment of the complex is evaluated. For complexes involving one non-dipolar component measurements are usually made in the presence of a large excess of the non-dipolar component. This ensures maximum complex formation, but a correction for any residual free dipolar molecules is still necessary.

TABLE 7.1

Dielectric measurements[15] on the system chloranil–naphthalene in carbon tetrachloride solution at 293 K. (The subscripts A, D and DA refer to the acceptor, donor and complex respectively)

$10^4\, x_A$	$10^4\, x_D$	$10^4\, x_{DA}$	ε	ρ/g cm^{-3}	$10^6\, P_{DA}$/m^3 mol^{-1}
3.576	207.9	1.215	2.2466	1.5796	100.0 ± 16.5
3.365	262.2	1.521	2.2488	1.5755	115.7 ± 13.1
4.204	327.6	1.900	2.2525	1.5709	116.6 ± 10.5
3.602	321.3	2.124	2.2521	1.5713	115.7 ± 9.4
4.811	429.2	2.837	2.2581	1.5637	114.9 ± 7.0

The evaluation of the dipole moment involves a knowledge of the distortion polarizability ($\alpha_D = \alpha_E + \alpha_A$). Permittivity measurements in the far-infrared spectral region (near 200 cm^{-1}) enable accurate determinations of α_D. Few such measurements are available. The usual practice is to calculate the electronic polarizability from the molar refraction as measured at optical frequencies (e.g. using the sodium D line) and assume that the atomic polarizability contributes up to $0.15\alpha_E$ to the distortion polarizability. The magnitude of the distortion polarizability has an important effect on the calculated value of the dipole moment. This is especially so for weak complexes with low dipole moments. Consider a case with ${}_{\infty}P_2 = 10^{-4}\ m^3\ mol^{-1}$, and $\alpha_E = 3.68 \times 10^{-39}\ C\ m^2\ V^{-1}$. Substitution in equation 7.4 (with $\alpha_A = 0$) yields $\mu = 3.0 \times 10^{-30}$ C m. If the calculation is repeated with $\alpha_D = 1.15\alpha_E$ then $\mu = 1.46 \times 10^{-30}$ C m, a difference of 50% in the dipole moment. The relative importance of the distortion polarization contribution to the total molar polarization usually decreases with increasing dipole moment.

Dielectric absorption measurements (often combined with far-infrared studies) give a direct determination of the orientation polarization in polar complexes. The results of such measurements are discussed on page 569.

7.2.2. DIELECTRIC METHODS FOR THE DETERMINATION OF STOICHIOMETRY AND EQUILIBRIUM CONSTANT

Before discussing the experimental dipole moment determinations a brief account of the dielectric titration technique is relevant. Using this technique both the equilibrium constant and the stoichiometry of the complex may be determined.

The method involves measuring the change in the permittivity ($\Delta\varepsilon$) of a solution of one component on the addition of successive amounts of the other component. For a 1:1 complex the results of such a titration are schematically shown in figure 7.1. Consider the addition of a dipolar acceptor to a non-dipolar donor producing a complex with a dipole moment much larger than the dipole moment of the acceptor (for example, see reference 16). The addition of a small amount of acceptor generates an equivalent amount of complex. The permittivity increases on successive addition of the acceptor and continues to increase until equivalent quantities of the donor and acceptor are present. This situation is schematically shown by line 0f of figure 7.1. Further addition of the acceptor increases the permittivity by an amount depending on the acceptor dipole moment (line fg in figure 7.1.) A break is observed in the graph at the donor-acceptor ratio corresponding to the ratio existing in the complex. In practice, dissociation of the complex causes departure from the line 0fg. The magnitude of this departure enables the equilibrium constant of the reaction to be determined[17,18,81]. The slope of the initial portion (0f) of the graph

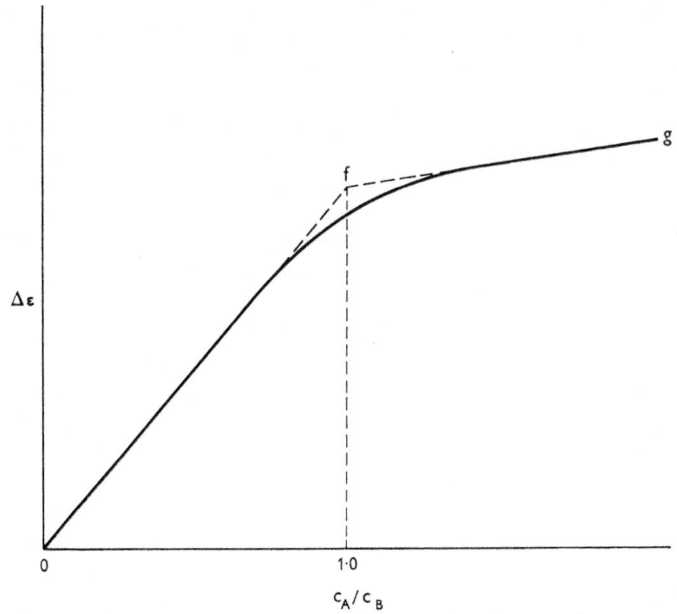

FIG 7.1 Dielectric titration of a non-polar donor and a polar acceptor. The permittivity change ($\Delta\varepsilon$) is plotted against the acceptor: donor concentration ratio (C_A/C_D).

gives the dipole moment of the complex; the slope of fg gives the dipole moment of the polar acceptor.

A more complex case is shown in figure 7.2 which illustrates the results obtained on titrating tin(IV) chloride against dibutylsulphide in benzene solution. The first maximum shows the presence of a highly polar 1:1 complex between tin(IV) chloride and dibutylsulphide. A 1:2 complex is also present, but has a lower dipole moment than does the 1:1 complex. The slope of the line following the 1:2 complex corresponds to the dipole moment of the added sulphide.

This technique is particularly valuable in the study of hydrogen-bonded complexes (see, for example, reference 20). The determination of equilibrium constants and of dipole moments of hydrogen-bonded complexes is often carried out by the Few and Smith[21] procedure. Maryott[22] investigated the equilibrium between picric acid (C_6H_2-$(NO_2)_3OH$) and tribenzylamine ($(C_6H_5)_3N$) in benzene solution. Assuming the permittivity change ($\Delta\varepsilon$) observed on mixing the acid and base is a linear function of the acid, base and complex concentrations then:

$$\frac{\Delta\varepsilon}{x_c} = \frac{\varepsilon_{12} - \varepsilon_1}{x_c} = k_A\alpha + k_B\alpha + k_c(1 - \alpha) \qquad 7.9$$

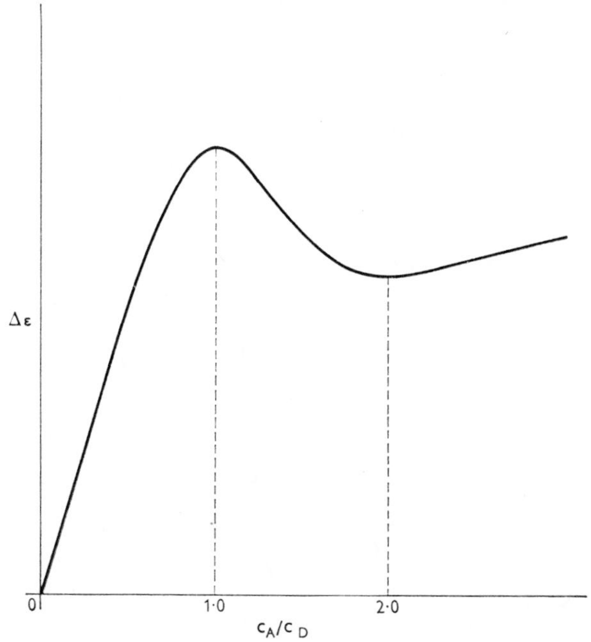

FIG 7.2 Dielectric titration (schematic) of tin(IV) chloride with a thioether in benzene solution (adapted from references 18 and 19). The permittivity change ($\Delta\varepsilon$) is plotted against the acceptor: donor concentration rate (C_A/C_D).

where subscripts A, B and C refer to the acid, base and complex respectively, k is a proportionality constant, x_c the mole fraction of the complex and α the fraction of the complex dissociated. Both k_A and k_B are determined from data obtained from solutions of the pure acid and pure base. The equilibrium constant K_x is then:

$$K_x = \frac{((\Delta\varepsilon/x_c) - k_A - k_B)(k_c - k_A - k_B)}{(k_c - (\Delta\varepsilon/x_c))^2 x_c} \qquad 7.10$$

The equation may be solved for K_x and k_c by plotting $\Delta\varepsilon/x_c$ against $\{(\Delta\varepsilon/x_c - k_A - k_B)/x_c\}^{1/2}$. For the equilibrium,

$$(C_6H_5)_3N + HOC_6H_2(NO_2)_3 \leftrightharpoons (C_6H_5)_3NH^+\text{—}O^-C_6H_2(NO_2)_3,$$

Maryott deduced $K_x = 0.13 \times 10^3$ at 303 K, $K_x = 0.07 \times 10^3$ at 313 K and a dipole moment of 38.1×10^{-30} C m for the complex which obviously exists chiefly in the ionic form.

7.2.3. EXPERIMENTAL RESULTS

The dipole moment of some molecular complexes are given in table 7.2 p. 541 and in the appendix. The accuracy of these dipole moments is often no better than $\pm 0.06 \times 10^{-30}$ C m but may be $\pm 0.4 \times 10^{-30}$ C m for low dipole moments of about 2×10^{-30} C m (*i.e.* 0.6 Debye). In a large number of determinations the non-dipolar donor also acts as solvent. This ensures maximum complex formation but it must be remembered that for weak complexes only about sixty to seventy per cent of the acceptor is complexed.

It is sometimes suggested[68] that dipole moment measurements in the pure donor as solvent do not represent the dipole moment of the 1:1 complex, and that more realistic values are obtained from measurements in inert solvents. There is insufficient experimental evidence available to support this statement. Some solvent dependence of the dipole moment is normally observed (see references 8–14), but this does not usually amount to more than a few per cent. In the pure donor solvent systems the possibility exists of 1:n complexes, but their effect on the observed dipole moment is not known. In weak complexes, the lifetime of the complex is of the order of 10^{-12} s (see section 7.3.3) and it is reasonable to assume that the lifetime of any other complex formed (*e.g.* 1:2, etc.) would be much shorter and also have a much lower effective concentration. The measured dipole moment is a weighted mean value for all the configurations possible in solution; the 1:1 complex (or a 'sticky' binary collision) is the most favoured statistically. A selection of dipole moments for 1:1 complexes is given in the appendix.

The dipole moment gives a measure of the strength of the interaction between the donor and acceptor molecules. In complexes formed between non-dipolar acceptors and donors the measured dipole moment is a direct result of this interaction; in complexes containing a dipolar component the dipole moment resulting from complex formation must be evaluated after due allowance is made for the polarities of the component molecules. The overall dipole moment ($\vec{\mu}_c$) of the complex is the resultant of three contributions: $\vec{\mu}_c = \vec{\mu}_D + \vec{\mu}_A + \vec{\mu}_{DA}$ where the subscripts D and A refer to the donor and to the acceptor, and $\vec{\mu}_{DA}$ is the dipole moment arising from donor-acceptor interaction.

Consider the pyridine–acetic acid and the pyridine–trichloroacetic acid complexes[69]. The configuration of the complexes is assumed as in figure 7.3 where the corresponding group moments are also indicated.

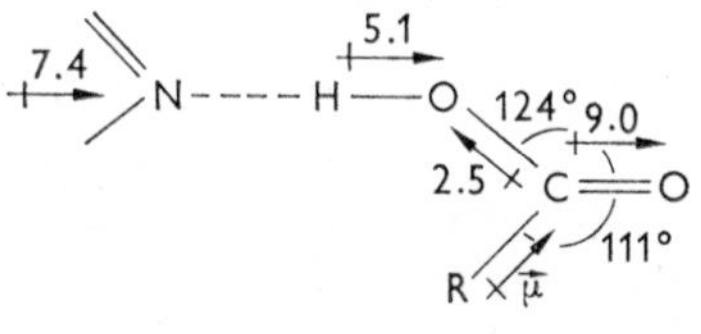

FIG 7.3 Assumed configuration and group dipole moments ($\times 10^{30}$ C m) for the pyridine–acetic acid and pyridine–trichloroacetic acid complexes (if R = H then μ = 1.3; if R = CCl_3 then μ = −5.3).

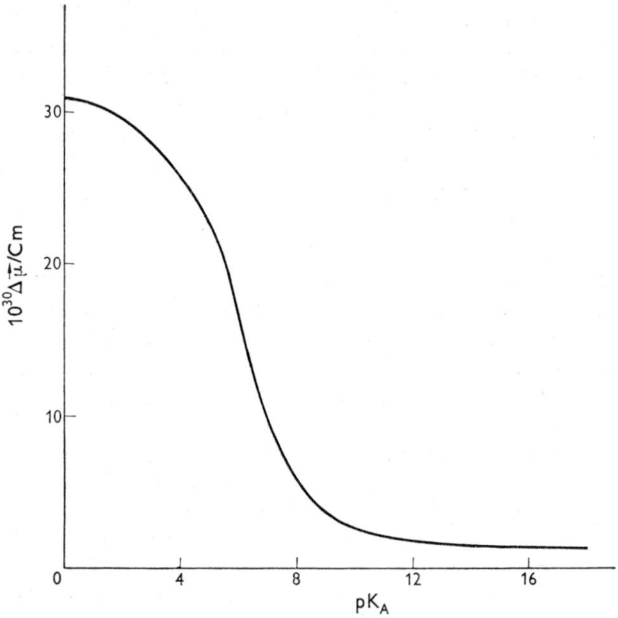

FIG 7.4 Variation of the dipole moment ($\Delta\mu$) of the N—H—O group in triethylamine-acid complexes with acid strength (pK_A) (adapted from reference 75).

The observed dipole moment of the pyridine–acetic acid complex is reproduced only when the moment along the ⩾N—H—O group is increased from 12.5×10^{-30} C m (as in figure 7.3) to 15.7×10^{-30} C m. Assuming the simplest model for the O—H group dipole moment in the free acid (r(OH)) = 0.096 nm) the effective charge at the O and H centres in the free acid is 5.3×10^{-20} C. An adequate estimate of r(OH) in the complex is 0.102 nm and leads to an effective charge of 8.0×10^{-20} C on the H and O centres. In the pyridine–trichloroacetic acid complex the total moment along the N—H—O line is 28.8×10^{-30} C m. If r(OH) remains 0.102 nm the effective charge at the O and H centres is 28.2×10^{-20} C (*i.e.* in excess of the electronic charge). This is sufficient to establish the essentially ionic character of the pyridine–trichloroacetic acid complex. If r(OH) = 0.134 nm the ⩾N—H—O moment of 28.8×10^{-30} C m is accommodated without change in the pyridine dipole moment. The change of the donor-acceptor interaction from a partly covalent to ionic bond is well illustrated in figure 7.4 where the variation of the group dipole moment calculated (as above) for the ⩾N—H—O group for a series of triethylamine complexes with acids of different strengths is shown. With increasing acid strength the polarity of the ⩾N—H—O group initially increases gradually, and is followed by a much more rapid increase to a constant value. The low polarity in complexes with weak acids reflects the predominance of the essentially covalent structure ⩾N—H—OR, but with strong acids the complex becomes essentially ionic and proton transfer occurs to give the structure ⩾NH⁺—OR⁻. (see chapter 4 for a further discussion of this phenomenon and section 1.3 for details of how complex dipole moments are obtained from molecular orbital calculations.)

The calculation of μ_{DA} in other complexes may be performed in a similar manner but some care must be exercised when assigning group moments (and their orientations in the complex). Configurational changes often accompany complex formation. For example, the C—O—C bond angle in dioxan increases from 109.5° to 116.6° on complexing with bromine,[78] the tetrahedral trimethyltin chloride adopts a pyramidal configuration on complexing with pyridine,[7] and an octahedral configuration appears in the 2:1 complexes of tin compounds[79,80]. These configurational changes also involve a change in the group moments. The Sn–Cl group moment depends on the hybridization at the tin atom and will change when the configuration around the tin atom changes from tetrahedral to pyrimidal. The dependence of group moment on the state of hybridization of the atoms is reasonably well established for C—C bonds[14], but is not quantified for other groups.

The stronger the complex the greater the dipole moment (μ_{DA}) arising from donor-acceptor interaction. The Mulliken 'charge-transfer' model of complex formation predicts[24,106] an increasing dipole moment with decreasing donor ionization potential; the electrostatic model (Mantione[95,97], Dewar and Thompson[92]) for complex formation predicts an increasing dipole moment with increasing donor polarizability. Both these predictions are obeyed for the dipole moment of tetracyanoethylene complexes with various methyl benzenes[48] (see table 7.2). No distinction is possible between these models without detailed calculations of the electrostatic contributions (see p 545). Attempts to evaluate such contributions are described in sections 1.3 and 4.7).

In a series of *n-v* complexes (tin and titanium tetrachlorides and tetrabromides with organic sulphides, ethers and nitriles) Gol'dshtein, Bur'yanova and Kharlamova[59] assumed that the main contribution to μ_{DA} is the transfer of charge from donor to the acceptor. The degree of charge transfer being $\mu_{DA}/\mathbf{e}r$ where **e** is the electronic charge and r the intermolecular separation. The stronger the complex the greater is the degree of charge transfer and the greater is the enthalpy of complex formation ($\Delta H^{\ominus}$). A linear relation $\Delta H^{\ominus}$ = const. $\mu_{DA}/\mathbf{e}r$ was found for the *n-v* complexes studied.

Very large μ_{DA} values indicate substantial interactions and possibly ionic forms. The electrical conductance of pyridine–iodine complexes

TABLE 7.2

The dipole moment (μ_{DA}) and enthalpy of formation ($\Delta H^{\ominus}$) of tetracyanoethylene–methyl benzene complexes in carbon tetrachloride solutions.[48] The donor first ionization energy (I_D) and polarizability (α) are also listed

Donor	I_D^a/ kJ mol^{-1}	$10^{39}\,\alpha$/ C m^2 V^{-1}	$-\Delta H^{\ominus}$/ kJ mol^{-1}	$10^{30}\mu_{DA}$/ C m
C_6H_6	904	1.16	11.3	2.5
$C_6H_5CH_3$	887	1.37	12.6	2.8
1,4-$(CH_3)_2C_6H_4$	863	1.59	13.8	3.2
1,3,5-$(CH_3)_3C_6H_3$	846	1.80	17.1	3.7
1,2,4,5-$(CH_3)_4C_6H_2$	824	2.00	20.1	4.4
$C_6(CH_3)_6$	792	2.39	26.8	5.5

a. 1 kJ mol^{-1} = 1.038 eV.

in pyridine solution emphasises the ionic character of the complexes. Hamilton and Sutton[61] found a temperature dependent dipole moment for the triethylamine–iodine complex. At 233 K the dipole moment is 23.0×10^{-30} C m. This dipole moment increases to 29.7×10^{-30} C m at room temperature. The room temperature ultraviolet spectrum of the complex showed the two time-dependent absorptions at 294 and 363 nm characteristic of the I_3^- ion. A chemical reaction must occur between the amine and the iodine:

$$R_3N + 2I_2 \rightleftharpoons R_3N \cdots I - I + I_2 \rightleftharpoons R_3NI^+ + I_3^-$$

and the high dipole moment of the complex reflects its ionic character.

7.2.4. ORIGIN OF THE DIPOLE MOMENT PRODUCED ON COMPLEX FORMATION

Mulliken[82] writes the wavefunction (Ψ_N) for the ground state of a molecular complex as:

$$\Psi_N = a\Psi_0(D, A) + b\Psi_1(D^+{-}A^-) + \cdots\cdots$$

where a and b are the relative weighting coefficients, $\Psi_0(D, A)$ is the wavefunction for the 'no-bond' state and $\Psi_1(D^+ - A^-)$ is the wavefunction for the 'dative' state (see section 1.1.2 p 7). In the dative state electron transfer has occurred between the donor and acceptor molecules and is accompanied by the formation of a (weak) covalent bond involving the odd electrons in A^- and D^+. The 'no-bond' wavefunction includes contributions from intermolecular interactions (e.g. dipole-induced dipole, dipole-dipole, dispersion etc.) between the donor and the acceptor molecules. $\Psi_0(D, A)$ is usually constructed from the unperturbed wavefunctions of the individual donor and acceptor molecules, and for complexes involving non-dipolar donors and non-dipolar acceptors it is often wrongly assumed that the 'no-bond' form is non-dipolar (see section 1.1.4 for further discussion of this point). The dipolar properties of the complex are thus often assumed to arise entirely from the dative state. The weighting coefficients (a and b) calculated in this way from dipole moment determinations thus seriously over-estimate the contribution of the dative state since purely electrostatic interactions between the donor and the acceptor molecules ensures some induced polar character in the 'no-bond' state. The polar properties of the 'no-bond' state must be considered before the relative contribution of the dative state (and, hence, the degree of charge transfer) can be evaluated.

Most calculations of the dipole moment of the 'no-bond' state have been evaluated for $b\pi$-$a\pi$ complexes. The molecular geometry in the ground state is often well established in the crystal. The donor and acceptor molecules usually lie 0.32–0.35 nm apart with their molecular planes parallel to each other. It is assumed that this configuration is retained in solution. The resultant dipole moment is the vector sum of (1) the moments of individual donor and acceptor (2) the induced-dipole components and (3) any contribution from charge transfer. The donor and acceptor molecules are sufficiently close to each other that the electrostatic interactions are calculated using the point charge (monopole) or point dipole approximations. The molecular polarizability is also reduced to its anisotropic bond components acting at the bond centres. The electric field strength (E) produced at a point P at a distance r from a dipole $\vec{\mu}$ is:

$$\vec{E} = \frac{\vec{\mu}}{4\pi\epsilon r^3}(3\cos^2\theta + 1)^{1/2} \qquad 7.11$$

where ϵ is the permittivity of vacuum and θ is the angle made by the line joining P to the centre of the dipole and the dipole direction. This electric field produces an induced moment $\vec{\mu}_{IND} = \alpha\vec{E}$ (where α is the bond polarizability component) in a polarizable bond at P. The total induced moment is then the vector sum of the moments induced in the individual bonds. The naphthalene-nitrobenzene complex illustrates the dipole components involved (figure 7.5). The nitrobenzene dipole ($\vec{\mu}_A$) induces a dipole moment ($\vec{\mu}_{IND}$) in the plane of the naphthalene molecule and any dipole ($\vec{\mu}_{CT}$) arising from charge transfer is perpendicular to the molecular planes. The resultant dipole moment is

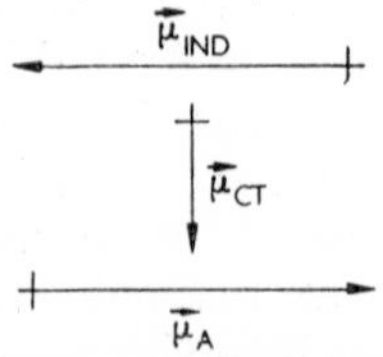

FIG 7.5 Representation of the dipole moment components in a $b\pi$ – $a\pi$ complex formed between a polar acceptor and a non-polar donor.

$\vec{\mu} = (\vec{\mu}_A + \vec{\mu}_{IND} + \vec{\mu}_{CT})$. The gas phase dipole moment of nitrobenzene is 14.1×10^{-30} C m and the dipole moment of the naphthalene-nitrobenzene complex is 12.6×10^{-30} C m. A calculation[95] of this induced dipole and hence the resultant dipole moment of the complex (neglecting the charge-transfer contribution) yields a value of (12.5 to 12.7) $\times 10^{-30}$ C m in very good agreement with the observed value. Charge transfer alone would produce a dipole moment for the complex which is greater than the dipole moment of the isolated molecule. Generally the dipole moment of complexes involving a dipolar component is found to be smaller than the gas-phase dipole moment of the dipolar component or its dipole moment in an inert solvent. The dipole moment of maleic anhydride is 12.9×10^{-30} C m in carbon tetrachloride; in benzene solution the dipole moment decreases to 11.8×10^{-30} C m.

Generally, dielectric studies in solutions produce dipole moments which appear to be solvent dependent even in non-interacting solvents. The dipole moment of acetonitrile is 13.2×10^{-30} C m in the gas phase, 11.5×10^{-30} C m in benzene solution, 11.4×10^{-30} C m in carbon tetrachloride solution and 11.1×10^{-30} C m in n-hexane solution. There is no completely satisfactory means of calculating this 'solvent effect' (chapter 1), but two approaches to the problem deserve mention. Both involve considerations of the molecular shape of the polar molecule and the interaction between the molecule and the surroundings. (The Debye evaluation assumes a spherical cavity for the molecule and neglects intermolecular interaction). Higasi[83] expressed the dipole moment in solution as $\vec{\mu} = \vec{\mu}(\text{gas}) + \Sigma\vec{\mu}_i$, where $\Sigma\vec{\mu}_i$ is the sum of the dipole moments induced in the solvent molecules. A qualitative picture of the sign of the 'solvent effect' in three simple cases is immediately available. A spherical molecule with its dipole at the centre of a spherical cavity exhibits little or no solvent effect. The observed dipole moment of t-butyl chloride is 7.15×10^{-30} C m in the gas, 7.12×10^{-30} C m in benzene and 7.20×10^{-30} C m in n-hexane. A dipole directed along the major axis of an ellipsoidal molecule induces opposing dipoles in the surrounding molecules and thus a negative solvent effect (figure 7.6(a)), while a dipole directed along the minor axis of an ellipsoidal molecule induces dipoles in the solvent molecules which reinforce the parent dipole (figure 8.6(b)). This gives a positive solvent effect. These predictions are reasonably well obeyed, and the influence of molecular shape on the apparent dipole moments observed in solutions is established. Buckingham[84] and Buckingham and Raab[85] developed an expression for the concentration variation of the polarization in solution. The orientation polarization P_O is expressed in virial form in terms of the concentration (c):

$$P_O = \left\{\frac{\varepsilon_{12} - 1}{\varepsilon_{12} + 2} - \frac{n_{12}^2 - 1}{n_{12}^2 + 2}\right\}\frac{1}{c} = A + Bc + Cc^2 + \ldots \qquad 7.12$$

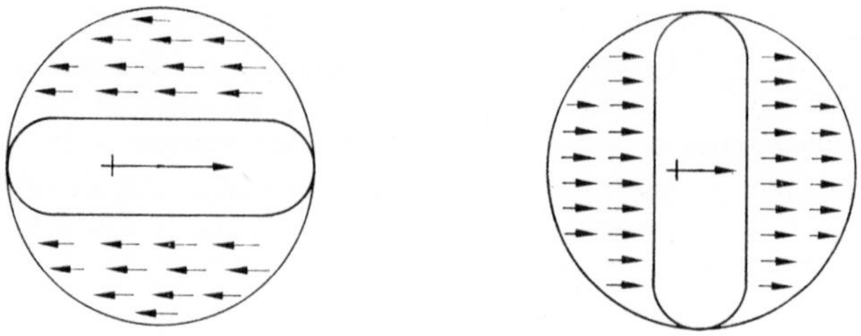

FIG 7.6

The coefficients A, B, C etc. are concentration independent. A is the Debye term (see equation 7.4), B the orientation polarization contribution from polar molecules paired by collision (or other means) for times longer than the period of the applied field, and C the contribution from molecules grouped in triplets, etc. Each molecular grouping is also assumed to be solvated. The apparent dipole moment of the solvated single molecule is calculated from the coefficient A and the Debye equation. The apparent solute dipole moment (μ_s) of the single molecule immersed in a continuum of permittivity ε_{12} is related to the dipole moment in the gas (μ_g) by:

$$\mu_s = \frac{3\varepsilon_{12}\{1 - (n_2^2 - 1)F\}}{(\varepsilon_{12} + 2)\{\varepsilon_{12} + (n_2^2 - \varepsilon_{12})F\}}\mu_g \tag{7.13}$$

where F is a form factor depending on the molecular shape and the direction of the dipole moment within the molecular ellipsoid (F may be evaluated using tables provided by Ross and Sack[86].) The subscripts 12 and 2 refer to the solution and solute respectively. Some typical values of μ_s/μ_g are shown in table 7.3. It is important to note that, even for systems where charge-transfer interaction is postulated between solvent and solute (*e.g.* benzene–sulphur dioxide), reasonable agreement is observed between the calculated and observed values of μ_s/μ_g. The calculated values are obtained by consideration of electrostatic interactions without considering charge transfer.

TABLE 7.3

μ_s/μ_g calculated according to equation 7.13 compared with some experimental determinations in benzene solution

Solute	CH_3Cl	SO_2	$C_6H_5NO_2$
μ_s/μ_g (obs)	0.91	1.00	0.93
μ_s/μ_g (calc)	0.94	0.99	0.97

It is therefore essential to allow for these solvent and molecular shape factors when interpreting dipole moment differences between free dipoles (as in the gas phase) and the values obtained in solution. For *bπ-aπ* complexes involving dipolar components deductions of the 'degree of charge transfer' (for example F_{1N}—see chapter 1, p 8) based on dipole moment measurements are quite uncertain at present (see section 1.1.4 for some of the values obtained and their estimated uncertainties.

A polarization greater than expected from mixture laws is observed in solutions of complexes formed between non-dipolar donors and non-dipolar acceptors. This excess polarization was examined by Briegleb[31,33,87–91] in a series of papers published in the early 1930's and

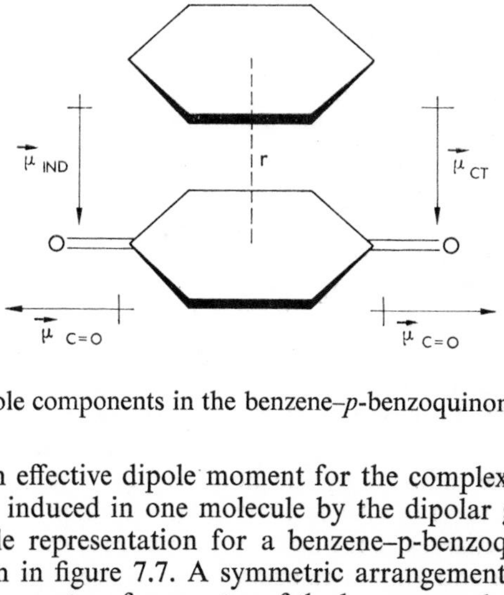

FIG 7.7 Dipole components in the benzene–*p*-benzoquinone complex.

evaluated as an effective dipole moment for the complex which arises from moments induced in one molecule by the dipolar groups in the other. A simple representation for a benzene–p-benzoquinone interaction is shown in figure 7.7. A symmetric arrangement of the polar groups about the centre of symmetry of the benzene molecule produces a zero resultant induced moment parallel to the molecular planes, but a finite induced moment perpendicular to the molecular planes. This is also the direction of any dipole moment produced by charge transfer between the donor and the acceptor molecules. Calculations of the 'degree of charge transfer' in *bπ-aπ* complexes have invariably neglected this induced moment. It is only recently that this contribution has been re-examined and one is forced to conclude that for *bπ-aπ* complexes the charge-transfer contribution to the stabilization of the complex is much smaller than previously estimated. Dewar and Thompson[92] and Malrieu and Claverie[93] point out that charge-transfer interactions in the ground-state can be evaluated, along with the other contributions to complex stability, using a second-order perturbation treatment. They do not regard the charge-transfer contribution as special or significant for these complexes.

Before considering the calculation of the dipole moment of the molecular complexes it is necessary to describe the variety of interactions between the donor and acceptor molecules in the ground state. Murrell, Randic and Williams[94] developed a perturbation theory for treating intermolecular forces in the region of small overlap and factorized the stabilization energy into terms identifiable with each possible intermolecular interaction. The stabilization energy (U) of the molecular complex may be written as a series of terms:

$$U = U_C + U_{ID} + U_R + U_{CT} + \dots \qquad 7.14$$

(cf. equation 1.4) where U_C is the coulomb energy, U_{ID} is the induction and dispersion energy, U_R the exchange repulsion and U_{CT} the charge-transfer energy. A knowledge of the exact wavefunctions for the molecules is required before a proper solution of the stabilization energy is available. These exact wavefunctions are not available for large molecules but approximate solutions are possible (see section 1.1 for a fuller description of this approach and its use in vibrational spectroscopy, in particular).

(a) The Coulomb Energy (U_C)

The coulomb energy arises from electrostatic interactions between the two molecules. The charge distribution within a molecule may be represented as a set of point dipoles located at the mid-point of a particular bond (cf. Briegleb[31,34,87–91], Le Fèvre[101]) or as point charges (monopoles) and multipoles located at the atomic centres (cf. Mantione[95], Hanna[96], Le Fèvre *et al.*[101]). Le Fèvre assumed that the point charges at the atomic centres are adequately represented by dividing the bond dipole moment by the bond length. Thus, for tetracyanoethylene ($\vec{\mu}(C{\equiv}N) = 12.7 \times 10^{-30}$ C m, $\vec{\mu}(C(sp^2) - C(sp)) = 4 \times 10^{-30}$ C m) the charges at the atomic centres are shown in figure 7.8(a). Both Mantione and Hanna calculated the charges from the electronic wavefunctions. Contributions from both the σ and the π electrons were considered, but the resulting charge distribution was very dependent

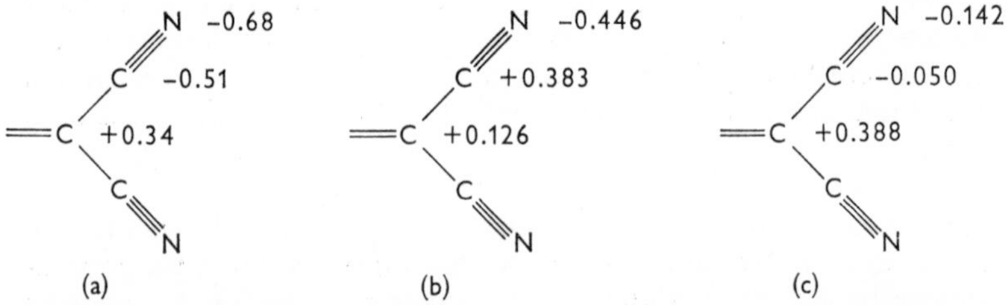

FIG 7.8 Monopole distribution in tetracyanoethylene as calculated after (a) Le Fèvre[109], (b) Mantione[95] and (c) Hanna[96]. The numbers represent the fractional electronic charge at the nucleus.

on the approximations used and the wavefunctions chosen. The charge distribution in tetracyanoethylene as calculated by Mantione and by Hanna is shown in figures 7.8(b) and 7.8(c). These charge distributions are quite different from each other but for the *p*-xylene–tetracyanoethylene the monopole interaction energy was calculated as -11.3 kJ mol^{-1} (Mantione) and -10.5 kJ mol^{-1} (Hanna), and are in surprisingly good agreement with each other despite the different charge distributions. Hanna[96] identified an additional contribution to the charge distribution in conjugated systems. The π-type atomic orbitals produce an effective quadrupole moment which may be located as a point quadrupole at the atomic centres. Consideration of the monopole-quadrupole and the quadrupole-quadrupole interactions increases the Coulomb energy, as calculated by Hanna, to -17.3 kJ mol^{-1}.

(b) Induction and Dispersion Energy (U_{ID})

The induction (or polarization) energy arises from the polarization of one molecule by the electric field produced by another molecule. The induction energy U_{ID} is $\frac{1}{2}\alpha E^2$ where α is the polarizability at a point where the electric field strength is E.

TABLE 7.4

Variation of the enthalpy of formation ($\Delta H^{\ominus}$) of molecular complexes formed between trinitrobenzene and aromatic hydrocarbons with the mean polarizability (α) of the hydrocarbon (see reference 88 and also table 7.2)

Donor	$-\Delta H^{\ominus}$/kJ mol^{-1}	10^{39} α/C m^2 V^{-1}
Benzene	2.0	1.16
Diphenyl	5.8	2.37
Naphthalene	14.2	2.03
Phenanthrene	16.7	2.94
Anthracene	18.4	3.43

Briegleb[31,34,87–91] calculated the induction energy contribution to the stabilization of molecular compounds formed between aromatic nitro-compounds and aromatic hydrocarbons. He showed that the enthalpy of complex formation decreased with the decreasing mean polarizability of the aromatic hydrocarbon (table 7.4, see also table 7.2). A calculation of the induction energy in various $b\pi$-$a\pi$ complexes gave values in good agreement with the observed enthalpy of formation (table 7.5). In this calculation Briegleb assumed point dipoles on the nitro-compound and separated the polarizability of the aromatic hydrocarbon into components arising from the σ- and the π-electrons. Good agreement is observed between the calculated interaction energy and the

TABLE 7.5

Comparison of the enthalpy of formation ($\Delta H^{\ominus}$) of some complexes and the calculated induction energies (U_{IND}) (see reference 91). $U_{IND} = U_\sigma + U_\pi$, where U_π is the contribution from the inductive effect on the π-electrons, and U_σ is the inductive effect on the σ-bonds

Complex	$-\Delta H^{\ominus}$/ kJ mol^{-1}	$-U_\pi$/ kJ mol^{-1}	$-U_\sigma$/ kJ mol^{-1}	$-U_{IND}$/ kJ mol^{-1}
trinitrobenzene–benzene	<RT	3.8	3.8	7.6
trinitrobenzene–styrene	7.5	5.4	5.4	10.8
trinitrobenzene–naphthalene	14.2	7.5	5.4	12.9
trinitrobenzene–anthracene	18.4	12.5	6.7	19.2
p-dinitrobenzene–anthracene	5.0	1.7	4.4	6.1

enthalpy of formation. The Coulomb energy does not appear in Briegleb's calculation. The group dipoles on the aromatic hydrocarbon were assumed negligibly small compared with group dipoles on the nitro-compound.

The Mantione[95,97,98] monopole calculation for the induction energy between *p*-xylene and tetracyanothylene gave an energy of -5.8 kJ mol^{-1} at an intermolecular distance of 0.35 nm (-7.5 kJ mol^{-1} at 0.33 nm). These compare reasonably well with the induction energy of -4.8 kJ mol^{-1} (intermolecular separation of 0.35 nm) calculated by Hanna[96,98,100] (assuming point monopoles and multipoles at the atomic centres). A rather more straight forward calculation of the charges at the atomic centres was adopted by Le Fèvre, Radford and Stiles.[101] The charges at the atomic centres were calculated by dividing the bond moment by the bond length. For the hexamethylbenzene complex with trinitrobenzene and with tetracyanoethylene Le Fèvre[101] calculated an induction energy of -12.6 kJ mol^{-1} and -14.7 kJ mol^{-1} respectively. The measured enthalpies of formation of these complexes in solution are -19.7 kJ mol^{-1} and -32.5 kJ mol^{-1}.

The above calculations were performed on molecules containing groups with large dipole moments. Hanna[99] showed that the electric field produced within about 0.4 nm of a benzene molecule is sufficient to account for a very large part of the enthalpy of formation of the benzene-halogen complexes. The electric field at a point along the six-fold axis of the benzene molecule arises from the quadrupole moment of the molecule and may be approximated to a π electron contribution and a σ-bond dipole contribution (the C—H bond with a dipole moment 2.1×10^{-30} C m with a positive hydrogen atom). The electric field strength variation with distance from the plane of the benzene ring is shown in figure 7.9. The σ-bond dipole contribution is approximately 0.6 of that due to the π-electrons. An induction energy of -1.5 kJ mol^{-1} was calculated for the benzene–iodine complex

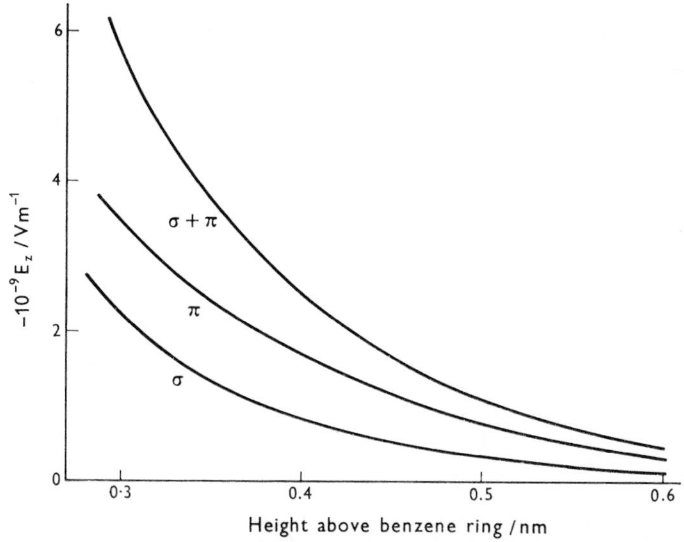

FIG 7.9 Variation of the electric field strength (E_z) with distance along the hexagonal axis of the benzene ring. σ is the contribution from the σ-polar bonds, π is the contribution from the π-electrons and $\sigma + \pi$ is the total field. (Reproduced from *J. Amer. Chem. Soc.*, **90,** 285 (1968), by permission of the American Chemical Society.)

assuming that the polarizability of the iodine atom is concentrated at the mid-point of the bond. This calculated energy increased to -4.9 kJ mol^{-1} if it was assumed that the molecular polarizability may be divided into two equal parts located at the individual iodine atoms. The gas-phase enthalpy of formation of the benzene–iodine complex was -8.4 kJ mol^{-1} [102]. The induction energy calculated for a cyclohexane-iodine interaction was about -16 J mol^{-1} [99], much smaller than **RT** at room temperature. (See section 1.2.5 for some further comments on these calculations on benzene–halogen complexes.)

In the gas phase London dispersion forces between the donor and acceptor molecules play an important part in the stabilization of the complex. The dispersion energy is usually written as:

$$U_D = -\frac{3}{2}\frac{I_A I_D}{I_A + I_D}\frac{\alpha_A \alpha_D}{r^6} \qquad 7.15$$

where I and α represent the ionization potential and polarizability respectively, the subscripts A and D refer to the acceptor and donor and r is the intermolecular separation. The dispersion energy between a *p*-xylene and tetracyanoethylene molecule separated by 0.33 nm is -19.2 kJ mol^{-1} [97]. In solution the significance of the dispersion energy contribution to the overall stabilization energy is uncertain (see section 1.1.2). The dispersion interactions are often assumed to cancel each other[96]; the interaction between the isolated donor and the isolated acceptor with the solvent molecules are of the same order of magnitude as the interaction between the donor and acceptor molecules[96]. It may well be more reasonable to assume that the solvent interactions cancel each other leaving a finite contribution from the interaction between the donor and acceptor molecules. A decision on this point is not possible until data become available on the ionization potentials and polarizabilities of molecular complexes (see section 1.1.2 p 19 and section 1.2.9, p 70 for a further discussion of the solvent effects on these interactions).

Claverie[103] and Mantione[97] suggested that the London equation (equation 7.15) underestimates the dispersion energy by a factor of approximately 3. They calculated the dispersion energy using a semi-empirical treatment developed by Kitaygorodsky[104]. The dispersion energy was represented as:

$$U_D = -C_1\left(\frac{r_0}{r}\right)^6 \qquad 7.16$$

where r_0 is the 'contact distance' (evaluated from Pauling's atomic radii), r is the intermolecular separation and C_1 (a constant) is 15.9 kJ mol^{-1}. The dispersion energy calculated for the p-xylene–tetracyanoethylene complex is −42.7 kJ mol^{-1} at an intermolecular separation of 0.35 nm. This is about twice the dispersion energy calculated by Hanna using equation 7.15.

(c) Charge-Transfer Energy (U_{CT})

The transfer of an electron from the donor to the acceptor contributes to the stabilization energy of the complex. This energy may be regarded as arising from the interaction of the charge distribution on the acceptor molecule after receiving an electron with the charge distribution on the donor molecule after losing an electron. A large number of calculations on molecular complexes in the ground state assume that this is the major contributor to the stabilization energy. Often this is probably not so but it should be realized that electron-donor-acceptor complexes show a very wide range of stabilities. In weak complexes (e.g. *bπ-aπ* complexes) van der Waals forces make important contributions to the stabilization energy (section 1.3) but some charge transfer must occur in strong complexes (e.g. tin(IV) chloride–ether system) where configurational changes often accompany complex formation. The electron transfer between sodium and chlorine atoms producing an ionic species is one reaction involving complete charge transfer. At the other extreme is the weak interaction described in electron-donor-acceptor terminology[106] as 'contact charge transfer' (see sections 1.2.10, 3.3.4).

Mulliken[24,105] derived an approximate expression for the charge-transfer energy,

$$U_{CT} = \sum_m \sum_n \frac{-|kS_{mn}|^2}{h\nu_{mn}} \qquad 7.17$$

where $h\nu_{mn}$ is the spectroscopic charge-transfer excitation energy for a transition between filled orbitals of one molecule to the unfilled orbitals of another molecule, S_{mn} is the overlap integral, and k a constant determined by the degree of charge transfer, F_{IN}. Normally k is evaluated from the dipole moment of the complex or from the extinction coefficient of the charge-transfer band. These procedures grossly overestimate k when the van der Waals interactions are neglected. There is no satisfactory method available at present to accurately evaluate k nor, therefore, the value of U_{CT}. Hanna[96] choses k in such a way that the calculated dipole moments bracket the *experimental* values. This is an over-estimate (the experimental dipole moment is uncorrected for induced moments) and leads to a charge-transfer stabilization energy of -10.4 kJ mol^{-1} for the *p*-xylene–tetracyanoethylene complex.

(d) Repulsion Energy, (U_R)

The complete calculation of the stabilization energy must consider the repulsive effects between the molecules at short intermolecular separations. These are difficult to evaluate. Mantione[97], used a semi-empirical equation developed by Kitaygorodsky[104], to evaluate this repulsion energy:

$$U_R = C_2 \exp\left(-\frac{\beta r}{r_0}\right) \qquad 7.18$$

where r and r_0 have the same significance as in equation 7.16, C_2 is 12.6×10^4 kJ mol^{-1} and β is 11. The value of β is chosen to give a minimum in the potential energy curve at an intermolecular separation of 0.35 nm.

A different approach was proposed by Murrell, Randic and Williams[94]. The repulsion energy arises from quantum mechanical considerations of the repulsion between electrons of the same spin in overlapping charge distributions. The contribution is zero for electrons of opposite spin. A good approximation for this repulsion is:

$$U_R = cS_{AB}^2/r \qquad 7.19$$

where S_{AB} is the overlap integral between the two molecules at intermolecular separation r. For two large molecules Hanna evaluated this repulsion energy using the pairwise sum over the atomic orbitals:

$$U_R = \sum_i \sum_j c_{ij} S_{ij}^2/r \qquad 7.20$$

where the sums over i and j are over all atomic orbitals on molecules A and B. c_{ij} may be evaluated from the intermolecular distance in the complex (where the potential energy is a minimum). It should be noted that both the charge-transfer energy (equation 7.17) and this repulsion energy are functions of S^2 and in weak complexes these contributions are sometimes said to approximately cancel[96]. The greater the overlap between the donor and acceptor electron orbitals the greater is the charge-transfer energy and the greater, also, is the repulsion energy.

TABLE 7.6
Summary of some complex stabilization energy calculations (all energies in kJ mol^{-1})

Complex	Coulomb energy ($-U_C$)	Induction energy ($-U_I$)	Dispersion energy ($-U_D$)	Charge-transfer energy ($-U_{CT}$)	Repulsion energy (U_R)	Total energy ($-U$)	Notes	$-\Delta H^\ominus$ (observed)	Reference
benzene–TNB		7.5				7.5	Briegleb—point dipole		91
naphthalene–TNB		10.9				10.9	approximation.	14.2 (s)	91
anthracene–TNB		19.2				19.2		18.4 (s)	91
anthracene-p-dinitrobenzene		6.1				6.1		5.0 (s)	91
p-xylene–TCNE	11.3	5.8	42.7		21.9	37.9	Mantione–monopole	31.0 (g)	97
mesitylene–TCNE	11.5	6.3	48.1		25.0	41.0	approximation. Dispersion	38.5 (g)	97
							and repulsion energies		
durene–TCNE	12.5	7.0	54.1		30.1	43.5	using equations 7.16 and 7.18	42.2 (g)	97
benzene–TCNE	16.8	4.4	15.0	7.5	17.7	26.0	Hanna—monopole point	9.6 (s)	96
							quadrupole approximation.	30.8 (g)	96
p-xylene–TCNE	17.3	4.8	15.1	10.5	21.4	26.2	Dispersion and repulsion	14.1 (s)	
benzene–I_2	17.8	1.7	4.5				energies using equations	8.4 (g)	96
benzene–Br_2	16.2	1.7	4.5	3.4	10.1	15.7	7.15 and 7.17.	4.6 (s)	96
HMB–hexafluorobenzene		4.1				4.1	Le Fèvre—monopole		101
HMB–chloranil		8.2				8.2	approximation	22.4 (s)	101
HMB–TNB		10.9				10.9		19.7 (s)	101
HMB·TCNE		12.6				12.6		32.4 (s)	101

TNB = trinitrobenzene: TCNE = tetracyanoethylene: HMB = hexamethylbenzene: (s) in solution: (g) in gas phase.

On the other hand, there is some reason to believe (section 1.1.2, p 15) that the polarisation and repulsion energies may cancel. The situation is therefore unclear at the present time.

A summary of the results of various calculations on the stabilization energies is given in table 7.6. Clearly from table 7.6 more data on the energies of complex formation in the gas phase are required before quantitative conclusions may be reached on the nature of the interaction energy. Nevertheless, the van der Waals interactions are of a major factor in the stabilization of the complexes listed. A rather good agreement between the observed enthalpy of formation in the gas phase and the calculated stabilization energy is obtained by Mantione for the tetracyanoethylene complexes. The calculation does not include any charge-transfer interaction. Comparison of Mantione's and Hanna's calculations on the *p*-xylene–tetracyanoethylene complex shows that they both obtained a similar overall stabilization energy, but very different dispersion energies. Hanna calculated a charge-transfer contribution which was of the same order of magnitude as the electrostatic contributions. On the basis of these calculations we do not believe it is possible to determine, with any great certainty, the charge-transfer contribution to the stabilization of the complexes (see section 1.1.2 for more evidence to support this conclusion).

The above discussion of the stabilization energies has been essentially restricted to complexes formed from non-dipolar components. Calculations reported[106] for complexes formed between ammonia and iodine and between trimethylamine and iodine yielded a polarization contribution of -6.6 and -0.5 kJ mol^{-1} respectively to the total energy. A point dipole located at the mid-point of the N—H or the C—N bond was assumed in the calculations. It is, however, not clear whether these calculations included the contribution of the dipole moment located on the lone-pair of electrons of the nitrogen atom. This atomic dipole has a dipole moment of at least 3×10^{-30} C m and is in very close proximity to the acceptor molecule. The neglect of this contribution would cause the interaction energy to be underestimated. Hydrogen bonding is another phenomenon where the nature of the bonding is under discussion. Electrostatic forces must contribute to the bond formation, but again their magnitude is uncertain, and good calculation of the electrostatic energy requires a better knowledge of wave-functions than is at present available. For reviews on the theories of hydrogen bond formation readers are referred to references 107–110 (see also chapter 4).

The dipole moment increment on complex formation arises from the induction interaction and from any charge-transfer present. The latter cannot be properly evaluated unless the induction contribution is known. Calculations of the induced dipole moment suffer from the same approximations as are used in the calculations of the induction energies. Table 7.7 summarises the results of some induced dipole moment calculations

TABLE 7.7
Comparison of the calculated induced dipole moment (μ_{IND}) and observed dipole moment (μ_{OBS}) for some molecular complexes

Complex	Notes	Intermolecular separation/ nm	$10^{30}\,\mu_{IND}$/ C m	Reference	$10^{30}\,\mu_{OBS}$/ C m	Reference
durene–TCNE	Mantione—monopole approx.	0.33	4.1	97	4.2	42
		0.33	3.7			
pyrene–TCNE	Mantione—monopole approx. (Calculation for various relative orientations of the components)	0.34	5.7–5.2	95	6.7	44
naphthalene–TCNE	ditto	0.34	4.9–4.4	95	4.3	42
naphthalene–nitrobenzene	ditto	0.34	12.7–12.5	95	12.6	95
HMB–hexafluoro-benzene	Le Fèvre—monopole approx.	0.34	1.0	101		
	point dipole approx.	0.34	1.0	101		

HMB–chloranil	Le Fèvre—monopole approx.	0.34	1.7	101	3.3	68
	point dipole approx.	0.34	1.7	101		
HMB–TNB	Le Fèvre—monopole approx.	0.34	3.3	101	2.9	68
		0.32	3.4	101		
HMB–TCNE	Le Fèvre—monopole approx.	0.34	2.3	101	4.5	68
		0.32	2.3	101		
benzene–TCNE	Hanna—monopole, point quadrupole approximation	0.35	0.4	96	3.6	43
					2.7	43
p-xylene–TCNE	ditto	0.35	0.5	96		
benzene–iodine	Hanna—monopole, point	0.48	4.6 (a)	99	6.0	25
	quadrupole approximation. (a) iodine polarizability located at mid point of halogen bond, (b) iodine polarizability divided into two equal parts located at each atom.	0.48	3.1 (b)	99	2.3	26
					2.4	23, 24

TCNE = tetracyanoethylene: HMB = hexamethylbenzene: TNB = trinitrobenzene.

and compares them with the observed dipole moments. The induced dipole moment contribution is at least 50% of the observed dipole moment. Mantione's calculations of the dipole moment produce good agreement between the induced and the observed dipole moments without considering any charge-transfer contributions, as do Hanna's calculations on the benzene-iodine complex. Calculations of the percentage charge-transfer should therefore be revised to take account of the induced moment. Writing the ground state wavefunction for the molecular complex as:

$$\Psi_N = a\Psi_0(D, A) + b\Psi_1(D^+ - A^-) \qquad 7.21$$

with the normalizing condition:

$$a^2 + b^2 + 2abS_{01} = 1 \qquad 7.22$$

(where S_{01} is the overlap integral usually assumed as 0.1 for weak complexes[106]). The dipole moment of the complex is given by Person (section 1.1.3):

$$\vec{\mu}_N = a^2\vec{\mu}(D, A) + b^2\vec{\mu}(D^+ - A^-) + 2ab\vec{\mu}_{01} \qquad 7.23$$

where

$$\vec{\mu}_{01} \approx \tfrac{1}{2}S_{01}\vec{\mu}(D^+ - A^-)$$

(see ref. 106, p 335). $\vec{\mu}(D^+ - A^-)$ is 50.1×10^{-30} C m for an intermolecular separation of 0.32 nm. Numerous calculations of the percentage charge-transfer $(100b^2/(a^2 + b^2))$ assuming $\vec{\mu}(D, A) = 0$ are available; a 6.7% charge-transfer is calculated in the tetracyanoethylene–hexamethylbenzene complex using the above assumption. Le Fèvre estimates that the induced dipole moment in this complex is 2.3×10^{-30} C m $(= \vec{\mu}(D, A))$ and the percentage charge-transfer is correspondingly reduced to about 3%. For the hexamethylbenzene–trinitrobenzene complex Le Fèvre calculated an induced dipole moment of 3.3×10^{-30} C m (the observed value is 2.9×10^{-30} C m) and, hence, a zero contribution from charge-transfer effects.

Clearly, at present, the precise determination of the extent of charge transfer in the ground state of molecular complexes is very difficult. However, it seems likely that electrostatic interactions play a much more important role in the stabilization of complexes (especially "weak" complexes) than previously assumed (see sections 1.1 and 1.3 for further discussion).

It was assumed in these calculations that there is a preferred geometry for the complexes. This is often well-established in the solid state but (generally) not in solution. The lifetime of some weak complexes in solution is of the order of 10^{-11} s and 'complexes' of such short lifetimes are more appropriately regarded as existing during 'sticky' collisions which may not adopt a preferred geometry. (See below and chapter 2 for further discussion of the consequences of such short lifetimes).

7.3. MICROWAVE AND FAR INFRA-RED MEASUREMENTS

7.3.1. INTRODUCTION TO DIELECTRIC RELAXATION[10,13,14]

A rigid dipole tends to align itself in the direction of an applied electric field. In fluids this tendency is opposed by Brownian motion. At low frequencies the dipole follows the variation of the applied field and provides the maximum orientation polarization contribution to the total polarization. Electromagnetic energy (the displacement current in the dielectric) is conveyed through the dielectric by the reorientation

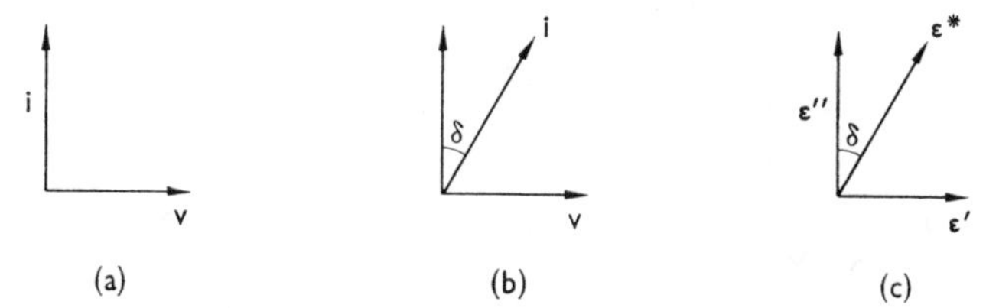

FIG 7.10. Phase relations in a dielectric; (a) current (i) and voltage (V) at low (static) frequency, (b) current and voltage at high frequency, (c) the complex permittivity.

of the dipoles. At low frequencies there is no lag between the orientation of the molecules and the variation of the alternating applied voltage. The displacement current is 90° out of phase with the voltage and is ahead of the latter (figure 7.10). The Joule heating in the dielectric is zero. On increasing the frequency of the applied voltage into the microwave region the dipoles begin to lag behind the variations of the alternating voltage. The current through the dielectric acquires a component in the direction of the applied field and Joule heating occurs in the dielectric (dielectric loss). The inability of the dipoles to follow the variations of the applied field produces a decrease in the permittivity of the dielectric (dispersion of the permittivity). Under these conditions the permittivity is expressed by a complex quantity, ε^* (see figure 7.10(c)) given by,

$$\varepsilon^* = \varepsilon' - i\varepsilon'' \tag{7.24}$$

where $i = (-1)^{1/2}$, ε' is the real part of the permittivity and ε'' the dielectric loss factor. The energy dissipation factor is $\tan \delta = \varepsilon''/\varepsilon'$. At still higher frequencies (around 10^{13} Hz) the dipole ceases to follow the variation of the applied field. The orientation polarization is now zero and the remaining polarization arises from the atomic and the electronic contributions.

The theory of dielectric relaxation was first evaluated by Debye[3]. The rotation of a molecule of moment of inertia, I, under the influence of an applied couple, C, is described by the equation of motion:

$$I\ddot{\theta} + \zeta\dot{\theta} = c \qquad 7.25$$

where $\ddot{\theta}$ is the angular momentum and $\zeta\dot{\theta}$ the resistive couple related to the effective viscosity in the medium. Debye neglected the inertial term and showed that, when the relaxation of the orientation polarization is a simple exponential decay process:

$$P_0(t) = P_0(0) \exp(-kt) = P_0(0) \exp\left(-\frac{t}{\tau}\right) \qquad 7.26$$

where $P_0(t)$ and $P_0(0)$ are the orientation polarizations at time t = t and t = 0 respectively, and τ (the reciprocal rate coefficient, k^{-1}) is the relaxation time, then the real and imaginary parts of the complex permittivity are expressed as:

$$\varepsilon - \varepsilon_\infty = \frac{(\varepsilon_0 - \varepsilon_\infty)}{1 + \omega^2\tau^2} \qquad 7.27$$

and

$$\varepsilon'' = (\varepsilon_0 - \varepsilon_\infty)\frac{\omega\tau}{1 + \omega^2\tau^2} \qquad 7.28$$

where ε_0 is the limiting permittivity at low frequency, ε_∞ the limiting permittivity at high frequency (obtained by extrapolation using the Debye equation) and $\omega = 2\pi f$ (f = frequency). These equations are illustrated in figure 7.11. The dielectric loss factor reaches a maximum

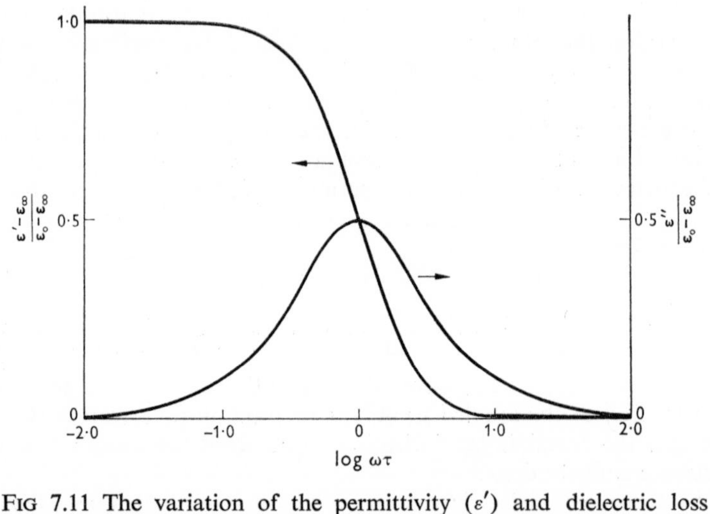

FIG 7.11 The variation of the permittivity (ε') and dielectric loss factor (ε'') with angular frequency (ω) (Debye single relaxation time).

TABLE 7.8
The relaxation times (298 K) of some rigid molecules in the liquid state and in solution[114]. (η is the liquid viscosity)

Solute	Solvent	$10^3\ \eta$/ N m^{-2} s	$10^{12}\ \tau$/ s	$f_m = (2\pi\tau)^{-1}$/ GHz
chlorbenzene	liquid	0.8	10.0	15.9
	benzene	0.6	8.3	19.1
bromobenzene	benzene	0.6	10.6	15.0
nitrobenzene	benzene	0.6	11.4	13.9
1-chloronaphthalene	*n*-heptane	0.4	10.6	15.0
1-chloronapthahlene	nujol	1970	63.7	2.5
chloroform	liquid	0.6	7.5	21.2
	carbon tetrachloride	1.0	5.0	31.8
	benzene	0.6	7.1	22.4
	n-heptane	0.4	3.6	44.1

($\varepsilon''_m = (\varepsilon_0 - \varepsilon_\infty)/2$) at $\omega\tau = 1$. The orientation polarization is proportional to $(\varepsilon_0 - \varepsilon_\infty)$ which is related to the dipole moment in dilute solution by a modified form[111] of equation 7.4

$$(\varepsilon_0 - \varepsilon_\infty) = 2\varepsilon''_m = \frac{(\varepsilon_0 + 2)^2 \mathrm{Nc}\mu^2}{2700\mathbf{kT}\epsilon} \qquad 7.29$$

where c is the solute concentration. Evaluation of the dipole moment using equation 7.29 does not require a knowledge of the distortion polarization.

Debye related the relaxation time of a spherical molecule to its molecular dimensions and to the viscosity of the medium. The relation was later modified[112] for ellipsoidal molecules with semi-axes a, b and c to give:

$$\tau = \frac{4\pi\eta \mathrm{abc}}{\mathbf{k}\mathrm{T}}\,\mathrm{s} \qquad 7.30$$

where η is the microscopic viscosity and s is a form factor depending on the relative direction of the molecular dipole moment with respect to the principal ellipsoidal axes. A table of s values for various values of a, b and c is available[113]. No satisfactory formulation of the microscopic viscosity is available and it is sometimes equated to the macroscopic viscosity. The relaxation times of some rigid molecules are listed in table 7.8. The relaxation times increases with increasing molecular volume and with increasing liquid viscosity. Equation 7.30 predicts a linear relation between the relaxation time and microscopic viscosity. The results shown in table 7.8 demonstrate that the macroscopic viscosity is an inadequate substitute for the microscopic viscosity.

Nevertheless, equation 7.30 is often used to predict relaxation times in solution provided the relaxation time of a reference molecule of known volume is available in the same solvent. Poor agreement between the calculated relaxation time and the experimentally observed quantity often establishes the presence of a second relaxation process.

The width of the Debye absorption (equation 7.28) at one half peak height corresponds to a log. frequency displacement ($\Delta \log f$) of 1.14 and is characteristic of a relaxation process described by a single relaxation time. Often the experimentally observed absorption is broader than expected. The dispersion process is no longer defined in terms of a single exponential decay function and the relaxation behaviour is better described in terms of the Fuoss-Kirkwood[115] or the Cole-Cole[116] relations.

The Fuoss-Kirkwood empirical relation:

$$\cosh^{-1}\left(\frac{\varepsilon''_m}{\varepsilon''}\right) = \beta \ln\left(\frac{f_m}{f}\right) \tag{7.31}$$

describes a broadened absorption in terms of the parameter β ($0 < \beta \leqslant 1$). When $\beta = 1$ equation 7.31 reduces to the Debye equation. In most liquids the experimentally observed values of β range from 0.95 to 1. The Cole-Cole relation:

$$\varepsilon^* - \varepsilon_\infty = \frac{\varepsilon_0 - \varepsilon_\infty}{1 + (i\omega\tau)^{1-h}} \tag{7.32}$$

gives a depressed semicircular arc when ε'' is plotted against ε' (figure 7.12). The breadth of the absorption is described again by an empirical factor h ($0 \leqslant h < 1$). When $h = 0$ equation 7.32 reduces to the Debye equation. Again in most liquids the experimentally observed value of h usually lies between about 0.04 and 0.

The dielectric relaxation behaviour of molecular complexes in solution is determined by the rate of molecular reorientation and also by the

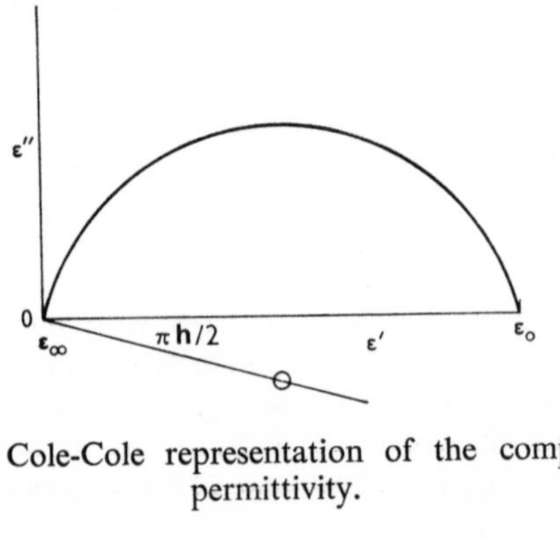

FIG 7.12 The Cole-Cole representation of the complex dielectric permittivity.

rate of formation and decomposition of the complex. The situation may be exemplified by considering a dipolar molecular complex whose components are non-polar. The relaxation behaviour depends on the rate of molecular reorientation of the complex (*cf.* that of a polar molecule) and also by the rate of decomposition of the complex. The latter process produces a change in the total orientation polarization by the decomposition of the polar complex into its non-polar components. The two processes involved may be classified as dipole relaxation and chemical (or exchange) relaxation respectively. Usually the observed dielectric behaviour is determined by the *faster* of these two mechanisms.

References to table 7.8 shows that the maximum dielectric absorption in normal liquids occurs at frequencies around 15 GHz. The width of the absorption at a half and at a tenth peak height corresponds to log frequency displacements ($\Delta \log f$) of 1.14 and 2.60 respectively. Thus to adequately cover the absorption process measurements are required over two or three decades of frequency centred around about 15 GHz. In this frequency region waveguide techniques are required. These operate at essentially single frequencies and measurements at frequencies separated by factors of two (*i.e.* $\Delta \log f = 0.3$) usually suffice to establish the absorption contour. Detailed descriptions of the experimental techniques available in the microwave region are available in reference 14. Recently, a method has been developed which allows of dielectric measurements as a continuous function of frequency up to about 18 GHz[117].

Microwave dielectric loss (ε'') measurements are often accurate to $\pm 2\%$ or $\pm 0.5 \times 10^{-3}$, whichever is the greater, and dielectric relaxation times evaluated from measurements made over a sufficiently wide frequency range are accurate to better than $\pm 5\%$. The uncertainty in the evaluated relaxation time increases significantly when the measured dielectric loss factor (ε'') is of the order of 1 to 2×10^{-3}. This situation frequently occurs in systems where the molecular complex is formed from non-polar components.

7.3.2. THE DIELECTRIC PROPERTIES OF COMPLEXES FORMED BETWEEN POLAR COMPONENTS

Very long lived molecular complexes show relaxation times characteristic of the molecular reorientation of the complex. The relaxation times of some such typically strong complexes (see table 7.9) are, as expected, much larger than the relaxation time of the individual components. Thus for pyridine (in benzene solution) $\tau = 5$ ps; for triethylamine $\tau = 8$ ps. The relaxation time reflects the molecular volume of the complex which increases with increasing size of the complex. Thus in the pyridine-substituted acetic acid complexes the relaxation time increases with increased chlorination of the acid. The

TABLE 7.9
The relaxation times (τ) of some molecular complexes in solution at 293 K

Complex	Solvent	τ/ps	Reference
triethylamine–trinitrobenzene	dioxan (308 K)	53	118
tributylamine–trinitrobenzene	dioxan (308 K)	75	118
pyridine–acetic acid	benzene	36	69
pyridine–monochloroacetic acid	benzene	45	69
pyridine–dichloroacetic acid	benzene	83	70
pyridine–trichloroacetic acid	benzene	83	70

identical relaxation times reported for the dichloro- and the trichloroacetic acid complexes probably reflects the change in character of the hydrogen bonding on going from the dichloro- to the trichloroacetic acid complex. The former appears to be primarily covalent, and the latter ionic, in character[69,70]. Probably the increased ionic character of the bonding draws the components closer together and offsets the increased volume caused by the addition of the chlorine atom.

In systems where the polar complex exists in equilibrium with a significant concentration of a polar component two relaxation times are expected when the chemical relaxation process is slow. One relaxation time corresponds to the relaxation of the donor, the other (and lower frequency process) to the relaxation of the complex. The resulting dielectric absorption is written as:

$$\frac{\varepsilon''}{\varepsilon_0 - \varepsilon_\infty} = \sum \frac{c_i \omega \tau_i}{1 + \omega^2 \tau_i^2} \qquad 7.33$$

where c_i is an intensity factor proportional to the square of the dipole moment and to the concentration of the i'th species. Only when the relaxation times differ by a factor of at least 6 are the individual absorptions sufficiently well separated on a log frequency scale so that the absorption maximum may be distinguished for each process. Otherwise a single absorption maximum is observed and the absorption contour is broader than the Debye contour. The resulting absorption is described by the Fuoss-Kirkwood equation (7.31). The situation is illustrated in figure 7.13(a) where two relaxation processes of equal intensity but with relaxation times differing by a factor of 3 are combined to give the illustrated resultant absorption. The Fuoss-Kirkwood plot is shown in figure 7.13(b) from which a distribution parameter (β) of 0.91 is deduced. The observed relaxation time is intermediate between that of

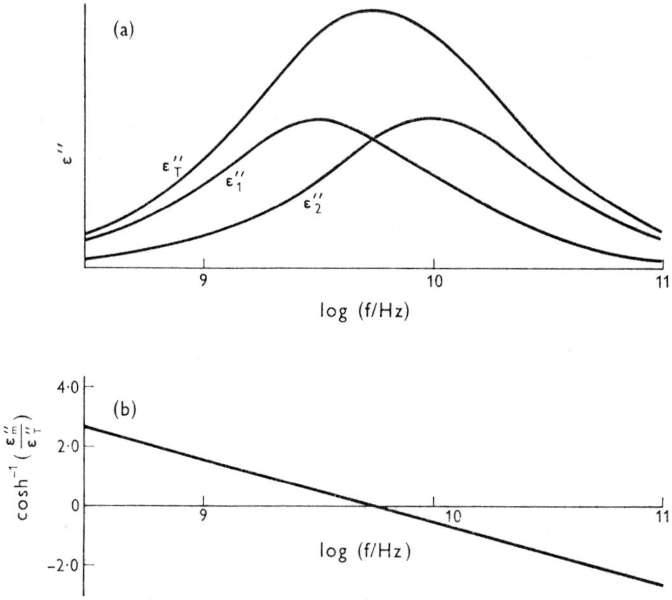

FIG 7.13 (a) The variation of the dielectric absorption ($\varepsilon''_T = \varepsilon''_1 + \varepsilon''_2$) with log frequency computed for two absorption processes (ε''_1 and ε''_2) having relaxation times differing by a factor of 3. (b) Fuoss-Kirkwood plot for ε''_T (frequencies measured in Hz).

the two components. In any relaxation mechanism involving overlap between a molecular reorientation and a faster chemical relaxation the effective relaxation time is smaller than that for the molecular reorientation. Care must be exercised in the interpretation of relaxation times obtained by the resolution of broad dielectric absorption curves into a number of individual Debye type absorptions. Often distribution parameters (β) of about 0.96 are obtained for liquids (particularly viscous liquids) containing only one relaxing species.

Debye type absorption is also observed when there is rapid chemical exchange between the complex and its components. This situation has been treated by Kauzmann[121] and by Anderson[119,120]. An example of this situation occurs with maleic anhydride and with phthalic anhydride in aromatic solvents. Complex formation occurs between the anhydride and the solvent, but because the chemical exchange is a fast process (compared to the time of molecular reorientation) the observed relaxation time is an averaged value of the relaxation time of the complexed and of the uncomplexed anhydride. The relaxation times observed for solutions of maleic anhydride and for phthalic anhydride in various solvents are listed in table 7.10. The relaxation times in the aromatic solvents are all greater than the relaxation times measured in

TABLE 7.10

Dielectric relaxation time (τ) of maleic and of phthalic anhydrides in various solvents[47] at 298 K

Solute	Solvent	Solute viscosity $10^3\ \eta$/N s m^{-2}	τ/ps
maleic anhydride	carbon tetrachloride	0.97	5.8
	benzene	0.66	9.4
	p-xylene	0.63	11.3
	mesitylene	0.71	12.6
phthalic anhydride	carbon tetrachloride	0.97	14.0
	benzene	0.66	18.4
	p-xylene	0.63	22.0
	mesitylene	0.71	28.1

carbon tetrachloride. Viscosity considerations alone predict τ (CCl_4 solution) $>$ τ (aromatic solutions) as is observed in a very large number of instances (*e.g.* for chlorobenzene in carbon tetrachloride solution $\tau = 8.6$ ps, in benzene solution $\tau = 8.2$ ps; for nitrobenzene in carbon tetrachloride solution $\tau = 14.7$ ps, in benzene solution $\tau = 12.8$ ps). Complex formation occurs in the aromatic solvents and the relaxation time increases with increasing volume of the complex. Furthermore, the observed relaxation time for maleic anhydride dissolved in mesitylene-carbon tetrachloride mixtures increases with increasing mesitylene concentration despite the decreasing solution viscosity (table 7.11). The greater the percentage complex formation the longer the relaxation time. These results follow the pattern predicted by Anderson for the behaviour of systems where rapid chemical relaxation occurs between the complex and its components. In an equilibrium (with rapid exchange) involving two polar molecules (*e.g.* maleic anhydride

TABLE 7.11

Relaxation time (τ) and solution viscosity (η) for maleic anhydride dissolved in mesitylene/carbon tetrachloride mixtures[47] at 293 K

Mole fraction mesitylene	Conc. maleic anhydride/mol dm^{-3}	τ/ps	$10^3\ \eta$/ N s m^{-2}
0.0	0.0223	5.8	0.97
0.102	0.0430	9.1	0.90
0.187	0.0557	11.0	0.88
0.409	0.0599	12.3	0.83
1.00	0.189	12.6	0.71

(A) + benzene ⇌ complex (C)) the relaxation behaviour is characterised by an effective relaxation time (τ) given by:

$$\tau^{-1} = \alpha\tau_c^{-1} + (1 - \alpha)\tau_A^{-1} \qquad 7.33$$

where τ_c and $\tau_A(\tau_c > \tau_A)$ are the relaxation times of the complex and the maleic anhydride respectively and α is the equilibrium fraction of complexed maleic anhydride. The greater the degree of complexing the larger is the effective relaxation time.

No direct evaluation of the lifetime of the complex is possible from these data. However since the chemical exchange process is more rapid than the molecular reorientation process the lifetime of the complex cannot be greater than about 10^{-12} s.

7.3.3. COMPLEXES FORMED BETWEEN NON-POLAR COMPONENTS

The relaxation behaviour in these cases has been evaluated by Schwarz[122], Williams[123] and by Anderson[119]. The dielectric properties of most complexes are usually investigated in the presence of excess donor and the equilibrium is then represented as:

$$A \underset{k_{-1}}{\overset{k_1D}{\rightleftharpoons}} DA \qquad 7.34$$

where k_1 and k_{-1} are the respective rate coefficients and D is the equilibrium donor concentration. The formation of the complex is a pseudo-first order reaction. Both Schwarz and Williams predict a dielectric behaviour described by:

$$\frac{\varepsilon^* - \varepsilon_\infty}{\varepsilon_0 - \varepsilon_\infty} = \frac{K_1}{1 + K_1}\frac{1}{1 + i\omega\tau_{DA}} + \frac{1}{1 + K_1}\frac{1}{1 + i\omega\tau_2} \qquad 7.35$$

where K_1 is the equilibrium constant k_1D/k_{-1}, τ_{DA} the relaxation time for the complex (effectively behaving as a stable, long lived entity) and

$$\tau_2^{-1} = \tau_{DA}^{-1} + \tau_{Ch}^{-1} \qquad 7.36$$

where τ_{Ch} is the chemical relaxation time given by:

$$\tau_{Ch} = (k_1D + k_{-1})^{-1} \qquad 7.37$$

Two limiting conditions are distinguished. When $\tau_{Ch} \ll \tau_{DA}$ (*i.e.* fast chemical relaxation) then $\tau_2 \sim \tau_{Ch}$ and two relaxation processes may be detectable. The lower frequency process arising from the reorientation of the 'stable' complex and a higher frequency process due to chemical relaxation. This situation corresponds to weak, short lived complexes with low equilibrium constants. The ratio of the intensity of the two processes is K:1 and if K is small the contribution from the molecular orientation may be too weak to be detected. The second limiting case occurs when $\tau_{Ch} \gg \tau_{DA}$ (*i.e.* slow chemical relaxation).

Equation 7.35 then becomes:

$$\frac{\varepsilon^* - \varepsilon_\infty}{\varepsilon_0 - \varepsilon_\infty} = \frac{1}{1 + i\omega\tau_{DA}} \qquad 7.38$$

i.e. a single process describing the reorientation of the complex. This situation exists for long-lived, 'stable' complexes. In the intermediate case bimodal relaxation behaviour is predicted. (This bimodal behaviour may well be indistinguishable from that of a single relaxation process if τ_{DA} and τ_2 are not very different from each other.) The experimental observations show that the properties of the systems listed in table 7.12

TABLE 7.12

Dielectric relaxation time (τ) and dipole moment (μ) of some molecular complexes dissolved in excess of the donor

Donor	Acceptor	τ/ps	$10^{30}\ \mu$/C m	Reference
mesitylene	tetracyanoethylene	13	2.7	39
mesitylene	tetrachloro-*p*-benzoquinone	11	2.7	39
mesitylene	2,5-dichloro-*p*-benzoquinone	9	1.4	39
mesitylene	*p*-benzoquinone	8	1.3	39
benzene	tin(IV)bromide	12	2.5	53
p-xylene	tin(IV)bromide	32	1.8	53
mesitylene	tin(IV)bromide	40	1.3	53
p-dioxan	tin(IV)chloride	190	17.0	52
mesitylene	tin(IV)chloride	30	—	53

are adequately described by single relaxation times and no bimodal behaviour has yet been reported. The relaxation times for the mesitylene complexes with tetracyanoethylene and with the substituted quinones are lower than expected for the molecular reorientation of a rigid entity. The estimated relaxation times for a complex behaving as a stable entity may be calculated using equation 7.30 and the results of such a calculation (estimated as being significant to about ± 20 to 30 %) are listed (as τ_{DA}) in table 7.13. These values are greater than the observed relaxation times (τ (obs)) suggesting that the relaxation mechanism involves a chemical relaxation occurring at a faster rate than molecular relaxation. These data allow one (in conjunction with equation 7.36 and 7.37) to evaluate the order of magnitude of the rate coefficients involved in equation 7.34 when the equilibrium constant $K(= [DA]/[D][A])$ is known. The results and the calculated rate coefficients are listed in table 7.13. The strength of the complex may be estimated by considering the dipole-induced dipole forces between the polar bonds of one molecule and the polarisable bonds of the second

TABLE 7.13

Rate parameters for complex formation in mesitylene solution at 298 K

Acceptor	τ (obs)/ ps	τ_{DA}/ ps	K^a/ $dm^3\ mol^{-1}$	$10^{-9}\ k_1$/ $mol^{-1}\ dm^3\ s^{-1}$	$10^{-9}\ k_{-1}$/ s^{-1}
tetracyanoethylene	13	20	17.3	3.4	0.2
tetrachloro-*p*-benzoquinone	11	21	5.9	5.9	1
2,5-dichloro-*p*-benzoquinone	9	18	0.6	6	10
p-benzoquinone	8	14	0.24	4.8	20

a. Values from references 124–126.

molecule. One expects that the lifetime of the complex ($(k_{-1})^{-1}$) is related to this interaction energy U by $\log k_{-1} = -U/2.30\ \mathbf{R}T +$ constant. The interaction energies (U) for the complexes mentioned in table 7.13 are listed in table 7.14. A plot of $\log k_{-1}$ against U is linear within the experimental uncertainty and gives a gradient of -1.9×10^{-4} J mol^{-1} (for an intermolecular separation of 0.32 nm). The expected gradient (*i.e.* $-(2.3\ \mathbf{R}T)^{-1}$) is -1.74×10^{-4} J mol^{-1} in good agreement with the observed value. If nothing else, this calculation again emphasizes the importance of electrostatic interactions in the stabilization of these complexes.

Table 7.13 shows that the rate coefficient for the formation of the complex does not vary significantly from complex to complex. The rate coefficient deduced from binary molecular collisions is about 10^{11} mol^{-1} dm^3 s^{-1} and is much larger than the observed value of about 10^9 mol^{-1} dm^3 s^{-1}. Better agreement is obtained when one considers the rate for a diffusion controlled reaction. The rate coefficient[127] for such a

TABLE 7.14

Electrostatic interaction energy (U) in mesitylene complexes (data from reference 39)

Acceptor	$-U$/kJ mol^{-1} at 0.32 nm	$-U$/kJ mol^{-1} at 0.35 nm
tetracyanoethylene	13.2	9.0
tetrachloro-*p*-benzoquinone	6.8	4.6
2,5-dichloro-*p*-benzoquinone	4.2	2.8
p-benzoquinone	2.0	1.4

reaction is:

$$k = \frac{kT}{3\eta} \frac{(r_A + r_B)^2}{r_A r_A} N \qquad 7.39$$

where η is the viscosity of the liquid and r is the molecular radius of the molecular species A and B. For mesitylene solutions ($\eta = 0.7 \times 10^{-3}$ N s m^{-2}) and $r_A = r_B = 0.5$ nm the calculated value of k is 4.7×10^9 mol^{-1} dm^3 s^{-1} in good agreement with the observed k_1 values. Similar diffusion controlled rate coefficients have been reported for complex formation between tetracyanoethylene and hexamethylbenzene in 1-chlorobutane solution[128].

The effective relaxation time in solutions of iodine in benzene has been reported[26] as 3 ps. This result was deduced from dielectric absorption measurements made over a limited frequency range up to 36 GHz. An extension[129] of these measurements at frequencies up to 6 THz (200 cm^{-1}) shows that the absorption reaches a maximum at about 80 cm^{-1} and not in the microwave region as previously suggested. The variation of the dielectric loss factor with frequency for a solution of iodine in benzene and in dioxan is shown in figure 7.14. The continuity between the microwave and far infra-red spectral region is clearly

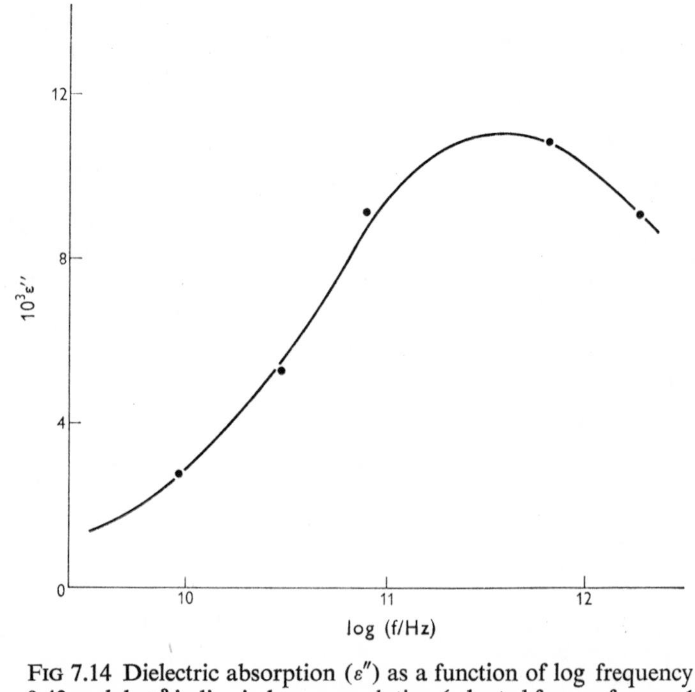

FIG 7.14 Dielectric absorption (ε'') as a function of log frequency for 0.43 mol dm^{-3} iodine in benzene solution (adapted from reference 129).

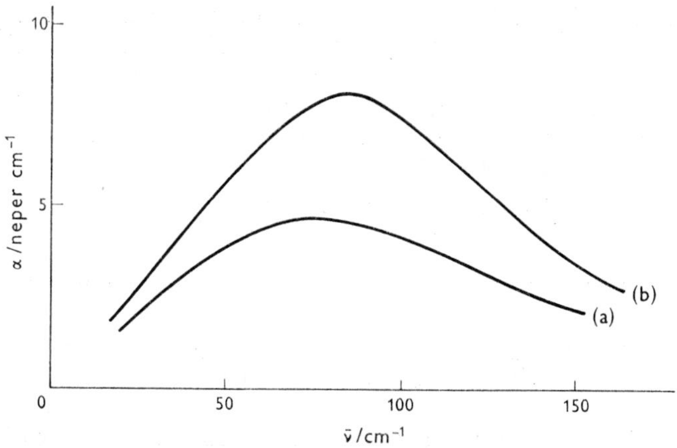

FIG 7.15 Far infra-red absorption in (a) benzene and (b) 0.5 mol dm^{-3} iodine in benzene at 329 K. (Reproduced from *Trans. Faraday Soc.*, II, **68,** 1306 (1972), by permission of the Chemical Society.)

established. The suggested structure of the benzene-iodine 'complex' is one where the iodine is perpendicular to the benzene ring and lies along the six fold axis. Molecular volume considerations predict a relaxation time (τ_{DA}) for this 'stable complex' of 18 ps corresponding to a frequency of maximum absorption of 9 GHz. Such a maximum is not observed.

In 1950 Whiffen[130] established the presence of a weak microwave absorption in non-polar molecules such as benzene and carbon tetrachloride. This absorption is proportional to the frequency up to about 36 GHz (1.2 cm^{-1}) and shows that the absorption not due to a permanent dipole moment. Such a process would show an absorption maximum in this region. Whiffen suggested the presence of transient collision-induced dipoles in the liquid having an effective relaxation time ($\sim 10^{-12}$ s) determined by the time between molecular collisions. In carbon tetrachloride an induced dipole moment of about 0.3×10^{-30} C m adequately accounts for the intensity of the absorption. The general features of this absorption become evident from measurements in the far-infrared spectral region (10 to 200 cm^{-1})[14,131,132,133]. In this region the absorption intensity for pure liquids is usually expressed in terms of the absorption coefficient α (with units of neper cm^{-1}) which is related to the dielectric loss factor by:

$$\alpha = l^{-1} \ln (T_0/T) = 2\pi\varepsilon''\tilde{\nu}/n \qquad 7.40$$

where T_0 and T are the apparent incident and transmitted radiation intensities and l is the sample thickness. $\tilde{\nu}$ is the wavenumber of the radiation and n the sample refractive index at wavenumber $\tilde{\nu}$. The absorption spectrum for benzene is shown in figure 7.15. The absorption in non-polar liquids is usually quite broad (the width at half peak intensity is often about 100 cm^{-1}) and the intensity becomes a maximum in the region 50 to 100 cm^{-1}. The effective dipole moment ($\vec{\mu}_{eff}$) for the absorption may be calculated using the Kramers-Krönig relation to relate the experimental integrated intensity to a dispersion amplitude according to:

$$\int \frac{\varepsilon''}{\bar{\nu}}\, d\bar{\nu} = \frac{n}{2\pi} \int \frac{\alpha}{\bar{\nu}^2}\, d\bar{\nu} = \frac{\pi}{2}(\varepsilon_0 - \varepsilon_\infty) \qquad 7.41$$

The Debye equation relates $(\varepsilon_0 - \varepsilon_\infty)$ to the effective dipole moment giving:

$$\mu_{eff}^2 = \frac{27\mathbf{k}T}{\pi^2} \frac{M\epsilon}{\rho N} \frac{n}{(n^2+2)^2} \int \frac{\alpha}{\bar{\nu}^2}\, d\bar{\nu} \qquad 7.42$$

where M is the molar mass and ρ the liquid density. The absorption in benzene is adequately accounted for by an effective dipole moment of 0.6×10^{-30} C m, the absorption in carbon disulphide and in carbon tetrachloride by an effective dipole moment of 0.4×10^{-30} C m. These dipoles are produced during intermolecular collisions and persist only for the duration of the collision. During collision the molecular symmetry may be disturbed and this gives rise to a dipole moment; local charge asymmetry also provides an electric field which induces local dipole moments in neighbouring molecules. Both these processes may be described in terms of an effective quadrupole moment. The effect of binary collisions between quadrupolar molecules in the gas phase has been evaluated[134,135] and applied[133] to solutions of non-dipolar substances. The effective dipole moment deduced (equation 7.42) from the far-infrared absorption in non-polar liquids is related to the molecular quadrupole moment (θ) by:

$$\mu_{eff}^2 = \frac{3\alpha^2\theta^2 s\langle R^{-8}\rangle}{(4\pi\epsilon)^2} \qquad 7.43$$

where α is the mean molecular polarizability, s the number of molecules per unit volume and $\langle R^{-8}\rangle$ is the radial average of the distance between interacting molecules. For a Lennard-Jones 12–6 potential:

$$\langle R^{-8}\rangle = \tfrac{1}{3}\pi y^{-4}\sigma^{-5}H_8(y) \qquad 7.44$$

where $y = (\varepsilon/\mathbf{k}T)^{1/2}$, ε and σ are the force parameters in the Lennard-Jones potential and the function $H_8(y)$ is tabulated[136]. Some quadrupole moments calculated from equation 7.43 are listed in table 7.16 and agree with the literature values (where available). Thus it seems reasonable to use quadrupole moments evaluated in the above way to calculate molecular interaction parameters in liquids despite the limitations involved in applying expressions derived for binary collisions in the gas phase to data obtained in the liquid phase.

TABLE 7.15

Effective molecular parameters obtained from far-infrared spectra. The dipole moment (μ_{eff}) is calculated using equation 7.42 and the quadrupole moment[b] (θ) using equation 7.43

Molecule	$10^{30}\ \mu_{eff}$/ C m	$10^{40}\ \theta$/ C m^2	$10^{40}\ \theta$ (lit)/ C m^2	Reference
benzene	0.55	8.6	12.0	137, 138
carbon disulphide	0.36	3.5	6.0, 4.0, 2.7	137–139
iodine	1.34[a]	32	42	96
bromine	1.18[a]	10		

a. calculated from the data reported by Wagner[140].
b. literature values from reference 129.

In figure 7.15 the absorption spectrum of benzene and of a solution of iodine in benzene are shown. There is a close similarity between the general features found in both spectra. In binary mixtures the effective dipole moment is given by:

$$\mu^2 = x_1^2\mu_{11}^2 + 2x_1x_2\mu_{12}^2 + x_2^2\mu_{22}^2 \qquad 7.45$$

where x_1 and x_2 are the mole fractions of components 1 and 2, μ_{11} and μ_{22} are the dipole moments produced on binary collision between similar molecules and μ_{12} is the dipole moment produced on binary collision between molecules 1 and 2. Both μ_{11} and μ_{22} are calculated from the experimental data and equation 7.42 and μ_{12} is calculated using:

$$\mu_{12}^2 = \tfrac{3}{2}(\alpha_1^2\theta_2^2 + \alpha_2^2\theta_1^2)\,\frac{s\langle R^{-8}\rangle}{(4\pi\epsilon)^2} \qquad 7.46$$

with $\langle R^{-8}\rangle$ evaluated assuming $\sigma_{12} = \frac{1}{2}(\sigma_1 + \sigma_2)$ and $\varepsilon_{12} = (\varepsilon_1\varepsilon_2)^{1/2}$ and s is the average number of molecules per unit volume. The results of some such calculations for solutions of iodine in benzene, iodine in carbon disulphide and bromine in carbon disulphide are shown in table 7.16. The effective dipole moment evaluated from the experimentally observed spectra and equation 7.42 are shown in column 5 and the calculated dipole moment from equation 7.45 (using the data in table 7.15) are shown in column 6. The agreement between the observed and calculated dipole moments is very good and the absorptions are adequately accounted for without considering charge-transfer effects. The very similar behaviour observed for the benzene–iodine system and for the two carbon disulphide systems is also very striking. The benzene-iodine system is the 'classic' example of a charge-transfer complex, while carbon disulphide is usually regarded as in 'inert' solvent and no ultra-violet charge-transfer bands have been reported for systems involving this substance. It appears that a similar mechanism explains

TABLE 7.16
The effective dipole moment (μ_{eff}) of iodine and bromine solutions compared with the calculated value (μ_{calc}) from equation 7.45 (reference 129)

Solvent	Solute	T/K	Solute concentration/ mol dm^{-3}	10^{30} μ_{eff}/ C m	10^{30} μ_{calc}/ C m
benzene	iodine	294	0.20	0.62	0.62
		294	0.43	0.69	0.69
		329	0.30	0.66	0.69
		329	0.50	0.74	0.78
		329	0.69	0.79	0.85
carbon disulphide	iodine	294	0.21	0.42	0.49
		294	0.41	0.43	0.59
		294	0.60	0.46	0.67
carbon disulphide	bromine	294	0.37	0.47	0.40
		294	0.60	0.48	0.42

the origin of the far infra-red spectrum in both these systems. (see section 3.4.3 and 2.3.6 for further discussion of this far-infrared work).

Dielectric relaxation studies establish the very wide range of stabilities exhibited by molecular complexes. Complexes between *e.g.* acids and bases or amines and iodine exist for very long times and may effectively be regarded as stable entities. At the other extreme lie the weak interactions (as found in benzene–iodine) which persist only for times of the order of 10^{-12} s, a time sufficiently long to allow for observations in the ultraviolet and at least part of the infrared spectral regions.

APPENDIX

The dipole moment of some molecular complexes. μ_D and μ_A are the dipole moments of the pure donor and acceptor molecules as reported[27] in the gas phase or in a "non-complexing" solvent

Donor	Acceptor	$10^{30}\ \mu_D$/C m	$10^{30}\ \mu_A$/C m	Solvent	$10^{30}\ \mu_{DA}$/C m	Reference
			bπ–aσ complexes			
benzene	iodine	0	0	benzene	2.4	23, 24
				benzene	2.3	26
				benzene	1.9	29
				cyclohexane	6.0	25
				decalin	2.4	30
benzene	ICN	0	12.3	benzene	12.5	28, 27
p-xylene	iodine	0	0	*p*-xylene	3.0	23
				p-xylene	2.6	29
mesitylene	iodine	0	0	mesitylene	3.3	26
				mesitylene	4.6	29
				decalin	4.4	30
naphthalene	iodine	0	0	cyclohexane	8.7	25
				decalin	5.5	30
toluene	iodine	1.3	0	toluene	2.6	29
				decalin	3.6	30
di-iso-butylene	iodine	0	0	*di-iso*-butylene	5.0	23
di-iso-butylene	ICN	0	12.3	*di-iso*-butylene	13.3	28
anthracene	iodine	0	0	decalin	8.6	30
phenanthrene	iodine	0	0	decalin	5.0	30
			bπ–aπ complexes			
benzene	1,3,5-trinitrobenzene	0	0	benzene	2.0	31
		0	0	benzene	2.6	32

Appendix (*Cont.*)

Donor	Acceptor	$10^{30}\ \mu_D$/C m	$10^{30}\ \mu_A$/C m	Solvent	$10^{30}\ \mu_{DA}$/C m	Reference
naphthalene	1,3,5-trinitrobenzene	0	0	naphthalene	2.9	33
		0	0	carbon tetrachloride	2.3	34
hexamethylbenzene	1,3,5-trinitrobenzene	0	0	carbon tetrachloride	2.9	34
durene	1,3,5-trinitrobenzene	0	0	carbon tetrachloride	1.8	34
benzene	*p*-dinitrobenzene	0	0	benzene	0.6	35
		0	0	benzene	2.3	32
		0	0	benzene	2.7	38
benzene	nitrobenzene	0	14.3	benzene	13.1	36, 27
naphthalene	nitrobenzene	0	14.3	naphthalene	12.4	36
p-xylene	nitrobenzene	0	14.3	*p*-xylene	14.2	27
benzene	*p*-benzoquinone	0	0	benzene	2.0	37
				carbon tetrachloride	2.5	37
				benzene	0.3	35
mesitylene	*p*-benzoquinone	0	0	mesitylene	1.3	39
hexamethylbenzene	*p*-benzoquinone	0	0	carbon tetrachloride	3.3	40
mesitylene	2,5-dichloro-*p*-benzoquinone	0	0	mesitylene	1.4	39
mesitylene	chloranil	0	0	chloranil	2.7	39
hexamethylbenzene	chloranil	0	0	carbon tetrachloride	3.3	40
benzene	chloranil	0	0	benzene	0	41
				benzene	0.6	35
naphthalene	chloranil	0	0	carbon tetrachloride	3.0	42
durene	chloranil	0	0	carbon tetrachloride	3.0	42
benzene	1,4-naphthaquinone	0	0	benzene	4.0	41

benzene	tetracyanoethylene	0	0	benzene	3.6	43
				benzene	2.7	43
				carbon tetrachloride	2.5	48
mesitylene	tetracyanoethylene	0	0	mesitylene	2.7	39
				carbon tetrachloride	3.7	48
naphthalene	tetracyanoethylene	0	0	carbon tetrachloride	4.3	42
durene	tetracyanoethylene	0	0	carbon tetrachloride	4.2	42
hexamethylbenzene	tetracyanoethylene	0	0	carbon tetrachloride	4.5	42
				carbon tetrachloride	5.5	48
pyrene	tetracyanoethylene	0	0	carbon tetrachloride	6.7	44
benzene	hexafluorobenzene	0	0	benzene	0	45
				benzene	1.1	46
p-xylene	hexafluorobenzene	0	0	*p*-xylene	1.3	46
mesitylene	hexafluorobenzene	0	0	mesitylene	1.3	46
benzene	maleic anhydride	0	12.9	benzene	11.8	47
p-xylene	maleic anhydride	0	12.9	*p*-xylene	12.3	47
mesitylene	maleic anhydride	0	12.9	mesitylene	12.0	47
benzene	phthalic anhydride	0	19.6	benzene	15.7	47
p-xylene	phthalic anhydride	0	19.6	*p*-xylene	15.9	47
mesitylene	phthalic anhydride	0	19.6	mesitylene	16.1	47
benzene	sulphur dioxide	0	5.3	benzene	5.1	51
mesitylene	sulphur dioxide	0	5.3	mesitylene	5.5	51
		$b\pi$–v complexes				
benzene	mercuric chloride	0	0	benzene	4.1	49
benzene	mercuric bromide	0	0	benzene	3.2	49
benzene	diphenyl mercury	0	0	benzene	0.6	50
benzene	diphenyl cadmium	0	0	benzene	2.2	50
benzene	diphenyl zinc	0	0	benzene	2.8	50
benzene	antimony trichloride	0	8.5	benzene	13.1	51

Appendix (*Cont.*)

Donor	Acceptor	$10^{30}\,\mu_D$/C m	$10^{30}\,\mu_A$/C m	Solvent	$10^{30}\,\mu_{DA}$/C m	Reference
p-xylene	antimony trichloride	0	8.5	*p*-xylene	13.4	51
mesitylene	antimony trichloride	0	8.5	mesitylene	13.8	51
benzene	tin tetrachloride	0	0	benzene	2.6	52
				benzene	2.9	54
benzene	tin tetrachloride	0	0	benzene	2.5	53
p-xylene	tin tetrachloride	0	0	*p*-xylene	1.8	53
mesitylene	tin tetrachloride	0	0	mesitylene	1.3	53
benzene	aluminium bromide	0		benzene	3	55
benzene	boron tribromide	0	0	benzene	0.7	56
			n–aσ complexes			
dioxan	iodine	0.6	0	dioxan	4.3	23
				dioxan	3.3	26
				dioxan	3.2	57
				cyclohexane	10.0	25
dioxan	bromine	0.6	0	dioxan	4.3	57
dioxan	ICN	0.6	12.3	dioxan	14.7	28
dioxan	IBr	0.6	4.0	dioxan	9.2	57
dioxan	ICl	0.6	2.0	dioxan	13.3	57
diethylether	iodine	3.8	0		6.3	18, 59
diethylthioether	iodine	5.3	0		15.4	18, 59
triethylamine	iodine	2.2	0	dioxan	37.7	60
				toluene	23.0 (233 K)	61
					29.7 (289 K)	61
trimethylamine	iodine	2.1	0	toluene	21.7	61

pyridine	iodine	7.4	0	cyclohexane	15.0	25
					19.0	18, 59
				carbon tetrachloride	20.8	62
				benzene	13.9	57
		n–aπ complexes				
dioxan	maleic anhydride	0.6	12.9	dioxan	13.1	47
dioxan	phthalic anhydride	0.6	19.6	dioxan	16.2	47
dioxan	tetracyanoethylene	0.6	0	dioxan	4.7	43
				dioxan	3.8	43
				dioxan	2.5	39
dioxan	tetracyanoquino-dimethane	0.6	0	dioxan	4.7	43
dioxan	chloranil	0.6	0	dioxan	2.5	39
dioxan	2,5 dichloro-*p*-benzoquinone	0.6	0	dioxan	1.3	39
dioxan	*p*-benzoquinone	0.6	0	dioxan	1.0	39
dioxan	trinitrobenzene	0.6	0	dioxan	1.3	39
dioxan	pyrole	0.6	6.1	cyclohexane	7.2	63
triethylamine	pyrole	2.3	6.1	cyclohexane	10.4	63
				benzene	10.3	63
acetone	pyrole	8.6	6.1	cyclohexane	12.3	63
				carbon tetrachloride	12.3	63
dimethylaniline	chloranil	5.0	0	carbon tetrachloride	8.8	40
dimethylaniline	*p*-benzoquinone	5.0	0	carbon tetrachloride	5.1	40
diethylether	sulphur dioxide	3.8	5.3	diethylether	10.3	51
		n–v complexes				
dioxan	mercuric chloride	0.6	0	dioxan	4.4	64
dioxan	mercuric bromide	0.6	0	dioxan	3.5	64
dioxan	diphenyl mercury	0.6	0	dioxan	0.13	50

Appendix (*Cont.*)

Donor	Acceptor	$10^{30}\,\mu_D$/C m	$10^{30}\,\mu_A$/C m	Solvent	$10^{30}\,\mu_{DA}$/C m	Reference
dioxan	diphenyl cadmium	0.6	0	dioxan	4.8	50
dioxan	diphenyl zinc	0.6	0	dioxan	9.0	50
dioxan	diphenyl magnesium	0.6	0	dioxan	16.3	50
diethylether	antimony trichloride	3.8	8.5	diethylether	23.2	51
dioxan	tin tetrachloride	0.6	0	dioxan	17.0	52(a)
				dioxan	17.0	65
				dioxan	11.3	66
ammonia	trimethyl boron	4.6			13.8	18, 59
dimethylether	boron trifluoride	4.3	0		14.5	18, 59
pyridine	boron trifluoride	7.4	0		24.3	67
trimethylamine	boron trifluoride	2.1	0	benzene	25.7	67
acid–base complexes						
pyridine	acetic acid	7.40	5.83	benzene	9.78	69
pyridine	chloracetic acid	7.40	7.70	benzene	15.6	69
pyridine	dichloroacetic acid	7.40		benzene	18.2	70
pyridine	trichloroacetic acid	7.40		benzene	25.9	69
N-methylpiperidine	benzoic acid	2.4	5	benzene	12.9	71
N-methylpiperidine	trichloroacetic acid	2.4		benzene	29.3	71
trimethylamine	phenol	2.7	5.1	benzene	10.8	72, 73
trimethylamine	phenol	2.7	5.1	benzene	9.9	74, 75
triethylamine	p-chlorophenol	2.7	7.3	benzene	13.5	74
triethylamine	2,4,6-trichlorophenol	2.7	4.7	benzene	23.5	74
tri-n-butylamine	picric acid	2.5	5.0	dioxan	50.3	76
				dioxan	40.7	77
trimethylamine	picric acid	2.7	5.0	benzene	39.0	77

REFERENCES

1. O. F. Mossotti, *Memorie Mat. Fis. Modena*, **24**, 49 (1850); R. Clausius, Die Mechanische Warmtheorie, Vol. II (Braunschreig, 1897), p 62
2. H. A. Lorentz, *Ann. Physik*, **9**, 641 (1880); L. V. Lorenz, *Ann. Physik*, **11**, 70 (1880)
3. P. Debye, Polar Molecules, (Chemical Catalog Co., New York, 1929)
4. L. A. Onsager, *J. Amer. Chem. Soc.*, **58**, 1486 (1936)
5. G. Hedestrand, *Z. Phys. Chem.*, **B2**, 428 (1929)
6. I. F. Halverstadt and W. D. Kumler, *J. Amer. Chem. Soc.*, **64**, 2988 (1942)
7. E. A. Guggenheim, *Trans. Faraday Soc.*, **45**, 714 (1949)
8. R. J. W. Le Fèvre, Dipole Moments, (Methuen and Co. Ltd., London, 1953)
9. J. W. Smith, Electric Dipole Moments, (Butterworths Scientific Publications, London, 1955)
10. C. P. Smyth, Dielectric Behaviour and Structure, (McGraw-Hill Book Co. Inc., New York, 1955)
11. C. P. Smyth and J. G. Powles, Technique of Organic Chemistry, Vol. I, (Part III—Physical Methods), Ed., Weissberger (Interscience Pub. Inc., New York, 1960), chapter 38
12. C. P. Smyth, Technique of Organic Chemistry Vol. I, (Part III—Physical Methods), Ed., Weissberger (Interscience Pub. Inc., New York, 1960), chapter 39
13. Mansel Davies, Electrical and Optical Aspects of Molecular Behaviour, (Pergamon Press, Ltd., Oxford, 1965)
14. N. E. Hill, W. E. Vaughan, A. H. Price, Mansel Davies, Dielectric Properties and Molecular Behaviour, (van Nostrand Reinhold Co. Ltd., London, 1969)
15. G. Briegleb, J. Czekalla and G. Reuss, *Z. Phys. Chem. (Frankfurt)*, **30**, 333 (1961)
16. C. Abragall and R. Barre, *C.R. Acad. Sci.*, **253**, 439 (1961)
17. A. Findlay and A. N. Campell, Phase Rule, (Longmans, London 1940), p 113
18. E. N. Gur'yanova, *Russ. Chem. Revs.*, **37**, 863 (1968)
19. I. P. Gol'dshtein and E. N. Gur'yanova, *Dokl. Acad. Nauk. SSSR.*, **138**, 1099 (1961)
20. L. Sobczyk, A. Koll and H. Ratajczak, *Bull. Acad. Polon. Sci.*, **11**, 85 (1963)
21. A. V. Few and J. W. Smith, *J. Chem. Soc.*, 2781 (1949); R. J. Bishop and L. E. Sutton, *J. Chem. Soc.*, 6100 (1964)
22. A. A. Maryott, *J. Res. Nat. Bur. Stand.*, **41**, 1, 7 (1948)
23. F. Fairbrother, *J. Chem. Soc.*, 1051 (1948)
24. R. S. Mulliken, *J. Amer. Chem. Soc.*, **74**, 811 (1952)
25. G. Kortum and H. Walz, *Z. Elektrochem.*, **57**, 73 (1953)
26. A. H. Price and V. L. Brownsell, Molecular Relaxation Processes, Chemical Society Special Publication Number 20, (Academic Press, 1966), p 83
27. A. L. McClellan, Tables of Experimental Dipole Moments, (W. H. Freeman and Company, San Francisco, 1963)
28. F. Fairbrother, *J. Chem. Soc.*, 180 (1950)
29. J. Gerbier, *C.R. Acad. Sci.*, **261**, 5037 (1965)
30. A. S. Grishchenko, U. V. Kolodyaklunji and O. A. Osipov, *Zh. Obshch. Khim.*, **39**, 2165 (1969)
31. G. Briegleb and J. Kambeitz, *Z. Phys. Chem.*, **27B**, 11 (1934)
32. C. G. Le Fèvre and R. J. W. Le Fèvre, *J. Chem. Soc.*, 957 (1935); R. J. W. Le Fèvre, *Trans. Faraday Soc.*, **33**, 210 (1937)
33. G. Briegleb and J. Kambeitz, *Z. Phys. Chem.*, **25B**, 251 (1934)
34. G. Briegleb and J. Czekalla, *Z. Elektrochem.*, **59**, 184 (1955)

35. (a) E. N. DiCarlo and C. P. Smyth, *J. Amer. Chem. Soc.*, **84,** 1128 (1962); (b) E. N. DiCarlo, T. P. Logan and R. E. Stronski, *J. Phys. Chem.*, **72,** 1517 (1968)
36. G. Briegleb and J. Kambeitz, *Naturwiss.*, **22,** 105 (1934)
37. C. C. Meredith, L. Westland and G. F. Wright, *J. Amer. Chem. Soc.*, **79,** 2385 (1957)
38. P. Podleschka, L. Westland and G. F. Wright, *Can. J. Chem.*, **36,** 574 (1958)
39. R. A. Crump and A. H. Price, *Trans. Faraday Soc.*, **66,** 92 (1970)
40. G. Briegleb and J. Czekalla, *Z. Elektrochem.*, **58,** 249 (1954)
41. S. Soundararajan and M. J. Vold, *Trans. Faraday Soc.*, **54,** 1155 (1958)
42. G. Briegleb, J. Czekalla and G. Reuss, *Z. Phys. Chem.*, **30,** 333 (1961)
43. H. Huber and G. F. Wright, *Can. J. Chem.*, **42,** 1446 (1964)
44. H. Kuroda, T. Amano, I. Ikemoto and H. Akamatu, *J. Amer. Chem. Soc.*, **89,** 6056 (1967)
45. M. E. Baur, D. A. Horsma, C. M. Knobler and P. Perez, *J. Phys. Chem.*, **73,** 641 (1969)
46. W. A. Duncan, J. P. Sheridan and F. L. Swinton, *Trans. Faraday Soc.*, **62,** 1090 (1966)
47. R. A. Crump and A. H. Price, *Trans. Faraday Soc.*, **65,** 3195 (1969)
48. R. K. Chan and S. C. Liao, *Can. J. Chem.*, **48,** 299 (1970)
49. A. R. Tourky and H. A. Rizk, *Can. J. Chem.*, **35,** 630 (1957); see also I. Eliezer, *J. Chem. Phys.*, **41,** 3276 (1964)
50. W. Strohmeier, *Z. Elektrochem.*, **60,** 58 (1956)
51. J. Gerbier, *C.R. Acad. Sci.*, **262,** 685 (1966)
52. (a) R. A. Crump and A. H. Price, *Chem. Comm.*, 254 (1969); (b) D. J. Denney and A. H. Price, *J. Chem. Soc.*, **A,** 351 (1970)
53. J. P. Kettle and A. H. Price (unpublished results)
54. I. P. Gol'dshtein, E. N. Gur'yanova, D. E. Delinskaya and K. A. Kocheshkov, *Proc. Acad. Sci. U.S.S.R.*, **136,** 173 (1961)
55. I. P. Romm, E. N. Gur'yanova, *Zh. Obshch. Khim.*, **36,** 393 (1966)
56. A. Finch, P. N. Gates and D. Steele, *Trans. Faraday Soc.*, **61,** 2623 (1965)
57. Ya. K. Syrkin and K. M. Anisimova, *Dokl. Akad. Nauk, SSSR.*, **59,** 1457 (1948)
58. M. T. Rogers and W. K. Meyer, *J. Phys. Chem.*, **66,** 1397 (1962)
59. I. P. Gol'dshtein, E. N. Gur'yanova and E. N. Kharlamova, *Zh. Obshch. Khim.*, **38,** 1982 (1968)
60. H. Tsubomura and S. Nagakura, *J. Chem. Phys.*, **27,** 819 (1957)
61. A. J. Hamilton and L. E. Sutton, *Chem. Comm.*, 460 (1968)
62. L. Sobczyk and L. Budziszewski, *Rocz. Chem.*, **40,** 901 (1966)
63. D. M. Bertin and H. Lumbroso, *C.R. Acad. Sci.*, **263C,** 181 (1966)
64. A. R. Tourky, H. A. Rizk and Y. M. Girgis, *J. Phys. Chem.*, **64,** 565 (1960); see also I. Eliezer, *J. Chem. Phys.*, **41,** 3276 (1964)
65. O. A. Osipov and Yu. B. Kletenick, *Zhur. Neorg. Khim.*, **4,** 1494 (1959)
66. T. J. Lane, P. A. McCusker and B. C. Curran, *J. Chem. Soc.*, **64,** 2076 (1942)
67. M. Taillandier and E. Taillandier, *Spectrochim. Acta*, **25A,** 1807 (1969)
68. G. Briegleb, Elektronen-Donator-Acceptor-Komplexe, (Springer-Verlag, 1961), p 125
69. M. M. Davies and L. Sobczyk, *J. Chem. Soc.*, 3000 (1962)
70. S. R. Gough and A. H. Price, *J. Phys. Chem.*, **73,** 459 (1969)
71. L. Sobjczyk, Hydrogen Bonding, (Ed., D. Hadži), (Pergamon Press, London, 1959)
72. J. R. Hulett, J. A. Pegg and L. E. Sutton, *J. Chem. Soc.*, 3901 (1955)
73. J. W. Smith, *J. Chim. Phys.*, **61,** 125 (1964)
74. H. Ratajczak and L. Sobczyk, *Zh. Strukt. Khim.*, **6,** 262 (1965)
75. H. Ratajczak and L. Sobczyk, *Bull. Acad. Polon. Sci.*, **18,** 93 (1970)
76. Mansel Davies and G. Williams, *Trans. Faraday Soc.*, **56,** 1619 (1960)

77. A. A. Maryott, *J. Res. Nat. Bur. Stand.*, **41,** 1 (1948)
78. O. Hassel and J. Hvoslef, *Acta. Chem. Scand.*, **8,** 873 (1954)
79. I. R. Beattie, G. P. Mcquillan and R. Hulme, *Chem. and Ind.*, 1429 (1962)
80. F. P. Mullins, *Can. J. Chem.*, **48,** 1677 (1970)
81. I. G. Arzamanova, E. N. Gur'yanova and I. P. Gol'dshtein, *Dokl. Akad. Nauk. SSSR.*, **155,** 1391 (1964)
82. R. S. Mulliken, *J. Amer. Chem. Soc.*, **74,** 811 (1952)
83. K. Higasi, *Bull. Inst. Phys. Chem. Research* (*Tokyo*), **14,** 146 (1955); *Sci. Papers Inst. Phys. Chem. Research*, **28,** 284 (1936)
84. A. D. Buckingham, *Trans. Faraday Soc.*, **52,** 1551 (1956)
85. A. D. Buckingham and R. E. Raab, *Trans. Faraday Soc.*, **55,** 377 (1959)
86. I. G. Ross and R. A. Sack, *Proc. Phys. Soc.*, **63B,** 893 (1950)
87. G. Briegleb, *Z. Phys. Chem.*, **14B,** 97 (1931)
88. G. Briegleb and T. Schachowsky, *Z. Phys. Chem.*, **19B,** 255 (1932)
89. G. Briegleb, *Z. Phys. Chem.*, **23B,** 105 (1933)
90. G. Briegleb, *Z. Phys. Chem.*, **26B,** 63 (1934)
91. G. Briegleb, *Z. Phys. Chem.*, **31B,** 58 (1935)
92. M. J. S. Dewar and C. C. Thompson, *Tetrahedron Suppl.*, **7,** 97 (1966)
93. J. P. Malrieu and P. Claverie, *J. Chim. Phys.*, **65,** 735 (1968)
94. J. N. Murrell, M. Randic and D. R. Williams, *Proc. Roy Soc.*, **A284,** 566 (1965)
95. M-J. Mantione, *Theor. Chim. Acta*, **11,** 119 (1968)
96. J. L. Lippert, M. W. Hanna, P. J. Trotter, *J. Amer. Chem. Soc.*, **91,** 4035 (1969)
97. M-J. Mantione, *Theor. Chim. Acta*, **15,** 141 (1969)
98. M-J. Mantione, Molecular Associations in Biology, (Ed. B. Pullman) (Academic Press, New York, 1968), p 411
99. M. W. Hanna, *J. Amer. Chem. Soc.*, **90,** 285 (1968)
100. M. W. Hanna and D. E. Williams, *J. Amer. Chem. Soc.*, **90,** 5358 (1968)
101. R. J. W. Le Fèvre, D. V. Radford and P. J. Stiles, *J. Chem. Soc.*, **B,** 1297 (1968)
102. F. T. Lang and R. L. Strong, *J. Amer. Chem. Soc.*, **87,** 2345 (1965)
103. P. Claverie, Molecular Associations in Biology, (Ed. B. Pullman), (Academic Press, New York, 1968), p 115
104. A. I. Kitaygorodsky, *Tetrahedron*, **14,** 230 (1961)
105. R. S. Mulliken, *Rec. Trav. Chim.*, **75,** 845 (1956)
106. R. S. Mulliken and W. B. Person, Molecular Complexes (Wiley-Interscience, New York 1969), p 310
107. C. A. Coulson, Valence (Oxford University Press, Oxford 1961) p 344; *Research* (*London*), **10,** 149 (1957)
108. G. C. Pimentel and A. L. McClellan, The Hydrogen Bond (W. H. Freeman and Co., San Francisco, 1960), p 229
109. J. N. Murrell, *Chem. Brit.*, **5,** 107 (1969)
110. C. A. Coulson, Hydrogen Bonding, (Ed. D. Hadži), (Pergamon Press, London, 1959), p 339
111. C. J. F. Böttcher, Theory of Electric Polarization (Elsevier Publishing Co., Amsterdam 1952), p 374
112. F. Perrin, *J. Phys. Radium*, **5,** 497 (1934); E. Fischer, *Phys. Z.*, **60,** 645 (1939)
113. A. Budó, E. Fischer and S. Miyamoto, *Phys. Z.*, **40,** 337 (1939)
114. F. Buckley and A. A. Maryott, Tables of Dielectric Dispersion Data for Pure Liquids and Dilute Solutions, N.B.S. Circular 589 (1958)
115. R. M. Fuoss and J. G. Kirkwood, *J. Amer. Chem. Soc.*, **63,** 385 (1941)
116. K. S. Cole and R. H. Cole, *J. Chem. Phys.*, **9,** 341 (1941)
117. H. Fellner-Feldegg, *J. Phys. Chem.*, **73,** 616 (1969); H. W. Loeb, G. M. Young, P. A. Quickenden and A. Suggett, *Ber. Bunsenges Phys. Chem.*, **75,** 1155 (1971)
118. J. E. Anderson and C. P. Smyth, *J. Amer. Chem. Soc.*, **85,** 2904 (1963)
119. J. E. Anderson, Ford Motor Company Technical Report SR70-80, Michigan, U.S.A.
120. J. E. Anderson and R. Ullman, *J. Chem. Phys.*, **47,** 2178 (1967)

121. W. Kauzmann, *Rev. Mod. Phys.*, **14**, 12 (1942)
122. G. Schwarz, *J. Phys. Chem.*, **71**, 4021 (1967)
123. G. Williams, *Adv. Mol. Relaxation Processes*, **1**, 409 (1970); *Chem. Rev.*, **72**, 55 (1972)
124. R. E. Merrifield and W. D. Phillips, *J. Amer. Chem. Soc.*, **80**, 2778 (1958)
125. R. Foster, D. L. Hammick and B. N. Parsons, *J. Chem. Soc.*, 555 (1956)
126. R. Foster and A. C. Fyfe, *Trans. Faraday Soc.*, **62**, 1400 (1966)
127. For example, see E. A. Moelwyn-Hughes, Physical Chemistry, (Pergamon Press, Oxford 1965), pp 1211, 1275
128. E. F. Caldin, J. E. Cooks, D. O'Donnell, D. Smith and S. Toner, *J. Chem. Soc., Faraday*, **1**, **68**, 849 (1972)
129. J. P. Kettle and A. H. Price, *J. Chem. Soc., Faraday II*, **68**, 1306 (1972)
130. D. H. Whiffen, *Trans. Faraday Soc.*, **46**, 124 (1950)
131. Mansel Davies, G. W. F. Pardoe, J. Chamberlain and H. A. Gebbie, *Trans. Faraday Soc.*, **66**, 273 (1970)
132. G. W. F. Pardoe, *Trans. Faraday Soc.*, **66**, 2699 (1970)
133. S. K. Garg, J. E. Bertie, H. Kilp and C. P. Smyth, *J. Chem. Phys.*, **49**, 2551 (1968)
134. See, for example, J. O. Hischfelder, C. F. Curtis and R. B. Bird, Molecular Theory of Gases and Liquids, (John Wiley and Sons, Inc., New York, 1964)
135. A. D. Buckingham, *Quart. Rev.*, **13**, 183 (1959)
136. (a) J. A. Pople, *Proc. Roy. Soc.*, **A221**, 508 (1954); (b) A. D. Buckingham and J. A. Pople, *Trans. Faraday Soc.*, **51**, 1173 (1955)
137. D. E. Strogyn and A. P. Strogyn, *Mol. Phys.*, **11**, 371 (1966)
138. G. J. Davies, Thesis, University of Wales (1971)
139. C. R. Fischer and P. J. Kemmey, *Mol. Phys.*, **22**, 1133 (1971)
140. V. Wagner, *Z. Phys.*, **224**, 353 (1969)

Subject Index

Compound Index